Principles of Sedimentology and Stratigraphy

Principles of Sedimentology and Stratigraphy

Fourth Edition

Sam Boggs, Jr.
University of Oregon

PEARSON
Prentice
Hall

Upper Saddle River, New Jersey 07458

Library of Congress Cataloging-in-Publication Data

Boggs, Sam.
   Principles of sedimentology and stratigraphy / Sam Boggs, Jr.—4th ed.
     p. cm.
   Includes bibliographical references and index.
   ISBN 0-13-154728-3
    1. Sedimentation and deposition. 2. Geology, Stratigraphic. I. Title.

QE571.B66 2006
552'.5—dc22   2005047667

Executive Editor: *Patrick Lynch*
Project Manager: *Dorothy Marrero*
Executive Managing Editor: *Kathleen Schiaparelli*
Assistant Managing Editor: *Beth Sweeten*
Editorial Assistant: *Sean Hale*
Production Editor: *Kevin Bradley/GGS Book Services*
AV Editor: *Greg Dulles*
Art Studio: *Laserwords*
Manufacturing Buyer: *Alan Fischer*
Art Director: *Jayne Conte*
Cover Design: *Bruce Kenselaar*
Cover Photo: *Eroded Permian and Triassic sedimentary rocks viewed from the Green River Overlook, Canyonlands National Park, near Moab, Utah. Photograph courtesy of the U. S. Department of Interior.*

© 2006, 2001, 1995 by Pearson Education, Inc.
Pearson Prentice Hall
Pearson Education, Inc.
Upper Saddle River, NJ 07458

Earlier edition © 1987 by Merrill Publishing Comapny.

Pearson Prentice Hall™ is a trademark of Pearson Education, Inc.

Printed in the United States of America
10 9 8 7 6

ISBN 0-13-154728-3

Pearson Education Ltd., *London*
Pearson Education Australia Pty. Ltd., *Sydney*
Pearson Education Singapore, Pte. Ltd
Pearson Education North Asia Ltd., *Hong Kong*
Pearson Education Canada, Inc., *Toronto*
Pearson Educación de Mexico, S.A. de C.V.
Pearson Education—Japan, *Tokyo*
Pearson Education Malaysia, Pte. Ltd.

*Dedicated to my mother, Lucinda Caudill Boggs,
to whom I owe everything*

Contents

Preface

The roots of sedimentology and stratigraphy extend back to the 16th century; however, these disciplines are still growing and changing. Geologists continue to "fine tune" sedimentologic and stratigraphic concepts through a variety of research avenues and by using an array of increasingly sophisticated research tools. The result is a continuous outpouring of fresh data and new ideas. In fact, it is becoming increasingly difficult to keep abreast of the flood of new information appearing in the geological literature. A glance through recent issues of a well-known sedimentology journal reveals important new papers on sedimentation and tectonics, depositional systems, carbonates, biosedimentology, diageneis, provenance, geochemistry, sediment transport and sedimentary structures, stratigraphic architecture, chronostratigraphy, numerical modeling, paleoclimatology, sequence stratigraphy, and basin analysis—to name but a few research areas.

I make no claim that I have, in this fourth edition of *Principles of Sedimentology and Stratigraphy*, fully evaluated all of these new data or captured all of the new ideas and concepts that may have been put forward since publication of the third edition. I have, however, tried to weave important new information into the basic structure of previous editions and revise concepts that may have become outdated. In addition, I have reorganized some of the chapters, added numerous references to pertinent new research articles and books, and added a significant number of new photographs, line drawings, and tables. I hope that these changes increase both the readability of the book for students and also keep them abreast of recent developments in the fields of sedimentology and stratigraphy.

As mentioned in the preface to the third edition, career opportunities for geology students are shifting away from the more traditional avenues of petroleum and mining geology toward environmental geology and other disciplinary areas that deal with problems of society. To be prepared for these careers, students need to gain a solid foundation in the basic principles of sedimentology, stratigraphy, and related sciences, as well as to develop insight into innovative applications of these principles to areas of study such as environmental analysis, paleoclimate evaluation, groundwater resources, and marine pollution. I hope that this book provides a useful part of the basic background that students need to advance into these exciting career fields.

I want to thank the following people for reviewing the Third Edition. I used their reviews to guide me in the preparation of the Fourth Edition:

Edwin J. Anderson, Temple University; Janok P. Bhattacharya, University of Texas – Dallas; Charles W. Byers, University of Wisconsin; Beth Christensen, Georgia State University; Joachim Dorsch, Saint Lewis University; James E. Evans, Bowling Green State University; Larry T. Middleton, Northern Arizona University; Michael R. Owen, Saint Lawrence University; Bruce Selleck, Colgate University; and Mark A. Wilson, College of Wooster.

SAM BOGGS, JR.

Introduction

Kinds of Sedimentary Rocks

This book describes and discusses the physical, chemical, and biological characteristics of sedimentary rocks and the interpretations that we draw from these characteristics about the origin of sedimentary rocks. Geologists disagree somewhat about how the various kinds of sedimentary rocks should be classified; however, such rocks can conveniently be placed into three fundamental groups on the basis of composition and origin: siliciclastic, chemical/biochemical, and carbonaceous. **Siliciclastic** sedimentary rocks are composed dominantly of silicate minerals, such as quartz and feldspar, and rock fragments (clasts). These materials originate mainly by the chemical and physical breakdown (weathering) of igneous, metamorphic, or (older) sedimentary rock. Conglomerates, sandstones, and shales belong to this group. Silicate detritus, including silicate minerals, rock fragments, and glass shards, can also be generated by explosive volcanism. Siliciclastic sedimentary rocks that formed mainly from the products of explosive volcanism are called **volcaniclastic** rocks. **Chemical/biochemical** sedimentary rocks are composed of minerals precipitated mainly from ocean or lake water by inorganic (chemical) and/or organic (biogenic) processes. They include limestone, chert, evaporites such as gypsum, phosphorites, and iron-rich sedimentary rocks. Evaporites are probably precipitated entirely by inorganic processes resulting from evaporation of lake or seawater. Biogenic processes, as well as inorganic processes, play an important role in the formation of many limestones and likely play some role in the origin of chert, phosphorites, and iron-rich sedimentary rocks. **Carbonaceous** sedimentary rocks contain a substantial amount (>~15 %) of highly altered remains of the soft tissue of plants and animals, referred to as organic matter. The principal carbonaceous rocks are coal and oil shale. Carbonaceous sedimentary rocks make up only a small fraction of the total sedimentary record; however, these rocks (especially coals) have great economic importance as fossil fuels.

Distribution of Sedimentary Rocks in Time and Space

Sedimentary rocks are confined to Earth's outer crust, where they make up only 5–10 percent of the outer 10 miles (16 km) or so of the crust. On the other hand, they are the most common rocks at Earth's surface. Sedimentary rocks and sediments cover nearly three-fourths of Earth's land surface and most of the ocean floor. They range in age from Precambrian to modern. The first sedimentary rocks were deposited nearly four billion years ago, at which time most of Earth's surface was covered with volcanic rocks. The relative proportion of sedimentary rocks at Earth's surface has increased progressively with time, as weathering processes brought about decomposition of other kinds of rock and deposition of the decomposition products to form sedimentary rocks.

Sedimentology Versus Stratigraphy

The record of Earth history locked up in sedimentary rocks dates back almost four billion years. It is the study of this reservoir of Earth history that constitutes the sciences of sedimentology and stratigraphy. **Sedimentology** is the scientific study of the classification, origin, and interpretation of sediments and sedimentary rocks. It is often difficult to draw a sharp distinction between sedimentology and **stratigraphy**, which is defined simply and broadly as the science of rock strata. In general, however, sedimentology is concerned with the physical (textures, structures, mineralogy), chemical, and biologic (fossils) properties of sedimentary rocks and the processes by which these properties are generated. It is these properties that provide much of the basis for interpreting the physical features, climate, and environmental conditions of Earth in the geologic past. Stratigraphy, on the other hand, is concerned more with age relationships of strata, successions of beds, local and worldwide correlation of strata, and stratigraphic order and chronological arrangement of beds in the geologic column. Stratigraphy finds special applications in the study of plate reconstructions (plate tectonics) and in the unraveling of the intricate history of landward and seaward movements of ocean shorelines (transgressions and regressions) and rise and fall of sea level through time. Particularly exciting developments in stratigraphy have come about recently by applying the principles of seismology and paleomagnetism to stratigraphic problems.

Brief History of Sedimentology and Stratigraphy

Sedimentologic and stratigraphic study date back to about A.D. 1500 with the observations of Leonardo da Vinci on fossils in sedimentary rocks of the Italian Apennines. Since that time, a steady drumbeat of progress in understanding sedimentary rocks has taken place, punctuated at intervals by significant new developments in tools and techniques for studying sedimentary rocks and emergence of new concepts and ideas about their origin. Especially noteworthy among these seminal events were (1) initiation of the use of the microscope to study fossils by Robert Hooke in the latter part of the 17th century, (2) elucidation of the concept of uniformitarianism (loosely, the present is the key to the past) by James Hutton in the late 18th century, (3) the birth of biostratigraphy (study and interpretation of sedimentary rocks on the basis of the fossils they contain) by William Smith in the early 19th century, (4) application of the petrographic microscope to study of sedimentary rocks by Henry Clifton Sorby around 1850, (5) development of one of the most far-reaching concepts in geologic philosophy—seafloor spreading and global plate tectonics—in the early 1960s, and (6) emergence of the concepts of seismic stratigraphy (study of seismic data for the purpose of extracting stratigraphic information), sequence stratigraphy (application of the concept of depositional sequences to stratigraphic interpretation), and magnetostratigraphy (study of rock magnetism as a stratigraphic tool) in the 1960s and 1970s. The pace of new developments in sedimentology and stratigraphy continues to the present time, spurred by the availability of technologically advanced laboratory tools such as the scanning electron microscope and mass spectrometer and development of advanced field procedures such as the ability to drill deep holes in the ocean floor and recover sediment cores in water several thousand meters deep.

Why Study Sedimentary Rocks?

The sheer abundance of sedimentary rocks at Earth's surface provides a partial answer to a question frequently asked by students, "Why should we study sedimentary rocks; why bother?" In addition to their abundance, however, they are also important because of information they yield about Earth's history and because of the economic products they contain. All geologic study is aimed in one way or another at developing a better understanding of Earth's history. All rocks, whether sedimentary, igneous, or metamorphic, contain clues to some aspect of this history, but sedimentary rocks are unique with regard to the information they provide. From the composition, textures, structures, and fossils in sedimentary rocks, experienced geologists can decipher clues that provide insight into past climates, oceanic environments and ecosystems, the configurations of ancient land systems, and the locations and compositions of ancient mountain systems long since vanished. Thus, study of sedimentary rocks forms the primary basis for the sciences of paleoclimatology (study of climates throughout geologic time), paleogeography (study and description of the physical geography of Earth's past), paleoecology (study of the relationship between ancient organisms and their environment), and paleooceanography (study of the characteristics of ancient oceans). In addition, many sedimentary rocks have economic significance. Most of the world's oil and gas and all of its coal are contained in sedimentary rock successions. Iron-bearing minerals, uranium minerals, evaporite minerals, phosphate minerals, and many other economically valuable minerals also occur in these rocks.

Thus, the disciplines of sedimentology and stratigraphy, while having their roots in studies dating back to the early 16th century, are still vibrant, exciting, growing disciplines. I hope that this book will help students capture some of this sense of excitement: It provides an integrated view of sedimentology and stratigraphy. The first few chapters are devoted to description and discussion of the processes that form sedimentary rocks, the physical, chemical, and biological properties of rocks that result from these processes, and the principal kinds of sedimentary rocks. Succeeding chapters deal with sedimentary environment and their interpretation from the rock record; stratigraphic relationships revealed through study of lithology, seismic reflection characteristics, remanent magnetism, fossils, and radiometric ages; and basin analysis, which is the integrated sedimentological and stratigraphic study of sedimentary rocks.

Additional Sources of Information

Numerous references are made throughout this book to research papers that provide detailed information about particular topics. In addition, a list of pertinent monographs is provided at the end of each chapter. Readers should find these research papers and books a useful starting point for additional literature research. Finally, Appendix E furnishes an extended list of Web sites where online information about sedimentology and stratigraphy is available.

PART I

Origin and Transport of Sedimentary Materials

Sediment transport in the braided Kongakut River, Arctic National Wildlife Refuge, Alaska

Sedimentary rocks form through a complex set of processes that begins with **weathering**, the physical disintegration and chemical decomposition of older rock to produce solid particulate residues (resistant minerals and rock fragments) and dissolved chemical substances. Some solid products of weathering may accumulate *in situ* to form soils that can be preserved in the geologic record (paleosols). Ultimately, most weathering residues are removed from weathering sites by erosion and subsequently transported, possibly along with fragmental products of explosive volcanism, to more distant depositional sites.

Transport of siliciclastic detritus to depositional basins can involve a variety of processes. Mass-transport processes such as slumps, debris flows, and mud flows are important agents in the initial stages of sediment transport from weathering sites to valley floors. Fluid-flow processes, which include moving water, glacial ice, and wind, move sediment from valley floors to depositional basins at lower elevations. When transport processes are no longer capable of moving sediment, **deposition** of sand, gravel, and mud takes place, either subaerially (e.g., in desert dune fields) or subaqueously in river systems, lakes, or the marginal ocean. Sediment deposited at the ocean margin may be reentrained and retransported tens to hundreds of kilometers into deeper water by turbidity currents or other transport processes. Sediments deposited in basins are eventually buried and undergo physical and chemical changes (**diagenesis**) resulting from increased temperature, pressure, and the presence of chemically active fluids. Burial diagenetic processes convert siliciclastic sediments to lithified sedimentary rock: conglomerate, sandstone, shale.

Weathering processes also release from source rocks soluble constituents such as calcium, magnesium, and silica that make their way in surface water and groundwater to lakes or the ocean. When concentrations of these chemical elements become sufficiently high, they are removed from water by chemical and biochemical processes to form "chemical" sediments. Subsequent burial and diagenetic alteration of these sediments generates lithified sedimentary rock: limestone, chert, evaporites, and other chemical/biochemical sedimentary rocks.

In summary, the origin of sedimentary rocks involves weathering of older rock to generate the materials that make up sedimentary rock, erosion and transport of weathered debris and soluble constituents to depositional basins, deposition of this material in continental (terrigenous) or marine environments, and diagenetic alteration during burial to ultimately produce lithified sedimentary rock. Because weathering plays such a critical role in generating the solid particles and chemical constituents that make up sedimentary rocks, Chapter 1 focuses on the physical and chemical processes of weathering, the nature of the resulting weathering products, and a brief discussion of soils. Chapter 2 continues with a detailed discussion of the various processes by which sediment grains are transported from weathering sites to depositional basins. Other aspects of the origin of sedimentary rocks are introduced and discussed in succeeding chapters, as appropriate.

1

Weathering and Soils

1.1 INTRODUCTION

Weathering involves chemical, physical, and biological processes, although chemical processes are by far the most important. A brief summary of weathering processes is presented here to illustrate how weathering acts to decompose and disintegrate exposed rocks, producing particulate residues and dissolved constituents. These weathering products are the source materials of soils and sedimentary rocks; thus, weathering constitutes the first step in the chain of processes that produce sedimentary rocks.

It is important to understand how weathering attacks exposed source rocks and what remains after weathering to form soils and be transported as sediment and dissolved constituents to depositional basins. The ultimate composition of soil and terrigenous sedimentary rock bears a relationship to the composition of their source rock; however, study of residual soil profiles shows that both the mineral composition and the bulk chemical composition of soils may differ greatly from those of the bedrock on which they form. Some minerals in the source rock are destroyed completely during weathering, whereas more chemically resistant or stable minerals are loosened from the fabric of the decomposing and disintegrating rock and accumulate as residues. During this process, new minerals such as iron oxides and clay minerals may form *in situ* in the soils from chemical elements released during breakdown of the source rocks. Thus, soils are composed of survival assemblages of minerals and rock fragments derived from the parent rocks plus any new minerals formed at the weathering site. Soil composition is governed not only by the parent-rock composition but also by the nature, intensity, and duration of weathering and soil-forming processes. It follows from this premise that the composition of terrigenous sedimentary rocks such as sandstones, which are derived from soils and other weathered materials, is also controlled by parent-rock composition and weathering processes.

Most ancient soils were probably eroded and their constituents transported to furnish the materials of sedimentary rocks; however, some survived to become part of the geologic record. We call these ancient soils **paleosols**. Weathering and soil-forming processes are significantly influenced by climatic conditions. Geologists are greatly interested in the study of past climates, called paleoclimatology, because of this relationship and because paleoclimates also influenced past sea levels and sedimentation processes as well as the life forms on Earth at various times.

In this chapter, we examine the principal processes of subaerial weathering and discuss the nature of the particulate residues and dissolved constituents that result from weathering. We also consider the less important but highly interesting processes of submarine weathering. Submarine weathering includes both the interaction of seawater with hot oceanic rocks along mid-ocean ridges—a process that leaches important amounts of chemical constituents from hot crustal rocks—and low-temperature alteration of volcanic rocks and sediments on the ocean floor. Finally, we take a brief look at soils and paleosols and discuss important soil-forming processes and the factors, such as climate, that influence soil development.

1.2 SUBAERIAL WEATHERING PROCESSES

Physical Weathering

Physical (mechanical) weathering is the process by which rocks are broken into smaller fragments through a variety of causes, but without significant change in chemical or mineralogical composition. Except in extremely cold or very dry climates, physical and chemical weathering act together, and it is difficult to separate their effects.

Freeze-Thaw (Frost) Weathering

Disruption of rock fabrics owing to stresses generated by freezing and thawing of water in rock fractures is an important physical weathering process in climates where recurring, short-term changes from freezing to thawing temperatures take place. Water increases in volume by about 9 percent when it changes to ice, creating enough pressure in tortuous rock fractures to crack most types of rock. To be effective, water must be trapped (sealed by freezing) within the rock body, and repeated freezing and thawing are necessary to allow progressive disintegration of the rock, which occurs very slowly. Other processes, such as the movement of water into a freezing zone rather than conversion of water in place to ice, may also, or alternatively, cause freeze-thaw expansion of cracks (Bland and Rolls, 1998, p. 89).

Freeze-thaw weathering commonly produces large, angular blocks of rock (Fig. 1.1) but may also cause granular disintegration of coarse-grained rocks such as granites. The presence of microfractures and other microstructures exerts an important control on the sizes and shapes of shattered blocks. Mechanically weak rocks such as shales and sandstones tend to disintegrate more readily than do hard, more strongly cemented rocks such as quartzites and igneous rocks.

Insolation Weathering

Expansion of rock surfaces heated by the Sun (insolation) followed by contraction as the temperature falls can allegedly weaken bonds along grain boundaries and cause subsequent flaking off of rock fragments or dislodging of mineral grains. A thermal gradient is set up between the surface and interior of a rock that has been heated; the rock surface expands more than the interior, creating stresses. These stresses presumably lead to formation of small cracks and possibly granular disintegration (Ollier and Pain, 1996, p. 26). Once a small crack in a rock's surface expands with heating, silt or sand particles may sift into the crack and prevent it from closing when the rock cools. Repeated heating and cooling causes the crack to grow wider and wider, resulting in small-scale disruption of the rock surface. These kinds of physical changes are caused mainly by heating from sunshine but may also result from fires (e.g., Allison and Goudie, 1994). Although observations

Figure 1.1
Large, angular blocks of rock generated by freeze-thaw weathering of thin-bedded sandstones and mudstones of the Canning Formation (Paleocene) exposed along the Canning River, Arctic National Wildlife Refuge, Alaska. [Photograph by C. J. Schenk, U.S. Geological Survey Open File Report 98–34, The oil and gas resource potential of the Arctic National Wildlife Refuge 1002 Area, Alaska, 1999.]

in desert areas suggest that insolation weathering does occur, heating and cooling experiments in the laboratory have not yielded conclusive proof that insolation weathering is an important process. The concept remains controversial.

Salt Weathering

High temperatures in desert environments also tend to promote weathering caused by the crystallization of salts in pore spaces and fractures (Sperling and Cooke, 1980; Watson, 1992; Bland and Rolls, 1998). Evaporation of water concentrates dissolved salts in saline solutions that have access to rock fractures and pores. Growth of salt crystals generates internal pressures (crystallization pressures) that can force cracks apart or cause granular disintegration of weakly cemented rocks. Expansion pressures may also be generated when salts in fractures become hydrated (absorb water) and expand. Salt weathering is most common in semiarid regions but can occur also along seacoasts where salt spray is blown onto sea cliffs.

Wetting and Drying

Alternate wetting and drying of soft or poorly cemented rocks such as shales causes fairly rapid breakdown of the rocks, and most disintegration may occur during the drying cycle. The exact causes of disintegration are not well understood, but drying may lead to negative pore pressures and consequent tensile stresses (contraction) that tend to pull the rock apart. On the other hand, absorption of water during wetting phases creates "swelling" pressures that push cracks apart. Disintegration by wetting and drying appears to be particularly effective on well-exposed, steep cliff faces where loosened fragments fall off and expose fresh surfaces.

Stress-Release Weathering

A rock unit buried below a land surface experiences high compressional stresses because of the weight of the overlying rock. If some of the overlying rock is removed by erosion, compressional stresses on the rock unit are reduced and the rock unit "rebounds" upward. Expansion of the rock upward creates tensile stresses (pulls the rock apart), causing fractures to develop that are oriented nearly parallel to the topographic surface. These fractures divide the rock into a series of layers or sheets; hence, this process of crack formation is often called sheeting. These layers increase in thickness with depth and may exist for several tens of meters below Earth's surface. Sheeting is most conspicuous in homogeneous rocks such as granite but may occur also in layered rock, such as massive sandstone.

Other Physical Processes

Other factors that may contribute to mechanical weathering under certain conditions include volume increases caused by absorption of water (hydration) by clay minerals or other minerals; volume changes caused by alteration of minerals such as biotite and plagioclase to clay minerals; growth of plant roots in the cracks of rocks; plucking of mineral grains and rock fragments from rock surfaces by lichens as they expand and contract in response to wetting and drying; and burrowing and ingestion of soils and loosened rock materials by worms or other organisms.

Some physical weathering effects may be the result of two or more processes operating together. **Exfoliation**, the peeling off of large, curved sheets or slabs of rock from the weathered surfaces of an outcrop, is an apposite example. Stress release may create initial fractures, which then allow the entry of water that further widens fractures by freeze-thaw or other processes. **Spheroidal weathering** is smaller-scale weathering of roughly cubic rock masses, cut by intersecting joints, causing layers or "skins" to spall off to produce spheroidal cores (Fig. 1.2). The fractures that separate the weathering rinds may form in response to stress release or possibly thermal changes (Taylor and Eggleton, 2001, p. 166); entry of water into fractures promotes additional physical stresses arising from freeze-thaw or chemical processes such as those mentioned in the preceding paragraph.

Figure 1.2
Spheroidal weathering in granite. Note how successive, thin layers of weathered rock are spalled off to produce a spheroidal core.

Chemical Weathering

Chemical weathering involves changes that can alter both the chemical and the mineralogical composition of rocks. Minerals in the rocks are attacked by water and dissolved atmospheric gases (oxygen, carbon dioxide), causing some components of the minerals to dissolve and be removed in solution. Other mineral constituents recombine *in situ* and crystallize to form new mineral phases. These chemical changes, along with changes caused by physical weathering, disrupt the fabric of the weathered rock, producing a loose residue of resistant grains and secondary minerals. Water and dissolved gases play a dominant role in every aspect of chemical weathering. Because some water is present in almost every environment, chemical weathering processes are commonly far more important than physical weathering processes, even in arid climates. Nevertheless, owing to the low temperatures of the weathering environment ($<30°C$), chemical weathering occurs very slowly. The processes of chemical weathering are listed and briefly described in Table 1.1, along with selected examples of new minerals formed *in situ* during the weathering processes.

Major Chemical Weathering Processes

Simple Solution. Simple solution (*congruent dissolution*) occurs when a mineral goes into solution completely without precipitation of other substances (e.g., Birkland, 1999, p. 59). Simple solution of highly soluble minerals such as calcite, dolomite, gypsum, and halite, and even less soluble minerals such as quartz, occurs during exposure to meteoric water (rainwater). Chemical bonds between ions in the minerals are broken, destroying the minerals and releasing constituent ions into solution in surface and groundwaters. If carbon dioxide is dissolved in the rainwater through interaction with atmospheric or soil CO_2, the usual case in the weathering environment, the solubilizing ability of water is enhanced. Dissolution of CO_2 in water forms carbonic acid (H_2CO_3—this is what you consume in your soft drinks), which subsequently dissociates to produce hydrogen ions and carbonate ions ($CO_2 + H_2O \leftrightarrow H_2CO_3 \leftrightarrow H^+ + HCO_3^-$). Increase in H^+ ions, relative to OH^- ions, makes meteoric waters more acidic and thus more aggressive dissolution agents, particularly for carbonate minerals. Simple solution of this type is an important weathering process, particularly in moderately wet climates where carbonate rocks or evaporites are present near the surface or at the water table.

Hydrolysis. Hydrolysis is an extremely important chemical reaction between silicate minerals and acids that leads to the breakdown of the silicate minerals and release of metal cations and silica, but the reaction does not lead to complete dissolution of the minerals. In other words, the amount of ions from the mineral that are taken into solution during weathering does not correspond to the formula of the weathering mineral. This kind of incomplete dissolution is called

---

Box 1.1　pH

The acidity or alkalinity of a solution is expressed by its **pH**. The pH is defined as the negative logarithm to the base 10 of the approximate hydrogen-ion concentration in moles per liter. The pH scale extends from 0 to 14, corresponding to H^+ concentrations ranging from 10^0 to 10^{-14}. For example, a solution containing a H^+ concentration of 10^{-1} moles per liter has a pH of 1, an H^+ concentration of 10^{-7} yields a pH of 7, and so forth. Solutions with a pH of 7 are considered neutral. Acids have pH values lower than 7 and bases have values greater than 7.

Table 1.1 Principal processes of chemical weathering

Most important processes	Examples	Principal kinds of rock materials affected
Simple (congruent) Solution—Dissolution of soluble minerals in H_2O (direct solution) or in $H_2O + CO_2$ (carbonation) to yield cations and anions in solution	$SiO_2 + 2H_2O \rightarrow H_4SiO_4$ (direct solution) (quartz) (silicic acid) aq $CaCO_3 + H_2O + CO_2 \leftrightarrow Ca^{2+} + 2HCO_3^-$ (Carbonation) (calcite) aq aq	Highly soluble minerals (e.g., gypsum, halite), quartz Carbonate rocks
Hydrolysis (incongruent dissolution)—Reaction between H^+ and OH^- ions of water and the ions of silicate minerals, yielding soluble cations, silicic acid, and clay minerals (if Al present)	$2KAlSi_3O_8 + 2H^+ + 9H_2O \rightarrow H_4Al_2Si_2O_9 + 4H_4SiO_4 + 2K^+$ (orthoclase) aq (kaolinite) (silicic acid) aq $2NaAlSi_3O_8 + 2H^+ + 9H_2O \rightarrow H_4Al_2Si_2O_9 + 4H_4SiO_4 + 2Na^+$ (albite) aq (kaolinite) (silicic acid) aq	Silicate minerals
Oxidation—Loss of an electron from an element (commonly Fe or Mn) in a mineral, resulting in the formation of oxides or hydroxides (if water present)	$2FeS_2 + 15/2O_2 + 4H_2O \rightarrow Fe_2O_3 + 4SO_4^{2-} + 8H^+$ (pyrite) (hematite) aq aq $MnSiO_3 + 1/2O_2 + 2H_2O \rightarrow MnO_2 + H_4SiO_4$ (rhodonite) (pyrolusite) (silicic acid)	Iron- and manganese-bearing silicate minerals, iron sulfides
Other Processes		
Hydration and Dehydration—Gain (hydration) or loss (dehydration) of water molecules from a mineral, resulting in formation of a new mineral	$Fe_2O_3 + H_2O \leftrightarrow 2FeOOH$ (hydration) (hematite) (goethite) $CaSO_4 \cdot 2H_2O \leftrightarrow CaSO_4 + 2H_2O$ (dehydration) (gypsum) (anhydrite)	Ferric oxides Evaporites
Ion Exchange—Exchange of ions, principally cations, between solutions and minerals	K-clay + $Mg^{2+} \leftrightarrow Mg$-clay + K^+ Ca-zeolite + $Na^+ \leftrightarrow Na$-zeolite + Ca^{2+}	Clay minerals and zeolites
Chelation—Bonding of metal ions to organic molecules having ring structures	Metal ions (cations) + chelating agent (e.g., secreted by lichens) $\rightarrow H^+$ ions + chelate (metal ions/organic molecules in solution)	Silicate minerals

Note: aq = aqueous

incongruent dissolution. If aluminum is present in the minerals undergoing incongruent dissolution during weathering, clay minerals such as kaolinite, illite, and smectite may form as a by-product of hydrolysis. For example, orthoclase feldspar can break down to yield kaolinite or illite, albite (plagioclase feldspar) can decompose to kaolinite or smectite, and so on, as illustrated by the reactions in Table 1.1. As mentioned, the H^+ ions shown in Table 1.1 are commonly supplied by the dissociation of CO_2 in water. Thus, the more CO_2 that is dissolved in water, the more aggressive the hydrolysis reaction. Hydrolysis can also take place in water containing little or no dissolved CO_2, with H^+ ions being supplied either by clay minerals that have a high proportion of H^+ ions in cation exchange sites or by living plants, which create an acid environment. Most of the silica set free during hydrolysis goes into solution as silicic acid (H_4SiO_4); however, some of the silica may separate as colloidal or amorphous SiO_2 and be left behind during weathering to combine with aluminum to form clay minerals. Hydrolysis is the primary process by which silicate minerals decompose during weathering. A more rigorous and detailed discussion of this process is given by Nahon (1991, p. 7).

Oxidation and Reduction. Chemical alteration of iron and manganese in silicate minerals such as biotite and pyroxenes, caused by oxygen dissolved in water, is an important weathering process because of the abundance of iron in the common rock-forming silicate minerals. An electron is lost from iron during oxidation ($Fe^{2+} \rightarrow Fe^{3+} + e^-$, where e^- = electron transfer), which causes loss of other cations such as Si^{4+} from crystal lattices to maintain electrical neutrality. Cation loss leaves vacancies in the crystal lattice that either bring about the collapse of the lattice or make the mineral more susceptible to attack by other weathering processes. Oxidation of manganese minerals to form oxides and silicic acid or other soluble products is a less important but common weathering process. Another element that oxidizes during weathering is sulfur. For example, pyrite (FeS_2) is oxidized to form hematite (Fe_2O_3), with release of soluble sulfate ions. Under some conditions where material undergoing weathering is water saturated, oxygen supply may be low and oxygen demand by organisms high. These conditions can bring about **reduction** of iron (gain of an electron) from Fe^{3+} to Fe^{2+}. Ferrous iron (Fe^{2+}) is more soluble, and thus more mobile, than ferric iron (Fe^{3+}) and may be lost from the weathering system in solution.

Other Chemical Weathering Processes. Although simple solution, hydrolysis, and oxidation are the most important chemical weathering processes, under certain conditions several other processes can facilitate chemical weathering of minerals. **Hydration** is the process whereby water molecules are added to a mineral to form a new mineral. Common examples of hydration are the addition of water to hematite to form goethite, or to anhydrite to form gypsum. Hydration is accompanied by volume changes that may lead to physical disruption of rocks. Under some conditions, hydrated minerals may lose their water, a process called **dehydration**, and be converted to the anhydrous forms, with accompanying decrease in mineral volume. Dehydration is relatively uncommon in the weathering environment because some water is generally present.

Ion exchange is a process whereby ions in a mineral are exchanged with ions in solution; for example, the exchange of sodium for calcium. Most ion exchange takes place between cations (positively charged ions), but anion exchange also occurs. This reaction causes one mineral to be altered to another (new) mineral and, in the process, releases soluble ions into solution. Ion exchange is particularly important in alteration of one clay mineral to another (e.g., alteration of smectite to illite). Ion exchange also plays a role in alteration of one kind of zeolite to another (e.g., alteration of heulandite, a Ca-zeolite to analcime, a Na-zeolite).

Chelation involves the bonding of metal ions to organic substances to form organic molecules having a ring structure (e.g., Boggs, Livermore, and Seitz, 1985). During weathering, chelation (i.e., organic complexing) performs the dual role of removing cations from mineral lattices and also keeping the cations in solution until they are removed from the weathering site. Chelated metal ions will remain in solution under pH conditions and at concentrations at which nonchelated ions would normally be precipitated. The bonding of aluminum or iron with a complexing agent, and subsequent removal of these elements from a rock, is of particular importance. A good example of natural chelation is provided by lichens that increase the rate of chemical weathering on rock surfaces on which they grow by secreting organic chelating agents. In addition to their role as chelating agents, plants also enhance chemical weathering processes by retaining soil moisture and by acidifying waters by release of CO_2 and various types of organic acids during decay.

Weathering Rates

Determining the rate at which weathering takes place is a difficult and uncertain task. Various techniques are used to evaluate weathering rates: estimating the rate at which the landscape is lowered, estimating the rate at which bedrock is converted into soil, estimating the volume of solid detritus removed from weathering sites by streams, and making chemical mass-balance calculations to evaluate the amount of soluble material removed in surface water and groundwater. Weathering processes proceed at different rates depending upon the climate and the mineral composition and grain size of the rocks undergoing weathering. Physical weathering processes may be quite effective in moderately cold climates (freeze-thaw) or arid climates (salt weathering), whereas chemical weathering processes are accelerated in humid, hot climates. Average rainfall is known to be a controlling factor in the rate of chemical weathering (Nahon, 1991, p. 4); however, the influence of temperature on weathering rate is difficult to quantify although we know that the rate of chemical reactions is accelerated by increasing temperature. Slope of the land surface is also important. Weathering tends to be more effective on low to moderate slopes as compared to steep slopes. Water is more likely to be retained on low slopes, and material undergoing weathering remains for a longer time before being removed by erosion.

The rate of weathering of silicate rocks, such as granite and gneiss, of a given grain size may be related to the relative chemical stabilities of the common rock-forming silicate minerals. Table 1.2 shows the order of relative stability to weathering of the most important mafic and felsic minerals, as determined by Goldich (1938) through empirical study of sand- and silt-size particles in soil profiles. Readers will recognize this order as the same as that in which minerals crystallize in Bowen's reaction series. Minerals that crystallize at high temperatures (e.g., olivine) have the greatest degree of disequilibrium with surface weathering temperatures and thus tend to be less stable than minerals that crystallize at lower temperatures (e.g., quartz). Furthermore, the high-temperature minerals are bonded with weaker ionic or ionic-covalent bonds, whereas quartz is bonded with strong covalent bonds. Jackson (1968) suggests that the stability of very fine size (clay-size) particles may differ somewhat from that of larger particles (Table 1.2).

Rates of weathering must take into account both physical and chemical processes, and they are very likely to be site specific. Therefore, it is probably unwise to generalize too much about weathering rates. In particular, there is no rule of weathering susceptibility that can be applied generally to sedimentary rocks. Rates of weathering of these rocks are a function of the mineralogy, the amount and type of cement in the rocks, and the climate. For example, limestones weather rapidly by solution in wet climates and much more slowly in very arid or very cold climates. Quartz-rich sandstones cemented with silica cement weather very

Table 1.2 Relative stability of common sand-size minerals and various clay-size minerals under conditions of weathering

Sand- and silt-size minerals*		Clay-size minerals**
Mafic minerals	Felsic minerals	1. Gypsum, halite
Olivine		2. Calcite, dolomite, apatite
		3. Olivine, amphiboles, pyroxenes
	Ca plagioclase	4. Biotite
Pyroxene		5. Na plagioclase, Ca plagioclase, K-feldspar, volcanic glass
	Ca-Na plagioclase	6. Quartz
Amphibole	Na-Ca plagioclase	7. Muscovite
	Na plagioclase	8. Vermiculite (clay mineral)
		9. Smectite (clay mineral)
Biotite		10. Pedogenic (soil) chlorite
	K-feldspar,	11. Allophane (clay mineral)
	muscovite, quartz	12. Kaolinite, halloysite (clay minerals)
		13. Gibbsite, boehmite (clay minerals)
		14. Hematite, goethite, magnetite
▼(Increasing stability)		15. Anatase, titanite, rutile, ilmenite (all, titanium-bearing minerals), zircon

Source: *Goldich (1938); ** Jackson (1968).

slowly under most climatic conditions. Finally, it is likely that rates of weathering have varied throughout geologic time depending upon climatic conditions and vegetative cover. Prior to the development of land plants in early Paleozoic time, absence of plant cover to hold soil moisture and contribute organic acids probably slowed rates of chemical weathering while contributing to increased rates of physical erosion.

Products of Subaerial Weathering

Subaerial weathering generates three types of weathering products that are important to the formation of sedimentary rocks (Table 1.3): (1) source-rock residues consisting of chemically resistant minerals and rock fragments derived particularly from siliceous rocks such as granite, rhyolite, gneiss, and schist, (2) secondary minerals formed *in situ* by chemical recombination and crystallization, largely as a result of hydrolysis and oxidation, and (3) soluble constituents released from parent rocks mainly by hydrolysis and solution. Until they are removed by erosion, residues and secondary minerals accumulate at the weathering site to form a soil mantle composed of particles of various compositions and of grain sizes ranging from clay to gravel. Grain size and composition depend upon the grain size and composition of the parent rock and upon the nature and intensity of the weathering process. These characteristics of the weathering environment are in turn functions of climate, topography, and duration of the weathering process.

Source Rock Residues

The residual particles in young or immature soils developed on igneous or metamorphic rocks may include, in addition to rock fragments, assemblages of minerals with low chemical stability: e.g., biotite, pyroxenes, hornblende, and calcic plagioclase. Mature soils, developed after more prolonged or intensive weathering of these rocks, commonly contain only the most stable minerals: quartz, muscovite,

Table 1.3   Principal kinds of products formed by subaerial weathering processes and the types of sedimentary rocks ultimately formed from these products

Weathering process	Type of weathering product	Example	Ultimate depositional product
Physical weathering	Particulate residues	Silicate minerals such as quartz and feldspar; all types of rock fragments	Sandstones, conglomerates, mudrocks
Chemical weathering			
Hydrolysis	Soluble constituents	Silicic acid (H_4SiO_4); K^+, Na^+, Mg^{2+}, Ca^{2+}, etc.	Cherts, limestones, etc.
	Secondary minerals	Clay minerals	Mudrocks (shales)
Simple solution	Soluble constituents	Silicic acid; K^+, Na^+, Mg^{2+}, Ca^{2+}, HCO_3^-, SO_4^{2-}, etc.	Limestones, evaporites, chert, etc.
Oxidation	Secondary minerals	Ferric oxides (Fe_2OOH); manganese oxides (MnO_2)	Minor constituent in siliciclastic rocks
	Soluble constituents	Silicic acid; SO_4^{2-}	Chert, evaporites, etc.

and perhaps potassium feldspars. Because the silicate minerals that make up siliciclastic sedimentary rocks such as sandstones have already passed through a weathering cycle before the siliciclastic rocks were formed, the weathering products of these rocks tend to be depleted in easily weathered minerals. Thus, even young soils developed on siliciclastic sedimentary rocks may have assemblages of mature minerals. Weathering of limestones by solution produces thin soils composed of the fine-size insoluble silicate and iron oxide residues of these rocks.

Secondary Minerals

Secondary minerals developed at the weathering site are dominantly clay minerals, iron oxides or hydroxides, and aluminum hydroxides. The common secondary iron minerals include goethite, limonite, and hematite. The weathering products reflect both the nature and the intensity of the weathering process and the composition of the parent rock. Clay minerals formed in immature soils under only moderately intense chemical weathering conditions may be illites or smectites. More prolonged and intense leaching conditions lead to formation of kaolinite. Under extremely intense chemical weathering conditions, aluminum hydroxides such as gibbsite and diaspore are formed. These latter clay minerals are aluminum ores.

Comparing the chemical composition of unweathered silicate rocks with that of the weathering products of these rocks shows a net loss attributed to weathering of all major cations except aluminum and iron (e.g., Krauskopf, 1979). In the oxidized state, aluminum and ferric iron (Fe^{3+}) are both relatively insoluble. Although considerable silica is lost as soluble silicic acid during weathering, loss of Mg, Ca, Na, and K is comparatively much greater. Therefore, the relative abundance of silica, aluminum, and ferric iron in the particulate weathering residues of silicate rocks is greater than that in the parent source rocks.

Soluble Materials

Soluble materials extracted from parent rocks by chemical weathering are removed from the weathering site in surface water or soil groundwater more or less continuously throughout the weathering process. Ultimately these soluble products make their way into rivers and are carried to the ocean. The most abundant inorganic constituents of rivers, representing the principal soluble products of weathering, are, in order of decreasing abundance, HCO_3^- (bicarbonate), Ca^{2+}, H_4SiO_4 (silicic acid), SO_4^{2-} (sulfate), Cl^-, Na^+, Mg^{2+}, and K^+ (Garrels and McKenzie, 1971). These constituents are the raw materials from which chemically and biochemically deposited rocks such as limestones and cherts are formed in the oceans.

1.3 SUBMARINE WEATHERING PROCESSES AND PRODUCTS

Although we commonly think of weathering as being a subaerial process, an important kind of weathering also takes place on the ocean floor. Geologists have long recognized that sediments and rocks on the seafloor are altered by reaction with seawater, a process called **halmyrolysis** or submarine weathering. Halmyrolysis includes alteration of clay minerals of one type to another, formation of glauconite from feldspars and micas, and formation of phillipsite (a zeolite mineral) and palagonite (altered volcanic glass) from volcanic ash. Dissolution of the siliceous and calcareous tests of organisms may also be considered a type of submarine weathering. Prior to the 1970s, submarine weathering processes had not received a great deal of research, and it was not recognized that they might have a significant effect on the overall chemical composition of the oceans. Our concept of the importance of submarine weathering has changed dramatically since the mid-1970s because studies of volcanic rocks and weathering processes on the seafloor show that submarine weathering of basalts, particularly on mid-ocean ridges, is an extremely important chemical phenomenon. This process results in both widespread hydration and leaching of basalts as well as changes in composition of seawater owing to ion exchange during the reaction of seawater with basalt.

Alteration of oceanic rocks occurs both at low temperatures (less than 20°C) and at higher temperatures ranging to ~350°C. Low-temperature alteration takes place as seawater percolates through fractures and voids in the upper part of the ocean crust, perhaps extending to depths of 2–5 km. Olivine and interstitial glass in the basalts are replaced by smectite clay minerals, and further alteration may lead to formation of zeolite minerals and chlorite. As a result of these changes, chemical elements are exchanged between rock and water, and large volumes of seawater become fixed in the oceanic crust in hydrous clay minerals and zeolites.

The discovery in 1977 of submarine thermal springs along the Galapagos Rift (Corliss et al., 1979) led to the awareness that large-scale hydrothermal activity takes place in the ocean. Since that initial discovery, scientists using submersible vehicles and water-sampling techniques have located many additional hot springs along mid-ocean ridges in both the Pacific and Atlantic oceans, as well as along convergent plate margins, in back-arc basins, and even on mid-plate volcanoes in the Hawaiian chain (e.g., Karl et al., 1988; Parson, Walker, and Dixon, 1995). These hot springs originate where seawater enters the ocean crust along fractures or other voids and comes in contact with hot volcanic rock. The heated water then flows out into the ocean through vents on the ocean floor and mixes with the overlying water. The heated water rises as hydrothermal plumes 100–300 m above the vent field. Exceptional plumes rising to heights of 1000 m have also been reported (e.g., Cann and Strens, 1989).

At the sites of many oceanic hot springs, investigators have found spectacular vents composed of sulfide, sulfate, and oxide deposits up to 10 m or more tall that discharge plumes of hot solutions (Fig. 1.3). These vents or chimneys are called **black smokers** if they discharge water containing suspended, fine-grained,

Figure 1.3
A multiple-orifice black smoker, Faulty Towers complex, Mothra hydrothermal vent field, Endeavour Segment, Juan de Fuca ridge. The constructional chimneys in the foreground were built by precipitation of sulfides and other minerals from heated water issuing from the vents at temperatures exceeding 250°C. [Photograph courtesy of John R. Delaney and Deborah S. Kelley, University of Washington School of Oceanography.]

dark-colored minerals or **white smokers** if the water contains no suspended dark minerals (McDonald, Spiess, and Ballard, 1980). The temperature of the water when it emerges from the vents may exceed 350°C. When these hot solutions mix with seawater of ambient temperature, they precipitate various minerals, particularly pyrite (FeS_2) and chalcopyrite ($CuFeS_2$), to build sulfide deposits around the vents. The deposits of fossil hydrothermal systems have now been observed in ancient oceanic ophiolite complexes exposed on land (e.g., Cann and Strens, 1989).

Reactions between hot basalt and seawater play a role in regulating the chemical composition of seawater. Magnesium, sulfate, and sodium ions are removed from seawater during this exchange, whereas many other elements such as calcium, iron, manganese, silicon, potassium, lithium, and strontium are enriched in the seawater (Edmond et al., 1982; Palmer and Edmond, 1989; Von Damm, 1990). The entire ocean apparently circulates through ocean-floor hydrothermal systems on a time scale of 10^6–10^7 years, which has a significant impact on the budget of several elements, including silica (Kadko et al., 1995).

The magnitude of hydrothermal alteration of basalts along mid-ocean ridges and its effect on ocean chemistry is still being investigated and uncertainties remain; however, it now appears that circulation of ocean water through hydrothermal systems throughout geologic time has added significant quantities of certain ions to the ocean, while removing others. Thus, both seafloor hydrothermal reactions and continental weathering processes supply the ocean with ions that may eventually be extracted to form chemically deposited rocks such as limestones, iron-rich sedimentary rocks, and cherts. Stanley and Hardie (1999) argue that changes in spreading rates along mid-ocean ridges, where hydrothermal activity takes place, have exerted a major control on the calcium and magnesium content of seawater throughout geologic time. High spreading rates result in significant adsorption and loss of magnesium with concomitant increase in calcium, thus causing a decrease in the ratio of magnesium to calcium (Mg/Ca). Low spreading rates have the opposite effect of increasing the Mg/Ca ratio. As discussed in Chapters 6 and 11, these changes have important implications regarding the kinds of calcium-carbonate minerals deposited in the ocean.

1.4 SOILS

Considered from the standpoint of sedimentary-rock origin, we are perhaps more interested in the products of weathering than in the processes that bring about weathering, although it is useful for students to understand just how weathering processes operate to generate these products. The materials that make up sedimentary rocks are either siliciclastic grains derived from the land as a result of weathering (or explosive volcanism in some cases) or they are so-called "chemical" minerals that were precipitated from ocean or lake water. The elements that make up these chemical minerals were released from parent rocks by chemical weathering processes operating on land and in the ocean. Thus, it is quite reasonable to consider that the generation of both siliciclastic and chemical/biochemical sedimentary rocks begins with weathering.

Subaerial weathering products initially form soils of varied thickness over weathered bedrock. Throughout geologic time, most of these soils have ultimately been stripped away and transported as sediment to sedimentary basins; however, some soils are preserved to become part of the sedimentary record. Thus, because soils represent an incipient stage in the generation of siliciclastic sedimentary rocks and some are preserved in their own right, a discussion of soil-forming processes and the various kinds of soils that result from these processes is pertinent.

Soil-Forming Processes

Subaerial weathering processes generate a mantle of soil above bedrock. The characteristics and thickness of this soil mantle are a function of the bedrock lithology, the climate (rainfall, temperature), and the slope of the bedrock surface. These factors govern the intensity of weathering and determine which minerals survive to become part of the soil profile, what new minerals are created in the soil, and the length of time soil materials remain before being eroded and transported to depositional basins. On very steep slopes, for example, the weathered mantle may be removed so rapidly by erosion that little soil accumulates.

In addition to the chemical and physical weathering processes that cause the breakdown of bedrock to form soils, several other biologic and chemical processes operate within soils over time to modify their characteristics (e.g., Birkland, 1999, p. 105):

1. *Additions to the ground surface*—Precipitation of dissolved ions in rainwater; influx of solid particles such as windblown dust; addition of organic matter from surface vegetation
2. *Transformations*
   a. Decomposition of organic matter within soils to produce organic compounds
   b. Weathering of primary minerals; formation of secondary minerals, including iron oxides

   3. *Transfers*
      a. Movement of solid or suspended material downward from one soil horizon to a lower horizon by groundwater percolation (eluviation)
      b. Accumulation of soluble or suspended material in a lower horizon (illuviation)
      c. Transfer of ions upward by capillary movement of water and precipitation of ions in the soil profile
   4. *Removals*—Removal of substances still in solution to become part of the dissolved constituents in groundwater or surface water
   5. *Bioturbation of soil*—Soil disruption by animals (e.g., ants, termites) and plants.

This list of soil-forming processes is highly simplified. Buol et al. (1997, p. 112) recognize and define more than two dozen soil-forming processes. These processes generate distinct soil horizons, which are collectively referred to as the soil profile. Further details of soil-forming processes may be found in additional readings listed at the end of this chapter.

Soil Profiles and Soil Classification

Soils are classified on the basis of the characteristic horizontal layers or horizons that are visible in road cuts, pits, and so on. The thickness and nature of these **soil horizons** are determined by the various soil-forming processes mentioned and may vary widely. Soil profiles can be divided crudely into five *major* horizons: O, A, E, B, and C. The **O horizon** is the surface accumulation of mainly organic matter. The **A horizon**, which occurs at the surface or below the O horizon, consists of a dark-colored accumulation of organic matter (e.g., leaf litter) that is decaying and mixing with mineral soil. The **E horizon**, which underlies an O or A horizon, is a light colored eluvial horizon (a horizon from which material was removed by downward movement) characterized by less organic matter, fewer iron and aluminum compounds, and/or less clay than the underlying horizon. The **B horizon** underlies an O, A, or E horizon and may contain illuvial (added material derived from an upper horizon) concentrations of fine organic matter, clay, etc.; most of the original rock structures have been obliterated by soil-forming processes. The **C horizon,** which lies above bedrock, is partly altered bedrock that can be deeply weathered but is relatively unaffected by soil-forming processes. Studies of soil profiles show, however, that soil layers are commonly much more complex than indicated by this simple scheme. As many as twenty-four different kinds of soil horizons have been described (e.g., Birkland, 1999, p. 5).

Numerous systems for more detailed classification of soils are in existence: e.g., the Australian handbook classification, the U.S. Soil Taxonomy classification, and the FAO (UNESCO) world map classification (Eswaran et al., 2003). One of the more widely used soil classifications in the United States appears in *Soil taxonomy: A basic system for making independent soil surveys, 2nd ed.* (Soil Survey Staff, 1999), which recognizes twelve major classes or orders of soils with names such as aridosol (soils of arid regions) and ultisol (leached soils of warm, humid regions). These soil types are differentiated on the basis of a variety of complex criteria, such as the amount of contained organic material, the presence of clay layers, and the presence of oxic (iron-rich) horizons.

The factors that influence soil formation, and thus the kinds of soils that form, include the parent rock material, length of the soil-forming process, climate (e.g., wet or dry), topography (steep or gentle slopes), and organisms (vegetation cover and soil fauna such as earthworms). Climate plays a particularly important role in soil formation.

Paleosols

In the context of this book, we are concerned primarily with ancient soils, called paleosols, rather than modern soils. Paleosols, sometimes referred to as *fossil soils*, are buried soils or horizons of the geologic past. Most soil horizons that developed in the past on elevated landscapes were eventually destroyed as erosion lowered the landscape. Nonetheless, some soils, presumably those formed mainly in low-lying areas, escaped erosion to become part of the stratigraphic record. Quaternary soils that formed particularly on glacial or fluvial deposits are most common (e.g., Catt, 1986). Such soils that have not been buried are called relict soils. Many buried soils of Quaternary and much older age are also known. Old paleosols occur in the stratigraphic record at major unconformities, including unconformities in Precambrian rocks, where their presence may reflect the combined processes of soil formation, erosional landscape lowering, reorganization of preexisting soil horizons, and changing flow of groundwater (Retallack, 1990, p. 14). Paleosols are also present as interbeds in sedimentary successions, particularly in alluvial successions, that are at least as old as the Ordovician (e.g., Reinhardt and Sigleo, 1988). Geologists are becoming increasingly interested in paleosols as indicators of paleoenvironments and ancient climatic conditions.

Recognition of Paleosols

Because interbedded paleosols in sedimentary successions superficially resemble sediments or sedimentary rocks, many paleosols have unquestionably gone unrecognized in the past. Many of us have simply identified them as gray, red, or green mudstones. As awareness of paleosols has increased, however, more and more paleosols are being recognized. How can the ordinary geologist, not specifically trained in soil science, recognize paleosols in the field? Retallack (1988, 1997) suggests three principal kinds of diagnostic characteristics of paleosols that help distinguish them from sedimentary rocks: traces of life, soil horizons, and soil structure (Fig. 1.4).

Root traces are the most important **traces of life** preserved in paleosols. Root traces provide diagnostic evidence that rock was exposed to the atmosphere and colonized by plants, thus forming a soil. The top of a paleosol is the surface from which root traces emanate. Root traces mostly taper and branch downward (Fig. 1.5), which helps to distinguish them from burrows. On the other hand, some root traces spread laterally over hardpans in soils, and some kinds branch upward and out of the soil. Root traces are most easily recognized when their original organic matter is preserved, which occurs mostly in paleosols formed in waterlogged, anoxic lowland environments. Root traces in red, oxidized paleosols consist mainly of tubular features filled with material different from the surrounding paleosol matrix.

The presence of **soil horizons** is a second general feature of paleosols. The top of the uppermost horizon of a paleosol is commonly sharply truncated by an erosional surface, but soil horizons typically show gradational changes in texture, color, or mineral content downward into the parent material. Differences in grain size, color, reaction with weak hydrochloric acid (to test for the presence of carbonates), and the nature of the boundaries must all be examined to detect soil horizons (Retallack, 1988). Comparison with modern soil horizons aids in recognition.

Bioturbation (disruption) by plants and animals, wetting and drying, and other soil-forming processes cause paleosols to develop characteristic **soil structures** at the expense of the original bedding and structures in the parent rock. One of the characteristic kinds of soil structure is a network of irregular planes (called **cutans**) surrounded by more stable aggregates of soil material called **peds**. This structure gives a hackly appearance to the soil. Peds occur in a variety of sizes and

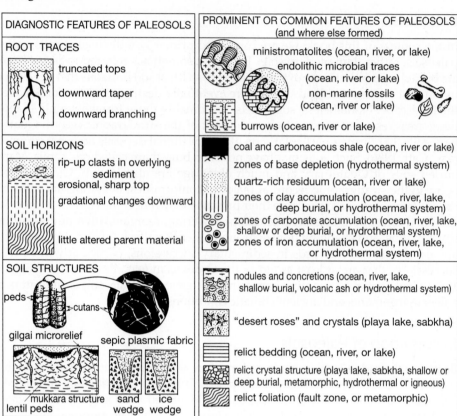

DIAGNOSTIC FEATURES OF PALEOSOLS	PROMINENT OR COMMON FEATURES OF PALEOSOLS (and where else formed)
ROOT TRACES truncated tops / downward taper / downward branching	ministromatolites (ocean, river, or lake) / endolithic microbial traces (ocean, river or lake) / non-marine fossils (ocean, river or lake) / burrows (ocean, river or lake)
SOIL HORIZONS rip-up clasts in overlying sediment / erosional, sharp top / gradational changes downward / little altered parent material	coal and carbonaceous shale (ocean, river or lake) / zones of base depletion (hydrothermal system) / quartz-rich residuum (ocean, river or lake) / zones of clay accumulation (ocean, river, lake, deep burial, or hydrothermal system) / zones of carbonate accumulation (ocean, river, lake, shallow or deep burial, or hydrothermal system) / zones of iron accumulation (ocean, river, lake, or hydrothermal system)
SOIL STRUCTURES peds / cutans / sepic plasmic fabric / gilgai microrelief / mukkara structure / lentil peds / sand wedge / ice wedge	nodules and concretions (ocean, river, lake, shallow burial, volcanic ash or hydrothermal system) / "desert roses" and crystals (playa lake, sabkha) / relict bedding (ocean, river, or lake) / relict crystal structure (playa lake, sabkha, shallow or deep burial, metamorphic, hydrothermal or igneous) / relict foliation (fault zone, or metamorphic)

Figure 1.4
Characteristic and common features useful in recognition of paleosols. [From Retallack, G. J., 1992, How to find a Precambrian paleosol, *in* Schidlowski, M., et al. (eds.), Early organic evolution: Implications for mineral and energy resources, Springer-Verlag, Berlin, Fig. 10, p. 27, reproduced by permission.]

Figure 1.5
An example of root traces in a paleosol. The original organic matter has been partially replaced by iron oxides. Early Miocene, Molalla Formation, western Oregon. [Photograph courtesy of G. J. Retallack.]

TYPE	PLATY	PRISMATIC	COLUMNAR	ANGULAR BLOCKY	SUBANGULAR BLOCKY	GRANULAR	CRUMB
SKETCH	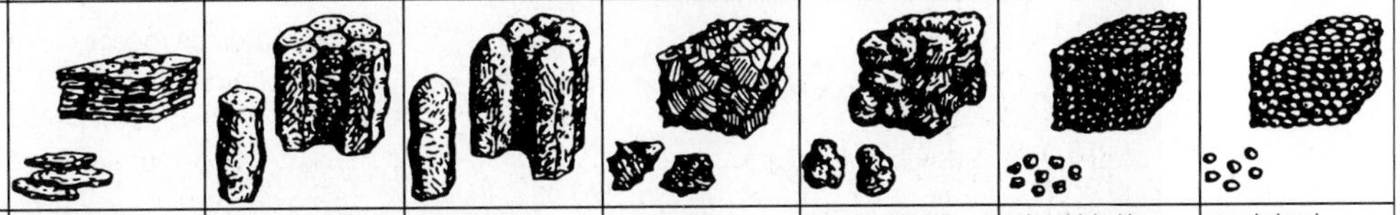						
DESCRIPTION	tabular and horizontal to land surface	elongate with flat top and vertical to land surface	elongate with domed top and vertical to surface	equant with sharp interlocking edges	equant with dull interlocking edges	spheroidal with slightly interlocking edges	rounded and spheroidal but not interlocking
USUAL HORIZON	E,Bs,K,C	Bt	Bn	Bt	Bt	A	A
MAIN LIKELY CAUSES	initial disruption of relict bedding; accretion of cementing material	swelling and shrinking on wetting and drying	as for prismatic, but with greater erosion by percolating water, and greater swelling of clay	cracking around roots and burrows, swelling and shrinking on wetting and drying	as for angular blocky, but with more erosion and deposition of material in cracks	active bioturbation and coating of soil with films of clay, sesquioxides, and organic matter	as for granular; including fecal pellets and relict soil clasts
SIZE CLASS	very thin < 1 mm	very fine < 1 cm	very fine < 1 cm	very fine < 0.5 cm	very fine < 0.5 cm	very fine < 1 mm	very fine < 1 mm
	thin 1 to 2 mm	fine 1 to 2 cm	fine 1 to 2 cm	fine 0.5 to 1 cm	fine 0.5 to 1 cm	fine 1 to 2 mm	fine 1 to 2 mm
	medium 2 to 5 mm	medium 2 to 5 cm	medium 2 to 5 cm	medium 1 to 2 cm	medium 1 to 2 cm	medium 2 to 5 mm	medium 2 to 5 mm
	thick 5 to 10 mm	coarse 5 to 10 cm	coarse 5 to 10 cm	coarse 2 to 5 cm	coarse 2 to 5 cm	coarse 5 to 10 mm	not found
	very thick > 10 mm	very coarse > 10 cm	very coarse > 10 cm	very coarse > 5 cm	very coarse > 5 cm	very coarse > 10 mm	not found

Figure 1.6
Characteristics of various kinds of soil peds. [From Retallack, G. J., 1988, *in* Reinhardt, J., and W. R. Sigleo (eds.), Field recognition of paleosols: Geol. Soc. America Spec. Paper 216, Fig. 9, p. 216. Reproduced by permission of Geol. Soc. America, Boulder, Colo.]

shapes (Fig. 1.6). Their recognition in the field depends upon recognition of the cutans that bound them, which commonly form clay skins around the peds. Other kinds of soil structure include concentrations of specific minerals that form hard, distinct, calcareous, ferruginous, or sideritic lumps called **glaebules** (a general term including nodules and concretions). More diffuse, irregular, or weakly mineralized concentrations are called mottles. Figure 1.7 shows the field appearance of some Miocene paleosols. These paleosols are red; however, paleosols can have a variety of colors and properties (Retallack, 1997).

Paleosols can be recognized to have characteristics similar to those of modern soils; thus, U.S. Soil Taxonomy names such as aridosol and ultisol can be applied to paleosols (e.g., Retallack, 1992). Because the characteristics of paleosols reflect the conditions under which they formed, including climatic conditions, the study of paleosols is an important tool in paleoenvironmental analysis. For example, aridosols suggest formation under desert conditions whereas ultisols reflect weathering under warm, moist conditions. Clearly, the processes of weathering that lead to generation of sedimentary particles and soil formation are intimately tied up with climatic conditions. Weathering did not begin on Earth until an atmosphere containing water vapor and carbon dioxide had accumulated sometime during the early Precambrian; subsequent addition of oxygen also had an important bearing on the weathering processes. Geologists are becoming increasingly aware of the need to study Earth's past climates (paleoclimatology).

This short, generalized description of paleosols is intended only to pique reader interest in fossil soils. Several of the books listed under Further Readings at the end of the chapter provide further details.

1.5 CONCLUDING REMARKS

The processes that form sedimentary rocks can be considered to begin with weathering, a process strongly influenced by climatic conditions. Weathering brings

Figure 1.7
Red paleosols exposed below bedded sandstones in the Middle Miocene, Chinji Formation, Siwalik Group, in a creek bed 3 km south of Khaur, Potwar Plateau, Pakistan. The hammer is 25 cm long. From G. J. Retallack, 1997, [A colour guide to paleosols.] Chichester

about the breakdown of older rocks exposed in upland areas to yield soluble ions, which are transported to the ocean in solution, and insoluble, chemically resistant minerals such as quartz that may accumulate at the weathering site for a time as soils. Soil formation, like weathering, is intimately related to climatic conditions. Some soils, called paleosols, are preserved to become part of the sedimentary record; however, most insoluble soil materials are removed by erosion and transported by gravity processes, water, glaciers, or wind to basins at lower elevations, where deposition takes place. Succeeding chapters of this book describe the processes of sediment transport, deposition, and burial that ultimately result in the generation of lithified sedimentary rocks.

FURTHER READINGS

Weathering

Bland, W., and D. Rolls, 1998, Weathering: An introduction to the scientific principles: Oxford University Press Inc., New York, 271 p.

Humphris, S. E., R. A. Zierenberg, L. S. Mullineaux, and R. E. Thompson (eds.), 1995, Seafloor hydrothermal systems, Geophysical Monograph 91, American Geophysical Union, Washington, D.C., 446 p.

Martini, I. P., and W. Chesworth (eds.), 1992, Weathering, soils & paleosols: Elsevier, Amsterdam, 618 p.

Nahon, D. B., 1991, Introduction to the petrology of soils and chemical weathering: John Wiley & Sons, New York, 313 p.

Parson, L. M., C. L. Walker, and D. R. Dixon (eds.), 1995, Hydrothermal vents and processes: The Geological Society, London, 411 p.

Robinson, D. A., and R. B. G. Williams (eds.), 1994, Rock weathering and landform evolution: John Wiley & Sons, Chichester, 519 p.

White, A. F., and S. L. Brantley (eds.), 1995, Chemical weathering rates of silicate minerals: Mineralogical Society of America Reviews in Mineralogy, v. 31, 583 p.

Soils and Paleosols

Birkland, P. W., 1999, Soils and geomorphology, 3rd ed.: Oxford University Press, New York, 430 p.

Bronger, A., and J. A. Catt (eds.), 1989, Paleopedology: Nature and application of paleosols: Catena Verlag, Destedt, Germany, 232 p.

Buol, S. W., F. D. Hole, R. J. McCracken, and R. J. Southard, 1997, Soil genesis and classification, 4th ed.: Iowa State University Press, Ames, 527 p.

Meyer, R., 1997, Paleoalterites and paleosols: A. A. Balkema, Rotterdam, 151 p.

Ollier, C., and C. Pain, 1996, Regolith, soils and landforms: John Wiley & Sons, Chichester, 316 p.

Paquet, H., and N. Clauer (eds.), 1997, Soils and sediments: Mineralogy and geochemistry: Springer-Verlag, Berlin, 369 p.

Reinhardt, J., and W. R. Sigleo (eds.), 1988, Paleosols and weathering through geologic time: Principles and applications: Geological Society of America Special Paper 216, 181 p.

Retallack, G. J., 1997, A colour guide to paleosols: John Wiley & Sons, Chichester, 175 p.

Retallack, G. J., 2001, Soils of the past: Blackwell Science, Oxford, 404 p.

2

Transport and Deposition of Siliciclastic Sediment

2.1 INTRODUCTION

Silicate minerals and rock fragments weathered from older rocks on land, together with pyroclastic particles generated by explosive volcanism, are the source materials of siliciclastic sedimentary rocks—conglomerates, sandstones, shales. These materials are eroded from highlands and transported to depositional basins at lower elevations, where they may undergo additional transport before final deposition. Mass-wasting processes such as slides and slumps commonly play an initial role in moving sediment short distances down steep slopes to sites where other transport processes take over. Subsequent transport may involve fluid flows (e.g., moving water) or sediment-gravity flows, such as mud flows, that may behave like fluids. Thus, study of sediment transport requires some understanding of the principles of fluid flow.

The fundamental laws of fluid dynamics are moderately complex when applied to fluid flow alone. These complexities are magnified when particles are entrained in the flow during sediment transport. Sediment transport can take place under a variety of conditions: subaerially by wind and certain kinds of sediment-gravity flows, and subaqueously in rivers, lakes, and the ocean by currents, waves, tides, and sediment-gravity flows.

In this chapter, we investigate sediment transport processes by first examining some of the properties of fluids and the basic concepts of fluid flow and sediment-gravity flow. We then consider the problems involved in entrainment and transport of particles by fluid- and sediment-gravity-flow processes. No attempt is made here to give a comprehensive review of fluid mechanics. Only those concepts of flow that are important to understanding sediment transport and deposition are discussed, and these concepts are presented in very simplified form. More rigorous treatment of fluid dynamics is available in numerous specialized books. Discussions of fluid mechanics particularly pertinent to the problems of sediment transport include those of Middleton and Southard (1984), Allen (1984), Carlin and Dawson (1996), and Leeder (1999). Details of sediment transport peculiar to various depositional environments (e.g., river systems, lakes, the ocean) are discussed in appropriate sections of subsequent chapters.

2.2 FUNDAMENTALS OF FLUID FLOW

Fluids are substances that change shape easily under their own weight. Air, water, and water containing various amounts of suspended sediment are the fluids of interest in sediment transport. The basic physical properties of these fluids are density and viscosity. Differences in these properties markedly affect the ability of fluids to erode and transport sediment.

Fluid **density,** commonly referred to as ρ (rho), is defined as mass per unit fluid volume. Density affects the magnitude of forces that act within a fluid and on the bed as well as the rate at which particles fall or settle through a fluid (slower in denser fluids). Density particularly influences the movement of fluids downslope under the influence of gravity. Density varies with different fluids and increases with decreasing temperature of a fluid. The density of water (0.998 g/mL at 20°C) is more than 700 times greater than that of air. This density difference influences the relative abilities of water and air to transport sediment; e.g., water can transport particles of much larger size than those transported by wind.

Fluid **viscosity** is a measure of the ability of fluids to flow. Put simply, fluids with low viscosity flow readily and fluids with high viscosity flow sluggishly. For example, air has very low viscosity and ice has very high viscosity. Water has low viscosity; honey has high viscosity. The viscosity of water at 20°C is almost 55 times greater than that of air (Blatt, Middleton, and Murray, 1980, p. 91). Like density, viscosity increases with decreasing temperature of the fluid. Viscosity has a particularly important influence on water turbulence. Increasing viscosity tends to suppress turbulence (random movement of water molecules), thereby slowing the rate at which particles settle through water—a significant factor in transport of suspended sediment. Decreased turbulence also reduces the ability of running water to erode and entrain sediment.

Box 2.1 Molecular (Dynamic) Viscosity

For a more rigorous examination of viscosity, consider a simple experiment in which a fluid is trapped between two parallel plates (Fig. 2.1.1). The lower plate is stationary, and the upper plate is moving over it with a constant velocity (V). The fluid can be thought of as forming parallel sheets between the plates. As the upper plate moves over the lower, the fluid in between is put in motion with a velocity that varies linearly from zero at the lower plate to velocity (V) at the upper plate.

The shear applied to the moving plate, which is being transmitted through the fluid to the static plate, is given by the relation:

$$\tau = \mu \frac{du}{dy} \tag{2.1.1}$$

where τ (tau) is the shear stress, u is the fluid velocity, y is the normal distance from the stationary plate, and μ is the molecular viscosity. Shear stress is the shearing force per unit area (e.g., dynes/cm^2) exerted across the shearing surface at some point in a fluid. It acts on the fluid parallel to the surface of the fluid body. Shear stress is generated at the boundary of two moving fluids, and it is a function of the extent to which a slower moving mass retards a faster moving one. Thus, as a faster moving layer moves over a slower moving layer, the shear stress is the force that produces a change in velocity (du) relative to height (dy), the velocity gradient (Fig. 2.1.1).

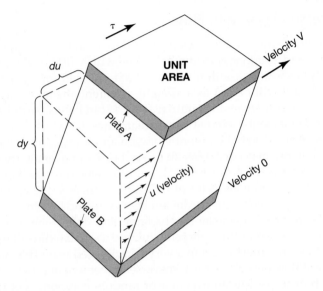

Figure 2.1.1
Geometric representation of the factors that determine fluid viscosity. A fluid is enclosed between two rigid plates, A and B. Plate A moves at a velocity (V) relative to
Plate B. A shear force (τ) acting parallel to the plates creates a steady-state velocity
profile, shown by the inclined line, where fluid velocity (u) is proportional to the length
of the arrows. The shear stress may be thought of as the force that produces a change
in velocity (du) relative to height (dy) as one fluid layer slides over another. The ratio of
shear stress to du/dy is the viscosity (μ).

Molecular (Dynamic) viscosity μ (mu) is the measure of resistance of a substance to change in shape taking place at finite speeds during flow. It is the pro-
portionality factor that links shear stress to the rate of strain, defined as the ratio
of shear stress (τ) to the rate of deformation (du/dy) sustained across the fluid:

$$\mu = \frac{\tau}{du/dy} \tag{2.1.2}$$

The shearing force per unit area needed to produce a given rate of shearing, or a
given velocity gradient normal to the shear planes, is determined by the viscosity—the greater the viscosity the greater the shear stress must be. Viscosity decreases with higher temperature; thus, a given fluid flows more readily at higher
temperatures. Equation 2.1.1 is the equation for a **Newtonian fluid**, a fluid that
does not undergo a change in viscosity as the shear rate increases, e.g., ordinary
water. Because both density and dynamic viscosity strongly affect fluid behavior,
fluid dynamicists commonly combine the two into a single parameter called
kinematic viscosity ν (nu), which is the ratio of dynamic viscosity to density:

$$\nu = \frac{\mu}{\rho} \tag{2.1.3}$$

Kinematic viscosity is an important factor in determining the extent to which
fluid flows exhibit turbulence.
    Note: The relation depicted in Fig. 2.1.1 is a specialized case of shearing in
a fluid confined between two plates. In natural systems, the rate of shearing
and the orientation of the shear planes are likely to vary from point to point;
that is, the velocity profile is curved rather than linear.

Laminar versus Turbulent Flow

Fluids in motion display two modes of flow depending upon the flow velocity, fluid viscosity, and roughness of the bed over which flow takes place. Experiments with dyes show that a thin stream of dye injected into a slowly moving, unidirectional fluid will persist as a straight, coherent stream of nearly constant width. This type of movement is **laminar flow**. It can be visualized as a series of parallel sheets or filaments, referred to as **streamlines**, by which movement occurs on a molecular scale owing to constant vibration and translation of the fluid molecules. The streamlines may curve over an object, but they never intertwine (Fig. 2.1). Laminar flow takes place only at very low fluid velocities over smooth beds. If flow velocity increases or viscosity of the fluid decreases, the dye stream is no longer maintained as a coherent stream but breaks up and becomes highly distorted. It moves as a series of constantly changing and deforming masses in which there is sizable transport of fluid perpendicular to the mean direction of flow; that is, the streamlines are intertwined in a very complicated way. This type of flow is called **turbulent** flow because of the transverse movement of these masses of fluid (Fig. 2.1). Turbulence is thus an irregular or random component of fluid motion. Highly turbulent water masses are referred to as **eddies**. Most flow of water and air under natural conditions is turbulent, although flow of ice and mud flows (Section 2.3) are essentially laminar.

The upward motion of water particles in turbulent water masses slows the fall rate of settling particles and, thus, decreases their settling velocity. Also, fluid turbulence tends to increase the effectiveness of fluid masses in eroding and entraining particles from a sediment bed. Because of the significance of turbulence in sediment transport, it is important to develop a fuller understanding of this property. Velocity measured over a period of time at a particular point in a laminar flow is constant. By contrast, velocity measured at a point in turbulent flow tends toward an average value when measured over a long period of time, but it varies from instant to instant about this average value. As we shall see, a calculated variable

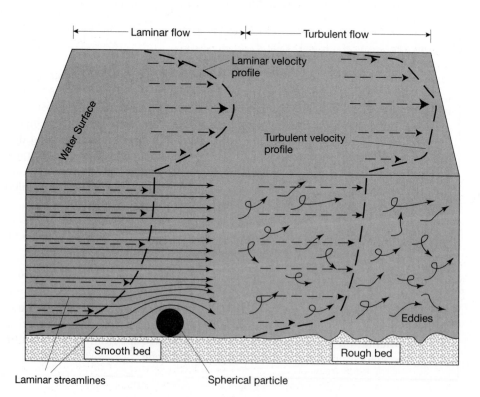

Figure 2.1
Schematic representation of laminar and turbulent fluid flow. Compare streamline flow under laminar flow conditions to the chaotic pattern of flow under turbulent conditions. Also, compare the shape of the laminar velocity profile to that of the turbulent velocity profile (heavy dashed lines).

called the Reynolds number can be used to predict the boundary conditions separating laminar and turbulent flow. Turbulent flow resists distortion to a much greater degree than does laminar flow. Thus, a fluid undergoing turbulent flow appears to have a higher viscosity than the same fluid undergoing laminar flow. As mentioned, this apparent viscosity, which varies with the character of the turbulence, is called **eddy viscosity**. Eddy viscosity results from turbulent momentum, and it is the rate of exchange of fluid mass between adjacent water bodies. It is necessary in dealing with fluids undergoing turbulence to rewrite the equation for shear stress to include a term for eddy viscosity. Thus, for laminar flow, shear stress is given by the relation shown in Equation 2.1.1; however, for turbulent flow the relationship becomes

$$\tau = (\mu + \eta)\frac{du}{dy} \qquad (2.1)$$

where η (eta) is eddy viscosity, which is commonly several orders of magnitude higher than dynamic viscosity. For a more rigorous discussion of turbulence, see Middleton and Wilcock (1994) and Williams (1996).

Reynolds Number

The fundamental differences in laminar and turbulent flow arise from the ratio of inertial forces to viscous forces. Inertial forces, which are related to the scale and velocity of fluids in motion, tend to cause fluid turbulence. Viscous forces, which increase with increasing viscosity of a fluid, resist deformation of a fluid and thus tend to suppress turbulence. The relationship of inertial to viscous forces can be shown mathematically by a dimensionless value called the **Reynolds number** (R_e), which is expressed as

$$R_e = \frac{UL\rho}{\mu} = \frac{UL}{\nu} \qquad (2.2)$$

where U is the mean velocity of flow, L is some length (commonly water depth) that characterizes the scale of flow, and ν is kinematic viscosity. When viscous forces dominate, as in highly concentrated mud flows, Reynolds numbers are small and flow is laminar. Very low flow velocity or shallow depth also produces low Reynolds numbers and laminar flow. When inertial forces dominate and flow velocity increases, as in the atmosphere and most flow in rivers, Reynolds numbers are large and flow is turbulent. Thus, as mentioned, most flow under natural conditions is turbulent. Note from Equation 2.2 that an increase in viscosity can have the same effects as a decrease in flow velocity or flow depth. The transition from laminar flow to turbulent flow takes place above a critical value of Reynolds number, which commonly lies between 500 and 2000 and which depends upon the boundary conditions such as channel depth and geometry. Thus, under a given set of boundary conditions, the Reynolds number can be used to predict whether flow will be laminar or turbulent and to derive some idea of the magnitude of turbulence. Because the Reynolds number is dimensionless, it is of particular value when used to compare scaled-down models of flow systems with natural flow systems, as in modeling natural systems.

Boundary Layers and Velocity Profiles

When a fluid flows over a solid surface (boundary) such as a streambed, flow in the immediate vicinity of the boundary is retarded by the frictional resistance of the boundary. This zone of retardation is called a **boundary layer**. The boundary layer is the region of fluid flow next to the boundary across which the fluid velocity grades from that of the boundary (commonly zero) to that of the unaffected part of the flow (see velocity profiles in Fig. 2.1). Boundary layers may be comparatively thin or they may extend all the way to the free surface of a flow. Also, flow within boundary layers may be laminar or turbulent or may grade from laminar to turbulent.

Because of the greater shear stress required to maintain a particular velocity gradient in turbulent flow, turbulent flow-velocity profiles, both vertically in the flow and horizontally across the flow, have different shapes than do laminar flow-velocity profiles (Fig. 2.1). Owing to variations in flow velocity during turbulent flow, the shape of the turbulent-flow vertical profile is determined by time-averaged values of velocity. Under conditions of turbulent flow, laminar or near-laminar flow occurs only very near the bed. The exact shape of the turbulent profile depends upon the nature of the bed over which the flow takes place. For smooth beds, there is a thin layer close to the bed boundary where molecular viscous forces dominate. Molecular adhesion causes the fluid immediately at the boundary to remain stationary. Successive overlying layers of fluid slide relative to those beneath at a rate dependent upon the fluid viscosity. Flow within this thin layer tends toward laminar, although it is characterized by streaks of faster and slower moving fluid and is not truly laminar. This layer is the **viscous sublayer**, or laminar sublayer.

If sediment particles on a streambed are so small (mud to fine-sand size) that they lie within the viscous sublayer, the near-bed flow is dominated by viscous forces and the flow is said to be **hydraulically smooth**. If the grains are so large that they exceed the thickness of the viscous sublayer and thus protrude into the turbulent part of the flow, the flow is **hydraulically rough**. Over a very rough or irregular bed such as coarse sand or gravel, the viscous sublayer is destroyed by these irregularities, which extend through the layer into the turbulent flow. The flow of fluid over a boundary is thus affected by the roughness of the boundary. Obstacles on the bed generate eddies at the boundary of a flow; the larger and more abundant the obstacles, the more turbulence is generated.

Most sediment transport takes place within boundary layers; turbulent boundary flow is much more effective in eroding and transporting sediment than is laminar flow. The presence or absence of a viscous sublayer may be an important factor in initiating grain movement. That is, extremely small grains that lie entirely within the viscous sublayer may be difficult to move.

Box 2.2 Boundary (Bed) Shear Stress

As a fluid flows across its bed, a stress that opposes the motion of the fluid exists at the bed surface. This stress, called the **boundary shear stress** (τ_0) to differentiate it from fluid shear stress (τ), is defined as force per unit area parallel to the bed, that is, the tangential force per unit area of surface (the mean shear stress acting over the wetted perimeter). It is a function of the density of the fluid, slope of the bed, and water depth. Boundary shear stress is expressed (Allen, 1994) as

$$\tau_0 = \gamma \rho g h s \qquad (2.2.1)$$

where γ is density of the fluid, ρ is fluid density, g is gravitational acceleration, h is flow depth, and s is the slope of the parallel bed and water surface (gradient). The boundary shear stress is also a function of velocity of flow, a complex mathematical relationship not shown here. It tends to increase as velocity increases, although not in a direct way. The bed shear stress increases linearly with depth and slope.

Because boundary shear stress is determined by the force that a flow is able to exert on the sediment bed and is related to flow velocity, it is an important variable in determining the erosion and transport of sediment on the bed below a flow. Equation 2.2.1 indicates that boundary shear stress increases directly with increasing density of the moving fluid, increasing diameter and depth of the stream channel, and increasing slope of the streambed. *Other factors being equal, we would thus expect to see greater boundary shear stress, and greater ability to erode and transport sediment, in water flows than in air flows, in larger stream channels than in smaller channels, and in high-gradient streams than in low-gradient streams.*

Froude Number

In addition to the effects of fluid viscosity and inertial forces, gravity also plays a role in fluid flow because gravity influences the way in which a fluid transmits surface waves. The velocity with which small gravity waves move in shallow water is given by the expression $\sqrt{gL}$, in which g is gravitational acceleration and L is water depth. The ratio between inertial and gravity forces is the **Froude number** (F_r), which is expressed as

$$F_r = \frac{U}{\sqrt{gL}} \qquad (2.3)$$

where U is again the mean velocity of flow and L is water depth, in the case of water flowing in an open channel. The Froude number, like the Reynolds number, is a dimensionless value and thus is very useful in modeling studies.

When the Froude number is less than 1, the velocity at which waves move is greater than flow velocity, and waves can travel upstream. That is, waves in a stream move upstream in the direction opposite to current flow. For example, if you threw a stone into a slow-moving stream having a Froude number less than 1, waves created by impact of the stone could spread upstream. Flow under these conditions is called tranquil, streaming, or subcritical. If the Froude number is greater than 1, waves cannot be propagated upstream, and flow is said to be rapid, shooting, or supercritical. Thus, the Froude number can be used to define the critical velocity of moving water at which flow at a given depth changes from tranquil to rapid (such as change from tranquil flow in a stream channel with a gentle slope to rapid flow where the channel becomes steeper) or vice versa. The Froude number also has a relationship to flow regimes, which are defined by characteristic bedforms, such as ripples, that develop during fluid flow over a sediment bed. This relationship is discussed further in Chapter 4.

2.3 PARTICLE TRANSPORT BY FLUIDS

Having established some of the fundamentals of fluid behavior during flow of fluids alone, we are now at a point where we can consider the more complicated processes of transport of sediment by fluid flow. Transport of sediment by fluid flow involves two fundamental steps: (1) erosion and entrainment of sediment from the bed and (2) subsequent, sustained downcurrent or downwind movement of sediment along or above the bed. The term **entrainment** refers to the processes involved in lifting resting grains from the bed or otherwise putting them into motion. More energy is commonly required to initiate particle movement than to keep particles in motion after entrainment. Thus, a great deal of experimental and theoretical work has been directed toward study of the conditions necessary for particle entrainment. Once particles are lifted from the sediment bed into the overlying water or air column, the rate at which they fall back to the bed (the settling velocity) is an important factor in determining how far the particles travel downcurrent before they again come to rest on the sediment bed. Like particle entrainment, the settling velocity of particles has been studied extensively. We will now examine some of these fundamental aspects of particle transport by fluids, beginning with a look at the factors involved in entrainment of sediments by a moving body of fluid.

Particle Entrainment by Currents

As the velocity and shear stress of a fluid moving over a sediment bed increase, a critical point is reached at which grains begin to move downcurrent. Typically, the smaller and lighter grains move first. As shear stress increases, larger grains are put into motion until finally grain motion is common everywhere on the bed. This **critical threshold** for grain movement is a direct function of several variables, including the boundary shear stress, fluid viscosity, and particle size, shape, and density. Indirectly, it is also a function of the velocity of flow, which varies as the logarithm of the distance above the bottom.

To understand the problems involved in lifting particles from the bed and initiating motion, let us consider the opposing forces that come into play as a fluid moves across its bed. As shown in Figure 2.2A, forces caused by gravity act downward to resist motion and hold particles against the bed. The gravity forces result from the mass of the particles and are aided in resisting grain movement by frictional resistance between particles. Fine, clay-size particles have added resistance to movement owing to cohesiveness that arises from electrochemical bonds between these small grains. The motive forces that must be generated by fluid flow to overcome the resistance to movement imposed by these retarding factors include a **drag force**, which acts parallel to the bed and is related to the boundary shear stress, and a **lift force** caused by the Bernoulli effect of fluid flow over projecting grains. The drag force (F_D) depends upon the boundary shear stress

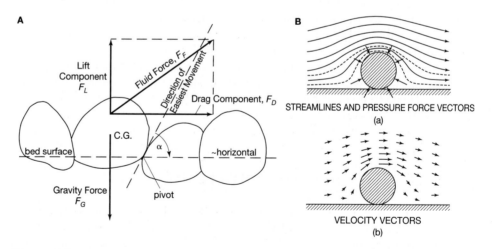

Figure 2.2

A. Forces acting during fluid flow on a grain resting on a bed of similar grains. B. Flow pattern of fluid moving over a grain, illustrating the lift forces generated owing to the Bernoulli effect: (a) streamlines and the relative magnitude of pressure acting on the surface of the grain. (b) direction and relative velocity of velocity vectors; higher velocities occur where streamlines are closer together. [A., after Middleton, G. V., and J. B. Southard, 1978, Mechanisms of sediment movement: Eastern Section, Soc. Econ. Paleontologists and Mineralogists Short Course No. 3, Fig. 6.1, p. 6., reprinted by permisssion of SEPM, Tulsa, Okla. B. Streamlines and velocity vectors, from Blatt, H., G. V. Middleton, and R. Murray, 1980, Origin of sedimentary rocks, 2nd ed., Fig. 4.9, p. 107, reprinted by permission of Prentice-Hall, Englewood Cliffs, N.J.]

(τ_0) and the drag exerted on each grain exposed to this stress. The hydraulic lift force (F_L) known as the Bernoulli effect is caused by the convergence of fluid streamlines over a projecting grain. The Bernoulli effect results from an increase in flow velocity in the zone where the streamlines converge over the grain. This velocity increase causes pressure to decrease above the grain. Hydrostatic pressure from below then tends to push the grain up off the bed into this low-pressure zone (Fig. 2.2B) for the same reason that lift is created when air flows over the curved wing of an airplane. The drag and lift forces combine to produce the total fluid force, represented by the fluid-force vector (F_F) in Figure 2.2A. For grain movement to occur, the fluid force must be large enough to overcome the gravity and frictional forces.

The preceding discussion is greatly simplified and generalized, and a number of factors complicate calculation of critical thresholds of grain movement under natural conditions. These factors include variations in shape, size, and sorting of grains; bed roughness, which controls the presence or absence of a viscous sublayer; and cohesion of small particles. Because of these complicating factors, the critical conditions for particle entrainment cannot be calculated and must be determined experimentally. The simplest plot that shows an experimentally derived threshold graph for initiation of grain movement is the **Hjulström diagram**.

In the Hjulström diagram (Fig. 2.3), the velocity at which grain movement begins as flow velocity increases above the bed is plotted against mean grain size (grain diameter). This diagram shows the critical velocity for movement of quartz grains on a plane bed at a water depth of 1 m. The curve separates the graph into two fields. Points above the graph indicate the conditions under which grains are in motion, and points below indicate no motion. Note from this figure that critical entrainment velocity for grains larger than about 0.5 mm increases gradually with increasing mean grain size, whereas the entrainment velocity for consolidated

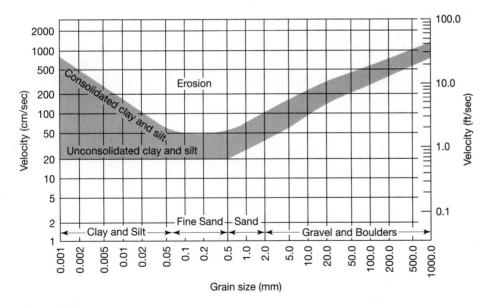

Figure 2.3
The Hjulström diagram, as modified from Sundborg, showing the critical current velocity required to move quartz grains on a plane bed at a water depth of 1 m. The shaded area indicates the scatter of experimental data, and the increased width of this area in the finer grain sizes shows the effect of sediment cohesion and consolidation on the critical velocity required for sediment entrainment. [After Sundborg, A., 1956, The River Klarälven, a study of fluvial processes: Geografiska Annaler, Ser A., v. 38, p. 197, reprinted by permission.]

clay and silt grains smaller than 0.05 mm increases with decreasing grain size. This seemingly anomalous behavior at smaller grain sizes is apparently due mainly to increasing cohesion of finer size particles, making them more difficult to erode than larger, noncohesive particles. Also, extremely small grains may lie within the viscous sublayer, where little grain movement takes place.

Box 2.3 The Shields Diagram

The Hjulström diagram is somewhat limited in its application because it is strictly valid only at a water depth of 1 m on a plane bed and only in water in which fluid and grain densities and dynamic viscosity are constant, as in freshwater streams in a given season during average flow. The Shields diagram (Fig. 2.3.1) is also a threshold graph for initiation of sediment grain movement that is widely used by sedimentologists and is well established by experimental work. It has more rigor than the Hjulström diagram and it has a more general application. For example, it can be used for wind as well as water and for a variety of conditions in water. It is, however, more complex and difficult to understand than is the Hjulström diagram because it involves two dimensionless relationships. In Figure 2.3.1, dimensionless shear stress (θ_τ) (called β by some workers) is used instead of flow velocity as a measure of critical shear, and the mean grain size parameter used in the Hjulström diagram is replaced by the **grain Reynolds number** (R_{eg}), another dimensionless quantity. The dimensionless bed shear stress (θ_t) is given by

$$\theta_t = \frac{\tau_t}{(\rho_s - \rho)gD} \tag{2.3.1}$$

where τ_t is boundary shear stress, ρ_s is density of the particles, ρ is density of the fluid, g is gravitational acceleration, and D is particle diameter. The value of dimensionless shear stress increases with increasing bed shear stress and increasing velocity; it decreases with increasing density and size of the particles. Instead of velocity in Hjulström's diagram, the dimensionless shear stress thus incorporates shear stress (velocity), grain size, and grain and fluid densities into a single term, which is plotted on the vertical axis of Shields' diagram. An increase in dimensionless shear stress indicates either an increase in flow velocity and shear stress or a decrease in grain size or grain density.

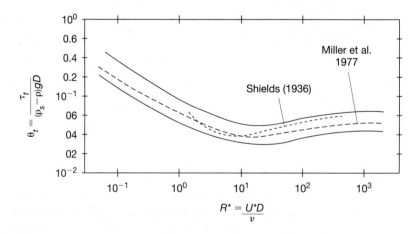

Figure 2.3.1
The Shields diagram as modified by Miller. (After Miller, M.C., et al., 1977, Threshold of sediment motion under unidirectional currents: Sedimentology v. 24, Fig. 2, p. 511, reproduced by permission.)

The grain Reynolds number (R_{eg}) differs from the ordinary Reynolds number previously discussed. The length or water depth value (L) of the ordinary Reynolds number is replaced by particle diameter (d) and the flow velocity (U) by shear velocity (U^*). The grain Reynolds number is a measure of turbulence at the grain-fluid boundary. It is thus expressed as

$$R_{eg} = \frac{U^* d}{\nu} \qquad (2.3.2)$$

The grain Reynolds number, plotted on the horizontal axis of Shields' diagram, is clearly not the same thing as mean grain size; however, it can be seen from Figure 2.3.1 that the grain Reynolds number increases with increasing grain size if shear velocity and kinematic viscosity remain constant. Thus, an increase in grain Reynolds number means an increase in grain size, an increase in shear velocity and turbulence, or a decrease in kinematic viscosity.

The Shields diagram is less intuitive than Hjulström's diagram in which fluid velocity is plotted against grain size; however, points above the curve indicate that noncohesive grains on the bed are fully in motion and points below indicate no motion, as in the Hjulström diagram. Beginning of motion is determined by the dimensionless shear stress, which increases with increasing bed shear stress under a given set of conditions for grain density, fluid density, and grain size. The critical dimensionless shear stress required to initiate grain movement thus depends upon the grain Reynolds number, which in turn is a function of grain size, kinematic viscosity, and turbulence. Note from the Shields diagram in Figure 2.3.1 that the dimensionless bed shear stress increases slightly with increasing grain Reynolds number above about 5 to 10, although it remains mainly between 0.03 and 0.05. At lower Reynolds numbers, the value increases steadily up to a value of 0.1 or higher. This greater rate of increase at lower Reynolds numbers is related to the presence of the viscous sublayer. When the bed is composed of small particles on the order of fine sand or smaller, a smooth boundary and hydraulically smooth flow result; the particles lie entirely within the viscous sublayer, where flow is essentially nonturbulent and instantaneous velocity variations are less than in the lower part of the overlying turbulent boundary layer. For coarser particles, the viscous sublayer is so thin that the grains project through the layer into the turbulent flow.

Several complicating factors not covered in the Hjulström and Shields diagrams make prediction of the onset of grain movement difficult. Instantaneous fluctuations in boundary shear stress may arise from local eddies or from wave action superimposed on current flow, and these fluctuations may cause some particles to move before the general onset of grain movement. Fine muds and silts may not erode to yield individual grains owing to the tendency of such cohesive materials to be removed as chunks or aggregates of grains.

Many of the processes involved in sediment entrainment and transport can be effectively modeled by computer simulation. An excellent book on this subject, *Simulating Clastic Sedimentation* (Tetzlaff and Harbaugh, 1989), takes the reader through the computations and reasoning of erosion, transportation, and deposition of clastic sediment. The book further provides detailed instructions on the techniques and computer programs used to simulate clastic sedimentation. For further insight into simulating sediment transport, see Slingerland, Harbaugh, and Furlong (1994).

Role of Particle Settling Velocity in Transport

As soon as grains are lifted above the bed during the entrainment process, they begin to fall back to the bed. The distance that they travel downcurrent before again coming to rest on the bed depends upon the drag and lift forces exerted by the current, including turbulence, and the settling velocity of the particles. A particle initially accelerates as it falls through a fluid, but acceleration gradually decreases until a steady rate of fall, called the **terminal fall velocity,** is achieved. For small particles, terminal fall velocity is reached very quickly. The rate at which particles settle after reaching fall velocity is a function of the viscosity of the fluid and the size, shape, and density of the particles. The settling rate is determined by the interaction of upwardly directed forces—owing to buoyancy of the fluid and viscous resistance (drag) to fall of the particles through the fluid—and downwardly directed forces arising from gravity. The drag force exerted by the fluid on a falling spherical grain is proportional to the density of the fluid (ρ_f), the diameter (d) of the grains, and the fall velocity (V), as given by the relationship

$$\text{Drag force} = C_D \pi \frac{d^2}{4} \frac{\rho_f V^2}{2} \tag{2.4}$$

where C_D is a drag coefficient that depends upon the grain Reynolds number and the particle shape. The upward force resulting from buoyancy of the fluid is given by

$$F_{upward} = \frac{4}{3}\pi \left(\frac{d}{2}\right)^3 \rho_f g \tag{2.5}$$

where ρ_f is fluid density and g is gravitational acceleration. [*Note:* $4/3\pi(d/2)^3$ is the volume of a sphere.] The downward force owing to gravity is given by

$$F_g = \frac{4}{3}\pi \left(\frac{d}{2}\right)^3 \rho_s g \tag{2.6}$$

where ρ_s is particle density. As the particle stops accelerating and achieves fall velocity, the drag force of the liquid on the falling particle is equal to the downward force due to gravity minus the upward force resulting from buoyancy of the liquid. Thus

$$C_D \pi \frac{d^2}{4} \frac{\rho_f V^2}{2} = \frac{4}{3}\pi \left(\frac{d}{2}\right)^3 \rho_s g - \frac{4}{3}\pi \left(\frac{d}{2}\right)^3 \rho_f g \tag{2.7}$$

Rearranging terms, we can explain this relationship in terms of fall velocity (V) as

$$V^2 = \frac{4gd}{3C_D} \frac{(\rho_s - \rho_f)}{\rho_f} \tag{2.8}$$

For slow laminar flow at low concentrations of particles and low grain Reynolds numbers (R_{eg}, see Box 2.3), C_D has been determined to equal $24/R_{eg}$ (Rouse and Howe, 1953, p. 182). Substituting this value ($24/U^*d/\mu/\rho_f$) for C_D yields

$$V = \frac{1}{18} \frac{(\rho_s - \rho_f)gd^2}{\mu} \tag{2.9}$$

which is **Stokes's Law** of settling, with particle size expressed as diameter in centimeters. This law, formulated by Stokes in 1845, is often simplified to

$$V = CD^2 \text{ (in cm/sec)} \tag{2.10}$$

where C is a constant equaling $(\rho_s-\rho_f)g/18\,\mu$ and D is the diameter of particles (spheres) expressed in centimeters. Values of C have been calculated for a range of common laboratory temperatures (e.g., Galehouse, 1971); thus, settling velocity (V) can be determined quickly for any value of particle diameter (D). Note that the

Reynolds number is a distinguishing factor in grain settling behavior just as it is in laminar and turbulent flow.

Experimental determination of particle fall velocity shows that Stokes's Law accurately predicts settling velocity of particles in water only for particles less than about 0.1 to 0.2 mm in diameter. Larger particles have fall velocities lower than those predicted by Stokes's Law, apparently owing to inertial (turbulent) effects caused by the increased rates of fall of these larger grains. Thus, the Stokes equation cannot be used for determining the settling velocity of sand, a very important component of most sediment. The fall velocity is also decreased by decrease in temperature (which increases viscosity), decrease in particle density, and decrease in the sphericity (the degree to which the shape of a particle approaches the shape of a sphere) of the particles. Most natural particles are not spheres, and departure from spherical shape decreases fall velocity. Fall velocity is also decreased by increasing concentration of suspended sediment in the fluid, which increases the apparent viscosity and density of the fluid, and by turbulence.

Sediment Loads and Transport Paths

Once sediment has been eroded and put into motion, the transport path that it takes during further sustained downcurrent movement is a function of the settling velocity of the particle and the magnitude of the current velocity and turbulence. Under a given set of conditions, the sediment load may consist entirely of very coarse particles, entirely of very fine particles, or of mixtures of coarse and fine particles. Coarse sediment such as sand and gravel moves on or very close to the bed during transport and is considered to constitute the **bed load** (Fig. 2.4). Finer material carried higher up in the main flow above the bed makes up the **suspended load**. If the shear velocity (U^*) is greater than the settling velocity (V), material will remain in suspension.

Bed Load Transport

Particles larger than sand size are commonly transported as part of the bed load in essentially continuous contact with the bed. This type of transport, called **traction** transport, may include rolling of large or elongated grains, sliding of grains over or past each other, and creep. Creep results from grains being pushed a short distance along the bed in a downcurrent direction owing to impact of other moving grains. **Saltation** is a type of bedload transport in which grains, particularly sand-size grains, tend to move in intermittent contact with the bed. Saltating grains move by a series of jumps or hops, rising off the bed at a steep angle (~45°) to a height of a few grain diameters and then falling back along a shallow descent path of about 10°. This asymmetric saltation path may be interrupted by turbulence or

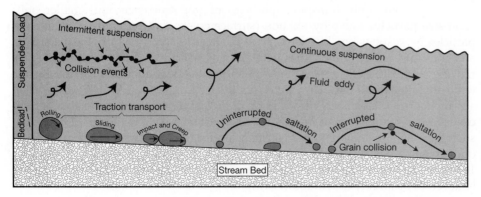

Figure 2.4
Schematic illustration of grain paths during bedload, suspension, and saltation transport. [Based in part on Leeder, M. R., 1979, Bedload dynamics: Grain interactions in water flows: Earth surface processes, v. 4, Fig. 5, p. 237, John Wiley & Sons.]

by collision with another grain (Fig. 2.4). Saltation transport may be thought of as intermediate between traction transport and suspension transport, but it is described here as part of bedload transport because most saltating grains remain relatively close to the bed during movement.

Suspended Load Transport

As the flow strength of a current increases, the intensity of turbulence increases close to the bed. Particle trajectories become longer, more irregular, and higher up from the bed than the trajectories of saltating grains. Upward components of fluid motion resulting from turbulence increase to the point that they balance downward gravitational forces on the particles, allowing the particles to stay suspended above the bed far longer than could be predicted from their settling velocities in nonturbulent water. If the lift forces arising from turbulence are erratic and do not continuously maintain this balance, a common occurrence during transport of fine-to-medium sand, the grains may drop back from time to time onto the bed. This behavior is called **intermittent suspension** (Fig. 2.4). Intermittent suspension differs from saltation because the suspended particles tend to be carried higher above the bed and remain off the bed for longer periods of time. Smaller particles have settling velocities that may be so low that they remain in nearly **continuous suspension** and are carried along at almost the same velocity as the fluid flow.

Wash Load

Much of the sediment load undergoing continuous suspension transport is composed of fine, clay-size particles with very low settling velocities. In rivers, this sediment is derived either from upstream source areas or by erosion of the bank, rather than from the streambed, and is called the **wash load**. Rivers have the capacity to transport large wash loads even at very low velocities of flow. Because the wash load travels in continuous suspension at about the same velocity as the water, it is transported rapidly through river systems.

Transport by Wind

Wind can be considered a very low density, low viscosity "fluid" that is capable of flowing and bringing about sediment transport. The principles involved in entrainment and transport of particles by wind (eolian transport) are similar to those for water; however, wind's low density and viscosity cause the threshold values for wind entrainment and transport to be quite different (see discussion by Nickling, 1994). Entrainment of grains by wind action can be strongly affected by the impact of moving grains hitting the bed. At a value of wind velocity below the critical velocity needed to initiate grain movement, grain motion can be started and propagated downwind by throwing grains onto the bed, a process referred to as seeding. This lower threshold for grain movement is called the **impact threshold**.

Wind commonly transports particles of fine-sand size and smaller only. Sand-size particles move by traction (surface creep) and saltation and dust-size particles by suspension. Transport takes place at relatively high wind velocities, and the flow is commonly turbulent, characterized by eddies of various sizes moving with different speeds and directions. Suspended loads carried by the wind are called **dustloads**. Upward diffusion in unstable, buoyant air masses at an advancing front have been known to carry dust clouds rapidly to heights of hundreds or even thousands of meters during volcanic eruptions. Material carried to such great heights may remain in suspension for long periods of time and subsequently be spread over a very broad area, including the ocean basins (Prospero, 1981). In fact, the very fine grained component of deep-sea pelagic sediments is believed to be largely of windblown origin.

Like water transport, eolian transport results in deposition of bedforms ranging in size from ripples a few centimeters high to gigantic dunes hundreds of meters high. Eolian transport and depositional processes are discussed in much greater detail in Chapter 8.

Transport by Glacial Ice

The high viscosity of glacial ice causes it to flow very slowly. Glaciers are capable of entraining huge volumes of sediment by scraping and plucking the underlying bedrock and adjacent valley walls and from rockfall onto the glacier from above. As laminar flow of ice takes place within the glacier, sediment of all sizes is carried along with the moving ice in contact with the bed and suspended at various heights above the base of the glacier. When the front of a glacier melts, the sediment load is dumped as unsorted, poorly layered glacial moraine. Details of sediment transport by ice are given in Chapter 8.

Deposits of Fluid Flows

Water and air are responsible for most sediment transport by fluid flow; however, ice may account for local transport of large volumes of sediment and particles of very large size, as mentioned. Sediment entrainment and transport by moving water and wind stop, and deposition occurs, when local hydrologic or wind conditions change sufficiently to decrease bed shear stress to the point that it is no longer adequate to initiate and sustain particle movement. This decrease in bed shear stress is caused fundamentally by decrease in flow velocity. Flow velocity and shear stress may decrease below the critical level required for sediment transport owing to a variety of causes. In the case of water transport, these causes include decrease in the slope of the bed, increase in bed roughness, and loss of water volume. Decrease in wind velocity may also result from increase in bed roughness or from changes in surface topography and weather conditions. Deposition from glaciers is brought about on land when the glaciers either become stagnant or retreat owing to decrease in snow accumulation rates or increase in melting rates. Glaciers that run out to sea and calve to form icebergs eventually melt and drop their load on the seafloor.

Sediment deposition may be temporary or permanent. For example, sediment deposited in river channels and point bars, beach environments, and other very nearshore environments can be reentrained and subjected to continued transport as seasonal or longer range changes in the hydrologic regimen take place. In fact, river sediment may be deposited and reentrained numerous times before finally reaching a depositional basin in the ocean. On the other hand, some river sediment, lake sediment, and wind-transported sediment may become deposited in continental settings and be preserved for long periods of time to become part of the geologic record. The great bulk of sediment undergoing transportation ultimately finds its way into ocean basins, where it is eventually deposited below wave base and more or less permanently immobilized until buried.

Sediments deposited by fluid flow of water or wind are commonly characterized by layers or beds of various thickness, scarcity of vertical grain-size grading, grain-size sorting ranging from poor to excellent depending upon depositional conditions, and a variety of sedimentary structures (Fig. 2.5). Sediments deposited from traction currents commonly preserve sedimentary structures such as cross-beds, ripple marks, and pebble imbrication (pebbles overlap like shingles on a roof) that display directional features from which the direction of the ancient fluid flow can be determined. Sediments deposited from suspension lack these flow structures and are commonly characterized instead by fine laminations. Wind is competent to transport and deposit particles in the size range of

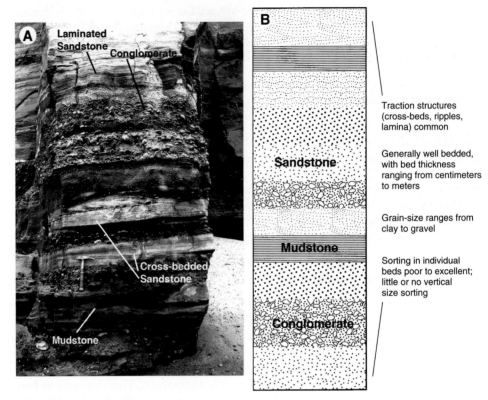

Figure 2.5
A. Photograph of well-bedded fluid-flow deposits, Miocene, Blacklock Point, southern Oregon coast.
B. Schematic representation of typical characteristics of fluid-flow deposits.

Traction structures (cross-beds, ripples, lamina) common

Generally well bedded, with bed thickness ranging from centimeters to meters

Grain-size ranges from clay to gravel

Sorting in individual beds poor to excellent; little or no vertical size sorting

sand to dust (clay) only. By contrast, the grain size of sediment deposited by water may range from clay size to cobbles or boulders tens to hundreds of centimeters in diameter. These variations in grain size reflect the wide range of energy conditions of wind and water that prevail under natural conditions and the variations in relative competence of wind and water to initiate and sustain sediment transport. Because of its high viscosity, ice is capable of transporting particles of enormous size as well as particles of the smallest sizes. Sediments deposited by glaciers are characteristically poorly layered and extremely poorly sorted, with particles ranging from meter-size boulders to clay-size grains.

This brief description of fluid-flow deposits is given here to illustrate the relationship between flow processes and the characteristics of the resulting sedimentary deposits. The textures and structures of sedimentary rocks are discussed in detail in Chapters 3 and 4; the characteristics of sediments deposited by fluid flows in different sedimentary environments are described in Chapters 8 through 11.

2.4 PARTICLE TRANSPORT BY SEDIMENT GRAVITY FLOWS

In the preceding section, we examined sediment transport resulting from the interaction of moving fluids and sediment. During fluid-flow transport, the fluids (water, wind, ice) move in various ways under the action of gravity and the sediment is simply carried along with and by the fluid. Sediment can also be transported essentially independently of fluid by the effect of gravity acting directly on the sediment. In this type of transport, fluids may play a role in reducing internal friction and supporting grains, but they are not primarily responsible for downslope movement of the sediment. Movement of sediment under the influence of gravity creates the flow, and flow stops when the sediment load is deposited.

Sediment can be transported by the direct action of gravity in both subaerial and subaqueous environments. Gravity transport under submarine conditions has the greater sedimentological significance. A spectrum of gravity movements exists, ranging from those in which sediment is moved en masse and fluids act mainly to reduce internal friction by lubricating the grains, to those in which transport is on a grain-by-grain basis and fluids play an important role in supporting the sediment during transport. Gravity mass movements can be grouped into rock falls, slides, and sediment gravity flows, as shown in Table 2.1. **Rock fall** involves free fall of blocks or clasts from cliffs or steep slopes. **Slides** are en-masse movements of rock or sediment owing to shear failure that take place with little accompanying internal deformation of the mass. **Sediment gravity flows** are more "fluid" types of movement in which breakdown in grain packing occurs and internal deformation of the sediment mass is intense.

Sediment gravity flows are of particular interest because they are capable of rapidly transporting large quantities of sediment, including very coarse sediment, even into very deep water in the oceans. Gravity flows that occur in subaerial environments can be considered, in a broad sense, to include snow avalanches, pyroclastic flows and base surges resulting from volcanic eruptions, grain flow of dry sand down the slip face of sand dunes, and both volcanic and nonvolcanic debris flows and mud flows, in which large particles are transported in a slurrylike matrix of finer material. Subaqueous sediment gravity flows also include grain flows and debris flows, as well as turbidity currents and liquefied sediment flows.

Sediment gravity flows can occur only when grains become separated and dispersed to the point that internal friction and cohesiveness are sufficiently reduced to

Table 2.1 Major types of mass-transport processes, their mechanical behavior, and transport and sediment support mechanisms

Mass transport processes			Mechanical behavior	Transport mechanism and sediment support
Rock fall			Elastic	Free fall and subordinate rolling of individual blocks or clasts along steep slopes
Slide	Glide		Elastic	Shear failure among discrete shear planes with little internal deformation or rotation
	Slump		Elastic	Shear failure accompanied by rotation along discrete shear surfaces with little internal deformation
Sediment gravity flow — Mass flow	Debris flow / Mud flow	Plastic limit / Plastic	Shear distributed throughout sediment mass; strength principally from cohesion due to clay content; additional matrix support possibly from buoyancy
	Grain flows: Inertial, Viscous	Liquid limit	Cohesionless sediment supported by dispersive pressure; flow in inertial (high-concentration) or viscous (low-concentration) regime; steep slopes usually required
Sediment gravity flow — Fluidal flow	Liquefied flow	Viscous fluid	Cohesionless sediment supported by upward displacement of fluid (dilatance) as loosely packed structure collapses, settling into a more tightly packed framework; slopes in excess of 3° required
	Fluidized flow	Viscous fluid	Cohesionless sediment supported by the forced upward motion of escaping pore fluid; thin (<10 cm) and short-lived
	Turbidity current	Viscous fluid	Supported by fluid turbulence

Source: Nardin, T. R., F. J. Hein, D. S. Gorsline, and B. D. Edwards, 1979. A review of mass movement processes, sediments, and acoustical characteristics, and contrasts in slope and base-of-slope systems versus canyon-fan-basin flow systems, in L. J. Doyle and O. R. Pilkey (eds.), Geology of continental slopes: Soc. *Econ. Paleontologists and Mineralogists Spec. Pub. 27.* Table 1, p. 64, reprinted by permission of SEPM, Tulsa, Okla.

Figure 2.6
The principal kinds of sediment-gravity flows and the relationship of flow type to grain-support mechanisms and fluid types. [Based on Middleton, G. V., and M. A. Hampton, 1976, Subaqueous sediment transport and deposition by sediment gravity flows, *in* Stanley, D. H., and D. J. P. Swift, (eds.), Marine sediment transport and environmental management: John Wiley & Sons, Inc., Fig. 1, p. 198.]

Observed Type of flow	Turbidity Current	Liquefied Flow	Grain Flow	Mud Flow; Debris Flow
Support Mechanism	Turbulent Fluid	Upward escape of intergranular fluid	Grain interaction (dispersive pressure)	Strength of matrix
	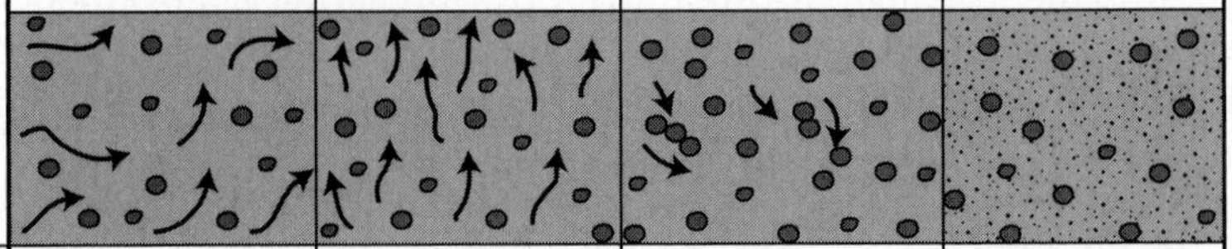			
Type of Fluid	Turbulent fluid	Newtonian fluid (high viscosity)	Non-Newtonian fluid	Bingham plastic

lower the strength of the sediment mass below the critical point required for gravity to initiate movement. Four theoretical types of dispersive and support flow mechanisms that can achieve this reduction in internal strength have been identified: turbulent flow (turbulence), upward escape of intergranular fluid, grain interaction (dispersive pressure), and support by a cohesive matrix (Fig. 2.6). Four observed flow types can be identified that correspond to these theoretical support mechanisms: turbidity currents, liquefied flow, grain flow, and mud and debris flow. These four mechanisms of gravity transport, described in greater detail below, can be thought of as members of a spectrum of sediment-gravity-flow processes. One type may grade into another under some conditions, as when a submarine mud flow changes into a turbidity current downslope with additional mixing and dilution by water.

Turbidity Currents

Flow Mechanisms and Characteristics

A turbidity current is a kind of density current that flows downslope along the bottom of an ocean or lake because of density contrasts with the surrounding (ambient) water arising from sediment that becomes suspended in the water owing to turbulence. Turbidity currents can be generated experimentally in the laboratory by the sudden release of muddy, dense water into the end of a sloping flume filled with less dense, clear water. They have been observed to occur under natural conditions in lakes where muddy river water enters the lakes, and they are believed to have occurred throughout geologic time in the marine environment on continental margins. In this setting, they originate particularly in or near the heads of submarine canyons.

Turbidity currents can be generated by a variety of mechanisms, including sediment failure, storm-triggered flow of sand and mud into canyon heads, bedload inflow from rivers and glacial meltwater, and flows during eruption of airfall ash (Normark and Piper, 1991). They may move as surges or as steady, uniform flows.

Surges, or spasmodic turbidity currents, are initiated by some short-lived catastrophic event, such as earthquake-triggered massive sediment slumping or storm waves acting on a continental shelf. Such an event creates intense turbulence in the water overlying the seafloor, resulting in extensive erosion and entrainment of sediment, which is rapidly thrown into suspension. The sediment then remains suspended, supported in the water column by turbulence. This process generates a dense, turbid cloud that moves downslope, eroding and picking up more sediment as it increases in speed. Surge flows develop into three main parts as they move away from the source: the head, body, and tail.

Box 2.4 Flow Velocity of Turbidity Currents

The **head** of the surge is about twice as thick as the rest of the flow and is characterized by intense turbulence. The velocity, U_{head}, at which the head advances into still water is:

$$U_{head} = 0.7\sqrt{\frac{\Delta\rho}{\rho}gh} \qquad (2.4.1)$$

where $\Delta\rho$ is the density contrast between the turbidity current and the ambient water, ρ is the density of the ambient water, g is gravitational acceleration, and h is the height of the head (Middleton and Hampton, 1976). The head is overhanging and is divided transversely into lobes and clefts (Fig. 2.4.1 and 2.4.2).

Flow within the **body** of a surge-type turbidity current is nearly steady and uniform and the flow is almost uniform in thickness. The body moves at a velocity, U_{body}, that is:

$$U_{body} = \sqrt{\frac{8g}{f_0 + f_1}\left(\frac{\Delta\rho}{\rho}\right)hs} \qquad (2.4.2)$$

where h is the height or thickness of the flow, s is the slope of the bottom, f_0 is the frictional resistance at the bottom of the flow, and f_1 is the frictional resistance at the upper interface of the flow in contact with the overlying ambient water layer. The body flows at a velocity that is faster in deep water than that of the head. This difference in velocity causes the forward part of the body to consume itself within the head in the process of mixing with the ambient water (Allen, 1985). The **tail** of the flow thins abruptly away from the body and becomes more dilute.

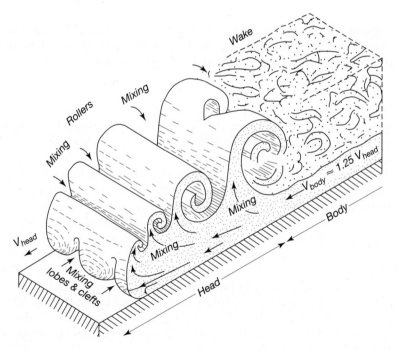

Figure 2.4.1
Postulated structure of the head and body of a turbidity current advancing into deep water. The tail is not shown. [From Allen, J. R. L., 1985, Loose boundary hydraulics and fluid mechanics: Selected advances since 1961, *in* Brenchley, P., and D. J. P. Swift, (eds.), Sedimentology: Recent developments and applied aspects. Fig. 8, p. 20, reprinted by permission of Blackwell Scientific Publications, Ltd., Oxford.]

Figure 2.4.2
The head of an experimentally-generated turbidity current advancing across the floor of a small flume. Note the lobes and clefts created by extreme turbulence in the head. Modified from "Gravity Deposits" (video), Institut Français du Petrole. Reproduced by permission.

Some turbidity currents are steady, uniform flows that lack a turbulent head. These flows move at velocities similar to those of the body of surge-type flows. Although the velocity is sensitive to the slope over which flow takes place, flow may occur on slopes as low as 1 degree (Kersey and Hsü, 1976). Steady, uniform flows have been observed along the sloping bottom of lakes where sediment-laden rivers run into the lakes. They may occur also on continental shelves where muddy rivers discharge; however, they are less likely in this setting because the density contrast between muddy river water and ocean water is less than that between muddy river water and fresh water.

Once sediment is suspended in a turbidity current, the turbidity current continues to flow for some time under the action of gravity and inertia. Flow will stop when the sediment-water mixture that produces the density contrast with the ambient water is exhausted by settling of the suspended load. Rapid deposition of coarser particles from suspension appears to occur in regions near the source owing to early decay of the extremely intense turbulence generated by the initial event. As the flow continues to move forward, the remaining coarser material will be progressively concentrated in the head of the flow; denser fluid must be continuously supplied to the head to replace that lost to eddies that break off from the head and rejoin the body of the flow. Owing to differences in turbulence in the head and body, the head may be a region of erosion while deposition is taking place from the body.

Theoretically, sediment remaining in suspension after initial deposition of coarse material in the proximal area can, during further transport, be maintained in suspension for a very long time in a state of dynamic equilibrium called **autosuspension** (Bagnold, 1962; Pantin, 1979; Parker, 1982). A condition of auto-suspension is presumably maintained because turbulence continues to be generated in the bottom of the flow owing to gravity-generated downslope flow of the turbidity current over the bed. Thus, loss of energy by friction of the flow with the bottom is compensated for by gravitational energy. The distance that turbidity currents can travel in the ocean is not known from unequivocal evidence. A presumed turbidity current triggered by the 1929 Grand Banks earthquake off Nova Scotia appears to have traveled south across the floor of the Atlantic for a distance of more than 300 km at velocities up to 67 km/hr (19 m/s), as timed by breaks in submarine telegraph cables (Piper, Shor, and Clark, 1988). Transport of sediment over this distance suggests that autosuspension may actually work; nonetheless, some geologists remain skeptical of the autosuspension process (e.g., review by Middleton, 1993).

The velocity of a turbidity current eventually diminishes owing to flattening of the canyon slope, overbank flow of the current along a submarine channel, or spreading of the flow over the flat ocean floor at the base of the slope. As the flow slows, turbulence generated along the sole of the flow also diminishes, and the current gradually becomes more dilute owing to mixing with ambient water around the head and along the upper interface. The remaining sediment carried in the head eventually settles out, causing the head to sink and dissipate. The exact process by which deposition takes place from various parts of a turbidity current is still not thoroughly understood, although it seems clear from experimental results that deposition does not occur in all parts of the current at the same time. As mentioned above, for example, the head may be a region of potential erosion at the same time that the body behind the head is depositing sediment. Sediment that is deposited very rapidly from some parts of the flow, such as the head, may undergo little or no subsequent traction transport before being quickly buried. On the other hand, in more distal parts of the flow or in areas where the head overflows the channel, a period of scouring by the head may be followed by slow deposition from the body and tail, during which additional traction transport of the deposited sediment takes place. Final deposition from the tail may take place after movement of the current is too weak to produce traction transport.

Depending upon position within the turbidity flow and the initial amount of sediment put into suspension by the flow, turbidity currents may contain either high or relatively low concentrations of sediment. Two principal types of turbidity currents, on the basis of suspended particle concentration, can be considered: **low-density flows,** containing less than about 20 to 30 percent grains, and **high-density flows,** containing greater concentrations (Lowe, 1982). Low-density flows are made up largely of clay, silt, and fine- to medium-grained sand-size particles that are supported in suspension entirely by turbulence. High-density flows may include coarse-grained sands and pebble- to cobble-size clasts as well as fine sediment. Support of coarse particles during flow is provided by turbulence aided by hindered settling resulting from their own high sediment concentrations and the buoyant lift provided by the interstitial mixture of water and fine sediment. (High-density flows differ from debris flows in that debris flows are not turbulent and are less fluid.) Note that the heads of turbidity currents may be high-density flows, whereas the tails may be dilute, low-density flows.

Turbidity Current Deposits

The deposits of turbidity currents, commonly called **turbidites,** are of two basic types. Turbidites deposited from high-density flows with high sediment concentrations tend to form thick-bedded turbidite successions containing coarse-grained sandstones or gravels. Individual flow units typically have relatively poor grading and few internal laminations, and basal scour marks are either poorly developed or absent. Some turbidites with thick, coarse-grained basal units may grade upward to finer grained deposits that display traction structures such as laminations and small-scale cross bedding (Fig. 2.7). In the uppermost part of the flow units, the sediments may consist of very fine grained, nearly homogeneous muds

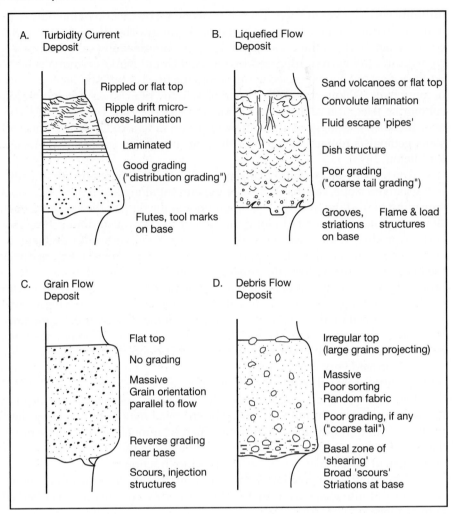

Figure 2.7
Comparison of sedimentary structures in different types of sediment gravity-flow deposits. [After Middleton, G. V., and M. A. Hampton, 1976, Subaqueous sediment transport and deposition by sediment gravity flows, *in* Stanley, D. H., and D. J. P. Swift, (eds.), Marine sediment transport and environmental management: John Wiley & Sons, Inc., Fig. 9, p. 213. Reproduced by permission.]

deposited from the tail of the flow. The deposits of more dilute, low-concentration turbidity current flows generally form thin-bedded turbidite successions. Individual flow units are fine grained at the base and display good vertical size grading, well-developed laminations, and small-scale cross-bedding. Scour marks may be present on the soles or bottoms of the beds.

Bouma (1962) proposed an ideal turbidite sequence, now commonly called a **Bouma sequence**. This ideal sequence consists of five structural units (Fig. 2.8-1) that include the characteristics of both types of turbidites. These structural subdivisions presumably record the decay of flow strength of a turbidity current with time and the progressive development of different sedimentary structures and bedforms in adjustment to different flow regimes (upper-flow regime to lower-flow regime) as current-flow velocity wanes. Most turbidites do not contain all of these structural units. Thick, coarse-grained turbidites tend to show well-developed A and B units, but C through E units are commonly poorly developed or absent. Thin, finer grained turbidites may display well-developed C through E units and poorly developed or absent A and B units. In fact, Hsü (1989, p. 116) claims that Bouma's D unit rarely occurs and that most turbidites can be divided into only two units: a lower horizontally laminated unit (unit A + B, Fig. 2.8-2 and -3) and an upper, cross-laminated unit (unit C). Unit E is a problem because it may consist of fine material deposited slowly from the water column, and thus it may not be part of a turbidite flow unit.

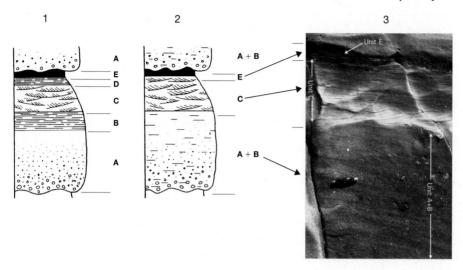

Figure 2.8
Ideal sequence of sedimentary structures in graded-bed units as proposed by Bouma (1) and Hsü (2). Note that in Hsü's model, Bouma units A and B are combined and unit D is omitted. (3) Photograph of a Bouma unit that is very similar to Hsü's model (Cretaceous, southern Oregon coast). [1 and 2 after Hsü, K.J., 1989, Physical principles of sedimentology, Fig. 7.8, p. 116, reprinted by permission of Springer-Verlag, Berlin.]

Turbidites laid down near the source, particularly within the main transport channel where suspended sediment concentrations are high, are generally the coarse-grained, massive, or poorly laminated type. Some very high concentration flows may also deposit coarse-grained turbidites within the main channel at considerable distances from the source. On the other hand, thin, fine-grained turbidites can also be deposited near the source where turbidity currents overflow the banks of a channel and become more dilute as they spread out over the seafloor, as well as in areas more distant from the source. The deposits of a single turbidity current flow typically display horizontal size grading in addition to vertical size grading. That is, thick, coarse-grained deposits generally grade laterally to thinner and finer grained sediments. For a more comprehensive treatment of turbidites, see Mutti (1992).

Liquefied Flows

Flow Mechanisms

Liquefied flows are concentrated dispersions of grains in which the sediment is supported either by the upward flow of pore water escaping from between the grains as they settle downward by gravity or by pore water that is forced upward by injection from below. Loosely packed, cohesionless sediment such as sand can become temporarily liquefied owing to a sudden shock, or series of shocks, that causes the grains to momentarily lose contact with each other and become suspended in their own pore fluid. Grain contact may also be lost if a fluid is introduced into the base of a mass or column of cohesionless sediment and injection is continued until the grains are pushed apart, with their weight being supported by the rising fluid. This process is called fluidization. Once the cohesionless sediment has become liquefied (or fluidized), it loses its strength and behaves like a high-viscosity fluid that can, nonetheless, flow quite rapidly down slopes as low as 3°.

Liquefied flow can occur only as long as grain dispersion is maintained. As soon as the grains settle out of the fluid and reestablish grain-to-grain contact, the flowing layer will "freeze up" and stop moving. "Freezing" begins at the base of the flow; a surface of settled grains rises up through the dispersion at a rate determined by the settling velocity of the particles (Fig. 2.9). The time required for settling to occur is on the order of hours for thick, fine-grained flows (Lowe, 1976); therefore, liquefied flows may travel short, though potentially important, distances before deposition. The upward movement of pore waters through the settling grains as deposition occurs leads to formation of a number of fluid escape

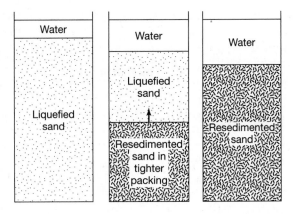

Figure 2.9
Schematic representation of grain settling and water expulsion during deposition of sand from a liquefied flow. [After Allen, J. R. L., and N. L. Banks, 1972, An interpretation and analysis of recumbent-folded deformed cross bedding: Sedimentology, v. 19, Fig. 3, p. 267, reprinted by permission of Elsevier Science Publishers, Amsterdam.]

structures, such as dish structures. Some liquefied flows may become turbulent as the flowing sediment mass is accelerated downslope and thus change into turbidity currents.

Liquidized-Flow Deposits

The deposits of liquefied flows are typically thick, poorly sorted sand units. They are characterized particularly by fluid escape structures (Chapter 4), such as the dish structures, pipes, and sand volcanoes shown in Figure 2.7B.

Grain Flows

Flow Mechanism

Grain flows are dispersions of cohesionless sediment in which the sediment is supported in air by dispersive pressures owing to direct grain-to-grain collisions and in water by collisions and close approaches. Sediment can flow rapidly under both subaerial and subaqueous conditions, especially on steep slopes that approach the angle of repose for the sediment. Grain flow results in the movement of cohesionless sediments down steep slopes owing to sudden loss of internal shear strength of the sediment. Grain flow begins when traction processes cause cohesionless sediment, commonly sand, to be piled up beyond the critical angle of repose. This angle is a function of grain packing and grain shape and tends to be greatest in deposits with angular grains of low sphericity. When the angle of repose for a particular sediment is exceeded, avalanching occurs; flow quickly begins when the internal shear stresses owing to gravity exceed the internal shear strength of the sediment. The **dispersive pressures** needed to force the grains apart and keep them suspended during flow are provided not by fluid but by grain-to-grain collisions in air and grain collisions and close encounters in water as the failed mass of sediments moves down a slope. During the interaction of grains, dispersive pressure is the force normal to the plane of shearing which tends to expand or disperse the grains in that direction. Bagnold (1956) suggested that the relation between the shear stress (T) acting on grains and the dispersive pressure (P) is

$$T/P = \tan a \qquad (2.11)$$

where a is the angle of internal friction. This formula suggests that the minimum slope on which sustained grain flow in air is possible is about 30°; under water, greater slopes may be required for flow to occur. Although dispersion or dilation of sand grains is achieved and maintained during flow primarily by grain collisions, dispersion may be aided under some conditions by upward flow of pore fluids as grains settle or possibly by buoyancy of a dense mud matrix. Grain flow is similar to liquefied flow in many respects and may, in fact, grade into these flows. In contrast to liquefied flows, grain flow can occur under subaerial conditions as well as subaqueous conditions.

Grain flow is common on the lee slopes of sand dunes. Flows of cohesionless sand have also been observed and photographed in the ocean as they moved down steep slopes in submarine canyons (Shepard, 1961; Dill, 1966; Shepard and Dill, 1966). Grain flows over the floors of Norwegian fjords are reported to be responsible for breaking submarine telephone cables. Grain flows may be of limited geological significance because of the steep slopes required to initiate flow, although it has been suggested that grain flow may accompany turbidity currents on less steep slopes, moving beneath but independently of the turbidity currents. Deposition of grain-flow sediment occurs quickly and en masse by sudden "freezing," primarily as a result of reduction of slope angle.

Grain-Flow Deposits

Grain-flow origin has been suggested by some workers for very thick, almost massive sandstone beds; however, Lowe (1976) concluded that the deposits of a single grain flow in any environment cannot be thicker than a few centimeters for sand-size grains. Reverse grading (that is, grading from fine size to coarse size upward, which occurs in some sandstones) has been attributed to grain-flow processes. Reverse grading is assumed to form during grain flow as a result of smaller particles filtering down through larger particles while they are in the dispersed state, a process called **kinetic sieving**.

Grain-flow deposits are massively bedded with little or no internal laminations and grading except possible reverse grading in the base (Fig. 2.7C). Deposits of a single grain flow are commonly less than about 5 cm thick.

Debris Flows and Mud Flows

Subaerial debris flows occur under many climatic conditions but are particularly common in arid and semiarid regions where they are usually initiated after heavy rainfalls. They are also common in volcanic regions where volcanic debris become water saturated during heavy rains that accompany eruptions or from melting ice and snow that accumulate on volcanic cones between eruptions. Debris flows are slurrylike flows composed of highly concentrated, poorly sorted mixtures of sediment and water that behave in a different manner than fluid flows.

Box 2.5 Types of Fluids

Depending upon the extent to which dynamic viscosity (μ) changes with shear or strain (deformation) rate, three general types of fluids can be defined. **Newtonian fluids** have no strength and do not undergo a change in viscosity as the shear rate increases (Fig. 2.5.1). Thus, ordinary water, which does not change viscosity as it is stirred or agitated, is a Newtonian fluid. **Non-Newtonian fluids** have no strength but show variable viscosity (μ) with change in shear or strain rate. Water that contains dispersions of sand in concentrations greater than about 30 percent by volume, or even lower concentrations of cohesive clay, behaves as a non-Newtonian fluid. Therefore, highly water-saturated, noncompacted muds may display non-Newtonian behavior. Such muds may flow sluggishly at low flow velocities but display less viscous flow at higher velocities.

Some extremely concentrated dispersions of sediment may behave as plastic substances, which have an initial strength that must be overcome before they yield. If the plastic material behaves as a substance with constant viscosity after the yield strength is exceeded, it is called a **Bingham plastic**. Debris flows in which large cobbles or boulders are supported in a matrix of interstitial fluid and fine sediment are examples of natural substances that behave as Bingham plastics. Water with dispersed sediment, and other plastic materials (such as ice), which behave as substances with variable viscosity after yield strength is exceeded and they start to flow, are called **pseudoplastics**. **Thixotropic substances,** a special type of pseudoplastic, have strength until sheared. Shearing destroys their strength; the substances then behave like a fluid (commonly non-Newtonian) until allowed to rest a short while, after which strength is regained. Freshly deposited muds commonly display thixotropic behavior. Shearing resulting from earthquake tremors, for example, can cause liquefaction and failure of such muds. Such momentary liquefaction may result in downslope movement of sediment that otherwise would not undergo transport. Differences in behavior of Newtonian fluids, non-Newtonian fluids, and plastic substances in response to shear stress are illustrated schematically in Figure 2.5.1.

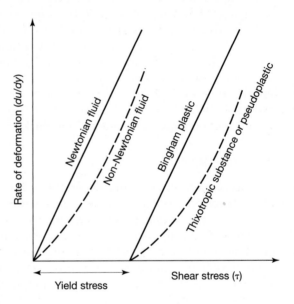

Figure 2.5.1
Types of fluids: rates of deformation vs. shear stress for fluids and plastics. [After Blatt, H., G. V. Middleton, and R. Murray, 1980, Origin of sedimentary rocks, 2nd ed., Fig. 5.26, p. 187. Reprinted by permission of Prentice-Hall, Englewood Cliffs, N.J.]

Debris-Flow Mechanisms

Debris flows behave as Bingham plastics; that is, they have a yield strength that must be overcome before flow begins. The cohesive mud matrix in debris flows has enough strength to support large grains, but cohesiveness is not great enough to prevent flow on an adequate slope. Debris flows are generally initiated on steep slopes (>10°), but they can flow considerable distances on gentle slopes of 5° or less; they occur in both subaerial and subaqueous environments. They consist of poorly sorted mixtures of particles, which may range to boulder-size, in a fine gravel, sand, or mud matrix. Those composed predominantly of mud-size grains are **mud flows** and those with a lower but substantial mud fraction (>5 percent by volume) are **muddy debris flows** (Middleton, 1991). The grains in these

Figure 2.10
Poorly sorted debris-flow deposits (Eocene), north-central Oregon. (Photograph courtesy of Abbas Seyedolali)

mud-bearing debris flows are supported in a matrix of mud and interstitial water that has enough cohesive strength to prevent larger particles from settling but not enough strength to prevent flow. Debris flows that have a matrix composed predominantly of cohesionless sand and gravel are **mud-free debris flows** (Middleton, 1991). The support mechanism for these mud-free debris flows is poorly understood.

After the yield strength of a debris flow has been overcome owing to water saturation, and movement begins, the flow may continue to move over slopes as low as 1° or 2° (Curray, 1966). Debris flows are believed to occur also in subaqueous environments, possibly as a result of mixing at the downslope ends of subaqueous slumps. As subaqueous debris flows move rapidly downslope and are diluted by mixing with more water, their strength is reduced, and they may pass into turbidity currents. Deposition of the entire mass of debris flows and mud flows occurs quickly. When the shear stress owing to gravity no longer exceeds the yield strength of the base of the flow, the mass "freezes" and stops moving.

Debris-Flow Deposits

Debris-flow deposits are thick, poorly sorted units that lack internal layering (Fig. 2.7D; Fig. 2.10). They typically consist of chaotic mixtures of particles that may range in size from clay to boulders. The large particles commonly show no preferred orientation. They are generally poorly graded, but if grading is present, it may be either normal or reverse.

FURTHER READING

Carling, P. A., and M. R. Dawson, 1996, Advances in fluvial dynamics and stratigraphy: John Wiley & Sons, Chichester, 530 p.

Clifford, N., J. R. French, and J. Hardisty (eds.), 1993, Turbulence: Perspectives on flow and sediment transport: John Wiley & Sons, Chichester, 360 p.

Edwards, D. A., 1993, Turbidity currents: Dynamics, deposits and reversals: Lecture Notes in Earth Sciences, Springer-Verlag, Berlin, 173 p.

Julien, P. Y., 1995, Erosion and sedimentation: Cambridge University Press, Cambridge, 280 p.

Leeder, M., 1999, Sedimentology and sedimentary basins: Blackwell Science Ltd., Oxford, 592 p.

Middleton, G. V., and J. B. Southard, 1984, Mechanics of sediment movement: Soc. Econ. Paleontologists and Mineralogists Short Course Notes No. 3, 2nd ed., 401 p.

Middleton, G. V., and P. R. Wilcock, 1994, Mechanics in the earth and environmental sciences: Cambridge University Press, Cambridge, 459 p.

Pye, K., 1994, Sediment transport and depositional processes: Blackwell Scientific Publications, Oxford, 397 p.

Yang, C. T., 1996, Sediment transport: Theory and practice: McGraw-Hill Companies, Inc., New York, 396 p.

PART II

Physical Properties of Sedimentary Rocks

Flute casts and groove casts on the base of a turbidite sandstone bed (Cretaceous), northern Klamath Mountains, California

The transport and depositional processes described in Chapter 2 generate a wide variety of sedimentary rocks, each characterized by distinctive textural and structural properties. **Sedimentary texture** refers to the features of sedimentary rocks that arise from the size, shape, and orientation of individual sediment grains. Geologists have long assumed that the texture of sedimentary rocks reflects the nature of transport and depositional processes and that characterization of texture can aid in interpreting ancient environmental settings and boundary conditions. An extensive literature has thus been published dealing with various aspects of sediment texture, particularly methods of measuring and expressing grain size and shape and interpretation of grain size and shape data. The textures of siliciclastic sedimentary rocks are produced primarily by physical processes of sedimentation and are considered to encompass grain **size, shape** (form, roundness, surface texture), and **fabric** (grain orientation and grain-to-grain relations). The interrelationship of these primary textural properties controls other, derived, textural properties such as **bulk density, porosity,** and **permeability**. The textures of some nonsiliciclastic sedimentary rocks such as certain limestones and evaporites are also generated partly or wholly by physical transport processes. The texture of others is principally caused by chemical or biochemical sedimentation processes. Extensive recrystallization or other diagenetic changes may destroy the original textures of nonsiliciclastic sedimentary rocks and produce crystalline textural fabrics that are largely of secondary origin. Obviously, the textural features of chemically or biochemically formed sedimentary rocks, and of rocks with strong diagenetic fabrics, have quite different genetic significance from those of unaltered siliciclastic sedimentary rocks.

Whereas the term "texture" applies mainly to the properties of individual sediment grains, **sedimentary structures**, such as cross-bedding and ripple marks, are features formed from aggregates of grains. These structures are generated by a variety of sedimentary processes, including fluid flow, sediment gravity flow, soft-sediment deformation, and biogenic activity. Because sedimentary structures reflect environmental conditions that prevailed at or very shortly after the time of deposition, they are of special interest to geologists as a tool for interpreting ancient depositional environments. We can use sedimentary structures to help evaluate such aspects of ancient sedimentary environments as sediment transport mechanisms, paleocurrent flow directions, relative water depths, and relative current velocities. Some sedimentary structures are also used to identify the tops and bottoms of beds and thus to determine if sedimentary successions are in depositional stratigraphic order or have been overturned by tectonic forces. Sedimentary structures are particularly abundant in coarse siliciclastic sedimentary rocks that originate through traction transport or turbidity current transport. They occur also in nonsiliciclastic sedimentary rocks such as limestones and evaporites.

3

Sedimentary Textures

3.1 INTRODUCTION

This chapter focuses primarily on the physically produced textures of silici-clastic sedimentary rocks. Some of the special textural features that are important to understanding the classification and genesis of limestones and other nonsiliciclastic sedimentary rocks are discussed in Chapters 6 and 7. In this chapter, we examine the characteristic textural properties of grain size and shape, particle surface texture, and grain fabric and discuss the genetic significance of these properties. Although the study of sedimentary textures may not be the most exciting aspect of sedimentology, it is nonetheless an important field of study. A thorough understanding of the nature and significance of sedimentary textures is fundamental to interpretation of ancient depositional environments and transport conditions, although much uncertainty still attends the genetic interpretation of textural data. Some long-standing ideas about the genetic significance of sediment textural data are now being challenged, while new ideas and techniques for studying and interpreting sediment texture continue to emerge. No textbook on sedimentology would be complete without some discussion of sediment texture and its genetic significance.

3.2 GRAIN SIZE

Grain size is a fundamental attribute of siliciclastic sedimentary rocks and thus one of the important descriptive properties of such rocks. The sizes of particles in a particular deposit reflect weathering and erosion processes, which generate particles of various sizes, and the nature of subsequent transport processes, as discussed in Chapter 2. Grains can range in size from clay-size particles that require a microscope for clear visualization to boulders several meters in diameter. Sedimentologists are particularly concerned with three aspects of particle size: (1) techniques for measuring grain size and expressing it in terms of some type of grain-size or grade scale, (2) methods for summarizing large amounts of grain-size data and presenting them in graphical or statistical form so that they can be more easily evaluated, and (3) the genetic (e.g., environmental) significance of these data. We will now examine each of these concerns.

Grain-Size Scales

As mentioned, particles in sediments and sedimentary rocks range in size from a few microns to a few meters. Because of this wide range of particle sizes, logarithmic or geometric scales are more useful for expressing size than are linear scales. In a geometric scale there is a succession of numbers such that a fixed ratio exists between successive elements of the series. The grain-size scale used almost universally by sedimentologists is the **Udden-Wentworth scale**. This scale, first proposed by Udden in 1898 and modified and extended by Wentworth in 1922, is a geometric scale in which each value in the scale is either twice as large as the preceding value or one-half as large, depending upon the sense of direction (Table 3.1). The scale extends from <1/256 mm (0.0039 mm) to >256 mm and is divided into four major size categories (clay, silt, sand, and gravel), which can be further subdivided (e.g., fine sand, medium sand, coarse sand). Blair and McPherson (1999) suggest that the coarse end of the Udden-Wentworth scale be divided into a greater number of subdivisions than those shown in Table 3.1 by adding *block* (4.1–65.5 m), *slab* (65.5 m–1.0 km), *monolith* (1.0–33.6 km), and *megalith* (>33.6 km).

A useful modification of the Udden-Wentworth scale is the logarithmic **phi scale**, which allows grain-size data to be expressed in units of equal value for the purpose of graphical plotting and statistical calculations. This scale, proposed by Krumbein in 1934, is based on the following relationship:

$$\phi = -\log_2 d \qquad (3.1)$$

where ϕ is phi size and d is the grain diameter in millimeters. For example, a grain 4 mm in diameter has a phi size of –2, which is the power required to raise the base (2) of the logarithm to 4 (i.e., 2^2). A grain 8 mm in size has a phi value of –3 (the base 2^3). Some equivalent phi and millimeter sizes are shown in Table 3.1. Note that the phi scale yields both positive and negative numbers. The real size of particles, expressed in millimeters, decreases with increasing positive phi values and increases with decreasing negative values. Because sand-size and smaller grains are the most abundant grains in sedimentary rocks, Krumbein chose the negative logarithm of the grain size in millimeters so that grains of this size will have positive phi values, avoiding the bother of constantly working with negative numbers. This usage is also consistent with the common practice of plotting coarse sizes to the left and fine sizes to the right in graphs.

Measuring Grain Size

The size of siliciclastic grains can be measured by several techniques (Table 3.2). The choice of methods is dictated by the purpose of the study, the range of grain sizes to be measured, and the degree of consolidation of sediment or sedimentary rock. Large particles (pebbles, cobbles, boulders) in either unconsolidated sediment or lithified sedimentary rock can be measured manually with a caliper or tape. Grain size is commonly expressed in terms of either the long dimension or the intermediate dimension of the particles. Granule- to silt-size particles in unconsolidated sediments or sedimentary rocks that can be disaggregated are commonly measured by sieving through a set of nested, wire-mesh screens (see Ingram, 1971). The sieve numbers of U.S. Standard Sieves that correspond to various millimeter and phi sizes are shown in Table 3.1. Sieving techniques measure the intermediate dimension of particles because the intermediate particle size generally determines whether or not a particle can go through a particular mesh.

Grain size of small, unconsolidated particles can also be measured by sedimentation techniques on the basis of the settling velocity of the particles. In these techniques, grains are allowed to settle through a column of water at a specified

Table 3.1 The Wentworth grain-size scale for sediments, showing Wentworth size classes, equivalent phi (ϕ) units, and sieve numbers of U.S. Standard Sieves corresponding to various millimeter and ϕ sizes

U.S. standard sieve mesh	Millimeters	Phi (ϕ) units	Wentworth size class
	4096	−12	Boulder
	1024	−10	Boulder
	256 / 256	−8	
	64 / 64	−6	Cobble
	16	−4	Pebble
5	4 / 4	−2	
6	3.36	−1.75	
7	2.83	−1.5	Granule
8	2.38	−1.25	
10	2.00 / 2	−1.0	
12	1.68	−0.75	
14	1.41	−0.5	Very coarse sand
16	1.19	−0.25	
18	1.00 / 1	0.0	
20	0.84	0.25	
25	0.71	0.5	Coarse sand
30	0.59	0.75	
35	0.50 / ½	1.0	
40	0.42	1.25	
45	0.35	1.5	Medium sand
50	0.30	1.75	
60	0.25 / ¼	2.0	
70	0.210	2.25	
80	0.177	2.5	Fine sand
100	0.149	2.75	
120	0.125 / ⅛	3.0	
140	0.105	3.25	
170	0.088	3.5	Very fine sand
200	0.074	3.75	
230	0.0625 / 1/16	4.0	
270	0.053	4.25	
325	0.044	4.5	Coarse silt
	0.037	4.75	
	0.031 / 1/32	5.0	
	0.0156 / 1/64	6.0	Medium silt
	0.0078 / 1/125	7.0	Fine silt
	0.0039 / 1/256	8.0	Very fine silt
	0.0020	9.0	
	0.00098	10.0	Clay
	0.00049	11.0	
	0.00024	12.0	
	0.00012	13.0	
	0.00006	14.0	

(GRAVEL, SAND, MUD — SILT, CLAY left-margin labels)

temperature in a settling tube, and the time required for the grains to settle is measured. For coarser particles (granules, sand, silt), the settling time of the particles is related empirically to a standard size-distribution curve (calibration curve) to obtain the equivalent millimeter or phi size (see, for example, Poppe, Eliason, and Fredricks, 1985). As mentioned in Chapter 2, settling velocity of particles is affected

Table 3.2 Methods of measuring sediment grain size

Type of sample	Sample grade	Method of analysis
Unconsolidated sediment and disaggregated sedimentary rock	Boulders Cobbles Pebbles	Manual measurement of individual clasts
	Granules Sand Silt ───	Sieving, settling-tube analysis, image analysis
	Clay	Pipette analysis, sedimentation balances, photohydrometer, Sedigraph, laser-diffractometer, electroresistance (e.g., Coulter counter)
Lithified sedimentary rock	Boulders Cobbles Pebbles	Manual measurement of individual clasts
	Granules Sand Silt ───	Thin-section measurement, image analysis
	Clay	Electron microscope

by particle shape. Spherical particles settle faster than nonspherical particles of the same mass. Therefore, determining the grain sizes of natural, nonspherical particles by sedimentation techniques may not yield exactly the same values as those determined by sieving.

The grain size of fine silt and clay particles can be determined by sedimentation methods based on Stokes's Law (Chapter 2). The standard sedimentation method for measuring the sizes of these small particles is **pipette analysis** (Galehouse, 1971). Pipette analysis is a laborious process because of the many operations involved. To simplify these procedures, automatic-recording settling tubes have now been developed that allow the sizes of both sand-size and clay-size sediment to be more easily and rapidly determined. Several other kinds of automated particle-size analyzers are also available, each based on a slightly different principle. A **photohydrometer** is a settling tube that empirically relates changes in intensity of a beam of light passed through a column of suspended sediment to particle settling velocities and thus to particle size (Jordan, Freyer, and Hemmen, 1971). The **Sedigraph** determines particle size by measuring the attenuation of a finely collimated X-ray beam as a function of time and height in a settling suspension (Stein, 1985; Jones, McCave, and Patel, 1988). A **laser-diffracter size analyzer** operates on the principle that particles of a given size diffract light through a given angle, which increases with decreasing particle size (McCave et al., 1986). **Electroresistance size analyzers**, such as the Coulter counter or Electrozone particle counter, measure grain size on the basis of the principle that a particle passing through an electrical field maintained in an electrolyte will displace its own volume of the electrolyte and thus cause a change in the field. These changes are scaled and counted as voltage pulses; the magnitude of each pulse is proportional to particle volume (Swift, Schubert, and Sheldon, 1972; Muerdter, Dauphin, and Steele, 1981). Semiautomated **image analysis** techniques use TV cameras to capture and digitize grain images from which, with the aid of appropriate computer software, grain-size diameters can be calculated (Kennedy and Mazzullo, 1991). For a comparison of some of these analytical techniques, see Singer et al. (1988). Additional information is available in *Principles, Methods, and Applications of Particle Size Analysis*, a monograph edited by Syvitski (1991).

The grain size of particles in consolidated sedimentary rocks that cannot be disaggregated must be measured by techniques other than sieving or sedimentation analysis. The size and sorting of sand- and silt-size particles can be estimated by using a reflected-light binocular microscope and a standard size-comparison set, which consists of grains of specific sizes mounted on a card. More accurate size determination can be made by measuring grains in thin sections of rock by use of a transmitted-light petrographic microscope fitted with an ocular micrometer or by the image analysis technique mentioned above. Both microscopic and image analysis techniques tend to yield grain sizes that are smaller than the maximum diameter of the grains because the plane of a thin section does not cut exactly through the centers of most grains. Grain sizes measured by these methods are commonly corrected mathematically in some way to make them agree more closely with sieve data (Burger and Skala, 1976; Piazzola and Cavaroc, 1991). Fine silt- and clay-size grains in consolidated rocks may be studied by use of an electron microscope, although the electron microscope is not commonly used for grain-size measurements.

Graphical and Mathematical Treatment of Grain-Size Data

Measurement of grain size by the techniques described generates large quantities of data that must be reduced to a more condensed form before they can be used. Tables of data showing the weights of grains in various size classes must be simplified to yield such average properties of grain populations as mean grain size and sorting. Both graphical and mathematical data-reduction methods are in common use. Graphical plots are simple to construct and provide a readily understandable visual representation of grain-size distributions. On the other hand, mathematical methods, some of which are based on initial graphical treatment of data, yield statistical grain-size parameters that may be useful in environmental studies.

Graphical Plots

Figure 3.1 illustrates three common graphical methods for presenting grain-size data. Figure 3.1A shows typical grain-size data obtained by sieve analysis. Raw sieve weights are first converted to individual weight percents by dividing the weight in each size class by the total weight. Cumulative weight percent may be calculated by adding the weight of each succeeding size class to the total of the preceding classes. Figure 3.1B shows how individual weight percent can be plotted as a function of grain size to yield a grain-size **histogram**—a bar diagram in which grain size is plotted along the abscissa of the graph and individual weight percent along the ordinate. Histograms provide a quick, easy, pictorial method for representing grain-size distributions because the approximate average grain size and the sorting—the spread of grain-size values around the average size—can be seen at a glance. Histograms have limited application, however, because the shape of the histogram is affected by the sieve interval used. Also, they cannot be used to obtain mathematical values for statistical calculations.

A **frequency curve** (Figure 3.1B) is essentially a histogram in which a smooth curve takes the place of a discontinuous bar graph. Connecting the midpoints of each size class in a histogram with a smooth curve gives the approximate shape of the frequency curve. A frequency curve constructed in this manner does not, however, accurately fix the position of the highest point on the curve; this point is important for determining the modal size, to be described. A grain-size histogram plotted from data obtained by sieving at exceedingly small sieve intervals would yield the approximate shape of a frequency curve, but such small sieve intervals are not practical. Accurate frequency curves can be derived from cumulative curves by special graphical methods described in detail by Folk (1974).

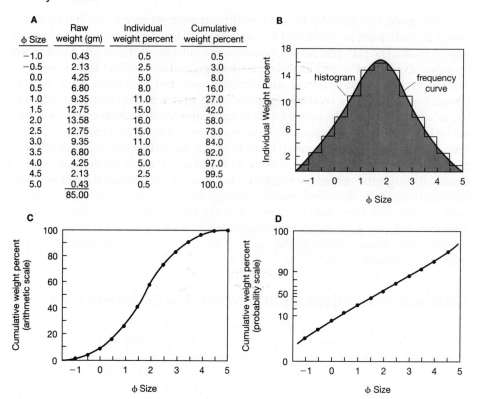

φ Size	Raw weight (gm)	Individual weight percent	Cumulative weight percent
−1.0	0.43	0.5	0.5
−0.5	2.13	2.5	3.0
0.0	4.25	5.0	8.0
0.5	6.80	8.0	16.0
1.0	9.35	11.0	27.0
1.5	12.75	15.0	42.0
2.0	13.58	16.0	58.0
2.5	12.75	15.0	73.0
3.0	9.35	11.0	84.0
3.5	6.80	8.0	92.0
4.0	4.25	5.0	97.0
4.5	2.13	2.5	99.5
5.0	0.43	0.5	100.0
	85.00		

Figure 3.1
Common visual methods of displaying grain-size data. A. Grain-size data table. B. Histogram and frequency curve plotted from data in A. C. Cumulative curve with an arithmetic ordinate scale. D. Cumulative curve with a probability ordinate scale.

A grain-size **cumulative curve** is generated by plotting grain size against cumulative weight percent frequency. The cumulative curve is the most useful of the grain-size plots. Although it does not give as good a pictorial representation of the grain-size distribution as does a histogram or frequency curve, its shape is virtually independent of the sieve interval used. Also, data that can be derived from the cumulative curve allow calculation of several important grain-size statistical parameters. A cumulative curve can be plotted on an arithmetic ordinate scale (Fig. 3.1C) or on a log probability scale in which the arithmetic ordinate is replaced by a log probability ordinate (Fig. 3.1D). When phi-size data are plotted on an arithmetic ordinate, the cumulative curve typically has the S-shape shown in Figure 3.1C. The slope of the central part of this curve reflects the sorting of the sample. A very steep slope indicates good sorting, and a very gentle slope poor sorting. If the cumulative curve is plotted on log probability paper, the shape of the curve will tend toward a straight line if the population of grains has a normal distribution (actually log-normal, as illustrated in Figure 3.1D). In a normal distribution, the values show an even distribution, or spread, about the average value. In conventional statistics, a normally distributed population of values yields a perfect bell-shaped curve when plotted as a frequency curve. Deviations from normality of a grain-size distribution can thus be easily detected on log probability plots by deviation of the cumulative curve from a straight line. Most natural populations of grains in siliciclastic sediments or sedimentary rocks do not have a normal (or log-normal) distribution; the nearly normal distribution shown in Figure 3.1B is not typical of natural sediments. Some investigators believe that the shape of the log probability curve reflects conditions of the sediment transport process and thus can be used as a tool in environmental interpretation. We shall return to this point subsequently.

Graphical plots permit quick, visual inspection of the grain-size characteristics of a given sample; however, comparison of graphical plots becomes cumbersome and inconvenient when large numbers of samples are involved. Also, average grain-size and grain-sorting characteristics cannot be determined very accurately by visual inspection of grain-size curves. To overcome these disadvantages, mathematical methods that permit statistical treatment of grain-size data can be used to derive parameters that describe grain-size distributions in mathematical language. These statistical measures allow both the average size and the average sorting characteristics of grain populations to be expressed mathematically. Mathematical values of size and sorting can be used to prepare a variety of graphs and charts that facilitate evaluation of grain-size data.

Mathematical Measures

Average Grain Size.   Three mathematical measures of average grain size are in common use. The **mode** is the most frequently occurring particle size in a population of grains. The diameter of the modal size corresponds to the diameter of grains represented by the steepest point (inflection point) on a cumulative curve or the highest point on a frequency curve. Siliciclastic sediments and sedimentary rocks tend to have a single modal size, but some sediments are bimodal, with one mode in the coarse end of the size distribution and one in the fine end. Some are even polymodal. The **median size** is the midpoint of the grain-size distribution. Half of the grains by weight are larger than the median size, and half are smaller. The median size corresponds to the 50th percentile diameter on the cumulative curve (Fig. 3.2). The **mean** size is the arithmetic average of all the particle sizes in a sample. The true arithmetic mean of most sediment samples cannot be determined because we cannot count the total number of grains in a sample or measure each small grain. An approximation of the arithmetic mean can be arrived at by picking selected percentile values from the cumulative curve and averaging these values. As shown in Figure 3.2 and Table 3.3, the 16th, 50th, and 84th percentile values are commonly used for this calculation. The mode, median, and mean sizes may or may not be the same, as subsequently discussed under the topic of skewness (e.g., Fig. 3.5).

Sorting.   The sorting of a grain population is a measure of the range of grain sizes present and the magnitude of the spread or scatter of these sizes around the mean size. Sorting can be estimated in the field or laboratory by use of a hand lens or

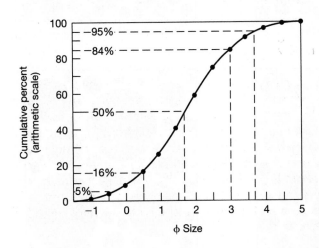

Figure 3.2
Method for calculating percentile values from the cumulative curve.

Table 3.3 Formulas for calculating grain-size statistical parameters by graphical methods

Graphic mean	$M_z = \dfrac{\phi_{16} + \phi_{50} + \phi_{84}}{3}$	(1)
Inclusive graphic standard deviation	$\sigma_i = \dfrac{\phi_{84} - \phi_{16}}{4} + \dfrac{\phi_{95} - \phi_{5}}{6.6}$	(2)
Inclusive graphic skewness	$SK_t = \dfrac{(\phi_{84} + \phi_{16} - 2\phi_{50})}{2(\phi_{84} - \phi_{16})} + \dfrac{(\phi_{95} + \phi_{5} - 2\phi_{50})}{2(\phi_{95} - \phi_{5})}$	(3)
Graphic kurtosis	$K_G = \dfrac{(\phi_{95} - \phi_{5})}{2.44(\phi_{75} - \phi_{25})}$	(4)

Source: Folk, R. L., and W. C. Ward, 1957, Brazos River bar: A study in the significance of grain-size parameters: *Jour. Sed. Petrology*, v. 27, p. 3–26.

microscope and reference to a visual estimation chart (Fig. 3.3). More accurate determination of sorting requires mathematical treatment of grain-size data. The mathematical expression of sorting is **standard deviation**. In conventional statistics, one standard deviation encompasses the central 68 percent of the area under the frequency curve (Fig. 3.4). That is, 68 percent of the grain-size values lie within plus or minus one standard deviation of the mean size. A formula for calculating standard deviation by graphical-statistical methods is shown in Table 3.3. Note that the standard deviation calculated by this formula is expressed in phi (ϕ) values and is called **phi standard deviation**. The symbol ϕ must always be

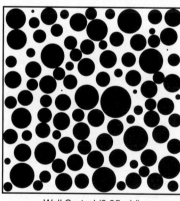

Well Sorted (0.35 phi)

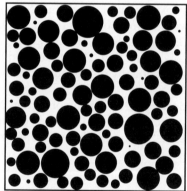

Moderately Well Sorted (0.50 phi)

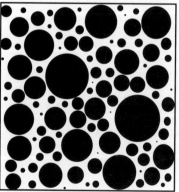

Poorly Sorted (1.00 phi)

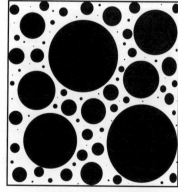

Very Poorly Sorted (2.00 phi)

Figure 3.3
Visual images for estimating grain-size sorting. [After Harrell, J., 1984, A visual comparator for degrees of sorting in thin and plane sections: Jour. Sed. Petrology, v. 54, Fig. 3, 4, 5, 6, p. 684, reproduced by permission of SEPM, Tulsa, Okla.]

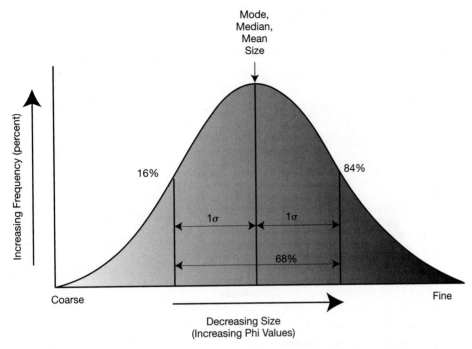

Figure 3.4
Grain-size frequency curve for a normal distribution of phi values, showing the relationship of standard deviation to the mean, mode, and median size. One standard deviation (1σ) on either side of the mean size accounts for 68 percent of the area under the frequency curve. [Based in part on Friedman and Sanders, 1978.]

attached to the standard deviation value. Verbal terms for sorting corresponding to various values of standard deviation are as follows (after Folk, 1974):

Standard Deviation	
$<0.35\phi$	very well sorted
$0.35–0.50\phi$	well sorted
$0.50–0.71\phi$	moderately well sorted
$0.71–1.00\phi$	moderately sorted
$1.00–2.00\phi$	poorly sorted
$2.00–4.00\phi$	very poorly sorted
$>4.00\phi$	extremely poorly sorted

Skewness and Kurtosis. As mentioned, most natural sediment grain-size populations do not exhibit a normal or log-normal grain-size distribution. The frequency curves of such nonnormal populations are not perfect bell-shaped curves, such as the example shown in Figure 3.4. Instead, they show some degree of asymmetry, or **skewness**. The mode, mean, and median in a skewed population of grains are all different, as illustrated in Figure 3.5. Skewness reflects sorting in the "tails" of a grain-size population. Populations that have a tail of excess fine particles (Fig. 3.5A) are said to be positively skewed or fine skewed, that is, skewed toward positive phi values. Populations with a tail of excess coarse particles (Fig. 3.5B) are negatively skewed or coarse skewed. Graphic skewness can be calculated by Equation 3 in Table 3.3. Verbal skewness is related to calculated values of skewness as follows (Folk, 1974):

Skewness	
$>+0.30$	strongly fine skewed
$+0.30$ to $+0.10$	fine skewed
$+0.10$ to -0.10	near symmetrical
-0.10 to -0.30	coarse skewed
<-0.30	strongly coarse skewed

A. Positively (fine) Skewed

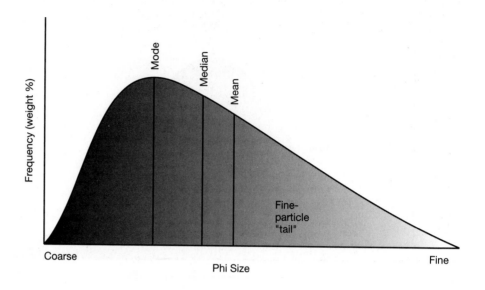

B. Negatively (coarse) Skewed

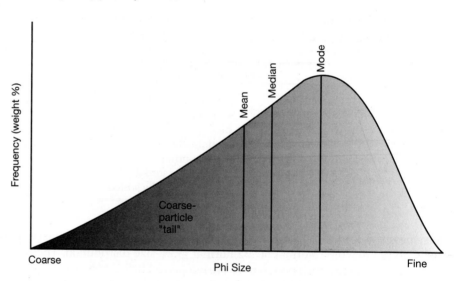

Figure 3.5
Skewed grain-size frequency curve, illustrating the difference between positive (fine) and negative (coarse) skewness. Note the difference between these skewed, asymmetrical curves and the normal frequency curve shown in Figure 3.4. [Based in part on Friedman and Sanders, 1978.]

Grain-size frequency curves can show various degrees of sharpness or peakedness. The degree of peakedness is called **kurtosis**. A formula for calculating kurtosis is shown in Table 3.3. Although kurtosis is commonly calculated along with other grain-size parameters, the geological significance of kurtosis is unknown, and it appears to have little value in interpretative grain-size studies.

Moments. Grain size statistical parameters can be calculated directly without reference to graphical plots by the mathematical **method of moments** (Krumbein and Pettijohn, 1938). The method had not been used extensively until comparatively recently because of the laborious calculations involved and because it had not been definitely proven that moment statistics are of greater value than graphical statistics in application to geologic problems. With the advent of modern computers, lengthy calculations no longer pose a problem, and moment statistics are now commonly used. The computations in moment statistics involve multiplying a weight (weight frequency in percent) by a distance (from the midpoint of each

Table 3.4 Formulas for calculating grain-size parameters by the moment method

Mean (1st moment)	$\bar{x}_\phi = \dfrac{\Sigma fm}{n}$	(1)
Standard deviation (2nd moment)	$\sigma_\phi = \sqrt{\dfrac{\Sigma f(m - \bar{x}_\phi)^2}{100}}$	(2)
Skewness (3rd moment)	$Sk_\phi = \dfrac{\Sigma f(m - \bar{x}_\phi)^3}{100\sigma_\phi^3}$	(3)
Kurtosis (4th moment)	$K_\phi = \dfrac{\Sigma f(m - \bar{x}_\phi)^4}{100\sigma_\phi^4}$	(4)

where f = weight percent (frequency) in each grain-size grade present
$\quad\quad m$ = midpoint of each grain-size grade in phi values
$\quad\quad n$ = total number in sample; 100 when f is in percent

size grade to the arbitrary origin of the abscissa). Formulas for computing moment statistics are given in Table 3.4, and a sample computation form using $1/2\,\phi$ size classes is given in Table 3.5. Millimeter values can also be used in computing moment statistics; that is, the size data need not be transformed to phi units. The formulas shown in Table 3.3 can easily be programmed into a computer or programmable calculator to calculate moment statistics from sieve data.

Application and Importance of Grain-Size Data

Grain size is a fundamental physical property of sedimentary rocks and, as such, is a useful descriptive property. Because grain size affects the related derived properties of porosity and permeability, the grain size of potential reservoir rocks is of considerable interest to petroleum geologists and hydrologists. For example, coarse-grained, well-sorted sandstones are better reservoir rocks for oil than are fine-grained, poorly sorted sandstones. Such well-sorted sandstones are also better aquifers for groundwater. Grain-size data are used in a variety of other ways (summarized by Syvitski, 1991):

1. To interpret coastal stratigraphy and sea-level fluctuations
2. To trace glacial sediment transport and the cycling of glacial sediments from land to sea
3. By marine geochemists to understand the fluxes, cycles, budgets, sources, and sinks of chemical elements in nature
4. To understand the mass physical (geotechnical) properties of seafloor sediment, i.e., the degree to which these sediments are likely to undergo slumping, sliding, or other deformation

Finally, because the size and sorting of sediment grains may reflect sedimentation mechanisms and depositional conditions, grain-size data are assumed to be useful for interpreting the depositional environments of ancient sedimentary rocks.

It is this assumption that grain-size characteristics reflect conditions of the depositional environment that has sparked most of the interest in grain-size analyses. Geologists have studied the grain-size properties of sediments and sedimentary rocks for more than a century; research efforts since the 1950s have

Table 3.5 Form for computing moment statistics using $1/2\phi$ classes

Class interval (ϕ)	m Midpoint (ϕ)	f Weight %	fm Product	$m - \bar{x}$ Deviation	$(m - \bar{x})^2$ Deviation squared	$f(m - \bar{x})^2$ Product	$(m - \bar{x})^3$ Deviation cubed	$f(m - \bar{x})^4$ Product	$(m - \bar{x})^4$ Deviation quadrupled	$f(m - \bar{x})^4$ Product
0–0.5	0.25	0.9	0.2	−2.13	4.54	4.09	−9.67	−8.70	20.60	18.54
0.5–1.0	0.75	2.9	2.2	−1.63	2.66	7.71	−4.34	−12.59	7.07	20.50
1.0–1.5	1.25	12.2	15.3	−1.13	1.28	15.62	−1.45	−17.69	1.63	19.89
1.5–2.0	1.75	13.7	24.0	−0.63	0.40	5.48	−0.25	−3.43	0.16	2.19
2.0–2.5	2.25	23.7	53.3	−0.13	0.02	0.47	0.00	0.00	0.00	0.00
2.5–3.0	2.75	26.8	73.7	0.37	0.13	3.48	0.05	1.34	0.02	0.54
3.0–3.5	3.25	12.2	39.7	0.87	0.76	9.27	0.66	8.05	0.57	6.95
3.5–4.0	3.75	5.6	21.0	1.37	1.88	10.53	2.57	14.39	3.52	19.71
>4.0	4.25	2.0	8.5	1.87	3.50	7.00	6.55	13.10	12.25	24.50
Total		100.0	237.9			63.65		−5.53		112.82

Source: McBride, E. F., Mathematical treatment of size distribution data, *in* R. E. Carver (ed.), *Procedures in sedimentary petrology*. © 1971 by John Wiley & Sons, Inc. Table 2, p. 119, reprinted by permission of John Wiley & Sons, Inc., New York.

focused particularly on statistical treatment of grain-size data. This prolonged period of grain-size research has generated hundreds of learned papers in geological journals. Thus, it would be logical to assume that by now the relationship between grain-size characteristics and depositional environments has been firmly established. Alas, that is not the case!

Many techniques for utilizing grain-size data to interpret depositional environments have been tried; however, little agreement exists regarding the reliability of these methods. For example, Friedman (1967, 1979) used two-component grain-size variation diagrams, in which one statistical parameter is plotted against another (e.g., skewness vs. standard deviation [sorting], or mean size vs. standard deviation). This method putatively allows separation of the plots into major environmental fields such as beach environments and river environments (e.g., Fig. 3.6). Passega (1964, 1977) developed a graphical approach to interpreting grain-size data that makes use of what he calls C-M and L-M diagrams, in which grain diameter of the coarsest grains in the deposit (C) is plotted against either the median grain diameter (M) or the percentile finer than 0.031 mm (L). Passega maintains that most samples from a given environment will fall within a specific environmental field in these diagrams. Visher (1969), Sagoe and Visher (1977), and Glaister and Nelson (1974) have all suggested that depositional environments can be interpreted on the basis of the shapes of grain-size cumulative curves plotted on log probability paper. Cumulative curves plotted in this way commonly display two or three straight-line segments rather than a single straight line (Fig. 3.7; see Visher, 1969, for additional examples). Each segment of the curve is interpreted to represent different subpopulations of grains that were transported simultaneously but by different transport modes, that is, by suspension, saltation, and bedload transport. Sediments from different environments (dune, fluvial, beach, tidal, nearshore, turbidite) can putatively be differentiated on the basis of the general shapes of the curves, the slopes of the curve segments, and the positions of the truncation points (breaks in slope) between the straight-line segments (e.g., Visher, 1969). Although success using these various techniques has been reported by some investigators, all of the techniques have been criticized for yielding incorrect results in an unacceptably high percentage of cases (e.g., Tucker and Vacher, 1980; Sedimentation Seminar, 1981; Vandenberghe, 1975; Reed, LeFever, and Moir, 1975).

More sophisticated multivariate statistical techniques, such as factor analysis and discriminant function analysis (e.g., Chambers and Upchurch, 1979; Taira and

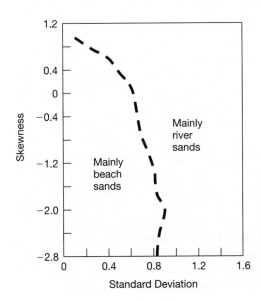

Figure 3.6
Grain-size bivariate plot of moment skewness vs. moment standard deviation showing the fields in which most beach and river sands plot. [Redrawn from Friedman, G. M., 1967, Dynamic processes and statistical parameters for size frequency distributions of beach and river sands: Jour. Sed. Petrology, v. 37, Fig. 5, p. 334, reproduced by permission of SEPM, Tulsa, Okla.]

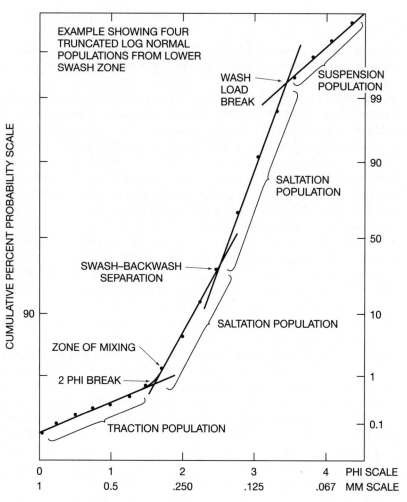

CUMULATIVE PERCENT PROBABILITY SCALE

EXAMPLE SHOWING FOUR
TRUNCATED LOG NORMAL
POPULATIONS FROM LOWER
SWASH ZONE

WASH
LOAD
BREAK

SUSPENSION
POPULATION

SALTATION
POPULATION

SWASH–BACKWASH
SEPARATION

SALTATION POPULATION

ZONE OF MIXING

2 PHI BREAK

TRACTION POPULATION

| 0 | 1 | 2 | 3 | 4 | PHI SCALE |
| 1 | 0.5 | .250 | .125 | .067 | MM SCALE |

Figure 3.7
Relation of sediment transport dynamics to populations and truncation points
in a grain-size distribution as revealed
by plotting grain-size data as a cumulative curve on log probability paper.
[After Visher, G. S., 1969, Grain size distributions and depositional processes:
Jour. Sed. Petrology, v. 39, Fig. 4, p.
1079, reprinted by permission of SEPM,
Tulsa, Okla.]

Scholle, 1979; Stokes, Nelson, and Healey, 1989) and the so-called **log-hyperbolic distribution** (Barndorff-Nielsen, 1977; Bagnold and Barndorff-Nielsen, 1980; Barndorff-Nielson et al., 1982), have also been used. These techniques have not yet been widely applied, however, and some of the assumptions used in these statistical methods have come under criticism (e.g., Forrest and Clark, 1989). Also, some investigators (e.g., Fieller, Gilbertson, and Olbricht, 1984; Wyrwoll and Smith, 1985) report no better results with log-hyperbolic distributions than with results based on the normal probability function. Thus, it appears that after several decades of intensive research into the techniques and significance of grain-size analysis, during which the techniques for interpreting grain-size data have come to demand increasingly sophisticated statistical applications, there is little consensus as to their reliability for environmental analysis. Such is science!

Reasons why grain-size techniques for identifying depositional environments of sediments fail to work consistently are probably related to variability in depositional conditions within major environmental settings. The energy conditions and sediment supply within river systems, for example, can differ considerably from one river to another and even within different parts of the same river system. Thus, in some cases, the grain-size characteristics of sediments may show as much variability within different parts of the same environmental setting as between different environments. Grain-size distributions reflect processes, not environments, and sediment transport processes are not unique to a particular environment. In any case, grain-size data should be considered as only one of the available tools for environmental interpretation and should not be used alone for this purpose.

3.3 PARTICLE SHAPE

The shapes of minerals and clasts (rock fragments) in sedimentary rocks are determined by a variety of factors: the original shapes of mineral grains in the source rocks; the orientation and spacing of fractures in bedrock that influence the shapes that clasts (rock fragments) take on when they weather from exposed rock; the nature and intensity of sediment transport, which can abrade grains and change original shapes; and sediment burial processes such as compaction, which can also change original shapes. Therefore, sedimentary particles may display a wide range of shapes, depending upon their history.

Particle shape is defined by three related but different aspects of grains. **Form** refers to the gross, overall configuration (outline) of particles and reflects variations in their proportions. The form of some particles resembles that of a sphere; other particles may have a platy (flattened) or rodlike form (Fig. 3.8A). Form should not be confused with **roundness**, which is a measure of the sharpness of grain corners. Well-rounded grains have smooth corners and edges; poorly rounded grains have sharp or angular corners and edges (Fig. 3.8B). **Surface texture** refers to small-scale, microrelief markings such as pits, scratches, and ridges that occur on the surface of grains (Fig. 3.8C). Form, roundness, and surface texture are independent properties, as illustrated in Figure 3.8, and each can theoretically vary without affecting the other. Actually, form and roundness tend to be positively correlated in sedimentary deposits; particles that are highly spherical in shape also tend to be well rounded. Surface texture can change without significantly changing form or roundness, but a change in form or roundness will affect surface texture because new surfaces are exposed. The three aspects of shape can be thought of as constituting a hierarchy, in which form is a first-order property, roundness a second-order property superimposed on form, and surface texture a third-order property superimposed on both the corners of a grain and the surfaces between corners (Barrett, 1980).

Particle Form (Sphericity)

Form reflects variations in the proportions of particles, that is, the relative lengths of their three major axes (long, intermediate, short). The term **sphericity** was introduced by Wadell (1932) to express this relationship (Appendix A). If all three axes have about the same length, a particle has high sphericity. If the axes differ markedly in length, the particle has low sphericity. A formula for expressing this relationship mathematically was developed by Krumbein (1941), which yields a mathematical value of 1 for a perfect sphere; less spherical particles have lower, fractional values (see Appendix A).

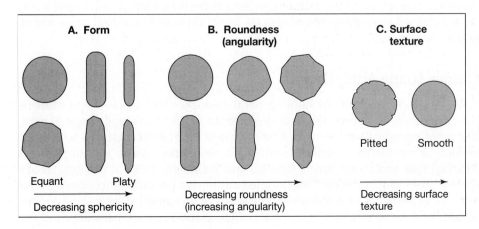

Figure 3.8
Schematic representation of the principal aspects of particle shape: form, roundness, and surface texture. Note that sphericity and roundness are independent properties. For example, a highly spherical (equant) particle can be either well rounded or poorly rounded, and a well-rounded particle can have either high or low sphericity.

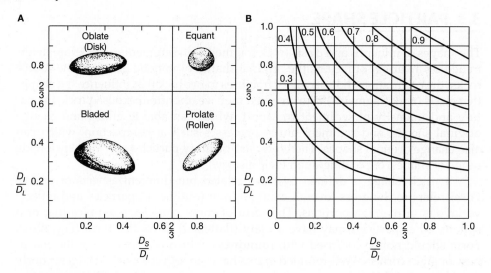

Figure 3.9
A. Classification of shapes of pebbles after Zingg (1935). B. Relationship between mathematical sphericity and Zingg shape fields. The curves represent lines of equal sphericity. [A, from Blatt, H., G. V. Middleton, and R. Murray, 1980, Origin of sedimentary rocks, 2nd ed., Fig. 3.20, p. 80, reprinted by permission of Prentice-Hall, Englewood Cliffs, N.J.; B, from Pettijohn, F. J., 1975, Sedimentary rocks, 3rd ed., Fig. 3.19, p. 34, Harper and Row Publishers, New York.]

Zingg (1935) took a somewhat different approach by proposing the use of two shape indices D_I/D_L and D_S/D_I (Appendix A) to define four shape fields on a bivariate plot: oblate (disc), equant (spheres), bladed, and prolate (rollers) (Fig. 3.9A). Lines of equal intercept sphericity (Krumbein) can be drawn on the Zingg shape fields (Fig. 3.9B), illustrating that particles of quite different form can have the same mathematical sphericity. See also Sneed and Folk (1958, 114–150).

Particle Roundness

Particle roundness refers to the degree of sharpness of the corners and edges of a grain. If the corners and edges are quite smooth, the grain is said to be well rounded (Fig. 3.8B). If the corners and edges are sharp and angular, the grain is poorly rounded. Wadell (1932) developed a formula that expresses mathematically the roundness of a particle (Appendix A). Like the sphericity formula, the roundness formula yields a mathematical value of 1 for perfectly rounded particles and smaller, fractional values for less well rounded particles. Because of the laborious process necessary to measure and express roundness mathematically, most workers are content to estimate roundness by comparison with a visual grain roundness scale (Fig. 3.10).

Fourier Shape Analysis

Sphericity and roundness are rather general descriptors of particle shape. More recently, attempts have been made to describe the two-dimensional shape of particles more rigorously by using a method based on Fourier analysis, which is a method of representing periodic mathematical functions as an infinite series of summed sine and cosine terms (Ehrlich and Weinberg, 1970). If the outline of a grain is "cut" and "unrolled," the unrolled outline is a periodic function somewhat resembling the shape of a sine wave. The shape of this unrolled grain can be represented by a series of terms called harmonics. Harmonics are periodic functions

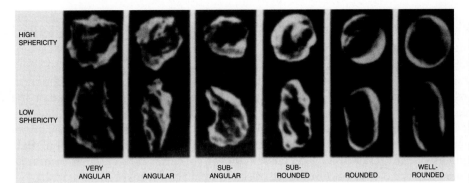

Figure 3.10
Powers' grain images for estimating roundness of sedimentary particles. [After Powers, M. C., 1953, A new roundness scale for sedimentary particles: Jour. Sed. Petrology, v. 23, Fig. 1, p. 118, reprinted by permission of Society of Economic Paleontologists and Mineralogists, Tulsa, Okla.]

(e.g., a sine wave) that can have various shapes owing to differences in wave amplitude and frequency.

By adding together (graphically) the shapes of various harmonics (done by computer), we can faithfully reproduce the shape of any periodic function. The method involves digitizing the periphery of grains by projecting grains onto a grid and recording, either manually or with an automatic digitizer, intercepts of the grain outline with the grid. Digitized data are reduced by computer to obtain the harmonics and produce computer-generated grain outlines. Presumably, the results of Fourier analysis reflect both sphericity and roundness of a particle. Fourier analysis has been used both to study the source of sediment grains and to characterize grains from particular depositional environments. See Boggs (1992, p. 52–60) for additional review of Fourier grain-size techniques.

Significance of Particle Shape

Sphericity

The sphericity of particles in sedimentary deposits is a function mainly of the original shapes of the grains, although the shapes of gravel-size particles can be modified somewhat by abrasion and breakage during transport. Sphericity affects the settling velocity of small particles (spherical particles settle faster than nonspherical particles) and the transportability of gravel-size particles, which move by traction (spheres and roller-shaped pebbles roll more readily than do pebbles of other shapes). Although sphericity is thus known to affect particle transport, it has not yet been demonstrated that the sphericity of particles can be used alone as a reliable tool for interpreting depositional environments. Although empirical studies show some relative differences in sphericity of grains from different environments, these differences have not proven to be sufficiently distinctive to permit environmental discrimination.

Particle Roundness

The roundness of grains in a sedimentary deposit is a function of grain composition, grain size, type of transport process, and distance of transport. Hard, resistant grains such as quartz and zircon are rounded less readily during transport than are weakly durable grains such as feldspars and pyroxenes. Pebble- to cobble-size grains commonly are more easily rounded by abrasion during transport than are sand-size grains. Resistant mineral grains smaller than 0.05–0.1 mm do not appear to become rounded by any transport process. Because of these factors, it is always necessary to work with particles of the same size and composition when doing roundness studies.

Experimental studies in flumes and wind tunnels of the effects of abrasion on transport of sand-size quartz grains show that transport by wind is 100 to 1000 times more effective in rounding these grains than transport by water (Kuenen, 1959, 1960). In fact, almost no rounding occurs in as much as 100 km of transport by water. Most roundness studies of small quartz grains in rivers have corroborated these experimental results. For example, Russell and Taylor (1937) observed no increase in rounding of quartz grains throughout a distance of 1100 mi (1775 km) in the Mississippi River between Cairo, Illinois, and the Gulf of Mexico. The effectiveness of surf action on beaches in rounding sand-size quartz grains is not well understood. In general, surf processes appear to be less effective in rounding grains than wind transport but more effective than river transport.

Once acquired, the roundness of quartz grains is not easily lost and may be preserved through several sedimentation cycles. Well-rounded quartz grains in an ancient sandstone may well indicate an episode of wind transport in its history, but it may be difficult or impossible to determine if rounding took place during the last episode of transport or during some previous cycle.

The roundness of transported pebbles is strongly related to pebble composition and size (Boggs, 1969). Soft pebbles such as shale and limestone become rounded much more readily than quartzite or chert pebbles, and large pebbles and cobbles are commonly better rounded than smaller pebbles. Although stream transport is relatively ineffective in rounding small quartz grains, pebble-size grains can become well rounded by stream transport. Depending upon composition and size, pebbles can become well rounded by stream transport in distances ranging from 11 km (7 mi) for limestones to 300 km (186 mi) for quartz (Pettijohn, 1975).

Well-rounded pebbles in ancient sedimentary rocks generally indicate fluvial transport. The degree of rounding cannot, however, be depended upon to give reliable estimates of the distance of transport. The greatest amount of rounding takes place in the early stages of transport, generally within the first few kilometers. Also, the roundness of pebbles is not an unequivocal indicator of fluvial environments because pebbles can also become rounded in beach environments and possibly on lake shores. Furthermore, rounded fluvial pebbles may eventually be transported into nearshore marine environments where they may be reentrained by turbidity currents and resedimented in deeper parts of the ocean.

Surface Texture

The surface of pebbles and mineral grains may be polished, frosted (dull, matte texture like frosted glass), or marked by a variety of small-scale, low-relief features such as pits, scratches, fractures, and ridges. These surface textures originate in diverse ways, including mechanical abrasion during sediment transport; tectonic polishing during deformation; and chemical corrosion, etching, and precipitation of authigenic growths on grain surfaces during diagenesis and weathering. Gross surface textural features such as polishing and frosting can be observed with an ordinary binocular or petrographic microscope; however, detailed study of surface texture requires high magnifications. Krinsley (1962) pioneered use of the electron microscope for studying grain surface texture at high magnifications.

Most investigators who study the surface texture of sediment grains carry out their studies on quartz grains because the physical hardness and chemical stability of quartz grains allow these particles to retain surface markings for geologically long periods of time. Through study of thousands of quartz grains, investigators have now been able to fingerprint the markings on grains from various modern depositional environments. More than twenty-five different surface textural features have been identified, including conchoidal fractures, straight and curved scratches and striations, upturned plates, meander ridges, chemically etched V's,

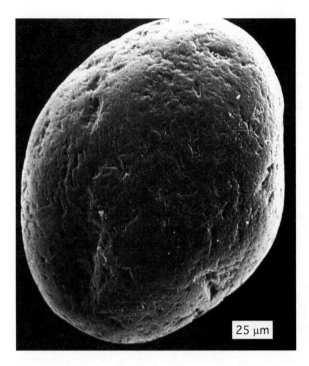

Figure 3.11
Electron micrograph of a quartz grain from unconsolidated Pliocene-Pleistocene sand, Louisiana salt dome edge, southern Louisiana, showing details of the surface texture. The grain has been well rounded by wind transport and contains tiny "upturned plates" characteristic of dune sands. Photograph courtesy of David Krinsley. Scale bar = 25 microns.

mechanically formed V's, and dish-shaped concavities (e.g., Bull, 1986). Examples of some of these markings are shown in Figure 3.11, illustrating the kinds of markings that can be seen at high magnification. Many other excellent electron micrographs of quartz surfaces may be found in Krinsley and Doornkamp's (1973) *Atlas of Quartz Sand Surface Textures*.

Surface texture appears to be more susceptible to change during sediment transport and deposition than do sphericity and roundness. Removal of old surface textural features and generation of new features are more likely to occur than marked changes in sphericity and roundness, and surface texture is more likely to record the last cycle of sediment transport or the last depositional environment. Therefore, geologists are interested in surface textural features as possible indicators of ancient transport conditions and depositional environment. The usefulness of surface texture in environmental analysis is limited, however, because similar types of surface markings can be produced in different environments. Also, the markings produced on grains in one environment may be retained on grains that are transported into another environment. Although less abrasion is required to remove surface markings from grains than is required to change roundness or sphericity, markings inherited from a previous environment may remain on grains for a long time before they are removed or replaced by different markings produced in the new environment. For example, grains on an arctic marine shelf may still retain surface microrelief features acquired during glacial transport of the grains to the shelf.

With care and the use of statistical methods, it has proven possible on the basis of surface textures to distinguish quartz grains from at least three major modern environmental settings: littoral (beach and nearshore), eolian (desert), and glacial. Quartz grains from littoral environments are characterized especially by V-shaped percussion marks and conchoidal breakage patterns. Grains deposited in eolian environments show surface smoothness and rounding, irregular upturned plates, and silica solution and precipitation features. Grains from glacial deposits have conchoidal fracture patterns and parallel to semiparallel striations. Techniques for studying the surface texture of quartz grains in modern environments have also been extended to study of ancient sedimentary deposits. Interpretation of pale-oenvironments on the basis of surface texture is complicated by the fact that surface microtextures can be changed during diagenesis by addition of cementing overgrowths or by chemical etching and solution. Krinsley and Trusty (1986) and Marshall (1987) provide additional details and discuss applications of the study of quartz grain surface textures by using the scanning electron microscope (SEM).

3.4 FABRIC

The fabric of sedimentary rocks is a function of grain orientation and packing and is thus a property of grain aggregates. Orientation and grain packing in turn control such physical properties of sedimentary rocks as bulk density, porosity, and permeability.

Grain Orientation

Particles in sedimentary rocks that have a platy (blade or disc) shape or an elongated (rod or roller) shape commonly show some degree of preferred orientation (Fig. 3.12). Platy particles tend to be aligned in planes that are roughly parallel to the bedding surfaces of the deposits. Elongated particles show a further tendency to be oriented with their long axes pointing roughly in the same direction. The preferred orientation of these particles is caused by transport and depositional processes and is related particularly to flow velocities and other hydraulic conditions at the depositional site. Most orientation studies have shown that sand-size particles deposited by fluid flows tend to become aligned parallel to the current direction (Fig. 3.12A; Parkash and Middleton, 1970), although a secondary mode of grains oriented normal to current flow (Fig. 3.12B) may be present. If the grains have a streamline or tear-drop shape, the blunt end of the grains commonly points upstream because this is the most stable orientation within a current. Sand grains can also show well-developed imbrication with long axes generally dipping up-current at angles less than about 20°. Imbrication refers to the overlapping arrangement of particles like that of shingles on a roof (Fig. 3.12C). Particles of

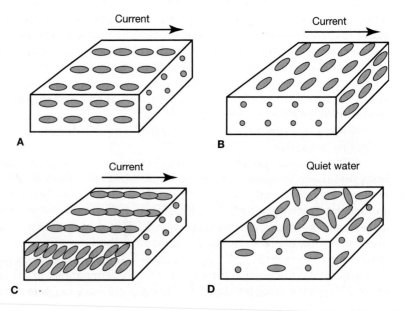

Figure 3.12
Schematic Illustration of the orientation of elongated particles in relation to current flow. A. Particles oriented parallel to current flow. B. Particles oriented perpendicular to current flow. C. Imbricated particles. D. Randomly oriented particles, characteristic of deposition in quiet water.

sandy sediment deposited by turbidity currents and grain-flow or sandy debris-flow processes also tend to be aligned parallel to flow directions and display up-stream imbrication at angles exceeding 20° (Hiscott and Middleton, 1980); however, in some gravity-flow deposits, orientation and imbrication directions can be variable or polymodal. Particles deposited under quiet water conditions may also show various orientations and a lack of imbrication (Fig. 3.12D). Fabric inconsistencies or bimodality appear to be related mainly to very rapid deposition from suspension or from sandy debris flow.

Pebbles in many gravel deposits and ancient conglomerates also display preferred orientation and imbrication. River-deposited pebbles are commonly oriented with their long axes normal to flow direction (Fig. 3.12C) and may display up-stream imbrication of up to 10 to 15° (Fig. 3.13). Orientation can also be parallel to flow, or even bimodal. Increasing flow intensity appears to favor orientation with long axes parallel to current flow rather than normal to flow (Johansson, 1976). Pebbles deposited by turbidity currents or other gravity-flow processes also become oriented with their long axes mainly parallel to flow direction, although orientation in some deposits can be random. Pebbles in glacial tills show preferred orientation parallel to flow and a minor mode oriented normal to flow.

Grain Packing, Grain-to-Grain Relations, and Porosity

Grain packing refers to the spacing or density patterns of grains in a sedimentary rock and is a function mainly of grain size, shape, and the degree of compaction of the sediment. Packing strongly affects the bulk density of the rocks as well as their porosity and permeability. The effects of packing on porosity can be illustrated by considering the change in porosity that takes place when even-size spheres are rearranged from loosest packing (cubic packing) to tightest packing (rhombohedral packing) as shown in Figure 3.14. Cubic packing yields porosity of 47.6 percent, whereas the porosity of rhombohedrally packed spheres is only 26.0 percent. The packing of natural particles is much more complex because of variations in size, shape, and sorting and is further complicated in lithified sedimentary rocks by the effects of compaction.

Poorly sorted sediments tend to have lower porosities and permeabilities than well-sorted sediments because grains are packed more tightly in these sediments owing to finer sediment filling pore spaces among larger grains. Petroleum and groundwater geologists are especially concerned with the porosity of sedimentary rocks because porosity determines the volume of fluids (oil, gas, groundwater) that can be held within a particular reservoir rock. Compaction causes major reduction in porosity. For example, a sandstone having an original porosity of about 40 percent may have porosity reduced during burial to less than 10 percent

Figure 3.13
Well-developed imbrication of river cobbles, Kiso River, Japan. Imbrication was produced by river currents flowing from left to right (arrow). Note hammer for scale.

Figure 3.14
Progressive decrease in porosity of spheres owing to increasingly tight packing. [After Graton, L. C., and H. J. Fraser, 1935, Systematic packing of spheres with particular relation to porosity and permeability: Jour. Geology, v. 43, Fig. 3, p. 796.]

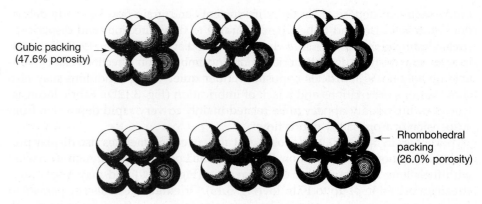

Cubic packing (47.6% porosity)

Rhombohedral packing (26.0% porosity)

owing to compaction resulting from the weight of overlying sediment. Compaction forces grains into closer contact and causes changes in the types of grain-to-grain contacts. Taylor (1950) identified four types of contacts between grains that can be observed in thin-sections: **tangential contacts**, or point contacts; **long contacts**, appearing as a straight line in the plane of a thin-section; **concavo-convex contacts**, appearing as a curved line in the plane of a thin section; and **sutured contacts**, caused by mutual stylolitic interpenetration of two or more grains (Fig. 3.15). In very loosely packed fabrics, some grains may not make contact with other grains in the plane of the thin-section and are referred to as "floating grains." Contact types are related to both particle shape and packing. Tangential contacts occur only in loosely packed sediments or sedimentary rocks, whereas

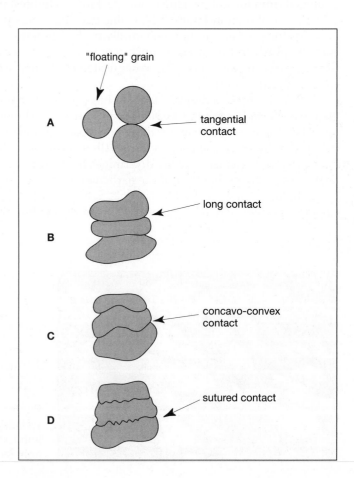

"floating" grain

A   tangential contact

B   long contact

C   concavo-convex contact

D   sutured contact

Figure 3.15
Diagrammatic illustration of principal kinds of grain contacts. A. Tangential. B. Long. C. Concavo-convex. D. Sutured. [Based on Taylor, J. M., 1950, Pore-space reduction in sandstones: Am. Assoc. Petroleum Geologists Bull., v. 34, p. 701–716.]

concavo-convex contacts and sutured contacts occur in rocks that have undergone considerable compaction during burial. The relative abundance of these various types of contacts can be used as a rough measure of the degree of compaction and packing and thus the depth of burial of sandstones. Several other methods for expressing the packing of sediment have been proposed; see Boggs (1992, p. 68–69) for details.

The sand-size grains in sandstones are commonly in continuous grain-to-grain contact when considered in three dimensions; thus, they form a **grain-supported** fabric. Conglomerates deposited by fluid flows also generally have a grain-supported fabric. On the other hand, conglomerates in glacial deposits, mud-flow deposits, and debris-flow deposits commonly have a **matrix-supported** fabric. In this type of fabric, the pebbles are not in grain-to-grain contact but "float" in a matrix of sand or mud. Matrix-supported conglomerates indicate deposition under conditions where fine sediment is abundant and deposition occurs by mass-transport processes or by processes that cause little reworking at the depositional site.

FURTHER READING

Bunge, H. J., S. Siegsmund, W. Skrotzki, and K. Weber (eds.), 1994, Textures of geological materials: Deutche Gesellschaft fur Materialkunde, 400 p.

Carver, R. E. (ed.), 1971, Procedures in sedimentary petrology: John Wiley & Sons, New York, 653 p.

Folk, R. L., 1974, Petrology of sedimentary rocks: Hemphill, Austin, Tex., 182 p.

Lewis, D. W., 1984, Practical sedimentology, p. 58–108: Hutchinson Ross, Stroudsburg, Pa., 229 p.

Syvitski, J. P. M., 1991, Principles, methods, and applications of particle size analysis: Cambridge University Press, Cambridge, 368 p.

4

Sedimentary Structures

4.1 INTRODUCTION

Sedimentary structures are large-scale features of sedimentary rocks such as parallel bedding, cross bedding, ripples, and mudcracks that are best studied in the field. They are generated by a variety of sedimentary processes, including fluid flow, sediment-gravity flow, soft-sediment deformation, and biogenic activity. Because they reflect environmental conditions that prevailed at, or very shortly after, the time of deposition, they are of special interest to geologists as a tool for interpreting such aspects of ancient sedimentary environments as sediment transport mechanisms, paleocurrent flow directions, relative water depth, and relative current velocity. Some sedimentary structures can also be used to identify the tops and bottoms of beds and thus to determine if sedimentary sequences are in depositional stratigraphic order or have been overturned by tectonic forces. Sedimentary structures are particularly abundant in siliciclastic sedimentary rocks, but they occur also in nonsiliciclastic sedimentary rocks such as limestones and evaporites.

This chapter describes and discusses the more important sedimentary structures. The discussions are brief, but they include a summary of current ideas on mechanisms of formation and, where appropriate, an analysis of the usefulness and limitations of the structures in environmental interpretation. The chapter focuses on primary sedimentary structures, formed essentially contemporaneously with sediment deposition, because these structures are most useful in environmental analysis. Sedimentary structures that form some time after deposition, during burial diagenesis, are secondary structures. A short discussion of a few common secondary sedimentary structures is provided also near the end of the chapter.

A very large body of literature on sedimentary structures has developed since the 1950s owing to their potential usefulness in environmental interpretation and paleocurrent analysis. These publications include several important monographs that contain excellent photographs and drawings illustrating a large variety of mainly primary sedimentary structures. Some of the more useful, and recent, books are listed under "Further Reading" at the end of this chapter.

4.2 KINDS OF PRIMARY SEDIMENTARY STRUCTURES

The most common and abundant primary sedimentary structures are listed in Table 4.1. This listing is fundamentally descriptive; that is, it is based primarily upon observable properties. Sedimentary structures are grouped into three broad categories: stratification structures and bedforms, bedding-plane markings, and other structures. Stratification structures and bedforms are further subdivided into four descriptive categories: bedding and lamination, bedforms, cross-stratification, and irregular stratification. Primary sedimentary structures are generated by four fundamental kinds of processes: (1) mainly deposition (depositional structures), (2) processes that involve an episode of erosion followed by deposition (erosional structures), (3) deposition followed by physical soft-sediment deformation (deformation structures), and (4) biogenically mediated deposition or nonbiogeneic deposition followed by biogenic modification (biogenic structures). The inset in Table 4.1 shows the relationship between sedimentary processes and the kinds of sedimentary structures formed.

Table 4.1 Principal primary sedimentary structures

STRATIFICATION AND BEDFORMS
  Planar bedding and lamination
    Laminated bedding[1,2,3]
    Graded bedding[1,12]
    Massive (structureless) bedding[1,12]
  Bedforms
    Ripples[1,2]
    Dunes[1,2]
    Antidunes[1]
  Cross-stratification
    Cross-bedding[1,2]
    Ripple cross-lamination[1,2]
    Flaser and lenticular bedding[1]
    Hummocky cross-stratification[1]
Irregular stratification
  Convolute bedding and lamination[7]
  Flame structures[7]
  Ball and pillow structures[7]
  Synsedimentary folds and faults[6]
  Dish and pillar structures[9]
  Channels[4]
  Scour-and-fill structures[4]
  Mottled bedding[12]
  Stromatolites[13]

BEDDING-PLANE MARKINGS
  Groove casts; striations; bounce, brush, prod, and roll marks[5]
  Flute casts[4]
  Parting lineation[1]
  Load casts[7]
  Tracks, trails, burrows[12]
  Mudcracks and syneresis cracks[10]
  Pits and small impressions[11]
  Rill and swash marks[1]

OTHER STRUCTURES
  Sedimentary sills and dikes[8]

Depositional Structures
1. Suspension settling and current- and wave-formed structures
2. Wind-formed structures
3. Chemically and biochemically formed structures

Erosional Structures
4. Scour marks
5. Tool marks

Deformation Structures
6. Slump structures
7. Load and founder structures
8. Injection (fluidization) structures
9. Fluid-escape structures
10. Desiccation structures
11. Impact structures (rain, hail, spray)

Biogenic Structures
12. Bioturbation structures
13. Biostratification structures

4.3 STRATIFICATION AND BEDFORMS

Bedding and Lamination

Nature of Bedding

Bedding is a fundamental characteristic of sedimentary rocks. **Beds** are tabular or lenticular layers of sedimentary rock that have lithologic, textural, or structural unity that clearly distinguishes them from strata above and below. The upper and lower surfaces of beds are known as **bedding planes** or **bounding planes**. Otto (1938) regarded beds as sedimentation units; that is, as the thickness of sediment deposited under essentially constant physical conditions. It is not always possible, however, to identify individual **sedimentation units**. Many beds defined by the criteria stated may contain several true sedimentation units. Beds are defined as strata thicker than 1 cm (McKee and Weir, 1953); layers less than 1 cm thick are called **laminae**. Terms used for describing the thickness of beds and laminae are shown in Figure 4.1.

Beds can be differentiated internally into a number of informal units (Fig. 4.2). A single bed may contain **subdivisions** arising from distinctive associations of sedimentary structures, such as plane laminae or ripple laminae. Also, thin units of different composition, texture, cementation, or color, such as a **lens** of pebbles or a **band** of chert, may be present. A marked discontinuity (commonly an erosional surface) between two beds of similar composition is called an **amalgamation surface,** and beds separated by such surfaces are called **amalgamated beds**. The term **layer** is used in a loose, informal sense for any bed or stratum of rock. The term has also been used in a more formal sense but usage is not consistent. For example, Blatt, Middleton, and Murray (1980) suggest that layers are parts of a bed thicker than laminae that are separated by minor but distinct discontinuities in texture or composition. On the other hand, Ricci Lucchi (1995) uses the term in the sense of a sedimentation unit (Otto, 1938); thus, according to Ricci Lucchi, a layer can include two or more beds.

Beds are separated by bedding planes or bedding surfaces, most of which represent planes of nondeposition, an abrupt change in composition (which reflect changes in depositional conditions), or an erosion surface (Campbell, 1967). Some bedding surfaces may be postdepositional features formed by diagenetic processes

Figure 4.1
Terms used for describing the thickness of beds and laminae. [Modified from McKee, E. D., and G. W. Weir, 1953, Terminology for stratification and cross-stratification in sedimentary rocks: Geol. Soc. America Bull., v. 64, Table 2, p. 383; and Ingram, R. L., 1954, Terminology for the thickness of stratification and parting units in sedimentary rocks: Geol. Soc. America Bull., v. 65, Fig. 1, p. 937.]

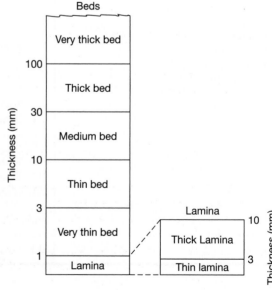

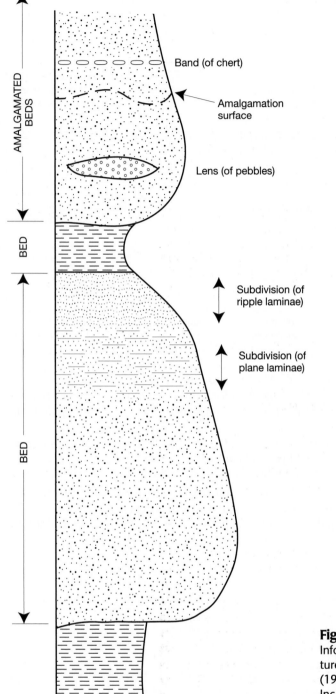

Figure 4.2
Informal subdivision of beds on the basis of internal structures. Based in part on Blatt, Middleton, and Murray (1980, Origin of sedimentary rocks, 2nd ed.: Prentice-Hall, Inc., Englewood Cliffs, Fig. 5.1, p. 130).

or weathering. The gross geometry of a bed depends upon the relationship between bedding surfaces. The bottom and top surfaces of beds are commonly parallel to each other; however, some bedding surfaces are nonparallel (Fig. 4.3). The bedding surfaces themselves may be even, wavy, or curved. Depending upon the combination of these characteristics, beds can have a variety of geometric forms such as uniform-tabular, tabular-lenticular, curved-tabular, wedge-shaped, and irregular. The interior of beds (the interval between bedding planes) may contain layers and laminae that are essentially parallel to the bedding planes; that is, beds may display internal planar stratification or laminated bedding. Layers and laminae that

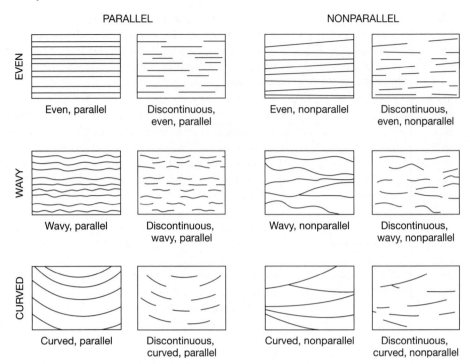

PARALLEL   NONPARALLEL

EVEN

Even, parallel   Discontinuous,   Even, nonparallel   Discontinuous,
                      even, parallel   even, nonparallel

WAVY

Wavy, parallel   Discontinuous,   Wavy, nonparallel   Discontinuous,
                      wavy, parallel   wavy, nonparallel

CURVED

Curved, parallel   Discontinuous,   Curved, nonparallel   Discontinuous,
                      curved, parallel   curved, nonparallel

Figure 4.3
Descriptive terms used for the configuration of bedding surfaces. [From Campbell, C. V., 1967, Lamina, laminaset, bed and bedset: Sedimentology, v. 8, Fig. 2, p. 18, reprinted by permission of Elsevier Science Publishers, Amsterdam.]

make up the internal structure of some beds are deposited at an angle to the bounding surfaces of the bed and are, therefore, called **cross-strata** or **cross-lamina**. Beds composed of cross-stratified or cross-laminated units are called **cross-beds**. The bounding surfaces of cross-beds may be either parallel or nonparallel.

Groups of similar beds or cross-beds are called **bedsets**. A **simple bedset** consists of two or more superimposed beds characterized by similar composition, texture, and internal structures. A bedset is bounded above and below by bedset (bedding) surfaces. A **composite bedset** refers to a group of beds differing in composition, texture, and internal structures but associated genetically, representing a common type of deposited succession (Reineck and Singh, 1980). The terminology of bedsets is illustrated in Figure 4.4.

Figure 4.4
Diagram illustrating the terminology of bedsets. [From Collinson, J. D., and D. B. Thompson, 1982, Sedimentary structures: George Allen & Unwin, London, Fig. 2.2, p. 8.]

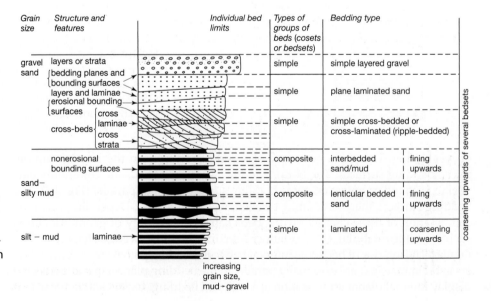

Many beds are characterized by lateral continuity, and some beds can be traced for many kilometers. Others may terminate within a single outcrop. Beds terminate laterally by (1) convergence and merging of upper and lower bounding surfaces (pinch-out), (2) lateral gradation of the composition of a bed into another bed of different composition so that the bounding bed surfaces die out, and (3) meeting a cross-cutting feature such as a channel, fault, or unconformity.

Individual beds are produced under essentially constant physical, chemical, or biological conditions. Many beds must have been produced very rapidly by a single event such as a flood that lasted only a few hours or days. Even more rapid deposition lasting perhaps only seconds or minutes, such as deposition of sand laminae by grain flow down the slip face of a sand dune, occurs in some environments. On the other hand, suspension deposition of very fine clay could take months or years to produce a bed.

The true bedding planes or bounding surfaces between beds represent periods of nondeposition, erosion, or changes to completely different depositional conditions. Many beds are not preserved to become part of the geologic record but are destroyed by succeeding erosional episodes. The preservation potential appears to be greater for those beds deposited by an event of great magnitude, such as a very large flood, than for those formed by very small scale events.

Kinds of Planar Bedding

The term **planar bed** is used here to differentiate beds that do not contain internal dipping laminae (cross-laminae) and which are bounded by nearly planar bedding surfaces that are essentially parallel to each other (e.g., Fig. 4.5). Planar bedding can be further differentiated, on the basis of internal structures, into laminated bedding, graded bedding, and massive bedding. Planar beds that contain internal parallel, or nearly parallel, laminae are (plane) laminated beds. Those that display vertical size grading are called graded beds. Beds that appear to contain no internal structures are called massive beds, particularly if the beds are very thick. Planar beds may range in thickness from a centimeter or so to meters.

Laminated Bedding. Parallel laminae (e.g., Fig. 4.6) are produced by less severe, or shorter lived, fluctuations in sedimentation conditions than those that generate beds. They result from changing depositional conditions that cause variations in grain size, content of clay and organic material, mineral composition, or microfossil content of sediments. They can form both by settling of fine-size particles from suspension (e.g., slow settling of clay in lakes) and traction transport of sand in water under some conditions. Examples of traction deposition include formation of heavy- and light-mineral laminae by swash and backwash on beaches and transport of sand in rivers at high flow velocities (e.g., Harms and Fahnestock,

Figure 4.5
Planar bedding in the Helena Formation (Precambrian), Glacier National Park, Montana.

Figure 4.6
Parallel laminae (arrows) in fine-grained sandstone, Elkton Formation (Eocene), southern Oregon coast. The dark, rounded objects in the lower right corner of the photograph are concretions.

1965; Allen, 1984; Bridge and Best, 1997). They can also form by wind transport (e.g., McKee, Douglass, and Rittenhouse, 1971; Hunter, 1977); however, wind-formed parallel laminae are not common. Thus, laminated bedding can develop in a variety of environments, and its presence is not a unique environmental indicator.

Once formed, parallel laminae are commonly preserved unless they are deposited in an environment in which sediment is being actively reworked by organisms. The burrowing and feeding activities of organisms in many environments can quickly destroy lamination. Laminae have the greatest potential for preservation in reducing or toxic environments, where organic activity is minimal, or in environments where deposition is so rapid that the sediment is buried below the depth of active organic reworking before organisms can destroy stratification.

Graded Bedding.   Graded beds are sedimentation units characterized by distinct vertical gradations in grain size. They range in thickness from a few centimeters to a few meters or more and commonly have sharp basal contacts. Beds that show gradation from coarser particles at the base to finer particles at the top are said to have **normal grading** (Fig. 4.7; 2.8). Normal graded bedding can form by several processes, e.g., sedimentation from suspension clouds generated by storm activity on the shelf or deposition in the last phases of a heavy flood, but the origin of most such graded beds in the geologic record has been attributed to turbidity currents. Differences in the rate at which particles of different sizes settle from suspension during the waning stages of turbidity current flow appear to account for the grading, but the exact manner in which the grading process operates is not well understood. The graded materials may be mud, sand, or, more rarely, gravel. As discussed in Chapter 2, some graded turbidite units display an ideal sequence of sedimentary structures, called a Bouma sequence (Fig. 2.8), but more commonly the sequence is truncated at the top or bottom. The basal A division may be present, but some or all of the overlying divisions may be absent; or the A division itself may be missing. Normally graded turbidite beds commonly occur in thin, repetitious successions referred to as rhythmic bedding.

Reverse grading can also occur but is much less common than normal grading. Reverse grading has been attributed to two types of mechanisms: (1) dispersive pressures and (2) kinetic sieving. Dispersive pressures (Chapter 2) are believed to be proportional to grain size. In a sediment of mixed grain size, the higher dispersive pressures acting on the larger particles tend to force them up into the zone of least shear. Alternatively, reverse grading may be explained by a kinetic sieve mechanism. In a mixture of grains undergoing agitation, the smaller grains presumably fall down through the larger grains as grain motion opens up spaces between the larger particles. Overall, reverse grading is a relatively rare phenomenon, and its origin is still poorly understood.

Figure 4.7
Normal graded bedding in the Cretaceous Hudspeth Formation, north-central Oregon. Four graded beds (arrows) are visible.

Massive (Structureless) Bedding.   The term **massive bedding** is used to describe beds that appear to be homogeneous and lacking in internal structures (e.g., Fig. 4.8). Use of X-radiography techniques (Hamblin, 1965) or etching and staining methods often reveals that such beds are not truly massive but rather that they contain very faintly developed structures. Nonetheless, one occasionally finds beds, particularly thick sandstone beds, in which internal structures cannot be recognized even with the aid of X-ray or staining techniques. Such beds are rare, which is fortunate for us because they are very difficult to explain. Reported occurrences of massive beds include both graded bed units in some turbidites, which may lack internal structures other than size grading, and certain thick, nongraded sandstones such as those in Figure 4.8. The beds in Figure 4.8 appear massive on a large scale but do contain some internal layers.

Liquefaction of sediment owing to sudden shock or other mechanisms shortly after deposition has been suggested as a means of destroying original stratification to produce massive bedding. Otherwise, it is assumed that lack of stratification is a primary feature that occurs in the absence of traction transport and results from very rapid deposition from suspension or deposition from very highly concentrated sediment dispersions during sediment-gravity flows. It has commonly been suggested that the sediment is dumped very rapidly without subsequent reworking to form a more or less homogeneous mass. Kneller and Branney (1995) propose, however, that massive turbidite deposits could also form by gradual aggradation of sand beneath sustained steady or quasi-steady, high-density turbidity currents.

Bedforms

Anyone who has examined the sandy bed of a clear, shallow stream has certainly noticed that the bed is rarely perfectly flat and even. Instead, it is commonly

Figure 4.8
Thick, massive-bedded Jurassic/Triassic sandstones exposed along the Green River, Utah. Bedding planes (arrows) divide the lower part of the section into several massive-bedded units. The vertical stripes in the upper half of the photograph are weathering stains. Note the river raft for scale.

marked by ripples and similar bedforms of various sizes. Such bedforms also occur in eolian and submarine environments where they range in size from small ripples a few centimeters long and a fraction of a centimeter high to gigantic eolian sand dunes and undersea sand waves tens to hundreds of meters long and several meters to several tens of meters high. If we carefully dissect a ripple exposed on the dry bed of a stream to reveal its internal structure, we almost invariably find internal fine-scale cross-laminae that dip in a downcurrent direction. Clearly, there is a close genetic relationship between fluid-flow mechanisms, ripple bedforms, and cross-lamination.

The preservation potential of ripples is relatively low; therefore, they are not extremely common features on the bedding planes of ancient sedimentary rocks. On the other hand, cross-beds are exceedingly common in many ancient sandstone successions. In an attempt to better understand the origin of bedforms and cross-stratification, many investigators have turned to the study of sediment transport in flumes. Flumes are long, slightly sloping troughs fitted with glass sides to allow observation. Sand or other sediment is placed on the floor of the flume, and water is constrained to flow over the floor at various depths and velocities.

Numerous flume experiments have established that under unidirectional fluid flow, small ripples begin to develop in sandy sediment as soon as the critical entrainment velocity for the sediment is reached. The exact sequence of other kinds of bedforms that develop with increasing velocity depends upon the grain size of the material. If flow is over a bed of sediment ranging in size from about 0.25 mm to 0.7 mm (medium to coarse sand), for example, the succession of bedforms illustrated in Figure 4.9 is generated, beginning with ripples. Ripples are the smallest bedform, ranging in length from about 5 to 20 cm and in height from about 0.5 to 3 cm. Thus, they have a ripple index (ratio of ripple length/ripple height) ranging from about 8 for coarse sand to 20 for fine sand. They form in sediment

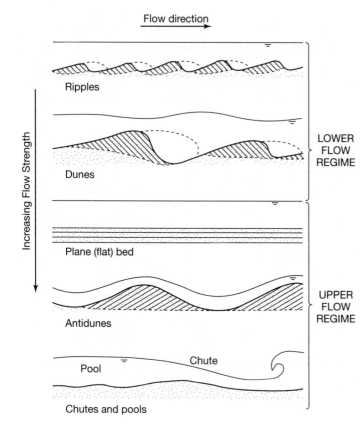

Flow direction →

Ripples

LOWER
FLOW
REGIME

Dunes

Increasing Flow Strength

Plane (flat) bed

UPPER
FLOW
REGIME

Antidunes

Pool   Chute

Chutes and pools

Figure 4.9
The succession of bedforms that develops during unidirectional flow of sandy sediment (0.25–0.7 mm) in shallow water as flow velocity increases. [Modified slightly from Blatt, H., G. V. Middleton, and R. Murray, 1980, Origin of sedimentary rocks, 2nd ed., Fig. 5.3, p. 137. Reprinted by permission of Prentice-Hall, Englewood Cliffs, N.J.]

ranging in size from silt (0.06 mm) to sand as coarse as 0.7 mm. Larger bedforms with spacing, or wave length, ranging from under 1 m to over 1000 m are called **dunes** (Ashley, 1990). Dunes are similar in general appearance to ripples except for size. They form at higher flow velocities in sediment ranging in grain size from fine sand to gravel. The ripple index of dunes ranges from about 5 in finer sands to 50 in coarser sediment. In the lower part of the dune stability field, ripples may be superimposed on the backs of dunes.

The hydraulic conditions that generate ripples and dunes take place at Froude numbers $<\sim 1$. Under these flow conditions, either the water surface shows little disturbance or the water waves are out of phase with bedforms, and flow is said to be in the **lower flow regime** (Simons and Richardson, 1961). Downstream migration of ripples and dunes leads to formation of cross-lamination that dips downstream at angles of up to about 30°. With further increase in flow velocity, dunes are destroyed and give way to an **upper flow regime** stage of flow, which takes place at Froude numbers $>\sim 1$. Sheetlike, rapid flow of water takes place, which generates surface water waves that are in phase with bedforms. Intense sediment transport results, over an initially relatively flat bed, during what is referred to as the plane-bed stage of flow. Plane-bed flow gives rise to internal planar lamination in which individual laminae range in thickness from a few millimeters to a few centimeters. At still higher velocities of flow, plane beds give way to **antidunes**, which are low, undulating bedforms up to 5 m in length. Antidunes form in very fast, shallow flows. They migrate upstream during flow, giving rise to low-angle ($<10°$) cross-bedding directed upstream.

The characteristics of bedforms that develop under unidirectional flow are summarized in Table 4.2. Two-dimensional dunes are generally straight-crested dunes whose shapes can be adequately described in a two-dimensional plane oriented parallel to the flow direction. Three-dimensional dunes are characterized by curved faces and scour pits, and their shapes must be described in three dimensions.

Table 4.2   Characteristics of bedforms developed under unidirectional flow

	Ripples	2D dunes	3D dunes	Lower plane bed	Upper plane bed	Antidunes
Length (spacing)	0.1–0.2 m	a few 10s of cm to 100s of m	a few 10s of cm to 10s of m (or more?)	—	—	10s of cm–m
Height	a few cm	cm to a few 10s of m	10s of cm to a few m (or more?)	—	—	cm–10s of cm
Ripple index (length/height)	relatively low	relatively high	relatively low	—	—	relatively high
Plan geometry	strongly irregular/ short-crested	straight/ sinuous and long-crested	strongly irregular/ short-crested	—	—	long-crested and short-crested
Characteristic flow velocity	low	low/moderate	moderate/high	low	high	high
Characteristic flow depth	> a few cm	> a few dm	> a few cm	all	all	shallow lows
Characteristic sediment size	0.03–0.6 mm	>0.3 mm	>0.2 mm	>0.6 mm	all	all

Source: Modified from Harms, J. C., J. B. Southard, and R. G. Walker, 1982, Structures and sequences in clastic rocks: *Soc. Econ. Paleontologists and Mineralogists Short Course No. 9.* Table 2-1, p. 2–11, reprinted by permission of SEPM, Tulsa, Okla.

2D = Two dimensional

3D = Three dimensional

Effects of Grain Size and Water Depth on Bedform Development

Experimental studies show that the succession of bedforms that develops at a given water depth during fluid flow depends not only upon flow velocity but also upon grain size; therefore, the succession of bedforms shown in Figure 4.9 does not occur in sediment of all particle sizes. Figure 4.10 shows the relationship of bedforms to flow velocity and grain size at a water depth ranging from 0.25 to 0.40 m.

Figure 4.10
Plot of mean flow velocity against median sediment size showing the stability fields of bed phases. Note that the recommended terminology for bedforms is (1) lower plane bed, (2) ripples, (3) dunes (all large-scale ripples), (4) upper plane bed, and (5) antidunes. F_r = Froude number. [After Southard, J. B., and L. A. Boguchwal, 1990, Bed configurations in steady unidirectional water flows. Part 2. Synthesis of flume data: Jour. Sed. Petrology, v. 60, Fig. 3, p. 664, reprinted by permission of Society for Sedimentary Petrology, Tulsa, Okla.]

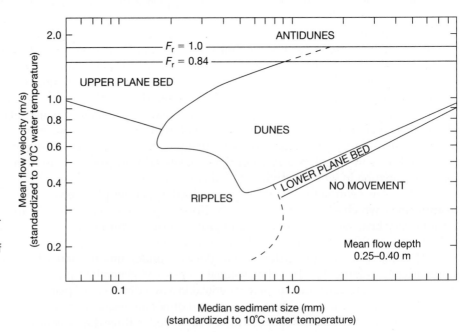

If flow takes place over sediment coarser than about 0.9 mm, for example, the ripple phase does not develop. Instead, a lower plane-bed phase forms just prior to the formation of dunes. Note also from Figure 4.10 that below a grain size of about 0.15 mm, dunes do not form. The ripple phase is succeeded abruptly by the upper plane-bed phase. For details of these plots as well other data on bed configurations, see Southard and Boguchwal (1990).

Most studies of bedforms have been carried out in laboratory flumes or under shallow-water conditions in natural environments. Therefore, most available sediment-size/velocity data pertain to the formation of bedforms under shallow-water conditions (commonly less than about 1 m). Much less is known about the development of bedforms under deeper water conditions. Based on limited available information, Harms, Southard, and Walker (1982) suggest that the nature of small ripples is approximately the same in deep-water flows as in shallow-water flows; however, the larger bedforms (dunes) can grow much larger in deep-water flows. The hydraulic relationships in deep water are the same as for shallow water; that is, dunes form at higher velocities than ripples and at lower velocities than plane beds and antidunes. The exact relationship between grain size, flow velocity, and bedform phase is not well documented for deeper water, but a generalized relationship is shown in Figure 4.11. Note from Figure 4.11 that exceedingly high velocities are required to produce antidunes at a water depth greater than a few meters; therefore, it appears that antidunes are unlikely to occur under natural conditions in deep water, except perhaps under some turbidity currents.

Nature of Flow over Bedforms

The mechanisms of sediment transport that are responsible for formation of the different bedforms are very complex. In general, the formation of transverse bedforms is related to a phenomenon called **flow separation**. Sediment is transported in suspension or by traction up the stoss side of the bedform to the brink or crest. At the brink, the flow separates from the bed to form a zone of reverse circulation

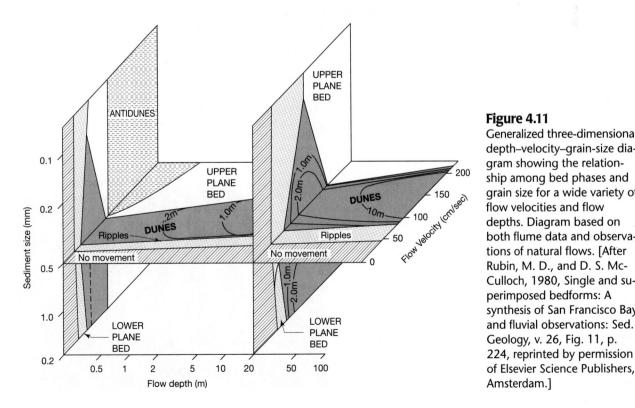

Figure 4.11
Generalized three-dimensional depth–velocity–grain-size diagram showing the relationship among bed phases and grain size for a wide variety of flow velocities and flow depths. Diagram based on both flume data and observations of natural flows. [After Rubin, M. D., and D. S. McCulloch, 1980, Single and superimposed bedforms: A synthesis of San Francisco Bay and fluvial observations: Sed. Geology, v. 26, Fig. 11, p. 224, reprinted by permission of Elsevier Science Publishers, Amsterdam.]

Figure 4.12
The terminology used to describe asymmetric ripples. [From Tucker, M. E., 1981, Sedimentary petrology, an introduction: John Wiley & Sons, Inc., Fig. 2.18, p. 28. Reproduced by permission of Blackwell Science.]

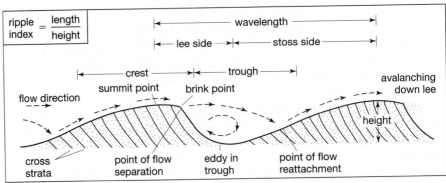

or backflow, producing a separation eddy (Figure 4.12). A zone of diffusion is present between the zone of backflow and the main flow above, owing to turbulent mixing with the main flow. Downstream from the point of separation, at a distance several times the height of the bedform, the flow becomes **reattached** to the bottom. Flow separation causes separation of the transported sediment into bedload and suspended load fractions. The bedload fraction accumulates at the ripple crest until the lee slope exceeds the angle of repose and avalanching takes place. The suspended load fraction is transported downcurrent where the coarser particles in the suspended load settle through the zone of diffusion into the zone of backflow and are deposited in the lee of the ripple. It is these processes that cause development and movement of the bedforms.

Classification of Ripples

Ripples are common sedimentary structures in modern environments, where they occur in both siliciclastic and carbonate sediments. They can form by both water and wind transport. Ripples can develop in granular material under either unidirectional current flow or oscillatory flow (wave action). Ripples that develop in response to unidirectional flow are asymmetrical in shape, and the steep or lee side faces downstream in the direction of current flow (Fig. 4.13). Asymmetrical ripples formed in this fashion are called **current ripples**. Under natural conditions they form by river and stream flow, by backwash on beaches, and by longshore currents, tidal currents, and deep-ocean bottom currents. In plan view, the crests of current ripples and dunes have a variety of shapes: **straight, sinuous, catenary, linguoid**, and **lunate** (Fig. 4.14). The plan-view shape of ripples and dunes is apparently related to water depth and velocity (Allen, 1968); however, the factors

Figure 4.13
Asymmetrical current ripples formed on a river bar, Rogue River, southern Oregon. The current flowed from left to right.

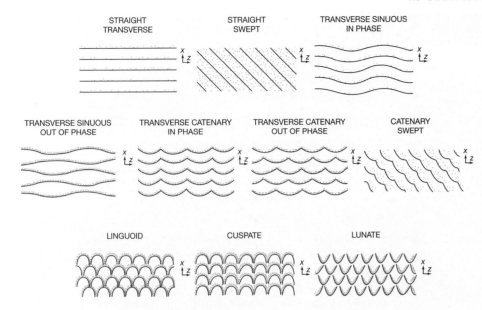

Figure 4.14
Idealized classification of current ripples and dunes on the basis of plan-view shape. Flow is from the bottom to the top in each case. [After Allen, J. R. L., 1968, Current ripples: Their relation to patterns of water motion: North Holland Pub., Amsterdam, Fig. 4.6, p. 65, reprinted by permission of Elsevier Science Publishers, Amsterdam.]

that control their shapes are not well understood. It has been observed under natural conditions that the more complex forms tend to develop in shallower water and at higher velocities than the less complex forms; the order in which the succession of bedforms develops with decreasing water depth and velocity is straight to sinuous to symmetric linguoid to asymmetric linguoid for ripples and straight to sinuous to catenary to lunate for dunes. Ripples that form by wave action under oscillatory flow are called **oscillation ripples**. Oscillation ripples tend to be nearly symmetrical in shape and have fairly straight crests (e.g., Fig. 4.15).

Ripples are most common in shallow-water environments; however, they have been photographed on the floor of the modern ocean at depths of a few thousand meters. Ripples have relatively low preservation potential because they tend to be eroded and destroyed by current erosion before burial. Therefore, ancient ripples such as the modern ripples shown in Figure 4.13 are not extremely abundant in the sedimentary record. Dunes are even less commonly preserved; nonetheless, ancient dunes are present in some thick sandstone units (e.g., Fig. 4.16).

Cross-Stratification Structures

Cross-Bedding

Cross-bedding forms primarily by migration of ripples and dunes in water or air. Ripple or dune migration leads to formation of dipping foreset laminae owing to

Figure 4.15
Oscillation ripples on the surface of fine-grained marine sandstone, Elkton Siltstone (Eocene), southern Oregon coast. Note hammer (lower-left corner) for scale.

Figure 4.16
Large dunes on the surface of a sandstone bed, Tyee Formation (Tertiary), exposed along the Umpqua River, southern Oregon Coast Range. The dunes are about 15 cm high and 70 cm from crest to crest. [Photograph courtesy of Ewart Baldwin.]

avalanching or suspension settling in the zone of separation on the lee sides of these bedforms (Fig. 4.12). If most of the sediment is too coarse to be transported in suspension, avalanching of the bedload sediment down the lee side of the ripple will cause formation of laminae that are steep and straight. These inclined **foreset** laminae make contact with the nearly horizontal, thin **bottomset** laminae (deposited from suspension) at a distinct (nontangential) angle, which is approximately the same as the angle of repose. Roughly the same effect is achieved if the height of the lee slope is large compared to total flow depth, so that the suspended load falls mainly on the lee slope. If the suspended load is large, or if the height of the lee slope is small compared to flow depth, suspended sediment will pile up at the base of the lee slope rapidly enough to keep pace with growth of the avalanche deposits. This process causes the lower part of the foreset laminae to curve outward and approach the bottomset laminae asymptotically (Blatt, Middleton, and Murray, 1980). Thus, the cross-laminae are said to be tangential.

The preservation potential of cross-laminae is much higher than that of the bedforms themselves (because the tops of bedforms tend to be planed off by subsequent current or wind erosion); therefore, cross-bedding is a very common type of sedimentary structure in ancient sedimentary rocks. Cross-stratification can be formed also by filling of scour pits and channels, by deposition on the point bars of meandering streams, and by deposition on the inclined surface of beaches and marine bars. Cross-bedding formed under different environmental conditions can be very similar in appearance, and it is often difficult in field studies of ancient sedimentary rocks to differentiate cross-bedding formed in fluvial, eolian, and marine environments.

Cross-beds commonly occur in sets (Fig. 4.4). Cross-bedding in sets less than about 5 cm thick is called small-scale cross-bedding; that in sets thicker than 5 cm is large-scale cross-bedding. Because of their diverse origin, many types of cross-beds occur. Allen (1963) proposed a very elaborate classification of cross-bedding based upon such properties as grouping of cross-bed sets, scale, nature of the

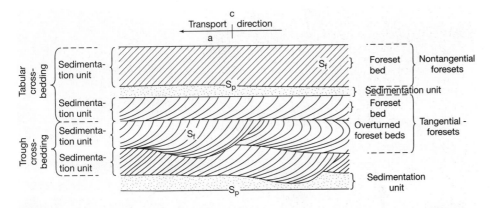

Figure 4.17
The terminology and defining characteristics of cross-bedding. Symbols: a, direction parallel to average sediment transport direction; c, direction perpendicular to (a) and the transport plane (bed) in which (a) lies; S_p, the principal bedding surface or bedding plane; S_f, the foreset surface of cross-bedding. [After Potter, P. E., and F. J. Pettijohn, 1977, Paleocurrent and basin analysis, 2nd ed., Fig. 4.1, p. 91, reprinted by permission of Springer-Verlag, Heidelberg.]

bounding surface of the beds, angular relation of cross-strata in a set or coset to the bounding surfaces, and degree of grain-size uniformity in different laminae. The much simpler scheme of McKee and Weir (1953), as modified by Potter and Pettijohn (1977), is adopted herein. Cross-beds are divided into two principal types, tabular and trough, on the basis of overall geometry and the nature of the bounding surfaces of the cross-bedded units (Fig. 4.17).

Tabular cross-bedding consists of cross-bedded units that are broad in lateral dimensions with respect to set thickness and that have essentially planar bounding surfaces (e.g., Fig. 4.18). The foreset laminae of tabular cross-beds are also commonly planar, but curved laminae that have a tangential relationship to the basal surface also occur. **Trough cross-bedding** consists of cross-bedded units in which one or both bounding surfaces are curved (e.g., Fig. 4.19). The units are trough-shaped sets consisting of an elongated scour filled with curved foreset laminae that commonly have a tangential relationship to the base of the set.

Tabular cross-bedding is formed mainly by the migration of large-scale, straight-crested ripples and dunes (Fig. 4.20A); thus, it forms during lower flow regime conditions. Individual beds range in thickness from a few tens of centimeters to a meter or more, but bed thicknesses up to 10 m have been observed (e.g., Harms et al., 1975). Trough cross-bedding can originate both by migration of small

Figure 4.18
Large-scale tabular cross-bedding in Permian sandstones, Canyon de Chelly National Monument, Arizona. Note that bounding surfaces of the cross-bedded units are planar and that the foreset laminae form mainly tangential contacts with these planar surfaces. [Photograph by James Stovall.]

Figure 4.19
Small-scale trough cross-bedding, Coaledo Formation (Eocene), southern Oregon coast. Notice that several episodes of scouring produced small, erosional troughs (arrows), which were subsequently filled with low-angle cross-laminae.

current ripples, which produces small-scale cross-bed sets, or by migration of large-scale, trough-shaped dunes (Fig. 4.20B). Trough cross-bedding formed by migration of large-scale ripples commonly ranges in thickness to as much as a few tens of centimeters and in width from less than 1 m to more than 4 m.

Ripple Cross-Lamination

Ripple cross-lamination (climbing ripples) forms when deposition takes place very rapidly during migration of current or wave ripples (McKee, 1965; Jopling

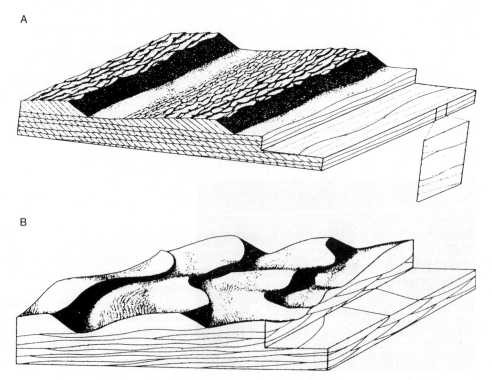

Figure 4.20
Diagram illustrating (A) large-scale tabular cross-bedding formed by migrating straight-crested dunes (with rippled surfaces) and (B) large-scale trough cross-bedding formed by migrating, trough-shaped dunes. Flow is from left to right in both A and B. [From Harms, J. C., J. B. Southard, and R. G. Walker, 1982, Structures and sequences in clastic rocks: Soc. Econ. Paleontologists and Mineralogists Short Course No. 9, Fig. 3-11, p. 3–23 and Fig. 3-10, p. 3–19, reprinted by permission of SEPM, Tulsa, Okla.]

A

B

and Walker, 1968). A series of cross-laminae are produced by superimposing migrating ripples. The ripples climb on one another such that the crests of vertically succeeding laminae are out of phase and appear to be advancing upslope. This process results in cross-bedded units that have the general appearance of waves (Fig. 4.21) in outcrop sections cut normal to the wave crests. In sections with other orientations, the laminae may appear horizontal or trough-shaped, depending upon the orientation and the shape of the ripples.

The formation of ripple cross-lamination appears to require an abundance of sediment, especially sediment in suspension, which quickly buries and preserves original rippled layers. Abundant suspended sediment supply must be combined with just enough traction transport to produce rippling of the bed, but not enough to cause complete erosion of laminae from the stoss side of ripples. Some ripple laminae may be in phase (one ripple crests lies directly above the other), indicating that the ripples did not migrate. In-phase ripple laminae form under conditions where a balance is achieved between traction transport and sediment supply, and are balanced so that the ripples do not migrate despite a growing sediment surface. Ripple cross-lamination occurs in sediments deposited in environments characterized by rapid sedimentation from suspension—fluvial flood plains, point bars, river deltas subject to periodic flooding, and environments of turbidite sedimentation. Figure 4.21 shows the sequence of bedforms developed in a river during waning flood stage. Laminae at the bottom of Figure 4.21 developed during a plane-bed phase of upper-flow regime transport at high flood velocity. As velocity waned into the lower-flow regime, ripple cross-lamination formed on top of the plane-bed laminae.

Flaser and Lenticular Bedding

Flaser bedding is a type of ripple bedding in which thin streaks of mud occur between sets of cross-laminated or ripple-laminated sandy or silty sediment (Fig. 4.22). Mud is concentrated mainly in the ripple troughs but may also partly cover the crests. Flaser bedding suggests deposition under fluctuating hydraulic conditions. Periods of current activity, when traction transport and deposition of rippled sand take place, alternate with periods of quiescence, when mud is deposited.

Figure 4.21

Ripple cross-lamination (below ballpoint pen) in flood deposits of the Illinois River, southwestern Oregon. Parallel laminae at the bottom of the photograph developed during a plane-bed phase of upper-flow-regime conditions; as current velocity diminished into the lower-flow regime, ripple cross-lamination formed on top of the laminae. A later flood pulse deposited upper-flow-regime plane beds on top of the ripple cross-lamination.

Figure 4.22
Flaser bedding in tidal-flat sediments of the North Sea. [After Reineck, H. E., 1967, Layered sediments of tidal flats, beaches, and shelf
bottoms of the North Sea, *in* Lauff, G. H. (ed.), Estuaries: Amer.
Assoc. for the Advancement of Science, Washington, D.C., Fig. 8,
p. 195.]

Repeated episodes of current activity erode previously deposited ripple crests, allowing new rippled sand to bury and preserve rippled beds with mud flasers in
the troughs (Reineck and Singh, 1980). **Lenticular bedding** is a structure formed
by interbedded mud and ripple cross-laminated sand in which the ripples or sand
lenses are discontinuous and isolated in both a vertical and a horizontal direction
(Fig. 4.23). Reineck and Singh (1980) suggest that flaser bedding is produced in environments in which conditions for deposition and preservation of sand are more
favorable than for mud, but that lenticular bedding is produced in environments
in which conditions favor deposition and preservation of mud over sand. Flaser
and lenticular bedding appear to form particularly on tidal flats and in subtidal
environments where conditions of current flow or wave action that cause sand deposition alternate with slack-water conditions when mud is deposited. They also
form in marine delta-front environments, where fluctuations in sediment supply
and current velocity are common; in lake environments in front of small deltas;
and possibly on the shallow marine shelf owing to storm-related transport of sand
into deeper water.

Hummocky Cross-Stratification

The name **hummocky cross-stratification** was introduced by Harms et al. in 1975,
although the structures had been recognized and described under different names
by earlier workers. Hummocky cross-stratification is characterized by undulating
sets of cross-laminae that are both concave-up (swales) and convex-up (hummocks) (Fig. 4.24). The cross-bed sets cut gently into each other with curved erosion surfaces (Fig. 4.25). Hummocky cross-bedding commonly occurs in sets 15 to
50 cm thick with wavy erosional bases and rippled, bioturbated tops (Harms et al.,
1975). Spacing of hummocks and swales is from 50 cm to several meters. The
lower bounding surface of a hummocky unit is sharp and is commonly an erosional surface. Current-formed sole marks may be present on the base. Hummocky

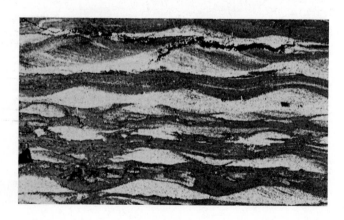

Figure 4.23
Lenticular bedding in tidal flats of the North Sea. Sand
lenses are wave ripples. [After Reineck, H. E., and I. B.
Singh, 1975, Depositional sedimentary environments,
Springer-Verlag, New York, Fig. 176, p. 103.]

HUMMOCKY CROSS STRATIFICATION - HCS

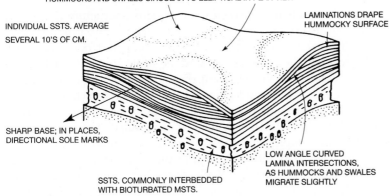

LONG WAVELENGTH, 1 - 5 M
LOW HEIGHT, FEW 10'S OF CM
HUMMOCKS AND SWALES CIRCULAR TO ELLIPTICAL IN PLAN VIEW

LAMINATIONS DRAPE
HUMMOCKY SURFACE

INDIVIDUAL SSTS. AVERAGE
SEVERAL 10'S OF CM.

SHARP BASE; IN PLACES,
DIRECTIONAL SOLE MARKS

LOW ANGLE CURVED
LAMINA INTERSECTIONS,
AS HUMMOCKS AND SWALES
MIGRATE SLIGHTLY

SSTS. COMMONLY INTERBEDDED
WITH BIOTURBATED MSTS.

HCS CHARACTERIZED BY –

1, UPWARD CURVATURE OF LAMINATIONS
2, LOW ANGLE, CURVED LAMINA INTERSECTIONS
3, VERY LONG WAVELENGTHS, LOW HEIGHTS;
   LAMINA DIPS NORMALLY LESS THAN 10°

Figure 4.24
Schematic diagram of hummocky cross-stratification, which typically occurs interbedded with bioturbated mudstone. MSTS, mudstones; SSTS, sandstones. [From Walker, R. G., 1984, Shelf and shallow marine sands, in Walker, R. G. (ed.), Facies models, 2nd ed.: Geoscience Canada Reprint Ser. 1, Fig. 11, p. 149, reproduced by permission.]

Figure 4.25
Hummocky cross-stratification in fine-grained sandstone, Elkton Siltstone (Eocene), southwestern Oregon. Note the clearly defined erosional surface with fine, laminated sand draped over the erosional hummock. The width of the area photographed is ~60 cm.

cross-stratification typically occurs in fine sandstone to coarse siltstone that commonly contains abundant mica and fine carbonaceous plant debris (Dott and Bourgeois, 1982).

Hummocky cross-stratification has not yet been produced in flumes or reported from modern environments, but it has been reported in ancient strata from numerous localities. Harms et al. (1975, 1982) suggest that this structure is formed by strong surges of varied direction (oscillatory flow) that are generated by relatively large storm waves in the ocean. Strong storm-wave action first erodes the seabed into low hummocks and swales that lack any significant orientation. This topography is then mantled by laminae of material swept over the hummocks and swales. More recently, Duke, Arnott, and Cheel (1991) and Cheel and Leckie (1993) suggest that hummocky cross-stratification originates by a combination of unidirectional and oscillatory flow related to storm activity. Although hummocky cross-stratification is commonly confined to shallow marine sedimentary rocks, Duke (1985) reports this structure in some lacustrine sedimentary rocks.

Irregular Stratification

Deformation Structures

Convolute Bedding and Lamination.   Convolute bedding is a structure formed by complex folding or intricate crumpling of beds or laminations into irregular, generally small-scale anticlines and synclines. It is commonly, but not necessarily, confined to a single sedimentation unit or bed, and the strata above and below this bed may show little evidence of deformation (Fig. 4.26). Convolute bedding is most common in fine sands or silty sands, and the laminae can typically be traced through the folds. Faulting generally does not occur, but the convolutions may be truncated by erosional surfaces which may also be convoluted. The convolutions increase in complexity and amplitude upward from undisturbed laminae in the lower part of the unit. They may either die out in the top part of the unit or be truncated by the upper bedding surface. Beds containing convolute lamination commonly range in thickness from about 3 to 25 cm (Potter and Pettijohn, 1977), but convoluted units up to several meters thick have been reported in both eolian and subaqueous deposits.

Convolute lamination is most common in turbidite successions. It also occurs in intertidal-flat sediments, river flood-plain and point-bar sediments, and deltaic deposits. The origin of convolute bedding is still not thoroughly understood, but it appears to be caused by plastic deformation of partially liquefied sediment soon after deposition. The axes of some convoluted folds have a preferred orientation which commonly coincides with the paleocurrent direction, suggesting that the process that produces convolutions occurs during deposition, at least in these cases. Liquefaction of sediment can be caused by such processes as differential overloading, earthquake shocks, and breaking waves.

Flame Structures.   Flame structures are wavy or flame-shaped tongues of mud that project upward into an overlying layer, which is commonly sandstone (Fig. 4.27). The crests of some flames are bent over or overturned; generally, overturned crests tend to all point in the same direction, but this is not always the case, as illustrated in Figure 4.27. Flame structures are commonly associated with other structures

Figure 4.26
Convolute lamination in fine-grained sandstone and shale, Coaledo Formation (Eocene), southwestern Oregon. Note absence of deformation in the underlying layers.

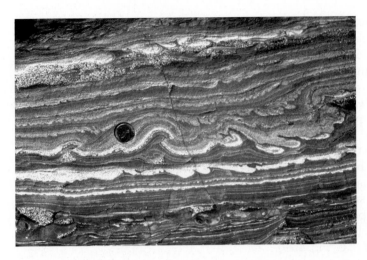

Figure 4.27
Flame structures in a thin-bedded, fine sandstone-shale succession, Elkton Siltstone, southwestern Oregon. The best developed flames are in the layer immediately below the coin.

caused by sediment loading. They are probably caused mainly by loading of water-saturated mud layers which are less dense than overlying sands and are consequently squeezed upward into the sand layers. The orientation of over-turned crests suggests that loading may be accompanied by some horizontal drag or movement between the mud and sand bed.

Ball and Pillow Structures. Ball and pillow structures are present in the lower part of sandstone beds, and less commonly in limestone beds, that overlie shales (Fig. 4.28). They consist of hemispherical or kidney-shaped masses of sandstone or limestone that show internal laminations. In some hemispheres, the laminae may be gently curved or deformed, particularly next to the outside edge of the hemispheres where they tend to conform to the shape of the edge. The balls and pillows may remain connected to the overlying bed, as in Figure 4.28, or they may be completely isolated from the bed and enclosed in the underlying mud. Ball and pillow structures are believed to form as a result of foundering and breakup of semiconsolidated sand, or limy sediment, owing to partial liquefaction of under-lying mud, possibly caused by shocking. Liquefaction of the mud causes the

Figure 4.28
Ball and pillow structures (arrows) on the base of a thin, steeply dipping sandstone bed. Lookingglass Formation (Eocene), near Illahe, southwestern Oregon.

overlying sand beds or limy sediment to deform into hemispherical masses that may subsequently break apart from the bed and sink into the mud. Kuenen (1958) experimentally produced structures that closely resemble natural ball and pillow structures by applying a shock to a layer of sand deposited over thixotropic clay.

Synsedimentary Folds and Faults.   The general term **slump structures** has been applied to structures produced by penecontemporaneous deformation resulting from movement and displacement of unconsolidated or semiconsolidated sediment, mainly under the influence of gravity. Potter and Pettijohn (1977) describe slump structures as being the products of either (1) pervasive movement involving the interior of the transported mass, producing a chaotic mixture of different types of sediments, such as broken mud layers embedded in sandy sediment, or (2) a décollement type of movement in which the lateral displacement is concentrated along a sole, thus producing beds that are tightly folded and piled into nappelike structures (e.g., Fig. 4.29).

Slump structures may involve many sedimentation units, and they are commonly faulted. Thicknesses of slump units have been reported to range from less than 1 m to more than 50 m. Slump units may be bounded above and below by strata that show no evidence of deformation. It may be difficult in some stratigraphic successions, however, to differentiate between slump units and incompetent beds such as shale that were deformed between competent sandstone or limestone beds during tectonic folding.

Slump structures typically occur in mudstones and sandy shales and less commonly in sandstones, limestones, and evaporites. They are generally found in units that were deposited rapidly, and they have been reported from a variety of environments where rapid sedimentation and oversteepened slopes lead to instability. They occur in glacial sediments, varved silts and clays of lacustrine origin, eolian dune sands, turbidites, delta and reef-front sediments, subaqueous dune sediments, and in sediments from the heads of submarine canyons, continental shelves, and the walls of deep-sea trenches.

Dish and Pillar Structures.   **Dish structures** are thin, dark-colored, subhorizontal, flat to concave-upward, clayey laminations (Fig. 4.30) that occur principally in sandstone and siltstone units (Lowe and LoPiccolo, 1974; Rautman and Dott, 1977). The laminations are commonly only a few millimeters thick, but individual dishes may range from 1 cm to more than 50 cm wide. They typically occur in thick beds where dish and pillar structures may be the only structures visible.

Figure 4.29
Small-scale décollement-type synsedimentary folds (arrows) in thin, fine-grained sandstone layers interbedded with shale, Elkton Siltstone (Eocene), southwestern Oregon.

0   4cm

Figure 4.30
Strongly curved to nearly flat dish structures (large arrow) and pillars (small arrows) formed by dewatering of siliciclastic sediment, Jackfork Group (Pennsylvanian), southeast Oklahoma. [From Lowe, D. R., 1975, Water escape structures in coarse-grained sediment: Sedimentology, v. 22, Fig. 8, p. 175, reprinted by permission of Elsevier Science Publishers, Amsterdam.]

They also occur in beds less than about 0.5 m thick, where they commonly cut across primary flat laminations and other laminations. **Pillar structures** generally occur in association with dish structures (Fig. 4.30). Pillars are vertical to near-vertical, cross-cutting columns and sheets of structureless or swirled sand that cut through either massive or laminated sands that also commonly contain dish structures and convolute laminations. They range in size from tubes a few millimeters in diameter to large structures greater than 1 m in diameter and several meters long. Pillars are not actually stratification structures. They are discussed here with dish structures because of their close association with these structures and because they form by a similar mechanism, discussed below.

Dish and pillar structures were first observed in sediment gravity-flow deposits (turbidites and liquefied flows) and are most abundant in such deposits; however, they have now also been reported in sediments from deltaic, alluvial, lacustrine, and shallow marine deposits, as well as from volcanic ash layers. They indicate rapid deposition and form by escape of water during consolidation of sediment. During gradual compaction and dewatering, semipermeable laminae act as partial barriers to upward-moving water carrying fine sediment. The fine particles are retarded by the laminae and are added to them, forming the dishes. Some of the water is forced horizontally beneath the laminations until it finds an easier escape route upward. This forceful upward escape of water forms the pillars. Therefore, both dish structures and pillars are dewatering structures.

Erosion Structures

Channels are structures that show a U-shape or V-shape in cross section and cut across earlier-formed bedding and lamination (Fig. 4.31). They are formed by erosion, principally by currents but in some cases by mass movements. Channels may be filled with sediment that is texturally different from the beds they truncate. Channels visible in outcrop range in width and depth from a few centimeters to many meters. Even larger channels may be definable by mapping or drilling. It is seldom possible to trace their length in outcrop, but they can presumably extend

Figure 4.31
Channel (arrows) incised into fluvial, volcaniclastic sediments of the Colestin Formation (Eocene), Oregon-California border. Width of the channel is ~4–5m.

Figure 4.32
Scour-and-fill structures in Miocene sandstone, Blacklock Point, southern Oregon coast. A small depression was scoured into underlying sand by currents, then filled with gravel and sand. Lens cap for scale.

for distances many times their width. Channels are very common in fluvial and tidal sediments. They also occur in turbidite sediments, where the long dimensions of the channels tend to be oriented parallel to current direction as shown by other directional structures.

Scour-and-fill structures are similar to channels but are commonly smaller (Fig. 4.32). They consist of small, filled asymmetrical troughs a few centimeters to a few meters in size, with long axes that point downcurrent and that commonly have a steep upcurrent slope and a more gentle downcurrent slope. They may be filled with either coarser grained or finer grained material than the substrate. These structures are most common in sandy sediment and are thought to form as a result of scour by currents and subsequent backfilling as current velocity decreases. In contrast to channels, several scour-and-fill structures may occur together closely spaced in a row. They are primarily structures of fluvial origin that can occur in river, alluvial-fan, or glacial outwash-plain environments.

4.4 BEDDING-PLANE MARKINGS

Markings Generated by Erosion and Deposition

Many bedding-plane markings occur on the underside of beds as positive-relief casts and irregular markings. Owing to their location on the bases or soles of beds, they are often referred to as **sole markings**. Sole markings are preserved particularly well on the undersides of sandstones and other coarser grained sedimentary rocks that overlie shale beds. Many sole markings show directional features that make them very useful for interpreting the flow directions of ancient currents.

These so-called erosional sole markings are actually formed by a two-stage process that involves both erosion and deposition. First, a cohesive, fine-sediment bottom is eroded by some mechanism to produce grooves or depressions. Because of the cohesiveness of the sediment, the depressions may be preserved long enough to be filled in and buried during subsequent deposition, typically by coarser-grained sediment than the bottom mud. This coarser sediment is probably deposited very shortly after erosion produced the depression, possibly in some cases by the same current that formed the depression. After burial and lithification, a positive relief feature is left attached to the base of the overlying bed. If the bed subsequently undergoes tectonic uplift, these structures may be exposed by weathering and subaerial erosion (Fig. 4.33). The initial erosional event that creates the depressions in a mud bottom can take the form of current scour, or the depression can result from the action of objects called **tools** that are carried by the current and intermittently or continuously make contact with the bottom. These tools can be pieces of wood, the shells of organisms, or any similar object that can be rolled or dragged along the bottom. Erosional structures may thus be classified genetically as either **current-formed structures** or **tool-formed structures.**

Erosional sole markings are most common on the soles of turbidite sandstones, but they are also present in sedimentary rocks deposited in other environments. They can form in any environments where the requisite conditions of an erosive event followed reasonably quickly by a depositional event are met. They have been reported in both fluvial and shelf deposits in addition to turbidites.

Groove Casts

Groove casts are elongate, nearly straight ridges that result from infilling of erosional relief produced as a result of a pebble, shell, piece of wood, or other object being dragged or rolled across the surface of cohesive sediment (Fig. 4.34). They typically range in width from a few millimeters to tens of centimeters and have a relief of a few millimeters to a centimeter or two; however, much larger groove casts also occur. Groove casts are greatly elongated in comparison to their width. They are directional features that are oriented parallel to the flow direction of the ancient currents that produced them; thus, they have paleocurrent significance. Groove casts on the same bed commonly have the same general orientation, although they may diverge at slight angles and even cross. Most groove casts do not have features that show the unique flow direction; that is, we cannot tell from them which direction was downcurrent and which upcurrent. **Chevrons** are a variety of

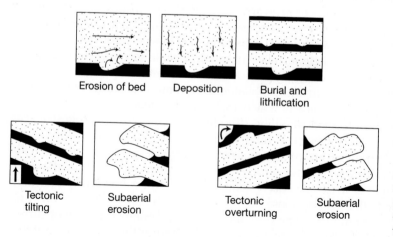

Erosion of bed   Deposition   Burial and lithification

Tectonic tilting   Subaerial erosion   Tectonic overturning   Subaerial erosion

Figure 4.33
Postulated stages of development of sole markings owing to erosion of a mud bottom followed by deposition of coarser sediment. The diagram also illustrates how the sole markings appear as positive-relief features on the base of the infilling bed after tectonic uplift and subaerial weathering; it suggests how sole markings can be used to tell top and bottom of overturned beds. [From Collinson, J. D., and D. B. Thompson, 1982, Sedimentary structures, Fig. 4.1, p. 37, reprinted by permission of George Allen & Unwin, London. After Ricci Lucchi, F., 1970, Sedimentografia: Zanchelli, Bologna, Italy.]

Figure 4.34
Large intersecting groove casts on the base of a
turbidite sandstone bed, Fluornoy Formation
(Eocene), Oregon Coast Range.

groove casts made up of continuous V-shaped crenulations in which the V points
in a downstream direction; thus, this type of groove cast can be used to determine
the true direction of flow. Dzulynski and Walton (1965) suggest that chevrons are
formed by tools moving just above the sediment surface but not touching the surface, causing rucking-up of the sides of the groove. Groove casts are especially
common on the soles of turbidite beds owing to shell fragments, pieces of wood,
or other tools that are carried in the base of turbidity current flows being dragged
across a mud bottom. They occur also on the soles of beds deposited in shallow-
water environments such as tidal flats and flood plains where floating tools may
touch bottom and leave grooves.

Bounce, Brush, Prod, Roll, and Skip Marks

Small gouge marks are produced by tools that make intermittent contact with the
bottom, creating small marks. **Brush** and **prod marks** are asymmetrical in cross-
sectional shape, and the deeper, broad part of the mark is oriented downcurrent.
Bounce marks are roughly symmetrical. **Roll** and **skip marks** are formed by a tool
bouncing up and down or rolling over the surface to produce a continuous track.
The genesis of these structures is illustrated in Figure 4.35.

Flute Casts

Flute casts are elongated welts or ridges that have a bulbous nose at one end that
flares out in the other direction and merges gradually with the surface of the bed
(Fig. 4.36). They occur singly or in swarms in which all of the flutes are oriented in
the same general direction. On a given sole, the flutes tend to be about the same
size; however, flute casts on different beds can range in width from a centimeter or
two to 20 cm or more, in height (relief) from a few centimeters to 10 cm or more,
and in length from a few centimeters to a meter or more. The plan-view shape of
flutes varies from nearly streamline, bilaterally symmetrical forms to more elongate and irregular forms, some of which are highly twisted.
       Flute casts are formed by filling of a depression scoured in cohesive sedi-
ment by current eddies created behind some obstacle, or by chance eddy scour.
This type of current scour produces asymmetrical depressions in which the
steepest and deepest part of the depression is oriented upstream or upslope (see
Fig. 4.33). Therefore, when such depressions are filled, the filling forms a positive-
relief structure with a bulbous nose oriented upstream. Flute casts thus make excellent paleocurrent indicators because they show the unique direction of current
flow. Flutes are particularly prevalent on the soles of turbidite sequences, but they

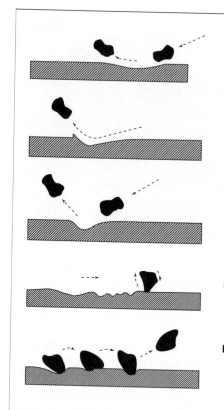

A   Bounce marks. The tool approaches the sediment surface at a low angle and immediately bounces back into the current.

B   Brush marks. The tool approaches the sediment surface at a very low angle, with the axis of the tool inclined upcurrent, and is then lifted away by the current, producing a ridge of mud downcurrent of the mark.

C   Prod marks. The tool reaches the sediment surface at a fairly high angle and is then lifted up and away by the current.

D   Roll marks. The tool rolls over the sediment surface, producing a continuous roll mark.

E   Skip marks. The tool travels downcurrent with a saltating movement, hitting the sediment surface at nearly regular intervals.

Figure 4.35
Development in a cohesive mud bottom of (A) bounce marks, (B) brush marks, (C) prod marks, (D) roll marks, and (E) skip marks by action of "tools" making contact with the bottom in various ways. These tool-formed depressions are subsequently filled with coarser sediment to produce positive-relief casts. [After Reineck, H. E., and I. B. Singh, 1980, Depositional sedimentary environments, 2nd ed., Fig. 127, 129, 125, 132, p. 82, 83, reprinted by permission of Springer-Verlag, Heidelberg.]

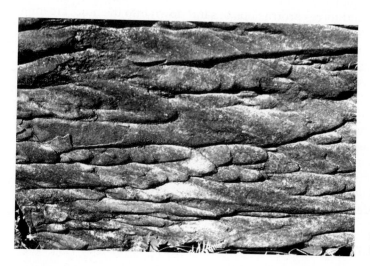

Figure 4.36
Flute casts on the base of a turbidite sandstone, Fluornoy Formation (Eocene), Oregon Coast Range. The bulbuous terminations of the flute casts indicate that paleocurrent flow was from right to left. [Photograph courtesy of Ewart Baldwin.]

are also present in sediments deposited in shallow marine and nonmarine environments. They have been reported on the soles of limestone beds as well as sandstone beds.

Markings Generated by Deformation: Load Casts

Load casts are described by Potter and Pettijohn (1977) as "swellings ranging from slight bulges, deep or shallow rounded sacks, knobby excrescences, or highly irregular protuberance." They commonly occur on the soles of sandstone beds that overlie mudstones or shales, and they tend to cover the entire bedding surface (Fig. 4.37). They range in diameter and relief from a few centimeters to a few tens

Figure 4.37
Irregularly shaped load casts on the base of a loose slab of Cretaceous sandstone, southern Oregon coast.

of centimeters. Load casts may superficially resemble flute casts; however, they can be distinguished from flutes by their greater irregularity of shape and their lack of definite upcurrent and downcurrent ends. Also, load casts do not display a preferred orientation with respect to current direction.

Although they are called casts, load casts are not true casts because they are not fillings of a preexisting cavity or mold. They are formed by deformation of uncompacted, hydroplastic mud beds owing to unequal loading by overlying sand layers. Uncompacted muds with excess fluid pore pressures, or muds liquefied by an externally generated shock, can be deformed by the weight of overlying sand, which may sink unequally into the incompetent mud. Loading owing to unequal weight of the sand forces protrusions of sand down into the mud, creating positive-relief features on the base of the sandstone beds that may resemble some erosional structures, as mentioned. Load casts are closely related genetically to ball and pillow structures and flame structures. Flute and groove casts may be modified by loading, which tends to exaggerate their relief and destroy their original shapes.

Load casts can form in any environment where water-saturated muds are quickly buried by sand before dewatering can take place. They do not indicate any particular environment, although they tend to be most common in turbidite sequences. Their presence on the bases of some beds and not on others seems to reflect the hydroplastic state of the underlying mud. They apparently will not form on the bases of sand beds deposited on muds that have already been compacted or dewatered prior to deposition of the sand.

Biogenic Structures

Trace Fossils

The burrowing, boring, feeding, and locomotion activities of organisms can produce a variety of trails, depressions, and open burrows and borings in mud or semiconsolidated sediment bottoms. Filling of these depressions and burrows with sediment of a different type or with different packing creates structures that may be either positive-relief features, such as trails on the bases of overlying beds, or features that show up as burrow or bore fillings on the tops of the underlying mud bed. Burrows and borings commonly extend down into beds; therefore, these structures are not exclusively bedding-plane structures.

Tracks, trails, burrows, borings, and other structures made by organisms on bedding surfaces or within beds are known collectively as **trace fossils**, also referred to as ichnofossils, or lebensspuren. Study of trace fossils constitutes the discipline of **ichnology**, which has become increasingly complex since the mid-1950s and has spawned a massive body of literature. Several of these books are listed under "Further Reading" at the end of this chapter.

Kinds of Trace Fossils.   Trace fossils are not true bodily preserved fossils; that is, they do not form by conversion of a skeleton into a body fossil. They are simply structures that originated through the activities of organisms. Interpreted broadly, biogenic structures can be considered to include the following: (1) bioturbation structures (burrows, tracks, trails, root penetration structures), (2) biostratification structures (algal stromatolites, graded bedding of biogenic origin), (3) bioerosion structures (borings, scrapings, bitings), and (4) excrement (coprolites, such as fecal pellets or fecal castings). Not all geologists regard biostratification structures as trace fossils, and these structures are not commonly included in published discussions of trace fossils.

Trace fossils are classified into **ichnogenera** on the basis of characteristics that relate to major behavioral traits of organisms and are given generic names such as *Ophiomorpha*. Distinctive but less important characteristics are used to identify **ichnospecies**, e.g., *Ophiomorpha nodosa*. Trace fossils are produced by a host of marine organisms such as crabs, flatfish, clams, molluscs, worms, shrimp, and eel. In nonmarine environments, organisms such as insects, spiders, worms, millipedes, snails, and lizards can produce a variety of burrows and tunnels; vertebrate organisms leave tracks; and plants leave root traces. The organisms that produce traces are rarely preserved with the traces; thus, the trace maker is commonly not known. Therefore, the names applied to ichnogenera and ichnospecies generally do not refer to the trace makers themselves.

Trace Fossil Assemblages.   From a sedimentological standpoint, study of trace-fossil assemblages has commonly proven to be more useful than study of individual ichnogenera or ichnospecies. A trace-fossil assemblage is a basic collective term that embraces all of the trace fossils present within a single unit of rock. Although various kinds of trace-fossil assemblages are recognized, grouping of trace fossils into **ichnofacies** has particular significance in paleoenvironmental studies. Seilacher (1964, 1967) introduced the concept of ichnofacies to describe associations of trace fossils that are recurrent in time and space and that reflect environmental conditions such as water depth (bathymetry), salinity, and the nature of the substrate in or on which they formed (e.g., mud vs. sand bottom). Fundamentally, ichnofacies are sedimentary facies defined on the basis of trace fossils, and each ichnofacies may include several ichnogenera.

Seilacher (1967) established six ichnofacies, which he named after characteristic ichnogenera. Four of these (*Skolithos, Cruziana, Zoophycos*, and *Nereites*) were based on the marine water depth at which they were interpreted to occur (Table 4.3; Fig. 4.38). The *Glossifungites* ichnofacies was established for traces that occur in firm to hard marine surfaces, and the *Scoyenia* ichnofacies characterized nonmarine environments. Subsequently, Frey and Seilacher (1980) established the *Trypanites* ichnofacies for hardgrounds and rockgrounds; Bromley, Pemberton, and Rahmani (1984) proposed the *Teredolites* ichnofacies for borings in wood (woodgrounds); and Frey and Pemberton (1987) established the *Psilonichus* ichnofacies for softgrounds in the marine to nonmarine environment. Several additional ichnofacies have also been proposed (e.g., Bromley, 1996, p. 241); however, the nine ichnofacies shown in Table 4.3 are most commonly used. Sedimentologists are particularly interested in the *Skolithos, Cruziana, Nereites*, and *Zoophycos* ichnofacies, which have the greatest potential for interpreting ancient marine environmental conditions.

Table 4.3 Principal ichnofacies

Ichnofacies	Substrate	Environment	Water depth	Water energy	Distinguishing characteristics
Teredolites	Woodground	Estuarine, nearshore marine	—	—	Club-shaped, stumpy to elongate, subcylindrical to subparallel borings
Trypanites	Rockground	Rocky coasts, reefs, hard-grounds	—	—	Cylindrical, tear-, or U-shaped, vertical to branching borings
Scoyenia	Firmground	Freshwater, terrestrial	—	—	Horizontal to curved or tortuous burrows; sinuous crawling traces; vertical cylindrical to branching shafts; tracks and trails
Glossifungites	Firmground	Marine to nonmarine	Various	Various	Vertical, cylindrical, U- or tear-shaped borings and/or densely branching burrows
Psilonichnus	Softground sand, mud	Marine to nonmarine	—	—	J-, Y-, or U-shaped burrows; vertical shafts and horizontal tunnels; tracks, trails, root traces
Skolithos	Softground sand	Marine	Beach	High	Vertical, cylindrical, or U-shaped burrows; very few horizontal burrows; low diversity
Cruziana	Softground sand, mud	Marine	Lagoon, shelf	Medium to low	Mixed association of vertical, inclined, and horizontal structures; high diversity of traces
Zoophycos	Softground mud	Marine	Slope-abyssal	Low	Simple to moderately complex grazing and feeding structures; horizontal to slightly inclined feeding or dwelling structures arranged in delicate sheets, ribbons, lobes, or spirals
Nereites	Softground sand, mud	Marine	Slope-abyssal	Turbidity current event	Complex horizontal, crawling, and grazing traces and patterned feeding/dwelling traces; low diversity

Data from: Bromley et al., 1984; Frey and Seilacher, 1980; Frey and Pemberton, 1987; Pemberton et al., 1992; Seilacher, 1967.

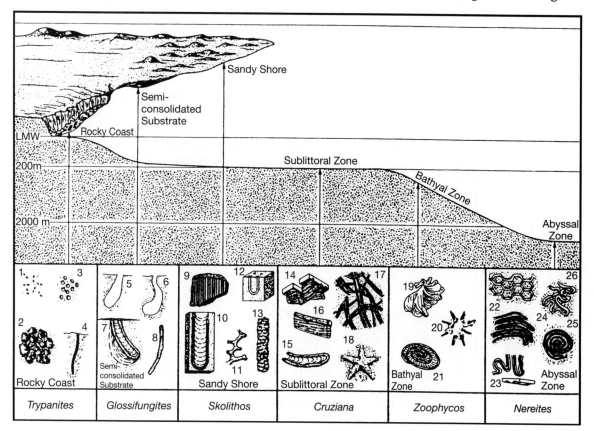

Figure 4.38

Schematic representation of the relationship of characteristic trace fossils to sedimentary facies and depth zones in the ocean. Borings of 1, *Polydora*; 2, *Entobia*; 3, echinoid borings; 4, *Trypanites*; 5,6, pholadid burrows; 7, *Diplocraterion*; 8, unlined crab burrow; 9, *Skolithos*; 10, *Diplocraterion*; 11, *Thalassinoides*; 12, *Arenicolites*; 13, *Ophiomorpha*; 14, *Phycodes*; 15, *Rhizocorallium*; 16, *Teichichnus*; 17, *Crossopodia*; 18; *Asteriacites*; 19; *Zoophycos*; 20, *Lorenzinia*; 21, *Zoophycos*; 22, *Paleodictyon*; 23, *Taphrhelminthopsis*; 24, *Helminthoida*; 25, *Spirohaphe*; 26, *Cosmorhaphe*. [From Ekdale, A. A., R. G. Bromley, and S. B. Pemberton, 1984, Ichnology: Trace fossils in sedimentology and stratigraphy: Soc. Econ. Paleontologists and Mineralogists Short Course No. 15, Fig. 15.2, p. 187, reprinted by permission of SEPM, Tulsa, Okla. Modified from Crimes, T. P., 1975, The stratigraphical significance of trace fossils, *in* Crimes, T. P., and J. C. Harper, eds., The study of trace fossils, Fig. 7.2, p. 118: Springer-Verlag, New York.]

Significance of Trace Fossils. Trace fossils are important paleoecological indicators; however, they are not infallible paleodepth indicators. In general, organisms in the littoral or intertidal zones adapt to harsh conditions resulting from high wave or current energy, desiccation, and large temperature and salinity fluctuations by burrowing into sand to escape. Thus, vertical and U-shaped dwelling burrows, some with protective linings, characterize the *Skolithos* ichnofacies of this zone. The neritic zone or subtidal zone extending from the low-tide zone to the edge of the continental shelf (at about 200 m water depth) is a less demanding environment, although erosive currents may be present. Vertical dwelling burrows and protected, U-shaped burrows are less common in this zone. Burrows tend to be shorter, and surface markings made by organisms such as crustaceans (or trilobites during early Paleozoic time) are more common. In the deeper part of the neritic zone, organic matter becomes abundant enough for sediment feeders to become established and produce feeding burrows. In these deeper waters, vertical

Box 4.1 Important Ichnofacies

SKOLITHOS ICHNOFACIES

Trace fossils of this association are characterized especially by vertical, cylindrical, or U-shaped burrows (e.g., *Ophiomorpha, Diplocraterion,* and *Skolithos,* Fig. 4.4.1). Overall diversity of ichnogenera is low and few horizontal structures are present. This ichnofacies is developed primarily in sandy sediment where relatively high levels of wave or current energy are typical. Organisms in this environment construct deep burrows to protect against desiccation or unfavorable temperature or salinity changes during low tide, and as a means of escaping the shifting substrate of the surface (Pemberton, MacEachern, and Frey, 1992). The *Skolithos* ichnofacies is typical of sandy shoreline environment (Fig. 4.38), but it may grade seaward into shallow shelf environments. It has also been reported from some deeper-water environments, such as deep-sea fans and bathyal slopes.

CRUZIANA ICHNOFACIES

The *Cruziana* ichnofacies commonly occurs in somewhat deeper water than the *Skolithos* ichnofacies within subtidal zones below fair-weather wave base but above storm wave base (Frey and Seilacher, 1980), typical of the middle and outer shelf. It may also be present in sediments from some nearshore environments. It is characterized by a mixed association of traces that may include nearly vertical burrows, inclined U-burrows (*Rhizocorallium*), horizontal structures (*Cruziana*), the traces of organisms that move about on or near the

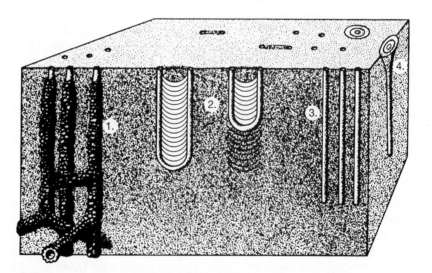

Figure 4.4.1
Trace fossil association characteristic of the *Skolithos* ichnofacies. 1) *Ophiomorpha,* 2) *Diplocraterion,* 3) *Skolithos,* 4) *Moncraterion.* [From Pemberton, S. G., J. A. MacEachern, and R. W. Frey, 1992, Trace fossil facies models: Environmental and allostratigraphic significance, *in* Facies models, Walker, R. G., and N. P. James (eds.), Geological Association of Canada, Fig. 9, p. 53. Reproduced by permission.]

sediment surface (*Thalassinoides*), and other odd traces having star shapes (*Asteriacites*) or C-shapes (*Arenicolites*) (Fig. 4.4.2). Note: Some authors (e.g., Bromley, 1996, p. 249) recognize a separate *Arenicolites* ichnofacies. The *Cruziana* ichnofacies commonly has high diversity and abundance of traces (Fig. 4.4.2); in fact, a profusion of burrows may be present. It is typically developed in well-sorted silts and sands but may be present in muddy sands or silts.

ZOOPHYCOS ICHNOFACIES

This ichnofacies appears most typical of quiet-water environments with moderately low oxygen levels and muddy bottoms but can occur in other substrates. It is characterized by traces that range from simple to moderately complex, such as *Spirophyton* (Fig. 4.4.3). Individual traces may be abundant, but overall diversity is low. Sediments of the Zoophycos ichnofacies may be totally bioturbated (Bromley, 1996, p. 250). Although commonly considered to indicate deeper water (Fig. 4.38), it is known to occur also in shallow water. Thus, its value as a paleodepth indicator is problematical. Its distribution appears to be tied more closely to oxygen levels and bottom sediment type than water depth.

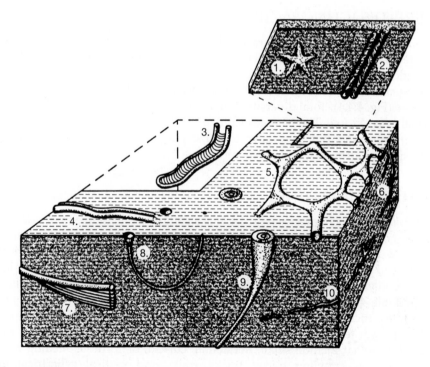

Figure 4.4.2
Trace fossil association characteristics of the *Cruziana* ichnofacies. 1) *Asteriacites*, 2) *Cruziana*, 3) *Rhizocorallium*, 4) *Aulichnites*, 5) *Thalassinoides*, 6) *Chondrites*, 7) *Teichichnus*, 8) *Arenicolites*, 9) *Rossella*, 10) *Planolites*. [From Pemberton, S. G., J. A. MacEachern, and R. W. Frey, 1992, Trace fossil facies models: Environmental and allostratigraphic significance, *in* Facies models, Walker, R. G., and N. P. James (eds.), Geological Association of Canada, Fig. 10, p. 54. Reproduced by permission.]

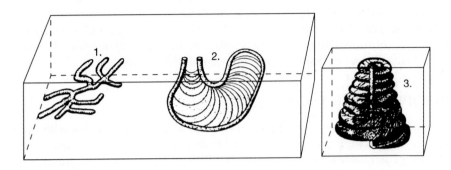

Figure 4.4.3
Trace fossil association characteristic of the *Zoophycos* ichnofacies. 1) *Phycosiphon*, 2) *Zoophycos*, 3) *Spirophyton*. [From Pemberton, S. G., J. A. MacEachern, and R. W. Frey, 1992, Trace fossil facies models: Environmental and allostratigraphic significance, *in* Facies models, Walker, R. G., and N. P. James (eds.), Geological Association of Canada, Fig. 11, p. 55. Reproduced by permission.]

NEREITES ICHNOFACIES

The *Nereites* ichnofacies is characteristic of deep water and is apparently restricted to turbidite deposits. It is distinguished by complex horizontal crawling and grazing traces and patterned feeding or dwelling structures. The ichnogenera are ornate and complicated, such as *Paleodictyon, Spirorhaphe*, and *Nereites* (Fig. 4.4.4). Total diversity of traces is high, but the abundance of individual traces is low. The Nereites ichnofacies develops initially in sandy (turbidite) substrates but may later colonize parts of some muddy (pelagic) deposits that form on top of sandy turbidites.

OTHER ICHNOFACIES

The *Psilonichnus* ichnofacies (Fig. 4.4.5) is a softground ichnofacies developed under nonmarine to very shallow marine or quasi-marine conditions. It is characterized by J-, Y-, or U-shaped burrows of marine organisms, vertical shafts, and horizontal tunnels of insects and tetrapods; tracks and trails of insects, reptiles, birds, and mammals; and root traces. The other ichnofacies listed in Table 4.3 are distinguished by development in firm but uncemented substrates, rocky substrates, or woody material. *Scoyenia* ichnofacies, which occur in both terrestrial and aquatic environments, are characterized by diverse traces that include small, horizontal, curved, or tortuous feeding burrows, sinuous crawling traces, tracks, trails, and vertical cylindrical to irregular shafts (see Hasiotis, 2002). The *Trypanites* ichnofacies develops in fully lithified marine substrates (beachrock, rocky coasts, hardgrounds, reefs). Traces include cylindrical, tear-, or U-shaped borings, commonly vertical to branching, most of which are dwelling structures for suspension-feeding organisms (Fig. 4.38, 1–4). Other structures in this ichnofacies include rasping and scraping traces made by feeding organisms, holes drilled by predatory gastropods, and microborings made by algae and fungi. The *Glossifungites* ichnofacies develops in a variety of marine environments in firm, but unlithified,

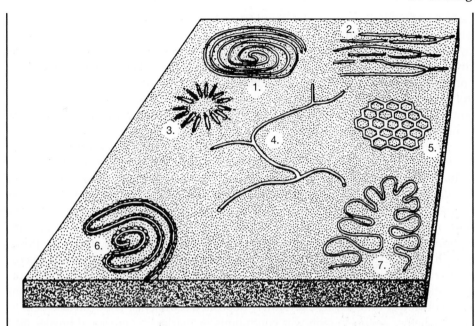

Figure 4.4.4
Trace fossil association characteristics of the *Nereites* ichnofacies. 1) *Spirorhaphe*, 2) *Uroheiminthoidea*, 3) *Lorenzinia*, 4) *Megagrapton*, 5) *Paleodictyon*, 6) *Nereites*, 7) *Cosmorhaphe*. [From Pemberton, S. G., J. A. MacEachern, and R. W. Frey, 1992, Trace fossil facies models: Environmental and allostratigraphic significance, *in* Facies models, Walker, R. G., and N. P. James (eds.), Geological Association of Canada, Fig. 12, p. 56. Reproduced by permission.]

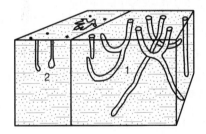

Figure 4.4.5
Trace fossil association characteristic of the *Psilonichnus* ichnofacies. 1) *Psilonichnus*, 2) *Macanopsis*. [After Pemberton, S. G., J. A. MacEachern, and R. W. Frey, 1992, Trace fossil facies models: Environmental and allostratigraphic significance, *in* Facies models, Walker, R. G., and N. P. James (eds.), Geological Association of Canada, Fig. 8, p. 53. Reproduced by permission.]

substrates that typically consist of dewatered, cohesive muds. It is characterized by vertical, cylindrical, U- or tear-shaped borings and/or densely branching burrows of suspension feeders or carnivores such as shrimp, crabs, worms, and pholadid bivalves (Fig. 4.38, 5–8). Individual structures may be abundant but diversity is low. *Teredolites* ichnofacies are restricted to woody substrates (so-called woodground) commonly in estuarine or very nearshore environments where substantial amounts of woody material can accumulate on the bottom. The traces consist of profuse club-shaped borings that may be stumpy to elongate and subcylindrical to subparallel.

escape or dwelling burrows thus tend to give way to horizontal feeding burrows. This zone of the ocean is distinguished by the *Cruziana* ichnofacies, characterized by such traces as those shown in Figure 4.38, 14–18 and Fig. 4.4.2. The deep bathyal and abyssal zones of the ocean exist below wave base where low-energy conditions generally prevail, although erosion and deposition can occur in these zones owing to turbidity currents or deep-bottom currents. Complex feeding burrows, such as those of the *Nereites* ichnofacies (Fig. 4.38, 22–26; Fig. 4.4.4), are particularly common in these zones.

Although each of these marine ichnofacies tends to be characteristic of a particular bathymetric zone of the ocean, as shown in Figure 4.38, we now know that individual trace fossils can overlap depth zones. No single biogenic structure is an infallible indicator of depth and environment. The basic controls on the formation of trace fossils are not simply depth but include nature of the substrate, water energy, rates of deposition, water turbidity, oxygen and salinity levels, toxic substances, and quantity of available food (Pemberton, MacEachern, and Frey, 1992).

In addition to their role as environmental indicators, trace fossils are also useful in several other ways. They may, for example, serve as an indicator of relative sedimentation rates based on the assumption that rapidly deposited sediments contain relatively fewer trace fossils than slowly deposited sediments. They can also help to show whether sedimentation was continuous or marked by erosional breaks, and they provide a record of the behavior patterns of extinct organisms. They may even be useful in paleocurrent analysis; study of the orientation of resting marks of organisms that may have preferred to face into the current while resting establishes the paleocurrent-flow direction. Some trace fossils such as U-shaped burrows, which opened upward when formed, can be used to tell the top and bottom orientation of beds. Trace fossils also have biostratigraphic and chronostratigraphic significance for zoning and correlation, and they may be useful for recognition of bounding discontinuities between stratigraphic successions (Pemberton, MacEachern, and Frey, 1992; also see Frey and Pemberton, 1985, and Frey and Wheatcroft, 1989).

Stromatolites

Stromatolites are organically formed, laminated structures composed of fine silt- or clay-size sediment or, more rarely, sand-size sediment. Most ancient stromatolites occur in limestones; however, stromatolites have also been reported in siliciclastic sediments. Stromatolitic bedding ranges from nearly flat laminations that may be difficult to differentiate from sedimentary laminations of other origins to hemispherical forms in which the laminae are crinkled or deformed to various degrees (Fig. 4.39). The hemispherical forms range in shape from biscuit- and cabbage-like forms to

Figure 4.39
Stromatolites in limestones of the Helena Formation (Precambrian), Glacier National Park, Montana.

columns. Logan, Rezak, and Ginsburg (1964) classified these hemispherical stromatolites into three basic types: (1) laterally linked hemispheroids, (2) discrete, vertically stacked hemispheroids, and (3) discrete spheroids, or spheroidal structures (Fig. 4.40). Laterally linked hemispheroids and discrete, vertically stacked hemispheroids can combine in various ways to create several different kinds of compound stromatolites. The term **thrombolite** was proposed by Aitken (1967) for structures that resemble stromatolites in external form and size but lack distinct laminations. The laminations of stromatolites are generally less than 1 mm thick and are caused by concentrations of fine calcium carbonate minerals, fine organic matter, and detrital clay and silt. Stromatolites composed of quartz grains have also been reported (Davis, 1968).

Stromatolites were considered true body fossils by early workers, but they are now known to be organosedimentary structures formed largely by the trapping and binding activities of blue-green algae (cyanobacteria). They are forming today in many localities where they occur mainly in the shallow subtidal, intertidal, and

Types	Description	Vertical section of stromatolite structure
Laterally linked hemispheroids	Space-linked hemispheroids with close-linked hemispheroids as a microstructure in the constituent laminae	
Discrete, vertically stacked hemispheroids	Discrete, vertically stacked hemispheroids composed of close-linked hemispheroidal laminae on a microscale	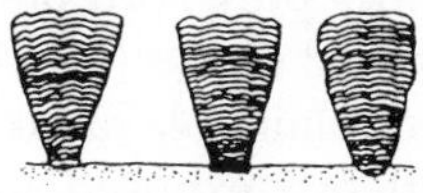
Discrete spheroids	Spheroidal structures consisting of inverted, stacked hemispheroids	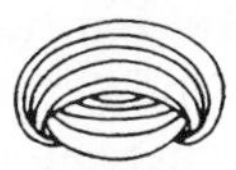
	Spheroidal structures consisting of concentrically stacked hemispheroids	
	Spheroidal structures consisting of randomly stacked hemispheroids	
Combination forms	Initial space-linked hemispheroids passing into discrete, vertically stacked hemispheroids with upward growth of structures	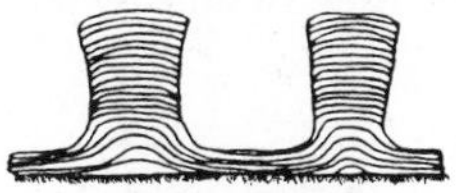
	Initial discrete, vertically stacked hemispheroids passing into close-linked hemispheroids by upward growth	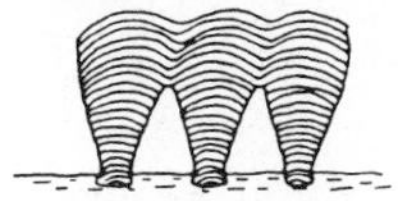
	Alternation of discrete, vertically stacked hemispheroids and space-linked hemispheroids due to periodic sediment infilling of interstructure spaces	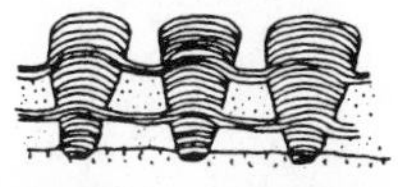
	Initial space-linked hemispheroids passing into discrete, vertically stacked hemispheroids; both with laminae of close-linked hemispheroids	
	Initial discrete, vertically stacked hemispheroids passing into close-linked hemispheroids; both with laminae of close-linked hemispheroids	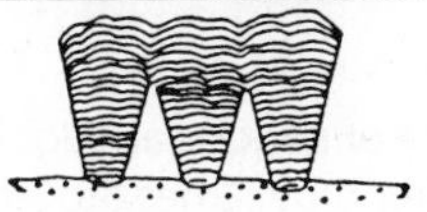

Figure 4.40
Structures of hemispherical stromatolites showing examples of laterally linked hemispheroids, vertically stacked hemispheroids, and discrete spheroids. [After Logan, B. W., R. Rezak, and R. N. Ginsburg, 1964, Classification and environmental significance of algal stromatolites: Jour. Geology, v. 72, Fig. 4, p. 76, and Fig. 5, p. 78, reprinted by permission of University of Chicago Press.]

supratidal zones of the ocean. They have also been found in lacustrine environments. Because they are related to the activities of blue-green algae, which carry out photosynthesis, they are restricted to water depths and environments where enough light is available for photosynthesis. The laminated structure forms because fine sediment is trapped in the very fine filaments of algal mats. Once a thin layer of sediment covers the mat, the algal filaments grow up and around sediment grains to form a new mat that traps another thin layer of sediment. This successive growth of mats produces the laminated structure. The shapes of the hemispheres are related to water energy and scouring effects in the depositional environment. Laterally linked hemispheroids tend to form in low-energy environments where scouring effects are minimal. In higher-energy environments, scouring by currents prevents linking of the stromatolite heads; thus, vertically stacked or discrete hemispheroids form. Stromatolites are forming in the world's oceans at the present time, and they have been reported in ancient rocks as old as 3.45 billion years (e.g., Hofmann et al., 1999).

Bedding-Plane Markings of Miscellaneous Origin

Mudcracks in modern sediment are downward-tapering, V-shaped fractures that display a crudely polygonal pattern in plan view. The area between the cracks is commonly curved upward into a concave shape. Mudcracks form in both siliciclastic and carbonate mud owing to desiccation. Subsequent sedimentation over a cracked surface fills the cracks. In ancient sedimentary rocks, mudcracks are commonly preserved on the tops of bedding surfaces as positive-relief fillings of the original cracks (Fig. 4.41). Mudcracks occur in estuarine, lagoonal, tidal-flat, river floodplain, playa lake, and other environments where muddy sediment is intermittently exposed and allowed to dry. They may be associated with raindrop or hailstone imprints, bubble imprints and foam impressions, flat-topped ripple marks, and vertebrate tracks (Plummer and Gostin, 1981).

Syneresis cracks are subaqueously formed markings on bedding surfaces that superficially resemble mudcracks; however, they are discontinuous and vary in shape from polygonal to spindle-shaped or sinuous (Plummer and Gostin, 1981). They commonly occur in thin mudstones interbedded with sandstones as either positive-relief features on the base of the sandstones or negative-relief features on the top of the mudstones. Syneresis cracks are subaqueous shrinkage cracks that form in clayey sediment by loss of pore water from clays that have flocculated rapidly or that have undergone shrinkage of swelling-clay mineral lattices because of changes in salinity of surrounding water (Burst, 1965).

Figure 4.41
Ancient mudcracks on the surface of a rock slab (age?), Death Valley, California. [Photograph by James Stovall.]

Small craterlike pits with slightly raised rims commonly occur together with mudcracks and are thought to be impressions made by the impact of rain (**raindrop imprints**) or hail (**hailstone imprints**). They are commonly only a few millimeters deep and less than 1 cm in diameter, and they may occur as either widely scattered pits or very closely spaced impressions. When they can be unambiguously recognized, their presence indicates subaerial exposure; however, small circular depressions created by bubbles breaking on the surface of sediment (**bubble imprints**), escaping gas, and some types of organic markings can be confused with raindrop or hailstone imprints.

Rill marks are small dendritic channels or grooves that form on beaches by the discharge of pore waters at low tide, or by small streams debouching onto a sand or mud flat. They have very low preservation potential and are seldom found in ancient sedimentary rocks. **Swash marks** are very thin, arcuate lines or small ridges on a beach formed by concentrations of fine sediment and organic debris. They are caused by wave swash and mark the farthest advance of wave uprush. They likewise have low preservation potential, but when found and recognized in ancient sedimentary rocks, they indicate either a beach or a lakeshore environment.

Parting lineation, sometimes called current lineation, forms on the bedding surfaces of parallel laminated sandstones. It consists of subparallel ridges and grooves a few millimeters wide and many centimeters long (Fig. 4.42). Relief on the ridges and hollows is commonly on the order of the diameter of the sandstone grains. The grains in the sandstone generally have a mean orientation of their long axes parallel to the lineation. The lineation is oriented parallel to current flow, and thus its presence in ancient sandstones is useful in paleocurrent studies, although it shows only that the current flowed parallel to the parting lineations and does not show which of the two diametrically opposed directions was the flow direction. Parting lineation occurs in newly deposited sands on beaches and in fluvial environments. It is most common in ancient deposits in thin, evenly bedded sandstones. Its origin is obviously related to current flow and grain orientation, probably owing to flow over upper-flow regime plane beds, but the exact mechanism by which parting lineation forms is poorly understood.

Figure 4.42
Parting lineation in sandstone, Haymond Formation, Texas. Paleocurrent flow was parallel to the lineation. [Photograph courtesy of E. F. McBride.]

4.5 OTHER STRUCTURES

Sandstone dikes and sills are tabular bodies of massive sandstone that fill fractures in any type of host rock. They range in thickness from a few centimeters to more than 10 m. They lack internal structures except for oriented mica flakes and other elongated particles that are commonly aligned parallel to the dike walls. Sandstone dikes are formed by forceful injection of liquefied sand into fractures, commonly in overlying rock; however, injection appears to have been downward in some rocks. Sandstone sills are similar features that formed by injection parallel to bedding. These sills may be difficult or impossible to distinguish from normally deposited sandstone beds unless they can be traced into sandstone dikes or be traced far enough to show a cross-cutting relationship with other beds. Suggested causes of liquefaction of sand include shocks owing to earthquakes or triggering effects related to slumps, slides, or rapid emplacement of sediment by mass flow.

Secondary sedimentary structures are structures that form sometime after deposition during sediment burial. These structures are largely of chemical origin, formed by precipitation of mineral substances in the pores of semiconsolidated or consolidated sedimentary rock or by chemical replacement processes. **Concretions** are probably the most common kind of secondary structure. Most concretions are composed of calcite, but concretions composed of dolomite, hematite, siderite, chert, pyrite, and gypsum are also known. They form by precipitation of mineral matter around some kind of nucleus, such as a shell fragment, and gradually build up a globular mass (e.g., the dark, rounded features in Fig. 4.6), which may or may not display concentric layering. Shapes of these masses range from spherical to disc-shaped, cone-shaped, and pipe-shaped, and they may range in size from less than 1 cm to as much as 3 m. Concretions are especially common in sandstones and shales but can occur in other sedimentary rocks.

Stylolites are suturelike seams of clay or other insoluble material that commonly occur in limestones owing to pressure solution (discussed in Chapter 6). Less common secondary structures include sand crystals (large crystals of calcite, barite, or gypsum filled with sand inclusions) and cone-in-cone structures (nested sets of small concentric cones composed of carbonate minerals). See Boggs (1992, p. 119–124) for additional descriptions of secondary structures.

4.6 PALEOCURRENT ANALYSIS FROM SEDIMENTARY STRUCTURES

As mentioned, many sedimentary structures yield directional data that show the direction ancient current flowed at the time of deposition. The dip direction of cross-bed foresets; the asymmetry and orientation of the crests of current ripples; and the orientation of flute casts, groove casts, and current lineation are all examples of directional data that can be obtained from sedimentary structures. Cross-bedding is one of the most useful sedimentary structures for determining paleocurrent direction. Because the foreset laminae in cross-beds are generated by avalanching on the downcurrent (lee) side of ripples, the foresets dip in the downcurrent direction. To measure paleocurrent direction from cross-beds requires that they be exposed in a three-dimensional outcrop. The strike of the foreset laminae is determined first; the dip direction is 90° to the strike. If cross-beds have been tilted by tectonic uplift after deposition, a correction must be made for this tilt (e.g., Collinson and Thompson, 1989, p. 200).

The orientation of directional sedimentary structures is determined in the field with a Brunton compass by taking measurements from as many different outcrops and individual beds as possible and practical. The orientation of directional structures determined from a particular bed or stratigraphic unit commonly shows considerable scatter. Therefore, directional data must be treated statistically in some manner to reveal primary and secondary directional trends. For example, the dip direction of cross-bed foresets in the ancient deposits of a meandering river system may range from N 20° W to N 20° E because of variations in flow direction of the stream in different parts of the meandering river system. By examining the orientation data statistically, we may be able to determine that the primary flow direction of the stream was approximately due north. Because all of the cross-bed foresets in this example indicate flow in the same general direction, in spite of some scatter, we say that flow was **unidirectional**. By contrast, the cross-bed foresets in sandy deposits of marine tidal channels may display two opposing dip directions owing to formation of cross-beds during both incoming and outgoing tides. This type of opposing flow is referred to as **bidirectional**. In some environments, such as the eolian environment, depositing currents may flow in several directions (**polydirectional**) at various times during deposition of a particular sedimentary unit.

The paleocurrent data collected from stratigraphic units that have undergone little or no tectonic deformation or tilting can be compiled and summarized directly. If the rocks have undergone considerable tilting, it is necessary to correct the measured orientation by restoring directions to their original attitude before tilting. A simple procedure using a stereogram can be used to reorient directional data collected from tilted stratigraphic units (Collinson and Thompson, 1989, p. 200). After any reorientation of data has been done, the data are commonly plotted as a circular histogram or "rose diagram" (Fig. 4.43). (Commercial computer software programs are available for plotting rose diagrams, e.g., *Rose*, from Rockware, Wheat Ridge, Colorado.) Such diagrams show the principal direction of paleocurrent flow and any secondary or tertiary modes of flow. If the paleocurrent flow as revealed by the rose diagram is dominantly in a single direction, the paleocurrent vector is said to be **unimodal**. If two principal directions of flow are indicated, it is **bimodal**, and if three or more directions of flow are revealed by the directional data, the paleocurrent flow is called **polymodal**.

Local paleocurrent directions may have environmental significance. For example, sediments from alluvial and deltaic environments tend to have unimodal paleocurrent vector patterns, whereas bimodal paleocurrent patterns are more common in shoreline and shelf sediments. Paleocurrent data have their greatest usefulness when plotted on a regional scale to reveal regional paleocurrent patterns (discussed in Chapter 16).

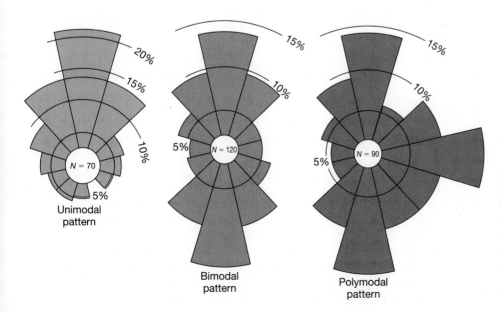

Figure 4.43
Hypothetical paleocurrent data plotted as rose diagrams: unimodal, bimodal, and polymodal patterns of paleoflow directions are shown. *N* = number of measurements (directions) taken in the field.

FURTHER READING

Mainly Physically Formed Structures

Allen, J. R. L., 1982, Sedimentary structures: Their character and physical basis, v. 1-2: Elsevier, Amsterdam, 664 p.

Collinson, J. D., and D. B. Thompson, 1989, Sedimentary structures, 2nd ed.: Chapman and Hall, London, 207 p.

Harms, J. C., J. B. Southard, and R. G. Walker, 1982, Structures and sequences in clastic rocks: Soc. Econ. Paleontologists and Mineralogists Short Course No. 9, Tulsa, Okla.

Pettijohn, F. J. , and P. E. Potter, 1964, Atlas and glossary of primary sedimentary structures: Springer-Verlag, New York, 370 p.

Ricci Lucchi, F., 1995, Sedimentographica: A photographic atlas of sedimentary structures, 2nd ed.: Columbia University Press, New York, 255 p.

Biogenic Structures

Bromley, R. G., 1996, Trace fossils: Biology, taphonomy and applications: Chapman and Hall, London, 361 p.

Curran, H. A. (ed.), 1985, Biogenic structures: Their use in interpreting depositional environments: Soc. Econ. Paleontologists and Mineralogists Special Publication No. 35, Tulsa, Okla., 347 p.

Donovan, S. K., 1994, The paleobiology of trace fossils: The Johns Hopkins University Press, Baltimore, 308 p.

Ekdale, A. A., R. G. Bromley, and S. G. Pemberton, 1984, Ichnology, trace fossils in sedimentology and stratigraphy: Soc. Econ. Paleontologists and Mineralogists Short Course No. 15, Tulsa, Okla., 317 p.

Hasiotis, S. T., 2002, Continental trace fossils: SEPM (Society for Sedimentary Geology), Tulsa, Okla., 130 p.

Maples, C. G., and R. R. West (eds.), 1992, Trace fossils: Paleontological Society Short Courses in Paleontology No. 5, 238 p.

Miller, M. F., A. A. Ekdale, and M. D. Picard (eds.), 1984, Trace fossils and paleoenvironments: Marine carbonate, marginal marine terrigenous and continental terrigenous settings: Jour. Paleontology, v. 58, p. 283–597.

Pemberton, S. G., J. A. MacEachern, and R. W. Frey, 1992, Trace fossil facies models: Environmental and allostratigraphic significance, in Walker, R. G., and N. P. James (eds.), Facies models: Response to sea level change: Geological Association of Canada, p. 47–72.

PART III

Composition, Classification, and Diagenesis of Sedimentary Rocks

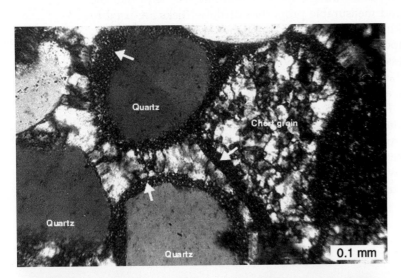

Top: Microquartz (arrows) cementing quartz and chert grains in sandstone, Jefferson City Formation (Ordovician), Missouri
Bottom: Ooids in limestone (Devonian), Canada

In Chapters 3 and 4 we examined sedimentary textures and structures and discussed some of the ways that these physical properties are used to interpret the genesis of sedimentary rocks. Particle and chemical composition are also fundamental properties of sedimentary rocks, properties that allow us to distinguish one kind of sedimentary rock from another and that provide additional information about the history of the rocks. Geologists commonly use the term **mineralogy** to refer to the identity of all the particles or grains in rocks. When we talk about the mineralogy of a sandstone, for example, we mean quartz, feldspar, micas, and other particulate mineral grains that make up the sandstone. Bulk **chemistry** is another aspect of the overall composition of sedimentary rocks and is directly related to the mineralogy of the rocks. Traditionally, sedimentologists have been interested more in mineralogy than in bulk chemistry because they consider mineralogy to be more useful than chemical composition in characterizing and classifying sedimentary rocks and in interpreting their geologic history. That perception is changing, and we are now seeing an increasing number of published papers that are concerned in some way with sedimentary geochemistry.

The next three chapters deal with the composition of siliciclastic sedimentary rocks (sandstones, conglomerates, shales), carbonate rocks (limestones and dolomites), other biochemical/chemical sedimentary rocks (evaporites, cherts, iron-rich sedimentary rocks, phosphorites), and carbonaceous sedimentary rocks (oil shales and coals). Among other things, we will examine the ways that the characteristic, observable, and measurable properties of these rocks are used to classify them into different kinds. We will also examine the various ways that mineralogy and chemical composition are used to interpret sediment provenance (source), depositional process, depositional environments, and diagenesis (burial alteration). Many of the biochemical/chemical sedimentary rocks, such as dolomite, iron-rich sedimentary rocks, and phosphorites, are enigmatic rocks whose origins are still poorly understood. This section discusses ideas concerning the origin of these rocks and analyzes controversial and conflicting hypotheses. Study of the mineralogy of sedimentary rocks is commonly referred to as **sedimentary petrography**. The next three chapters provide an introduction to sedimentary petrography. Additional details are available in a host of papers and monographs, several of which are listed under "Further Reading" at the ends of the chapters.

5

Siliciclastic Sedimentary Rocks

5.1 INTRODUCTION

As discussed in Chapter 1, sedimentary rocks composed mainly of silicate particles derived by the weathering breakdown of older rocks and by pyroclastic volcanism are called siliciclastic sedimentary rocks. Sandstones, conglomerates, and shales compose the members of this group. The siliciclastic sedimentary rocks make up roughly three-fourths of all sedimentary rocks in the geologic record, and they are present in sedimentary successions ranging in age from Precambrian to Holocene. They are of special interest to geologists as indicators of Earth history. Many of the textures and structures discussed in Chapters 3 and 4 are particularly well developed in these rocks. These textures and structures provide important information about ancient sediment transport and depositional conditions. In addition, the minerals and rock fragments in siliciclastic sedimentary rocks furnish our most definitive clues to the nature and location of vanished ancient mountain systems such as the ancestral Rocky Mountains and Appalachian Mountains of the United States. Petroleum geologists are especially interested in sandstones because more than half of the world's reserves of oil and gas occur in these rocks. Shales are likewise of great interest because the organic matter contained in shales is believed to be the source material of oil and gas.

In this chapter, we focus on the particle composition (kinds of minerals and rock fragments) of sandstones, shales, and conglomerates and explore ways that composition can be used to classify these rocks and interpret aspects of their origin. We will also briefly examine changes in composition and texture that take place owing to burial diagenesis.

5.2 SANDSTONES

Sandstones make up 20–25 percent of all sedimentary rocks. They are common rocks in geologic systems of all ages, and they are distributed throughout the continents of Earth. Sandstones consist mainly of silicate grains ranging in size from 1/16 to 2 mm. These particles make up the **framework fraction** of the sandstones. Sandstones may also contain various amounts of cement and very fine size (<~0.03 mm) material called **matrix**, which are present within interstitial pore space among the framework grains. Because of their coarse size (relative to the

119

sizes of particles in shales), the framework mineralogy of sandstones can general-ly be determined with reasonable accuracy with a standard petrographic micro-scope or by backscattered electron microscopy (e.g., Krinsley et al., 1998). Bulk chemical composition can be measured by instrumental techniques such as X-ray fluorescence and inductively coupled argon plasma emission spectrometry (ICP). The chemistry of individual mineral grains is commonly determined by use of an electron probe microanalyzer or an energy dispersive X-ray detector (EDX) at-tached to a scanning electron microscope.

In this section, we examine the mineralogy and chemical composition of sandstones, discuss the classification of sandstones on the basis of mineral compo-sition, and evaluate the usefulness of particle and chemical composition in inter-preting the genesis of sandstones.

Framework Mineralogy

The particles that make up sandstones are mainly sand-size and coarse silt-size sil-icate minerals and rock fragments referred to as framework grains, as mentioned. Only a few principal kinds of minerals make up the bulk of all sandstones. These common minerals and rock fragments are shown in Table 5.1 and are discussed in greater detail in the following paragraphs.

Quartz

Quartz (SiO_2) is the dominant mineral in most sandstones, making up on average about 50–60 percent of the framework fraction. It is a comparatively easy mineral to identify, both megascopically in hand specimens and by petrographic examina-tion in thin sections, although it can be confused with feldspars. Because of its su-perior hardness and chemical stability, quartz can survive multiple recycling. The quartz grains in many sandstones display some degree of rounding acquired by abrasion during one or more episodes of transport, particularly by wind.

Quartz can occur as single (monocrystalline) grains (Fig. 5.1A) or as compos-ite (polycrystalline) grains (Fig. 5.1B). When examined under crossed polarizing prisms with a petrographic microscope, many quartz grains display sweeping patterns of extinction as the stage is rotated. This property is called **undulatory ex-tinction**. Some authors (Folk, 1974; Basu et al., 1975) suggest that the properties of polycrystallinity and undulatory extinction can be used to distinguish quartz de-rived from different sources. Quartz is derived from plutonic rock, particularly felsic plutonic rocks such as granites, metamorphic rocks, and older sandstones.

Feldspars

Feldspar minerals make up about 10–20 percent of the framework grains of aver-age sandstones. They are the second most abundant mineral in most sandstones. Several varieties of feldspars are recognized on the basis of differences in chemical composition and optical properties. They are divided into two broad groups: alka-li feldspars and plagioclase feldspars.

Alkali feldspars constitute a group of minerals in which chemical composi-tion can range through a complete solid solution series from $KAlSi_3O_8$ through $(K, Na)AlSi_3O_8$ to $NaAlSi_3O_8$. Because potassium-rich feldspars are such com-mon members of this group, it has become widespread practice to call the alkali feldspars **potassium feldspars**, often shortened to simply **K-spars**. Common members of the potassium-feldspar group include orthoclase, microcline (Fig. 5.1C), and sanidine. **Plagioclase feldspars** form a complex solid solution series ranging in composition from $NaAlSi_3O_8$ (albite) through $CaAl_2Si_2O_8$ (anorthite). A general formula for the series is $(Na, Ca)(Al, Si)Si_2O_8$. Plagioclase feldspars can

Table 5.1 Common minerals and rock fragments in siliciclastic sedimentary rocks

Major minerals (abundance >~1–2%)
    Stable minerals (greatest resistance to chemical decomposition):
        Quartz—makes up approximately 65% of framework grains in average sandstone, 30% of minerals in average shale.
    Less stable minerals:
        Feldspars—include K-feldspars (orthoclase, microcline, sanidine, anorthoclase) and plagioclase feldspars (albite, oligoclase, andesine, labradorite, bytownite, anorthite); make up about 10–15% of framework grains in average sandstone, 5% of average shale.
        Clay minerals and fine micas—clay minerals include the kaolinite group, illite group, smectite group (montmorillonite a principal variety), and chlorite group; fine micas are principally muscovite (sericite) and biotite; minor abundance in sandstones, where they occur as matrix minerals, but compose >60% of the minerals in average shale.

Accessory minerals (abundances <~1–2%)
    Coarse micas—principally muscovite and biotite.
    Heavy minerals (specific gravity >~2.9):
        Stable nonopaque minerals—zircon, tourmaline, rutile, anatase.
        Metastable nonopaque minerals—amphiboles, pyroxenes, chlorite, garnet, apatite, staurolite, epidote, olivine, sphene, zoisite, clinozoisite, topaz, monazite, plus about 100 others of minor importance volumetrically.
        Stable opaque minerals—hematite, limonite.
        Metastable opaque minerals—magnetite, ilmenite, leucoxene.

Rock Fragments (make up about 10–15% of the siliciclastic grains in average sandstone and most of the gravel-size particles in conglomerates; shales contain few rock fragments)
    Igneous rock fragments—clasts of any kind of igneous rock possible in conglomerate; however, fragments of fine-crystalline volcanic rock are most common in sandstones.
    Metamorphic rock fragments—may include clasts of any kind of metamorphic rock; however, metaquartzite, schist, phyllite, slate, and argillite clasts are most common in sandstones.
    Sedimentary rock fragments—clasts of any kind of sedimentary rock possible in conglomerates; clasts of fine sandstone, siltstone, shale, and chert are most common in sandstones; limestone clasts are comparatively rare in sandstones.

Chemical Cements (abundance variable)
    Silicate minerals—predominantly quartz; others may include microquartz (chert), opal, feldspars, and zeolites.
    Carbonate minerals—principally calcite; less commonly aragonite, dolomite, siderite.
    Iron oxide minerals—hematite, limonite, goethite.
    Sulfate minerals—anhydrite, gypsum, barite.

Note: The terms stable and metastable refer to chemical stability.

commonly be distinguished from potassium feldspars on the basis of optical properties such as twinning (compare Fig. 5.1D with Fig. 5.1C) by examination with a petrographic microscope. Unfortunately, some K-feldspars (e.g., orthoclase and sanidine) and some plagioclase are untwinned, making them difficult to differentiate from each other and from quartz. Potassium feldspars are generally considered to be somewhat more abundant overall in sedimentary rocks than plagioclase feldspars; however, plagioclase is more abundant in sandstones derived from volcanic rocks.

Figure 5.1
Principal kinds of framework grains in sandstones. A. Monocrystalline quartz, Roubidoux Fm. (Ordovician), Missouri. B. Polycrystalline quartz, Miocene sandstone, Japan Sea. C. Potassium feldspar (microcline), Bateman Fm. (Eocene), Oregon. D. Plagioclase feldspar, Bateman Fm. (Eocene), Oregon. E. Mica (muscovite), Bateman Fm. (Eocene), Oregon. F. Heavy minerals (Z = Zircon; M = magnetite; P = pyroxene), India. Crossed nicol photomicrographs.

Feldspars are chemically less stable than quartz and are more susceptible to chemical destruction during weathering and diagenesis. Because they are also softer than quartz, feldspars become more readily rounded during transport. They also appear to be somewhat more prone to mechanical shattering and breakup owing to their cleavage. Feldspars are less likely than quartz to survive several episodes of recycling, although they can survive more than one cycle if weathering occurs in a moderately arid or cold climate. Owing to this possibility of recycling, the presence of a few feldspar grains in a sedimentary rock does not necessarily mean that the rock is composed of first-cycle sediments derived directly from crystalline igneous or metamorphic rocks. On the other hand, a high content of feldspars, particularly on the order of 25 percent or more, probably indicates derivation directly from crystalline source rocks.

Accessory Framework Minerals

Minerals that have an average abundance in sedimentary rocks less than about 1–2 percent are called accessory minerals. These minerals include the common **micas**, muscovite (white mica) and biotite (dark mica), and a large number of so-called heavy minerals, which are denser than quartz.

The average abundance of coarse micas in siliciclastic sedimentary rocks is less than about 0.5 percent, although some sandstones may contain 2–3 percent. Micas are distinguished from other minerals by their platy or flaky habit (Fig. 5.1E). Muscovite is chemically more stable than biotite and is commonly much more abundant in sandstones than biotite. Micas are derived particularly from metamorphic source rocks as well as from some plutonic igneous rocks.

Minerals that have a specific gravity greater than about 2.9 are called **heavy minerals**. These minerals include both chemically stable and unstable (labile) varieties as shown in Table 5.1. Stable heavy minerals such as zircon (Fig. 5.1F) and rutile can survive multiple recycling episodes and are commonly rounded, indicating that the last source was sedimentary. Less stable minerals, such as magnetite, pyroxenes, and amphiboles, are less likely to survive recycling. They are commonly first-cycle sediments that reflect the composition of proximate source rocks. Thus, heavy minerals are useful indicators of sediment source rocks because different types of source rocks yield different suites of heavy minerals. Heavy minerals are derived from a variety of igneous, metamorphic, and sedimentary rocks.

Because of their low abundance in sandstones, heavy minerals are commonly concentrated for study by separating them from the light mineral fraction by using heavy liquids such as bromoform or sodium polytungstate (e.g., Lindholm, 1987, p. 214). In this separation process, disaggregated sediment is stirred into a heavy liquid contained in a funnel placed inside a fume hood. (Many heavy liquids are toxic and must be handled with extreme care.) The light minerals float on the surface of the heavy liquid, but the heavy minerals gradually sink into the stem of the funnel where they can be drawn off and separated from the light fraction. After the heavy liquid has been washed off in a suitable solvent, these heavy mineral concentrates can then be mounted on a glass microscope slide (Fig. 5.1F) and studied with a petrographic microscope.

Rock Fragments

Pieces of ancient source rocks that have not yet disintegrated to yield individual mineral grains are called rock fragments or clasts. Rock fragments make up about 15–20 percent of the framework grains in the average sandstone; however, the rock-fragment content of sandstones is highly variable and ranges from zero to more than 95 percent. Fragments of any kind of igneous, metamorphic, or sedimentary rock can occur in sandstones (Fig. 5.2); however, clasts of fine-grained source rocks are most likely to be preserved as sand-size fragments (Boggs, 1968). Very coarse grained source rocks such as granites typically yield clasts of coarse-sand size or larger (e.g. Fig. 5.2A); however, such clasts are not common because granites commonly disintegrate to yield individual minerals rather than clasts. The most common rock fragments in sandstones are clasts of volcanic rocks (Fig. 5.2B), volcanic glass (in younger rocks), and fine-grained metamorphic rocks such as slate, phyllite, schist (Fig. 5.2C), and quartzite (Fig. 5.2D). Sand-size fragments of silica-cemented siltstone, fine-grained sandstone (Fig. 5.2E), and shale are less

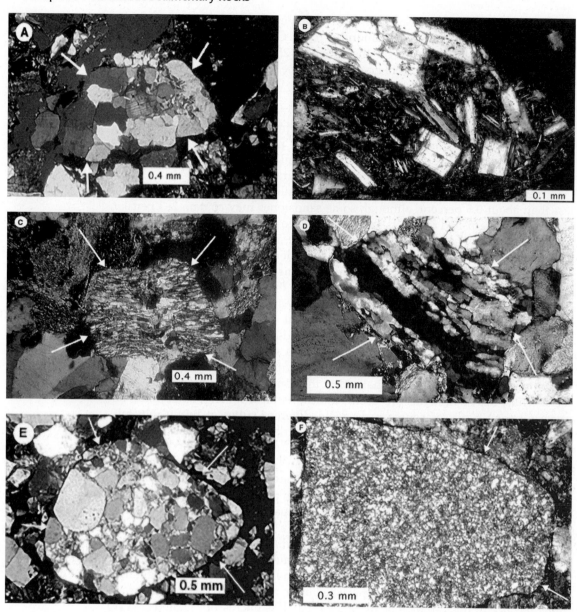

Figure 5.2
Common kinds of rock fragments in sandstones. A. Plutonic (granite; arrows), Bateman Fm. (Eocene), Oregon. B. Volcanic, Miocene sandstone, Japan Sea. C. Metamorphic (schist; arrows), Pottsville Group (Pennsylvanian), eastern Pennsylvania. D. Metamorphic (quartzite; arrows) Pottsville Group (Pennsylvanian), eastern Pennsylvania. E. Sedimentary (sandstone; arrows; note rounding of grains within this clast), Bateman Fm. (Eocene), Oregon. F. Sedimentary (chert; arrows), Otter Point Fm. (Jurassic), Oregon. Crossed nicol photomicrographs.

common. Clasts of limestone or other carbonate rocks are also less common, probably in part because they do not survive weathering and transport. Composite quartz grains that consist of exceedingly small quartz crystals, referred to as microcrystalline quartz, are called **chert**. Chert grains are actually rock fragments, derived by weathering of bedded chert or chert nodules in limestone, and can be abundant in some sandstones (Fig. 5.2F).

Rock fragments are particularly important in studies of sediment source rocks. They are moderately easily identified, and they are more reliable indicators of source rock types than are individual minerals such as quartz or feldspar, which can be derived from different types of source rocks.

Mineral Cements

The framework grains in most siliciclastic sedimentary rocks are bound together by some type of mineral cement. These cementing materials may be either silicate minerals such as quartz and opal or nonsilicate minerals such as calcite and dolomite. Quartz is the most common silicate mineral that acts as a cement. In most sandstones, the quartz cement is chemically attached to the crystal lattice of existing quartz grains, forming rims of cement called **overgrowths** (Fig. 5.3A). Such overgrowths that retain crystallographic continuity of a grain are said to be **syntaxial**. Because syntaxial overgrowths are optically continuous with the original grain, they go to extinction in the same position as the original grain when rotated on the stage of a polarizing microscope. Overgrowths can be recognized by a line of impurities or bubbles that mark the surface of the original grain. Quartz overgrowths are particularly common in quartz-rich sandstones. Less commonly, quartz cement is present as microcrystalline quartz, which has a fine-grained, crystalline texture similar to that of chert. When silica cement is deposited as microcrystalline quartz, it forms a mosaic of very tiny quartz crystals that fill the interstitial spaces among framework silicate grains (Fig. 5.3B). Not uncommonly, the

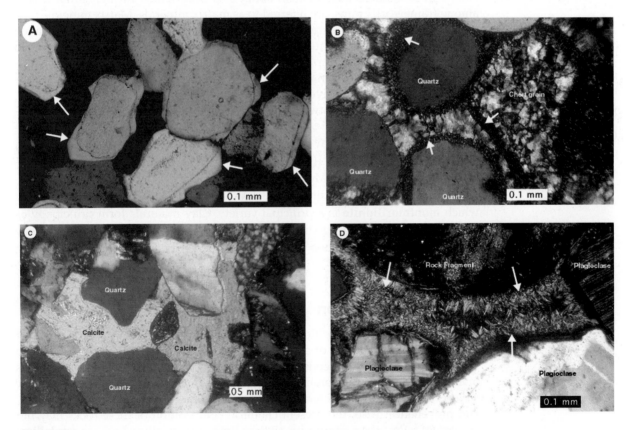

Figure 5.3

Common cements in sandstones. A. Quartz overgrowths (arrows), Lamotte Sandstone (Cambrian), Missouri. B. Microquartz (chert; arrows) cementing quartz and chert grains, Jefferson City Fm. (Ordovician), Missouri. C. Calcite, Kayenta Fm. (Triassic), Utah. D. Clay mineral (chlorite; arrows), Miocene sandstone, Japan Sea. Crossed nicol photomicrographs.

crystals next to the framework grains are small, slightly elongated, and are oriented normal to the surfaces of the framework grains. More rarely, opal (an isotropic mineral) occurs as a cement in sandstones, particularly in sandstones rich in volcanogenic materials. Like quartz and microcrystalline quartz (chert), opal is also composed of SiO_2, but, unlike these minerals, opal contains some water and lacks a definite crystal structure. Thus, it is said to be amorphous. Opal is metastable and crystallizes in time to microcrystalline quartz.

Carbonate minerals are the most abundant nonsilicate mineral cements in siliciclastic sedimentary rocks. Calcite is a particularly common carbonate cement. It is precipitated in the pore spaces among framework grains, typically forming a mosaic of smaller crystals (Fig. 5.3C). These crystals adhere to the larger framework grains and bind them together. Less common carbonate cements are dolomite and siderite (iron carbonate). Other minerals that act as cements in sandstones include the iron oxide minerals hematite and limonite, feldspars, anhydrite, gypsum, barite, clay minerals (Fig. 5.3D), and zeolite minerals. Zeolites are hydrous aluminosilicate minerals that occur as cements primarily in volcaniclastic sedimentary rocks (discussed in a subsequent section).

All cements are secondary minerals that form in sandstones after deposition and during burial. Details of the cementation process are given in Section 5.5 (Diagenesis).

Matrix Minerals

Grains in sandstones smaller than about 0.03 mm, which fill interstitial spaces among framework grains, are referred to as matrix minerals. Matrix minerals may include fine-size micas, quartz, and feldspars; however, clay minerals make up the bulk of matrix grains. Because of their small size, clay minerals are difficult to identify by routine petrographic microscopy. They must be identified by X-ray diffraction techniques, electron microscopy, or other nonoptical methods. Clay minerals are compositionally diverse. They belong to the phyllosilicate mineral group, which is characterized by two-dimensional layer structures arranged in indefinitely extending sheets.

The most common clay mineral groups are **illite** [$K_2(Si_6Al_2)Al_4O_{20}(OH)_4$], **smectite** (montmorillonite) [$(Al, Mg)_8(Si_4O_{10})_3(OH)_{10} \cdot 12H_2O$], **kaolinite** [$Al_2Si_2O_5(OH)_4$], and **chlorite** [$(Mg, Fe)_5(Al, Fe^{3+})_2Si_3O_{10}(OH)_8$]. Kaolinite is a two-layer clay; the others are three-layer clays. Smectite is a clay-mineral group, of which **montmorillonite** is a principal variety. Clay minerals form principally as secondary minerals during subaerial weathering and hydrolysis, although they can also form by subaqueous weathering in the marine environment and during burial diagenesis.

Chemical Composition

Sedimentologists have traditionally placed relatively little emphasis on study of the chemical composition of siliciclastic sedimentary rocks, in sharp contrast to geologists who study igneous and metamorphic rocks. This lack of interest is mainly attributed to the common belief that chemical composition of siliciclastic rocks is less useful than mineral composition for interpreting depositional history and provenance. Also, the present chemical composition of these rocks may not accurately reflect their composition at the time of deposition, because crystallization of new minerals during sediment burial and diagenesis can change the original chemical composition. The formerly high cost of doing chemical analyses is an additional factor that helped to discourage extensive chemical studies of sedimentary rocks. Several new tools, such as the electron probe microanalyzer and X-ray fluorescence equipment, now enable rapid and comparatively inexpensive chemical analyses of rock composition. These new tools, as well as changing attitudes regarding the significance of chemical composition, are causing sedimentologists to develop a deeper interest in the chemistry of sedimentary rocks. For example, chemical data, both bulk chemical composition and the trace-element composition of individual minerals, have been applied to provenance studies, that is, studies linking sedimentary materials to their source rocks (e.g., Morton et al., 1991; Johnsson and Basu, 1993).

Because most grains in siliciclastic sedimentary rocks are derived from various types of igneous, metamorphic, and sedimentary rocks, the mineralogy and chemical composition of siliciclastic rocks are clearly a function of parent rock composition. Nonetheless, sedimentary rocks display distinct chemical differences from parent source rocks owing to chemical changes that occur during weathering and diagenesis. For example, they tend to be enriched in silica and depleted in iron, magnesium, calcium, sodium, and potassium compared to the parent rocks. Enrichment in silica occurs because siliceous minerals resist chemical weathering; thus, silica is concentrated in weathering residues with respect to more soluble cations. Also, the superior chemical stability of SiO_2 minerals such as quartz and chert causes sedimentary rocks to become progressively enriched in these minerals during multiple recycling events. Thus, overall, silica content increases at the expense of less stable iron- and magnesium-rich minerals.

The average chemical composition of some sandstones reported in the literature is shown in Table 5.2. I stress that this table shows average composition of a few sandstones. Specimens of sandstone from other formations may have chemical compositions that deviate considerably from these average values. Silicon, expressed as SiO_2, is the most abundant chemical constituent in all types of sandstones because of the abundance of quartz in sandstones and the presence of silicon in all silicate minerals. Aluminum (Al_2O_3) is moderately abundant in sandstones containing abundant feldspars or in rock-fragment-rich sandstones that contain a matrix of clay minerals. It is much less abundant in quartz-rich sandstones, which commonly do not have a clay matrix. On average, iron, magnesium, calcium, sodium, and potassium are all less abundant in sandstones than is aluminum. Relative concentrations of these elements vary as a function of the mineralogy of the sand-size grains and the types of matrix clay minerals and diagenetic cements in the rock. For example, sandstones with abundant calcium carbonate cement or carbonate fossils may have anomalously high calcium content.

Classification of Sandstones

Descriptive classification of sandstones is based fundamentally on framework mineralogy, although the relative abundance of matrix plays a role in some classifications. Although mineralogy is the principal basis for classifying sandstones, finding a classification that is suitable for all types of sandstones and acceptable to most geologists has proven to be an elusive goal. In fact, more than fifty different classifications for sandstones have been proposed (Friedman and Sanders, 1978), but none has received widespread acceptance. Classifications that are all-inclusive tend to be too complicated and unwieldy for general use, and classifications that are oversimplified may convey too little useful information.

Textural Nomenclature of Mixed Sediments

Unconsolidated siliciclastic sediment is called gravel (dominance of >2 mm-size grains), sand (1/16–2 mm), or mud (<1/16 mm), depending upon grain size. The lithified rock equivalents of these sediments are conglomerate, sandstone, and shale (mudrock). Because many siliciclastic sedimentary rocks are composed of

Table 5.2 Average chemical composition (weight percent) of sandstones from some North American formations*

n =	(1) 11	(2) 23	(3) 30	(4) 16	(5) 18	(6) 12	(7) 119	(8) 12	(9) 59
SiO_2	86.5	67.8	65.6	56.9	56.2	68.4	70.6	37.3	50.3
TiO_2	0.53	0.95	0.91	1.42	0.89	0.69	0.64	0.34	0.64
Al_2O_3	5.71	15.4	15.1	12.3	15.3	13.5	12.6	7.91	14.0
$Fe_2O_3(t)$	2.69	6.46	6.09	6.18	6.48	5.30	4.97	3.18	6.40
MnO	0.02	0.07	0.15	0.11	0.07	0.09	0.08	0.10	0.13
MgO	0.69	1.73	1.82	4.20	2.35	1.68	1.51	1.07	3.25
CaO	0.05	0.42	1.94	5.82	5.74	2.38	1.61	26.0	9.90
Na_2O	0.02	1.07	0.87	1.92	1.28	3.15	2.76	0.92	—
K_2O	1.55	2.74	3.03	1.90	2.80	2.62	2.20	0.51	2.09
P_2O_5	0.02	0.16	0.17	0.17	0.17	0.18	0.02	0.10	0.21
V (ppm)	51	123	159	100	126	71	79	103	—
Cr	55	82	88	225	71	55	44	31	—
Ni	19	231	58	130	49	30	8	5	49
Zn	29	52	104	84	114	69	—	66	91
Rb	60	123	133	72	125	93	—	10	79
Sr	29	134	113	233	168	310	110	879	267
Y	17	31	40	21	35	36	37	15	29
Zr	417	238	260	191	187	333	413	58	118

Source: Argast and Donnelly, 1987, *Jour. Sed. Petrology*, v. 57, p. 813–823: Society Economic Mineralogists and Paleontologists, Tulsa, Okla.

Note: Iron is reported as total Fe_2O_3; n is the number of samples in each average; t indicates total; a dash indicates no reported value.

* (1) Shawangunk Formation near Ellenville, New York (quartz arenite)
  (2) Millport Member of the Rhinestreet Formation, Elmira, New York (lithic arenite/wacke)
  (3) Oneota Formation, Unadilla, New York (lithic arenite/wacke)
  (4) Cloridorme Formation, St. Yvon and Gros Morne, Quebec (lithic arenite/wacke)
  (5) Austin Glen Member of Normanskill Formation, Poughkeepsie, New York (lithic arenite/wacke)
  (6) Renessalaer Member of the Nassau Formation, near Grafton, New York (feldspathic arenite/wacke)
  (7) Renesselaer Member, averages of analyses from Ondrick and Griffiths (1969) (feldspathic arenite/wacke)
  (8) Rio Culebrinas Formation, La Tosca, Puerto Rico (fossiliferous volcaniclastics)
  (9) Turbidites from DSDP site 379A (lithic arenites/wacke)

grains of mixed sizes, it is not always easy to decide whether a rock should be called a conglomerate, a sandstone, or a shale. It might be difficult to decide, for example, whether a sedimentary rock composed of nearly equal portions of sand-size and mud-size particles should be called a sandstone or a shale. Various classification schemes have been devised for naming texturally mixed sediments and sedimentary rocks, most of which make use of triangular texture diagrams such as those shown in Figure 5.4.

The textural classification illustrated in Figure 5.4A includes particles ranging in size from mud (clay and fine silt) to gravel. Note that the textural boundaries in this classification scheme are not entirely symmetrical. Ideally, we might expect the boundary between gravel and mud-sand to be set at 50 percent; however, this is not always the case, as Figure 5.4A shows. Because particles of gravel size are commonly less abundant than sand and mud particles, many geologists consider a sediment with as little as 30 percent gravel-size fragments to be a gravel. If sediments contain only particles of sand size and smaller, a textural classification scheme such as Figure 5.4B or 5.4C that uses sand, silt, and clay as end members of the classification is more appropriate. Once the textural nomenclature

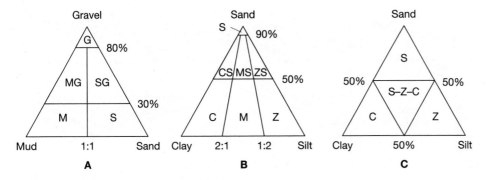

Figure 5.4
Nomenclature of mixed sediments. A, B. Simplified from Folk. C. After Robinson. C = clay, CS = clayey sand, G = gravel, M = mud, MG = muddy gravel, MS = muddy sand, S = sand, SG = sandy gravel, Z = silt, Zs = silty sand. [A and B after Folk, R. L., 1954, The distinction between grain size and mineral composition in sedimentary rock nomenclature: Jour. Geology, v. 62, Fig. 1a, p. 346, and 1b, p. 349, reprinted by permission of University of Chicago Press. C after Robinson, G. W., 1949, Soils, their origin, constitution, and classification, 3rd ed.; Murby, London.]

of siliciclastic sediment or sedimentary rocks has been established, rocks within each textural group can be further classified on the basis of composition.

Mineralogical Classification

Most sandstones are made up of mixtures of a very small number of dominant framework components. Quartz, feldspars, and rock fragments such as chert and volcanic clasts are the only framework constituents that are commonly abundant enough to be important in sandstone classification. In addition to framework grains, matrix may be present in interstitial spaces among these grains. In spite of the very simple composition of sandstones, geologists have not been able to agree on a single, acceptable sandstone classification. Published classifications range from those that have a strong genetic orientation to those based strictly on observable, descriptive properties of sandstones. Most authors of sandstone classifications use a classification scheme that involves a QFR or QFL plot. These plots are triangular diagrams on which quartz (Q), feldspars (F), and rock fragments (R or L) are plotted as end members at the poles of the classification triangle. There are numerous possible ways that such a triangle can be subdivided into classification fields, and geologists have explored the full range of these possibilities (see reviews by Klein, 1963; Boggs, 1967b; Okada, 1971; Yanov, 1978).

One of the simplest and easiest classifications to use is that of Gilbert (Williams, Turner, and Gilbert, 1982), shown in Figure 5.5, which is based on an earlier classification by Dott (1964). In this classification, sandstones that are effectively free of matrix (<5 percent) are classified as **quartz arenites, feldspathic arenites,** or **lithic arenites** depending upon the relative abundance of QFL constituents. If matrix can be recognized (at least 5 percent), the terms **quartz wacke, feldspathic wacke,** and **lithic wacke** are used instead. A principal difference between Gilbert's (1982) and Dott's (1964) classification is that Dott sets the boundary between arenites and wackes at 15 percent matrix. Sandstone classifications that include more classification "pigeonholes" than Gilbert's, and that have been rather widely used by American geologists, include those of McBride (1963) and Folk, Andrews, and Lewis (1970). These classifications do not include matrix as part of the classification scheme.

The name **arkose** is often used informally by geologists for any feldspathic arenite that is particularly rich (>~25 percent) in feldspars. Another term in

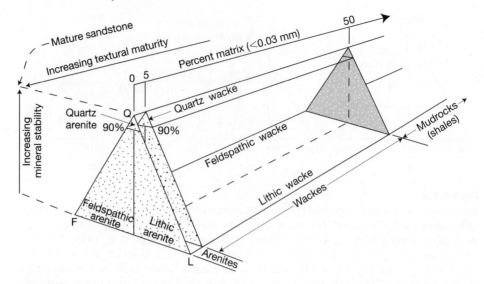

Figure 5.5
Classification of sandstones on the basis of three mineral components: Q = quartz, chert, quartzite fragments; F = feldspars; L = unstable, lithic grains (rock fragments). Points within the triangles represent relative proportions of Q, F, and L end members. Percentage of argillaceous matrix is represented by a vector extending toward the rear of the diagram. The term arenite is restricted to sandstones containing less than about 5 percent matrix; sandstones containing more matrix are wackes. [After Williams, H. F., F. J. Turner, and C. M. Gilbert, 1982, Petrography, an introduction to the study of rocks in thin section, 2nd ed., W. H. Freeman and Co., San Francisco, Fig. 13.1, p. 327. Modified from Dott, R. H., Jr., 1964, Wacke, graywacke, and matrix—what approach to immature sandstone classification: Jour. Sed. Petrology, v. 34, Fig. 3, p. 629, reprinted by permission of SEPM, Tulsa, Okla.]

general use is **graywacke.** This name is commonly applied to matrix-rich sandstones of any composition that have undergone deep burial, have a chloritic matrix, and are dark gray to dark green, very hard, and dense. This term has been much misused, and its continued used is controversial (see Boggs, 1992, p. 185–186). Some geologists think that the term should be abandoned entirely and that we should substitute the word wacke for graywacke. That is probably good advice. In any case, the name is best restricted to field use and should not be used as a petrographic term.

Sandstone Maturity

The term maturity is applied to sandstones in two different ways. **Compositional maturity** refers to the relative abundance of stable and unstable framework grains in a sandstone. A sandstone composed mainly of quartz is considered compositionally mature, whereas a sandstone that contains abundant unstable minerals (e.g., feldspars) or unstable rock fragments is compositionally immature. **Textural maturity** is determined by the relative abundance of matrix and the degree of rounding and sorting of framework grains, as illustrated in Figure 5.6. Textural maturity can range from immature (much clay, framework grains poorly sorted and poorly rounded) to supermature (little or no clay, framework grains well sorted and well rounded). Textural maturity allegedly reflects the degree of sediment transport and reworking; however, it may also be affected by diagenetic processes (i.e., clay minerals may form in pore spaces during burial diagenesis).

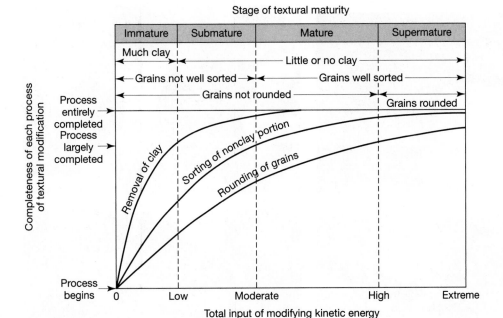

Figure 5.6
Textural maturity classification of Folk. Textural maturity of sands is shown as a function of input of kinetic energy. [From Folk, R. L., 1951, Stages of textural maturity in sedimentary rocks: Jour. Sed. Petrology, v. 21, Fig. 1, p. 128, reprinted by permission of SEPM, Tulsa, Okla.]

General Characteristics of Major Classes of Sandstones

The preceding discussion indicates that sandstones can be divided on the basis of framework mineralogy into three major groups: quartz arenites, feldspathic arenites, and lithic arenites (these groups include wackes). Some general characteristics of each of these major sandstone clans are discussed below.

Quartz Arenites

Quartz arenites are composed of more than 90 percent siliceous grains that may include quartz, chert, and quartzose rock fragments (Fig. 5.7A). They are commonly white or light gray but may be stained red, pink, yellow, or brown by iron oxides. They are generally well lithified and well cemented with silica or carbonate cement; however, some are porous and friable. Quartz arenites typically occur in association with assemblages of rocks deposited in stable cratonic environments such as eolian, beach, and shelf environments. Thus, they tend to be interbedded with shallow-water carbonates and, in some cases, with feldspathic sandstones. Most quartz arenites are texturally mature to supermature (Fig. 5.6); quartz wackes are uncommon. Cross-bedding is particularly characteristic of these sandstones, and ripple marks are moderately common. Fossils are rarely abundant, possibly owing to poor preservation or to the eolian origin of some quartz arenites, but fossils may be present. Also, trace fossils such as burrows of the *Skolithos* facies may be locally abundant in some shallow-marine quartz arenites. Quartz arenites are common in the geologic record. Pettijohn (1963) estimates that they make up about one-third of all sandstones.

Quartz arenites can originate as first-cycle deposits derived from primary crystalline or metamorphic rocks, but they are more likely to be the product of

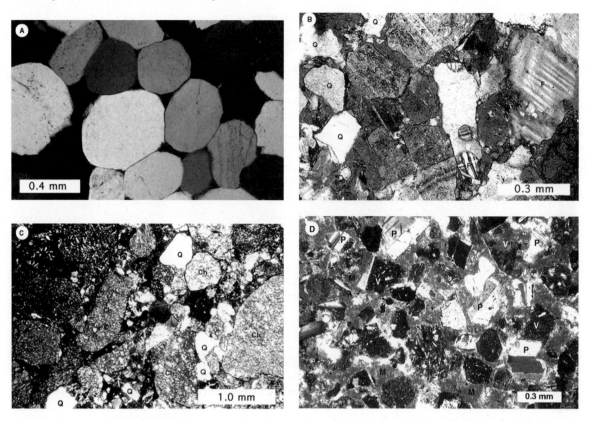

Figure 5.7
Representative photomicrographs illustrating major classes of sandstones. A. Quartz arenite (>95% quartz), Roubidoux Fm. (Ordovician), Missouri. B. Feldspathic arenite (F = feldspar; Q = quartz), Belt Group (Precambrian), Montana. C. Lithic arenite (V = volcanic rock fragment; Q = quartz; Ch = chert—some apparent chert grains may actually be silicified volcanic rock fragments), Otter Point Fm. (Jurassic), Oregon. D. Volcaniclastic sandstone (V = volcanic rock fragment; P = plagioclase; M = matrix), Miocene sandstone, Japan Sea. Crossed nicols.

multiple recycling of quartz grains from sedimentary source rocks. If they are first-cycle deposits, they must have formed under weathering, transport, and depositional conditions so vigorous that most grains chemically less stable than quartz were eliminated (e.g., Johnsson, Stallard, and Meade, 1988). Conceivably, extensive chemical leaching under hot, humid, low-relief weathering conditions; prolonged transport by wind; intensive reworking in the surf zone; or a combination of these processes might be adequate to generate a first-cycle quartz arenite. Most quartz arenites are probably polycyclic, and their history may have included at least one episode of eolian transport, although not necessarily during the last depositional cycle.

Quartz arenites are very common rocks in the geologic record, particularly in Mesozoic and Paleozoic stratigraphic successions. Some well-known examples of quartz arenites in North America include the Ordovician St. Peter Sandstone in the midcontinent United States, the Jurassic Navajo Sandstone of the Colorado Plateau, the Ordovician Eureka Quartzite in Nevada and California, parts of the Cretaceous Dakota Sandstone in the Colorado Plateau and Great Plains, and many Cambro-Ordovician sandstones in the Upper Mississippi Valley. Numerous examples of quartz arenites are also known from other continents. Pettijohn, Potter, and Siever (1987, p. 179–184) list additional examples of quartz arenites from North America, Europe, and other parts of the world.

Feldspathic Arenites

Feldspathic arenites (Fig. 5.7B) contain less than 90 percent quartz, more feldspar than unstable rock fragments, and minor amounts of other minerals such as micas and heavy minerals. Some feldspathic arenites are colored pink or red because of the presence of potassium feldspars or iron oxides; others are light gray to white. They are typically medium to coarse grained and may contain high percentages of subangular to angular grains. Matrix content may range from trace amounts to more than 15 percent, and sorting of framework grains can range from moderately well sorted to poorly sorted. Thus, feldspathic sandstones are commonly texturally immature or submature, i.e., wackes.

Feldspathic arenites are not characterized by any particular kinds of sedimentary structures. Bedding may range from essentially structureless to parallel laminated or cross laminated. Fossils may be present, especially in marine beds. Feldspathic arenites typically occur in cratonic or stable shelf settings, where they may be associated with conglomerates, shallow-water quartz arenites or lithic arenites, carbonate rocks, or evaporites. Less typically, they occur in sedimentary successions that were deposited in unstable basins or other deeper water, mobile-belt settings. Feldspathic arenites of the latter types, which are commonly matrix rich and well indurated owing to deep burial, are often called feldspathic graywackes. The abundance of feldspathic arenites in the geologic record is not well established. Pettijohn (1963) estimates that arkoses make up about 15 percent of all sandstones. If feldspathic graywackes are included, feldspathic arenites are probably more abundant than 15 percent.

Some arkoses originate essentially *in situ* when granite and related rocks disintegrate to produce a granular sediment called **grus**. These residual arkosic materials may be shifted a short distance downslope and deposited as fans or aprons of waste material, commonly referred to as clastic wedges. These fans may extend into basins and become intercalated or interbedded with better stratified and better sorted sediments. Other feldspathic arenites undergo considerable transport and reworking by rivers or the sea before they are deposited. These reworked sandstones commonly contain less feldspar than do residual arkoses, and they are better sorted and grains are better rounded.

Most feldspathic sandstones are derived from granitic-type primary crystalline rocks, such as coarse granite or metasomatic rocks containing abundant potassium feldspar. Feldspathic arenites containing feldspars that are dominantly plagioclase, derived from igneous rocks such as quartz diorites or from volcanic rocks, are also known. The preservation of large quantities of feldspars during weathering appears to require that feldspathic arenites originate either (1) in very cold or very arid climates, where chemical weathering processes are inhibited, or (2) in warmer, more humid climates where marked relief of local uplifts allows rapid erosion of feldspars before they can be decomposed. Although some feldspars may survive recycling from a sedimentary source, it appears unlikely that sedimentary source rocks can furnish enough feldspar to produce a feldspathic arenite or arkose.

Feldspathic arenites occur in sedimentary successions of all ages, although they appear to be particularly abundant in Mesozoic and Paleozoic strata. Some common examples include the Old Red Sandstone (Carboniferous) in Scotland, the Triassic Newark Group in the New Jersey area, the Pennsylvanian Fountain and Lyons formations of the Colorado Front Range, and the Paleocene Swauk Formation of Washington. The Swauk Formation is particularly interesting because it is a plagioclase arkose.

Lithic Arenites

Lithic arenites are an extremely diverse group of rocks that are characterized by generally high content of unstable rock fragments such as volcanic and metamorphic clasts; however, lithic arenites may contain some stable clasts such as chert (e.g., Fig. 5.7C). They contain less than 90 percent quartzose grains and contain more unstable rock fragments than feldspars. Colors may range from light gray, salt-and-pepper to uniform medium to dark gray. Many lithic arenites are poorly sorted; however, sorting ranges from well sorted to very poorly sorted. Quartz and many other framework grains are generally poorly rounded. Lithic arenites tend to contain substantial amounts of matrix, much of which may be of secondary origin. Thus, most lithic sandstones are texturally immature to submature (lithic wackes). Lithic arenites may range from irregularly bedded, laterally restricted, cross-stratified fluvial units to evenly bedded, laterally extensive, graded, marine turbidite units. They may occur in association with fluvial conglomerates and other fluvial deposits or in association with deeper water marine conglomerates, pelagic shales, cherts, and submarine basalts. Lithic arenites include sandstones that many geologists continue to refer to as graywackes. Graywackes differ from "normal" lithic arenites in that they are dark gray to dark green, are well indurated or lithified, and commonly have a matrix consisting of secondary chlorite. The term graywacke is used so loosely, however, that it might be best to simply drop it, as mentioned. Pettijohn (1963) estimates that lithic arenites and graywackes together make up nearly one-half of all sandstones.

Lithic arenites are typically compositionally immature sandstones that originate under conditions favoring the production and deposition of large volumes of relatively unstable materials. The mechanically weak character of many of the lithic fragments in these sandstones suggests that they are probably derived from rugged, high-relief source areas. Lithic arenites may be deposited in nonmarine settings in proximal alluvial fans or other fluvial environments. Alternatively, they may be deposited in marine foreland basins (Chapter 16) adjacent to fold-thrust belts, or they may be transported by large rivers off the continent into deltaic or shallow shelf environments. Lithic sediments deposited in coastal areas may be retransported into deeper water by turbidity currents or by other sediment gravity-flow mechanisms. These deeper-water sediments are particularly likely to undergo deep burial and incipient metamorphism, leading to development of characteristics generally ascribed to graywackes.

Common examples of lithic sandstones include the Paleozoic sandstone successions of the central Appalachians in the eastern United States (e.g., Ordovician Juniata Formation, Mississippian Pocono Formation, Pennsylvanian Pottsville Formation); many sandstones associated with the Coal Measures throughout the world; many Jurassic and Cretaceous sandstones of the U.S. and Canadian Rocky Mountains and the U.S. West Coast (e.g., Cretaceous Belly River Sandstone of Canada, Jurassic Franciscan Formation of California); and Tertiary sandstones of the Gulf Coast, the West Coast, and the Alps.

Volcaniclastic sandstones are a special kind of lithic arenite composed primarily of volcanic detritus (Fig. 5.7D). Volcaniclastic sandstones may be made up largely of pyroclastic materials that have been transported and reworked, or they may contain volcanic detritus derived by weathering of older volcanic rocks. They are especially characterized by the presence of euhedral feldspars, pumice fragments, glass shards, and volcanic rock fragments, and they generally have a very low quartz content (e.g., Boggs, 1992, p. 197–209).

Other Sandstones

The sandstones discussed above are composed of constituents derived primarily by weathering of preexisting rocks or by explosive volcanism. A few less abundant types of "sandstones" are known whose constituents formed largely within the depositional basin by chemical or biochemical processes. These rocks, called hybrid sandstones by some authors, include such uncommon varieties of sandstones as greensands (glauconitic sands), phosphatic sandstones, and calcarenaceous sandstones (composed of sand-size carbonate grains). These rocks are not true sandstones (siliciclastic rocks) but rather are chemical/biochemical sedimentary rocks (Chapters 6 and 7).

5.3 CONGLOMERATES

The term conglomerates is used in this book as a general class name for sedimentary rocks that contain a substantial fraction (at least 30 percent) of gravel-size (>2 mm) particles (Fig. 5.8A). Breccias (Fig. 5.8B), which are composed of very angular, gravel-size fragments, are not distinguished from conglomerates in the succeeding discussion. Conglomerates are common in stratigraphic successions of all ages but probably make up less than 1 percent by weight of the total sedimentary rock mass (Garrels and McKenzie, 1971, p. 40). They are closely related to sandstones in terms of origin and depositional mechanisms, and they contain some of the same kinds of sedimentary structures (e.g., tabular and trough crossbedding, graded bedding).

Particle Composition

Conglomerates may contain gravel-size pieces of individual minerals such as quartz; however, most of the gravel-size framework grains are rock fragments (clasts). Individual sand- or mud-size mineral grains are commonly present as a

Figure 5.8
A. Clast-supported conglomerate underlying laminated sands. Terrace deposits (Holocene) of the Umpqua River, southwest Oregon. B. Large breccia clasts (dark) cemented with calcite. Nevada Limestone (Devonian), Treasure Peak, Nevada. Photograph courtesy of Walter Youngquist.

matrix. Any kind of igneous, metamorphic, or sedimentary rock may be present in a conglomerate, depending upon source rocks and depositional conditions. Some conglomerates are composed of only the most stable and durable kinds of clasts (quartzite, chert, vein-quartz). Stable conglomerates composed mainly of a single clast type are referred to by Pettijohn (1975) as **oligomict conglomerates**. Most oligomict conglomerates were probably derived from mixed parent-rock sources that included less stable rock types. Continued recycling of mixed ultrastable and unstable clasts through several generations of conglomerates ultimately led to selective destruction of the less stable clasts and concentration of stable clasts. Conglomerates that contain an assortment of many kinds of clasts are **polymict conglomerates**. Polymict conglomerates that are made up of a mixture of largely unstable or metastable clasts such as basalt, limestone, shale, and metamorphic phyllite are commonly called **petromict conglomerates** (Pettijohn, 1975). Almost any combination of these clast types is possible in a petromict conglomerate. The matrix of conglomerates commonly consists of various kinds of clay minerals and fine micas and/or silt- or sand-size quartz, feldspars, rock fragments, and heavy minerals. The matrix may be cemented with quartz, calcite, hematite, clay, or other cements.

Classification

Conglomerates can originate by several processes, as shown in Table 5.3. We are interested most in epiclastic conglomerates, which form by breakdown of older rocks through the processes of weathering and erosion. Epiclastic conglomerates that are so rich in gravel-size framework grains that the gravel-size grains touch

Table 5.3 Fundamental genetic types of conglomerates and breccias

Major types	Subtypes	Origin of clasts
Epiclastic conglomerate and breccia	Extraformational conglomerate and breccia	Breakdown of older rocks of any kind through the processes of weathering and erosion; deposition by fluid flows (water, ice) and sediment gravity flows
	Intraformational conglomerate and breccia	Penecontemporaneous fragmentation of weakly consolidated sedimentary beds; deposition by fluid flows and sediment gravity flows
Volcanic breccia	Pyroclastic breccia	Explosive volcanic eruptions, either magmatic or phreatic (steam) eruptions; deposited by air-falls or pyroclastic flows
	Autobreccia	Breakup of viscous, partially congealed lava owing to continued movement of the lava
	Hyaloclastic breccia	Shattering of hot, coherent magma into glassy fragments owing to contact with water, snow, or water-saturated sediment (quench fragmentation)
Cataclastic breccia	Landslide and slump breccia	Breakup of rock owing to tensile stresses and impact during sliding and slumping of rock masses
	Tectonic breccia: fault, fold, crush breccia	Breakage of brittle rock as a result of crustal movements
	Collapse breccia	Breakage of brittle rock owing to collapse into an opening created by solution or other processes
Solution breccia		Insoluble fragments that remain after solution of more soluble material; e.g., chert clasts concentrated by solution of limestone
Meteorite impact breccia		Shattering of rock owing to meteorite impact

Source: Modified from Pettijohn, F. J., 1975, *Sedimentary rocks*, 3rd ed., Harper & Row, New York, p. 165.

and form a supporting framework are called **clast-supported** conglomerates. Clast-poor conglomerates that consist of sparse gravels supported in a mud/sand matrix are called **matrix-supported** conglomerates. Boggs (1992, p. 212) suggests that clast-supported conglomerates be referred to simply as conglomerates and that matrix-supported conglomerates be called **diamictites**. Although the term diamictite is often used for poorly sorted glacial deposits, it is actually a nongenetic term that can be applied to nonsorted or poorly sorted siliclastic sedimentary rocks that contain larger particles of any size in a muddy matrix.

Conglomerates and diamictites can be further divided on the basis of clast stability into **quartzose** (oligomict) conglomerate/diamictite and **petromict** conglomerate/diamictite on the basis of relative abundance of these clast types (Table 5.4). Further classification on the basis of clast type (igneous, metamorphic, sedimentary) can be made if desired (Fig. 5.9); however, such classification may not be necessary in many cases.

Origin and Occurrence of Conglomerates

Quartzose (oligomict) conglomerates are derived from metasedimentary rocks containing quartzite beds, igneous rocks containing quartz-filled veins, and sedimentary successions, particularly limestones, containing chert beds. As mentioned, less stable rock types must have been destroyed by weathering, erosion, and sediment transport, perhaps through several cycles of transport, to produce a residuum of stable clasts. Because quartzose clasts represent only a small fraction of a much larger original body of rock, the total volume of quartzose conglomerates is small. They tend to occur as thin, pebbly layers or lenses of pebbles in dominantly sandstone units. They may be either clast-supported or matrix-supported. Although their overall volume is small, quartzose conglomerates are common in the geologic record ranging from the Precambrian to the Tertiary. Most quartzose conglomerates appear to be of fluvial origin (Chapter 8) and are probably deposited mainly in braided streams. Marine, wave-worked quartzose conglomerates that were deposited in the littoral (beach) environment are also known.

Most **petromict conglomerates** are polymict conglomerates that consist of a variety of metastable clasts. They can be derived from many kinds of plutonic igneous, volcanic, metamorphic, and sedimentary rocks, although the clasts in a particular conglomerate may be dominantly one or another of these rock types. Thus, a particular conglomerate may be a limestone conglomerate, a basalt conglomerate, a schist conglomerate, and so on. Conglomerates composed dominantly of plutonic igneous clasts appear to be uncommon, probably because plutonic rocks such as granites tend to disintegrate into sand-size fragments rather than form larger blocks. The volume of ancient petromict conglomerates is far greater than that of quartzose conglomerates. They form the truly great conglomerate bodies of the geologic record and may reach thicknesses of thousands of meters. Preservation of such great thickness of conglomerate implies rapid erosion of sharply elevated highlands or areas of active volcanism. Petromict conglomerates may be transported by fluid-flow and sediment gravity-flow mechanisms; they

Table 5.4 Classification of conglomerates and diamictites on the basis of clast stability and fabric support

Percentage of ultrastable clasts	Type of fabric support	
	Clast-supported	Matrix-supported
>90	Quartzose conglomerate	Quartzose diamictite
<90	Petromict conglomerate	Petromict diamictite

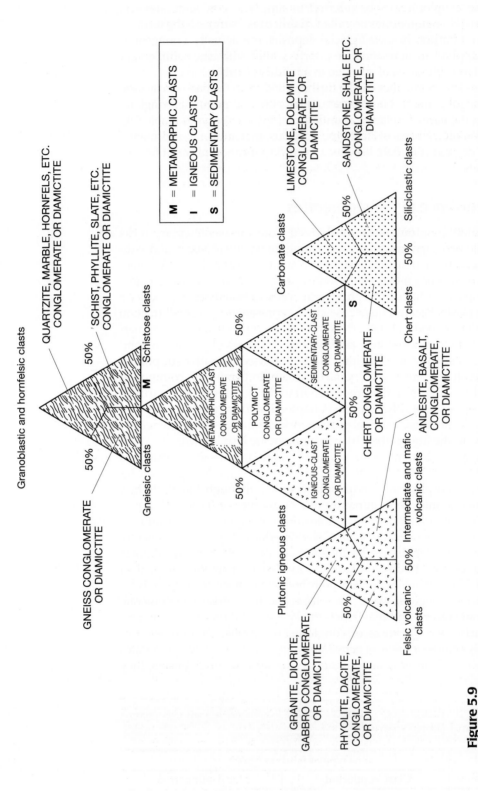

Figure 5.9

Classification of conglomerate on the basis of clast lithology and fabric support. [From Boggs, 1992, Petrology of sedimentary rocks, Fig. 6.3, p. 220: Macmillan Publishing Co., New York, reproduced by permission of Prentice Hall.]

are deposited in environments ranging from fluvial through shallow marine to deep marine. Deep-marine conglomerates are so-called **resedimented conglomerates** that were retransported from nearshore areas by turbidity currents or other sediment gravity-flow processes. The bulk of the truly thick conglomerate bodies (>20 m) were probably deposited in nonmarine (alluvial fan/braided river) settings or deep-sea fan settings.

Intraformational conglomerates are composed of clasts of sediments believed to have formed within depositional basins, in contrast to the clasts of extraformational conglomerates that are derived from outside the depositional basin. Intraformational conglomerates originate by penecontemporaneous deformation of semiconsolidated sediment and redeposition of the fragments fairly close to the site of deformation. Penecontemporaneous breakup of sediment to form clasts may take place subaerially, such as by drying out of mud on a tidal flat, or under water. Subaqueous rip-ups of semiconsolidated muds by tidal currents, storm waves, or sediment-gravity flows are possible causes. In any case, sedimentation is interrupted only a short time during this process. The most common types of fragments found in intraformational conglomerates are siliciclastic mud clasts and lime clasts. The clasts are commonly angular or only slightly rounded, suggesting little transport. In some beds, flattened clasts are stacked virtually on edge, apparently owing to unusually strong wave or current agitation, to form what is called **edgewise conglomerates** (Pettijohn, 1975, p. 184).

Intraformational conglomerates commonly form thin beds, a few centimeters to a meter in thickness, that may be laterally extensive. Although much less abundant than extraformational conglomerates, they nonetheless occur in rocks of many ages. So-called flat-pebble conglomerates composed of carbonate or limy siltstone clasts are particularly common in Cambrian-age rocks in various parts of North America. They also occur in many other early Paleozoic limestones of the Appalachian region. Intraformational conglomerates composed of shale rip-up clasts embedded in the basal part of sandstone units are very common in sedimentary successions deposited by sediment gravity-flow processes.

5.4 SHALES (MUDROCKS)

Shales are fine-grained, siliciclastic sedimentary rocks, that is, rocks that contain more than 50 percent siliciclastic grain less than 0.062 (1/256) mm. Thus, they are made up dominantly of silt-size (1/16–1/256 mm) and clay-size (<1/256 mm) particles. Shale is an historically accepted class name for this group of rocks (Tourtelot, 1960), equivalent to the class name sandstone, a usage accepted by Potter, Maynard, and Pryor (1980, p. 12–15). These authors use the term shale as the class name for all fine-grained siliciclastic sedimentary rocks, but they divide shales into several kinds, such as mudstones and mudshales, depending upon the percentage of clay-size constituents and the presence or absence of lamination (discussed subsequently under classification).

On the other hand, some authors prefer to use the class name **mudrock,** rather than shale, for all fine-grained rocks (e.g., Blatt, Middleton, and Murray, 1980, p. 382). They divide mudrocks into **shales** (if laminated) or **mudstones** (if nonlaminated). Thus, they restrict the usage of shale to fine-grained rocks, such as those in Figure 5.10A, that display lamination or fissility (the ability to split easily into thin layers). Fine-grained, nonlaminated rocks such as those shown in Figure 5.10B are, according to this usage, mudstones. In this book, I follow the usage of Potter, Maynard, and Pryor and apply the general class name shale to all fine-grained siliciclastic sedimentary rocks. Clearly, however, the usage of shale or mudrock as a class name for fine-grained siliciclastic rocks is a matter of personal preference.

Figure 5.10
A. Laminated (red) shale (pre-Mississippian), Arctic National Wildlife Refuge, Alaska. Note binoculars for scale. B. Lacustrine mudstones (nonlaminated), Furnace Creek Formation (Miocene/Pliocene), Death Valley, California. Photograph by James Stovall.

Shales are abundant in sedimentary successions, making up roughly 50 percent of all the sedimentary rocks in the geologic record. Historically, shales have been an understudied group of rocks, mainly because their fine grain size makes them difficult to study with an ordinary petrographic microscope. This perspective is changing, however, as instruments are developed such as the scanning electron microscope and electron probe microanalyzer that allow study of fine-size grains at high magnification (e.g., Fig. 5.11).

Composition

Mineralogy

Shales are composed primarily of clay minerals and fine-size quartz and feldspars (Table 5.5). They also contain various amounts of other minerals, including carbonate minerals (calcite, dolomite, siderite), sulfides (pyrite, marcasite), iron oxides (goethite), and heavy minerals, as well as a small amount of organic carbon. Figure 5.11 is a high-magnification photograph, taken by use of backscattered scanning electron microscopy, which allows both the mineralogy and texture of this laminated shale to be examined. [See Krinsley et al., 1998, for discussion of the use of backscattered electron microscopy in the study of sedimentary rocks.] The

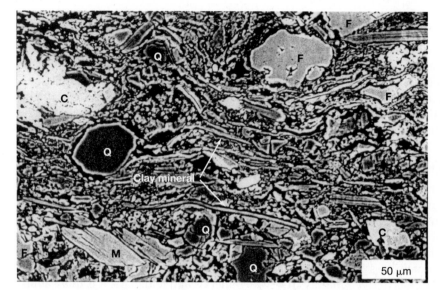

Figure 5.11
Backscattered scanning electron microscope (BSE) photograph of a laminated shale. The elongated, platy, or flaky minerals are clay minerals (illite) and fine micas. Other coarser minerals are quartz (Q), feldspar (F), calcite (C), mica (M), and pyrite (very bright mineral). Note that orientation of clay minerals and fine micas creates the lamination. Whitby Mudstone (Jurassic), Yorkshire, England. Scale bar = 50 μm. Photograph courtesy of David Krinsley.

data in Table 5.5 show mineral composition as a function of age. No discernible trend of mineralogy vs. age is evident from this table, except possibly a slight trend of decreasing feldspar with increasing age. Many factors affect the composition of shales, including tectonic setting and provenance (source), depositional environments, grain size, and burial diagenesis. Some minerals, such as carbonate minerals and sulfides, form in the shales during burial as cements or replacement minerals. Quartz, feldspars, and clay minerals are mainly detrital (terrigenous)

Table 5.5 Average percent mineral composition of shales of different ages

Age	Number of analyses	Clay minerals	Quartz	Potassium feldspar	Plagioclase feldspar	Calcite	Dolomite	Siderite	Pyrite	Other minerals	Organic carbon
Quaternary	5	29.9	42.3	12.4	—	6.6	2.4	—	5.6	—	0.9
Pliocene	4	56.5	14.6	5.7	11.9	3.2	—	2.9	1.8	<1.0	2.6
Miocene	9	25.3	34.1	7.4	11.7	14.6	1.2	—	1.9	2.4	1.4
Oligocene	4	33.7	53.5	3.0	—	5.5	—	—	—	4.0	0.4
Eocene	11	40.2	34.6	2.0	8.1	3.8	4.6	1.7	1.6	—	3.5
Cretaceous	9	27.4	52.9	3.6	1.6	2.9	7.9	0.1	1.6	—	2.0
Jurassic	10	34.7	21.9	0.6	4.4	14.6	1.6	0.4	10.9	—	10.9
Triassic	9	29.4	45.9	10.7	0.7	3.7	4.1	5.1	—	—	0.3
Permian	1	17.0	28.0	4.0	8.0	—	1.0	—	—	42.0	0.2
Pennsylvanian	7	48.9	32.6	0.8	6.2	1.4	2.1	3.4	3.5	—	1.0
Mississippian	3	57.2	29.1	0.4	2.9	—	—	0.6	5.1	—	4.7
Devonian	22	41.8	47.1	0.6	—	2.0	1.3	0.3	3.3	—	3.7
Ordovician	2	44.9	32.2	<1.0	6.3	9.8	0.5	0.5	3.4	—	1.5
Misc. ages	29	47.8	33.1	<1.0	5.5	5.2	2.3	0.8	3.1	—	4.5

Source: O'Brien, N. R., and R. M. Slatt, 1990, *Argillaceous rock atlas*, Springer-Verlag, New York, Table 1, p. 124–125.

Note: Values adjusted to 100% for shales of each age.

minerals, although some fraction of these minerals may also form during burial diagenesis. In particular, clay minerals appear to be strongly affected by diagenetic processes. As shown in Table 5.1, the principal clay mineral groups are kaolinite, illite, smectite, and chlorite. The relative proportions of these clay-mineral groups have been reported to change systematically with age (Fig. 5.12). With time, particularly in rocks older than the Mesozoic, the proportion of illite and chlorite increases at the expense of kaolinite and smectite. These trends are attributed to the diagenetic alteration of kaolinite and smectite to form illite and chlorite.

Chemical Composition

The chemical composition of shales is a direct function of their mineral composition. Compositions of some average North American and Russian shales are shown in Table 5.6. SiO_2 is the most abundant chemical constituent in these shales (57–68 percent), followed by Al_2O_3 (16–19 percent). The SiO_2 content of shales is affected by all silicate minerals present but particularly by quartz. Thus, shales tend to contain less SiO_2 than do sandstones, which commonly are enriched in quartz. Al_2O_3 is derived mainly from clay minerals and feldspars. It is more abundant in shales than in sandstones because of the greater clay mineral content of

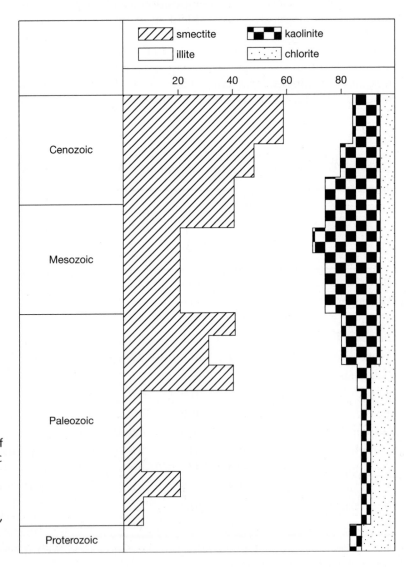

Figure 5.12
Systematic changes in relative abundance of major clay minerals as a function of geologic age. [After Singer, A., and G. Müller, 1983, Diagenesis in argillaceous sediments, *in* Larson, G., and Chilingarian, G. V. (eds.), Diagenesis in sediments and sedimentary rocks, v. 2, Fig. 3.28, p. 176, reproduced by permission of Elsevier Science Publishers, New York.]

Table 5.6   Average chemical composition of selected shales reported in the literature

	1	2	3	4	5	6	7	8	9	10	11	12	13
SiO_2	60.65	64.80	59.75	56.78	67.78	64.09	66.90	63.04	62.13	65.47	64.21	64.10	63.31
Al_2O_3	17.53	16.90	17.79	16.89	16.59	16.65	16.67	18.63	18.11	16.11	17.02	17.70	17.22
Fe_2O_3	7.11	—	—	—	—	—	—	—	—	—	—	2.70	0.82
FeO	—	5.66	5.59	6.56	4.11	6.03	5.87	7.66	7.33	5.85	6.71	4.05	5.45
MgO	2.04	2.86	4.02	4.56	3.38	2.54	2.59	2.60	3.57	2.50	2.70	2.65	3.00
CaO	0.52	3.63	6.10	8.91	3.91	5.65	0.53	1.31	2.22	4.10	3.44	1.88	3.52
Na_2O	1.47	1.14	0.72	0.77	0.98	1.27	1.50	1.02	2.68	2.80	1.44	1.91	1.48
K_2O	3.28	3.97	4.82	4.38	2.44	2.73	4.97	4.57	2.92	2.37	3.58	3.60	3.64
TiO_2	0.97	0.70	0.98	0.92	0.70	0.82	0.78	0.94	0.78	0.49	0.72	0.86	0.81
P_2O_5	0.13	0.13	0.12	0.13	0.10	0.12	0.14	0.10	0.17	—	—	—	0.10
MnO	0.10	0.06		0.08		0.07	0.06	0.12	1.10	0.07	0.05	—	0.06

Source:

 (1) Moore, 1978 (Pennsylvanian shale, Illinois Basin)
 (2) Gromet et al., 1984 (North American shale composite)
 (3) Ronov and Migdisov, 1971 (average North American Paleozoic shale)
 (4) Ronov and Migdisov, 1971 (average Russian Paleozoic shale)
 (5) Ronov and Migdisov, 1971 (average North American Mesozoic shale)
 (6) Ronov and Migdisov, 1971 (average Russian Mesozoic shale)
 (7) Cameron and Garrels, 1980 (average Canadian Proterozoic shale)
 (8) Ronov and Migdisov, 1971 (average Russian Proterozoic shale)
 (9) Cameron and Garrels, 1980 (average Canadian Archean shale)
(10) Ronov and Migdisov, 1971 (average Archean shale)
(11) Clarke, 1924 (average shale)
(12) Shaw, 1956 (compilation of 155 analyses of shale)
(13) Average of values in columns 1 through 12

shales. Fe in shales is supplied by iron oxide minerals (hematite, goethite), biotite, and a few other minerals such as siderite, ankerite, and smectite clay minerals. K_2O and MgO abundance is related mainly to clay mineral abundance, although some Mg may be supplied by dolomite and K is present in some feldspars. Na abundance is related to the presence of clay minerals (e.g., smectites) and sodium plagioclase. Ca is supplied by calcium-rich plagioclase and carbonate minerals (calcite, dolomite).

Classification

Because special analytical techniques are required to determine the mineral composition of shales, and because such techniques are time consuming and expensive, many geologists do not routinely determine the mineral composition of shales (mudrocks). Therefore, most classifications that have been proposed for shales have not been based on mineral composition, or at least not entirely on mineral composition. These classifications, none of which has been widely accepted, commonly emphasize the relative amounts of silt and clay, the hardness or degree of induration of the rocks, and the presence or absence of fissile lamination. (Fissility is defined as the property of a rock to split easily along thin, closely spaced, approximately parallel layers.) Exceptions to this general practice of classification are the classifications of Picard (1971), which emphasizes mineral composition of the silt-size grains in mudrocks (shales), and the classification of Lewan (1978), which requires semiquantitative X-ray diffraction analysis to determine mineralogy.

The classification of Potter, Maynard, and Pryor (1980), shown in Table 5.7, is based on grain size, lamination, and degree of induration. It is similar to the field classification of shales proposed by Lundegard and Samuels (1980). This classification emphasizes the importance of clay-size constituents and bedding thickness, that is, whether bedded or laminated. Depending upon these variables, shales can be divided into **mudstone** (33–65% clay-size constituents and bedded) or **mudshale** (33–65% clay-size constituents and laminated), and **claystone** (66–100% clay-size constituents and bedded) or **clayshale** (66–100% clay-size constituents and laminated). Fine-grained siliciclastic rocks that contain less than 33 percent clay-size constituents are **siltstones**.

Additional informal terms can be used with this classification to provide further information about the properties of the shales. These may include terms that express color, type of cementation (calcareous, or limy; ferruginous, or iron-rich; siliceous); degree of induration (hard, soft); mineralogy if known (e.g., quartzose, feldspathic, micaceous); fossil content (e.g., fossiliferous, foram-rich); organic matter content (e.g., carbonaceous, kerogen-rich, coaly); type of fracturing (conchoidal, hackly, blocky); or nature of bedding (e.g., wavy, lenticular, parallel).

Origin and Occurrence of Shales (Mudrocks)

Shales form under any environmental conditions in which fine sediment is abundant and water energy is sufficiently low to allow settling of suspended fine silt

Table 5.7   Classification of shales and siltstone (>50% grains <0.062 mm)

	Percentage clay-size constituents		0–32	33–65	66–100
	Field adjective		Gritty	Loamy	Fat or slick
NONINDURATED	Beds >10 mm		Bedded silt	Bedded mud	Bedded claymud
	Laminae <10 mm		Laminated silt	Laminated mud	Laminated claymud
INDURATED	Beds >10 mm		Bedded siltstone	Mudstone	Claystone
	Laminae <10 mm		Laminated siltstone	Mudshale	Clayshale
METAMORPHOSED	Degree of metamorphism	Low	Quartz Argillite	Argillite	
			Quartz Slate	Slate	
		High	Phyllite and/or Mica Schist		

Source: Potter, P. E., J. B. Maynard, and W. A. Pryor, 1980, Sedimentology of shales: Springer-Verlag, New York, Table 1.2, p. 14.

and clay. Shales are particularly characteristic of marine environments adjacent to major continents where the seafloor lies below the storm wave base, but they can form also in lakes and quiet-water parts of rivers, and in lagoonal, tidal-flat, and deltaic environments. The fine-grained siliciclastic products of weathering greatly exceed coarser particles; thus, fine sediment is abundant in many sedimentary systems. Because fine sediment is so abundant and can be deposited in a variety of quiet-water environments, shales are by far the most abundant type of sedimentary rock. They make up roughly 50 percent of the total sedimentary rock record. They commonly occur interbedded with sandstones or limestones in units ranging in thickness from a few millimeters to several meters or tens of meters. Nearly pure shale units hundreds of meters thick also occur. Shale units in marine successions tend to be laterally extensive.

A few shales that are particularly well known owing to their thickness, widespread areal extent, stratigraphic position, or fossil content include the Cambrian Burgess Shale of western Canada, which is famous for its well-preserved imprints of soft-bodied animals, the Eocene Green River (oil) Shale of Colorado; the Cretaceous Mancos Shale of western North America, which forms a thick, eastward-thinning wedge stretching from New Mexico to Saskatchewan and Alberta; the Devonian-Mississippian Chattanooga Shale and equivalent formations that cover much of North America and whose widespread extent is still poorly explained; the Silurian Gothlandian shales of western Europe, northern Africa, and the Persian Gulf region, which contain a pelecypod and graptolite faunal association, and the Precambrian Figtree Formation of South Africa, well known for its early fossils. The origin and occurrence of shales are discussed in detail by Potter, Maynard, and Pryor (1980) and Schieber, Zimmerle, and Sethi (1998).

5.5 DIAGENESIS OF SILICICLASTIC SEDIMENTARY ROCKS

Siliciclastic sedimentary rocks form initially as unconsolidated deposits of gravels, sand, or mud. The mineral and chemical compositions of these deposits are functions of a complex system of conditions and processes, including source-rock lithology, sediment transport, and environmental conditions (e.g., Johnsson, 1993). Newly deposited sediments are characterized by loosely packed, uncemented fabrics; high porosities; and high interstitial water content. As sedimentation continues in subsiding basins, older sediments are progressively buried by younger sediments to depths that may reach tens of kilometers. Sediment burial is accompanied by physical and chemical changes that take place in the sediments in response to increase in pressure from the weight of overlying sediment, downward increase in temperature, and changes in pore-water composition. These changes act in concert to bring about compaction and **lithification** of sediment, ultimately converting it into consolidated sedimentary rock. Thus, unconsolidated gravel is eventually lithified to conglomerate, sand is lithified to sandstone, and siliciclastic mud is hardened into shale (mudrock).

The process of lithification is accompanied by physical, mineralogical, and chemical changes. Loose grain packing gives way with burial to more tightly packed fabrics having greatly reduced porosity. Porosity may be further reduced by precipitation of cements into pore spaces. Minerals that were chemically stable at low surface temperatures and in the presence of environmental pore waters become altered at higher burial temperatures and changed pore-water compositions. Minerals may be completely dissolved or may be partially or completely replaced by other minerals.

Thus, porosity, mineralogy, and chemical composition may all be changed to various degrees during burial diagenesis. Diagenesis, the final stage in the process of forming conglomerates, sandstones, and shales, is a process that begins with weathering of source rocks and continues through sediment transport, deposition, and burial. To properly interpret the provenance, transport, and depositional history of sedimentary rocks, we must recognize and distinguish among features of sediment that were present at the time of deposition and those features of sedimentary rocks that resulted from burial alteration. Diagenesis also has economic significance because it can adversely affect the ability of siliciclastic rocks to store and transmit fluids, a subject of considerable interest to petroleum and groundwater geologists (e.g., Stonecipher, 2000). A short description of diagenetic processes and the physical and chemical effects of these processes is included here.

Stages and Realms of Diagenesis

Diagenesis takes place at temperatures and pressures higher than those of the weathering environment but below those that produce metamorphism. There is no clear boundary between the realms of diagenesis and metamorphism; however, we commonly consider diagenesis to occur at temperatures below about 250°C (Fig. 5.13). Diagenesis can begin almost immediately after deposition, while sediment is still on the ocean or other basin floor, and may continue through deep burial and eventual uplift. Burial subjects sediments to conditions of pressure and temperature markedly different from those that exist in the depositional environment. Increases in pressure, and temperature as a function of depth are shown in Figure 5.13. Pore-fluid composition changes also. There is both a general increase in salinity of pore waters with increasing burial depth and a change in pore-water chemistry (e.g., Heydari, 1997). Changes in pore-water chemistry are difficult to generalize, and they differ from basin to basin, but they include variations in abundance of such important mineral-forming ions as

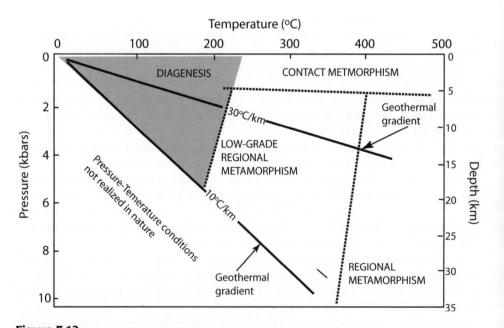

Figure 5.13
Pressure-temperature diagram relating diagenesis to metamorphic regimes and typical pressure-temperature, gradients in Earth's crust. The 10°C/km geothermal gradient is typical of stable cratons; the 30°C/km gradient is typical of rifted sedimentary basins. [Modified from Worden, R. H., and S. D. Burley, Sandstone diagenesis: The evolution of sand to stone, *in* Burley, S. D., and R. H. Worden, 2003, Sandstone diagenesis: Recent and ancient: Blackwell Pub., Malden, Mass. Fig. 1, p. 3. Reproduced by permission.]

Si^{4+}, Al^{3+}, Ca^{2+}, K^+, Mg^{2+}, Na^+, and HCO_3^- (bicarbonate). Many of these ions increase in abundance with increasing burial depth, concomitant with increase in salinity. For a recent look at fluids in depositional basins and their role in diagenesis, see Kyser (2000).

Various authors have suggested that sediments go through three to six stages of diagenesis. Perhaps the most widely accepted stages of diagenesis are those proposed by Choquette and Pray (1970). **Eodiagenesis** refers to the earliest stage of diagenesis, which takes place at very shallow depths (a few meters to tens of meters) largely under the conditions of the depositional environment. **Mesodiagenesis** is diagenesis that takes place during deeper burial, under conditions of increasing temperature and pressure and changed pore-water compositions. **Telodiagenesis** refers to late-stage diagenesis that accompanies or follows uplift of previously buried sediments into the regime of meteoric waters. Sedimentary rocks that are still deeply buried in depositional basins have not, of course, undergone telodiagenesis. Some authors now refer to these stages simply as **eogenesis**, **mesogenesis**, and **telogenesis** (e.g., Worden and Burley, 2003; Fig. 5.14). The most important diagenetic processes that take place in each of these diagenetic regimes, and the effects of these processes, are summarized in Table 5.8. These processes and effects are discussed in greater detail below.

Major Diagenetic Processes and Effects

Shallow Burial (Eogenesis)

The principal diagenetic changes that take place in the eodiagenetic regime include reworking of sediments by organisms (bioturbation), minor compaction and grain repacking, and mineralogical changes. Organisms rework sediment at or near the depositional interface through various crawling, burrowing, and sediment-ingesting activities. Bioturbation can destroy primary sedimentary structures such

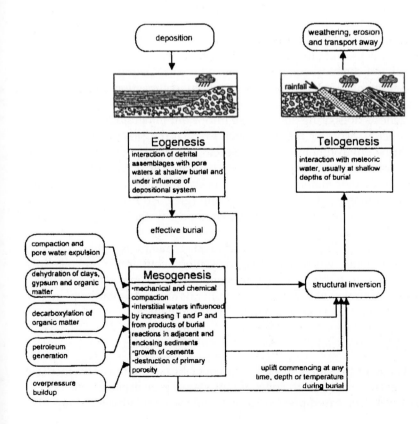

Figure 5.14

Flow chart illustrating the links between the regimes of diagenesis. Structural inversion refers to uplift. [From Worden, R. H., and S. D. Burley, Sandstone diagenesis: The evolution of sand to stone, *in* Burley, S. D., and R. H. Worden, 2003, Sandstone diagenesis: Recent and ancient: Blackwell Pub., Malden, Mass. Fig. 4, p. 7. Reproduced by permission.]

Table 5.8   Principal diagenetic processes and changes that occur in siliciclastic sedimentary rocks during burial

	Diagenetic stage	Diagenetic process	Result
Burial ↓	Eogenesis	Organic reworking (bioturbation)	Destruction of primary sedimentary structures; formation of mottled bedding and other traces
		Cementation and replacement	Formation of pyrite (reducing environments) or iron oxides (oxidizing environments); precipitation of quartz and feldspar overgrowths, carbonate cements, kaolinite, or chlorite
	Mesogenesis	Physical compaction	Tighter grain packing; porosity reduction and bed thinning
		Chemical compaction (pressure solution)	Partial dissolution of silicate grains; porosity reduction and bed thinning
		Cementation	Precipitation of carbonate (calcite) and silica (quartz) cements with accompanying porosity reduction
		Dissolution by pore fluids	Solution removal of carbonate cements and silicate framework grains; creation of new (secondary) porosity by preferential destruction of less stable minerals
		Mineral replacement	Partial to complete replacement of some silicate grains and clay matrix by new minerals (e.g., replacement of feldspars by calcite)
		Clay mineral authigenesis	Alteration of one kind of clay mineral to another (e.g., smectite to illite or chlorite, kaolinite to illite)
Uplift ↑	Telogenesis	Dissolution, replacement, oxidation	Solution of carbonate cements, alteration of feldspars to clay minerals, oxidation of iron carbonate minerals to iron oxides, oxidation of pyrite to gypsum, solution of less stable minerals (e.g., pyroxenes, amphiboles)

as lamination and create in their place a variety of traces that may include mottled bedding, burrows, tracks, and trails. Organic reworking commonly has little effect on the mineralogical and chemical composition of sediments. Owing to very shallow burial depth, sediments undergo only very slight compaction and grain rearrangement during early diagenesis.

Early diagenesis does bring about some important mineralogical changes in siliciclastic sediments. Most of these changes involve the precipitation of new minerals. In marine environments where reducing (low-oxygen) conditions can prevail, the formation of pyrite is particularly characteristic. Pyrite may form cement or may replace other materials such as woody fragments. Other important reactions include formation of chlorite, glauconite (greenish iron-silicate grains), illite/smectite clays, and iron oxides in oxygenated pore waters (e.g., red clays on the deep ocean floor); and precipitation of potassium feldspar overgrowths, quartz overgrowths (e.g., Fig. 5.3A), and carbonate cements (e.g., Fig. 5.3C). In nonmarine environments, where oxidizing conditions commonly prevail, little pyrite forms. Instead, iron oxides (goethite, hematite) are commonly produced, creating redbeds. Formation of kaolinitic clay minerals and precipitation of quartz and calcite cements may take place also in this environment.

Deep Burial (Mesogenesis)

Compaction.   The load pressures caused by deeper burial significantly increase the tightness of grain packing with concomitant loss of porosity (e.g., Fig. 5.15A) and thinning of beds. Increased pressure at the contact point between grains also

Figure 5.15
Fabrics in sandstones created by diagenetic processes: A. Physical compaction (note prevalence of concavo-convex and long contacts), Tuscarora Sandstone (Silurian), Pennsylvania. B. Chemical compaction owing to pressure solution (note irregular sutured contact indicated by arrows), Oriskany Quartzite (Devonian), Pennsylvania. C. Cementation by microquartz (chert), Jefferson City Fm. (Ordovician), Missouri. D. Replacement of quartz by calcite, creating "nibbled" contacts (arrows), Mauch Chunk Group (Mississippian), Pennsylvania. Crossed nicol photomicrographs.

increases the solubility of the grains at the contact, leading to partial dissolution of the grains. This process is referred to as **pressure solution** or **chemical compaction** (e.g., Fig. 5.15B). Chemical compaction further reduces porosity and increases bed thinning. Thus, under the influence of physical and chemical compaction, aided by cementation (below), the primary porosity of both sands and muds is reduced dramatically during deep burial (Fig. 5.16). Compaction also causes bending of flexible grains such as micas and squeezing of soft grains such as rock fragments (Fig. 5.17). Stone and Siever (1996) report that mechanical compaction and pressure solution cause porosity loss in quartzose sandstones mainly at burial depths less than about 2 km (Fig. 5.18) because the combined effects of compaction, pressure solution, and a small amount of quartz cement produce stable grain-packing arrangements. According to these authors, porosity loss at greater depths is primarily the result of quartz cementation. Worden and Burley (2003) suggest that some porosity loss owing to compaction can continue to depths of at least 5 km.

Chemical Processes and Changes. An increase in temperature of 10°C during burial can cause chemical reaction rates to double or triple. Thus, mineral phases that were stable in the depositional environment may become unstable during deep

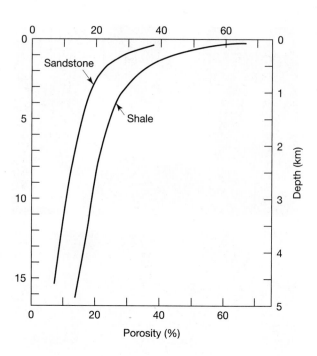

Figure 5.16

Approximate best-fit curves showing changes in porosity of sediments related to burial compaction and cementation in some California (sandstone) and Louisiana (shale) basins. [Sandstone curve based on Wilson, J. C., and E. F. McBride, 1988, Compaction and porosity evolution of Pliocene sandstones, Ventura Basin, California: Am. Assoc. Petroleum Geologists Bull., v. 72, Fig. 4, p. 669; shale curve based on Dzevanshir, R. D., et al., 1986, Sed. Geology, v. 46, Fig. 1, p. 170.]

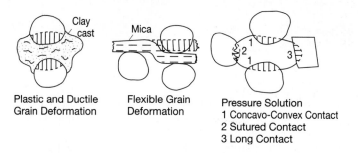

Figure 5.17

Schematic representation of textural criteria used to estimate volume loss in sandstones owing to compaction. The hachured areas indicate rock volume lost by grain deformation and pressure solution. [From Wilson, J. C., and E. F. McBride, 1988, Compaction and porosity evolution of Pliocene sandstones, Ventura Basin, California: Am. Assoc. Petroleum Geologists Bull., v. 72, Fig. 10, p. 679, reprinted by permission of AAPG, Tulsa, Okla.]

burial. Increasing temperature favors the formation of denser, less hydrous minerals and also causes an increase in solubility of most common minerals except the carbonate minerals. Thus, silicate minerals show an increasing tendency to dissolve with greater burial depths (and temperatures), whereas carbonate minerals such as calcite are more likely to precipitate. On the other hand, decrease in pH (increase in acidity) of pore waters with depth may bring about dissolution of carbonates. For example, organic materials may decompose during deep burial diagenesis to release CO_2. Increase in the CO_2 content of pore waters results in a decrease in pH (increase in acidity) that can bring about dissolution of carbonate minerals. As discussed above, increased pressure during deep burial causes an increase in solubility of minerals at point contacts, resulting in partial dissolution of the minerals. This process, which releases silica into pore waters, is an important mechanism for furnishing silica that can later precipitate as new silicate minerals. Several kinds of chemical/mineralogical diagenetic processes take place in siliciclastic sedimentary rocks during deep burial. The most important of these processes are cementation, dissolution, replacement, and clay-mineral authigenesis.

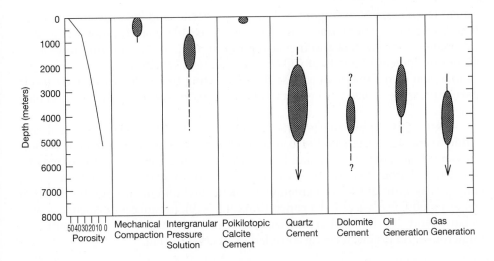

Figure 5.18
Summary diagram showing depth ranges at which mechanical compaction, pressure solution, and cementation reduce porosity in quartzose sandstones. Note that porosity is reduced from approximately 50 percent at the surface to virtually zero at a burial depth of about 5000 m. This diagram also shows the approximate depths at which oil and gas are generated in the subsurface (Chapter 7). [From Stone, W. N., and R. Siever, 1996, Quantifying compaction, pressure solution and quartz cementation in moderately- and deeply-buried quartzose sandstones from the greater Green River Basin, Wyoming, *in* Crossey, L. J., R. Loucks, and M. W. Totten, eds., 1996, Siliciclastic diagenesis and fluid flow: SEPM Special Publication No. 55, Fig. 5, p. 134.]

Box 5.1 Estimating Diagenetic Paleotemperatures

Because temperature has a particularly significant effect on diagenetic processes, geologists are greatly interested in estimating the temperatures at which particular diagenetic reactions take place. Considerable research has been carried out to develop reliable techniques for **paleotemperature** analysis. Tools used for determining paleotemperatures are called **geothermometers**. The principal techniques now in use for determining diagenetic paleotemperatures include methods based on (1) conodont color alteration, (2) vitrinite reflectance, (3) graphitization levels in kerogen, (4) clay mineral assemblages, (5) zeolite mineral assemblages, (6) fluid inclusions, and (7) oxygen isotope ratios. The methods have various degrees of reliability, and none can be considered an infallible estimator of the paleotemperatures of diagenesis; conodont color alteration and vitrinite reflectance are generally regarded to be the most useful methods. Methods based on analyses of mineral assemblages tend to be less sensitive and more equivocal. The formation of zeolite minerals, for example, depends upon pressure and the salinity and chemical composition of sediment pore waters, as well as temperature. Two or three different methods, which generally include examination of conodont color alteration and vitrinite reflectance, are commonly used together as a cross check on reliability. Appendix B provides additional details.

Cementation refers to the precipitation of minerals into the pore space of sediment, thereby reducing porosity and bringing about lithification of the sediment. Carbonate and silica cements are most common; however, feldspars, iron oxides, pyrite, anhydrite, zeolites, and many other minerals can also form as cements. As mentioned in the discussion of sandstones, calcite is the dominant carbonate cement (e.g., Fig. 5.3C); aragonite, dolomite, siderite, and ankerite are less common. Carbonate cementation is favored by increasing concentration of calcium carbonate in pore waters and increasing burial temperature. Precipitation is inhibited by increased levels of CO_2 in pore waters, which may result from decomposition of organic matter in sediments during burial. Increased CO_2 levels (partial pressure) cause pore waters to become acidic and corrosive to carbonate minerals.

Figure 5.3C shows calcite cement that is restricted to a relatively small area within a sandstone. Cementation can be much more extensive, and cement eventually can fill most of the pore space in the sandstone. In other cases, cement may be concentrated around some object, such as a fossil or fossil fragment, which apparently acts as a nucleus for cementation. Cement can build up around this object to create a globular mass called a **concretion** (Fig. 5.19). In rare cases, calcite, as well as barite and gypsum, can crystallize (precipitate) as large crystals that envelop numerous sand grains, forming so-called **sand crystals** (Fig. 5.20).

Figure 5.19
Large concretions weathering out on the surface of a laminated sandstone bed, Coaledo Formation (Eocene), southern Oregon coast. Calcite, precipitated around some kind of nucleus, filled pore spaces in the sandstone, gradually building up the globular masses. Note the sandstone lamination preserved in the concretions. (Photograph courtesy of Robert Q. Oaks, Jr.)

Figure 5.20
Sand crystals, Miocene sandstone, Badlands, South Dakota. The length of the specimen is about 16 cm. [From Pettijohn, F. J., 1975, Sedimentary rocks, 3rd ed., Harper and Row, Publishers, Inc., New York, Fig. 1.2, p. 467.]

Quartz precipitated as overgrowths around existing detrital quartz grains (e.g., Fig. 5.3A) is the most common kind of silica cement. Quartz overgrowth cements are particularly abundant in many quartz arenites. Less commonly, silica precipitates as microcrystalline quartz (chert) cement (e.g., Fig. 5.15C) or opal. Quartz cementation is favored by high concentrations of silica in pore waters and by low temperatures. Some silica is supplied locally by pressure solution or by dissolution of the siliceous skeletons of fossil organisms such as diatoms and radiolarians. Silica may also be imported from other areas of a basin during episodes of fluid flow related to deep-basin mineral dehydration or tectonic activity (Stone and Siever, 1996). Quartz cementation is particularly likely to occur in sedimentary basins where waters that circulated downward deeply into the basin, and dissolved silica at higher temperatures, rise upward and cool along basin edges.

Dissolution of framework silicate grains and previously formed carbonate cements may occur during deep burial under conditions that are essentially the opposite of those required for cementation. For example, carbonate minerals are dissolved in cooler pore waters with high carbon dioxide partial pressures. Rock fragments and low-stability silicate minerals, such as plagioclase feldspars, pyroxenes, and amphiboles, may dissolve as a result of increasing burial temperatures and the presence of organic acids in pore waters. The selective dissolution of less stable framework grains or parts of grain during diagenesis is called **intrastratal solution**. Dissolution of framework grains and cements increases porosity, particularly in sandstones. Petroleum geologists, who are especially interested in the porosity of sandstones, now believe that much of the porosity that exists in sandstones below a burial depth of about 3 km is secondary porosity, created by dissolution processes.

Mineral **replacement** refers to the process whereby one mineral dissolves and another is precipitated in its place essentially simultaneously. Replacement appears to take place without any volume change between the replaced and replacing mineral. Thus, delicate textures present in the original mineral may, in some cases, be faithfully preserved in the replacement mineral. Well-known examples of such preserved textures can be found in petrified wood and carbonate fossils replaced by chert.

Common replacement events include replacement of carbonate minerals by microcrystalline quartz (chert), replacement of chert by carbonate minerals, replacement of feldspars and quartz by carbonate minerals (e.g., Fig. 5.15D), replacement of feldspars by clay minerals, replacement of clay matrix by carbonate minerals, replacement of calcium-rich plagioclase by sodium-rich plagioclase (albitization), and replacement of feldspars and volcanic rock fragments by clay or zeolite minerals. Replacement may be partial or complete. Complete replacement destroys the identity of the original minerals or rock fragments and thereby gives a biased view of the original mineralogy of a rock. Porosity may also be affected by replacement, particularly replacement of framework grains by clay minerals, which tend to plug pore space and reduce porosity. Much of the clay matrix in sandstones may be produced diagenetically by alteration of unstable framework grains to clay minerals.

In addition to these common replacement processes, one kind of clay mineral may alter to another during diagenesis. For example, smectite clays may alter to illite at temperatures ranging from about 55–200°C, with concomitant release of water. This process is particularly common in shales and is referred to as shale **dewatering**. Smectite may also alter to chlorite within about the same temperature range, and kaolinite typically alters to illite at temperatures between about 120 and 150°C. It is these diagenetic processes that are believed to account for the trend of changing clay-mineral relative abundance with age shown in Fig. 5.12.

Telogenesis

Sedimentary rocks that have undergone deep burial diagenesis may subsequently be uplifted by mountain-building activities and unroofed by erosion. These processes bring mineral assemblages, including new minerals formed during mesogenesis, into an environment of lower temperature and pressure and in which mesogenetic pore waters are flushed and replaced by oxygen-rich, acidic meteoric (rain) waters of low salinity. Under these changed conditions, previously formed cements and framework grains may dissolve (creating secondary porosity) or framework grains may alter to clay minerals, e.g., potassium feldspar to kaolinite (reducing porosity). Alternatively, depending upon the nature of the pore waters, silica or carbonate cements can be precipitated. Other changes may include oxidation of iron carbonate minerals and other iron-bearing minerals to form iron oxides (goethite and hematite), oxidation of sulfides (pyrite) to form sulfate minerals (gypsum) if calcium is present in pore waters, and dissolution of less stable minerals such as pyroxenes and amphiboles. The processes of telogenesis grade into those of subaerial weathering as sedimentary rocks are exposed at Earth's surface.

5.6 PROVENANCE SIGNIFICANCE OF MINERAL COMPOSITION

The silicate mineralogy and rock-fragment composition of siliciclastic sedimentary rocks are fundamental properties of these rocks that set them apart from other sedimentary rocks. Mineralogy is a particularly important property for studying the origin of siliciclastic sedimentary rocks because it provides almost the only available clue to the nature of vanished source areas, that is, ancient mountain systems. The kinds of siliciclastic minerals and rock fragments preserved in sedimentary rocks furnish important evidence of the lithology of the source rocks. Rock fragments provide the most direct lithologic evidence: Volcanic rock fragments indicate volcanic source rocks, metamorphic rock fragments indicate metamorphic source rocks, etc. Feldspars and other minerals are also important source-rock indicators. For example, potassium feldspars suggest derivation mainly from alkaline plutonic igneous or metamorphic rocks, whereas sodic plagioclase is derived principally from alkaline volcanic rocks and calcic plagioclase comes mainly from basic volcanic rocks. Suites of heavy minerals are also used for source-rock determination. A suite of heavy minerals consisting of apatite, biotite, hornblende, monazite, rutile, titanite, pink tourmaline, and zircon indicates alkaline igneous source rocks. A suite consisting of augite, chromite, diopside, hypersthene, ilmenite, magnetite, and olivine suggests derivation from basic igneous rocks. Andalusite, garnet, staurolite, topaz, kyanite, sillimanite, and staurolite constitute a mineral suite diagnostic of metamorphic rocks, whereas a suite of heavy minerals consisting of barite, iron ores, leucoxene, rounded tourmaline, and rounded zircon suggests a recycled sediment source. The trace element composition of individual varieties of heavy minerals, such as the Ti and Fe content of ilmenite, has significance also as a provenance indicator (e.g., Darby and Tsang, 1987). See Morton and Hallsorth (1999) for an extended discussion of heavy minerals and provenance.

Quartz also has value as a provenance indicator. For example, Basu et al. (1975) suggest that a high percentage of quartz grains with undulose extinction greater than 5° combined with a high percentage of polycrystalline grains containing more than three crystal units per grain are typical of low-rank metamorphic source rocks. By contrast, nonundulose quartz and polycrystalline quartz containing less than three crystal units per grain indicate derivation from high-rank metamorphic or plutonic igneous source rocks. Seyedolali et al. (1997) demonstrated that provenance of quartz can also be determined by scanning electron microscope (SEM)-cathodoluminescence fabric analysis. Quartz grains from plutonic, volcanic, and metamorphic rocks display distinctively different patterns of cathodoluminescence when excited by an electron beam in the SEM, which provides reliable provenance interpretation (e.g., Kwon and Boggs, 2002).

In addition to providing information about source-rock lithology, the relative chemical stabilities and the degree of weathering and alteration of certain minerals can be used as a tool for interpreting the climate and relief of source areas (e.g., Folk, 1974, p. 85). For example, the presence of large, fresh, angular feldspars in a sandstone suggests derivation from a high-relief source area where grains were eroded rapidly before extensive weathering. Alternatively, they may have been derived from a source area having a very arid or extremely cold climate that retarded chemical weathering. Small, rounded, highly weathered feldspar grains indicate a source area of low relief and/or a warm, humid climate where chemical weathering was moderately intense. Absence of feldspars may indicate either that weathering was so intense that all feldspars were destroyed or that no feldspars were present in the source rocks. Such analyses of mineral constituents provide only tentative conclusions about climate and relief. Also, they are subject to misinterpretations owing to diagenetic alteration or destruction of source-rock minerals.

Geologists are also interested in the tectonic setting of source areas and associated depositional sites. With development of the theory of seafloor spreading and plate tectonics, this interest has focused on interpreting the tectonic setting in terms of plate tectonic provinces (Dickinson and Suczek, 1979; Dickinson, 1982; Dickinson et al., 1983). In other words, geologists want to know if a particular deposit was derived from source rocks located within a continent, in a volcanic arc associated with a subduction zone, or in other tectonic settings. Three principal types of tectonic settings, or **provenances**, as they are called, have been identified: (1) continental block provenances, (2) magmatic arc provenances, and (3) recycled orogen provenances (Fig. 5.21).

Continental block provenances (Fig 5.21A) are located within continental masses, which may be bordered on one side by a passive continental margin and on the other by an orogenic belt or zone of plate convergence. Source rocks consist of plutonic igneous, metamorphic, and sedimentary rocks but include few volcanic rocks. Sediment eroded from these sources typically consists of quartzose

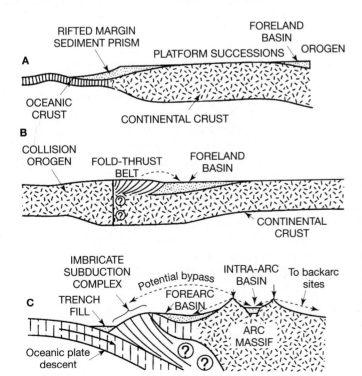

Figure 5.21
Schematic representation of the principal tectonic settings of sediment source areas. A. Continental block provenances. B. Recycled orogen provenances. C. Magmatic arcs. The dashed lines with arrows indicate sediment transport paths. (After Dickinson, W. R., and C. A. Suczek, 1979, Plate tectonics and sandstone composition: American Association Petroleum Geologists Bull., v. 63, Fig. 5, p. 2174, Fig. 6. p. 2175, Fig. 7, p. 2177, reprinted by permission of AAPG, Tulsa, Okla.).

sand, feldspars with high ratios of potassium feldspar to plagioclase feldspar, and metamorphic and sedimentary rock fragments. Sediment eroded from continental sources may be transported off the continent into adjacent marginal ocean basins, or it may be deposited in local basins within the continent.

Recycled orogen provenances (Fig. 5.21B) are zones of plate convergence, where collision of major plates creates uplifted source areas along the collision suture belt. Where two continental masses collide, source rocks in the collision uplifts are typically sedimentary and metamorphic rocks that were present along the continental margins prior to their collision. Detritus stripped from these source rocks commonly consists of abundant sedimentary-metasedimentary rock fragments, moderate quartz, and a high ratio of quartz to feldspars. Where a continental mass collides with a magmatic arc complex, uplifted source rocks may include deformed ultramafic rocks, basalts, and other oceanic rocks, and a variety of other rock types such as greenstone (weakly metamorphosed basic igneous rock), chert, argillite (weakly metamorphosed shale), lithic sandstones, and limestones. Sediment derived from these sources may contain many types of rock fragments, quartz, feldspars, and chert. Chert is a particularly abundant constituent of sediments derived from this provenance.

Magmatic arc provenances (Fig. 5.21C) are located in zones of plate convergence where sediment is eroded mainly from volcanic arc sources consisting of volcanogenic highlands (undissected arcs). Volcaniclastic debris shed from these highlands consists largely of volcanic lithic fragments and plagioclase feldspars. Quartz and potassium feldspars are commonly very sparse except where the volcanic cover is dissected by erosion to expose underlying plutonic rocks (dissected arcs). Sediment shed from volcanic highlands may be transported to an adjacent trench or deposited in fore-arc and back-arc basins.

To differentiate sediment derived from these three major tectonic provenances, Dickinson and Suczek (1979) and Dickinson et al. (1983) suggest the use of triangular composition diagrams showing framework proportions of monocrystalline quartz, polycrystalline quartz, potassium feldspars and plagioclase feldspars, and volcanic and sedimentary-metasedimentary rock fragments. Through study of sandstone compositions from many parts of the world, they generated the provenance diagrams shown in Figure 5.22. To use these diagrams as a guide to provenance determination of other sandstones, one determines the compositions of the sand-size grains in a sandstone and plots them on one or both of the diagrams shown in Figure 5.22. The field in which most of the plotted points fall (e.g., craton interior, recycled orogen) is the putative tectonic setting of the source rocks.

Dickinson's provenance model has been criticized because not every individual sand or sandstone plots where it should according to its tectonic setting. The model is valid, however, for *average* values of large data sets taken from large-scale sampling of various tectonic settings (Raymond Ingersoll, personal communication, 2004). Also, compositional data must be generated by using the so-called Gazzi-Dickinson point-counting method (see Ingersoll et al., 1984). Marsaglia and Ingersoll (1992) modified Dickinson's provenance triangle on the basis of contrasts between intraoceanic magmatic arcs and continental-margin magmatic arcs (their Figure 8), a useful refinement of the basic provenance model.

A comparatively recent aspect of provenance analysis of siliciclastic sedimentary rocks is estimation of ages of single mineral grains, such as apatite and zircon, by using various radiometric techniques. Determining the ages of single mineral grains in sedimentary rocks provides ages of the source rocks from which these grains were derived. Such analysis makes possible linking of the mineral grains to specific source areas of known ages (e.g., Bernet and Spiegel, 2004b). This kind of evaluation is known as **detrital thermochronology**. See Appendix B for details.

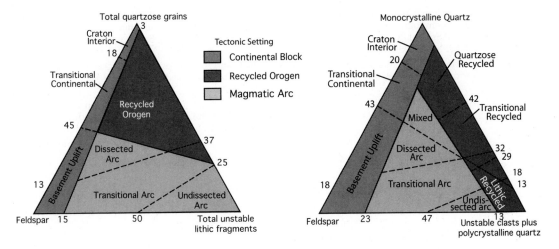

Figure 5.22
Relationship between framework composition of sandstones and tectonic setting. [After Dickinson, R. W., et al., 1983, Provenance of North American Phanerozoic sandstones in relation to tectonic setting: Geol. Soc. America Bull., v. 94, Fig. 1, p. 223.]

The discussion above provides only the barest introduction to the topic of provenance interpretation. The application of provenance study to basin analysis is explored further in Chapter 16. For additional information on this important subject, including discussion of the provenance of conglomerates and shales, see Boggs (1992, Chapter 8) and the volumes listed under "Further Reading–Provenance" at the end of this chapter.

FURTHER READING

Composition (sedimentary petrology)

Adams, A. E., W. S. Mackenzie, and C. Guilford, 1984, Atlas of sedimentary rocks under the microscope: John Wiley & Sons, New York, 104 p.

Boggs, S., Jr., 1992, Petrology of sedimentary rocks: Merrill/Macmillan, New York, 707 p.

Carozzi, A. V., 1993, Sedimentary petrography: PTR Prentice Hall, Englewood Cliffs, New Jersey, 263 p.

Folk, R. L., 1974, Petrology of sedimentary socks: Hemphill, Austin, Tex., 182 p.

Johnsson, M. J., 1993, The system controlling the composition of clastic sediments, *in* Johnsson, M. J., and A. Basu (eds.), Processes controlling the composition of clastic sediments: Geol. Soc. America Spec. Paper 284, p. 1–19.

Scholle, P. A., 1979, A color illustrated guide to constituents, textures, cements, and porosities of sandstones and associated Rocks: Am. Assoc. Petroleum Geologists Mem. 28, Tulsa, Okla., 201 p.

Sandstones and Conglomerates

Koster, E. H., and R. H. Steel (eds.), 1984, Sedimentology of gravels and conglomerates: Canadian Soc. of Petroleum Geologists Mem. 10, 441 p.

Mutti, E., 1992, Turbidite sandstones: Agip, Instituto di Geologia, Universitá di Parma, Milan, 275 p.

Pettijohn, F. J., P. E. Potter, and R. Siever, 1987, Sand and sandstone, 2nd ed.: Springer-Verlag, New York, 618 p.

Shales

Bennett, R. H., W. R. Bryant, and M. H. Hulbert, 1991, Microstructures of fine-grained sediments: From mud to shale: Springer-Verlag, New York, 582 p.

Krinsley, D. H., K. Pye, S. Boggs, Jr., and N. K. Tovey, 1998, Backscattered scanning electron microscopy and image analysis of sediments and sedimentary rocks: Cambridge University Press, Cambridge, 193 p.

O'Brien, N. R., and R. M. Slatt, 1990, Argillaceous rock atlas: Springer-Verlag, New York, 141 p.

Potter, P. E., J. B. Maynard, and W. A. Pryor, 1980, Sedimentology of shale: Springer-Verlag, New York, 553 p.

Schieber, J., W. Zimmerle, and P. S. Sethi (eds.), 1998, Shales and mudstones: E. Schweizerbartsche Verlagsbuchhandlung, Stuttgart, Vol I, 384 p., Vol II, 296 p.

Weaver, C. E., 1989, Clays, muds, and shales: Elsevier, Amsterdam, 819 p.

Diagenesis

Burley, S. D., and R. H. Worden, 2003, Sandstone diagenesis: Recent and ancient: Blackwell Pub., Malden, Mass. 649 p.

Crossey, L. J., R. Loucks, and M. W. Totten (eds.), 1996, Siliciclastic diagenesis and fluid flow: SEPM Special Publication No. 55, Society for Sedimentary Geology, Tulsa, Okla. 222 p.

McDonald, D. A., and R. C. Surdam (eds.), 1984, Clastic diagenesis: Am. Assoc. Petroleum Geologists Mem. 37, 434 p.

Montañez, I. P., J. M. Gregg, and K. L. Shelton (eds.), 1997, Basin-wide diagenetic patterns: integrated petrologic, geochemical, and hydrologic considerations: SEPM Special Publication No. 57, Society for Sedimentary Geology, Tulsa, Okla., 302.

Morad, S., 1998, Carbonate cementation in sandstones: distribution patterns and geochemical evolution: Blackwell Science, Malden, Mass., 511 p.

Wolf, K. H., and G. V. Chilingarian (eds.), 1992, Diagenesis III, Elsevier, Amsterdam, 671 p.

Provenance

Bahlburg, H., and P. A. Floyd, 1999, Advanced techniques in provenance analysis of sedimentary rocks: Special Issue, Sedimentary Geology, Vol. 124, 224 p.

Bernet, M., and C. Spiegel (eds.), 2004b, Detrital thermochronology: Provenance analysis, exhumation, and landscape evolution of mountain belts: Geol. Soc. America Special Paper 378, 126 p.

Johnsson, M. J., and A. Basu, eds., 1993, Processes controlling the composition of clastic sediments: Geol. Soc. America Spec. Paper 284, 342 p.

Morton, A. C., S. P. Todd, and P. D. W. Haughton (eds.), 1991, Developments in sedimentary provenance studies: Geological Society Spec. Publ. 57, Geological Society London, 370 p.

Zuffa, G. G. (ed.), 1984, Provenance of arenites: D. Reidel, Dordrecht, 408 p.

6

Carbonate Sedimentary Rocks

6.1 INTRODUCTION

Chemical/biochemical sedimentary rocks originate by precipitation of minerals from water through various chemical or biochemical processes. They are distinguished from siliciclastic sedimentary rocks by their chemistry, mineralogy, and texture. They can be divided on the basis of mineralogy and chemistry into five fundamental types: (1) carbonates, (2) evaporites, (3) siliceous sedimentary rocks (cherts), (4) iron-rich sedimentary rocks, and (5) phosphorites. Carbonaceous sedimentary rocks, such as coals and oil shales, make up a further special group of rocks that contain abundant nonskeletal organic matter in addition to various amounts of siliciclastic or chemical (e.g., carbonate) constituents.

The carbonate rocks, by far the most abundant kind of chemical/biochemical sedimentary rock, are described in this chapter. Other chemical/biochemical and carbonaceous sedimentary rocks are discussed in Chapter 7. Carbonate rocks can be divided on the basis of mineralogy into limestones and dolomites (dolostones). Limestones are composed mainly of the mineral calcite, and dolomites are composed mainly of the mineral dolomite. Carbonate sedimentary rocks make up 20 to 25 percent of all sedimentary rocks in the geologic record. They are present in many Precambrian assemblages and in all geologic systems from the Cambrian to the Quaternary. Precambrian and Paleozoic carbonate successions include abundant dolomite, whereas Mesozoic and Cenozoic carbonates are mainly limestone. Limestones contain richly varied textures, structures, and fossils that yield important information about ancient marine environments, paleoecological conditions, and the evolution of life forms, particularly marine organisms, through time. Carbonate sedimentary rocks are also an economically important group of rocks because limestones and dolomites are useful for agricultural and industrial purposes, they make good building stones, and, most important, they act as reservoir rocks for more than one-third of the world's petroleum reserves. Because of their environmental and economic significance, they have been extensively studied and their mineralogy, chemistry, and textural characteristics are described in hundreds of research papers. The characteristic properties of carbonate rocks have also been summarized in several books; see "Further Reading" at the end of this chapter.

6.2 CHEMISTRY AND MINERALOGY

The elemental chemistry of carbonate rocks is dominated by calcium (Ca^{2+}), magnesium (Mg^{2+}), and carbonate (CO_3^{2-}) ions. Calcium and magnesium are present in both limestones and dolomites; however, magnesium is a particularly important constituent of dolomites. Expressed as oxides, CaO, MgO, and CO_2 make up more than 90 percent of the average carbonate rock. Numerous other elements are present in carbonate rocks in minor or trace amounts. Many of the elements that occur in minor concentrations are contained in noncarbonate impurities. For example, Si, Al, K, Na, and Fe occur mainly in silicate minerals such as quartz, feldspars, and clay minerals that are present in minor amounts in most carbonate rocks. Trace elements that are common in carbonate rocks include B, Be, Ba, Sr, Br, Cl, Co, Cr, Cu, Ga, Ge, and Li. The concentration of these trace elements is controlled not only by the mineralogy of the rocks but also by the type and relative abundance of fossil skeletal grains in the rock. Many organisms concentrate and incorporate trace elements such as Ba, Sr, and Mg into their skeletal structures.

The chemistry and structure of the principal carbonate minerals, only a few of which are important components of limestones and dolomites, are shown in Table 6.1. A more detailed analysis of the crystal chemistry of the carbonates is given by Reeder (1983) and Tucker and Wright (1990, p. 284). Modern carbonate sediments are composed mainly of aragonite, but they also include calcite (especially in deep-sea calcareous ooze) and dolomite. Calcite ($CaCO_3$) can contain several percent magnesium in its formula because magnesium can readily substitute for calcium in the rhombohedral lattice of calcite crystals, owing to the fact that magnesium ions and calcium ions are similar in size and charge. Thus, we

Table 6.1   Principal carbonate minerals

Mineral	Crystal system	Formula	Remarks
Calcite group			
*Calcite	Rhombohedral	$CaCO_3$	Dominant mineral of limestones, especially in rocks older than the Tertiary
Magnesite	Rhombohedral	$MgCO_3$	Uncommon in sedimentary rocks but occurs in some evaporite deposits
Rhodochrosite	Rhombohedral	$MnCO_3$	Uncommon in sedimentary rocks; may occur in Mn-rich sediments associated with siderite and Fe-silicates
Siderite	Rhombohedral	$FeCO_3$	Occurs as cements and concretions in shales and sandstones; common in ironstone deposits; also in carbonate rocks altered by Fe-bearing solutions
Smithsonite	Rhombohedral	$ZnCO_3$	Uncommon in sedimentary rocks; occurs in association with Zn ores in limestones
Dolomite group			
*Dolomite	Rhombohedral	$CaMg(CO_3)_2$	Dominant mineral in dolomites; commonly associated with calcite or evaporite minerals
Ankerite	Rhombohedral	$Ca(Mg,Fe,Mn)(CO_3)_2$	Much less common than dolomite; occurs in Fe-rich sediments as disseminated grains or concretions
Aragonite group			
*Aragonite	Orthorhombic	$CaCO_3$	Common mineral in recent carbonate sediments; alters readily to calcite
Cerussite	Orthorhombic	$PbCO_3$	Occurs in supergene lead ores
Strontionite	Orthorhombic	$SrCO_3$	Occurs in veins in some limestones
Witherite	Orthorhombic	$BaCO_3$	Occurs in veins associated with galena ore

*Important minerals in limestones and dolomites.

recognize both **low-magnesian calcite** (called simply calcite) containing less than about 4 percent $MgCO_3$ and **high-magnesian calcite** containing more than 4 percent $MgCO_3$. High-magnesian calcite still retains the crystal structure of calcite in spite of the presence of Mg ions, which randomly substitute for Ca ions in the calcite crystal lattice. Note: Mg^{2+} ions commonly do not substitute for Ca^{2+} ions in the larger spaces available in the more open orthorhombic lattice of aragonite. In contrast to high-magnesian calcite, true dolomite, so-called **stoichiometric dolomite**, is a totally different mineral in which Mg ions occupy half of the cation sites in the crystal lattice and are arranged in well-ordered planes that alternate with planes of CO_3 ions and Ca ions. Dolomite occurs in a few restricted modern environments, particularly in certain supratidal environments and freshwater lakes, but it is much less abundant in modern carbonate environments than aragonite and calcite. Other carbonate minerals such as magnesite, ankerite, and siderite are even less common in modern sediments.

Although precipitation of aragonite and, to a lesser extent, high-magnesian calcite, is favored in the modern ocean, this preference for aragonite precipitation has not always been the case. It now appears likely that the oceans during early Paleozoic and middle to late Cenozoic time favored precipitation of calcite, probably because of a lower ratio of magnesium to calcium during these times.

The mineralogy and chemistry of carbonate sediments can be strongly influenced by the composition of calcareous fossil organisms present in the sediments (e.g., Scholle, 1978, p. xi; Jones and Desrochers, 1992). For example, many molluscs such as pelecypods, gastropods, pteropods, chitons, and cephalopods, as well as calcareous green algae, stromatoporoids, scleractinian corals, and annelids build skeletons of aragonite. Echinoids, crinoids, bottom-dwelling (benthonic) forams, and coralline red algae are composed mainly of high-magnesian calcite. Some carbonate-secreting organisms, e.g., planktonic (floating) forams, coccoliths, and brachiopods, have low-magnesian calcite shells or tests.

In contrast to the dominance of aragonite in modern shallow-water carbonate sediments, ancient carbonate rocks older than about the Cretaceous contain little aragonite. Aragonite is the metastable polymorph (having the same chemical composition but different crystal structure) of $CaCO_3$ and is converted fairly rapidly under aqueous conditions to calcite. Thus, aragonite deposited during earlier times, such as the late Paleozoic and early Cenozoic, has subsequently dissolved and been replaced by calcite. The ratio of dolomite to calcite is much greater in ancient carbonate rocks than in modern carbonate sediments, presumably because $CaCO_3$ minerals exposed to magnesium-rich interstitial waters during burial and diagenesis are converted to dolomite by replacement.

The stable isotope composition of carbonate rocks is also of considerable interest in paleoenvironmental studies and for the purpose of time-stratigraphic correlation. Isotopes of oxygen (^{18}O and ^{16}O) are particularly useful for these purposes; however, carbon, sulfur, and strontium isotopes also have significant utility. Stable-isotope studies commonly involve comparison of the ratios of stable isotopes (e.g., $^{18}O/^{16}O$) in a sample to those of a standard. The applications of such studies to sedimentological and stratigraphic problems are considered further in Chapter 15.

6.3 LIMESTONE TEXTURES

As discussed, ancient limestones are composed mainly of calcite. Calcite can be present in at least three distinct textural forms: (1) **carbonate grains,** such as ooids and skeletal grains, which are silt-size or larger aggregates of calcite crystals, (2) **microcrystalline calcite,** or carbonate mud, which is texturally analogous to the mud in siliciclastic sedimentary rocks but which is composed of extremely fine size calcite crystals, and (3) **sparry calcite**, consisting of much coarser grained calcite crystals that appear clear to translucent in plane (nonpolarized) light.

Carbonate Grains

Early geologists tended to regard limestones as simply crystalline rocks that commonly contained fossils and that presumably formed largely by passive precipitation from seawater. We now know that many, and perhaps most, carbonate rocks are not simple crystalline precipitates. Instead, they are composed in part of aggregate particles or grains that may have undergone mechanical transport before deposition. Folk (1959) suggested use of the general term **allochems** for these carbonate grains to emphasize that they are not normal chemical precipitates. Carbonate grains typically range in size from coarse silt (0.02 mm) to sand (up to 2 mm), but larger particles such as fossil shells also occur. They can be divided into five basic types, each characterized by distinct differences in shape, internal structure, and mode of origin: carbonate clasts, skeletal particles, ooids, peloids, and aggregate grains. Scholle and Ulmer-Scholle (2003) provide an outstanding collection of color photomicrographs illustrating all of the major kinds of carbonate grains.

Carbonate Clasts (Lithoclasts)

Carbonate clasts are rock fragments that were derived either by erosion of ancient limestones exposed on land or by erosion of partially or completely lithified carbonate sediments within a depositional basin. If carbonate clasts are derived from older limestones present in land sources located outside the depositional basin, they are called **extraclasts**. If they are derived from within the basin by erosion of semiconsolidated carbonate sediments from the seafloor, adjacent tidal flats, or a carbonate beach (beach rock), they are called **intraclasts**. The distinction between extraclasts and intraclasts has important implications for interpreting the transport and depositional history of limestones. Extraclasts may have iron-stained rims resulting from weathering, may contain recrystallized veins inherited from the parent rock, or may display other properties that distinguish them from intraclasts (Boggs, 1992, p. 425). Nonetheless, the distinction between fragments of ancient, weathered limestones and penecontemporaneously produced intraclasts is often difficult to make. **Lithoclast** (or limeclast) is a nonspecific term that can be used for carbonate clasts when this distinction cannot be made.

Lithoclasts range in size from very fine sand to gravel, although sand-size fragments are most common. They generally show some degree of rounding (Fig. 6.1A), indicative of transport, but subangular or even angular clasts (Fig. 6.1B) are not unusual. Some clasts display internal textures or structures such as lamination, older clasts, siliciclastic grains, fossils, ooids, or pellets, but others are internally homogeneous. A limestone composed mainly of gravel-size limeclasts is a kind of intraformational conglomerate. Clasts are not the most abundant type of carbonate grain in ancient limestones, but they occur with sufficient frequency in the geologic record to show that the clast-forming mechanism was a common process.

Skeletal Particles

Skeletal fragments occur in limestones as whole microfossils, whole larger fossils, or broken fragments of larger fossils. They are by far the most common kind of grain in carbonate rocks, and they are so abundant in some limestones that they make up most of the rock. Fossils representing all of the major phyla of calcareous marine invertebrates are present in limestones. The specific kinds of skeletal particles that occur depend upon both the age of the rocks and the paleoenvironmental

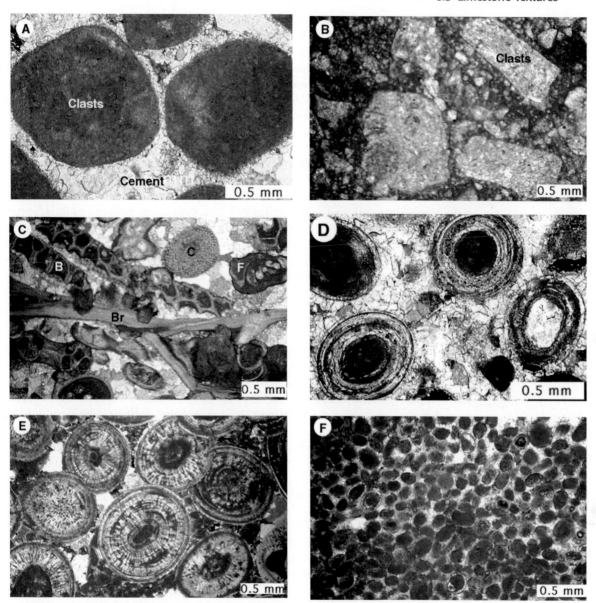

Figure 6.1
Fundamental kinds of carbonate grains (allochems) in limestones: A. Rounded clasts cemented with sparry calcite cement, Devonian limestone, Canada. B. Angular to subangular clasts in a micrite (dark) matrix, Calville Limestone (Permian), Nevada. C. Mixed skeletal grains (B = bryozoan, Br = brachiopod, C = crinoid, F = foraminifer) cemented with sparry calcite, Salem Formation (Mississippian), Missouri. D. Normal ooids cemented with sparry calcite (white), Miami Oolite (Pleistocene), Florida. E. Radial ooids cemented with sparry calcite (white) and micrite (dark); note relict concentric layering, Devonian limestone, Canada. F. Pellets cemented with sparry calcite, Quaternary-Pleistocene limestone, Grand Bahama Banks. Crossed nicols.

conditions under which they were deposited. Because of evolutionary changes in fossil assemblages through time, different kinds of fossil remains dominate rocks of different ages. For example, trilobite skeletal remains characterize early Paleozoic rocks, but they do not occur in Cenozoic rocks, which instead commonly contain abundant foraminifers. Likewise, certain kinds of skeletal particles characterize limestones formed in different environments. To illustrate, the remains of colonial corals, which build rigid, wave-resistant skeletal structures, are commonly restricted to limestones deposited in shallow-water, high-energy environments where the water was well agitated and oxygen levels were high. By contrast, branching types of bryozoa are fragile organisms that cannot withstand the rigors of high wave energy environments. Thus, their remains are found mainly in limestones deposited under quiet-water conditions.

Depending upon paleoenvironmental conditions, skeletal remains in a given specimen of limestone may consist entirely or almost entirely of one species of organism; however, they commonly include several species. An example of a mixed assemblage of skeletal particles is shown in Figure 6.1C. The serious student of carbonate rocks must learn to identify the many kinds of fossils and fossil fragments that occur in limestones because fossils have special significance for paleoenvironmental and paleoecological interpretation. Several photographic atlases illustrating whole fossils and fossil fragments as they appear in microscope thin sections are available (e.g., Horowitz and Potter, 1971; Adams and Mackenzie, 1998; Scholle and Ulmer-Scholle, 2003). By using these atlases, students should be able to identify many of the kinds of fossil remains commonly present in limestones.

Ooids

The term **ooid** is applied as a general name to coated carbonate grains that contain a nucleus of some kind—a shell fragment, pellet, or quartz grain—surrounded by one or more thin layers or coatings (the cortex) consisting of fine calcite or aragonite crystals. (In some ooids, the nucleus may be too small to be easily seen.) These coated grains are sometimes referred to also as **ooliths**; however, the term ooid is preferred. Carbonate rocks formed mainly of ooids are called **oolites**. Spherical to subspherical ooids that exhibit several internal concentric layers with a total thickness greater than that of the nucleus are called normal or mature ooids (Fig. 6.1D). Ooids form where strong bottom currents and agitated-water conditions exist and where saturation levels of calcium bicarbonate are high (Section 6.7). The coatings on modern ooids are composed mainly of aragonite, whereas ancient ooids are composed principally of calcite. Many of these ancient ooids were composed originally of aragonite that later transformed to calcite; however, petrographic evidence suggests that other ancient ooids originated as calcite. Precipitation of ancient calcitic ooids appears to have been particularly important during middle Paleozoic and middle Mesozoic time (Morse and Mackenzie, 1990, p. 538). Variations in ooid mineralogy appear to be related to sea levels. High stands of the sea apparently favor formation of calcite ooids because CO_2 levels tend to be higher and Mg/Ca ratios lower during such times; low stands favor aragonite ooids because of lowered CO_2 levels and elevated Mg/Ca ratios (Wilkinson, Owen, and Carroll, 1985). As discussed in Section 6.7, high Mg/Ca ratios favor precipitation of aragonite rather than calcite because Mg ions inhibit crystallization of calcite.

Although most ooids display an internal structure consisting of concentric layers, some ooids show a radial internal structure (Fig. 6.1E). Radial ooids that also display concentric layers, such as those shown in Figure 6.1E, probably formed by recrystallization of normal ooids; however, radial ooids may form also by primary sedimentation processes. The coating on some ooids consists only of one or two very thin layers, which have a total thickness less than that of the nucleus. Such ooids have been called **superficial ooids** or **pseudo-ooids**. Coated grains that have an internal structure similar to that of ooids but that are much larger—that is, greater than 2 mm—are called **pisoids** (a rock composed of pisoids is a pisolite). Pisoids are generally less spherical than ooids and are commonly crenulated. Some pisoids may be of algal origin, formed by the trapping and binding activities of blue-green algae (cyanobacteria) in the same way that stromatolites are formed (Chapter 4). Spheroidal stromatolites that reach a size exceeding 1 to 2 cm are called **oncoids**.

Peloids

Peloid is a nongenetic term for carbonate grains that are composed of microcrystalline or cryptocrystalline calcite or aragonite and that do not display distinctive internal structures (Fig. 6.1F). Peloids are smaller than ooids and are generally of silt to fine-sand size (0.03–0.1 mm), although some may be larger. The most common kind of peloids are fecal pellets, produced by organisms that ingest calcium carbonate muds and extrude undigested mud as pellets. Fecal pellets tend to be small, oval to rounded, and uniform in size. They commonly contain enough fine organic matter to make them appear opaque or dark colored. Pellets can be differentiated from ooids by their lack of concentric or radial internal structure and from rounded intraclasts by their uniformity of shape, good sorting, and small size. Because they are produced by organisms, their sizes and shapes are not related to current transport, although pellets may be transported by currents and redeposited after initial deposition by organisms.

Peloids may also be produced by other processes, such as micritization of small ooids or rounded skeletal fragments caused by the boring activities of certain organisms, particularly endolithic (boring) algae. These boring activities convert the original grains into a nearly uniform, homogeneous mass of microcrystalline calcite. Some marine peloids may form by precipitation around active clumps of bacteria (e.g., Chafetz, 1986). Other peloids may simply be very small, well-rounded intraclasts formed by reworking of semiconsolidated mud or mud aggregates.

Aggregate Grains

Aggregate grains are irregularly shaped carbonate grains that consist of two or more carbonate fragments (pellets, ooids, fossil fragments) joined together by a carbonate-mud matrix that is generally dark colored and rich in organic matter. Because the shapes of the aggregate grains in some modern carbonate-forming environments, such as the Bahama Banks, resemble a bunch of grapes, they are commonly called **grapestones** (Illing, 1954). Other aggregate grains with a somewhat smoother appearance have been referred to by the rather inelegant name of **lumps**. Tucker and Wright (1990, p. 12) suggest that lumps evolve from grapestones by continued cementation and micritization of the grains (Fig. 6.2). Aggregate grains in modern carbonate environments are composed mainly of aragonite,

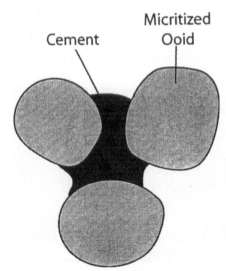

Cement

Micritized Ooid

Figure 6.2
Schematic representation of lumps formed by micritization and cementation of ooids.

but such grains in ancient limestones are dominantly calcite. Aggregate grains in modern environments can commonly be recognized by their botryoidal shapes and lack of internal structures; however, they can be confused with intraclasts. In fact, they are considered a type of intraclast by some geologists (e.g., Scholle and Ulmer-Scholle, 2003, p. 246). Aggregate grains are only rarely reported in ancient limestones, possibly because their shapes become distorted beyond recognition owing to compaction during diagenesis.

Microcrystalline Calcite

Carbonate mud composed of very fine size calcite crystals is present in many ancient limestones in addition to sand-size carbonate grains. Carbonate mud or lime mud occurs also in modern environments where it consists dominantly of needle-shaped crystals of aragonite about 1 to 5 microns (0.001–0.005 mm) long. The carbonate mud in ancient limestones is composed of similar-size crystals of calcite. Lime muds may also contain small amounts of fine-grained detrital minerals such as clay minerals, quartz, feldspar, and fine-size organic matter. They have a grayish to brownish, subtranslucent appearance under the microscope (Fig. 6.3), and they are easily distinguished from carbonate grains and sparry calcite crystals (discussed below) by their extremely small crystal size. Folk (1959) proposed the contraction **micrite** for microcrystalline calcite, a term that has been universally adopted to signify very fine grained carbonate sediments.

Micrite may be present as matrix among carbonate grains, or it may make up most or all of a limestone. A limestone composed mostly of micrite is analogous texturally to a siliciclastic mudrock or shale. The presence of micrite in an ancient limestone is commonly interpreted to indicate deposition under quiet-water conditions where little winnowing of fine mud took place. By contrast, carbonate sediments deposited in environments where bottom currents or wave energy are strong are commonly mud-free because carbonate mud is selectively removed in these environments. On the basis of purely chemical considerations, carbonate mud or micrite can theoretically form by inorganic precipitation of aragonite, later converted to calcite, from surface waters supersaturated with calcium bicarbonate. Geologists are uncertain, however, about how much aragonite is actually being generated by inorganic processes in the modern ocean. Much modern carbonate mud appears to originate through organic processes (Section 6.7). These processes include breakdown of calcareous algae in shallow water to yield aragonite mud, and deposition of carbonate nannofossils (<35 mm in size) such as coccoliths in deeper water to yield calcite muds (chalks).

Sparry Calcite

Many limestones contain large crystals of calcite, commonly on the order of 0.02 to 0.1 mm, that appear clear or white when viewed with a hand lens or in plane light

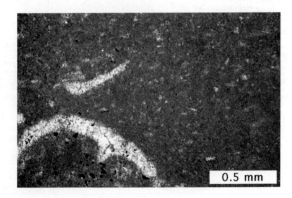

Figure 6.3
Micrite-rich limestone containing a few skeletal grains. Plattin Limestone (Ordovician), Missouri. Crossed nicols.

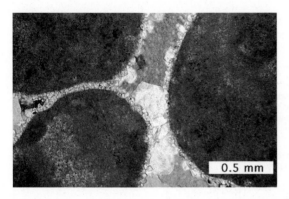

Figure 6.4
Sparry calcite cementing rounded intraclasts. Note that the cement displays drusy texture: small calcite crystals, oriented with their long dimensions perpendicular to the clast surfaces, grade outward from the margins of the clasts into larger, randomly oriented calcite crystals. Devonian limestone, Canada. Crossed nicols.

under a polarizing microscope. Such crystals are called sparry calcite. They are distinguished from micrite by their larger size and clarity and from carbonate grains by their crystal shapes and lack of internal texture. Some sparry calcite can be seen under the microscope to fill interstitial pore spaces among grains or to fill solution cavities as a cement (Fig. 6.4). The presence of sparry calcite cement in intergranular pore spaces indicates that grain framework voids were empty of lime mud at the time of deposition, suggesting deposition under agitated-water conditions that removed fine mud, as mentioned.

Sparry calcite can also form in ancient limestones by recrystallization of primary depositional grains and micrite during diagenesis (Section 6.8). Sparry calcite formed by recrystallization may be very difficult in some cases to differentiate from sparry calcite cement (Boggs, 1992, p. 549). It is important to distinguish between the two types of sparry calcite because incorrectly identifying recrystallized spar as sparry calcite cement can cause errors in both environmental interpretation and limestone classification.

6.4 DOLOMITE TEXTURES

Dolomite (dolostone) is composed mainly of the mineral dolomite [$CaMg(CO_3)_2$]. Unlike limestone, which is characterized by the presence of grains, micrite, and/or sparry cement, dolomite has a largely crystalline (granular) texture. On the basis of crystal shape, two kinds of dolomite are recognized. **Planar** (or idiotopic) dolomite (Fig. 6.5A) consists of rhombic, euhedral (well-formed) to anhedral (poorly

Figure 6.5
Dolomite crystals: A. Planar dolomite exhibiting euhedral crystals with planar faces (arrow), Bonneterre Formation (Cambrian), Missouri. B. Nonplanar dolomite, with curved or irregular faces, Davis formation (Cambrian), Missouri. Crossed nicols.

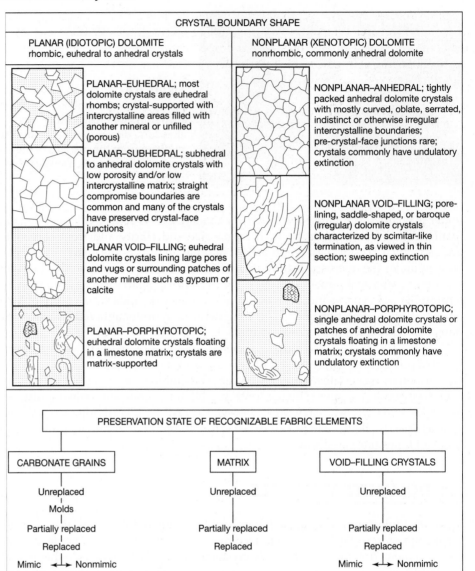

Figure 6.6
Classification of dolomite textures. [After Gregg, J. M., and D. F. Sibley, 1984, Epigenetic dolomitization and the origin of xenotopic dolomite texture: Jour. Sed. Petrology, v. 54, Fig. 6, p. 913, and Sibley, D. F., and Gregg, J. M., 1987, Classification of dolomite rock textures: Jour. Sed. Petrology, v. 57, Fig. 1, p. 968; figures reprinted by permission of Society of Economic Paleontologists and Mineralogists, Tulsa, Okla.]

formed) crystals. **Nonplanar** (or xenotopic) dolomite (Fig. 6.5B) is made of nonplanar, commonly anhedral crystals (Sibley and Gregg, 1987). Each of these major kinds of dolomite can be divided into subtypes as shown in Figure 6.6. Many dolomites form by replacement of a precursor limestone. Original limestone textures may be preserved in such dolomites to various degrees, ranging from virtually unreplaced to totally replaced (Fig. 6.6). That is, the replacing dolomite may preserve the original texture as a "ghost" (mimicking replacement) or the original texture may be completely destroyed (nonmimicking replacement).

6.5 STRUCTURES IN CARBONATE ROCKS

Carbonate rocks contain many of the same kinds of sedimentary structures as those present in siliciclastic rocks (Chapter 4). These structures include cross-stratification, laminated bedding, lenticular bedding, convolute lamination, flame structures, load casts, flute casts, groove casts, and mudcracks, as well as trace fossils such as tracks, trails, and burrows (Demicco and Hardie, 1994). They may also contain stromatolites and cryptoalgal structures and less common structures such as teepee structures (arched or upturned polygons), solution cavities, and stromatactis (irregularly-shaped masses of sparry calcite and internal sediments ranging in shape from elongate to globose).

6.6 CLASSIFICATION OF CARBONATE ROCKS

Attempts to classify carbonate rocks date back to at least 1904 with the publication of Grabeau's classic textbook on the classification of sedimentary rocks. Additional classifications were proposed by other authors in the 1930s, 1940s, and 1950s. Most of these early classifications were basically genetic schemes in which names such as "fore-reef talus limestone" or "low-energy limestone" were used to identify limestones according to their presumed environment of deposition (Ham and Pray, 1962). These classifications failed to recognize the clear distinction between carbonate grains and carbonate mud or to exploit difference in identity of the various kinds of carbonate grains. Publication in 1959 of Folk's largely descriptive "Practical Petrographic Classification of Limestones" marked the beginning of the modern period of limestone classification. In 1962, several additional classifications appeared (Ham, 1962) which, with one exception, are mainly descriptive classifications. Unlike the confusion attending the proliferation of sandstone classifications, the appearance of several descriptive limestone classifications seems to have had a largely positive effect because it forced geologists to become more keenly aware of the varied constituents that make up limestones, as well as the environmental significance of these constituents.

Mineralogy plays only a small role in classification of carbonate rocks because most carbonate rocks are essentially monomineralic. Mineralogy is used primarily to differentiate dolomite from limestone or carbonate rocks from noncarbonate rocks. The principal constituents or parameters used in carbonate classification are the types of carbonate grains or allochems and the grain/micrite ratio. The nature of the grain packing or fabric is also used in some classifications, in which the fabric is referred to as either grain-supported or mud-supported. A grain-supported fabric is one in which grains are in contact, creating an intact grain framework in which voids may or may not be filled with mud (matrix). In a mud-supported fabric, most grains do not touch, and they appear to "float" in the carbonate mud.

Folk's (1959, 1962) classification has probably been the most widely accepted limestone classification because of its applicability to a wide range of carbonate rock types and the ease with which its terms can be understood and utilized. The classification is based on the relative abundance of three major types of constituents: (1) carbonate grains or allochems, (2) microcrystalline carbonate mud (micrite), and (3) sparry calcite cement. As illustrated in Table 6.2, classification is made by first determining the relative abundance of total allochems vs. micrite plus sparry calcite cement. Further subdivision is then made on the basis of the relative abundance of the various types of carbonate grains (Fig. 6.7) and the relative abundance of micrite compared to sparry calcite cement. This classification approach yields a bipartite name that reflects both the major type of carbonate grain in the limestone and the relative abundance of micrite and sparry calcite cement. Thus, an **oosparite** is an ooid-rich rock cemented with sparry calcite that contains little micrite, whereas an **oomicrite** is an ooid-rich limestone in which micrite is abundant and sparry calcite is subordinate. Additional textural information can be added by use of the textural maturity terms shown in Figure 6.8. Thus, a **packed oomicrite** indicates a grain-supported oolitic limestone, and a **sparse oomicrite** is an oolitic rock with a mud-supported fabric. Note that Folk's classification can also be used to classify dolomite rock, if "ghosts" of the original allochems are still identifiable in the dolomite.

Table 6.2 Classification of carbonate rocks

Volumetric allochem composition	Limestones, partly dolomitized limestones — >10% Allochems (Allochemical rocks) — Sparry calcite cement > microcrystalline ooze matrix (Sparry allochemical rocks)	>10% Allochems — Microcrystalline ooze matrix > sparry calcite cement (Microcrystalline allochemical rocks)	<10% Allochems (Microcrystalline rocks) — 1%–10% Allochems (Most abundant allochem)	<1% Allochems	Undisturbed bioherm rocks	Replacement dolomites — Allochem ghosts (Evident allochem)	No allochem ghosts
>25% Intraclasts	Intrasparrudite* / Intrasparite	Intramicrudite* / Intramicrite*	Intraclasts: intraclast-bearing micrite*	Micrite if disturbed, dismicrite; if primary, dolomite, dolomicrite	Biolithite	Finely crystalline intraclastic dolomite, etc.	Medium crystalline dolomite
<25% Intraclasts, >25% Ooids	Oosparrudite* / Oosparite	Oomicrudite* / Oomicrite*	Ooids: ooid-bearing micrite*			Coarsely crystalline oolitic dolomite, etc.	Finely crystalline dolomite
<25% Ooids — Volume ratio of fossils to pellets >3:1	Biosparrudite / Biosparite	Biomicrudite / Biomicrite	Fossils: fossiliferous micrite			Aphanocrystalline biogenic dolomite, etc.	etc.
3:1–1:3	Biopelsparite	Biopelmicrite					
<1:3	Pelsparite	Pelmicrite	Pellets: pelletiferous micrite			Very finely crystalline pellet dolomite, etc.	

Source: Folk, R. L., 1962, Spectral subdivision of limestone types, in W. E. Ham (ed.), Classification of carbonate rocks: *Am. Assoc. Petroleum Geologists Mem.* 1. Table 1, p. 70, reprinted by permission of AAPG, Tulsa, Okla.

Note: Names and symbols in the body of the table refer to limestones. If the rock contains more than 10 percent replacement dolomite, prefix the term "dolomitized" to the rock name. The upper name in each box refers to calcirudites (median allochem size larger than 1.0 mm); the lower name refers to all rocks with median allochem size smaller than 1.0 mm. Grain size and quantity of ooze matrix, cements, or terrigenous grains are ignored.

* Designates rare rock types.

Allochemical Rocks
Orthochemical Rocks
I
II
III
SPARRY CALCITE CEMENT
MICROCRYSTALLINE CALCITE MATRIX
MICROCRYSTALLINE CALCITE LACKING ALLOCHEMS
Allochem Composition
INTRACLASTS (i)
INTRASPARITE(Ii)
INTRAMICRITE(IIi)
MICRITE(IIIm)
OOIDS (o)
OOSPARITE(Io)
OOMICRITE(IIo)
DISMICRITE(IIImX)
FOSSILS (b)
BIOSPARITE(Ib)
BIOMICRITE(IIb)
Autochthonous Reef Rocks
IV
PELLETS (p)
PELSPARITE(Ip)
PELMICRITE(IIp)
BIOLITHITE(IV)
Sparry calcite
Microcrystalline Calcite

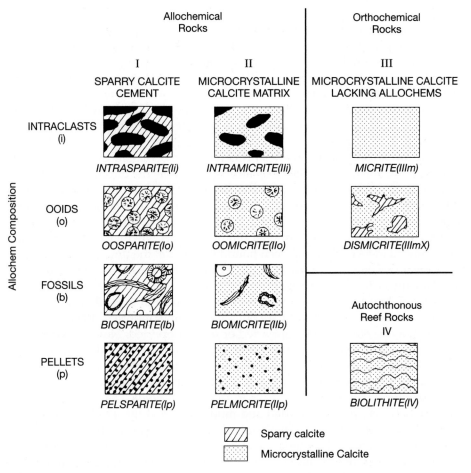

Figure 6.7
Schematic representation of the constituents that form the basis for Folk's classification of carbonate rocks (Table 6.2). [After Folk, R. L., 1962, Spectral subdivision of limestone types, *in* Ham, W. E. (ed.), Classification of carbonate rocks: Am. Assoc. Petroleum Geologists Memoir 1, Fig. 3, p. 71, reprinted by permission of AAPG, Tulsa, Okla.]

OVER 2/3 LIME MUD MATRIX
SUBEQUAL SPAR & LIME MUD
OVER 2/3 SPAR CEMENT
Percent Allochems
0 – 1%
1 – 10%
10 – 50%
OVER 50%
SORTING POOR
SORTING GOOD
ROUNDED & ABRADED
Representative Rock Terms
MICRITE & DISMICRITE
FOSSILI-FEROUS MICRITE
SPARSE BIOMICRITE
PACKED BIOMICRITE
POORLY WASHED BIOSPARITE
UNSORTED BIOSPARITE
SORTED BIOSPARITE
ROUNDED BIOSPARITE
Terminology
Micrite & Dismicrite
Fossiliferous Micrite
Biomicrite
Biosparite
Terrigenous Analogues
Claystone
Sandy Claystone
Clayey or Immature Sandstone
Submature Sandstone
Mature Sandstone
Supermature Sandstone
LIME MUD MATRIX
SPARRY CALCITE CEMENT

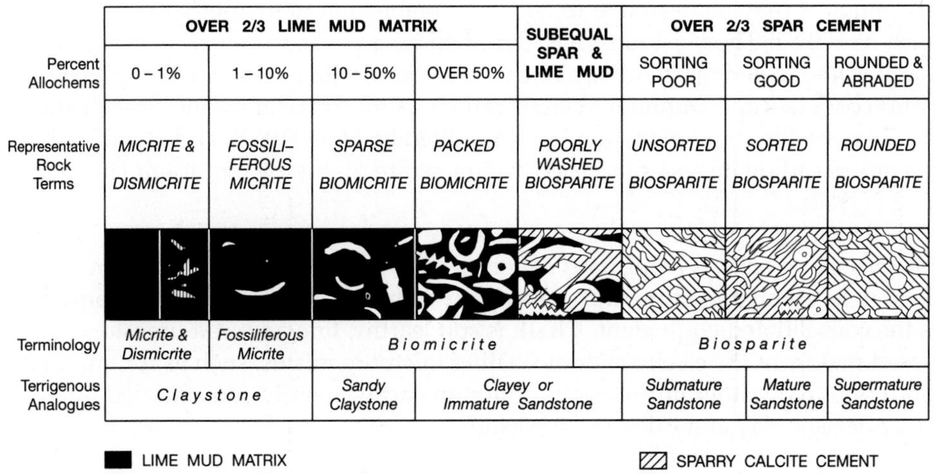

Figure 6.8
Textural classification of carbonate sediments on the basis of relative abundance of lime mud matrix and sparry calcite cement and on the abundance and sorting of carbonate grains (allochems). [After Folk, R. L., 1962, Spectral subdivision of limestone types, *in* Ham, W. E. (ed.), Classification of carbonate rocks: Am. Assoc. Petroleum Geologists Memoir 1, Fig. 4, p. 76, reprinted by permission of AAPG, Tulsa, Okla.]

The terms used by Folk (1959, 1962) to differentiate depositional textures are purely descriptive; however, they also have environmental significance. The term **biomicrite**, for example, conveys an interpretation of deposition under quiet-water conditions where micrite is abundant and winnowing of the lime mud is minimal. Thus, micrite accumulates along with skeletal particles. On the other hand, the term **biosparite** suggests deposition in a wave-agitated environment where micrite is removed by winnowing currents, allowing mud-free carbonate grains to accumulate. These grains are subsequently cemented with sparry calcite during burial (diagenesis).

Dunham (1962) has a somewhat different type of classification (Table 6.3A) that stresses the relative abundance of allochems and micrite but does not consider the identity of different kinds of carbonate grains. Dunham's classification is based solely upon depositional texture and considers two aspects of texture: (1) grain packing and the relative abundance of grains to micrite and (2) depositional binding of grains. Depositional binding means whether or not carbonate grains show evidence of having been bound together at the time of deposition, as in a colonial reef complex, a stromatolite (cyanobacteria) bed, or calcareous algae mat. Dunham's classification separates components that were not bound together at the time of deposition into those that lack lime mud and those that contain lime mud. Rocks that contain no mud are obviously grain-supported. Rocks that contain mud may be either grain-supported or mud-supported. Note, however, that grain support does not depend upon the absolute grain-to-mud ratio, because grain support is also a function of shapes of the carbonate grains. Platy or elongate grains such as bivalve shells may form a grain-supported fabric at much lower grain abundances than more spherical particles such as ooids. Therefore, Dunham's boundary between grain-supported and mud-supported limestones is not based on a fixed grain/micrite ratio. Dunham's classification was modified, with addition of two new names (floatstone, rudstone), by Embry and Klovan (1972) to better reflect the presence of gravel-size (>2 mm) carbonate grains (Table 6.3B). These authors also divided Dunham's boundstone into three types (framestone, bindstone, bafflestone) on the basis of the presumed kinds of organisms that bound the sediment together. Wright (1992) offers a classification with names derived mainly from the Embry-Klovan classification but which also takes into consideration features resulting from diagenetic processes.

Because Dunham's classification does not consider the identity of the carbonate grains, it may be desirable to use it in conjunction with another classification such as Folk's. Thus, a limestone identified as a packed oomicrite using Folk's classification could alternatively be called an **oomicrite packstone** using a combination of Folk's and Dunham's classifications. Additional limestone classifications are discussed in the symposium volume edited by Ham (1962), and classifications for mixed carbonate and siliciclastic sediment are discussed by Mount (1985) and Zuffa (1980).

The terms coquina, chalk, and marl are commonly used as informal names for carbonate rocks. A **coquina** is a mechanically sorted and abraded, poorly consolidated carbonate sediment consisting predominantly of fossil debris; **coquinite** is the consolidated equivalent. **Chalk** is soft, earthy, fine-textured limestone composed mainly of the calcite tests of floating microorganisms, such as foraminifers. **Marl** is an old, rather imprecise, term for an earthy, loosely consolidated mixture of siliciclastic clay and calcium carbonate.

6.7 ORIGIN OF CARBONATE ROCKS

The chemical weathering processes discussed in Chapter 1 release chemical ions from source rocks that eventually make their way, dissolved in groundwater and surface water, to lakes and the ocean. Bicarbonate ions (HCO_3^-) may be added

Table 6.3 Classification of limestones according to depositional textures

A

DEPOSITIONAL TEXTURE RECOGNIZABLE				DEPOSITIONAL TEXTURE NOT RECOGNIZABLE
Original components not bound together during deposition			Original components were bound together during deposition . . . as shown by intergrown skeletal matter, lamination contrary to gravity, or sediment-floored cavities that are roofed over by organic or questionably organic matter and are too large to be interstices.	CRYSTALLINE CARBONATE
Contains mud (particles of clay and fine silt size)		Lacks mud and is grain-supported		(Subdivide according to classifications designed to bear on physical texture or diagenesis.)
Mud-supported	Grain-supported			
Less than 10% grains / More than 10% grains				
MUDSTONE / WACKESTONE	PACKSTONE	GRAINSTONE	BOUNDSTONE	

B

ALLOCHTHONOUS LIMESTONE — Original components not organically bound during deposition					AUTOCHTHONOUS LIMESTONE — Original components organically bound during deposition		
Less than 10% >2 mm components				Greater than 10% >2 mm components	By organisms that build a rigid framework	By organisms that encrust and bind	By organisms that act as baffles
Contains lime mud (<0.03 mm)			No lime mud			BOUNDSTONE	
Mud-supported		Grain-supported	Grain-supported	Matrix-supported			
Less than 10% grains >0.03 mm <2 mm	Greater than 10% grains				>2 mm component-supported		
MUD-STONE	WACKE-STONE	PACK-STONE	GRAIN-STONE	FLOAT-STONE / RUD-STONE	FRAME-STONE	BIND-STONE	BAFFLE-STONE

Source: A, After Dunham, R. J., 1962, Classification of carbonate rocks according to depositional textures, in Ham, W. E., ed., Classification of carbonate rocks: *Am. Assoc. Petroleum Geologists Mem.* 1. Table 1, p. 117, reprinted by permission of AAPG, Tulsa, Okla. B, after Dunham, R. J., 1962, as modified by Embry, E. F., III and J. E. Klovan, 1972, Absolute water depth limits of late Devonian paleoecological zones: Geol. Rundschau, v. 61. Fig. 5, p. 676, reprinted by permission.

also by interaction of water with atmospheric and soil CO_2. Most dissolved ions end up in the ocean, where they remain dissolved in seawater for periods ranging from hundreds to millions of years. The average time that a particular chemical element remains in solution in the ocean before precipitating is called its **residence time**. Calcium ions (Ca^{2+}) have short residence times ($\sim$1,000,000 years) and carbonate ions ($CaCO_3^{2-}$) have even shorter residence times ($\sim$110,000 years).

The importance of residence times is indicated by comparing the abundance of calcium and carbonate ions in river water to that of ocean water. Bicarbonate ions (HCO_3^-) and carbonate ions (CO_3^{2-}) together make up almost 49 percent of the total dissolved solids in average river water but less than 1 percent of the dissolved solids in ocean water (Livingston, 1963; Mason, 1966). Calcium ions (Ca^{2+}) make up about 12 percent of dissolved solids in river water but only about 1 percent of the dissolved solids in ocean water. The explanation for these differences in relative abundance of carbonate, bicarbonate, and calcium ions in river water and ocean water is provided by examining the relative abundance of various kinds of nonsiliciclastic sedimentary rocks preserved in the geologic record. Measurements of the thickness and volume of different types of sedimentary rocks show that the chemically and biochemically deposited sedimentary rocks make up nearly one-fourth of all sedimentary rocks. The majority of these rocks are carbonate rocks, composed of calcium carbonate and calcium-magnesium carbonate minerals. Carbonate ions, bicarbonate ions, and calcium ions have been preferentially removed from the oceans throughout geologic time to form carbonate rocks, thus accounting for the low relative abundance of these ions in the modern ocean and for their short residence times. In this section, we explore the chemical/biochemical processes that control the removal of dissolved ions from the ocean to form carbonate rocks.

Limestones

Chemistry of Calcium Carbonate Deposition

The dissolution and precipitation of the calcium carbonate ($CaCO_3$) minerals calcite and aragonite are controlled chiefly by pH, which in turn is largely controlled by the partial pressure of dissolved carbon dioxide in water. When carbon dioxide is dissolved in water, the following reactions can occur:

$$CO_2 + H_2O \longleftrightarrow H_2CO_3 \quad \text{(carbonic acid)} \tag{6.1}$$

$$H_2CO_3 \longleftrightarrow H^+ + HCO_3^- \text{ (bicarbonate ion)} \tag{6.2}$$

$$HCO_3^- \longleftrightarrow H^+ + CO_3^{2-} \text{ (carbonate ion)} \tag{6.3}$$

These reactions show that the dissociation of carbonic acid to hydrogen ions and bicarbonate ions (Equation 6.2) and the further dissociation of bicarbonate ions to hydrogen ions and carbonate ions (Equation 6.3) release free hydrogen ions, thus lowering the pH of the solution. The reaction shown in Equation 6.3 suggests that adding CO_2 to water increases the amount of CO_3^{2-} in the water. The reaction in Equation 6.2 generates H^+ ions at a rate so much faster than the reaction indicated by Equation 6.3 generates carbonate ions, however, that adding CO_2 actually causes dissolution of carbonate ions by lowering pH (increasing acidity).

If calcite or aragonite crystals are allowed to react with a carbonic acid solution, these minerals are readily dissolved. This reaction can be summarized as

$$H_2O + CO_2 + CaCO_3 \longleftrightarrow Ca^{2+} + 2\,HCO_3^- \tag{6.4}$$
$$\text{(calcite or}$$
$$\text{aragonite)}$$

Note by the presence of double arrows that this reaction is reversible. If equilibrium conditions are disturbed by loss of carbon dioxide, the concentration of hydrogen ions decreases and the pH increases. The reaction shifts toward the left, resulting in precipitation of solid $CaCO_3$. The partial pressure of carbon dioxide thus clearly exerts a major control on calcium carbonate precipitation. Anything that causes loss of carbon dioxide (Table 6.4) should theoretically trigger the onset of precipitation because dissolved carbon dioxide increases the acidity of water by releasing H^+ ions, as mentioned, although we shall see subsequently that inorganic precipitation of calcium carbonate caused by loss of CO_2 may not be as important under natural conditions in the open ocean as suggested by Equation 6.4.

Two principal inorganic mechanisms that can cause loss of carbon dioxide from water are (1) increase in temperature and (2) decrease in water pressure. An increase in temperature causes a decrease in the solubility of carbon dioxide (and other gases) in water; that is, an increase in temperature reduces the capacity of water to dissolve and retain carbon dioxide, resulting in escape of carbon dioxide. Decrease in water pressure can also allow carbon dioxide to escape. Under natural conditions, pressure may be lowered and more CO_2 absorbed by the atmosphere, by wave agitation caused by storm activity or breaking of waves in the surf zone or over shallow banks. Circulation of deep, pressurized waters to the surface can likewise release carbon dioxide, and even lowering of atmospheric pressure may cause slight loss of carbon dioxide from ocean water.

In addition to its effect on CO_2 solubility, increase in temperature causes a decrease in the solubility of calcium carbonate minerals; that is, the calcium carbonate solubility product decreases with increasing temperature. Decrease in solubility means that a mineral will be more likely to precipitate under a given set of conditions. Thus, calcium carbonate deposition is favored in the more tropical areas of the ocean, where surface water temperatures may reach almost 30°C, compared to about 0°C in the polar regions.

The solubility of chemical constituents is affected also by salinity and the ionic strength of water (Table 6.4). Ionic strength is a function of the concentration of ions in solution and the charges on these ions; thus, ionic strength increases as salinity increases. The solubility of calcium carbonate minerals is markedly decreased at lower values of salinity because decrease in ionic strength decreases the concentration of foreign ions (e.g., Mg^{2+}) other than Ca^{2+} and CO_3^-. Foreign ions interfere with the formation of the calcium carbonate crystal structure, making it more difficult for calcite or aragonite minerals to grow and precipitate. Therefore, the solubility of calcium carbonate is several orders of magnitude lower in fresh water than in seawater (e.g., Degens, 1965), meaning that calcium carbonate precipitates more readily in fresh water than in seawater. On the other hand, the influence of salinity on the solubility of calcium carbonate in the surface waters of the open ocean may be slight because these waters range in salinities only from about thirty-two to thirty-six parts per thousand (o/oo).

This elementary discussion of carbonate solubility relationships is intended only to provide a very basic understanding of carbonate solubility and the factors

Table 6.4   Principal factors that affect inorganic precipitation of $CaCO_3$ in seawater or freshwater

Water condition	Direction of change	Directed effect	Effect on $CaCO_3$ solubility	Kind of $CaCO_3$ precipitated
Temperature	Increase	Loss of CO_2, increase in pH	Decrease*	Micrite or ooids
Pressure	Decrease	Loss of CO_2, increase in pH	Decrease	Micrite or ooids
Salinity	Decrease	Decrease in activity of "foreign" cations	Decrease	Micrite or ooids

*Decrease in $CaCO_3$ solubility = increase in tendency to precipitate.

that govern the precipitation of carbonate minerals. See Morse and Mackenzie (1990) for a more rigorous discussion of carbonate geochemistry.

Importance of Inorganic Precipitation

According to the theoretical considerations illustrated in Equation 6.4, significant loss of carbon dioxide by any mechanism should lead to precipitation of calcium carbonate minerals. Does loss of carbon dioxide cause precipitation of calcium carbonate on an important scale in the open ocean environment today? Available data show that near-surface water in the modern ocean is oversaturated by more than six times with respect to calcite and more than four times with respect to aragonite (Morse and Mackenzie, 1990, p. 217). Such gross oversaturation indicates a reluctance of calcium carbonate minerals to precipitate. Why? There appear to be at least two reasons why calcium carbonate minerals may not precipitate in the modern ocean as readily as suggested by the reaction in Equation 6.4.

First, the magnitude of the pH changes that occur in the open ocean owing to loss of carbon dioxide is relatively small because seawater is a well-buffered solution. Buffering occurs because a considerable portion of the carbon dioxide dissolved in seawater forms undissociated H_2CO_3 rather than dissociating to H^+ ions, HCO_3^- ions, and CO_3^{2-} ions as predicted by Equations 6.2 and 6.3. This buffering reaction is caused by the high alkalinity of ocean water; that is, the high concentrations of bicarbonate and carbonate ions already present in surface waters of the ocean inhibit breakdown of H_2CO_3 to form still more of these ions. Therefore, the actual change in pH in seawater caused by either gain or loss of carbon dioxide is comparatively small, and the pH values of seawater in the open ocean rarely fall outside the range of 7.8 to 8.3 (Bathurst, 1975).

Second, the presence of Mg^{2+} ions at the concentrations found in seawater has been shown experimentally to strongly inhibit the precipitation of calcite ($CaCO_3$). Experiments by Berner (1975) show that Mg^{2+} is readily adsorbed onto the surface of calcite crystals and incorporated into their crystal structure. This nonequilibrium incorporation of Mg^{2+} into growing calcite crystals was interpreted by Berner as decreasing their stability, resulting in an increase in calcite solubility. Thus, calcite crystals do not readily nucleate and grow in the presence of Mg^{2+} in seawater concentrations. See also Mucci and Morse (1983). Aragonite is also composed of $CaCO_3$ but has a different crystal structure (orthorhombic) from that of calcite (rhombohedral). Mg^{2+} ions appear to be less prone to sorb to aragonite nuclei and disrupt crystal growth. Therefore, aragonite is less affected by Mg^{2+} and has a tendency in the presence of Mg^{2+} to precipitate in preference to calcite. Nonetheless, aragonite does not precipitate completely freely in ocean water, even in surface waters supersaturated with respect to calcium carbonate, possibly owing to formation of thin organophosphatic coatings on aragonite seed nuclei that inhibit their growth (Berner et al., 1978).

Although calcite does not precipitate freely in the modern ocean owing to the presence of abundant Mg^{2+} ions, accumulating evidence suggests that calcite was precipitated in preference to aragonite at times in the geologic past (referred to as "calcite seas") when the concentration of Mg^{2+} ions in the ocean was low (Sandberg, 1983; Stanley and Hardie, 1999). Stanley and Hardie link these times of calcite precipitation to high rates of seafloor spreading, which increases removal of Mg^{2+} from seawater by absorption into hot seafloor basalts. Thus, skeletal and nonskeletal carbonates deposited during early Cambrian–middle Mississippian and middle Jurassic–late Tertiary were dominantly low-magnesium calcite, whereas those deposited during middle Mississippian–middle Jurassic and late Tertiary–Quaternary were dominantly aragonite and high-magnesium calcite (Fig. 6.9). Seas that preferentially precipitate aragonite are referred to as "aragonite seas" (Stanley and Hardie, 1999).

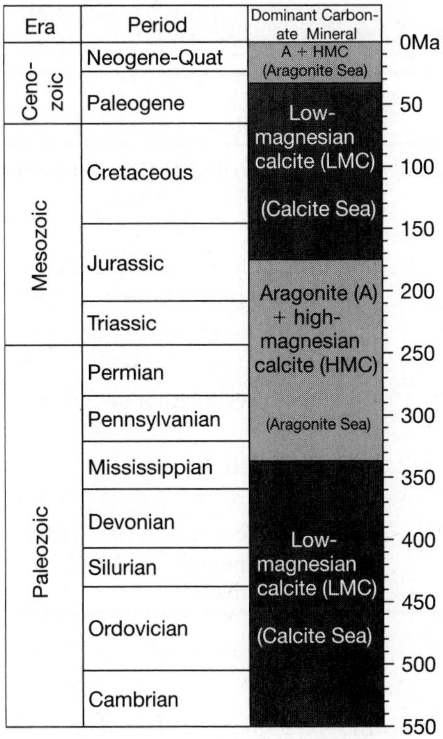

Era	Period	Dominant Carbonate Mineral	Ma
Cenozoic	Neogene-Quat	A + HMC (Aragonite Sea)	0Ma
Cenozoic	Paleogene	Low-magnesian calcite (LMC) (Calcite Sea)	50
Mesozoic	Cretaceous		100
Mesozoic			150
Mesozoic	Jurassic	Aragonite (A) + high-magnesian calcite (HMC) (Aragonite Sea)	200
Mesozoic	Triassic		250
Paleozoic	Permian		
Paleozoic	Pennsylvanian		300
Paleozoic	Mississippian		350
Paleozoic	Devonian	Low-magnesian calcite (LMC) (Calcite Sea)	400
Paleozoic	Silurian		
Paleozoic	Ordovician		450
Paleozoic			500
Paleozoic	Cambrian		550

Figure 6.9
The most favorable times for precipitating aragonite + high-magnesian calcite skeletal and nonskeletal carbonates, and low-magnesian calcite skeletal and nonskeletal carbonates in the Phanerozoic (post-Precambrian) ocean. The ocean during times of dominantly aragonite precipitation is referred to as an "aragonite sea" and during times of dominantly low-magnesian calcite precipitation as a "calcite sea." [Based on Stanley and Hardie, 1999, and Sandberg, 1983.]

The generation of ooids provides an example of what appears to be largely inorganic precipitation of $CaCO_3$. Ooids in modern environments consist mainly of aragonite, whereas many ancient ooids may have precipitated as calcite. Ooids form mainly under high-energy, agitated-water conditions in warm waters that are supersaturated with calcium carbonate. Warming and evaporation of cold ocean water driven onto shallow banks by tidal currents result in supersaturation of the water. Currents and waves keep the grains moving and intermittently suspended, allowing more or less even precipitation of calcium carbonate on all sides of the grains. Both supersaturation of the water and intermittent burial and resuspension of the ooids owing to agitation appear to be necessary for most ooids to form, although some ooids are known to form in quiet water. Cyanobacteria or other microorganisms may influence the formation of ooids—possibly by trapping carbonate grains on organic films or by mediating carbonate precipitation through removal of CO_2. The quantitative importance of organic influences on the formation of ooids is not well understood (Tucker and Wright, 1990, p. 6).

The Role of Organisms in Precipitation of Calcium Carbonate

Precipitation of minerals from water is fundamentally a chemical process; however, chemical processes can be aided in a variety of ways by organisms. Although purely inorganic precipitation of calcium carbonate minerals from normal-salinity seawater or fresh water apparently can occur, it may be less common today than precipitation aided in some way by organic processes (Table 6.5). Furthermore, geologic evidence suggests that organisms may have played a significant role in carbonate sedimentation throughout most of Phanerozoic (post-Precambrian) time, and some organisms (e.g., bacteria) may also have mediated carbonate precipitation during Precambrian time. The various ways that organisms may influence carbonate sedimentation are discussed below.

Direct Extraction of $CaCO_3$ from Water to form Skeletal Elements.   The most important role that organisms play in forming carbonate sediment is probably the direct

Table 6.5   Effects of organic activity on $CaCO_3$ precipitation

Kind of organic activity	Immediate effect	Ultimate effect
Extraction of $CaCO_3$ from seawater or fresh water	Promotes skeletal growth: Shells or tests	Forms silt- to gravel-size allochems upon death of organism
	Internal "stiffeners"	Forms micrite (carbonate mud) upon death of organism
Photosynthesis	Removes CO_2 from water; pH increases	Promotes precipitation of micrite or ooids
Decay of soft tissue	May increase alkalinity; pH increases	Promotes precipitation of $CaCO_3$
Feeding, sediment ingestion	Reshapes sediment	Generates pellets
Bacterial activity	Promotes $CaCO_3$ precipitation	Promotes precipitation of micrite, generates peloids, calcifies microbial mats

removal of dissolved carbonate constituents to build skeletal structures. The exact mechanisms by which organisms remove dissolved substances to build their shells or tests is not well understood, but the process is very common. Marine invertebrates that build protective shells or other skeletal structures of calcium carbonate range from freely drifting, planktonic species such as foraminifers and pteropods (winged marine snails) to bottom-dwelling benthonic organisms such as calcareous algae, corals, molluscs, and echinoderms. They can remove $CaCO_3$ not only from calcium-carbonate-saturated surface waters in tropical regions, but also from less saturated waters in temperate and colder regions. For example, shell sands and gravels are important deposits of the modern seafloor in shallow, cool water at high latitudes (e.g., Farrow, Allen, and Akpan, 1984; James and Clarke, 1997). Some organisms build skeletal materials from (low-magnesian) calcite, whereas others build skeletal material from high-magnesian calcite or aragonite. The importance of biologic removal of calcium carbonate from the oceans is demonstrated by the fact that most Phanerozoic limestones contain some recognizable calcium carbonate fossils, and many are composed dominantly of such remains. Also, large areas of the modern ocean floor are covered by calcareous oozes composed dominantly of the tests of foraminifers, one-celled algal coccolithophores, and pteropods.

Because calcium-carbonate-secreting organisms exist in huge numbers in some parts of the ocean, disintegration of their skeletal remains after death has the potential to supply large quantities of carbonate sediment of various sizes to the ocean floor. Some skeletal material consists of the shells of large invertebrate organisms such as pelecypods, gastropods, brachiopods, and corals. These large shells can become fragmented into sand-size or smaller pieces owing to biogenic activity or physical breakage. The calcareous tests of some invertebrates such as foraminifers and pteropods are sand size, and the tests of nannofossils such as coccoliths are fine-silt size. Some calcareous algae disintegrate to form sand-size carbonate grains (e.g., Hillis, 1991; Hudson, 1985), whereas others yield mud-size grains. For example, some red and green algae, such as some species of *Halimeda*, *Penicillis*, and *Udotea*, have calcareous skeletal elements composed of tiny, needle-like aragonite crystals that are deposited within intercelluar spaces (e.g., Macintyre and Reid, 1995) and act as stiffeners for soft tissue. When these organisms die, bacterial and chemical decomposition of the binding tissue releases the skeletal particles. This decay process yields a fine lime mud (micrite) composed of elongated aragonite crystals, 3–10 mm long, and very small (<1 mm) equant crystals (Macintyre and Reid, 1992, 1995).

Early quantitative studies of the rate of production of aragonite by disintegration of calcareous algae in the Florida Reef Tract (Stockman, Ginsburg, and Shinn, l967) and in the Bahamas (Neumann and Land, 1975) led to the conclusion that much or all of the aragonite mud deposited in these areas in the recent geologic past could have been supplied by skeletal disintegration of calcareous algae. On the other hand, subsequent observations by Shinn et al. (1989), suggest that only 10–20 percent of the carbonate mud in the Bahamas is algal carbonate, a suggestion supported by observations on crystal shape by Macintyre and Reid (1992) and the biochemical studies of Robbin and Blackwelder (1992). Nonetheless, many workers agree that disintegration of the skeletal elements of calcareous algae into fine detritus constitutes a major process for forming carbonate sediments in lagoons, reefs, and fore-reef slopes (e.g., Hillis, 1991; Hudson, 1985; Multer, 1988). Unfortunately, it does not appear possible at this time to provide quantitative estimates of the relative importance, in the modern ocean as a whole, of algal disintegration vs. precipitation of $CaCO_3$ owing to CO_2 loss. Such estimates become even more tenuous when we consider ancient carbonate rocks in the stratigraphic record.

Removal of CO_2 from Water by Photosynthesis.   Another type of organic activity that may be important to the formation of carbonate rocks is removal of carbon dioxide from water by photosynthesizing plants. As mentioned, any process that removes carbon dioxide from the water facilitates carbonate precipitation by increasing the pH. Aquatic plants remove carbon dioxide from water during photosynthesis as shown by the following relationship:

$$6H_2O + 6CO_2 \longrightarrow C_6H_{12}O_6 + 6O_2$$

$$(\text{water} + \text{carbon dioxide} \longrightarrow \text{carbohydrates} + \text{oxygen}) \qquad (6.5)$$

Blue-green algae (cyanobacteria), photosynthesizing bacteria, and small phytoplankton such as diatoms, dinoflagellates, and coccoliths are the most important users of carbon dioxide in the marine realm. The activities of photosynthesizing organisms are at a peak in sunlight and at a minimum in the dark; therefore, the carbon dioxide content of water in which active photosynthesis is taking place can vary measurably from day to night. Removal of CO_2 by organisms thus decreases the acidity of the water (increases pH).

Bacterial Mediation of Precipitation.   Bacteria may play an indirect role in precipitation of some carbonate sediment. For example, Chafetz (1986) suggests that some marine peloids originated as a fine-grained precipitate of high-magnesian calcite within and around active clumps of bacteria. Bacteria may also promote precipitation of calcium carbonate on dead cyanobacteria, leading to lithification of microbial mats to stromatolites (Buczynski and Chafetz, 1993; Chafetz, 1994). Microbial-mediated calcium carbonate precipitation is related to photosynthesis and ion transport through cell walls. Calcification occurs just outside cell walls in an alkaline microenvironment, which is generated as Ca^{2+} is exported from the cell in exchange for uptake of $2H^+$. Calcification results from either uptake of CO_2 (microalgae) or HCO_3^- (cyanobacteria). Excess inorganic carbon within the cell wall can be absorbed from the cell into the microalkaline environment, providing an additional source of carbon for calcification (Yates and Robbins, 2001). The overall importance of bacterial mediation of carbonate production throughout geologic time is not known; however, some microbiologists (e.g., Castainer, Le Métayer-Levrel, and Perthuisot, 1997; 1999) suggest that it may have been significant (see also Camoin, 1999).

Decay of Dead Organisms.   Decay of dead organisms also affects pH. Decay can release various organic acids and carbon dioxide to the water, causing acidity to increase (pH decreases). On the other hand, some decay products can be alkaline (pH increases). Alkalinity may be increased because organic matter is degraded through sulfate reduction by bacteria (e.g., Bernasconi, 1994). Increase in alkalinity favors $CaCO_3$ precipitation.

Generation of Pellets.   As mentioned in Section 6.4, many carbonate peloids are fecal pellets, generated by organisms such as sea cucumbers, mollusks, and worms. These organisms ingest calcium carbonate muds to obtain nutrients and extrude the remains as pellets. This process does not generate new carbonate sediments; it merely reshapes the sediment into a few form.

Relative Importance of Inorganic and Organic Precipitation of Calcium Carbonate

The organic production of sand- and gravel-size skeletal debris has unquestionably made a significant contribution to the overall budget of carbonate sediment throughout Phanerozoic time. The most controversial carbonate deposits, however, are the huge volumes of nonfossiliferous carbonate mud (micrites) present in both the Precambrian and Phanerozoic stratigraphic record. Did these thick successions of carbonate muds originate through inorganic processes or did organisms participate in some way?

An interesting phenomenon that may have a bearing on this question is the formation of **whitings** in such warm-water areas as the Bahamas, the Persian Gulf, and the Dead Sea. The sudden appearance of these whitings, which are milky patches of surface and near-surface water caused by dense concentrations of suspended aragonite crystals, has been suggested to result from spontaneous, large-scale, instantaneous physico-chemical nucleation of aragonite crystals in waters supersaturated with calcium bicarbonate, i.e., inorganic precipitation. This view has been challenged by other workers who propose that mechanisms such as re-suspension of aragonite mud from the shallow seafloor by wave action, turbulent tidal flow, turbulent boundary flow, or stirring up of mud by bottom-feeding fish are responsible for whitings, rather than spontaneous nucleation and precipitation of aragonite. Recent isotopic studies by Shinn, Holmes, and Marot (2000) indicate, however, that whitings are probably not the result of resuspension mechanisms, leaving open the question of exactly how they do form. The weight of opinion appears to be shifting toward microbially mediated precipitation by photosynthesizing microalgae or cyanobacteria as the likely origin of whitings (e.g., Robbins, Tao, and Evans, 1997; Yates and Robbins, 2001).

Does this mean that most lime mud precipitation was organically mediated? Let's consider carbonate deposition during the Precambrian. Judging from the abundance of calcareous skeletal fragments and whole fossils in Phanerozoic limestones, the removal of calcium carbonate from seawater owing to some aspect of organic activity may have been an important mechanism for forming carbonate sediments since at least early Paleozoic time—although the relative importance of biotic and abiotic precipitation of carbonates may have varied throughout this time. We have a more difficult time explaining the formation of Precambrian limestones. The Precambrian record contains impressive thicknesses of carbonate rocks (e.g., as much as 1000 m in Glacier National Park, Montana and Alberta) that, as far as we know, were deposited before the widespread appearance of calcium-carbonate-secreting organisms. Thus, it does not seem likely, on the basis of available evidence, that shelled organisms were directly responsible for deposition of large volumes of Precambrian limestone. Few, if any, Precambrian organisms could extract $CaCO_3$ to build skeletal elements. Blue-green algae (cyanobacteria) and other photosynthesizing bacteria may have played an indirect role in the precipitation of calcium carbonate through photosynthetic removal of carbon dioxide and by trapping and binding of fine carbonate sediment to form stromatolites. Cyanobacteria appear to have been particularly abundant in Precambrian time, possibly owing to fewer numbers of grazing organisms that fed on the algal mats.

The problem of Precambrian carbonate deposition is considered in detail in a recent monograph entitled "Carbonate sedimentation and diagenesis in the evolving Precambrian world," edited by Grotzinger and James (2000a). Early Precambrian (Archean) carbonate deposition was particularly characterized by precipitation of aragonite and high-magnesian calcite directly onto the seafloor as encrustations of both inorganic and microbial origin. Carbonate facies include large (up to meter-scale), upward-divergent "crystal fans" of calcite and dolomite, which replace original aragonite and high-magnesian calcite. Other facies include carbonate muds, stromatolites, and ooid-intraclast grainstones. Grotzinger and James (2000b) suggest that the common precipitation of aragonite and calcite directly on the seafloor in Archean time took place because the Precambrian surface seawater was substantially oversaturated with respect to calcium carbonate, well above the factor of 2–5 that is typical of the oceans today. Abiotic (inorganic) precipitation appears to have tapered off in later Precambrian (Proterozoic) time, as microbially mediated precipitation increased, suggesting a gradual depletion of the highly oversaturated Archean seawater.

Physical Processes in Carbonate Deposition

Calcium carbonate fossils, skeletal fragments, ooids, and other carbonate grains are subject to the same physical transport processes in the ocean as terrigenous grains. Thus, ultimate deposition of most limestones occurs through fluid-flow and sediment gravity-flow processes. Limestones may, therefore, display many of the same bedding characteristics and sedimentary structures as terrigenous sedimentary rocks.

Depth Control of Calcium Carbonate Production

When precipitation of a mineral phase just equals dissolution, the solution is in equilibrium with the solid and is said to be **saturated** with this mineral phase. A solution that precipitates a mineral is **supersaturated**, and a solution that dissolves the mineral is **undersaturated**. Much of the warm surface water of the modern ocean is supersaturated with calcium carbonate. Little inorganic $CaCO_3$ may actually be precipitating in these waters, however, owing to Mg inhibition or other factors discussed. This condition of supersaturation changes rapidly with depth. The degree of calcium carbonate saturation drops off abruptly in waters below the surface layer; at depths greater than a few hundred meters, seawater is undersaturated. The saturation of surface waters likewise decreases in colder waters of high latitudes.

The undersaturated state of deeper waters is a function of several factors, although increase in carbon dioxide partial pressure is one of the most important variables. In shallower water, CO_2 production is increased closer to the ocean floor by the respiration of benthonic organisms. Oxidation of organic matter on the seafloor in both shallow and deeper water also increases CO_2 production. Furthermore, colder water found at depth can contain more dissolved CO_2 than warmer surface waters. Both decrease in temperature and increase in hydrostatic pressure with depth cause an increase in the solubility of calcium carbonate and thus the corrosiveness of seawater.

Because of decreasing calcium carbonate saturation of seawater with depth, calcium carbonate production is confined mainly to the very shallow water areas of the ocean and to the supersaturated surface waters of the deeper ocean. These are the waters in which most calcium-carbonate-secreting organisms live. Calcium carbonate dissolution prevails in the deeper, undersaturated waters. The rate of dissolution does not, however, increase in a linear fashion with depth. Experiments in which calcite spheres were suspended on moorings at different depths in the ocean have demonstrated only slight corrosion of the spheres above a depth of about 3500 m, but solution of the spheres increased abruptly at that depth (Peterson, 1966). Effective solution of calcium carbonate thus occurs only at relatively great depths in the ocean (and to some extent in very cold, high-latitude surface waters). The particular depth at any locality at which the rate of dissolution of calcium carbonate equals the rate of supply of calcium carbonate to the seafloor, so that no net accumulation of carbonate takes place, is called the **calcium carbonate compensation depth (CCD)**. The position of the calcium carbonate compensation depth has been compared to the snowline of mountain ranges. Where biogenic oozes are accumulating in the modern ocean, white carbonate oozes cover elevated areas of the seafloor above the CCD but give way to brown or gray pelagic clays or siliceous oozes below. In different parts of the modern ocean, the CCD ranges in depth from about 3500 to 5500 m, because of differences in rates of production of $CaCO_3$ in surface waters and variations in the factors that control carbonate saturation. The average depth of the CCD in today's ocean is about 4500 m. (The average depth of the modern ocean is about 3800 m; depth ranges to slightly more than 11,000 m.)

Dolomite

General Statement

Dolomites are calcium carbonate rocks composed of more than 50 percent of the mineral dolomite [$CaMg(CO_3)_2$]. To differentiate the rock from the mineral, dolomites are sometimes referred to as **dolostones** or dolomite rock. They are abundant and widely distributed in the geologic record, ranging in age from Precambrian to Holocene, although the greatest volumes of dolomites are Paleozoic and older. Dolomites occur in close association with limestones and in many stratigraphic units as interbeds in the limestones; they are also commonly associated with evaporites.

Because dolomites recur so frequently in the stratigraphic record, they must have formed under environmental conditions that were relatively common and that were repeated again and again in various localities. Dolomites have been studied very extensively; therefore, in theory, we ought to understand their origin quite well. On the contrary, the origin of dolomites remains one of the most thoroughly researched but poorly understood problems in sedimentary geology. Although it is clear from the presence of relict limestone textures and structures that many coarsely crystalline dolomites are secondary rocks, formed by diagenetic replacement of older limestones, many fine-crystalline dolomites lack such textural evidence of replacement and cannot be proven to have originated by diagenetic alteration of limestones. It is these fine-crystalline dolomites that have created the so-called **dolomite problem**, which geologists have not been able to satisfactorily solve since dolomites were first recognized by the French naturalist Deodat de Dolomieu more than 200 years ago in 1791 (Zenger, Bourrouilh-Le Jan, and Carozzi, 1994; see also review by Warren, 2000).

The dolomite problem arises from the fact that scientists have not yet been successful in the laboratory in precipitating perfectly ordered dolomite at the normal temperatures (~25°C) and pressure (~1 atm) that occur at Earth's surface. Perfectly ordered dolomite has 50 percent of the cation sites filled by Mg and 50 percent filled by Ca (stoichiometric dolomite). Elevated temperatures, exceeding 60°C, are required to produce stoichiometric dolomite in the laboratory (e.g., Usdowski, 1994). In laboratory experiments carried out at the normal temperatures found in natural environments, only a dolomite-like material called **protodolomite** forms. Protodolomite contains excess $CaCO_3$ in its structure and is not a true (stoichiometric) dolomite; see also discussion by Lumsden and Lloyd (1997). Thus, geochemists have been unable to determine directly from low-temperature experimental work what geochemical conditions favor the precipitation of dolomite in natural environments. Nonetheless, geologic evidence suggests that dolomite does form naturally at, or near, near-surface temperatures.

Since the mid-1940s, minor modern dolomite sediments have been reported from numerous localities, including some in Russia, South Australia, the Persian Gulf, the Bahamas, Bonaire Island off the Venezuela mainland, the Florida Keys, the Canary Islands, and the Netherlands Antilles. Ages of these dolomites are estimated by radiocarbon methods to range from a few years to about 4000 years. Most are not perfectly ordered dolomites; mole percent $MgCO_3$ ranges from about 30 to 50 percent but falls mainly between 40 and 46 percent. Discovery of dolomite in modern environments was initially hailed as evidence by some workers that dolomite can be precipitated naturally as a primary deposit. Others suggested that these modern dolomites formed by replacement, that is, rapid alteration of an initial precipitate of $CaCO_3$ to dolomite—a process called **dolomitization**. Subsequent research has failed to establish unequivocally the relative importance of dolomite precipitation versus dolomite replacement in the origin of these early-formed, or **penecontemporaneous**, dolomites. Both mechanisms remain viable.

Early-formed dolomites include all those formed at or near the surface in the unconsolidated state as opposed to diagenetic dolomites that formed during burial and uplift by replacement of older, consolidated limestones. The volume of dolomite in the modern environment is small. Thus, an additional aspect of the dolomite problem has to do with the following question: Can the processes responsible for generation of early-formed dolomites account for the vast dolomite deposits of the past? That question is still being debated, and you will not find the answer here; however, it may be useful to examine some of the conditions that appear to favor the early formation of dolomite in modern environments. By extension, we may be able to gain some insight into the formation of ancient dolomite deposits.

Requirements for Dolomite Formation

The chemical reactions of interest with respect to formation of dolomite are as follows:

$$Ca^{2+}(aq) + Mg^{2+}(aq) + 2CO_3^{2-}(aq) = CaMg(CO_3)_2 \text{ (solid)} \qquad (6.6)$$

$$2CaCO_3 \text{ (solid)} + Mg^{2+}(aq) = CaMg(CO_3)_2 \text{ (solid)} + Ca^{2+} \text{ (aq)} \qquad (6.7)$$

Equation 6.6 illustrates the direct precipitation of dolomite from aqueous solution; Equation 6.7 illustrates replacement of calcite or aragonite by dolomite. As suggested at the beginning of this discussion, the problem with the reaction shown in Equation 6.6 is that this reaction requires temperatures far in excess of normal surface temperatures. The reasons why such high temperatures are necessary are far from well understood, but the problem is certainly related to kinetics (reaction rates). For example, it has been pointed out by a number of workers, such as Gains (1980), that the Mg^{2+} ion is strongly bound by water (hydrated) in solution and must be separated from the attached water before it can be incorporated into the solid dolomite crystal lattice. At low temperatures, Ca^{2+} ions, which are much less strongly bound by water, are more likely to enter the lattice and form $CaCO_3$ minerals. At elevated temperatures, Mg^{2+} ions are less strongly hydrated and thus more easily desolvated, allowing the naked Mg^{2+} ion to enter into the crystal lattice to form dolomite. The highly ordered state of dolomite also creates a kinetics problem at low temperatures. The nucleation and growth of the highly ordered dolomite lattice in a solution saturated in calcium bicarbonate are so slow that, in competition for calcium ions and carbonate ions, well-ordered dolomite is prevented from forming, and minerals such as aragonite or cation-disordered, high-magnesium calcites form instead. For additional discussion of the kinetics of dolomite formation, see Machel and Mountjoy (1986).

Models for Early-Formed Dolomite

As mentioned, the relative importance of dolomite precipitation vs. dolomitization (replacement) is not known; nonetheless, many geologists clearly believe that dolomitization was an important process in forming ancient dolomites. Much of the recent and current research on dolomites has thus focused on attempts to understand the mechanisms of dolomitization. Theoretical considerations suggest that dolomite formation is favored kinetically by high Mg^{2+}/Ca^{2+} ratios, low Ca^{2+}/CO_3^{2-} ratios, and low salinity (Machel and Mountjoy, 1986). It is favored also by higher temperatures, as mentioned. In fact, at temperatures exceeding about 100°C, most kinetic inhibitors, such as Mg^{2+} hydration, become ineffective.

By examining the various conditions under which dolomites in modern environments are forming, three principal models that meet, in one way or another, the conditions favorable for dolomite formation, particularly dolomitization have been proposed: (1) the hypersaline (sabkha, evaporation, reflux) model, (2) the mixed-water (mixing-zone) model, and (3) the seawater (shallow-subtidal) model (Fig. 6.10). Models 1 and 3 invoke seawater, concentrated by evaporation in the case of Model 1, as the dolomitizing fluid. Model 2 requires mixing of seawater and fresh (meteoric) water.

Hypersaline Model.   Many known occurrences of modern or Holocene dolomite are in hypersaline environments such as the sabkhas (coastal plains characterized by the presence of evaporites) of the Persian Gulf and the supratidal zones of arid climates (e.g., Fig. 6.11). Under strongly evaporative conditions, where rates of evaporation exceed rates of precipitation, seawater beneath the sediment surface becomes concentrated by evaporation. This concentration process leads to precipitation of aragonite and gypsum, which preferentially removes Ca^{2+} from the water and increases the Mg/Ca ratio. The Mg/Ca ratio in normal seawater is about 5:1. When this ratio rises to sufficiently high levels, possibly in excess of

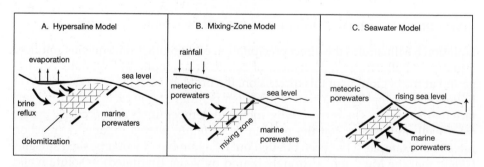

Figure 6.10
Dolomite models. Schematic representation of various conditions under which dolomitization may occur: A. Intense evaporation and brine reflux. B. Mixing of meteoric and marine porewaters. C. Pumping or flushing normal seawater through carbonate sediments. Arrows indicate the principal direction of fluid flow.

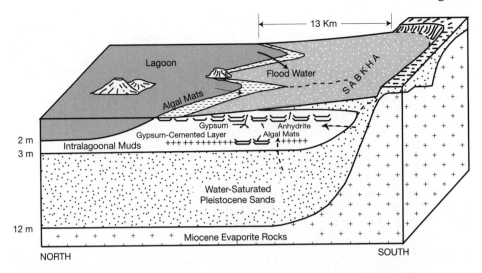

Figure 6.11
Abu Dhabi sabkha, Persian Gulf. Schematic representation of a typical sabkha environment in which penecontemporaneous dolomite forms, commonly in association with gypsum and anhydrite, owing to evaporitic concentration of Mg. [After Butler, G. P., 1969, Modern evaporite deposition and geochemistry of coexisting brines, the sabkha, Trucial coast, Arabian Gulf: Jour. Sed. Petrology, v. 39, Fig. 2, p. 72, reprinted by permission.]

10:1, dolomite is believed to form. One mechanism by which brines are concentrated involves evaporation of capillary water in the sediments of the sabkhas. Upward flow of water from the saturated groundwater zone replaces the water lost by capillary evaporation, a process called **evaporative pumping**. Brines may also be concentrated in surface ponds or bays by surface evaporation of water. These concentrated brines have higher density than that of normal seawater, causing them to sink downward. Flushing of large volumes of Mg-rich brine downward through calcium carbonate sediment can putatively bring about dolomitization, a process referred to as **seepage refluxion** (Fig. 6.10A).

The overall volume of dolomite that forms in sabkha environments is believed to be relatively small, and there is still considerable controversy with regard to the exact mechanism by which the dolomite forms. It is not definitely known if it forms by replacement of aragonite or high-magnesian calcite (dolomitization) or if it forms as a primary precipitate of disordered protodolomite, which presumably later develops better ordering to become true dolomite. See Hardie (1987) and Purser, Tucker, and Zenger (1994a) for additional discussion of this topic.

Mixing-Zone Model.   Several studies published since the early 1970s (e.g., Hanshaw, Back, and Deike, 1971; Badiozamani, 1973; Folk and Land, 1975) have suggested that brackish groundwaters produced by mixing of seawater with meteoric water could be saturated with respect to dolomite at Mg^{2+}/Ca^{2+} ratios much lower than those required under hypersaline conditions. Mixing of fresh water and saline water in environments such as the subsurface zones of coastal areas where meteoric waters come in contact with seawater (Fig. 6.10B) is suggested to lower salinities sufficiently so that dolomites can form at Mg^{2+}/Ca^{2+} ratios ranging from normal seawater values of about 5:1 to as low as 1:1 (Fig. 6.12). Presumably, dolomite can form at lower Mg^{2+}/Ca^{2+} ratios in these mixed waters compared to seawater because of less competition by other ions in the less saline water. The mixing-zone model, or variations thereof, has been referred to also as the **Dorag model** (Badiozamani, 1973) and the **schizohaline model** (Folk and Land, 1975).

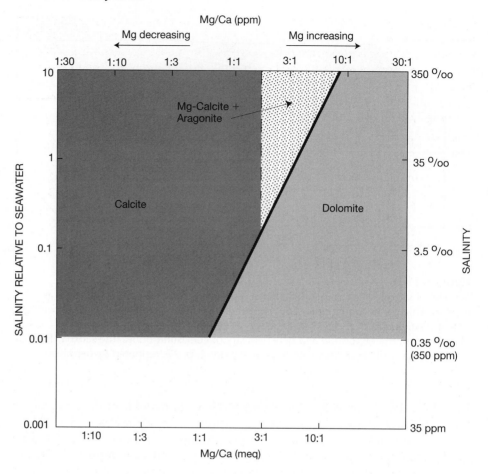

Figure 6.12
The preferred fields of occurrence of dolomite, calcite, magnesian-calcite, and
aragonite plotted as a function of salinity and Ca/Mg
ratios. Note that dolomite
can putatively form at progressively smaller Mg/Ca
ratios with decreasing
salinity owing to slower
crystallization rates and relative scarcity of competing
foreign ions at low salinities. [After Folk, R. L., and L.
S. Land, 1975, Mg/Ca ratio
and salinity: Two controls
over crystallization of
dolomite: Am. Assoc. Petroleum Geologists Bull., v. 59,
Fig. 1, p. 61, reprinted by
permission.]

Although the mixing-zone dolomite model has attracted many proponents, it has also come under some fairly devastating attacks. For example, Hardie (1987) points out that Badiozamani, in his original (1973) calculations for the model, used the solubility values of less soluble ordered dolomite when he should have used the values of more soluble, less ordered, Ca-rich dolomite (which is the kind of dolomite that actually forms under surface temperatures). Hardie also maintains that there is no actual documentation that dolomite can form at Mg/Ca ratios of 1:1, nor is there hard evidence that demonstrates the special power of low-salinity conditions to produce cation-ordered dolomite. Machel and Mountjoy (1986) further point out that dolomite does not form in most modern freshwater/seawater mixing zones, and where it does form, the volume of dolomite is small. Purser, Tucker, and Zenger (1994b) observe that mixed waters are potentially capable of dolomitization; however, the true significance of the mixing zone may be more in its role of inducing fluid movements in the marine groundwater below.

Seawater (Shallow Subtidal) Model. In the hypersaline model, seawater modified by evaporation processes is required for dolomitization. A few workers have proposed that early dolomitization can also take place in normal, unmodified seawater. According to the concept embodied in this model, dolomitization can occur in normal seawater if a sufficient volume of seawater is forced through the sediment so that each pore volume of water in the sediment is constantly being renewed with new seawater (e.g., Carballo, Land, and Miser, 1987; Land, 1991). Thus, new Mg^{2+} is constantly being supplied while replaced Ca^{2+} ions and other ions that might "poison" the dolomite crystal structure are removed. As an example, Carballo, Land,

and Miser (1987) report an area of Sugarloaf Key, Florida, where seawater is forced upward and downward through Holocene carbonate mud during rise and fall of seawater accompanying spring tides, a process they call **tidal pumping**. Owing to the large volume of seawater driven through the sediment by this mechanism, large quantities of Mg^{2+} are imported into the sediment, and pore fluids are constantly being replaced by new fluids. Under these conditions, dolomite is forming in the sediment even though little or no evaporation of the seawater has occurred. Carballo, Land, and Miser suggest that dolomite forms both by precipitation as a cement and by later replacement of preexisting crystallites. Dolomitization might also occur during a sea level rise (Fig. 6.10C) as marine porewaters move landward within a platform (e.g., Tucker, 1993). Although the seawater model has some problems, many geologists are apparently convinced that normal seawater has been a major dolomitizing medium in the past (Purser, Tucker, and Zenger, 1994b).

Other Factors Affecting Early Dolomitization

Experimental work on the formation of dolomite at 200°C by Baker and Kastner (1981) demonstrated that the presence of dissolved SO_4^{2-} inhibits the formation of dolomite. Extrapolating their experimental results to lower temperatures, they suggest that the reason for the scarcity of dolomite in open-marine environments is dissolved SO_4^{2-} in seawater. Dissolved SO_4^{2-} ions can allegedly inhibit the dolomitization of calcite at SO_4^{2-} values as low as 5 percent of their seawater value. Thus, according to these authors, any process that removes SO_4^{2-} from seawater (e.g., bacterial reduction of SO_4^{2-}; precipitation of calcium sulfate $[CaSO_4 \cdot 2H_2O]$) favors the formation of dolomite. Subsequent experimental work by Morrow and Abercrombie (1994) confirms that dissolved sulfate at a concentration of 0.005M retards, but does not prevent, dolomitization of calcite at high temperatures. These authors suggest that the observed rates of dolomitization may be due to dissolution of calcite at a more rapid rate in a sulfate-free environment at high temperatures because of its greater degree of undersaturation.

Bacteria may play a role in precipitation of dolomite under some conditions (e.g., Bernasconi, 1994; Gournay, Folk, and Kirkland, 1997; Vasconcelos and McKenzie, 1997; Wright, 2000). For example, Vasconcelos and McKenzie (1997) report precipitation of dolomite at normal earth-surface temperatures in black, organic-rich sediments in a shallow-water coastal lagoon (Lagoa Vermelha) near Rio de Janeiro, Brazil. They attribute precipitation to the activities of sulfate-reducing anaerobic bacteria. Precipitation apparently occurs owing to the release of excess Mg along with other by-products of sulfate reduction. Saturation of Mg on the submicron scale in microenvironments around the cell bodies creates conditions favorable for preferential precipitation of dolomite. The precipitate is a Ca-rich dolomite that undergoes ageing with time to increase ordering. In addition to observations in Lagoa Vermelha, dolomite was produced in the laboratory by using sulfate-reducing bacteria cultured from Lagoa Vermelha (Vasconcelos and McKenzie, 1995; Warthmann et al., 2000; Van Lith et al., 2003).

Changes in Climate and Ocean Chemistry

As discussed, the concentration of Mg^{2+} ions in the ocean was higher during periods of "aragonite seas," when rates of seafloor spreading and sea levels were low, than during periods of "calcite seas" when substantial amounts of Mg^{2+} were being absorbed onto hot seafloor basalts. Thus, dolomite precipitation may have been favored in aragonite seas.

Also, ocean temperature appears to have an effect on dolomite precipitation. During times of rapid seafloor spreading (and high sea level), CO_2 levels in the atmosphere are high owing to increased rates of CO_2 outgassing related to high rates of seafloor spreading. High concentrations of CO_2 in the atmosphere generate a "greenhouse" (hothouse) effect because CO_2 prevents heat loss from Earth and leads to global warming. Lower concentrations are present during low rates of seafloor spreading and lower sea level, producing so-called "icehouse" conditions. Some earlier observers (e.g. Givens and Wilkinson, 1987) reported that dolomite is more common in rocks deposited during greenhouse conditions than during icehouse conditions. Arvidson, Mackenzie, and Guidry (2000) suggest that increased atmospheric temperature results in increased rates of terrestrial weathering of siliciclastic and carbonate rocks and correspondingly increases rates of transfer of dissolved carbon, mainly as bicarbonate (HCO_3^-), to the ocean. This increase in ocean bicarbonate apparently causes greater supersaturation of ocean water with respect to dolomite than to calcite. Thus, dolomite precipitation is favored during warm, greenhouse conditions.

Subsurface (Burial) Dolomite

As mentioned, much dolomite in the geologic record has relict textures that indicate the dolomite was formed by replacement (dolomitization) of a precursor limestone. Such dolomitization appears to have taken place in the subsurface much later (perhaps millions to hundreds of millions of years later) than the time of formation of penecontemporaneous dolomite. For example, Mountjoy and Amthor (1994) report that massive replacement dolomites form 50–90 percent of all Devonian dolomites in the Western Canada Sedimentary Basin and that dolomitization took place both in the intermediate (500–1500 m) to deep (1500–3000 m) subsurface.

Reasoning from the concepts presented in the seawater model above, the problem of understanding late-stage, large-scale subsurface dolomitization may reduce mainly to finding a mechanism for circulating large volumes of Mg-rich water (normal seawater, modified seawater, basin brines, evaporite brines) deep into the subsurface. At the higher temperatures present in the subsurface, dolomitization can apparently take place readily in any buried limestone that has sufficient porosity and permeability to allow circulation of large volumes of Mg-bearing water. Several mechanisms have been suggested to drive circulation of fluids down or up through buried limestones in a basin, including (1) gravity-driven flow owing to the presence of a hydraulic head, the magnitude of which is determined by the elevation of the meteoric recharge area for the subsurface formations (e.g., Garven and Freeze, 1984), (2) thermal convection resulting from a geothermal heat source below the basin (e.g., Kohout, Henry, and Banks, 1977), and (3) buoyant circulation caused by circulation within freshwater lenses along the mixing zone with saline waters; the resulting discharge of brackish water at the coast causes a compensating inflow of saline waters at depth (e.g., Whitaker and Smart, 1990).

6.8 DIAGENESIS

Most carbonate sediments are deposited under marine conditions, although carbonate rocks can also form under some nonmarine conditions (Chapter 11). After deposition, carbonate sediments are subjected to a variety of diagenetic processes that bring about changes in porosity, mineralogy, and chemistry. Carbonate minerals are generally more susceptible to dissolution, recrystallization, and replacement than are most silicate minerals. Thus, the mineralogy of carbonate sediments may be pervasively altered. For example, an original aragonitic mud may alter entirely to calcite during early diagenesis or burial. In turn, the calcite may be replaced completely or nearly completely by dolomite at a later time. Such changes may also destroy or modify original depositional textures such as carbonate grains and micrite. Porosity of carbonate sediments may be either reduced by compaction and cementation or enhanced by dissolution.

Regimes of Carbonate Diagenesis

Carbonate sediments may go through the same general stages of diagenesis as siliciclastic sediments, that is, shallow burial (eogenesis), deep burial (mesogenesis), and uplift and unroofing (telogenesis). Diagenesis takes place in three major regimes or realms (Fig. 6.13): the marine, the meteoric, and the subsurface.

The **marine realm** includes the seafloor and the very shallow marine subsurface. The diagenetic environment here is characterized by seawater temperatures and marine waters of normal salinity. The principal diagenetic processes in this environment involve bioturbation of sediments, modification of carbonate shells and other grains by boring organisms, and cementation of grains in warm-water areas, particularly in reefs, platform-margin sand shoals, and carbonate beach deposits (beachrock).

Marine carbonate sediments may be brought from the seafloor realm into the **meteoric realm** in two ways: by falling sea level and by progressive sediment filling of a shallow carbonate basin. Older carbonate rock can also be brought into the meteoric realm by late-stage uplift and unroofing of a deeply buried carbonate complex (telogenesis). The meteoric realm is characterized by the presence of freshwater; it includes the unsaturated (sediment pores not filled with water) vadose zone above the water table and the phreatic zone, or water-saturated zone, below the water table. Meteoric waters are typically highly charged with CO_2; thus, they are chemically very aggressive (acidic). Because aragonite and high-magnesian calcite are more soluble than calcite, they dissolve readily in these corrosive waters. On the other hand, dissolution of aragonite and high-magnesian calcite may saturate the waters in calcium carbonate with respect to calcite, causing calcite to precipitate (a process called **calcitization)**. This dissolution reprecipitation process causes less stable aragonite and high-magnesian calcite to be replaced by more stable calcite. Calcite may also precipitate into open spaces as a cement. Thus, dissolution, alteration of aragonite and high-magnesian calcite to

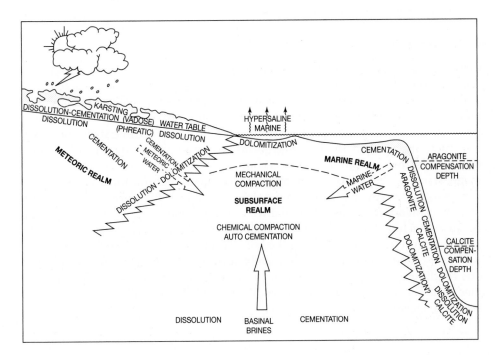

Figure 6.13
The principal environments in which postdepositional modification of carbonate sediments occurs. The dominant diagenetic processes that occur in each of the major diagenetic realms are also indicated. See text for details. [From Moore, C. H., 1989, Carbonate diagenesis and porosity. Fig. 3.1, p. 44, reprinted by permission of Elsevier Science Publishers, Amsterdam.]

calcite, and calcite cementation are the principal diagenetic processes in the meteoric realm.

After an initial period of diagenesis on the seafloor, and possibly in the meteoric realm, carbonate sediments are gradually buried and subjected to increased pressures, higher temperatures, and compositionally changed pore fluids in the **subsurface realm**. Under these changed conditions, carbonate sediments may undergo physical compaction, chemical compaction (dissolution at grain boundaries), and additional chemical or mineralogical changes that may include dissolution, cementation, aragonite-to-calcite transformation, and replacement of calcite by another mineral such as dolomite. The exact nature of the changes that take place during deep subsurface diagenesis depends upon the specific conditions (temperature, pore-fluid composition, pH) of the burial environment.

Major Diagenetic Processes and Changes

Biogenic Alteration

Organisms in carbonate depositional environments rework sediment by boring, burrowing, and sediment-ingesting activities, just as they do in siliciclastic environments. These activities may destroy primary sedimentary structures (e.g., Demicco and Hardie, 1994) in carbonate sediment and leave behind mottled bedding and various kinds of organic traces. In addition, many kinds of small organisms, such as fungi, bacteria, and algae, create microborings in skeletal fragments and other carbonate grains. Fine-grained (micritic) aragonite or high-magnesian calcite may then precipitate into these holes. This boring and micrite-precipitation process may be so intensive in some warm-water environments that carbonate grains are reduced almost completely to micrite, a process called **micritization**. If boring is less intensive, only a thin **micrite rim,** or micrite envelope, may be produced around the grain (Fig. 6.14A). Bacteria are suggested to affect carbonate diagenesis in a variety of other ways, such as mediating precipitation of carbonate cements and diagenetic formation of micrite (e.g., Camoin and Arnaud-Vanneau, 1997). Larger organisms, such as sponges and molluscs, create macroborings in skeletal grains and carbonate substrate, and other organisms, such as fish, sea cucumbers, and gastropods, may break down carbonate grains in various ways to smaller pieces.

Cementation

Cementation is an important process in all diagenetic realms. On the ocean floor, cementation takes place mainly in warm-water areas within the pore spaces of grain-rich sediments or in cavities. Reefs, carbonate sand shoals on the margins of platforms, and carbonate beach sands are favored areas for early cementation. Areas of the seafloor along the platform margin where sediments become well cemented are referred to as **hardgrounds**. Cemented carbonate beach sand is called **beachrock**. Seafloor cement is commonly aragonite, less commonly high-magnesian calcite. Seafloor cement can take several textural forms, as shown in Figure 6.15. Beachrock may contain meniscus cements that form where water is held by capillary forces as interstitial water drains from beaches during low tide. Because beach sediments are not constantly bathed in water, pendant cements may also form in beachrock along the bottoms of grains where drops of water are held. Isopachous rinds, which completely surround grains, form under subaqueous conditions where grains are constantly surrounded by water. Aragonite cements may also occur as a mesh of needles or as fibrous radial crystals that have a botryoidal form.

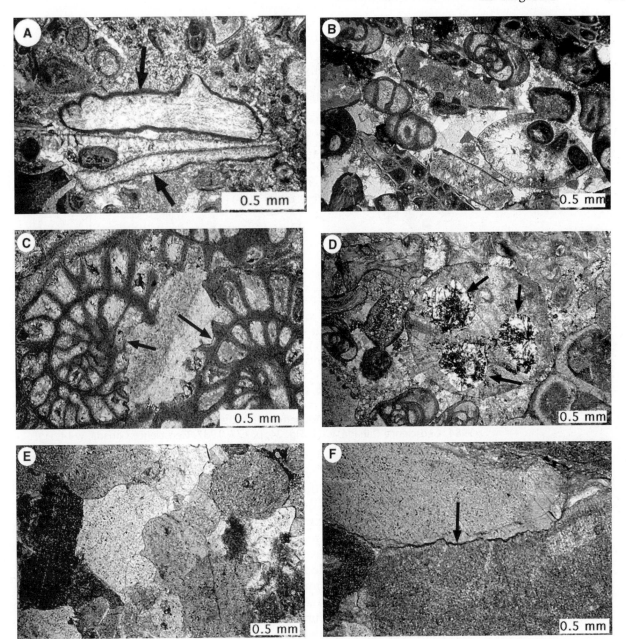

Figure 6.14

Diagenetic fabrics in limestones: A. Dark, micrite rims or envelopes (arrows) around fossil fragments, Renault Formation (Mississippian), Missouri. B. Sparry calcite (white) cementing fossils and fossil fragments, Salem Formation (Mississippian), Missouri. C. Recrystallization fabric that has partially destroyed fossil fusulinid foraminifers (arrows), Morgan Formation (Pennsylvanian), Colorado. D. Patchy replacement of a fossil fragment by chert (arrows), Salem Formation (Mississippian), Missouri. E. Grain fabric that is tightly packed owing to physical compaction of echinoderm (crinoid) fragments, Kimswick Limestone (Ordovician), Missouri. F. Irregular boundary (arrow) between two echinoderm fragments formed as a result of pressure solution (chemical compaction), middle Mississippian limestone, Oklahoma.

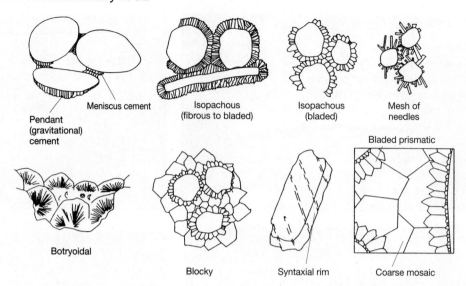

Figure 6.15
Principal kinds of cements that form in carbonate rocks during diagenesis. **Seafloor diagenetic environments** are characterized particularly by aragonitic meniscus and pendant cements (in beachrock), isopachous cement, needle cement, and botryoidal cement. **Meteoric-realm** cements are composed dominantly of calcite and include meniscus and pendant cements in the **vadose zone** and isopachous, blocky, and syntaxial rim cements in the **phreatic zone**. Cements of the **subsurface burial realm** are also mainly calcite and include syntaxial rims, bladed prismatic, and coarse mosaic types. [Modified from James, N. P., and P. W. Choquette, 1983, Geoscience Canada, v. 10, Fig. 3, p. 165; 1984, Geoscience Canada, v. 11, Fig. 24, p. 177; 1987, Geoscience Canada, v. 14, Fig. 21, p. 16.]

In the meteoric realm, dissolution is a more important process than cementation; however, cementation does occur. The cement is almost exclusively calcite. As mentioned, the calcium carbonate that forms this cement is derived by dissolution of less stable aragonite and high-magnesian calcite. In the (water-) unsaturated vadose zone, calcite cements are commonly meniscus and pendant cements. In the water-saturated phreatic zone, they are isopachous, blocky, or syntaxial rim cements. Syntaxial rims form by precipitation of optically continuous calcite around single-crystal fossil echinoderm fragments, in much the same way that cement overgrowths form around quartz grains.

Calcite cementation may also take place during deep burial, although the conditions that control cementation at depth are poorly understood. Factors that have been cited to favor carbonate cementation during deep burial include unstable mineralogy (aragonite and high-magnesian calcite favors solution and reprecrpitation); pore waters highly oversaturated in calcium carbonate; high porosity and permeability (which enable high rates of fluid flow); increase in temperature; and decrease in carbon dioxide partial pressure. The calcium carbonate needed for cementation at depth may be supplied, at least in part, by pressure solution of carbonate sediment in much the same way that pressure solution of quartz grains supplies silica to pore waters in siliciclastic sediment. Coarse mosaic calcite and bladed prismatic calcite (Fig. 6.15) are common kinds of deep-burial cements. The combination of bladed prismatic and coarse mosaic cement shown in Figure 6.15 is called **drusy** cement (see Fig. 6.1A). These calcite cements are commonly coarse grained and clear or white in appearance. They are usually referred to as sparry calcite cement. Figure 6.14B provides an additional example of a skeletal limestone cemented by sparry calcite cement.

Dissolution

Cementation is a very common diagenetic process in carbonate rocks, yet, somewhat paradoxically, so is dissolution. Dissolution of carbonate minerals requires conditions essentially opposite to those that lead to cementation. Dissolution is favored by unstable mineralogy (presence of aragonite or high-magnesian calcite), cool temperatures, and low pH (acidic) pore waters that are undersaturated with calcium carbonate. Dissolution takes place particularly in chemically aggressive pore waters highly charged with CO_2 and/or organic acids. Dissolution is relatively unimportant on the seafloor but is particularly prevalent in the meteoric realm where chemically aggressive meteoric waters percolate or flow down through the vadose zone into the phreatic zone. Extensive dissolution of aragonite and high-magnesian calcite takes place in this environment and even calcite may be dissolved if pore waters are sufficiently aggressive. Dissolution tends to be concentrated particularly along the water table (the boundary between the vadose and phreatic zones), which accounts for the common presence of caves in carbonate rocks at the level of the water table. Dissolution is less intensive in the deep-burial (subsurface) realm than in the meteoric realm for two reasons. First, most aragonite and high-magnesian calcite may already have been converted to more stable calcite in the meteoric realm (see "Neomorphism" below). Second, increasing temperature at depth decreases the solubility of all carbonate minerals. Dissolution may occur at depth if enough CO_2 is added to pore waters as a result of burial decay of organic matter (decarboxylation) to overcome the decrease in solubility resulting from increased temperature. Likewise, mixing of subsurface waters at depth may produce fluids that are undersaturated with respect to calcite, thus promoting destruction of carbonate cements or other carbonate elements (Morse, Hanor, and He, 1997). Buried carbonate sediments that are brought back into the meteoric zone after uplift may undergo extensive dissolution of both previously formed cements and other carbonate minerals under the influence of chemically aggressive, CO_2-charged meteoric waters.

Neomorphism

Neomorphism is a term used by Folk (1965) to cover the combined processes of inversion (e.g., transformation of aragonite to calcite) and recrystallization. **Inversion** refers to the change of one mineral to its polymorph, such as aragonite to calcite. Strictly speaking, inversion takes place only in the solid (dry) state. When the transformation of aragonite to calcite takes place in the presence of water, it occurs by means of dissolution of the less stable aragonite and nearly simultaneous precipitation (replacement) by more stable calcite. Many geologists refer to this process as **calcitization**, as mentioned. During diagenesis, most aragonite is eventually calcitized. **Recrystallization** indicates a change in size or shape of a crystal, with little or no change in chemical composition or mineralogy. Calcitization and recrystallization commonly go hand in hand.

Neomorphism may occur in all three diagenetic realms but is particularly important in the meteoric and subsurface diagenetic environments. Neomorphism may affect both carbonate grains and micrite and commonly increases crystal size. This process destroys original textures and fabrics and, when pervasive, may cause the entire rock to become recrystallized. Thus, a fine-grained (micritic) limestone can be converted into a coarse-grained sparry rock. On a smaller scale, recrystallization results in the formation of large, clear crystals of calcite that closely resemble sparry calcite cement. In fact, one of the most difficult problems in the microscopic study of carbonate rocks is to differentiate between sparry calcite cement and neomorphic spar. Figure 6.14C shows an example of neomorphic spar.

Replacement

As described under "Siliciclastic Diagenesis," replacement involves the dissolution of one mineral and the nearly simultaneous precipitation of another mineral of different composition in its place. Replacement of calcium carbonate minerals by other minerals is a common diagenetic process. Dolomitization of $CaCO_3$ sediment is one kind of replacement process. In addition, many other kinds of noncarbonate minerals may replace carbonate minerals during diagenesis, including microcrystalline quartz (chert; Fig. 6.14D), pyrite (iron sulfide), hematite (iron oxide), apatite (calcium phosphate), and anhydrite (calcium sulfate). Replacement can occur in all diagenetic environments. We have already discussed the replacement of $CaCO_3$ by dolomite in seafloor and burial environments. In carbonate-evaporite sequences, replacement of carbonate minerals by anhydrite at depth is a common process. Replacement of carbonates by microcrystalline quartz (chert) is also common in the meteoric and deep-burial environments. For example, Maliva and Siever (1989) report replacement of Paleozoic carbonates by chert at burial depths ranging from 30 to 1000 m. Replacement of carbonate minerals by silica may be very selective, with silica replacing fossils and other carbonate grains in preference to micrite, as in Figure 6.14D.

Physical and Chemical Compaction

Newly deposited, watery carbonate sediments have initial porosities ranging from 40 to 80 percent. As burial into the subsurface proceeds, the pressure of overlying sediments brings about grain reorientation and tighter packing (Fig. 6.14E). As with siliciclastic sediments, compaction results in loss of porosity and thinning of beds at fairly shallow burial depth. At deeper burial to depths of about 1000 ft (305 m) and at progressively higher overburden pressures, grains may also deform

Figure 6.16
Well-developed sutured stylolites in Cretaceous limestones, Calcare Massicio, Tuscany, Italy. [Photograph courtesy of E. F. McBride.]

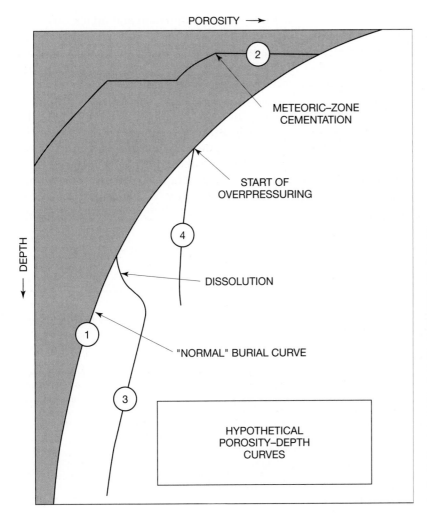

Figure 6.17
Hypothetical curves illustrating (1) a "normal" porosity-depth relationship for fine-grained sediments with marine pore waters; (2) cementation in the meteoric zone (horizontal segments) alternating with burial in marine pore waters; (3) reversal of normal-porosity depth trend owing to dissolution in the deep subsurface, followed by resumption of normal burial; and (4) arrested porosity reduction owing to abnormally high pore pressure. [From Choquette, P. W., and N. P. James, 1987, Diagenesis 12. Diagenesis in limestones—3. The deep-burial environment: Geoscience Canada, v. 14, Fig. 33, p. 23, reprinted by permission of Geological Association of Canada.]

by brittle fracturing and breaking and by plastic or ductile squeezing. Even at burial depths as shallow as 100 m, compaction can reduce the depositional thickness of carbonate sediments by as much as one-half, with accompanying porosity losses of 50 to 60 percent of original pore volumes (Shinn and Robbin, 1983).

At burial depths ranging from about 200 to 1500 m, chemical compaction of carbonate sediments is also initiated. As discussed under "Siliciclastic Diagenesis," pressure solution at grain-to-grain contacts can result in interpenetrating or sutured contacts between grains (Fig. 6.14F). On a larger scale, pressure-solution seams called **stylolites** develop. Stylolites are particularly common in carbonate rocks. The stylolite seams are marked by the presence of clay minerals and other fine-size noncarbonate minerals (commonly referred to as an insoluble residue) that accumulate as carbonate minerals dissolve. Stylolites range in size from microstylolites between grains, in which the amplitude of the interpenetrating contacts between grains is less than 0.25 mm, to stylolites with amplitudes exceeding 1 cm (e.g., Fig. 6.16). Pressure solution, with accompanying stylolite formation, causes significant loss of porosity (perhaps as much as 30 percent of original pore volume) and thinning of beds.

Summary Results of Carbonate Diagenesis

The combined effects of organic, physical, and chemical diagenesis produce significant changes in the depositional characteristics of carbonate sediments. Organisms destroy primary sedimentary structures such as lamination and attack carbonate grains by boring and sediment ingestion. Physical and chemical compaction resulting from overburden pressure cause extensive porosity reduction and bed thinning. Cementation further reduces porosity. On the other hand, dissolution processes in the meteoric realm, and to a lesser extent the subsurface realm, create new porosity. Even previously deposited cements may be dissolved during telogenesis. The various ways by which diagenetic processes can combine to affect the final porosity of carbonate sediments are illustrated in Figure 6.17. Finally, calcitization (neomorphism) converts most aragonite and high-magnesian calcite to calcite, and subsurface dolomitization can bring about pervasive replacement of carbonate minerals by dolomite. The photomicrographs offered in this chapter illustrate many of the common diagenetic textures of carbonate rocks. See Scholle and Ulmer-Scholle (2003) for numerous additional examples of diagenetic textures.

FURTHER READING

Adams, A. E., and W. S. MacKenzie, 1998, A color atlas of carbonate sediments and rocks under the microscope: John Wiley & Sons, New York, 180 p.

Bathurst, R. G. C., 1975, Carbonate sediments and their diagenesis, 2nd ed.: Elseveir, Amsterdam, 658 p.

Carozzi, A., 1993, Sedimentary petrography: PTR Prentice Hall, Englewood Cliffs, New Jersey, 263 p.

Demicco, R. V., and L. A. Hardie, 1994, Sedimentary structures and early diagenetic features of shallow marine carbonate deposits: SEPM Atlas Series No. 1, Society for Sedimentary Geology, Tulsa, Okla., 265 p.

Grotzinger, J. P., and N. P. James (eds.), 2000, Carbonate sedimentation and diagenesis in the evolving Precambrian world: SEPM Special Publication No. 67, Tulsa, Okla., 364 p.

James, N. P., and J. A. D. Clarke (eds.), 1997, Cool-water carbonates: SEPM Special Publication No. 56, Society for Sedimentary Geology, Tulsa, Okla., 440 p.

Montañez, I. P., J. M. Gregg, and K. L. Shelton (eds.), 1997, Basinwide diagenetic patterns: Integrated petrologic, geochemical, and hydrologic considerations: SEPM Special Publication No. 57, Society for Sedimentary Geology, Tulsa, Okla., 302 p.

Moore, C. H., 1989, Carbonate diagenesis and porosity: Elsevier, Amsterdam, 338 p.

Morse, J. W., and F. T. Mackenzie, 1990, Geochemistry of sedimentary carbonates: Elsevier, Amsterdam, 707 p.

Purser, B., M. Tucker, and D. Zenger (eds.), 1994, Dolomites: A volume in honor of Dolomieu: International Association of Sedimentologists, Special Publication No. 21, Blackwell Scientific Publications, Oxford, 451 p.

Rezak, R., and D. L. Lavoie (eds.), 1993, Carbonate microfabrics: Springer-Verlag, New York, 313 p.

Scholle, P. A., and D. S. Ulmer-Scholle, 2003, A color guide to the petrography of carbonate rocks: Grains, textures, porosity, diagenesis: AAPG Memoir 77, American Association of Petroleum Geologists, Tulsa, Okla., 474 p.

Scoffin, T. P., 1987, Carbonate sediments and rocks: Blackie, Glasgow, 274 p.

Stanley, S. M., and L. A. Hardie, 1999, Hypercalcification: Paleontology links plate tectonics and geochemistry to sedimentology: GSA Today, v. 9, p. 1–7.

Tucker, M. E., and V. P. Wright, 1990, Carbonate sedimentology: Blackwell Scientific Publications, Oxford, 482 p.

7

Other Chemical/Biochemical and Carbonaceous Sedimentary Rocks

7.1 INTRODUCTION

In addition to the carbonate rocks discussed in Chapter 6, the chemical/biochemical sedimentary rocks include a diverse group of rocks of which some (evaporites and iron-rich sedimentary rocks) form mainly by chemical processes and others (cherts and phosphorites) form largely through biogenic processes. Although volumetrically less significant than carbonate rocks, these sedimentary rocks are extremely important both as economic resources and as indicators of specialized paleoenvironments. Evaporite deposits such as gypsum, halite (rock salt), and trona (sodium carbonate) are mined for industrial and agricultural purposes. The iron-rich sedimentary rocks are iron ores that have enormous worldwide economic significance as a source for most of our iron. Sedimentary phosphorites are the major source of commercial phosphates used for fertilizers and chemical purposes. Even the siliceous sedimentary rocks may have some economic value in the semiconductor industry.

Aside from their economic value, these chemical/biochemical rocks are extremely interesting because they indicate past environmental conditions that appear to be uncommon on Earth today—or perhaps we simply don't recognize the modern counterparts of these ancient environments. For example, we know of no place on Earth where massive sedimentary iron deposits, of the type common in late Precambrian rocks, are forming today. Nor do we fully understand the mechanism of iron deposition or the source of the enormous amount of iron present in iron-rich sedimentary rocks. Likewise, the mechanisms by which low levels of phosphorus in ocean water become concentrated a millionfold to form phosphorite deposits is only partially understood. In this chapter, we look at the characteristics of these enigmatic sedimentary rocks and discuss some of the more interesting aspects of their origin.

We also briefly examine the characteristics and origins of the carbonaceous sedimentary rocks—rocks that contain significant amounts ($>\sim10$ percent) of organic carbon. Together with petroleum and natural gas, these organic-rich rocks, which include coals and oil shales, are the source of our fossil fuels. Fossil fuels currently supply most of the world's energy needs; therefore, commercial interest in exploiting carbonaceous sedimentary rocks is understandably high. Geologists are further interested in the origin of these rocks, particularly the processes and conditions that allow preservation of such high levels of organic matter. The average organic content of other sedimentary rocks is only about 1.5 percent.

7.2 EVAPORITES

Introduction

The term **evaporites** is used for all deposits, such as salt deposits, that are composed of minerals that originally precipitated from saline solutions concentrated by solar evaporation. Because of their generally high solubility and their susceptibility to deformation, many of these evaporites have subsequently been diagenetically or secondarily altered in various ways during burial. Thus, there are few completely "primary" evaporite beds older than about twenty-five million years (Warren, 1999, p. 1). Evaporites occur in rocks of most ages, including the Precambrian, but they are particularly common in Cambrian, Permian, Jurassic, and Miocene successions (Ronov et al., 1980). Although the total volume of evaporites in the geologic record is much less than that of carbonate rocks, some individual evaporite deposits, such as the Miocene Messinian of the Mediterranean region, reach thicknesses exceeding 1 km. Evaporites form under both marine and nonmarine conditions; however, marine evaporites tend to be thicker and more laterally extensive than nonmarine evaporites and are of greater geologic interest.

Evaporite deposits are composed dominantly of various proportions of halite (rock salt), anhydrite, and gypsum. Although approximately eighty minerals have been reported from evaporite deposits (Stewart, 1963; Warren, 1999), only about a dozen of these minerals are common enough to be considered important evaporite rock formers. Evaporite minerals are commonly classified into those of marine origin and those of nonmarine origin, although Hardie (1991) suggests that identifying the marine or nonmarine origin of evaporites on the basis of their mineralogy and chemistry may not be entirely justified. If carbonate minerals, most of which are not evaporites, are excluded, the most common minerals generally considered to characterize marine evaporites are the calcium sulfate minerals gypsum and anhydrite. Halite is next in abundance, followed by the potash salts, sylvite, carnellite, langbeinite, polyhalite, and kainite, and the magnesium sulfate, kieserite (Table 7.1). The marine evaporite minerals can be grouped by chemical composition into chlorides, sulfates, and carbonates. **Marine** evaporites commonly contain mixtures of minerals, although gypsum (or anhydrite) and halite predominate in most horizons. Deposits may range from those that are composed almost entirely of anhydrite or gypsum to those that are mainly halite. Gypsum is more abundant than anhydrite in modern evaporite deposits, but anhydrite is more abundant in ancient deposits, owing to diagenetic alteration of gypsum to anhydrite. Marine evaporite deposits may also contain various amounts of impurities such as clay minerals, quartz, feldspar, and sulfur.

Nonmarine evaporites are characterized by evaporite minerals that are not common in marine evaporites because the water from which nonmarine evaporites precipitate generally has proportions of chemical elements different from those of marine water (e.g., more bicarbonate and magnesium and little or no chlorine). These nonmarine minerals may include bloedite ($Na_2SO_4 \cdot MgSO_4 \cdot 4H_2O$), borax [$Na_2B_4O_5(OH)_4 \cdot 8H_2O$], epsomite ($MgSO_4 \cdot 7H_2O$), gaylussite ($Na_2CO_3 \cdot CaCO_3 \cdot 5H_2O$), glauberite [$Na_2Ca(SO_4)$], magadiite ($NaSi_7O_{13}(OH)_3 \cdot 3H_2O$), mirabilite ($Na_2SO_4 \cdot 10H_2O$), thenardite ($NaSO_4$), and trona [$Na_3H(CO_3)_2 \cdot 2H_2O$]. Nonmarine deposits may also contain anhydrite, gyspum, and halite and may even be dominated by these minerals.

Evaporites may be classified as chlorides, sulfates, or carbonates on the basis of their chemical composition (Table 7.1); however, few rock names have

Table 7.1 Classification of marine evaporites on the basis of mineral composition

Mineral class	Mineral name	Chemical composition	Rock name
Chlorides	Halite	NaCl	Halite; rock salt
	Sylvite	KCl	Potash salts
	Carnallite	$KMgCl_3 \cdot 6H_2O$	
	Langbeinite	$K_2Mg_2(SO_4)_3$	
	Polyhalite	$K_2Ca_2Mg(SO_4)_6 \cdot H_2O$	
	Kainite	$KMg(SO_4)Cl \cdot 3H_2O$	
Sulfates	Anhydrite	$CaSO_4$	Anhydrite
	Gypsum	$CaSO_4 \cdot 2H_2O$	Gypsum
	Kieserite	$MgSO_4 \cdot H_2O$	—
Carbonates	Calcite	$CaCO_3$	Limestone
	Magnesite	$MgCO_3$	—
	Dolomite	$CaMg(CO_3)_2$	Dolomite; dolostone

been applied to evaporite deposits. Rocks composed predominantly of the mineral halite are called halite or **rock salt**. Rocks made up predominantly of gypsum or anhydrite are simply called gypsum or anhydrite, although some geologists use the names **rock gypsum** or **rock anhydrite**. Few evaporite beds are composed predominantly of minerals other than the calcium sulfates and halite. Specific names have not been proposed for rocks enriched in other evaporite minerals, although the term **potash salts** is used informally for potassium-rich evaporites.

Evaporite deposits can display textures that range from primary depositional features to diagenetically produced features, depending upon their ages and deformation histories. Primary features can include a variety of crystal textures and bedding features that reflect chemical precipitation processes (e.g., crystal settling, bottom nucleation). Also, structures such as cross or graded bedding and ripple marks may reflect physical processes that indicate traction-current or turbidity-current transport. As mentioned, many ancient (subsurface) evaporites have undergone physical and chemical diagenetic modifications that have altered primary textures to secondary textures such as nodules and crystal pseudomorphs. Most buried gypsum and anhydrite deposits provide good examples of such diagenetically altered features.

Kinds of Evaporites

Gypsum and Anhydrite

Calcium sulfates are deposited dominantly as gypsum. Gypsum can be altered into, and be pseudomorphed by, anhydrite while the sediments are still in their general depositional environments. Gypsum is also dehydrated to anhydrite upon burial to a few hundred meters, and this loss of water is accompanied by a 38 percent decrease in solid volume of the gypsum. Because of this rapid dehydration with burial, most ancient calcium sulfate deposits are composed of anhydrite. Anhydrite can be hydrated back to gypsum after uplift and exposure to low-salinity surface waters, with an accompanying increase in volume. These volume changes resulting from dehydration and hydration can distort original depositional structures and textures, and many calcium sulfate deposits are characterized by distorted fabrics. Three fundamental structural groups of anhydrite

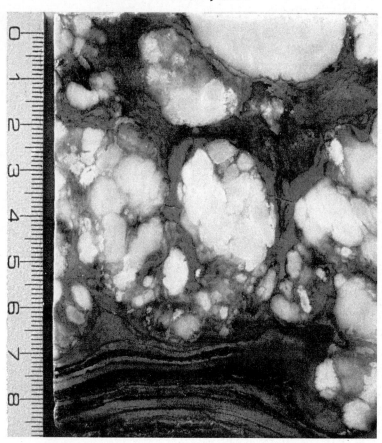

Figure 7.1
Nodular anhydrite with an abundant calcareous (dolomicrite) matrix. Falces Formation (Oligocene), Ebro Basin, Spain. Photograph courtesy of Joseph M. Salvany, Universitat Politécnica de Catalunya, Barcelona.

are recognized on the basis of fabric, bedding, and the presence or absence of distortion: nodular anhydrites, laminated anhydrites, and massive anhydrites.

Nodular anhydrites are irregularly shaped lumps of anhydrite that are partly or completely separated from each other by a salt or carbonate matrix (Fig. 7.1). The term **chickenwire structure** is used for a particular type of nodular anhydrite that consists of slightly elongated, irregular polygonal masses of anhydrite separated by thin dark stringers of other minerals such as carbonate or clay minerals (Fig. 7.2).

The formation of nodular anhydrite is initiated by displacive growth of gypsum in carbonate or clayey sediments. Gypsum crystals subsequently alter to anhydrite pseudomorphs, which continue to enlarge by addition of Ca^{2+} and SO_4^{2-} from an external source and ultimately grow displacively into anhydrite nodules. Chickenwire anhydrite forms when, with increasing size, the nodules ultimately coalesce and interfere. Most of the enclosing sediment is pushed aside and what remains forms thin stringers between the nodules.

Nodular anhydrites have been observed in many modern coastal sabkha environments (see Fig. 6.11). Nodular anhydrites can also form in deeper water environments. In fact, all that is needed for formation of nodular anhydrite is growth of crystals in mud in contact with highly saline brines, which can occur in deep or shallow standing water as well as in a sabkha environment.

Laminated anhydrites consist of thin, nearly white, anhydrite or gypsum laminations that alternate with dark gray or black laminae rich in dolomite or organic matter (Fig. 7.3). These laminae are commonly only a few millimeters thick and rarely reach 1 cm. Many of the thin laminae are remarkably uniform, with sharp planar contacts. Many laminae can be traced laterally for long distances—some more than 100 km (Dean and Anderson, 1978). Also, they may compose vertical successions hundreds of meters thick in which hundreds of thousands of laminae may be

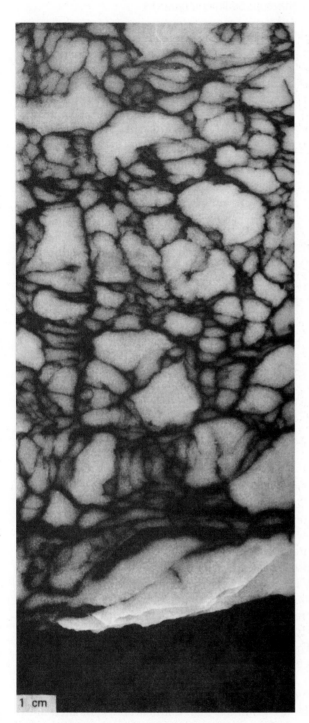

1 cm

Figure 7.2
Chickenwire structure in anhydrite. Evaporite series of the Lower Lias (Jurassic), Aquitaine Basin, southwest France. [From Bouroullec, J., 1981, Sequential study of the top of the evaporitic series of the Lower Lias in a well in the Aquitaine Basin (Auch 1), southwestern France, *in* Chambre Sydnical de la Recherche et de la Production due Pétrole et du Gaz Naturel (eds.), Evaporite deposits: Illustration and interpretation of some environmental sequences, Pl 36, p. 157, reprinted by permission of Editions Technip, Paris, and Gulf Publishing Co., Houston, Tex. [Photograph courtesy of J. Bouroullec.]

present (e.g., the Permian Castile Formation, southeast New Mexico and west Texas). Pairs of alternating light and dark bands have been suggested to be annual varves resulting from seasonal changes in water chemistry and temperature; however, they might equally well represent cyclic changes or disturbances of longer duration. Laminae of anhydrite can also alternate with thicker layers of halite, producing laminated halite.

Because of the lateral persistence of laminated evaporites, which indicates uniform depositional conditions over a wide area, laminites are commonly interpreted to form by precipitation of evaporites in quiet water below wave base. They could presumably form either in a shallow-water area protected in some manner from

Figure 7.3
Laminated anhydrite from the Prairie Evaporite (Devonian), Canada.

strong bottom currents and wave agitation or in a deeper water environment. Laminated anhydrites in ancient sedimentary successions such as the Permian Castile Formation indicate that these deposits have largely escaped physical deformation, although original gypsum minerals have been altered to anhydrite.

Some laminated anhydrite may form by coalescing of anhydrite nodules, which by continued growth in a lateral direction merge into one another to produce a layer. Layers formed by this mechanism are thought to be thicker and less distinct and continuous than laminae formed by precipitation. A special type of contorted layering that has resulted from coalescing nodules has been observed in some modern sabkha deposits where continued growth of nodules creates a demand for space. The lateral pressures that result from this demand cause the layers to become contorted, forming ropy bedding or **enterolithic structures** (Fig. 7.4).

Massive anhydrite is anhydrite that lacks perceptible internal structures. True massive anhydrite appears to be less common than nodular and laminated anhydrite, and little information is available regarding its formation. Presumably, it represents sustained, uniform conditions of deposition. Haney and Briggs (1964) suggest that massive anhydrite forms by evaporation at brine salinities of approximately 200 to 275 $\frac{0}{00}$ (parts per thousand), just below the salinities at which halite begins to precipitate. (Seawater has an average salinity of 35 $\frac{0}{00}$.)

Halite

Halite forms as crusts, presumably in shallow water, and as very finely laminated deposits (deep water?) that may reach thicknesses of as much as 1000 m. Laminated halite deposits commonly include anhydrite-carbonate laminae. Anhydrite along with other minerals such as dolomite, calcite, quartz, and clay minerals may also

Figure 7.4
Enterolithic structure in nodular gypsum of the Grenada Basin, southern Spain. [Photograph courtesy of J. M. Rouchy.]

be present in halite as inclusions. Dark, inclusion-rich laminae may alternate with light, inclusion-poor or inclusion-free laminae. Originally formed halite crystals can be chevron, coronet, or hopper shaped; however, halite that has been recrystallized and subjected to movement (salt flowage) may display highly interpenetrating, centimeter-size crystals. Halite deposits may also display sedimentary structures such as ripples and cross-bedding.

Origin of Evaporite Deposits

Evaporation Sequence

When ocean water is evaporated in the laboratory, evaporite minerals are precipitated in a definite sequence that was first demonstrated by Usiglio in 1848 (reported in Clarke, 1924). Minor quantities of carbonate minerals begin to form when the original volume of seawater is reduced by evaporation to about one-half. Gypsum appears when the original volume has been reduced to about 20 percent, and halite forms when the water volume reaches approximately 10 percent of the original volume (see also discussion in Kendall and Harwood, 1996). As discussed in Chapter 6, the precipitation of gypsum increases the Mg/Ca ratio in remaining water, which favors the process of dolomitization. Dolomite occurs in association with evaporites in many ancient sedimentary successions as well as in some modern environments. Magnesium and potassium salts are deposited when less than about 5 percent of the original volume of seawater remains. The same general sequence of evaporite minerals occurs in natural evaporite deposits, although many discrepancies exist between the theoretical sequences predicted on the basis of laboratory experiments and the sequences actually observed in the rock record. In general, the proportion of $CaSO_4$ (gypsum and anhydrite) is greater and the proportion of Na-Mg sulfates is less in natural deposits than predicted from theoretical considerations (Borchert and Muir, 1964).

Depositional Models for Evaporites

Modern evaporite deposits accumulate in a variety of subaerial and shallow subaqueous environments, as illustrated in Figure 7.5. Subaerial environments include both coastal and continental sabkhas, or salt flats, and interdune environments. Shallow subaqueous environments are present mainly in saline coastal lakes called **salinas**. With the possible exception of the Dead Sea in the Middle East, no modern examples of a deep-water evaporite basin exists; however, geologists

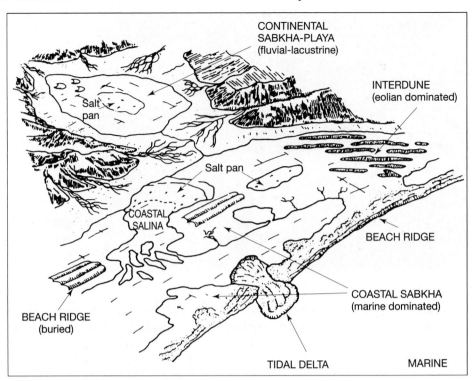

Figure 7.5
Principal settings in which modern evaporite deposits are accumulating. [From Kendall, A. C., 1984, Evaporites, in R. G. Walker (ed.), Facies models: Geoscience Canada Reprint Ser. 1, Fig. 1, p. 260, as modified slightly by Warren, 1989; reprinted by permission of Geological Association of Canada.]

believe that many of the thick, laterally extensive ancient evaporite deposits did accumulate in deep-water basins.

Some ancient evaporite deposits such as the Permian Zechstein of the North Sea area exceed 2 km in thickness, yet evaporation of a column of seawater 1000 m thick will produce only about 15 m of evaporites. Evaporating all the water in the Mediterranean Sea, for example, would yield a mean thickness of evaporites of only about 60 m. Obviously, special geologic conditions operating over a long period of time are required to deposit thick successions of natural evaporites. The basic requirements for deposition of marine evaporites are a relatively arid climate, where rates of evaporation exceed rates of precipitation, and partial isolation of the depositional basin from the open ocean. Isolation is achieved by means of some type of barrier that restricts free circulation of ocean water into and out of the basin. Under these restricted conditions, the brines formed by evaporation are prevented from returning to the open ocean, causing them to become concentrated to the point where evaporite minerals are precipitated.

Although geologists agree on these general requirements for formation of evaporites, considerable controversy still exists regarding deep-water vs. shallow-water depositional mechanisms for many ancient evaporite deposits. Figure 7.6 shows three possible models for deposition of thick successions of marine evaporites. The deep-water, deep-basin model assumes existence of a deep basin separated from the open ocean by some type of topographic sill. The sill acts as a barrier to prevent free interchange of water in the basin with water in the open ocean, but it allows enough water into the basin to replenish that lost by evaporation. Seaward escape of some brine allows a particular concentration of brine to be maintained for a long time, leading to thick deposits of certain evaporite minerals such as gypsum. The shallow-water, shallow-basin model assumes concentration of brines in a shallow, silled basin, but it allows for accumulation of great thicknesses of evaporites caused by continued subsidence of the floor of the basin. The shallow-water, deep-basin model requires that the brine level in the basin be reduced well below the level of the sill, a process called **evaporative drawdown**; recharge of water

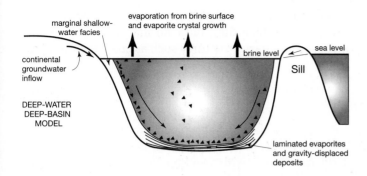

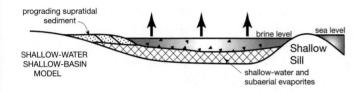

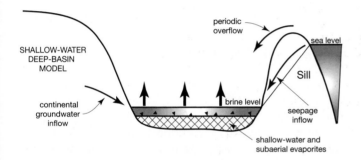

Figure 7.6
Schematic diagram illustrating three models for deposition of marine evaporites in basins where water circulation is restricted by the presence of a sill. [Modified from Kendall, A. C., 1979, Subaqueous evaporites, in R. G. Walker (ed.), Facies models: Geoscience Canada Reprint Ser. 1, Fig. 17, p. 170, reprinted by permission of Geological Association of Canada.]

from the open ocean takes place only by seepage through the sill or by periodic overflow of the sill. Total desiccation of the floors of such basins could presumably occur periodically, allowing the evaporative process to go to completion and thereby deposit a complete evaporite sequence, including magnesium and potassium salts. Thick evaporite deposits could accumulate under these conditions because of continued subsidence.

Applying these models to ancient evaporite deposits is a challenging task, and geologists have not always agreed upon the environmental interpretation of ancient evaporite deposits. Through time, the concept of evaporite deposition has swung from deep-water to shallow-water deposition; then it swung away from the tidal sabkha regime to very moderate water depths (Sonnenfeld and Kendall, 1989). See further discussion of evaporite environments in Chapter 11 and, for additional insight, Busson and Schreiber (1997), Kendall and Harwood (1996), Melvin (1991), Schreiber, Tucker, and Till (1986), Warren and Kendall (1985), and Warren (1989, 1999).

Physical Processes in Deposition of Evaporites

Although we tend to regard evaporite deposits as simply the products of chemical precipitation owing to evaporation, many evaporite deposits are not just passive chemical precipitates. The evaporite minerals have, in fact, been transported and reworked in the same way as the constituents of siliciclastic and carbonate deposits. Transport can occur by normal fluid-flow processes or by mass-transport processes such as slumps and turbidity currents. Turbidity current transport mechanisms may have been particularly important in deposition of ancient deep-water evaporite deposits (Schreiber, Tucker, and Till, 1986). Therefore, evaporite deposits may display clastic textures, including both normal and reverse size grading and various types of sedimentary structures such as cross-bedding and ripple marks.

Diagenesis of Evaporites

As mentioned, most buried evaporite deposits older than about twenty-five million years have undergone diagenetic alteration. Gypsum is converted to anhydrite with burial, particularly above 60°C, with a volume loss (from water loss) of about 38 percent. If exhumation subsequently occurs, anhydrite will hydrate to gypsum, with accompanying increase in volume. These volume changes are in part responsible for the formation of nodules and enterolithic structure. Also, evaporite deposits respond to burial and tectonic pressures by plastic deformation, which destroys original sedimentary structures. Further, deformation can result in folding and diapirism that create large-scale salt diapers (penetration structures) or salt domes that can rise through sediment for more than 5 km (e.g., Jackson, Roberts, and Snelson, 1996). During burial, evaporites may in addition undergo dissolution, cementation, replacement, and calcitization of sulfates (Schreiber, 1988; Warren, 1989).

7.3 SILICEOUS SEDIMENTARY ROCKS (CHERTS)

Introduction

Siliceous sedimentary rocks are fine-grained, dense, very hard rocks composed predominantly of the SiO_2 minerals quartz, chalcedony, and opal (in young rocks), with minor impurities such as siliciclastic grains and diagenetic minerals. **Chert** is the general term used for siliceous rocks as a group. Cherts are common rocks in geologic successions ranging in age from Precambrian to Tertiary; however, they make up only a minor fraction of all sedimentary rocks. They are particularly abundant in Jurassic to Neogene (Tertiary) rocks, moderately abundant in Devonian and Carboniferous rocks, and least abundant in Silurian and Cambrian deposits (Hein and Parrish, 1987). Geologists are particularly interested in cherts because of the information they provide about such aspects of Earth history as paleogeography, paleooceanographic circulation patterns, and plate tectonics. Cherts may also have minor economic significance. Silicon is used in the semiconductor and computer industries and for making glass and related products such as fire bricks, although much of this silica may come from quartz sand. Furthermore, siliceous deposits occur in association with important economic deposits of other minerals, such as Precambrian iron ores; uranium, manganese, and phosphorite deposits; and petroleum accumulations.

Chert is composed mainly of microcrystalline quartz, with minor chalcedony and perhaps opal, depending upon age. Cherts can be divided into three main textural types (Folk, 1974): **granular microquartz**, consisting of nearly equidimensional grains of quartz; average grain size is about 8–10 microns but grain size may range from <1 to 50 microns (Knauth, 1994); **chalcedony (fibrous silica)**, forming sheaflike bundles of radiating, extremely thin crystals about 0.1 mm long; and **megaquartz**, composed of equant to elongated grains commonly greater than 20 microns in size.

Figure 7.7 illustrates the texture of microquartz and megaquartz. The silica that makes up the tests of siliceous organisms is amorphous silica or opal, commonly called **opal-A**. Because the remains of siliceous organisms contribute to the formation of chert, opal A is present in some cherts, particularly those of Tertiary

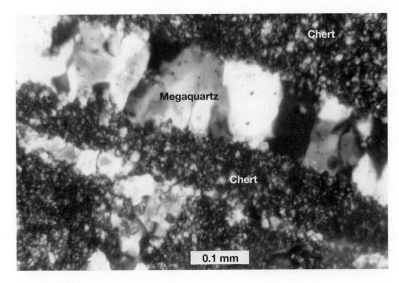

Figure 7.7
Fine-textured, nearly equigranular micro-quartz (chert) cut by a vein of much coarser megaquartz. Source of specimen unknown. Crossed nicols.

and younger age. Opal-A is metastable and crystallizes in time to opal-CT (discussed below) and finally to quartz (chert). Little opal is present in rocks older than about 60 Ma (Knauth, 1994). All gradations may be present in siliceous deposits, from nearly pure opal to nearly pure quartz chert, depending upon the age of the deposits and the conditions of burial.

Chert is composed dominantly of SiO_2 but may also include minor amounts of Al, Fe, Mn, Ca, Na, K, Mg, Ti, and a few other elements such as the rare earth elements cerium (Ce), europium (Eu), and lanthanum (La). Many of these additional elements are contained in impurities such as authigenic hematite and pyrite, detrital siliciclastic minerals, and pyroclastic particles. A few elements, including Fe, Mn, Ni, and Cu, may have precipitated from seawater or pore water as chert was formed. The amount of SiO_2 varies markedly in different types of cherts, ranging from more than 99 percent in very pure cherts such as the Arkansas Novaculite to less than 65 percent in some nodular cherts (Cressman, 1962). Aluminum is commonly the second most abundant element in cherts, followed by Fe and Mg or by K, Ca, and Na. See Jones and Murchey (1986) for additional details.

Varieties of Chert

Several informal names are applied to chert depending upon color, inclusions, and texture. **Flint** is a term used both as a synonym for chert and for a variety of chert, particularly chert nodules, that occurs in Cretaceous chalks. **Jasper** is a variety of chert colored red by impurities of disseminated hematite. Jasper interbedded with hematite in Precambrian iron formations is called **jaspilite**. **Novaculite** is a very dense, fine-grained, even-textured chert that occurs mainly in mid-Paleozoic rocks of the Arkansas, Oklahoma, and Texas region of the south central United States. **Porcelanite** is a term used for fine-grained siliceous rocks with a texture and a fracture resembling those of unglazed porcelain. **Siliceous sinter** is porous, low-density, light-colored siliceous rock deposited by waters of hot springs and geysers. Although most siliceous rocks consist predominantly of chert, some contain abundant detrital clays or micrite. These impure cherts grade into siliceous shales or siliceous limestones.

Cherts can be divided on the basis of gross morphology into two principal types: bedded cherts and nodular cherts. Bedded cherts are further distinguished by their content of siliceous organisms of various kinds. The principal distinguishing characteristics of bedded and nodular cherts are described below.

Bedded Chert

Bedded chert, also referred to as ribbon chert, consists of layers of nearly pure chert ranging to several centimeters thick that are commonly interbedded with millimeter-thick partings or laminae of siliceous shale (Fig. 7.8). Bedding may be even and uniform or may show pinching and swelling. Most chert beds lack internal sedimentary structures; however, graded bedding, cross-bedding, ripple marks, and sole markings have been reported in some cherts. These kinds of structures indicate that some mechanical transport was involved in the deposition of the sediment that composes these rocks. Bedded cherts are commonly associated with submarine volcanic rocks, pelagic limestones, and siliciclastic or carbonate turbidites.

Many bedded cherts are composed dominantly of the remains of siliceous organisms, which are commonly altered to some degree by solution and recrystallization. Bedded cherts can be subdivided on the basis of type and abundance of siliceous organic constituents into four principal kinds: (1) diatomaceous deposits, (2) radiolarian deposits, (3) siliceous spicule deposits, and (4) bedded cherts containing few or no siliceous skeletal remains.

Diatomaceous Deposits.   Diatomaceous deposits include both diatomites and diatomaceous cherts. **Diatomites** are light-colored, soft, friable siliceous rocks composed chiefly of the opaline (opal-A) frustules of diatoms, a unicellular aquatic algae (Fig. 7.9). Thus, they are fossil diatomaceous oozes. Diatomites of both marine and lacustrine (lake) origin are recognized. Marine diatomites are commonly associated with sandstones, volcanic tuffs, mudstones or clay shales, clayey limestones (marls), and, less commonly, gypsum. Lacustrine diatomites are almost invariably associated with volcanic rocks. **Diatomaceous chert** consists of beds and lenses of diatomite that have well-developed silica cement or groundmass that has converted the diatomite into dense, hard chert. Beds of marine diatomaceous chert comprising strata several hundred meters thick have been reported from sedimentary sequences such as the Miocene Monterey Formation of California

Figure 7.8
Thin, well-bedded chert in the Mino Belt Group (Triassic) near Inuyama, Honshu, Japan.

Figure 7.9
A loosely packed assemblage of diatoms, siliceous sponge spicules (rod shaped), and a few radiolarians. The diatoms are mainly *Coscinodiscus* sp. and *Arachnoidiscus* sp. Holocene sediments from the northeast Pacific ocean floor; water depth ~4000 m. Photograph courtesy of William N. Orr.

(Garrison et al., 1981) and occur in rocks as old as the Cretaceous. (Diatoms evolved in the Cretaceous.) Nonmarine diatomaceous deposits have been reported in rocks as old as the Eocene (Barron, 1987). When diatomaceous deposits are converted to quartz chert during diagenesis, the diatom tests are generally destroyed by dissolution and recrystallization.

Radiolarian Deposits.   Radiolarian deposits consist predominantly of the remains of radiolarians (e.g., Fig. 7.10), which are marine planktonic protozoans with a latticelike skeletal framework of opal. Radiolarian deposits can be divided into radiolarite and radiolarian chert. **Radiolarite** is the comparatively hard, fine-grained, chertlike equivalent of radiolarian ooze, that is, indurated radiolarian ooze.

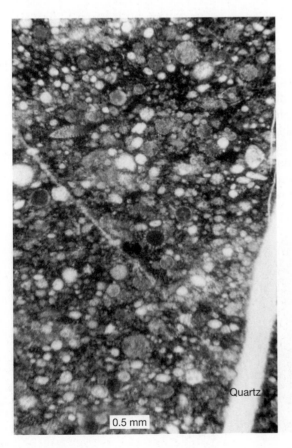

Figure 7.10
Radiolarian chert from the Otter Point Formation (Jurassic), southwestern Oregon. Most of the small, rounded bodies in this sample are radiolarians. The large fracture at the lower right is filled with silica (quartz) cement. Ordinary light photograph. [Photograph courtesy of Shelia A. Monroe.]

Radiolarian chert is well-bedded, microcrystalline radiolarite that has a well-developed siliceous cement or groundmass. Radiolarian cherts are commonly associated with tuffs, mafic volcanic rocks such as pillow basalts, pelagic limestones, and turbidite sandstones that indicate a deep-water origin. These bedded cherts, particularly those that exhibit a "pinch and swell" texture, are commonly called **ribbon cherts**. On the other hand, some radiolarian cherts are associated with micritic limestones and other rocks that suggest deposition in water perhaps as shallow as 200 m (Iijima, Inagaki, and Kakuwa, 1979). Radiolarians tend to survive silica diagenesis more effectively than do diatoms; therefore, they are common components of many quartz cherts (Hein, Yeh, and Barron, 1990).

Siliceous Spicule Deposits.   Spicularite (spiculite) is a siliceous rock composed principally of the siliceous spicules of invertebrate organisms, particularly sponges (see Fig. 7.9). Spicularite is loosely cemented in contrast to **spicular chert**, which is hard and dense. Spicular cherts are mainly marine in origin and are associated with glauconitic sandstones, black shales, dolomite, argillaceous (clayey) limestones, and phosphorites. They are not generally associated with volcanic rocks and are probably deposited mainly in relatively shallow water a few hundred meters deep.

Nonfossiliferous Cherts.   Many bedded chert deposits have been described that contain few or no recognizable remains of siliceous organisms. Some of these reported occurrences of fossil-barren cherts may simply be the result of inadequate microscopic examination of the cherts which, upon closer examination, might be found to contain siliceous organisms. Others have been examined closely and clearly contain few siliceous organisms. Cherts in this latter group include most cherts associated with Precambrian iron-formations, as well as some Phanerozoic cherts. The origin of these cherts is still poorly understood (see discussion under "Origin of Chert").

Nodular Chert

Nodular cherts are subspheroidal masses, lenses, or irregular layers or bodies that range in size from a few centimeters to several tens of centimeters (Fig. 7.11). They commonly lack internal structures, but some nodular cherts contain silicified fossils or relict structures such as bedding. Colors of these cherts vary from green to tan and black. Nodular cherts typically occur in shelf-type carbonate rocks where they tend to be concentrated along certain horizons parallel to bedding. They occur also in some sandstones, shales, deep-sea clays, lacustrine sediments, and evaporites. Nodular cherts originate by diagenetic replacement (see "Diagenesis of Chert"). Diagenetic origin is clearly demonstrated in many nodules by the presence of partly or wholly silicified remains of calcareous fossils or ooids.

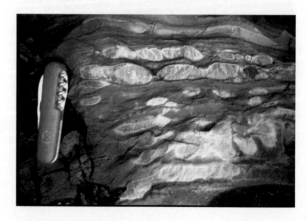

Figure 7.11
Nodular chert in limestones of the Helena Formation (Precambrian), Glacier National Park, Montana.

Origin of Chert

Two principal questions must be answered with respect to the origin of chert: What were the sources of the silica, and what mechanisms have operated in the past to extract silica from water, principally seawater, to form chert? The answers to these questions are known to a reasonable degree (see reviews by Carozzi, 1993; Hesse, 1990; Knauth, 1992, 1994); however, some aspects of the origin of chert, such as the mechanism responsible for deposition of Precambrian chert, remain unresolved.

Sources of Silica

Most bedded chert is present in marine sedimentary rocks. To understand the origin of chert requires that we have some knowledge of sources of silica and that we understand the mechanisms by which silica is extracted from seawater to form chert. In average river water, the concentration of silica in transport (from continental weathering sites) as H_4SiO_4 is about 13 ppm. In addition to silica transported to the oceans by rivers, silica is added to the oceans through reaction of seawater with hot volcanic rocks along mid-ocean ridges and by low-temperature alteration of oceanic basalts and detrital silicate particles on the seafloor (called halmyrolysis), as described in Chapter 1. Some silica may also escape from silica-enriched pore waters of pelagic sediments on the seafloor. These silica sources are summarized in Figure 7.12, which also depicts the pathway of silica diagenesis from opal-A to quartz chert (to be discussed). Despite contributions of silica from these various sources, the silica concentrations in different parts of the ocean range from less than 0.01 ppm to a maximum of about 11 ppm; the average dissolved silica content of the ocean is only 1 ppm (Heath, 1974). Clearly, silica is constantly being removed by some process and has a relatively short residence time in the ocean.

Silica Solubility

Solubility studies show that the solubility of silica in seawater differs for different silicate minerals. The solubility of SiO_2 at 25°C and normal ocean pH (~7.8–8.3) ranges from ~6 to 10 ppm for quartz to ~60 to 130 ppm for amorphous or noncrystalline varieties of silica such as opal (Krauskopf, 1959; Morey, Fournier, and

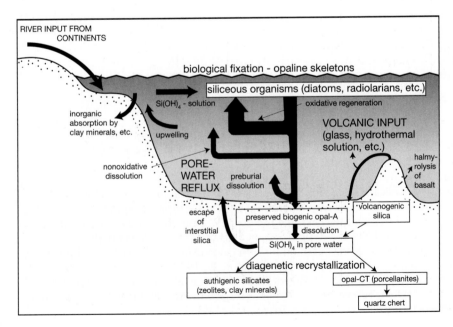

Figure 7.12

Sources of dissolved silica in seawater. [After Riech, V., and U. von Rad, 1979, Silica diagenesis in the Atlantic Ocean: Diagenetic potential and transformations, *in* Talwani, M., W. Hay, and W. B. F. Ryan (eds.), Deep drilling results in the Atlantic Ocean: Continental margins and paleoenvironment: Amer. Geophysical Union, Maurice Ewing Series, v. 3, Fig. 2, p. 322, modified from Heath, G. R., 1974, Dissolved silica and deep-sea sediments, *in* W. W. Hay (ed.), Studies in paleooceanography: Soc. Econ. Paleontologists and Mineralogists Spec. Pub. 20, Fig. 7, p. 81, reprinted by permission of SEPM, Tulsa, Okla.

Rowe, 1962, l964; Iller, 1979). Knauth (1994) suggests a lower limit of 4 ppm for quartz solubility. Therefore the ocean, which has an average dissolved silica content of only l ppm, is grossly undersaturated with respect to silica, in sharp contrast to the saturated state of the surface ocean with respect to calcium carbonate. This fact raises a very intriguing question. What mechanism (or mechanisms) is (are) capable of removing silica from highly undersaturated ocean water to form chert beds and maintain the low concentrations of dissolved silica in the ocean?

The solubility of silica is affected by both pH and temperature. Change in solubility of silica with pH is illustrated in Figure 7.13A. Solubility changes only slightly with increase in pH up to about 9, but it rises sharply at pH values above 9. Solubility increases with increasing temperature in essentially a linear fashion, and solubility at 100°C is nearly 3 to 4 times that at 25°C (Fig. 7.13B). Solubility also increases with increasing pressure. Clearly, the greater the solubility of silica under a given set of conditions, the less likely it will be to precipitate to form chert. See Dove and Rimstidt (1994) for an extended, and more rigorous, discussion of silica solubility.

Silica Extraction from Seawater

Chemical Extraction. In a laboratory experiment at 20°C lasting two years, Mackenzie and Gees (1971) precipitated quartz from seawater containing 4.4 ppm dissolved silica onto nucleation surfaces of cleaned quartz grains. Nonetheless, silica in solution under natural conditions of temperature and pH in the ocean apparently does not readily crystallize to form quartz, even from solutions that have

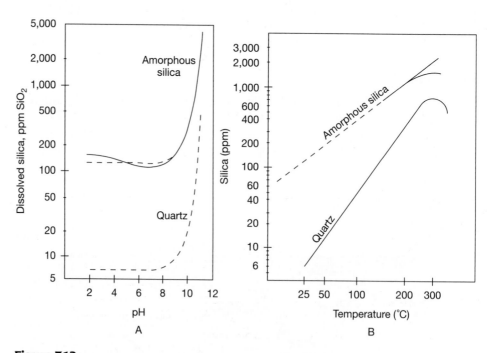

Figure 7.13
The solubility of silica as a function of (A) pH and (B) temperature. The solid line in A shows the variation in solubility of amorphous silica as determined experimentally. The upper dashed curve shows the calculated solubility of amorphous silica, based on an assumed constant solubility of 120 ppm SiO_2 at a pH below 8. The lower dashed line is the calculated solubility of quartz based on the approximate known solubility of 6 ppm SiO_2 in neutral and acid solutions. [A. after Krauskopf, K. B., 1979, Introduction to geochemistry, 2nd ed., Fig. 6.3, p. 133, McGraw-Hill, New York. B. after Fournier, R. O., 1970, Silica in thermal waters: Laboratory and field investigations: Proc. International Symposium on Hydrochemistry and Biochemistry, Tokyo, p. 122–139.]

silica concentrations exceeding the solubility of quartz. Therefore, it is unlikely that chert, which consists of microcrystalline quartz, can be precipitated by inorganic processes from highly undersaturated seawater. Chert might be precipitated in some local basins where waters are saturated with silica, owing perhaps to dissolution of volcanic ash or other processes related to volcanism (e.g., Hesse, 1989; Ledesma-Vázquez et al., 1997). Also, some silica may be removed from seawater in the open ocean by precipitation or adsorption onto clay minerals or other silicate particles, as in the Mackenzie and Gees (1971) experiment; however, such processes likely cannot account for the many bedded successions of nearly pure chert present in the geologic record.

Biogenic Extraction.   Removal of silica from ocean water by silica-secreting organisms to build opaline skeletal structures appears to be the only mechanism capable of large-scale silica extraction from undersaturated seawater. This biologic process has operated since at least early Paleozoic time to regulate the balance of silica in the ocean. **Radiolarians** (Cambrian/Ordovician–Holocene), **diatoms** (Cretaceous–Holocene), and **silicoflagellates** (Cretaceous–Holocene) are microplankton that build skeletons of opaline silica (opal-A). These siliceous microplankton (particularly diatoms and radiolarians) have apparently been abundant enough in the ocean during Phanerozoic time to extract most of the silica delivered to the oceans by rock weathering and other processes. Diatoms are probably responsible for the bulk of silica extraction from ocean waters in the modern ocean and during much of the past fifty million years (Calvert, 1983; Knauth, 1994); however, radiolarians were the important users of silica in the Phanerozoic oceans of Jurassic and older ages. Heath (1974) calculates that the residence time for dissolved silica in the ocean ranges from two hundred to three hundred years for biologic utilization to 11,000 to 16,000 years for incorporation into the geologic record—a very short time from a geologic point of view.

Box 7.1 Formation of Biogenic Chert

The mechanism by which the opaline tests of siliceous organisms are converted into chert is illustrated by pathway A in Figure 7.1.1. While silica-secreting organisms are alive, their siliceous (opal-A) skeletons undergo little dissolution in highly undersaturated and corrosive seawater. Cell walls are protected by some physicochemical system, such as armoring by metal ions, related to their vital activity (Lewin, 1961). After death, this protective system is disrupted and dissolution begins. In areas of the ocean where silicious organisms flourish, the rate of production of siliceous skeletons may be so high that they cannot all be dissolved as rapidly as they are produced. Under such conditions, a sufficient number of the siliceous skeletons may survive total dissolution to accumulate on the seafloor as siliceous oozes (sediments containing at least 30 percent siliceous skeletal material and commonly more than 60 percent). After burial by additional siliceous ooze or clayey sediment, these opaline skeletal materials continue to undergo solution; however, after burial much of the dissolving silica is trapped in the pore spaces of the sediment and does not escape back to the open ocean. The pore waters thus become increasingly enriched in silica, leading eventually to precipitation of chert.

During the process of transformation of biogenic opal to chert, opal-A may not convert directly to quartz chert, which is microcrystalline quartz, but commonly goes through an intermediate, metastable phase: **opal-CT** (Fig. 7.1.1). Although called opal-CT, this silica phase is composed mainly of low-temperature

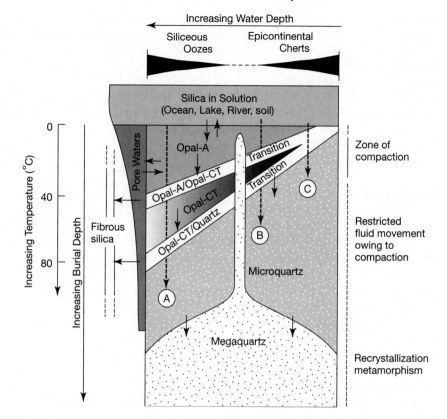

Figure 7.1.1

Schematic diagram showing major silica phases and their possible diagenetic transformations. Vertical dimension represents qualitative burial depth with associated increase in temperature and loss of porosity and permeability. Horizontal dimension represents qualitative water depth of the initial environment. In general, deep-sea siliceous oozes lie to the left of the diagram, whereas epicontinental deposits lie toward the right. Diagenetic path A represents silica initially deposited as opal-A (diatoms, radiolarians), which subsequently transforms to opal-CT and then microquartz by means of solution-reprecipitation steps. Path C represents early diagenetic cherts in which microquartz forms during shallow burial. Path B represents a possible pathway for chert formation under conditions intermediate between A and C. Megaquartz forms by metamorphic recrystallization of microquartz or by direct growth into voids (suggested by the "spike") at any stage of burial. Fibrous silica can grow in vugs and fractures at all burial depths. (After Knauth, L. P., 1994, Petrogenesis of chert, *in* Heaney, P. J., C. T. Prewitt, and G. V. Gibbs (eds.), Silica: Physical behavior, geochemistry and materials applications: Mineralogical Society of America Reviews in Mineralogy, v. 29, Fig. 4, p. 239, reproduced by permission.)

cristobalite disordered by interlayered tridymite lattices. Cristobalite and tridymite are metastable varieties of quartz that alter with time to quartz. Opal-CT may occur in open spaces in sediments as **lepispheres**, which are microcrystalline aggregates of blade-shaped crystals. It can also form as nonspherulitic blades, rim cements, and overgrowths, and as a massive cement (Maliva and Siever, 1988a). Note that the transformation of opal-A to opal-C is a solution-reprecipitation process; that is, opal-A dissolves to generate silica-rich pore waters from which opal-CT precipitates. In turn, opal-CT is converted into microquartz, also by solution-reprecipitation. Finally, microquartz is transformed to megaquartz as burial temperatures approach those of metamorphism. Megaquartz, as well as fibrous quartz, can also form at lower temperatures in small amounts as cavity and fracture fillings.

The rates of diagenetic evolution of silica from biogenic opal-A to opal-CT and finally to quartz chert are controlled by several physicochemical factors. Temperature is commonly considered to be a particularly important control, with increasing temperature promoting an increased rate of transformation (Siever, 1983). The transformation is fastest where rates of sedimentation and burial are high (sediment is rapidly buried to depths where high temperatures prevail) or where a high geothermal gradient exists in a region. In the deep ocean environment, conversion of opal-A to opal-CT apparently takes place at burial depths corresponding to a temperature of about 45°C and transformation of opal-CT to microquartz occurs at temperatures of about 80°C. Kastner and Gieskes (1983) demonstrated that the rate of transformation of opal-A to quartz chert depends also upon the nature of the opal starting material and the presence of magnesium hydroxide compounds, which serve as a nucleus for the crystallization of opal-CT. Williams, Parks, and Crerar (1985) conclude that increasing surface-to-volume ratio of siliceous particles, a ratio that increases with decreasing particle size, increases solubility of opal and the rate of transformation. Because these other factors also affect the rate of transformation of opal-A to opal-CT, Knauth (1994) suggests that some transformations may take place at shallower burial depths and lower temperatures (e.g., path B in Fig. 7.1.1). Bohrman et al. (1994) report the formation of opal-CT nodules and layers (porcellanites) at very shallow burial depths and low temperatures in cores of Antarctic deep-sea sediment. The porcellanites are present only in sediments rich in opal-A with extremely low levels of detrital minerals. This finding suggests that the absence of detrital impurities allows the transformation of opal-A to opal-CT to proceed faster, as previously reported by Isaacs (1982). It has also been suggested (e.g., Williams, Parks, and Crerar, 1985) that under some conditions opal-A may transform directly to quartz chert without going through an intermediate opal-CT stage (path C in Figure 7.1.1).

Origin of Nonfossiliferous Chert

The origin of chert that does not contain siliceous organic remains is poorly understood. Direct, inorganic precipitation of amorphous silica has been reported in some ephemeral Australian lakes (Peterson and von der Borch, 1965). Pleistocene cherts from alkaline Lake Magadi, Kenya, also apparently formed inorganically by alteration of sodium silicate precursors such as magadiite owing to removal of Na by meteoric waters; the residual silica crystallized to quartz chert (Schubel and Simonson, 1990). No similar occurrences have been reported in the open marine environment that could help explain the presence of nonfossiliferous bedded chert deposits. The scarcity of radiolarians and sponge spicules in Phanerozoic cherts does not preclude the possibility that these cherts were formed from the remains of siliceous organisms. They could have been derived from siliceous oozes that were subsequently almost completely dissolved and recrystallized, leaving few recognizable siliceous organic remains (Weaver and Wise, 1974). Murray, Jones, and Brink (1992) suggest that some bedded chert-shale couplets may form by diagenetic processes. According to these authors, biogenic SiO_2 in shales dissolves, migrates out of the shales, and then reprecipitates adjacent to the shale to form bedded chert.

Ledesma-Vázquez et al. (1997) describe a 14-m-thick, shallow-water chert (El Mono chert) from Pliocene deposits of Baja California that contains no recognizable fossil remains. They suggest that this chert formed when silica-rich geothermal solutions reacted with Pliocene sediments to deposit secondary silica in the form of Opal-A. Some reported nonfossiliferous chert may simply be the result of inadequate examination. Etching with hydrofluoric acid might reveal siliceous organisms in such cherts.

Precambrian Bedded Chert

Bedded cherts are also very common in stratigraphic successions of Precambrian age, associated with stromatolitic carbonates, iron formations, Archean greenstones, and silicic volcanic successions. These cherts do not contain unequivocal remains of silica-secreting organisms, although a few workers (e.g., LaBerge, Robbins, and Han, 1987) reported possible siliceous organic remains in some Precambrian cherts. The remains of cyanobacteria have also been reported in Precambrian cherts (e.g., Fairchild et al., 1996). Whether or not cyanobacteria or other bacteria may have aided in precipitation of silica gel (opal-A) remains to be established. Until the presence of silica-secreting Precambrian organisms is proven, or a definite link between bacteria and silica precipitation established, we must assume that deposition took place by inorganic processes, probably along pathway C in Figure 7.1.1. How these processes operated given the geochemical constraints discussed above is not understood, nor is the immediate source of the silica known. Perhaps the silica content of the Precambrian ocean was higher than that of the Phanerozoic ocean, possibly related to volcanism; or perhaps it was higher simply because silica concentrations could build in the absence of silica-secreting organisms. Thus, the origin of Precambrian cherts remains something of an enigma.

Nodular and Other Replacement Chert

In addition to occurrence as bedded chert, chert can occur in the form of small nodules, lenses, or thin, discontinuous beds, e.g., Fig. 7.11. Nodular cherts are especially common in limestones but may be present also in evaporites and siliciclastic sedimentary rocks, particularly shales. Relict textures in nodular cherts suggest that most are formed by diagenetic replacement. Some replacement cherts form in the open ocean where they replace carbonates and clays, as shown by preservation of burrows and other sedimentary structures (Hein and Karl, 1983). The silica that forms these so-called deep-sea cherts is furnished by dissolution of local siliceous organisms, especially diatoms (in Cenozoic occurrences). Nodular cherts are particularly common in shallow platform carbonate rocks. Silica is supplied in this environment by the dissolution of sponge spicules or other forms of biogenic opal-A within the sediment pile. This dissolution process causes the pore waters to become supersaturated in silica with respect to opal-CT and quartz. Chertification then occurs, possibly controlled by force-of-crystallization replacement of the host carbonate, in turn controlled by nonhydrostatic stresses resulting from opal-CT and quartz crystal growth that simultaneously causes calcite dissolution (Maliva and Siever, 1989). Silica cement can also accumulate in voids. The transformation from opal-A to quartz chert does not necessarily require an intermediate opal-CT stage and may be represented by path C in Figure 7.1.1. For additional insight into some of the geochemical conditions that favor the formation of replacement chert and the mechanisms of replacement, see Maliva and Siever (1988b) and Knauth (1994).

7.4 IRON-BEARING SEDIMENTARY ROCKS

Introduction

Some iron is present in almost all sedimentary rocks. The average iron content of siliciclastic shales, for example, is 4.8 percent. Sandstones contain 2.4 percent iron on average, and limestones contain about 0.4 percent (Blatt, 1982). The term **iron-rich** is reserved for sedimentary rocks that contain much higher iron content—at least 15 percent total iron. Most iron-rich sedimentary rocks were deposited during three time periods: the Precambrian, the early Paleozoic, and the middle to late Mesozoic (Jurassic-Cretaceous). They make up only a minor fraction of the total sedimentary record; however, they have great economic significance as iron ores. They constitute the major source of iron mined for commercial purposes. Economically important iron deposits are located on all major continents except Antarctica, and at least one deposit of sedimentary iron characterized as very large occurs in each of these continents: the Lake Superior region–Labrador Trough, North America; the Transvaal-Griquatown region, South Africa; the Hammersley Range, Australia; the Krivoy Rog-KMA, former USSR, Eurasia; and Minas Gerais, Brazil, South America.

Kinds of Iron-Rich Sedimentary Rocks

The major sedimentary iron deposits of the world were divided by James (1966) into two principal kinds: iron-formation (for dominantly Precambrian, cherty iron-rich sediments) and ironstone (for dominantly Phanerozoic noncherty iron-rich sediments). Some subsequent workers (e.g., Trendall, 1983; Young, 1989) have expressed dissatisfaction with this scheme, particularly labeling ironstones as solely Phanerozoic deposits. Kimberley (1994) suggests than an ironstone is any chemical sedimentary rock with >15 percent iron and an iron formation is a stratigraphic unit composed largely of ironstone. Thus, according to Kimberley, an iron formation can be either cherty or noncherty. More in keeping with previous usage, I have chosen in this book to retain the term **ironstone** for the mainly nonbanded, noncherty, commonly oolitic, iron-rich sedimentary rocks and the term **iron formation** for the mainly well-banded, cherty, iron-rich sedimentary rocks. Volumetrically, ironstones are much less important than iron formations. Other kinds of iron-rich sedimentary rocks of minor importance include iron-rich shales and miscellaneous iron-rich deposits such as bog iron ores and iron-rich laterites (e.g., Dimroth, 1979).

Iron Formations

Iron formations are iron-rich deposits that range in age from early Precambrian to Devonian, although they are primarily of Precambrian age (James and Trendall, 1982). They consist of distinctively banded successions (Fig. 7.14), 50–600 m thick, composed of layers enriched in iron alternating with layers rich in chert. Banding occurs on a scale ranging from millimeters to tens of meters. Cherty iron formations are associated with dolomite, quartz-rich sandstone, and black shale and can grade locally into chert or dolomite. Iron formations can have a variety of textures that may resemble those of limestones. Micritic, pelleted, intraclastic (rip-up clasts), peloidal, oolitic, pisolitic, and stromatolitic textures are recognized. Simonson (2003) suggests that the term **granular iron formation** (GIF) be used for iron formations that have (or originally had) coarse, granular textures and that the term **banded iron formation** (BIF) be reserved for iron formations that have

Figure 7.14
Banded iron formation, Helen Formation (Archean), Helen Mine, Algoma Steel Corp., Wawa, Ontario, Canada. The very dark layers are iron oxides with quartz (chert); lighter layers are iron carbonates (ankerite). [Photograph courtesy of Robert H. Hodder.]

much finer grained textures. Sedimentary structures reported from banded iron formations include cross-bedding, graded bedding, load casts, ripple marks, erosion channels, shrinkage cracks, and slump structures. These structures show that many of the particles that make up iron formations, especially the granular iron formations, have undergone mechanical transport and deposition.

Ironstones

Ironstones are dominantly Phanerozoic sedimentary deposits that occur on all continents. They are mainly early Paleozoic, Jurassic-Cretaceous, and early Cenozoic in age, but they range in age from middle Precambrian to Holocene (Petránek and Van Houten, 1997). They form bedded successions a few meters to a few tens of meters thick (e.g., Fig. 7.15), which are poorly banded or nonbanded, in sharp contrast to the much thicker, well-banded iron formations. They commonly have an oolitic texture (Fig. 7.16), and they may contain fossils that have been partly or completely replaced by iron minerals. Sedimentary structures that include cross-bedding, ripple marks, scour-and-fill structures, clasts, and burrows are present in many ironstones, indicating that mechanical transport of grains was involved in the origin of these rocks. Ironstones are commonly interbedded with carbonates, particularly limestones; shales; and fine-grained sandstones of shelf to shallow-marine origin. They may grade locally to siliciclastic

Figure 7.15
Oolitic ironstones of the Bell Island Group (Ordovician), at Wabana, Bell Island, Newfoundland. The succession shown also includes sandstones (light colored), dark shales, and minor carbonates. [Photograph courtesy of Robert H. Hodder.]

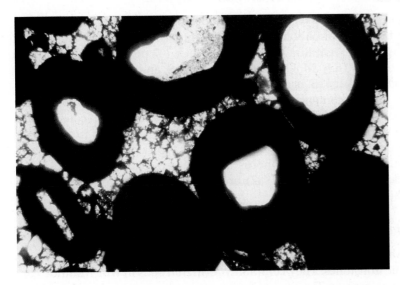

Figure 7.16
Ironstone ooids with quartz nuclei, cemented with sparry calcite cement, Clinton Formation (Silurian), New York. Ordinary light photograph. Scale bar = 0.5 mm.

sedimentary rock units. See Van Houten and Bhattacharyya (1982) and Young and Taylor (1989) for more details.

Mineralogy and Chemistry of Iron Formations and Ironstones

On the basis of relative abundance of major kinds of iron-bearing minerals, James (1966) defines four different mineral facies in iron-rich sedimentary rocks: oxides, silicates, carbonates, and sulfides. The principal minerals in each of these mineral classes are shown in Table 7.2. The oxides and silicates are commonly the most important iron-bearing minerals; however, sulfide minerals may constitute the major iron mineral in some thin beds.

Iron formations consist mainly of SiO_2 and Fe, but the chemical composition of these rocks ranges widely depending upon the type of deposit. It is difficult to

Table 7.2   Principal iron-bearing minerals in iron-rich sedimentary rocks

Mineral class	Mineral	Chemical formula
Oxides	Goethite*	FeOOH
	Hematite	Fe_2O_3
	Magnetite	Fe_3O_4
Silicates	Chamosite	$3(Fe, Mg)O \cdot (Al, Fe)_2O_3 \cdot 2SiO_2 \cdot nH_2O$
	Greenalite	$FeSiO_3 \cdot nH_2O$
	Glauconite	$KMg(Fe, Al)(SiO_3)_6 \cdot 3H_2O$
	Stilpnomelane	$2(Fe, Mg)O \cdot (Fe, Al)_2O_3 \cdot 5SiO_2 \cdot 3H_2O$
	Minnesotaite (iron talc)	$(OH)_2(Fe, Mg)_3Si_4O_{10}$
Sulfides	Pyrite	FeS_2
	Marcasite	FeS_2
Carbonates	Siderite	$FeCO_3$
	Ankerite	$Ca(Mg, Fe)(CO_3)_2$
	Dolomite	$CaMg(CO_3)_2$
	Calcite	$CaCO_3$

*Not found in Precambrian iron formations

establish a truly representative average composition; however, Gole and Klein (1981) suggest that iron formations commonly contain 40–50 percent SiO_2; 29–32 percent total Fe; 3–6 percent MgO; 2–7 percent CaO; 1–2 percent Al_2O_3; and less than 1 percent each of TiO_2, MnO, Na_2O, K_2O, P_2O_5, S, and C. Although iron, expressed as Fe_2O_3, FeO, or FeS, is the dominant chemical constituent in some iron-rich sediments, the iron content of many iron-rich sedimentary rocks is commonly exceeded by that of silica. Also, manganese concentration may reach considerable percentages in some iron formations. The average iron content of ironstones is similar that of iron formations; however, ironstones typically have a higher concentration of aluminum and phosphorus and a lower concentration of silicon. Appel and LaBerge (1987), Melnik (1982), and Trendall and Morris (1983) provide additional details.

Iron-Rich Shales

Pyritic black shales occur in association with both Precambrian iron formations and Phanerozoic ironstones. They commonly form thin beds in which sulfide content may range as high as 75 percent. Pyrite occurs disseminated in these black carbonaceous shales and in some limestones. It may be present also as nodules, as laminae, and as a replacement of fossil fragments and other iron minerals. Pyrite-rich layers have likewise been reported in some limestones. **Siderite-rich shales** (clay ironstones) occur primarily in association with other iron-rich deposits. They are present also in the coal measures of both Great Britain and the United States. Siderite (iron carbonate) occurs disseminated in the mudrocks or as flattened nodules and more or less continuous beds.

Miscellaneous Iron-Rich Sediments

Bog iron ores are minor accumulations of iron-rich sediments that occur particularly in small freshwater lakes of high altitude. They range from hard, oolitic, pisolitic, and concretionary forms to soft, earthy types. **Iron-rich laterites** are residual iron-rich deposits that form as a product of intense chemical weathering. They are basically highly weathered soils in which iron is enriched. **Manganese crusts and nodules** are widely distributed on the modern seafloor in deeper parts of the Pacific, Atlantic, and Indian oceans in areas where sedimentation rates are low. They have been reported also from ancient sedimentary deposits in association with such oceanic sediments as red shales, cherts, and pelagic limestones. Both iron-rich (15–20 percent Fe) and iron-poor ($<\sim6$ percent Fe) varieties of manganese nodules are known. These nodules contain various amounts of Cu, Co, Ni, Cr, and V in addition to manganese and iron oxide minerals. The presence of these valuable metals in manganese nodules has caused considerable interest in the possibility of mining them from the seafloor. Recovery vessels are already being planned and designed, and political negotiations have been underway for some time among the major nations of the world regarding undersea mining rights to these potentially valuable deposits. They are not likely to be mined, however, until cheaper land sources of iron are exhausted.

Iron-rich metalliferous sediments have been discovered in several oceanic settings, particularly near active mid-ocean spreading ridges. They form by precipitation from metal-rich hydrothermal fluids that have become enriched through contact and interaction with hot basaltic rocks. These sediments are enriched in Fe, Mn, Cu, Pb, Zn, Co, Ni, Cr, and V. Metal-enriched sediments have been reported also from some ancient sedimentary deposits in association with submarine pillow basalts and ophiolite sequences of ocean crustal rocks.

Heavy mineral placers are sedimentary deposits that form by mechanical concentration of mineral particles of high specific gravity, commonly in beach or alluvial environments. Magnetite, ilmenite, and hematite sands are common constituents of placers, particularly beach and marine placers. Placers are local accumulations, generally less than 1–2 m in thickness, that occur mainly in Pleistocene-Holocene sediments. Marine placer deposits containing about 5 percent iron ore were mined off the southern tip of Kyushu, Japan, for many years (Mero, 1965). Offshore placers containing up to 10 percent magnetite and ilmenite have been reported off the southeastern coast of Taiwan (Boggs, 1975). Beach placers containing ilmenite have been exploited commercially in Australia since about 1965 (Hails, 1976). "Fossil" placer deposits are comparatively rare, although thin, heavy-mineral laminae are common in some ancient beach deposits. Hails (1976) reports that outcrops of ilmenite- and magnetite-bearing placers of Cretaceous age are exposed discontinuously through New Mexico, Colorado, Wyoming, and Montana subparallel to the Rocky Mountains.

Origin of Iron Formations and Ironstones

Iron Deposition in Modern Environments

There are no modern counterparts to the ancient environments that presumably favored widespread deposition of iron-rich sediments to produce iron formations and ironstones. Iron-rich ooids and peloids have been reported off the northeast coast of Venezuela, in muddy deposits of the Orinoco Delta, and off the Amazon Delta (e.g., Van Houten, 2000). On the whole, however, modern examples of iron-rich sediments are comparatively rare and provide few clues to the depositional conditions that favored generation of ancient iron deposits.

Iron Deposition in Ancient Environments

The transport and deposition of iron are governed by both the Eh and the pH of the environment. Eh-pH diagrams such as Figure 7.17A can be used to predict the stability of iron-bearing minerals and serve to illustrate that Eh is commonly more important than pH in determining which iron-bearing mineral will be deposited. For example, hematite (Fe_2O_3) is precipitated under oxidizing conditions at the pHs commonly found in the ocean and most surface waters; siderite ($FeCO_3$) forms under moderately reducing conditions; and pyrite (FeS_2) forms under moderate to strong reducing conditions. Figure 7.17B shows the ranges of Eh and Ph in some natural environments.

Because the iron geochemistry of natural systems is far more complex than the simplified conditions assumed in constructing Eh-pH diagrams, such diagrams are of only limited use in interpreting the actual environment of iron deposition. Many problems are associated with the formation of sedimentary iron deposits, and the mechanisms by which iron was transported and deposited in the past to generate iron formations and ironstones are still poorly understood and very controversial. Three problems have to be addressed to account for the formation of iron-rich rocks: the source of the iron, transport of the iron to the depositional basin, and precipitation of the iron within the basin. Most workers now appear to agree that large iron formations were deposited in continental shelf to upper slope marine environments.

Many workers originally assumed that the iron was derived by subaerial weathering of iron silicate minerals; however, having the source on land creates a major problem with respect to transport of iron in solution. Iron in the oxidized or ferric (Fe^{3+}) state is much less soluble than iron in the reduced or ferrous (Fe^{2+}) state (Fig. 7.17). Ferric iron is soluble only at a pH less than about 4; such values rarely occur under natural conditions. Thus, under oxidizing conditions present in the weathering environment, iron tends to precipitate rather than undergo solution. How, then, can large quantities of iron be taken into solution and transported from subaerial weathering sites under the oxidizing conditions that commonly prevail in streams and rivers?

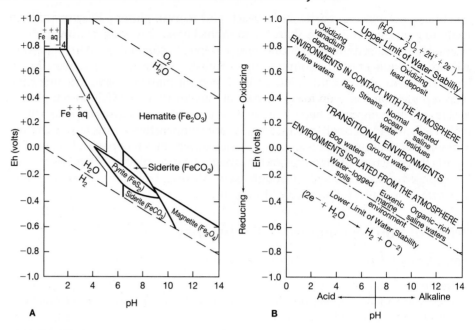

Figure 7.17
A. Eh-pH diagram showing the stability fields of the common iron minerals, sulfides, and carbonates in water at 25°C and 1 atm total pressure. Total dissolved sulfur = 10^{-6}; total dissolved carbonate = $10°$. B. Graph showing Eh and pH of waters in some natural environments. [A. from Garrels, R. M., and C. L. Crist, 1965, Solutions, minerals, and equilibria, Fig. 7.21, p. 224, reprinted by permission of Harper & Row, New York. B. after Blatt, H., G. V. Middleton, and R. Murray, 1980, Origin of sedimentary rocks, 2nd ed., Fig. 6.12, p. 241, reprinted by permission of Prentice-Hall, Englewood Cliffs, N.J., based on data from Baas Becking, L. G. M., et al., 1960, Limits of the natural environment in terms of pH and oxidation-reduction potentials: Jour. Geology, v. 68, p. 243–284.]

This problem of solution and transport of iron under oxidizing conditions prompted some workers (e.g., Lepp and Goldich, 1964; Cloud, 1973; Lepp, 1987) to postulate that the low atmospheric oxygen concentrations that apparently existed during the early Precambrian allowed great quantities of iron to be transported from land in the soluble, reduced (Fe^{2+}) state to marine basins. Lepp (1987) suggests that this transported ferrous iron was initially stored in basinal bottom waters in the ocean for long periods of time before finally precipitating. This argument of low oxygen concentrations cannot, however, be used to explain the solution and transport of iron during later Precambrian and Phanerozoic time when an oxidizing atmosphere existed. Some workers have suggested that iron may have been transported as colloids by physical processes rather than in true solution or that it was sorbed to clay particles or organic materials and transported along with these substances. It appears unlikely, however, that mechanisms such as colloidal transport can account for transport of the large quantities of iron that occur in iron formations or ironstones (Ewers, 1983). Reducing conditions seem to be required for transport of large amounts of iron in solution; hence the dilemma with respect to transport of iron by surface waters during late Precambrian and Phanerozoic time.

To get around this difficulty, many recent workers have now suggested that the source of the iron lay within the depositional basin itself. Some (e.g., Drever, 1974; Button et al., 1982) propose that dissolution of iron-bearing minerals in terrigenous siliciclastic bottom sediments or other submarine rocks provided iron to bottom waters. In other words, iron-bearing minerals were transported to the ocean where ferric iron was reduced by anoxic (low-oxygen) bottom waters and the resulting ferrous iron (Fe^{2+}) taken into solution. A considerable body of opinion now appears to be solidifying around the alternate possibility that the iron in major iron formations was derived by "exhalation" from oceanic rocks (e.g., Gross, 1980; Simonson, 1985, 2003; Kimberley, 1994; Isley, 1995), although not all geologists agree with this idea (e.g., Petránek and Van Houten, 1997). Submarine reaction of outpouring lava and hydrothermal activity from hot springs located along mid-ocean ridges furnished iron to ocean water (Chapter 1), or possibly iron-rich fluids were also exhaled from source regions too deep within the crust or mantle for seawater convection (Kimberley, 1994). Presumably the ocean at that time was stratified into an upper, oxygen-rich layer and intermediate-deep oxygen-poor (anoxic) layers (e.g., Simonson, 2003).

Deep, iron-rich, anoxic oceanic waters are postulated to move upward toward the surface along continental shelves because of (1) upwelling (e.g., Button et al., 1982), (2) spreading laterally as a plume from high-standing mid-ocean ridges (e.g., Isley, 1995), or (3) explosive exhalation through an ocean into the atmosphere and subsequent raining of atmospheric precipitates onto the shelf-slope (e.g., Kimberley, 1994). The relative importance of each of these postulated transport mechanisms is not known. Once iron-rich waters have moved into the upper slope-shelf environment, ferrous iron (Fe^{2+}) is oxidized to ferric iron (Fe^{3+}) and precipitation occurs. Oxidation can take place in the presence of molecular oxygen or, putatively, by a photochemical process caused by ultraviolet radiation in sunlight (e.g., Braterman et al., 1983; Anbar and Holland, 1992).

At this time, it does not appear possible to apply a single depositional model to all iron-rich sedimentary rocks of all ages. Although consensus seems to be emerging that a major source of the iron in iron formations probably lay within the ocean itself, some iron may have been derived from continental sources, particularly during the early Precambrian. Many puzzling aspects of the formation of sedimentary iron deposits, particularly banded iron formations, still remain. Why, for example, were chert and iron not deposited together as banded iron formations after the Precambrian? Presumably, the silica concentration of Precambrian seawater was much higher than in today's seawater, possibly as high as 60 ppm (Siever, 1992). But how was the silica precipitated? Coprecipitation with iron (Ewers, 1983), biogenic inducement (LaBerge, Robbins, and Han, 1987), or evaporative concentration and polymerization owing to electrolyte changes (Morris, 1993)? All of this is controversial. And what mechanism or mechanisms produced the banding? Evaporation of water in restricted basins (Garrels, 1987), periodic sea-level changes that affected the interface between underlying iron-rich seawater and overlying iron-poor seawater (Simonson and Hassler, 1996), or periodic explosive exhalations through an ocean that subsequently rained iron precipitates onto the shelf (Kimberley, 1994)? Again, these proposals are controversial. Also, did low forms of life such as bacteria and algae catalyze or initiate precipitation of iron in some manner? If so, how did they cause precipitation, and how important was such biologic activity? How did the iron-rich ooids that are common in ironstones form (e.g., Young, 1989)? The origin of iron-rich sedimentary deposits will likely remain controversial for some time!

7.5 SEDIMENTARY PHOSPHORITES

Introduction

The phosphorus content of rocks is usually expressed as percentage P_2O_5. The average sedimentary rock contains less than one percent P_2O_5 or one-half percent phosphorus. Sedimentary **phosphorites** are rocks that are significantly enriched in phosphorus as compared with other types of rocks. Significantly, in this context, is commonly taken to mean that they contain more than about 15 percent P_2O_5 or

6.5 percent phosphorus. These phosphorus-rich sedimentary rocks are called by a variety of other names—phosphate rock, rock phosphate, phosphates—in addition to phosphorites. Sedimentary rocks that contain less than about 15 percent P_2O_5 but considerably more than that in average sediment rocks are referred to as phosphatic, e.g., phosphatic shale. The total volume of sedimentary phosphates in the geologic record is small; however, like iron-rich sedimentary rocks, phosphorites have a special economic interest. They contribute more than 80 percent of the world's production of phosphate and make up about 96 percent of the world's total resources of phosphate rock. Total world resources of sedimentary phosphate rock are estimated to be about 158,000 million tons of all grades and types (Notholt, Sheldon, and Davidson, 1989).

Sedimentary phosphates occur in rocks of all ages from Precambrian to Holocene, but phosphorite deposition appears to have been particularly prevalent during the Precambrian and Cambrian in central and southeast Asia (China, USSR/MPR, Australia); the Permian in North America; the Jurassic and Early Cretaceous in eastern Europe; the Late Cretaceous to Eocene in the Tethyan province of the Middle East and North Africa; and the Miocene of southeastern North America (Fig. 7.18).

Phosphorite nodules and phosphatic sediments occur also on the present ocean floor at shallow depths in the vicinity of coastlines. They are particularly common off the coasts of Peru and Chile, southwest Africa, eastern United States, southern and Baja California, the continental margins of India, and on some seamounts and atolls in the Pacific (see Glen, Prévôt-Lucas, and Lucas, 2000). Many of these ocean-floor phosphate occurrences are older than the Holocene; however, modern phosphate nodules are present on the ocean floor in a few places.

Mineralogy and Chemistry

Sedimentary phosphorites are composed of calcium phosphate minerals, all of which are varieties of apatite. The principal varieties are fluorapatite [$Ca_5(PO_4)_3F$],

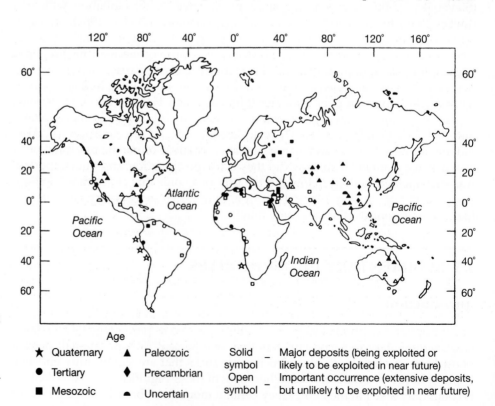

Figure 7.18
Worldwide distribution of major sedimentary phosphorite deposits. [After Cook, P. J., 1976, Sedimentary phosphate deposits, *in* Wolf, K. H. (ed.), Handbook of strata-bound and stratiform ore deposits, Fig. 1, p. 505, reprinted by permission of Elsevier Science Publishers, Amsterdam.]

Age

★ Quaternary   ▲ Paleozoic   Solid _ Major deposits (being exploited or
                                  symbol   likely to be exploited in near future)
● Tertiary   ◆ Precambrian   Open _ Important occurrence (extensive deposits,
                                  symbol   but unlikely to be exploited in near future)
■ Mesozoic   ▲ Uncertain

chlorapatite $[Ca_5(PO_4)_3Cl]$, and hydroxyapatite $[Ca_5(PO_4)_3OH]$. Most are carbonate hydroxyl fluorapatites in which up to 10 percent carbonate ions can be substituted for phosphate ions to yield the general formula $Ca_{10}(PO_4, CO_3)_6 F_{2-3}$. These carbonate hydroxyl fluorapatites are commonly called **francolite**. The wastebasket term **collophane** is often used for sedimentary apatites for which the exact chemical composition has not been determined. Detrital quartz, authigenic chert (microcrystalline quartz), opal-CT, calcite, and dolomite are also common constituents of many phosphorites. Glauconite, illite, montmorillonite, and zeolites may also be present in some deposits; moderately abundant organic matter is a characteristic constituent of many phosphorites (Nathan, 1984).

The chemistry of phosphorites is dominated by phosphorus, silicon (present in minerals other than apatite), and calcium. Slansky (1986, p. 70) shows that the abundance of these elements in 20 phosphorites ranging in age from Precambrian to Holocene is P_2O_5 = 22–39 percent; SiO_2 = <1–25 percent, and CaO = 43–53 percent. Other common constituents include Al_2O_3 (<1–5 percent); Fe_2O_3 (<1–4 percent);   MgO (<1–6 percent); Na_2O (<1 percent); K_2O (<1 percent); F (1–4 percent); Cl (<1 percent); SO_3 (0–11 percent); and organic carbon (0–2 percent). Many trace elements, such as Ag, Cd, Mo, Se, Sr, U, Yu, and Zn, as well as the rare earth elements may also be present in phosphorites in amounts exceeding their average compositions in seawater, the crust, and the average shale (Nathan, 1984). See also Notholt, Sheldon, and Davidson (1989) and McClellan and Van Kauwenbergh (1990).

Distinguishing Characteristics

Phosphate-rich sedimentary rocks may occur in layers ranging from thin laminae a few millimeters thick to beds a few meters thick. Some phosphate successions such as the Phosphoria Formation of the Idaho-Wyoming area may reach several hundred meters in thickness, although such successions are not composed entirely of phosphate-rich rocks. Phosphorites are generally interbedded with shales, cherts, limestones, dolomites, and, more rarely, sandstones. Phosphatic rocks commonly grade regionally into nonphosphatic sedimentary rocks of the same age.

Phosphorites have textures that resemble those in limestones. Thus, they may be made up of peloids, ooids, fossils (bioclasts), and clasts that are now composed of apatite. Some phosphorites lack distinctive granular textures and are composed instead of fine, micrite-like, textureless collophane. The phosphatic grains may contain inclusions of organic matter, clay minerals, silt-size detrital grains, and pyrite. Peloidal or pelletal phosphorites are particularly common; oolitic phosphorites are somewhat less so. Phosphatized fossils or fragments of original phosphatic shells are important constituents of some deposits. Most phosphorite grains are sand size, although particles greater than 2 mm may be present. These larger grains, referred to as nodules, can range in size to several tens of centimeters.

Because the textures of phosphorites have such close resemblance to those of limestones, some geologists suggest using modified limestone classifications to distinguish different kinds of phosphorites. For example, Slansky (1986) advocates using a classification system based to some extent on Folk's (1962) limestone classification, and Cook and Shergold (1986b) and Trappe (2001) suggest adapting Dunham's 1962 carbonate classification (modified by Embry and Klovan, 1971) for use in describing phosphorites. Using these modified classifications thus yields names such as wackestone phosphorite (Cook and Shergold) and phosclast wackestone (Trappe).

Principal Kinds of Phosphorite Deposits

Four kinds of phosphorite deposits are recognized: bedded, nodular, pebble bed, and guano. The major phosphorite deposits are mainly bedded marine deposits. **Bedded phosphorites** form distinct beds of variable thickness, commonly interbedded and interfingering with carbonaceous mudrocks, cherts, and carbonate rocks. The phosphorite in bedded deposits occurs as peloids, ooids, pisoids, phosphatized brachiopods and other skeletal fragments, micrite-like apatite mud, and cements. Perhaps the best-studied example of bedded phosphate deposits is the Permian Phosphoria Formation (Fig. 7.19). This formation has a total thickness of 420 m and extends throughout an area of about 350,000 km^2 in the Idaho-Wyoming area (McKelvey et al., 1959; Sheldon, 1989; Herring, 1995). Bedded marine phosphorites are also common in the Precambrian and Cambrian rocks of Australia, the Cretaceous-Tertiary rocks of North Africa, and many other parts of the world (Cook and Shergold, 1986a; Notholt, Sheldon, and Davidson, 1989; Burnett and Riggs, 1990; Soudry, 1992).

Bioclastic phosphorites are a special type of bedded phosphate deposit composed largely of vertebrate skeletal fragments such as fish bones, shark teeth, fish scales, and coprolites. The Rhaetic Bone Bed (Upper Triassic) of western England (Greensmith, 1989, p. 213) provides an example. Deposits composed mainly of invertebrate fossil remains such as phosphatized brachiopod shells are also known. These phosphate-bearing organic materials commonly become further enriched in P$_2$O$_5$ during diagenesis and may be cemented by phosphate minerals.

Nodular phosphorites are brownish to black, spherical to irregularly shaped nodules ranging in size from a few centimeters to a meter or more. Internal structure of phosphate nodules ranges from homogeneous (structureless) to layered or concentrically banded. Phosphatic grains, pellets, shark teeth, and other fossils may occur within the nodules. Nodular phosphorites are particularly common in many Neogene to Holocene phosphatic deposits of the world (Burnett and Riggs, 1990). Phosphate nodules are also forming today in zones of upwelling in the ocean, such as on the Peru continental margin (Burnett and Froelich, 1988). Many ancient nodular phosphorites may have had a similar origin under conditions of marine upwelling; however, some ancient phosphorite nodules may be of diagenetic origin.

Figure 7.19
Stratigraphic relations of the phosphatic Phosphoria Formation (Permian), Park City Formation (Permian), and Chugwater Formation (Triassic) of Idaho and Wyoming. Note in particular the Retort phosphate shale member and the Meade Peak phosphatic shale member. The section runs from west to east. [From Sheldon, R. P., 1986, Phosphorite deposits of the Phosphoria Formation, western United States, *in* Notholt, A. J. G., R. P. Sheldon, and D. F. Davidson (eds.), Phosphate deposits of the world, v. 2, Fig. 8.1, p. 54, Cambridge University Press, Cambridge.]

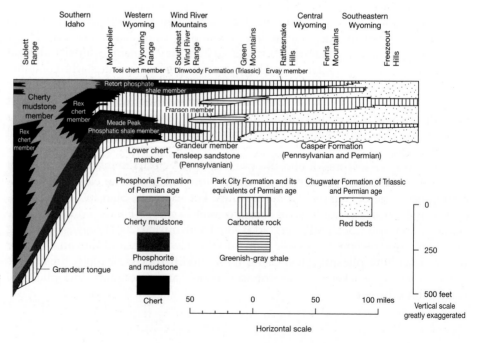

Pebble-bed phosphorites are composed of phosphatic nodules, phosphatized limestone fragments, or phosphatic fossils that have been mechanically concentrated by reworking of earlier formed phosphate deposits. The Miocene and Quaternary river-pebble and land-pebble deposits of Florida (Cathcart, 1989) provide a good example of this type of deposit.

Guano deposits are composed of bird and bat excrement that has been leached to form an insoluble residue of calcium phosphate. Guano occurs today on small oceanic islands in the eastern Pacific and the West Indies. Guano deposits are not important in the geologic record.

Origin of Phosphorites

Chemical/Biochemical Processes

As mentioned, the principal phosphate minerals in sedimentary rocks are various varieties of apatites, of which carbonate apatite $[Ca_{10}CO_3(PO_4)_6]$ is particularly important. Presumably, weathering of phosphorous-bearing rocks on land was the principal process that furnished phosphorus to the oceans, through river runoff, throughout geologic time. The average concentration of phosphorus in river water is 20 parts per billion (ppb), compared to 70 ppb in the ocean (Gulbrandsen and Roberson, 1973). Assuming approximately the same average content of phosphorus in the ancient ocean as in the modern ocean, how was the 70 ppb average phosphorus content of ancient oceans upgraded to form widespread deposits of carbonate apatite containing as much as 40 percent P_2O_5, an enrichment of up to two millionfold?

Although a variety of inorganic mechanisms for extracting phosphorus from ocean water have been considered by geologists, biologic utilization of phosphate to build soft body tissue appears to provide the most feasible answer to the problem of phosphate concentration in sediments. Modern phosphate nodules are forming in areas of oceanic upwelling where a steady supply of phosphate brought from the large, deep-ocean reservoir allows continuous growth of organisms in large numbers. After death, organisms and organic debris not consumed by scavengers pile up on the ocean floor under reducing conditions where decay is inhibited. These organic materials include the remains of phytoplankton and zooplankton, coprolites (feces), and the bones and scales of fish. All contain phosphorus; for example, phytoplankton contain about 0.4 percent phosphorus by dry weight (Gross, 1982, p. 326). Under the reducing conditions of the seafloor, some of the soft body tissue is thus preserved long enough to be buried and incorporated into accumulating sediment. Perhaps about 1 to 2 percent of the total phosphorus involved in primary productivity in upwelling zones is ultimately incorporated into the sediments in this way (Baturin, 1982).

Slow decay of body tissue after burial releases phosphorus to the interstitial waters of the sediment. Studies of the chemistry of interstitial waters in sediments where modern phosphate nodules are forming and in other areas of the seafloor where organic-rich sediments are accumulating under reducing conditions have reported phosphorus concentrations ranging from 1400 ppb to as much as 7500 ppb (Bentor, 1980; Froelich et al., 1988). At such high phosphorus concentrations, the interstitial waters are supersaturated with respect to calcium phosphate. The phosphate thus begins to precipitate on the surfaces of siliceous organisms, carbonate grains, particles of organic matter, fish scales and bones, siliciclastic mineral grains, or older phosphate particles (Baturin, 1982). Phosphate may also replace skeletal grains and carbonate grains, a process called **phosphatization**. Phosphorite grains thus form within the sediments by diagenetic reactions between organic-rich sediments and their phosphate-enriched interstitial waters. Studies of phosphorites on the Peru shelf indicate that some phosphorites form as thin (2–3 cm) crusts beneath a few centimeters of organic-rich sediment (Burnett et. al., 2000). Later exhumation may expose these crusts at the sediment surface where they are reworked by physical processes.

To allow phosphate precipitation to take place within sediments, Mg^{2+} ions (which inhibit phosphate precipitation, as they also do in carbonate precipitation) may be removed from pore waters owing to magnesium replacements of iron in clay minerals in the anoxic marine sediments (e.g., Drever, 1971). On the other hand, phosphate minerals are reported to precipitate from some sediments in which Mg concentrations are about that of seawater (e.g., Froelich et al., 1988). How phosphate precipitation under these conditions is possible is poorly understood. It may be related in some way to the presence of filamentous bacteria (cyanobacterial mats) and certain organic compounds within the pore waters (Glenn and Arthur, 1988; Schwennicke et al., 2000; Sisodia and Chauhan, 1990).

Physical Processes

The presence of clastic textures and primary depositional sedimentary structures in many ancient phosphorite deposits seems inconsistent with a diagenetic concentration mechanism. Therefore, Kolodny (1980) suggested a two-stage process for the origin of ancient phosphorite deposits. In the first stage, apatite forms diagenetically in reducing basins by mobilizing phosphorus in interstitial waters in the manner postulated for formation of modern phosphorites. The final stage involves reworking and enrichment of these diagenetically formed phosphorite grains by mechanical concentration processes under oxidizing conditions. Concentration presumably takes place in a high-energy environment, probably during lower stands of sea level. During this stage, the phosphate grains may be transported into a different depositional setting than that in which they formed. This final stage of phosphorite formation, during which the original diagenetically formed phosphorite sediments are mechanically reworked under shallow-water conditions, accounts for the clastic textures and primary sedimentary structures in many ancient phosphorites.

Summary of Phosphorite Deposition

In summary, most phosphorite workers propose that upwelling of phosphorus-rich waters from deeper parts of the ocean and biologic utilization of the phosphate in soft body tissue are the important factors in the origin of phosphorite deposits. Phosphorus is deposited on the seafloor in organic detritus and is buried with accumulating sediment. Phosphate becomes concentrated in the pore waters of sediment during slow decay of the phosphate-bearing, soft-bodied organisms and other organic detritus. Carbonate apatite precipitates diagenetically from these phosphate-enriched pore waters by some process not yet fully understood to form phosphate grains and cements. Apatite may also replace skeletal grains or other carbonate grains. Subsequently, these diagenetic deposits are reworked mechanically, possibly owing to lowered sea levels, allowing final concentration and deposition of phosphatic sediments by waves and currents. These processes are summarized diagrammatically in Figure 7.20.

This postulated multistage process for formation of phosphorite deposits has some limitations. It does not, for example, explain why phosphorites accumulated on a much vaster scale at some times in the geologic past than at present. A possible explanation for this phenomenon is that major episodes of phosphorite deposition were tied to climate and sea level changes. For example, a period of glaciation may produce a large volume of cold, nutrient-rich water that, after a long residence time, will eventually be circulated into shallower areas during rising sea level (transgression). Such an event would produce a major burst of

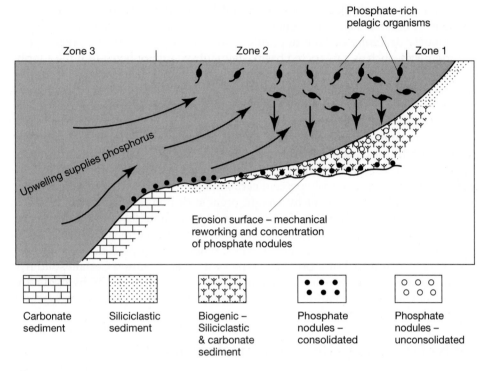

Figure 7.20
Schematic illustration of the formation of phosphorites in areas of upwelling on open ocean shelves. Near-shore, shallow-water siliciclastic deposits form in zone 1. Zone 2 is the zone where high contents of phosphate-rich biogenic detritus accumulate in sediments owing to rain-out of pelagic organisms; phosphate nodules form in this zone by diagenetic processes, followed by reworking of phosphate-rich sediments during lowered sea level. Zone 3 is a deeper water zone of carbonate sediments with local phosphate nodules. [After Baturin, G. N., 1982, Phosphorites on the sea floor: Origin, composition, and distribution, Fig. 5.4, p. 227, reprinted by permission of Elsevier Science Publishers.]

organic activity in the shallow zone (Cook and Shergold, 1986b), leading to increased phosphorite deposition. For additional discussion of the possible influence of climate on paleocirculation patterns (e.g., upwelling) and phosphorite deposition, see Parrish (1990).

To my knowledge, the only serious challenge to the upwelling (biologic-utilization) hypothesis for phosphogenesis has been mounted by Kimberley (1994). Kimberley suggests, as he suggested for iron-rich sedimentary rocks, that phosphorite deposition is brought about by precipitation from concentrated solutions, which acquired their phosphorus content by high-temperature dissolution of phosphorus-bearing minerals deep beneath the edge of a continental block. Thus, phosphorus is furnished to the ocean, according to Kimberley, by exhalative processes rather than by weathering and subsequent biologic-uptake.

7.6 CARBONACEOUS SEDIMENTARY ROCKS: COAL, OIL SHALE, BITUMENS

Introduction

Most sedimentary rocks, even some of Precambrian age, contain at least a small amount of organic matter that consists of the preserved residue of plant or animal tissue. When the tissue of organisms decays, particularly in an oxygen-deficient environment, organic degradation may not be complete; more decay-resistant fractions of organic substances such as cellulose, fats, resins, and waxes are not immediately decomposed. If a depositional basin happens to be an oxygen-poor environment—such as a restricted basin, stagnant swamp, or bog—or if the supply of organic matter is so great that it simply overwhelms all available oxidants, decay-resistant organic matter may be preserved long enough to become incorporated into accumulating sediment. Once buried, it may persist for hundreds of millions of years.

The average content of organic matter in sedimentary rocks is 2.1 weight percent in mudrocks, 0.29 percent in limestones, and 0.05 percent in sandstones (Degens, 1965). The average in all sedimentary rocks is about 1.5 percent. Organic matter contains about 50 to 60 percent carbon; therefore, the average sedimentary rock contains about 1 percent by weight organic carbon. A few special types of sedimentary rocks contain significantly more organic material than these average rocks. Black shales typically contain 3–10 percent organic matter. Oil shale or kerogen shale contains even higher percentages, ranging to 25 percent or more, and coals may be composed of more than 70 percent organic matter. Certain solid hydrocarbon accumulations—such as asphalt, formed from petroleum by oxidation and loss of volatiles—constitute another example of a sedimentary deposit greatly enriched in organic carbon.

Kinds of Organic Matter in Sedimentary Rocks

Three basic kinds of organic matter are accumulating in subaerial and subaqueous environments under present conditions: humus, peat, and sapropel. Soil **humus** is plant organic matter that accumulates in soils to form a number of decay products such as humic and fulvic acids (complex high molecular weight organic acids). Most soil humus is eventually oxidized and destroyed, and little is preserved in sedimentary rocks. **Peat** also consists of humic organic matter, but peat accumulates in freshwater or brackish-water swamps and bogs where stagnant, anaerobic conditions prevent total oxidation and bacterial decay. Therefore, some of the humus that accumulates under these reducing conditions can be preserved in sediments. **Sapropel** refers to fine organic matter that accumulates in lakes, lagoons, or marine basins where oxygen levels are low owing to poor water circulation or where the supply of organic remains is high enough to suppress oxygen concentration levels. It consists of the remains of phytoplankton, zooplankton, and spores and fragments of higher plants. Phytoplankton are tiny plants such as algae that drift about in the upper water column owing to currents; zooplankton are small drifting animals, such as foraminifers.

It is often difficult to differentiate accurately between the types of organic matter found in ancient sediments; however, both humic and sapropelic types are recognized. Humic organic matter is the chief constituent of most coals, although a few are formed of sapropel. The organic matter in oil shales and other carbonaceous mudrocks and limestones originated from sapropel, but it is so finely disseminated and altered that it is difficult to identify. This type of organic matter is called **kerogen** [see "Oil Shale (Kerogen Shale)" below].

Classification of Carbonaceous Sedimentary Rocks

The predominant organic constituents of carbonaceous sediments are thus humic and sapropelic organic matter. The nonorganic constituents are mainly either siliciclastic grains or carbonate materials. Carbonaceous sediments can be classified on the basis of relative abundance of nonorganic constituents and the kind of organic matter that composes the organic constituents (humic vs. sapropelic) into three basic types of organic-rich rocks: coal, oil shale, and asphaltic substances (Fig. 7.21). Each of these types of rocks contains at least 10 to 20 percent organic constituents.

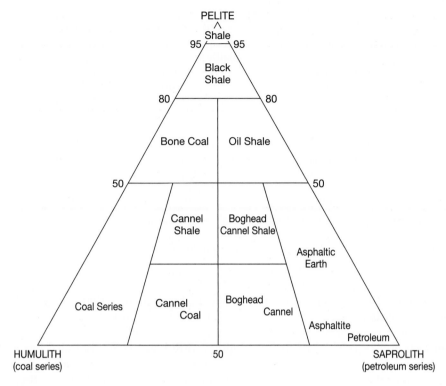

Figure 7.21
Classification and nomenclature of carbonaceous sediments on the basis of relative abundance of humic organic constituents (Humulith), sapropelic organic constituents (saprolith), and fine-grained terrigenous constituents (pelite). [From Pettijohn, F. J., 1957, Sedimentary rocks, 3rd ed., Fig. 11.37, p. 446. Copyright 1949, 1957 by Harper & Row, Publishers, N.Y. Copyright 1975 by F. J. Pettijohn. Reprinted by permisssion of Harper Collins Publishers, Inc.]

Coals

Coals are the best known kind of carbonaceous sediment. They are composed predominantly of combustible organic matter but contain various amounts of impurities (ash), which are largely siliciclastic materials. The amount of ash that coals can contain and still retain the name of coal is not precisely fixed. Some very impure coals (bone coals) may contain 70–80 percent ash, but most coals have less than 50 percent ash by weight. Most coals are humic, although a few are sapropelic coals made up mostly of spores, algae, and fine plant debris. Cannel coals and boghead coals (see below) are sapropelic coals. Coals are defined in various ways, but a commonly accepted definition is that of Schopf (1956, p. 527):

> Coal is a readily combustible rock containing more than 50 percent by weight and more than 70 percent by volume of carbonaceous material, formed from compaction or induration of variously altered plant remains similar to those of peaty deposits. Differences in the kinds of plant materials (type), in degree of metamorphism (rank), and range of impurities (grade), are characteristic of the varieties of coal.

Characteristics and Classification.   A common method of classifying coals is by **rank**, which is based on the degree of coalification or carbonification (increase in organic carbon content) attained by a given coal owing to burial and metamorphism (Table 7.3). Peat is included in Table 7.3 but is actually not a true coal. **Peat** consists of unconsolidated, semicarbonized plant remains with high moisture content. **Lignite** or brown coal is the lowest ranked coal. Lignites are brown to brownish black coals that have high moisture content and commonly retain many of the structures of the original woody plant fragments. They are dominantly Cretaceous or Tertiary in age. **Bituminous coals** are hard, black coals that contain fewer volatiles and less moisture than lignite and have a higher carbon content. They commonly display thin layers consisting of alternating bright and dull bands (Fig. 7.22). **Subbituminous coal** has properties intermediate between those of lignite and bituminous coal. **Anthracite** is a hard, black, dense coal commonly

Table 7.3   Classification of coal on the basis of rank

Class (rank steps)	Fixed carbon limits (wt. percent), dry, mineral- and matter-free basis	Volatile matter (wt. percent), dry, mineral- and matter-free basis	Calorific value limits (Btu/lb), moist, mineral- and matter-free basis
Anthracite	86–98	2–14	—
Bituminous	69–86	22–>31	10,500–14,000
Subbituminous	<69	>31	8,300–10,500
Lignite	<69	>31	6,300–8,300
Peat	low	high	low

Source: Data from American Society for Testing Materials (ASTM), 1981, *Annual book of ASTM standards*, Part 26, American Society for Testing Materials.

Note: Volatile matter is that part of coal that burns as a gas, mainly hydrogen.

containing more than 90 percent carbon. It is a bright, shiny rock that breaks with conchoidal fracture, such as the fractures in broken glass. Bituminous coals and anthracite are largely of Mississippian and Pennsylvanian (Carboniferous) age. **Cannel coal** and boghead coal are nonbanded, dull black coals that also break with conchoidal fracture; however, they have bituminous rank and much higher volatile content than anthracite. Cannel coal is composed mainly of spores. Boghead coals are composed dominantly of nonspore algal remains. **Bone coal** is very impure coal containing high ash content.

Coals are also classified on the basis of megascopic textural appearance and recognizable petrographic or microscopic constituents. Stopes (1919) recognized four types of coal, now called lithotypes, on the basis of megascopic appearance. Four lithotypes are recognized: vitrain, clarin, durain, and fusain. These lithotypes, which comprise millimeter-thick bands or layers of humic coal, are described in Table 7.4 and illustrated in Figure 7.23.

Figure 7.22
Layered and banded bituminous coal, Cedar Grove Seam (Pennsylvanian), Logan County, West Virginia. The thickness of the coal seam is about 2.3 m. [Photograph courtesy of Island Creek Coal Company.]

Table 7.4   Principal coal lithotypes and macerals

Lithotypes

Vitrain—brilliant, glossy, vitreous, black coal, bands 3–5 mm thick; breaks with a conchoidal fracture; clean to touch.

Clarin—smooth fracture with pronounced gloss; dull intercalations or striations; small-scale sublaminations within layers give surface a silky luster; the most common macroscopic constituent of humic coals.

Durain—occurs in bands a few cm thick; firm, somewhat granular texture; broken surfaces have a fine lumpy or matte texture; characterized by lack of luster, gray to brownish black color, and earthy appearance.

Fusain—soft, black; resembles common charcoal; occurs chiefly as irregular wedges; friable and porous if not mineralized.

Macerals

Vitrinites—originated as wood or bark; a major humic constituent of bright coals. Subtypes:

    Collinite—structureless or nearly structureless; commonly occurs as a matrix or impregnating material for fragments of other macerals.

    Tellinite—derived from cell-wall material of bark and wood and preserves some of the cellular texture.

Inertinites—composed of woody tissues, fungal remains, or fine organic debris of uncertain origin; relatively high carbon content. Subtypes:

    Fusunite—cell structures composed of carbonized or oxidized cell walls and hollow lumens (the space bounded by the wall of an organ) that are commonly mineral filled; characteristic of fusain.

    Semifusinite—a transitional state between fusinite and vitrinite.

    Schlerotinite—composed of the remains of fungal schlerotia (a hardened mass of tubular filaments or threads) or altered resins; characterized by oval shape and varying size.

    Micronite ($<$10 μm) and macronite (10–100 μm)—structureless, opaque, granular macerals derived from fine-grained organic detritus.

    Inertodetrinite—finely divided, structureless, clastic form of inertinite in which fragments of various kinds of inertinite macerals occur as dispersed particles.

Liptinites (exinites)—originate from spores, cuticles, resins, and algae; can be recognized from shapes and structures unless original constituents are compacted and squashed. Subtypes:

    Sporinite—composed of the remains of yellow, translucent bodies (spore exines) that are commonly flattened parallel to bedding.

    Cutinite—formed from macerated fragments of cuticles (layers covering the outer wall of a plant's epidermal cells).

    Resinite—the remains of plant resins and waxes; occur as isolated rounded to oval or spindle-shaped, reddish, translucent bodies, or as diffuse impregnations, or as fillings in cell cavities.

    Alginite—composed of the remains of algal bodies; serrated, oval shape; characteristic of bodhead coal.

Under the microscope, coal can be seen to consist of several kinds of organic units that are single fragments of plant debris or, in some cases, fragments consisting of more than one type of plant tissue. Stopes (1935) suggested the name maceral for these organic units as a parallel word for the term mineral used for the constituents of inorganic rocks. The starting materials for macerals are woody tissues, bark, fungi, spores, and so on; however, these materials are not always recognizable in coals. Macerals are divided into three major groups: vitrinite, inertinite, and liptinite (Table 7.4).

Coal macerals are identified on the basis of several characteristics: (1) reflectivity—the extent to which they reflect light, (2) degree of anisotopy (differences in reflectivity in different directions within a maceral) or isotopy as viewed under

Figure 7.23
Bituminous coal showing examples of three different lithotypes: V, vitrain; C, clarain; and D, durain. The small divisions
on the scale equal 1 cm. [From Bustin, R. M., et al., 1985, Coal
petrology, its principles, methods, and applications: Geol.
Assoc. Canada Short Course Notes, v. 3, Pl. 6A, p. 51, reprinted by permission.]

a petrographic microscope, (3) presence or absence of fluorescence when the
specimen is irradiated with blue (ultraviolet) light, (4) morphology (shape), (5)
relief, and (6) size. Study of macerals is referred to as coal petrology (e.g., Ting,
1982; Bustin et al., 1985; Ward, 1984).

Origin.   Coals occur in rocks ranging in age from Precambrian (algal coal) to Tertiary, and peat analogs of coal are present in Quaternary sediments. Coals originate in climates that promote plant growth under depositional conditions that
favor preservation of organic matter. Although ancient coals accumulated at all
latitudes from the equator to polar regions, most were deposited in middle latitudes (McCabe, 1984; see also, Cobb and Cecil, 1993). For coal to be preserved, the
rate of accumulation of organic matter must exceed the rate of decomposition
owing to microbial and chemical processes. Accumulating organic matter is most
likely to be preserved in depositional environments where oxidation of organic
matter is inhibited owing to rapid burial, such as in swampy areas in which the
water table is close to the peat surface. For thick coal deposits to form, these conditions must last for a geologically long period of time. Although land plants were
moderately well established by Devonian time, swampy environments large
enough to form major coal deposits have existed only since Carboniferous time
(Mississippian and Pennsylvanian Periods). Since that time, only the Triassic Period
appears to have been a time when coal-forming processes were at a minimum.

Compaction and loss of volatiles accompanying deep burial thins coal beds
by as much as 30 to 1 (Ryer and Langer, 1980); that is, 30 m of original peat may
produce only 1 m of coal. The rank of coal tends to increase with depth owing to
increase in temperature with depth. The formation of anthracite, for example, requires temperatures in excess of about 200°C (Daniels et al., 1990). Coals occur
predominantly in siliciclastic depositional sequences, although thin limestones
may be associated with some coals.

Oil Shale (Kerogen Shale)

Oil shale is fine-grained sedimentary rocks from which substantial quantities of
oil can be derived by heating. That is, sufficient oil is generated to produce more

energy than the energy required to produce the oil initially (Hutton, 1995). The term oil shale is actually a misnomer because relatively little free oil occurs in these rocks, although small blebs, pockets, or veins of asphaltic bitumins may be present. Oil shales are of particular interest because of their potential to generate oil when refined into fuel at sufficiently high temperatures. At least fifty countries of the world have reserves of oil shale that have the potential to be exploited as a fuel resource in the future. More than 80 percent of the organic matter in oil shales is present in the form of kerogen, which yields oil when heated to a temperature of about 350°C. The principal constituents of oil shales are shown in Figure 7.24. Organic constituents commonly do not exceed about 25 percent of the rock. Kerogen is disseminated organic matter that is insoluble in nonoxidizing acids, bases, or organic solvents (Durand, 1980; Horsfield, 1997). It consists of masses of almost completely macerated (disintegrated by biochemical or chemical processes) organic debris. On the basis of petrographic characteristics, this organic matter consists primarily of liptinite macerals (see Table 7.4). Vitrinite and inertinite macerals are commonly present in only minor amounts (Hutton, 1995). The organic matter in oil shale is derived from three primary sources: terrestrial plants, lacustrine (lake) algae, and marine organisms such as algae and dinoflagellates. Organic matter from terrestrial plants is commonly less important than that from lacustrine algae and marine organisms.

Not all so-called oil shales are actually shales. Some are organic-rich siltstones, limestones, and impure coals. **Carbonate-rich oil shales** are those in which the principal nonkerogen constituents are calcite, dolomite, ankerite, siderite, and various amounts of siliciclastic silt (Duncan, 1976). **Silica-rich oil shales** are shales in which the main constituents apart from kerogen are fine-grained quartz, feldspar, and clay minerals. They may also contain chert, opal, and phosphatic nodules. Siliceous oil shales are generally dark brown or black. **Cannel shale** is an oil shale that consists predominantly of organic matter that completely encloses other mineral grains. Cannel shales are sometimes classified as impure cannel coals. Many oil shales are characterized by distinct lamination caused by alternations of millimeter-thick organic laminae with either siliciclastic or carbonate laminae.

The amount of oil that can be extracted from oil shales through heating and retorting ranges between 10 and 150 gallons of oil per ton of rock (Duncan, 1976). The potential world supply of oil from oil shale is estimated to be 2000 trillion tons (Russell, 1990, p. 4). On the other hand, many technological problems exist with respect to mining, extracting, and refining oil shale. Oil shale is not now being processed commercially because it cannot compete economically with petroleum.

Oil shales form in environments where organic matter is abundant and anaerobic or reducing conditions prevent oxidation and total bacterial decomposition. They are deposited in both lacustrine (lake) and marine environments where the above conditions are met. The three principal environments are large lakes; shallow seas or continental platforms and continental shelves in areas where water

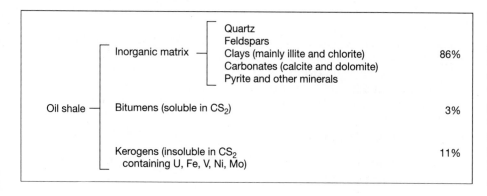

Figure 7.24
Principal constituents in oil shales. [After Yen, T. F., and G. V. Chilingarian, 1976, Introduction to oil shales, *in* Yen, T. F., and G. V. Chilingarian (eds.), Oil shales: Developments in petroleum science 5, Fig. 1.2, p. 3, reprinted by permission of Elsevier Science Publishers, Amsterdam.]

circulation was restricted and reducing or weakly oxidizing conditions existed; and small lakes, bogs, and lagoons associated with coal-producing swamps. Oil shales formed in lakes or swamps may be associated with impure cannel or boghead-type coal, tuffs and other volcanic rocks, or even evaporites. Many oil shales deposited in large lakes are carbonate-rich types and tend to have high oil yields, apparently owing to enhanced preservation potential of organic material in lake environments. Oil shales deposited in marine environments are characteristically the silica-rich type and have lower oil yields, although some Tertiary and Mesozoic siliceous oil shales have rich oil yields. Oil shales extend over wide geographic areas and are commonly associated with limestones, cherts, sandstones, and phosphatic deposits (Yen and Chilingarian, 1976).

Petroleum and Natural Bitumins

Petroleum.   Petroleum is not sedimentary rock but a carbon-rich, organic substance that accumulates predominantly in sandstones and carbonate rocks. For this reason, it is included here with the carbonaceous sedimentary rocks. Petroleum forms from plant and animal organic matter by a complex maturation process during burial that involves initial microbial alteration and subsequent thermal alteration that forms a complex organic substance called **kerogen**. Kerogen can subsequently undergo thermal degradation (cracking) at burial depths exceeding about 1000 m and temperatures of about 50° to 120°C to form liquid petroleum, a process called **catagenesis** (Hunt, 1996, p. 60; Tissot and Welte, 1984, p. 69). Liquid petroleum may subsequently be cracked at temperatures ranging from about 150° to 200°C to form natural gas (e.g., methane).

The source materials for the organic matter that eventually converts to petroleum are contained primarily in organic-rich shales and carbonate rocks. After petroleum has formed from organic materials in these fine-grained source rocks, at substantial burial depths, it migrates out of the source rocks (a process referred to as **primary migration**) into coarser grained, porous, and permeable sandstone or carbonate rocks. It then migrates through water-fill pore spaces in these rocks (**secondary migration**) to structurally higher sites, where it eventually accumulates in traps such as anticlines (Fig. 7.25). The rocks in which petroleum accumulates in

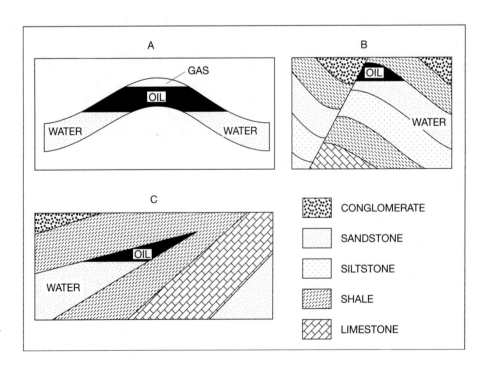

Figure 7.25
Schematic representation of three kinds of petroleum traps: A. Anticlinal trap. B. Fault trap, with an impermeable fault gouge or mineral seal along the fault. C. Stratigraphic (pinch-out) trap.

these traps, called **reservoir rocks**, are primarily porous sandstones, limestones, and dolomites.

Petroleum is composed dominantly of carbon (about 85 weight percent) and hydrogen (about 13 percent), with about 2 percent sulfur, nitrogen, and oxygen (Hunt, 1996). Despite its simple elemental chemical composition, the molecular structure of petroleum can be exceedingly complex. The molecules in petroleum range from the simple methane gas molecule (CH_4) with a molecular weight of 16 to larger molecules with molecular weights in the thousands. Several hundred different hydrocarbons have been recorded in natural crude oils; however, all hydrocarbons can be grouped into a few basic classes or series having common molecular structural form. These structural forms are complex and are not explained in detail here, but the main hydrocarbon series are as follows:

1. **paraffins (alkanes)**—open chain molecules with single covalent bonds between carbon atoms (Fig. 7.26)
2. **napthenes (cycloparaffins)**—closed ring molecules with single covalent bonds between carbon atoms (Fig. 7.27)
3. **aromatics (arenes)**—one or more benzene ring structures with double covalent bonds between some carbon atoms (Fig. 7.28)

Most natural gases as well as many liquid petroleums belong to the paraffin series of hydrocarbons. Most napthene hydrocarbons are liquid petroleums, although two occur as gases at normal temperatures. The aromatics, which are named for their strong aromatic odor, are liquid petroleums. They commonly make up only a small percentage of the petroleums in natural crude oils.

Natural Bitumens.   These substances are hydrocarbons such as natural asphalts and mineral waxes that occur in a semisolid or solid state. Most natural bitumens probably formed from liquid petroleums that were subjected to loss of volatiles, oxidation, and biologic degradation after seepage to the surface. Others may never have existed as light oils. Bitumins occur as seepages, surface accumulations, impregnations occupying the pore spaces of sandstones or other sedimentary rock (e.g., the Cretaceous Athabasca tar sands of Canada), and in veins and

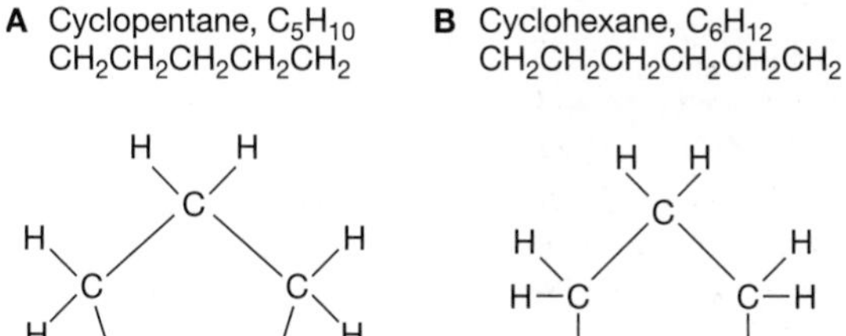

Figure 7.26
Schematic structure of paraffin hydrocarbons having the general formula C_nH_{2n+2} where n refers to the number of carbon or hydrogen atoms. A. Butane. B. Pentane.

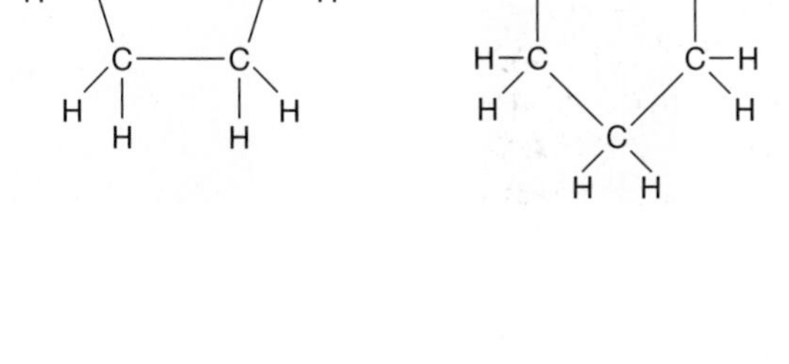

Figure 7.27
Schematic structure of naphthene (cycloparaffin) hydrocarbons having the general formula C_nH_{2n}. A. Cyclopentane. B. Cyclohexane.

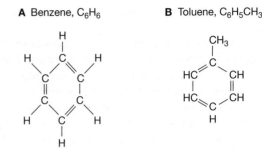

A Benzene, C_6H_6

B Toluene, $C_6H_5CH_3$

Figure 7.28
Schematic structure of aromatic hydrocarbons having the general formula C_nH_{2n-6}. A. Benzene. B. Toluene.

dikes. They are black or dark brown and have a characteristic odor of pitch or paraffin.

Natural bitumens have roughly the same elemental chemical composition as liquid petroleum, but the percentage of carbon and hydrogen tends to be somewhat lower and the content of sulfur, nitrogen, and oxygen somewhat higher. They are divided into two main types on the basis of solubility in carbon disulfide (CS_2), an organic solvent (Fig. 7.29): soluble bitumens and pyrobitumens. The soluble bitumens are further divided on the basis of ease of fusibility or melting into three groups: mineral waxes, natural asphalts, and asphaltites.

Mineral waxes are solid, waxy, light-colored substances that consist largely of paraffinic hydrocarbons of high molecular weight. Most represent the residuum of high-wax oils exposed at the surface. The most important native mineral wax is **ozocerite**, which consists of veinlike deposits of greenish or brown wax. Montan wax is an extract obtained from some kinds of brown coals or lignites.

Natural asphalts are soft, semisolid bitumins that occur as seeps, surface pools, or viscous impregnations in sediments (tar sands). They are dark colored, plastic to fairly hard, easily fusible, and soluble in carbon disulfide. Varietal names

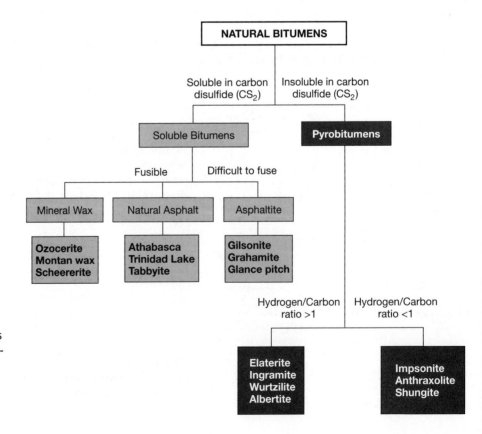

Figure 7.29
Terminology of principal kinds of naturally occurring solid hydrocarbons. Based on Rogers, McAlary, and Bailey, 1974; Hunt, 1979; Cornelius, 1987; Meyer and De Witt, 1990.

for asphalts from different areas are shown in Figure 7.29. Asphalts are commonly associated with active oil seeps.

Asphaltites occur primarily in dikes and veins that cut sediment beds. They are harder and denser than asphalts and melt at higher temperatures. They are largely soluble in carbon disulfide. Names applied to varieties of asphaltites that differ slightly in density, fusibility, and solubility are gilsonite, glance pitch, and grahamite.

Pyrobitumins, like asphaltites, occur in dikes and veins but are infusible and largely insoluble in carbon disulfide. Several varieties of pyrobitumins are recognized, which can be placed into two general groups on the basis of hydrogen/carbon ratio (Fig. 7.29). Those with H/C ratios >1 include **elaterite**, a soft elastic substance rather like India rubber, and **wurtzlite**, also a softer form. More indurated forms are **albertite**, a black, solid bitumin with a brilliant jetlike luster and conchoidal fracture, and **ingramite.** The metamorphosed pyrobitumins, **impsonite, anthraxolite,** and **shungite**, are indurated forms that have have H/C ratios <1.

The solid hydrocarbons are of interest to geologists because their presence at the surface is an indication of petroleum at depth in a region, and because study of their occurrence may help to solve the problems related to the origin and alteration of petroleum. Also, many of the solid hydrocarbons are of commercial value themselves. (See also Chilingarian and Yen, 1978; Cornelius, 1987; Meyer, 1987; and Meyer and De Witt, 1990.)

FURTHER READING

Evaporites

Busson, G., and Schreiber, B. C. (eds.), 1997, Sedimentary deposition in rift and foreland basins in France and Spain: Columbia University Press, New York, 479 p.

Melvin, J. L. (ed.), 1991, Evaporites, petroleum and mineral resources: Elsevier, Amsterdam, 555 p.

Schreiber, B. C. (ed.), 1988, Evaporites and hydrocarbons: Columbia University Press, New York, 475 p.

Warren, J. K., 1989, Evaporite sedimentology: Prentice Hall, Englewood Cliffs, N.J., 285 p.

Warren, J., 1999, Evaporites: Their evolution and economics: Blackwell Sciences Ltd., Oxford, 438 p.

Siliceous Sedimentary Rocks

Heaney, P. J., C. T. Prewitt, and G. V. Gibbs (eds.), 1994, Silica: Physical behavior, geochemistry and materials applications: Mineralogical Society of America Reviews in Mineralogy, v. 29, 606 p.

Hein, J. R. (ed.), 1987, Siliceous sedimentary rock-hosted ores and petroleum: Van Nostrand Reinhold, New York, 304 p.

Hein, J. R., and J. Obradovic (eds.), 1989, Siliceous deposits of the Tethys and Pacific regions: Springer-Verlag, New York, 244 p.

Iijima, A., J. R. Hein, and R. Siever (eds.), 1983, Siliceous deposits in the Pacific region: Elsevier, Amsterdam, 472 p.

Iron-rich Sedimentary Rocks

Appel, P. W. U., and G. L. LaBerge, 1987, Precambrian iron-formations: Theophrastus, S. A., Athens, Greece, 674 p.

Melnik, Y. P., 1982, Precambrian banded iron-formations: Developments in Precambrian Geology 5: Elsevier, Amsterdam, 310 p. (Translated from the Russian by Dorothy B. Vitaliano).

Petránek, J., and F. B. Van Houten, 1997, Phanerozoic ooidal ironstones: Czech Geological Survey Special Papers 7, Czech Geological Survey, Prague, 71 p.

Trendall, A. F., and R. C. Morris (eds.), 1983, Iron-formation facts and problems: Developments in Precambrian Geology 6: Elsevier, Amsterdam, 558 p.

Van Houten, F. B., and D. P. Bhattacharyya, 1982, Phanerozoic oolitic ironstone: Geologic record and facies models: Ann. Rev. Earth and Planet. Sci., v. 10, p. 441–457.

Young, T. P., and W. E. G. Taylor (eds.), 1989, Phanerozoic ironstones: Geol. Soc. Spec. Pub. 46, The Geological Society, London, 251 p.

Phosphorites

Baturin, G. N., l982, Phosphorites on the sea floor: Origin, composition and distribution: Developments in Sedimentology 33, Elsevier, Amsterdam, 343 p. (Translated from Russian by Dorothy B. Vitaliano.)

Bentor, Y. K. (ed.), 1980, Marine phosphorites—geochemistry, occurrence, genesis: Soc. of Econ. Paleontologists and Mineralogists Special Publication No. 29, Tulsa, Okla., 249 p.

Burnett, W. C., and S. R. Riggs (eds.), 1990, Phosphate deposits of the world: v. 3: Neogene to modern phosphorites: Cambridge University Press, Cambridge, 464 p.

Cook, P. J., and J. H. Shergold (eds.), 1986, Phosphate deposits of the world: v. 1: Proterozoic and Cambrian phosphorites: Cambridge University Press, Cambridge, 386 p.

Glen, C. R., L. Prévôt-Lucas, and J. Lucas (eds.), 2000, Marine authigenesis: From global to microbial: SEPM Spec. Pub. 66, 536 p.

Notholt, A. J. G., and I. Jarvis (eds.), 1990, Phosphorite, research and development: The Geological Society Special Publication 52, Bath, U.K., 326 p.

Notholt, A. J. G., R. P. Sheldon, and D. F. Davidson (eds.), 1989, Phosphate deposits of the world, v. 2: Phosphate rock resources: Cambridge University Press, Cambridge, 566 p.

Coal, Oil Shale, Bitumen

Bustin, R. M., A. R. Cameron, D. A. Grieve, and W. D. Kalkreuth, 1985, Coal petrology, its principles, methods, and applications: Geol. Assoc. Canada Short Course Notes, v. 3, 230 p.

Chilingarian, G. V., and T. F. Yen, 1978, Bitumins, asphalts and tar sands: Elsevier, New York, 331 p.

Cobb, J. C., and C. B. Cecil (eds.), 1993, Modern and ancient coal-forming environments: GSA Special Paper 286, 198 p.

Crelling, J. C., and R. R. Dutcher, 1980, Principles and applications of coal petrology: Soc. Econ. Paleontologists and Mineralogists Short Course Notes No. 8, Tulsa, Okla., 127 p.

Dissel, C. F., 1992, Coal-bearing depositional systems: Springer-Verlag, Berlin, 721 p.

Hunt, J. M., 1996, Petroleum geochemistry and geology, 2nd ed.: W. H. Freeman, New York, 743 p.

Meyers, R. A. (ed.), 1982, Coal structure: Academic Press, New York, 340 p.

Meyer, R. F. (ed.), 1987, Exploration for heavy crude oil and natural bitumens: AAPG Studies in Geology 25, Amer. Assoc. Petroleum Geologists, Tulsa, Okla., 731 p.

Russell, P. L., 1990, Oil shales of the world: Their origin, occurrence and exploitation: Pergamon Press, Oxford, 736 p.

Snape, C. (ed.), 1995, Composition, geochemistry and conversion of oil shales: Kluwer Academic Publishers, Netherlands, 505 p.

Stach, E., M.-Th. Mackowsky, M. Teichmüller, G. H. Taylor, D. Chandra, and R. Teichmüller, 1982, Handbook of coal petrology, 3rd ed.: Gebrüder Borntraeger, Berlin-Stuttgart, 535 p.

Thomas, L., 1992, Handbook of practical coal geology: John Wiley & Sons, Chichester, 338 p.

Tissot, B. P., and D. H. Welte, 1984, Petroleum formation and occurrence, 2nd ed.: Springer-Verlag, Berlin, 699 p.

Van Krevelen, D. W., 1993, Coal: Typology, physics, chemistry, composition: Elsevier, Amsterdam, 979 p.

Ward, C. R. (ed.), 1984, Coal geology and coal technology: Blackwell, Melbourne, 345 p.

Wignall, P. B., 1994, Black shales: Oxford University Press, New York, 127 p.

Yen, T. F., and G. V. Chilingarian (eds.), 1976, Oil shales: Elsevier, New York, 292 p.

PART *IV*

Depositional Environments

Thompson River north of Tok Junction, southeast Alaska.

241

The characteristic properties of sedimentary rocks are generated through the combined action of the various physical, chemical, and biological processes that make up the sedimentary cycle. Weathering, erosion, sediment transport, deposition, and diagenesis all leave their impress in some way on the final sedimentary rock product. The sedimentary processes and conditions that collectively constitute the depositional environment play the primary role in determining the textures, structures, bedding features, and stratigraphic characteristics of sedimentary rocks. The close genetic relationship between depositional process and rock properties provides a potentially powerful tool for interpreting ancient depositional environments. This linked set of reactions between environments and facies is commonly referred to as **process** and response (Fig. P4.1). If geologists can find ways to relate specific rock properties to particular depositional processes and conditions, they can work backwards to infer the ancient depositional processes and environmental conditions that created these particular rock properties.

Environmental analysis thus involves identifying response elements or properties that have environmental significance. These properties include sedimentary structures and textures (which reflect depositional processes such as current flow and suspension settling of grains), sedimentary facies associations (such as fining- and coarsening-upward successions of facies, which indicate shifts in environmental conditions), and fossils (which are useful indicators of salinity, temperature, water depths and water energy, and turbidity of ancient oceans). These properties can be used to construct facies models for each major depositional environment. A facies model is a general summary of the characteristics of a given depositional system. Such summary models act as a norm for the purposes of comparison and interpretation. They provide a kind of "mental picture" of the properties of rocks deposited in a given environment. Few of us have had enough field experience, have read enough books and papers, or have good enough memories

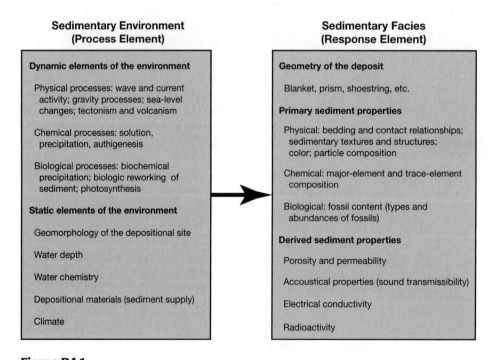

Figure P4.1
Process-response model illustrating the relationship between sedimentary environments and sedimentary facies.

to carry around a mental picture of each important depositional environment. Fortunately, we can draw on the experiences of many geologists through their published data and ideas to construct facies models that will provide the reference framework we need for interpreting ancient depositional environments.

Sedimentary rocks have been deposited through time in three major depositional settings: continental or terrestrial, i.e., on land; marginal-marine, the boundary between the sea and the land; and marine, the ocean proper. Each of these first-order depositional settings is divided into several major environments, which in turn are divided into subenvironments (Table P4.1). The following four chapters discuss the major environments in the continental, marginal-marine, and marine settings. Although not labeled specifically as such, facies models of various kinds are used throughout these chapters to summarize distinguishing characteristics of the principal sedimentary facies generated in these major sedimentary settings.

Table P4.1 Simplified classification of ancient depositional environments

Primary depositional setting		Major environment	Subenvironment
Continental	*Fluvial	*Alluvial fan	
		*Braided stream	
		*Meandering stream	
	*Desert		
	Lacustrine		
	*Glacial		
	*Deltaic	*Delta plain	
		*Delta front	
		*Prodelta	
Marginal-marine	*Beach/barrier island		
	*Estaurine/lagoonal		
	Tidal flat		
	Neritic	Continental shelf	
		**Organic reef	
Marine	Oceanic	Continental slope	
		Deep ocean floor	

*Dominantly siliciclastic deposition

**Dominantly carbonate deposition

Environments not marked by an asterisk(s) may be sites of siliciclastic, carbonate, evaporite, or mixed sediment deposition depending upon conditions.

8

Continental (Terrestrial) Environments

8.1 INTRODUCTION

We turn now to the study of continental or terrestrial depositional systems. Geologists recognize four major kinds of continental environments: fluvial (alluvial fans and rivers), desert, lacustrine (lake), and glacial. Although treated in this book as separate depositional systems, similar kinds of sediments can be generated in more than one of these environments. For example, eolian (windblown) sediments can accumulate both in desert environments and in some parts of glacial environments. Lacustrine sediments form in lakes in any environment, including deserts and glacial settings. Fluvial sediments are deposited mainly in river systems of humid regions, but they are generated also in rivers within desert areas and glacial environments.

Facies deposited in continental environments are dominantly siliciclastic sediments characterized by general scarcity of fossils and complete absence of marine fossils. Nonsiliciclastic sediments such as freshwater limestones and evaporites occur also in continental environments, but they are distinctly subordinate to siliciclastic deposits. Continental sedimentary rocks are less abundant overall than are marine and marginal marine sediments, but they nonetheless form an important part of the geologic record in some areas. Tertiary fluvial sediments of the Rocky Mountain–Great Plains region of the United States, Jurassic eolian sandstones of the Colorado Plateau, Tertiary lacustrine sediments (Green River Formation) of Wyoming and Colorado, and the late Paleozoic glacial deposits of South Africa and other parts of ancient Gondwanaland are all examples of continental deposits. Some terrestrial sediments have economic significance. They may contain important quantities of natural gas and petroleum, coal, oil shale, and uranium. We now examine, in turn, each of the major continental environments.

8.2 FLUVIAL SYSTEMS

Fluvial deposits, also referred to as alluvial deposits, encompass a wide spectrum of sediments generated by the activities of rivers, streams, and associated gravity-flow processes. Such deposits occur at the present time under a variety of climatic conditions and in various continental settings ranging from desert areas to humid and glacial regions. Although alluvial settings can be classified in many ways (e.g., Collinson, 1996) and many subenvironments of the fluvial system can be recognized, most ancient fluvial deposits can be assigned to one of two broad environmental settings: alluvial fan and river. These environments may be interrelated and overlapping.

Alluvial Fans

Definition and Depositional Setting

Alluvial fans are deposits with gross shapes approximating a segment of a cone and exhibiting convex-up cross-sectional profile (Fig. 8.1). Many have fairly steep depositional slopes. Sediments on alluvial fans are typically poorly sorted and include abundant gravel-size detritus. Modern alluvial fans are particularly common in areas of high relief, generally at the base of a mountain range, where an abundant supply of sediment is available. In many cases, they form downslope from major fault scarps. They occur both in sparsely vegetated arid or semiarid regions, where sediment transport occurs infrequently but with great violence during sudden cloudbursts, and in more-humid areas where rainfall is intense. In arid or semiarid settings, alluvial fans may pass downslope into desert-floor environments with internal drainage, including playa lake environments. In humid regions, they may merge downslope with alluvial or deltaic plains and beaches or tidal flats, or they may even build into lakes or the ocean. Fans that build into standing bodies of water are called **fan deltas** (Chapter 9). Along mountain fronts, alluvial fans developed in adjacent drainage systems may merge laterally to form an extensive piedmont, or bajada.

On the basis of depositional process, alluvial fans can be divided into debris-flow-dominated fans and stream-flow-dominated fans (Fig. 8.2). Although modern alluvial fans are common, the features that characterize alluvial fan deposits and that distinguish them from other fluvial deposits are controversial. Some authors (e.g., Blair and McPherson , 1994a) regard alluvial fans as relatively small scale features with steep slopes (1.5°–25°) that were deposited mainly by sediment-gravity flows, particularly debris flows, and upper-flow-regime fluid flows. According to this definition, many fluvial deposits originally considered to be fans are not true alluvial fans. Instead, they would be called distributary fluvial systems or braid

Figure 8.1
Aerial view of a debris-flow-dominated alluvial fan at the mouth of a canyon in the steep east wall of Death Valley, California. The highway gives the scale. [Photograph by John Shelton.]

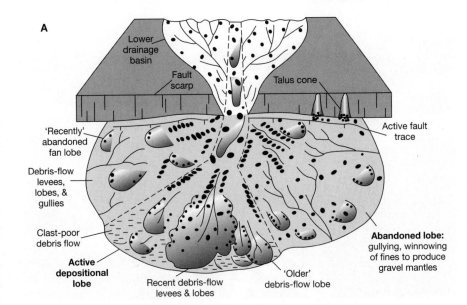

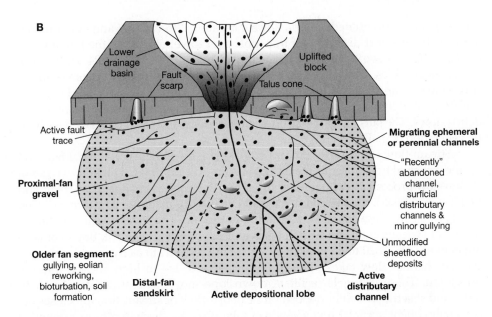

Figure 8.2
Schematic diagrams illustrating the depositional features of (A) debris-flow and (B) stream-flow-dominated alluvial fans adjacent to active normal faults. [Modified from Blair and McPherson, 1994, Alluvial fans and their natural distinction from rivers based on morphology, hydraulic process, sedimentary processes, and facies assemblages: Jour. Sedimentary Research, v. A34, Fig. 1, p. 455, reproduced by permission of the Society for Sedimentary Geology.]

deltas (see discussion by Miall, 1996, p. 246). Stanistreet and McCarthy (1993) propose a broader spectrum of fan types that include large fans with well-defined fluvial channels, such as the giant Kosi Fan of India and the huge Okavango Fan of Botswana (Africa), as well as smaller fans such as the Yana Fan of Alaska (Fig. 8.3).

Sedimentary Processes on Fans

As flows emerge from confined channels in a mountain front onto a fan, they are free to spread out, and water may infiltrate into the fan. Stream power is thus reduced, leading to deposition. Sediment-gravity flows, including debris flows and mudflows, are dominant transport and depositional processes on many fans in both arid-semiarid regions and humid setting. **Debris-flow deposits** (e.g., Fig. 2.10) are characteristically poorly sorted and lacking in sedimentary structures except possible reverse graded bedding in their basal parts. They may contain blocks of various sizes, including large boulders, and they are typically impermeable and nonporous owing to their high content of muddy matrix. Both clast-rich and clast-poor debris

Figure 8.3
Aerial view of the Yana outwash fan, Chugach Mountains, southeastern Alaska. [From J. C. Boothroyd and G. M. Ashley, 1975, Process bar morphology, and sedimentary structures on braided outwash fans, northeastern Gulf of Alaska, SEPM Special Pub. 23, Fig. 3B, p. 196, reproduced by permission.]

flows can be differentiated. Debris flows commonly "freeze up" and stop flowing after relatively short distances of transport over lower slopes on the fan; however, some flows have been reported to travel distances of up to 24 km (15 mi) (Sharp and Nobles, 1953). **Mudflows** are similar to debris flows but consist mainly of sand-size and finer sediments. **Landslides** are commonly associated with debris flows, and in many cases landslide deposits form a source of sediment for the debris flows. The surface of debris-flow-dominated fans tends to be steep with little vegetation (e.g., Fig. 8.1).

Stream-flow (fluid-flow) processes take place on all types of alluvial fans and are the principal transport mechanism on stream-flow dominated fans. Two types of stream-flow processes are operative: sheetflood and incised channel flow (Blair and McPherson, 1994b). **Sheetflood** is a broad expanse of unconfined, sediment-laden runoff water moving downslope, commonly produced by catastrophic discharge. Sediment concentration in water flows is typically about 20 percent; flows containing between about 20 to 45 percent sediment are referred to as hyperconcentrated. **Incised-channel flow** takes place through channels, 1–4 m high, incised into the upper fan. These channels facilitate downslope movement of sediment-gravity flows and sheetfloods. After deposition by debris-flow or stream-flow processes occurs, subsequent surficial reworking can take place by discharge from rainfall or snowmelt, eolian (wind) activity, and bioturbation by plants and animals.

Distinguishing Characteristics of Alluvial Fans

Alluvial fans are cone-shaped to arcuate in plan view, with network of branching distributary channels (Fig. 8.1, 8.2). The long profile, from fanhead to fantoe, is commonly concave upward; the greatest slope occurs at the fan apex and decreases down the fan. The transverse or cross-fan profile is generally convex upward. Alluvial fan sediments are dominated by gravelly deposits, which typically show down-fan decease in grain size and bed thickness and an increase in sediment sorting. Debris-flow-dominated fans are characterized by lobes of poorly sorted, coarse sediment, commonly with a muddy matrix. Stream-flow sediments constitute more sheetlike deposits of gravel, sand, and silt that may be moderately well sorted, cross-bedded, laminated, or nearly structureless.

Hooke (1967) suggested that runoff in the coarse deposits of the upper fan may percolate through the subsurface and rapidly deposit a gravel lobe as a **sieve deposit**. Presumably, highly permeable gravel deposits are generated that allow water to pass through rather than over the deposits, holding back only the coarser material. Sieve deposits have long been considered to be distinguishing features of alluvial fans; however, Blair and McPherson (1994b, p. 376) question the validity of the sieve concept, suggesting instead that most so-called sieve lobes are actually debris-flow deposits. Roger Hooke (personal communication, 2004) sees no reason to change his mind about the concept of sieve lobes, and maintains that the concept is still valid and useful.

Many individual beds in alluvial fans may display no detectable vertical grain-size trends; however, others may become either finer or coarser upward. Overall, alluvial-fan deposits tend to be characterized by strongly developed thickening- and coarsening-upward successions, caused by active fan progradation or outbuilding. Nonetheless, some fans display thinning- and fining-upward successions, which indicate relative inactivity of depositional processes or fan retrogradation (retreat) (Nilsen, 1982). The thickness of these fining- or coarsening-upward successions may be hundreds or even thousand of meters—for example, Miocene alluvial-fan deposits of the San Onofre Breccia near Dana Point, southern California, and Devonian alluvial-fan deposits along the northern margin of the Hornelen Basin, Norway. Alluvial fan deposits grade laterally into nonfan deposits such as fluvial-plain sediments, windblown deposits, or playa-lake sediments.

Ancient Alluvial-Fan Deposits

Alluvial fans may have been particularly important in Precambrian and early Paleozoic time, before the appearance of land plants that could provide an adequate vegetation cover to inhibit erosion; however, alluvial-fan deposits have been reported from stratigraphic successions of many other ages. Reported occurrences include alluvial-fan deposits in the Devonian-Hornelen Basin of Norway, the Devonian-Carboniferous of the Gaspé Peninsula, Canada, Permo-Carboniferous successions in England, the Triassic Mount Toby Conglomerate of Massachusetts, and the Jurassic Todos Santos Formation of New Mexico, as well as numerous Tertiary examples in the United States and other parts of the world (see listing by Blair and McPherson, 1994a).

The Cannes de Roche Formation (Carboniferous) of the Gaspé Peninsula, Canada, provides a Paleozoic example (Rust, 1981). A schematic depositional model for this formation is shown in Figure 8.4. The Lower Member of the formation, interpreted as alluvial-fan deposits, consists of coarse red breccia interbedded with silty sandstone and mudstone. The breccia clasts are predominately siliceous limestone. These coarse breccia units are interpreted as debris-flow deposits on the

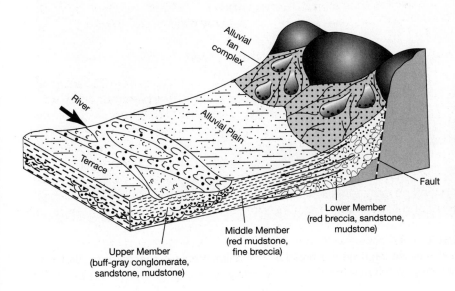

Upper Member
(buff-gray conglomerate,
sandstone, mudstone)

Middle Member
(red mudstone,
fine breccia)

Lower Member
(red breccia, sandstone,
mudstone)

Fault

Figure 8.4
Alluvial fan depositional model, the Cannes de Roche Formation (Carboniferous), Gaspé Peninsula, Canada. [Redrawn from Rust, 1981, Alluvial deposits and tectonic style: Devonian and Carboniferous successions in eastern Gaspé, *in* Miall, A. D. (ed.), Sedimentation and tectonics in alluvial basins: Geological Association of Canada Special Paper 23, Fig. 12, p. 65, reproduced by permission.]

basis of poor sorting and lack of stratification. Interbedded, horizontally stratified and cross-stratified red breccia, sandstone, and mudstone in the Lower Member, as well as the Middle Member, are interpreted to have formed by stream-flow processes. The Middle Member is the finer grained, down-fan equivalent of the coarse proximal deposits of the Lower Member. The Upper Member of the formation consists of buff to gray conglomerate with rounded cobbles, sandstone, and mudstone containing abundant plant fragments. This member is considered to be the deposits of a nearby river that flowed across the alluvial plain.

River Systems

River systems through time have been more important as sediment transport conduits to lakes and oceans than as sites of deposition. Nonetheless, rivers deposit sediment and some of this sediment is preserved under certain conditions to become part of the ancient sedimentary record. To recognize and understand the deposits of ancient river systems, it is useful to examine the channel shapes, sediment transport processes, and sediment characteristics of modern rivers.

Channel Form

According to Leeder (1999, p. 311), the channel form of rivers can be described in terms of the deviation of the channel from a straight path (*sinuosity*), the number of channels (*single* or *multiple*), the degree of channel subdivision by large bedforms (bars) and accreting islands around which channel reaches diverge and converge (*braiding*), and more permanent distributive channel subdivision into stationary smaller channels (separated by floodplains) that each contain their own channels and point bars (*anastomosing*). Some of these features, such as the size and shape of bars, vary as a function of river levels; that is, they may appear differently at low-water stage than at flood stage.

It has been common practice in the past to classify rivers into three main types on the basis of channel form: meandering (single-channel) (e.g., Fig. 8.5),

Figure 8.5
Two small meandering rivers in a broad alluvial valley, Brooks Range, northern Alaska. Note numerous cutoff meanders.

Figure 8.6
Braid bars in a braided lower reach of the Kongakut River, Arctic National Wildlife Refuge, northeastern Alaska.

braided (multiple-channel) (e.g., Fig. 8.6), and anastomosing (e.g., Fig. 8.7). Some geologists now suggest that such rigid classification is oversimplified and unsatisfactory because the different classes of channel patterns are not mutually exclusive (e.g., many rivers show combinations of sinuosity and braiding in different reaches of the river); also, different parameters are used to define the different patterns (e.g., Bridge, 2003, p. 147; Leeder, 1999, p. 311). Even so, geologists continue to refer to rivers by using these channel-form names.

The factors that influence channel sinuosity and braiding have been suggested to include the magnitude and variability of stream discharge, channel slope, grain size of sediment, bed roughnesss, the amount and kind of sediment load (bedload vs. suspended load), and the stability of the channel banks. These factors are complex, interrelated, and not fully understood. The exact causes of meandering and braiding remain somewhat obscure.

Box 8.1 Suggested Causes of Sinuosity and Braiding

Bridge (2003, p. 153) suggests that the geometry of alluvial rivers is mainly controlled by flow and sedimentary processes that operate during seasonal floods when discharge is maximal. Grain size of transported sediment is proportional to channel slope. In turn, grain size affects channel roughness, which increases with increasing grain size and stream power. The degree of braiding apparently increases as water discharge increases for a given slope and bed-sediment size, or as slope is increased for a given water discharge and bed-sediment size. Braiding occurs at lower slopes and/or discharge as bed material size decreases. Discharge variability has also been suggested to promote braiding; however, discharge variability may not actually be a critical factor. Many rivers with constant discharge display along-stream variations in channel pattern.

Sinuosity of channels increases with their width/depth for low-powered, single-channel streams, but decreases with width/depth for multiple-channel rivers. Sinuosity of single-channel rivers also increases with decreasing bed-material size for single channel rivers with a given discharge and slope.

It has also been suggested that rivers that transport large amounts of (coarse) bed load relative to suspended load tend to be associated with easily eroded banks of sand or gravel and that these rivers have large channel slopes and stream power. Such rivers have been assumed to be laterally unstable and thus prone to braiding. By contrast, large suspended loads were assumed to be characteristic of single-channel rivers of high sinuosity. Such rivers are allegedly associated with stable, cohesive muddy banks and low stream gradient and stream power. These generalities are not applicable in many cases. For example, Bridge (2003, p. 157) reports that many braided rivers are sandy and silty (e.g., Brahmaputra in Bangledesh, Yellow in China, Platte in Nebraska), and many single-channel, sinuous rivers are sandy and gravelly (Madison in Montana, South Esk in Scotland, Yukon in Alaska). Bridge also suggests that difficultly erodable banks, stabilized by vegetation or early cementation, may not have an important influence on the equilibrium channel pattern, as long as the *flood flow* is capable of eroding banks and transporting sediment.

Sediment Transport Processes in Rivers

Channel Transport.   Sediment transport (and erosion) in the higher gradient, proximal reaches of rivers occurs mainly within the river channels. Downstream flow of water around channel bends leads to helical spiraling of flow, out toward the surface and inward at the bed (Fig. 8.8). The channels are characterized by the presence of bars. **Point bars** (also referred to as side bars and lateral bars) are attached to the river bank (e.g., Fig. 8.5). The basic dynamics of flow around meanders leads to erosion on the outside parts of bends and deposition on the point bars. Helical flow transports sediment, eroded from the cut bank, across the stream along the bottom and deposits it by *lateral accretion* on the point bar. The resulting point-bar sediments are characterized by cross-bedding and general fining upward toward the top of the bar. In braided rivers, **braid bars** (also called channel, medial, longitudinal, and transverse bars, as well as sand flats) are present in midchannel position (e.g., Fig. 8.6). These braid bars can be thought of as double-sided point bars. As the current splits around the upstream end of the bar, helical flow causes lateral accretion on *both* sides of the bar. Because braid bars are free to move, in contrast to point bars, scouring and subsequent deltalike deposition takes place at the downstream end of the bar. Thus, braid bars can migrate downstream. On the other hand, some braid bars remain stable long enough to be colonized by vegetation, thus forming islands.

Floodplain Deposition.   Floodplains are strips of land adjacent to rivers that are commonly inundated during seasonal floods. Floodplains can be present along both braided and meandering rivers, although they appear to be particularly common along single-channel rivers. When the stream floods and overtops its banks, deposition of fine sediment occurs on natural levees, in adjacent flood-basins, and in oxbow lakes (Fig. 8.9). Deposition from overbank waters results in upbuilding of the sediment surface and is thus called vertical accretion, in contrast to the lateral accretion that takes place on point bars. **Natural-levee deposits** form primarily on the concave or steep-bank side of meander loops immediately adjacent to the channel as a result of sudden loss of competence, and they typically

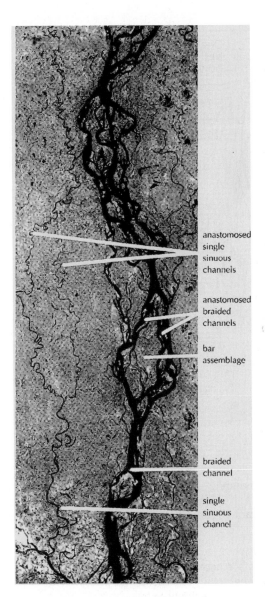

anastomosed
single
sinuous
channels

anastomosed
braided
channels

bar
assemblage

braided
channel

single
sinuous
channel

Figure 8.7
Landsat photograph of the Bramaputra River immediately north of its confluence with the Ganges, showing anastomosing, single-channel, and braided channel patterns. [From Bridge, J. S., 1993, The interaction between channel geometry, water flow, sediment transport and deposition in braided rivers, *in* Best, J. L., and C. S. Bristow (eds.), Braided rivers, Geological Society London Special Publication No. 75, Fig. 4, p. 21, reproduced by permission.]

contain horizontally stratified fine sands overlain by laminated mud. Floodplain deposits are fine-grained sediments that settle out of suspension from floodwaters carried into the floodbasin, which may be a broad, low-relief plain, a swamp, or even a shallow lake. These thin, fine-grained deposits commonly contain considerable plant debris and may be bioturbated by land-dwelling organisms or plant roots. **Crevasse-splay deposits** may also occur on floodplains where rising floodwaters breach natural levees (Fig. 8.9). Sedimentation from traction and suspension occurs rapidly after breaching as water containing both coarse bedload sediment and suspended sediment debouches suddenly onto the plain, resulting in graded deposits that may resemble a Bouma turbidite sequence (Walker and Cant, 1979). A river may also abandon its channel and move, relatively suddenly, to another position on the floodplain. This process is termed **avulsion**.

Characteristis of Fluvial Deposits

It is clear from the preceding discussion that sediments can be deposited in a variety of subenvironments within the fluvial system: on point bars and in channels of meandering rivers, in braid bars of braided rivers, and in natural levees, floodbasins,

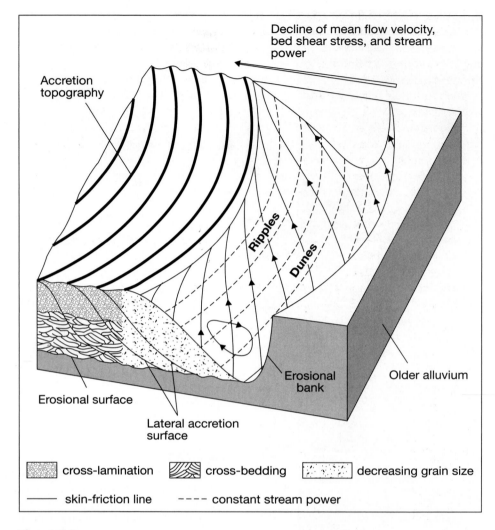

Decline of mean flow velocity, bed shear stress, and stream power

Accretion topography

Ripples

Dunes

Erosional bank

Older alluvium

Erosional surface

Lateral accretion surface

cross-lamination   cross-bedding   decreasing grain size

—— skin-friction line   ---- constant stream power

Figure 8.8
Helical flow in a meander bend, leading to lateral accretion of cross-bedded, fining-upward deposits. [From Leeder, M., 1999, Sedimentology and sedimentary basins, Fig. B17.1, p. 313. Reproduced by permission of Blackwell Science Ltd.]

Figure 8.9
The morphological elements of a meandering river system. Note: A thalweg is a line connecting the deepest points along a stream channel; it is commonly the line of maximum current velocity. [From Walker, R. G., and D. J. Cant, 1984, Sandy fluvial systems, *in* R. G. Walker (ed.), Facies models: Geoscience Canada Reprint Ser. 1, Fig. 1, p. 72, reprinted by permission of Geological Association of Canada.]

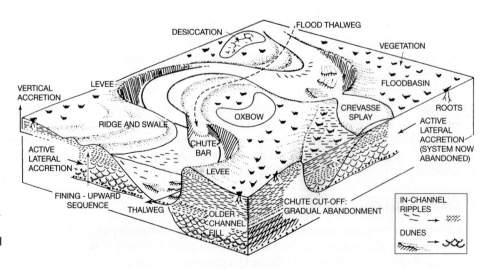

DESICCATION   FLOOD THALWEG   VEGETATION

LEVEE   FLOODBASIN

VERTICAL ACCRETION

RIDGE AND SWALE   OXBOW   CREVASSE SPLAY   ROOTS

ACTIVE LATERAL ACCRETION (SYSTEM NOW ABANDONED)

ACTIVE LATERAL ACCRETION   CHUTE BAR

LEVEE

FINING - UPWARD SEQUENCE   THALWEG   OLDER CHANNEL FILL   CHUTE CUT-OFF: GRADUAL ABANDONMENT

IN-CHANNEL RIPPLES

DUNES

and oxbow lakes of floodplains. Therefore, it is difficult to generalize about the characteristics of fluvial deposits. Nevertheless, fluvial sediments have some common properties. Most fluvial deposits consist of sand and gravel, although mud may be common in floodplain deposits of meandering streams. Some braided channels may also have been formed in muddy sediments on floodplains; however, the mud was probably transported as sand-sized pellets (Bridge, 2003, p. 157). Sorting of most fluvial sediments ranges from moderate to poor. The deposits of point bars and braid bars generally display fining-upward grain size owing to the helical nature of sediment transport on bars. Migration of meanders also produces a general fining-upward succession as channel lag deposits are overlain by fining-upward point bar deposits and, in turn, silty and muddy floodplain deposits (Allen, 1970b). Multiple episodes of channel shifting and bar migration in braided rivers produce vertical stacking of bar deposits, perhaps separated by thin mudstones (Fig. 8.10A). Multiple episodes of meander migration produce vertical stacking of fining-upward successions in meandering-river deposits (Fig. 8.10B). See Miall (1996) for additional examples of vertical profiles in fluvial sediments.

Fluvial deposits commonly display abundant traction structures, including planar and trough cross-bedding, upper-flow-regime planar bedding, and ripple-marked surfaces. Sedimentary structures yield unidirectional, downstream

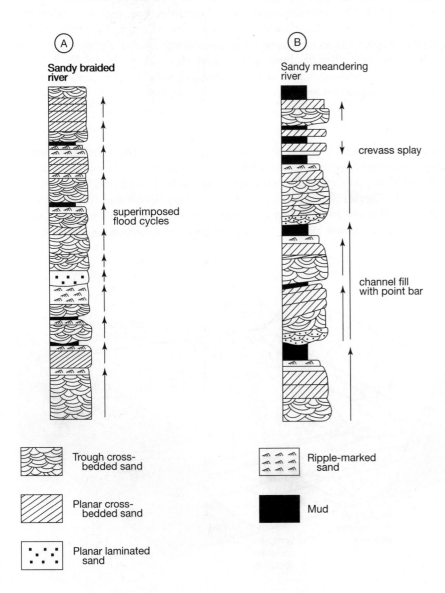

Figure 8.10
Examples of lithofacies and vertical profiles in sediment from a sandy braided river (A) and a sandy meandering river (B). [After Miall, A. D., 1996, The geology of fluvial deposits, Fig. 8.80, p. 205, and 8.8G, p. 204, Springer-Verlag, reproduced by permission.]

paleocurrent directions that tend to be more variable in meandering-river deposits than in braided-river deposits. Fluvial deposits may contain a variety of fossil hard parts of terrestrial animals as well as trace fossils created by both animals and plants (e.g., Bridge, 2003, p. 374).

Fluvial Architecture

Lateral migration of braided rivers leaves sheetlike or wedge-shaped deposits of channel and bar complexes (Cant, 1982). Lateral migration combined with aggradation leads to deposition of sheet sandstones or conglomerates that enclose very thin, nonpersistent shales within coarser sediments (Fig. 8.11). Migration of meandering streams, which are confined within narrow, sandy meander belts of stream floodplains, generate linear "shoestring" sand bodies oriented parallel to the river course. These shoestring sands are surrounded by finer grained, overbank floodplain sediments. Periodic stream avulsion may create new channels over time, leading to formation of several linear sand bodies within a major stream valley (Fig. 8.12).

The term fluvial (alluvial) architecture (Allen, 1978) refers to the three-dimensional geometry, proportion, and spatial distribution of the various types of alluvial deposits in sedimentary basins, as in Figures 8.11 and 8.12. Fluvial architecture concerns the large-scale, long-term aspects of alluvial erosion and deposition. The study of three-dimensional fluvial architecture requires extensive exposures (outcrops) or the availability of closely spaced sediment cores and/or seismic data (Chapter 13), as well as accurate age dating. Fluvial architecture is influenced by tectonics, climate, base levels, and channel types, which control processes such as subsidence rates, slope changes, channel incision and aggradation, and channel migration and avulsion (e.g., Leeder, 1993).

Ancient River Deposits

Many examples of ancient fluvial deposits, ranging in age from Precambrian to Holocene, have been cited in the literature. Numerous studies of fluvial deposits are documented in "Alluvial Sedimentation," edited by Marzo and Puigdefábregas (1993), and "Fluvial Sedimentology VI," edited by Smith and Rogers (1999). Many other studies are reported in geological journals; see, for example, references cited by Leeder (1999, p. 328). These published studies discuss a wide range

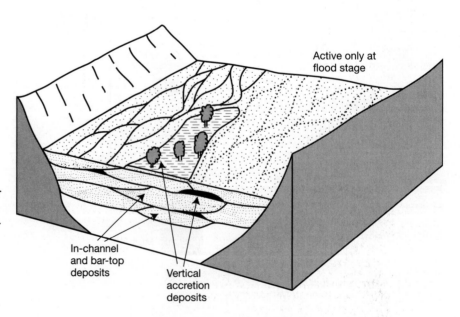

Figure 8.11
Schematic representation of the fluvial architecture of braided-river deposits. [After Walker, R. G., and D. J. Cant, 1984, Sandy fluvial systems, *in* R. G. Walker (ed.), Facies models: Geoscience Canada Reprint Ser. 1, Fig. 9, p. 77, reprinted by permission of Geological Association of Canada.]

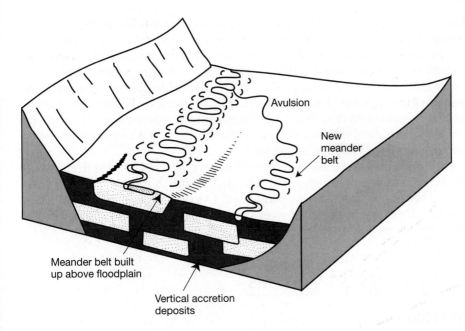

Figure 8.12
Schematic representation of the fluvial architecture of meandering-river deposits. [After Walker, R. G., and D. J. Cant, 1984, Sandy fluvial systems, *in* R. G. Walker (ed.), Facies models: Geoscience Canada Reprint Ser. 1, Fig. 9, p. 77, reprinted by permission of Geological Association of Canada.]

of fluvial sediments, such as meandering-river, braided-river, crevass-splay, avulsion, and floodplain deposits, as well as fluvial architecture.

The Triassic Buntsandstein facies in eastern Spain provides one example of an ancient river system that laid down deposits of both braided and meandering rivers (López-Gómez and Arche, 1993). The Buntsandstein of the southeastern Iberian Ranges consists of four continental red bed units (Fig. 8.13). The lowermost unit, which has very limited extent, is a basal breccia deposited by debris flows. The Boniches Formation overlying this basal unit consists of conglomerate with subrounded pebbles and abundant sandy matrix. These materials were deposited as longitudinal and lateral bars (braid bars) in shallow braided fluvial channels. The Alcotas Formation, which rests on the Boniches Formation, consists of red mudstone with many lenticular, multilateral, multistory sandstones and conglomerates. This unit was deposited as floodplain and channel-fill deposits in a fluvial system evolving from braided to high sinuosity channel pattern.

The Cañizar Formation, which ranges in thickness to about 210 m, rests above a scoured surface on the Alcotas Formation. It consists of pink to white,

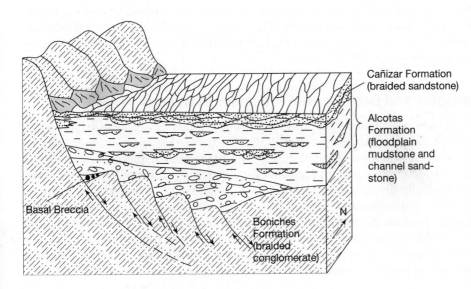

Figure 8.13
Schematic depositional model for the fluvial Buntsandstein facies (Triassic) southeast of the Iberian Ranges, eastern Spain. [Redrawn from López-Gómez, J., and A. Arche, 1993, Architecture of the Cañizar fluvial sheet sandstones, Early Triassic, Iberian ranges, eastern Spain, *in* Marzo and Puigdefábregas (eds.), Alluvial sedimentation: International Association of Sedimentologists Special Publ. No. 17: Blackwell Scientific Publ., Fig. 13, p. 377. Reproduced by permission.]

medium to fine sandstone with a few conglomerate and mudstone beds. The formation is made up of six multilateral, multistory sandstone sheet complexes, which are the main architectural units that build up the sheetlike sandstone formation. Most of the sandstone units display planar or trough cross-bedding. Some are marked by rippled surfaces, and a few are parallel laminated. Plant remains are scattered through the formation. The Cañizar Formation is interpreted as the deposits of a braided-river system with dominant bedload transport. The main facies formed as channel transverse bars, composite bars, and sandflat complexes. Lateral accretion was common on the bars.

The Buntsandstein facies as a whole displays gradation upward from coarse, braided-river deposits (Boniches Formation) through multistory, meandering-river sand bodies enclosed in floodplain muds (Alcotas Formation), to multistory sand sheets deposited in a braided-river system. The fluvial system flowed southeast through an asymmetrical graben (the Iberian Basin) in central Spain.

8.3 EOLIAN DESERT SYSTEMS

Introduction

Deserts cover broad areas of the world today, particularly within the latitudinal belts of about 10–30 degrees north and south of the equator, where dry, descending air masses create prevailing wind systems that sweep toward the equator. Deserts also lie in the interiors of continents and in the rain shadows of large mountain ranges where they are cut off from moisture from the oceans. Deserts are areas in which potential rates of evaporation greatly exceed rates of precipitation. They cover about 20–25 percent of the present land surface.

Because of their generally low rainfall, commonly less than about 25 cm/yr, we tend to think of deserts as extremely dry areas dominated by wind activity and covered by sand. In reality, a variety of subenvironments exist within deserts, such as alluvial fans; ephemeral streams that run intermittently in response to occasional rains; ephemeral saline lakes, also called playas or inland sabkhas; sand-dune fields; interdune areas covered by sediments, bare rocks, or deflation pavement; and areas around the fringe of deserts where windblown dust (loess) accumulates. Large areas of the desert environment may indeed be carpeted by windblown, or **eolian**, sand. Such areas that cover more than about 125 km^2 are called **sand seas** or **ergs** (Fig. 8.14); smaller areas are called **dune fields**. Ergs and dune fields cover about 20 percent of modern deserts or about 6 percent of the global land surface. The remaining areas of deserts are covered by eroding mountains, rocky areas, and desert flats. The largest desert in the world, the Sahara (7 million km^2), contains several ergs arranged in belts. The larger belts cover areas as extensive as 500,000 km^2.

Transport and Depositional Processes in Deserts

Most deserts are characterized by extreme fluctuations in temperature and wind, on both a daily and a seasonal basis. Rainfall rates are low, as mentioned, and the rains are very sporadic. Vegetation is generally extremely sparse. When rains do come, they tend, owing to the lack of vegetative cover, to create flash floods. Rainwater typically drains toward the centers of desert basins, where playas or inland sabkhas may develop and become sites of deposition of carbonate and evaporite minerals. Because periodic rains create flash floods and ephemeral streams and mobilize debris flows and mudflows, they are extremely important agents of sediment transport in deserts. Nonetheless, much of the time water plays a relatively small role in sediment transport in deserts. Most of time, wind is the dominant

Figure 8.14
Spaceborne radar image of part of the vast Namib Sand Sea on the west coast of southern Africa, just northeast of the city of Luderitz, Namibia, showing a variety of dune shapes and sizes. NASA image acquired by spaceborne imaging radar on board the space shuttle Endeavour, April 11, 1994. Downloaded from the Web 1/13/2000.

agent of sediment transport and deposition. Wind is much less effective than water as an agent of erosion, but it is an extremely effective medium of transport for loose sand and finer sediment. Not only does it account for the transport of vast quantities of siliciclastic sand in deserts, but it is also responsible for sediment transport in glacial environments, on river flood plains, and along many coastal areas, where both carbonate and siliciclastic sands may be transported inland. The windblown deposits of these latter environments are quite small compared to the sand seas of desert areas. Wind storms, or dust storms, may also carry silt and clay far from their sources and are responsible for transporting much of the pelagic sediment to deep ocean basins.

Wind transports sediment in much the same way as water, separating the sediment into three transport populations: traction, saltation, and suspension. Transport of grains by wind is initiated when wind strength rises to the fluid threshold and also when wind blowing at greater than threshold speed over an immobile surface encounters the leading edge of a deposit of loose, mobile material. Direct dislodgment by wind may also play a role in grain transport (Anderson, Sorenson, and Willets, 1991). Grain motion appears to cascade rapidly as those grains most susceptible to direct dislodgment collide (downwind) with and disturb less susceptible grains. The rapidity of the dislodgment depends upon the grain size, shape, sorting, and packing. At scattered locations, almost random, near-bed turbulence causes the wind flow to be seeded with low-energy ejected grains. Many of these grains translate downwind at a range of speeds, dislodging other grains as they go. A single flurry, therefore, tends to give rise to a translating and dispersing sequence of dislodgments. At a particular locality undergoing threshold wind flow, many such dislodgment sequences may be superimposed to produce overall entrainment and transport.

Wind effectively separates sediment finer than about 0.05 mm from coarser sediment and transports this fine sediment long distances in suspension. Except at unusually high wind velocities, coarser sediment travels by traction and saltation close to the ground. Saltation is a particularly important mode of wind transport, aided by downslope creep of grains owing to the impact of saltating grains as they strike the bed. Wind appears to be especially effective in transport of medium to fine sand and finer sediment, but coarse particles (up to 2 mm or somewhat larger) may also undergo transport by rolling and surface creep under high-velocity winds. The transporting and sorting action of wind tends to produce three kinds of deposits: dust (silt) deposits, sometimes referred to as **loess**, that commonly accumulate far from the source; sand deposits, which are commonly well sorted; and lag deposits, consisting of gravel-size particles that are too large to be transported by wind and that form a **deflation pavement.**

Wind transport and deposition generates many of the same kinds of bedforms and sedimentary structures—such as ripples, dunes, and cross-beds—as those produced by water transport. The bedforms that develop during wind transport range from ripples as small as 0.01 m long and a few millimeters in height to dunes 500–600 m long and 100 m high. Less commonly, gigantic bedforms called **draas** that may have wavelengths measured in kilometers (up to 5.5 km) and heights up to 400 m may also form by wind transport (Wilson, 1972; McKee, 1982). The wave length of wind-transported bedforms increases with increasing wind velocity, and wave height tends to increase with increasing grain size. Under a given set of conditions of grain size and wind velocity, ripples, dunes, and draas can coexist. Thus, dunes exist on the backs of draas, and ripples are created on the backs of dunes.

Bagnold's (1954) study dealing with the physics of blown sand remains the classic piece of research in the field of eolian sediment transport and deposition; however, more recent workers continue to investigate this subject (e.g., Barndorff-Nielsen and Willets, 1991; McEwan and Willets, 1993; Gilette, 1999). An interesting research trend is the use of computer modeling and simulation to generate data that can be compared to experimental observations from field and wind tunnel (e.g., McEwan and Willets, 1993).

Deposits of Modern Deserts

Eolian sediments accumulate in a variety of small-scale settings in deserts and even in shoreline environments; however, the major areas of accumulation are in **ergs** (sand seas). Ergs form under prevailing wind systems, primarily in arid regions, where copious supplies of fine sediment are present. Noteworthy present-day ergs include those of the Saharan and Arabian deserts of northern Africa, the Namib Desert of southern Africa (Fig. 8.14), the Mojave and Sonoran deserts of southwestern North America, and the Australian Desert of central Australia. Sediment supply, availability, and wind energy play major roles in determining the geomorphology of ergs. Dune patterns in sand seas are the product of (1) regional changes in wind regimes that promote the formation of dunes of different morphological types, and (2) temporal changes in sand supply, availability, and mobility that give rise to the generation of multiple episodes of dune formation (Lancaster, 1999).

The various environments of deserts can be grouped into three main subenvironments: dune, interdune, and sand sheet (Ahlbrandt and Fryberger, 1982; Fig. 8.15). The dune environment is primarily the site of wind transport and deposition of sand, which accumulates in a variety of dune forms, many having steeply dipping slip faces or avalanche faces. Interdune areas can receive both windblown sediment and sediment transported and deposited by ephemeral streams in stream floodplains or playa lakes. The sheet-sand environment exists around the margins of dune fields. The deposits of this environment form a transitional facies between dune and interdune deposits and deposits of other environments.

Dunes

Many types of dunes (e.g., Fig. 8.16) occur in the sand seas and dune fields of modern deserts, ranging from those with no slip faces to those with three or more

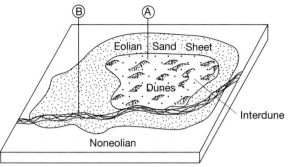

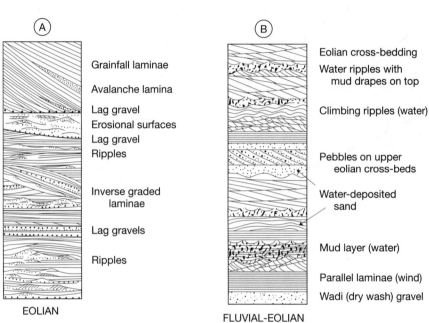

Grainfall laminae

Avalanche lamina

Lag gravel
Erosional surfaces
Lag gravel
Ripples

Inverse graded
laminae

Lag gravels

Ripples

EOLIAN

Eolian cross-bedding
Water ripples with
mud drapes on top

Climbing ripples (water)

Pebbles on upper
eolian cross-beds

Water-deposited
sand

Mud layer (water)

Parallel laminae (wind)
Wadi (dry wash) gravel

FLUVIAL-EOLIAN

Figure 8.15
Areal distribution and stratigraphic relationships of sheet sands and eolian dune sands. Column A shows cross-lamination and other typical bedding features, including lag gravels on erosional surfaces, in an eolian succession deposited in a dune environment. Column B depicts a fluvial-eolian succession formed in an eolian sand-sheet environment. [After Fryberger, S. G., T. S. Ahlbrandt, and S. Andrews, 1979, Origin, sedimentary features, and significance of low-angle eolian "sand sheet" deposits, Great Sand Dunes National Monument and vicinity, Colorado: Jour. Sed. Petrology, v. 49, Fig. 12, p. 745, reprinted by permission of Society of Economic Paleontologists and Mineralogists, Tulsa, Okla.]

Figure 8.16
Sand dunes near Stovepipe Wells, Death Valley, California. Note the sharply developed slip facies, indicating sand transport from left to right. Photograph by James Stovall.

slip faces (Fig. 8.17). Eolian bedforms range in scale from small ripples to transverse and longitudinal dunes 0.1 to 100 m high to complex dunes, called draas, with heights of 20 to 450 m. Dune morphology is determined by the availability of sand, wind intensity, and the variability of wind directions (e.g., Lancaster, 1999; Pye and Tsoar, 1990, Chapter 6).

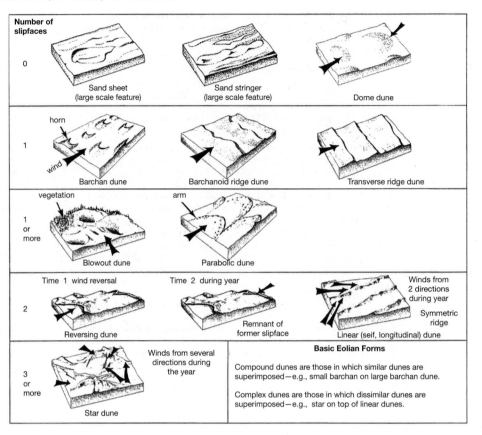

Figure 8.17
Basic eolian dune forms grouped by number of slip faces. [After Ahlbrandt, T. S., and S. G. Fryberger, 1982, Introduction to eolian deposits, in Scholle, P. A., and D. Spearing (eds.), Sandstone depositional environments: Am. Assoc. Petroleum Geologists Mem. 31, Fig. 3, p. 14, reprinted by permission of AAPG, Tulsa, Okla.]

Dune deposits commonly consist of texturally mature sands that are well sorted and well rounded; however, considerable textural variation can occur. They are also typically quartz rich, although many coastal dune deposits contain high concentrations of heavy minerals and unstable rock fragments. Coastal dunes in some tropical areas may consist largely of ooids, skeletal fragments, or other carbonate grains, and dunes composed of gypsum occur in some desert areas, such as White Sands, New Mexico. Eolian dunes are characterized particularly by large-scale cross-bedding (e.g., Fig. 4.18). Several kinds of small-scale internal structures may also be present, such as plane-bed laminae, rippleform laminae, ripple-foreset cross-laminae, climbing ripples, grainfall laminae, and sandflow cross-strata (e.g., Hunter, 1977). Migration of dunes generates a vertical succession of sandy facies that may display many of these structures (e.g., Fig. 8.15).

Owing to the variety of dune types that can form under different wind conditions, local paleocurrent vectors derived from eolian cross-bed data can range from unimodal to polymodal. Paleocurrent data may thus show a high degree of scatter that complicates calculation of ancient prevailing sediment transport directions. On a regional scale, eolian paleocurrent patterns are reported to swing over hundreds of miles around high-pressure wind systems.

Interdunes

Interdune areas occur between dunes and are bounded by dunes or other eolian deposits such as sand sheets (Fig. 8.15). Interdunes may be either deflationary (erosional) or depositional. Very little sediment accumulates in most deflationary interdunes except coarse, granule-size lag sediments that may show rippled surfaces and inverse grading. Deflationary interdunes are preserved in the rock record as a disconformity overlain by thin, discontinuous, winnowed lag deposits. Sediments deposited in depositional interdunes can include both subaqueous and subaerial deposits depending upon whether they are deposited in wet, dry, or evaporite interdunes (Ahlbrandt and Fryberger, 1981). All interdune deposits are characterized by low-angle stratification ($<\sim 10°$), because they are formed by processes other than dune migration, although many deposits may be almost structureless owing to secondary processes, largely bioturbation, that destroy stratification.

Dry interdunes or interdunes that are wetted only occasionally are most common. Deposits in dry interdunes are generated by ripple-related wind-transport processes, grainfall in the wind shadow in the lee of dunes, or sandflow (avalanching) from adjacent dunes. The deposits tend to be relatively coarse, bimodal, and poorly sorted, with gently dipping, poorly laminated layers. They are also commonly extensively bioturbated by both animals and plants.

Wet interdune areas are the sites of lakes or ponds where silts and clays are trapped by semipermanent standing bodies of water rather than being deflated and removed. These sediments may contain freshwater species of organisms such as gastropods, pelecypods, diatoms, and ostracods. They are also commonly bioturbated and may contain vertebrate footprints. Some wet interdune sediments become contorted owing to loading by dune sediments.

Evaporite interdunes, or inland sabkhas, occur where drying of shallow ephemeral lakes or evaporation of damp surfaces causes precipitation of carbonate minerals, gypsum, or anhydrite. Growth of carbonate minerals or gypsum in sandy sediment tends to disrupt and modify primary depositional features. Desiccation cracks, raindrop imprints, evaporite layers, and pseudomorphs may characterize these sediments (e.g., Lancaster and Teller, 1988).

Sheet Sands

Sheet sands are flat to gently undulating bodies of sand that commonly surround dune fields. They are typically characterized by low to moderately dipping (0–20°) cross-stratification and may be interbedded in some parts with ephemeral stream deposits (Fig. 8.15). Sheet-sand deposits may also contain gently dipping, curved, or irregular surfaces of erosion several meters in length; abundant bioturbation traces formed by insects and plants; small-scale cut-and-fill structures; gently dipping, poorly laminated layers resulting from adjacent grainfall deposition; discontinuous, thin layers of coarse sand intercalated with fine sand; and occasional intercalations of high-angle eolian deposits (e.g., Ahlbrandt and Fryberger, 1982; Kocurek and Nielson, 1986; Schwan, 1988).

Kinds of Eolian Systems

Desert systems can be characterized as wet, dry, or stabilized (Kocurek and Havholm, 1993; Kocurek, 1996). **Dry systems** are those in which the water table and its capillary fringe lie at depth below the depositional surface. Therefore, the water table has no stabilizing effect on the surface and near-surface sediment. The aerodynamic configuration or shape of the sediment surface (e.g., dune shape) alone determines whether sediment is deposited or simply moves across the surface (bypass) or, alternatively, if erosion of previously deposited sediment takes place. In **wet systems,** the water table or its capillary fringe is at or near the depositional surface. Therefore, deposition, bypass, and erosion along the substrate are controlled by the moisture content of the substrate as well as by its aerodynamic shape. **Stabilized systems** are those in which factors such as vegetation, surface cementation, or mud drapes play a significant stabilizing role and thus influence the behavior of the accumulating surface. Major eolian environments such as the Sahara may show a full range of these three kinds of eolian systems (Kocurek, 1996).

The extent to which eolian sediment is preserved to become part of the geologic record is strongly influenced by the kind of system in which it accumulates. The vertical space in which sediment accumulates is called its **accumulation space**; however, only that sediment which lies below the baseline of erosion is preserved (the **preservation space**). This baseline is affected mainly by subsidence (caused by tectonism, loading, and compaction) and the position of the water table. Not all of the sediment that accumulates in dry eolian systems may be preserved. Preservation can occur if subsidence brings the sediment below the erosional base level, the water table rises through the dry accumulation, or a combination of these two factors takes place (Kocurek, 1999; Fig. 8.18). In wet systems, the accumulation space is essentially also the preservation space because the water table is near the surface. In a stabilizing system, some preservation can occur above the regional baseline of erosion. Keep in mind, however, that as dunes migrate, the dune bedforms themselves (the shapes of the dunes) are not preserved. The depositional record that the migrating dunes leave behind is mainly the lower foresets only.

Bounding Surfaces in Eolian Deposits

As mentioned, bedforms are only rarely preserved in ancient eolian deposits. Instead, we see cross-bedding and other internal features (e.g., Fig. 8.15 A, B), mainly from the lowest parts of the original bedforms, that remain as a record of bedform migration across ancient deserts. In addition to foresets and other bedding features, several kinds of bounding surfaces may be generated within eolian successions as a record of complex depositional and erosional processes. Brookfield (1984) describes three kinds of bounding surfaces: flat, first-order surfaces related to migration of large bedforms; inclined, second-order surfaces that commonly slope downwind and enclose cosets of cross-strata deposited by smaller dunes superimposed on large bedforms; and third-order surfaces that are generated by erosional modification of the lee faces of migrating dunes.

Kocurek (1996) suggests that it is difficult to apply this hierarchical scheme of first-, second-, and third-order surfaces in surface sections and recommends abandoning the terminology. Instead, he proposes the following terminology for these surfaces: **reactivation or redefinition surfaces** (third-order), which occur because of periodic erosion of the lee faces of dunes; **superposition surfaces** (second-order), which form by migration of dunes or scour troughs superimposed

Figure 8.18
Examples of preservation of eolian sediment accumulations. A. New (increased) accumulation space created by a relative rise in the water table. B. Accumulation space created by subsidence below the baseline of erosion. [After Kocurek, G., and K. G. Havholm, 1993, Eolian sequence stratigraphy—A conceptual framework, in Weimer, P., and H. W. Posamentier (eds.), Siliciclastic sequence stratigraphy: Recent developments and applications: AAPG Memoir 58, Fig. 13, p. 405, reproduced by permission.]

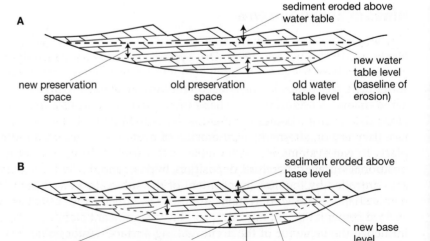

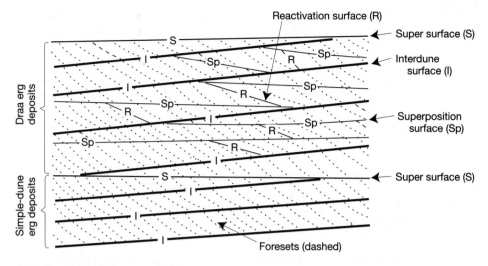

Figure 8.19

Bounding surfaces in eolian deposits. Note that superposition and reactivation surfaces are contained within cross-bed sets or cosets bounded by interdune surfaces, and that super surfaces (unconformities) can truncate interdune surfaces and entire erg successions. [Modified from Kocurek, G., 1988, First-order and super bounding surfaces in eolian sequences—Bounding surfaces revisited: Sedimentary Geology, v. 56, Fig. 2, p. 195, reproduced by permission of Elsevier Science Publishers.]

on the lee face of the main bedform; and **interdune surfaces** (first-order) formed between sets or cosets of cross-strata and that separate accumulations of the different bedforms (Fig. 8.19).

In addition to the above surfaces, regional unconformities may be present within eolian successions that mark the end of a major episode of eolian deposition. Such unconformities are referred to as **super surfaces** (Kocurek, 1996). They signify regional interruption of sand-sea deposition and develop in response to changes in sediment supply, climate, or possibly sea level in some cases. They may be surfaces of erosion, where sediment has deflated down to the water table, or surfaces of bypassing. On a bypassing surface, there is no net erosion or sediment accumulation. Sediment is simply transported across the surface to other areas.

Ancient Desert Deposits

Navajo/Nugget Sandstone

The Jurassic Navajo Formation of the southwestern United States is one of the thickest, most widespread, and best exposed ancient eolian (erg) depositional systems in the world (Kocurek, 2003). The Navajo, and its lateral equivalent Nugget Sandstone, reach nearly 700 m in thickness and extend over 265,000 km^2 over portions of five states (Fig. 8.20). The original extent of the Navajo sand seas was about 2.5 times as large as the present outcrop (Marzolf, 1988).

The Navajo has been suggested in the past to be a marine deposit; however, few geologists today doubt its eolian origin. Petrologically, it consists of fine- to medium-size quartz grains that are generally well rounded and commonly frosted. The most striking feature of the Navajo is the presence of huge tabular cross-bed sets that display sweeping foresets (Fig. 8.21). Dips of foresets commonly exceed 20°, and individual cross-bed sets range in thickness from about 5 m to almost 35 m. Freshwater invertebrate fossils (ostracods and crustaceans) have been reported from the Navajo, as well as dinosaur and pteropod tracks and skeletons of bipedal dinosaurs and early mammals. Slump structures such as contorted

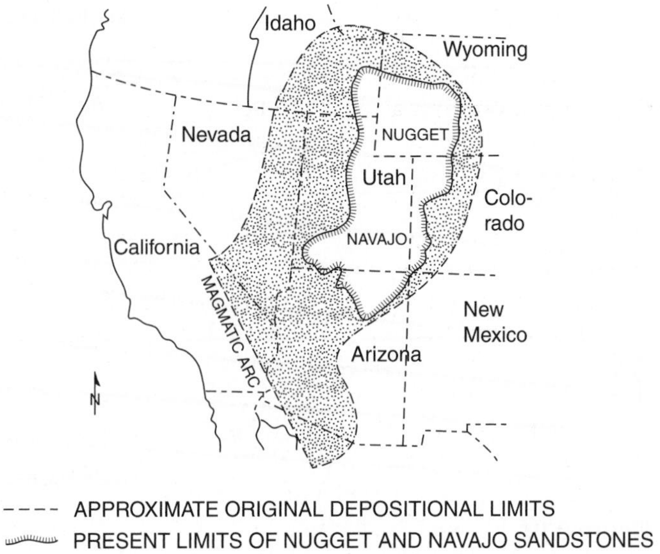

Figure 8.20
Estimates of the minimum and maximum areas of deposition of the Navajo Sandstone and its lateral equivalents. [After Marzolf, J. E., 1988, Controls on late Paleozoic and early Mesozoic deposition of the western United States: Sedimentary Geology, v. 56, Fig. 6, p. 179, reproduced by permission.]

- - - -   APPROXIMATE ORIGINAL DEPOSITIONAL LIMITS
〰〰〰〰 PRESENT LIMITS OF NUGGET AND NAVAJO SANDSTONES

Figure 8.21
Navajo Sandstone (Jurassic) in Zion National Park, Utah, showing large sets of cross strata generated by migration from left to right of ancient eolian sand dunes. Note the prominent interdune bounding surfaces (see Fig. 8.19) between cross-bed sets.

bedding, which are reported from modern dune sands that have been wetted, are also common.

As described in preceding paragraphs, sediment deposited during migration of dunes across a desert may be preserved in part, commonly the lower part of the foresets, owing to rise in water table, basin subsidence, or both. Succeeding, successive migrations across a subsiding basin result in vertical stacking of eolian facies separated by interdune bounding surfaces and super surfaces. At the margins of dune fields, the boundary between eolian and other (e.g., fluvial or marine) environments may shift back and forth, generating a vertical succession of facies in which eolian and noneolian sediments are interbedded. For example, the schematic illustration in Figure 8.15 shows an eolian succession in Column A and an interbedded fluvial-eolian succession in Column B.

The Navajo Sandstone provides a real example of this principle, as shown by intertonguing eolian deposits of the Navajo and fluvial deposits of the Kayenta Formation in northeastern Arizona (Fig. 8.22). Three fluvial to eolian drying-upward cycles are illustrated, each representing the advance of the Navajo erg across the Kayenta alluvial plain, probably in response to an increasingly arid climate. Return to wetter conditions terminated the advance of the erg, allowing fluvial deposits of the Kayenta to, in turn, advance over an erosional Navajo surface. Thus, in this succession, cross-bedded Navajo eolian dune deposits and flooded interdune deposits are interbedded vertically with floodplain and other fluvial deposits of the Kayenta Formation.

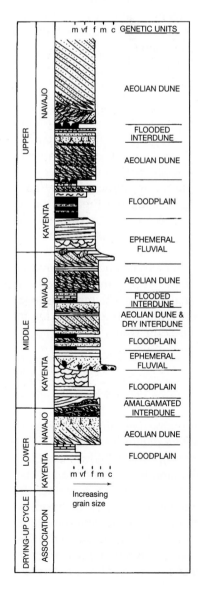

Figure 8.22
Representative intertonguing Jurassic eolian (Navajo) and fluvial (Kayenta) facies in northeastern Arizona. Total thickness of the column is about 100 m. Note three major drying-up cycles, indicating advance and retreat of the Navajo erg. [After Herries, R. D., 1993, Contrasting styles of fluvial-eolian interaction at a downwind erg margin: Jurassic Kayenta-Navajo transition, northeastern Arizona, *in* North, C. P., and D. J. Prosser (eds.), Characterization of fluvial and aeolian reservoirs, Geological Society London Special Publication No. 73, Fig. 17, p. 210, reproduced by permission.]

Other Ancient Desert Deposits

Ancient sandstones interpreted to be windblown deposits have been described from sedimentary successions as old as the Precambrian from many parts of the world. One of the most extensive and intensely studied eolian records is from the late Paleozoic and Mesozoic of the western interior of the United States (Blakey, Peterson, and Kocurek, 1988). In addition to the Navajo Sandstone, described above, eolian deposits are widespread from Montana to Arizona and include Pennsylvanian (e.g., Weber and Tensleep), Permian (e.g., Cedar Mesa and Coconino), Triassic (e.g., Jelm and Wingate), and Jurassic (Entrada) formations. This impressive eolian system consists of thick, extensive assemblages that represent deposition from different types of dunes and eolian complexes and interaction among eolian, fluvial, marine, and lacustrine environments.

Examples from other continents include the Permian Rotliegendes of northwestern Europe, the Jurassic-Cretaceous Botucatu Formation of the Parana Basin of Brazil, the Permian Lower Bunter Sandstone of Great Britain, the Permo-Triassic Hopeman Sandstone of Scotland, the Permian Corrie Sandstone of Scotland, and the Proterozoic (Precambrian) of India and northwestern Africa. The Rotliegendes has been particularly well studied (e.g., Glennie, 1986). It accumulated in a series of graben (fault) basins as interbedded eolian, fluvial, lacustrine (lake), and sabkha (evaporite) deposits, again illustrating the complex interaction of eolian and noneolian systems. Other examples of ancient eolian deposits can be found in "Further Reading—Eolian Systems" at the end of this chapter.

8.4 LACUSTRINE SYSTEMS

Lakes cover about 1–2 percent of Earth's surface. Because the world's continents are presently in a higher state of emergence than was typical of much of Phanerozoic time, lake sedimentation is more prevalent today than it was during much of the geologic past. In fact, ancient lake sediments appear to be of only minor importance volumetrically in the overall stratigraphic record, although they have been reported in stratigraphic successions ranging in age from Precambrian to Holocene. Although not abundant in the geologic record, lake sediments are nonetheless important. Lake chemistry is sensitive to climatic conditions, making lake sediments useful indicators of past climates. For example, several studies have shown that ancient episodes of wet and dry climates can be deciphered on the basis of lake sediment chemistry and mineralogy. Also, some lake deposits contain economically significant quantities of oil shales, evaporite minerals, coal, uranium, or iron. Many lake sediments also contain abundant fine organic matter that may act after burial as a source material for petroleum (Katz, 1990).

Origin and Size of Lakes

The basins, or depressions, in which lakes form can be created by a variety of mechanisms, including tectonic movements such as faulting and rifting; glacial processes such as ice scouring, ice damming, and moraine damming; landslides or other mass movements; volcanic activity such as lava damming or crater explosion and collapse; deflation by wind scour or damming by windblown sand; and fluvial activity such as the formation of oxbow lakes and levee lakes. Many existing lakes appear to have originated directly or indirectly by glacial processes (Picard and High, 1981) and thus may not be typical of ancient lakes, which formed predominantly by tectonic processes. On the other hand, we know that some large modern lakes also formed by tectonic processes (e.g., Lake Tanganyika in the East African rift system, Lake Baikal in the Baikel rift system in Siberia) and volcanic processes (e.g., Crater Lake, Oregon). Of the twenty-five largest lakes by surface area today, ten are of glacial origin, seven occupy cratonic depressions, and four are in rift valleys (Smith, 1990).

Modern lakes range in areal dimensions from a few tens of square meters to tens of thousands of square kilometers. The largest modern lake is the saline, inland Caspian Sea with a surface area of 436,000 km^2 (Van der Leeden, 1975). Other large lakes with surface areas ranging between 50,000 and 100,000 km^2 include Lake Superior, Lake Huron, and Lake Michigan in North America; Lake Victoria, located between Uganda and Kenya in east-central Africa; and Lake Aral east of the Caspian Sea. Water depths of modern lakes range from a few meters in small ponds to more than 1700 m in the world's deepest lake, Lake Baikal, Siberia. Water depth and surface area are not necessarily related; thus, some of the largest lakes have very shallow depths and vice versa. For example, Lake Victoria has a surface area of 68,000 km^2 but a maximum depth of only 79 m, whereas Crater Lake, Oregon, with a surface area of about 52 km^2 has a maximum depth of about 580 m.

Preserved lacustrine sediments show that ancient lakes also ranged in size from small ponds to large bodies of water exceeding 100,000 km^2. Three of the largest ancient lakes recognized are the Late Triassic Popo Agie Lake of Wyoming and Utah, which had a minimum areal extent, based on the preserved sediment record, of 130,000 km^2 (Picard and High, 1981); the Jurassic T'oo'dichi' Lake of the eastern Colorado Plateau, with an area of 150,000 km^2 (Turner and Fishman, 1991); and the Eocene Green River Basin, with an area of about 100,000 km^2 (Eugster and Hardie, 1978). According to Bohacs, Caroll, and Neal (2003), ancient lake strata in the Cretaceous system of the South Atlantic and eastern China and the Permian system of western China extend up to 300,000 km^2. Reported thickness of preserved ancient lake sediments ranges from less than 20 m to as much as 9000 m (e.g., Pliocene Ridge Basin Group, California; Link and Osborne, 1978). Lake size and character is a complex function of four main variables: basin-floor depth, sill height, water supply, and sediment supply (Bohacs, Carroll, and Neal, 2003).

Lake Settings and Principal Kinds of Lakes

Modern lakes occur in a variety of environmental settings, including glaciated inland plains and mountain valleys, nonglaciated inland plains and mountain regions, deserts, and coastal plains. They exist under a spectrum of climatic conditions ranging from very hot to very cold and from highly arid to very humid. Most lakes are filled with fresh water, but others, such as the Caspian Sea and many lakes in arid regions (e.g., Great Salt Lake, Utah) are highly saline. Many lakes are associated with other types of depositional systems, notably glacial, fluvial, eolian, and deltaic systems. The depositional processes that occur in lakes are influenced both by climatic conditions and by a variety of physical, chemical, and biological factors that include the chemistries of their waters and fluctuations in their shorelines and siliciclastic sediment supply. Some attributes of lacustrine depositional environment are similar to those of marine environments; however, important diffrences exist in terms of such factors as basin size, water chemistry, physical processes (e.g., no tides in lakes), and biologic processes (e.g., Gierlowski-Kordesch and Kelts, 1994b).

Open lakes are those that have an outflow of water and a relatively stable (fixed) shoreline and in which inflow and precipitation are approximately balanced by outflow and evaporation. Siliciclastic sedimentation commonly predominates in open lakes; however, chemical sedimentation can occur in open lakes that have a low supply of clastic sediment. **Closed lakes** do not have a major outflow and have fluctuating shorelines; inflow is commonly exceeded by evaporation and infiltration. These conditions lead to concentration of ions in lake water and a predominance of chemical sedimentation, although siliciclastic sediments may accumulate also.

Factors Controlling Lake Sedimentation

The kinds of sediments deposited in lakes are the result of a complicated balance among physical, chemical, and biological processes (Fig. 8.23). Climatic factors affect lake sedimentation in numerous ways. For example, the global distribution of lakes reflects global climate patterns. Water level in lakes is maintained by the balance between evaporation and precipitation. Climate can determine whether a lake is filled to overflowing (open) or acts as an internal drainage basin (closed). The kind of chemical sedimentation in lakes strongly reflects climatic conditions. For example, chemical sedimentation in lakes of arid regions is dominated by precipitation of gypsum, halite, and various other salts, but in humid climates chemical sedimentation is dominated by carbonate deposition. Sediment input to lakes is influenced by the vegetation cover in the drainage area of the lakes and is greatest in arid regions with low vegetation cover. In cold climates, seasonal drops in temperature lead to freezing of lakes, causing decrease in sediment input and cessation of wave activity, allowing deposition of fine-grained suspended sediment during these quiet-water conditions. Climate and the physiography of lake settings also determine the local weather conditions over lakes. Severe, localized storms with high winds can cause considerable shore erosion, coupled with sediment transport and deposition, during short periods of time.

Physical Processes

Physical processes that interact in lakes to bring about sediment transport and deposition include wind, river inflow, and atmospheric heating. Wind processes are of major importance because winds create waves and currents. River inflow may generate plumes of fine sediment that extend in surface waters far out into a lake (Fig. 8.23), or it may generate density underflows, or turbidity currents, that carry sediment along the bottom toward the basin center. River inflow can also create currents that flow along the margins of lakes. Other currents may be generated by flow-through of water along the lake bottom toward a point of lake discharge. Atmospheric heating, which is a function of climate, is responsible for density differences in lake water. These differences can cause stratification of water on the one hand (heating of surface water) or, under some conditions, generation of density

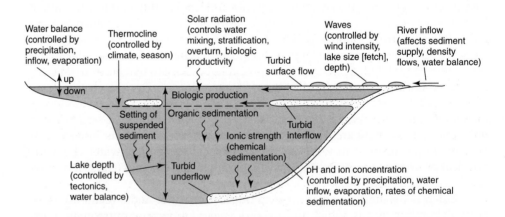

Figure 8.23
Sedimentary processes in lakes involve a balance between clastic input by rivers, offshore and longshore redistribution of clastics by waves and wave-produced currents, transport of fine clastics as turbid interflows, downslope movement of fine and coarse clastics by turbidity currents, and *in situ* production of biological and chemical sediment. The term thermocline refers to the boundary between warm, low-density surface water and colder, higher density deeper water. [Diagram based on numerous sources.]

currents (by cooling of surface water) that produce mixing and lake overturn. Also, temperature variations may cause alternate freezing and melting of lake surface waters, thereby affecting sediment transport within the lake.

Thus, a variety of sediment transport and depositional mechanisms operate in lakes. Deposition of siliciclastic sediment in the calmer, deeper portion of lakes can take place by settling of fine particles that were suspended in the water column owing to wave and current activity, or deposition may occur from turbidity currents generated where sediment-laden streams discharge into lakes. Sedimentation can occur also along the shallow shoreline of lakes from wind-generated traction currents or river-inflow currents deflected along the lake margin. One type of lake sedimentation process that appears to be particularly characteristic of cold-climate lakes is the formation of **varves**, which are very thin, alternating light- and dark-colored sediment layers. Thicker, light-colored, coarse-grained laminae accumulate suspension settling of fine sediment during summer conditions. Thinner, finer grained, organic-rich, dark laminae form by slow suspension settling during winter months when lakes are frozen.

Chemical Processes

Deposition of chemically formed sediment is particularly common in closed lakes. The chemistry of lake waters varies from lake to lake but is dominated by calcium, magnesium, sodium, potassium, carbonate, sulfate, and chloride ions. Thus, the most common chemical sediments in the lakes of humid regions are carbonates, although phosphates, sulfides, cherts, and iron and manganese oxides are present in some lakes. In arid regions, where rates of evaporation are high, chemical lake sediments are dominated by carbonates, sulfates, and chlorides. The evaporite deposits of lakes include many common marine evaporite minerals such as gypsum, anhydrite, halite, and sylvite, but they also include several minerals such as trona, borax, epsomite, and bloedite that are not common in marine evaporites (Chapter 7). The pH of lake waters commonly falls between 6 and 9; however, it can range from less than 2 (highly acidic) in some volcanic lakes to as much as 12 (highly alkaline) in some closed desert lakes. Although chemical sedimentation processes are most important in closed lakes, they may predominate also in some open lakes where the clastic sediment supply is low.

Biological Processes

Organisms play an important role in lake sedimentation by extracting chemical elements from lake water to build shells and the subsequent deposition of these shells, extraction of CO_2 during photosynthesis (thereby aiding precipitation of $CaCO_3$), contributing plant remains to form plant deposits, and bioturbation of sediments. Many kinds of organisms live in lakes and contribute their skeletal and nonskeletal remains to lake sediments. Siliceous diatoms are particularly widespread and noteworthy. Diatoms carry out photosynthesis and are the only important type of lake organism that produces siliceous tests. Their remains form important diatomite deposits in many Pleistocene lakes. Pelecypods, gastropods, calcareous algae, and ostracods also abound in many lakes and are important contributors of calcium carbonate sediments. Blue-green algae (cyanobacteria) carry on photosynthesis and also trap fine sediment to form stromatolites. Many different types of higher plants live in lakes. Under the reducing conditions and high sedimentation rates that exist in some lakes, the remains of higher plants may be partially preserved to eventually form peat and coal. Considering the small size of many lakes and their generally lower alkalinity and buffering capacity, compared to those of the open ocean, the assimilation of CO_2 by plants during photosynthesis is a much more important factor in controlling the pH of lakes than that of the ocean. Thus, increase in pH caused by photosynthetic removal of

CO_2 likely exerts a dominant control in facilitating carbonate sedimentation in lakes. Finally, organisms such as pelecypods, freshwater shrimp, and worms may burrow and rework lake sediments, destroying laminations and other primary sedimentary structures.

Characteristics of Lacustrine Deposits

Bohacs et al. (2000) suggests that lakes can be divided into three types on the basis of fill characteristics: overfill, balance-fill, and underfilled. **Overfilled** lake basins have persistently open hydrology, freshwater lake chemistry, progradational shoreline architecture, and commonly interbedded fluvial deposits. They occur when rate of supply of sediment + water consistently exceeds accomodation space (the space available in which sediments can accumulate). **Balance-fill** basins have intermittently open hydrology, fluctuating lake water chemistry, thermal and chemical stratification, mixed progradational and aggradational architecture, and varied interbedding of clastic and carbonate strata. This lake-basin type occurs when rates of sediment + water supply and accomodation are roughly in balance. **Underfilled** lake basins have persistently closed hydrology, characteristic chemical stratification, high solute content of lake waters, extensive desiccation (drying) features, highly contrasting lithologies, common association with evaporite deposits, and dominantly aggradational shoreline architecture. This basin type occurs when rates of accomodation consistently outstrip available water and sediment supply, resulting in closed basins with ephemeral lakes interspersed with playas or brine pools or both.

The sediment of most hydrologically open lakes are dominated by siliciclastic deposits, derived mainly from rivers—but possibly including windblown, ice-rafted, and volcanic detritus. Much of this sediment is deposited along the shores of lakes, particularly near river mouths. Gravelly sediment may be present in the toes of alluvial fans or fan deltas that extend to the lake edge or into the lake. Sand, likewise, accumulates mainly along the lakeshore in deltas, beaches, spits, or barriers. Sand may also be carried by turbidity currents into the middle of the lake (Fig. 8.23), however, deeper parts of the lake are characterized particularly by the presence of fine silt and clay. Some muddy sediment is transported into deeper water by surface overflows. In density-stratified lakes, muddy sediment may also be carried as a turbidity interflow above cold, denser lake water. Coarser particles in such interflows settle fairly quickly and accumulate as silt layers. Finer particles settle more slowly to form clay layers. Thus, the siliciclastic deposits of open lakes may consist of deltaic sands and muds (and possibly alluvial-fan gravels), turbidite sands and silts, and homogeneous to laminated muds.

In open lakes where the clastic sediment supply is low, chemical and biochemical processes predominate, resulting in deposition of largely chemical sediments. Primary inorganic carbonate precipitation (caused by loss of CO_2 through plant photosynthesis and/or increase in water temperature or mixing of water masses) and production of shells (by calcium carbonate- or silica-secreting organisms) account for most of the sedimentation. The principal types of invertebrate remains in lacustrine sediments include bivalves, ostracods, gastropods, diatoms, and charophytes and other algae. Chemical lake deposits consist mainly of carbonate sands and muds (less commonly siliceous diatom deposits). Stromatolites produced by blue-green algae (cyanobacteria) are common also in some lake deposits. Various amounts of noncarbonate organic matter and some siliciclastic sediment may be present. Plant life is commonly abundant in shallow water around lake margins, and plant deposits may become important during the late stages of lake filling. Carbonate sediments may interfinger along the lake margin with siliciclastic deltaic or alluvial deposits. Typical facies in an open lake with low siliciclastic sediment input are illustrated in Figure 8.24.

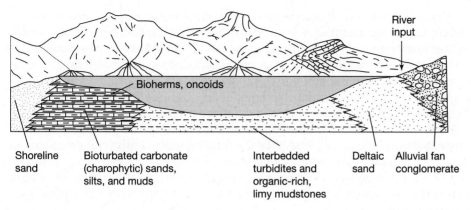

Figure 8.24

Sediment types in open lake characterized by low siliciclastic sediment input include both chemical/biochemical and siliciclastic sediment. By contrast, the deposits of open lakes with high clastic input consist predominantly of siliciclastic sediment. [Redrawn from Eugster, H. P., and K. Kelts, 1983, Lacustrine chemical sediments, in Goudie, J. J., and K. Pye (eds.), Chemical sediments and geomorphology: Academic Press, New York, Fig. 12.2, p. 333, reproduced by permission.]

Hydrologically closed lakes occur in regions of interior drainage where lake levels may experience considerable fluctuation owing to seasonal flooding. Alluvial fans are commonly present around the borders of such lakes, and the sandy aprons (sandflats) of such fans may extend into the lake. During high water, the edges of these sandflats can be reworked by wave action, resulting in redeposition of wave-rippled sandy sediment along the lake edge. Most sedimentation in closed lakes takes place by chemical/biochemical processes in waters made saline by high rates of evaporation. Two kinds of closed lakes are recognized. **Perennial** basins receive inflow from at least one perennial stream. They commonly do not dry up completely from year to year, although some may dry up occasionally. Most perennial lakes are saline, but some are dilute. The deposits of perennial lakes include carbonate muds, silts, and sands, commonly with intergrowths of evaporite minerals, and may include stromatolites (Fig. 8.25). Bedded evaporites may be present in the central part of the lake. **Ephemeral** salt-pan basins are fed by ephemeral runoff, springs, and groundwater and are generally dry through part of each year. Ephemeral salt-pan deposits may also contain carbonate sediments, including spring travertine or tufa, but bedded salt deposits are much more important. Saline deposits interfinger with siliciclastic sandflat deposits around the margin of the salt pan.

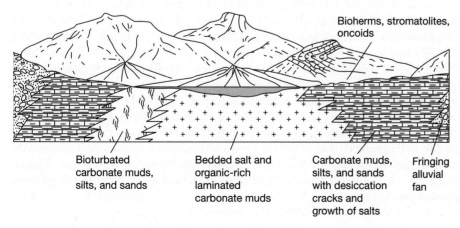

Figure 8.25

Depositional subenvironments and sediment types in a hydrologically closed, perennial saline lake basin. [Redrawn from Eugster, H. P., and K. Kelts, 1983, Lacustrine chemical sediments, in Goudie, J. J., and K. Pye (eds.), Chemical sediments and geomorphology: Academic Press, New York, Fig. 12.8 and 12.9, p. 351, reproduced by permission.]

Numerous kinds of sedimentary structures occur in lake sediments, including laminated bedding, varves, stromatolites, cross-bedding, ripple marks, parting lineations, graded bedding, groove casts, load casts, soft-sediment deformation structures, burrows and worm trails, raindrop and possible ice-crystal impressions, mudcracks, and vertebrate footprints. Varves are one of the more diagnostic characteristic of lake sediments, but light and dark laminae resembling varves have also been reported in nonlacustrine sediments (e.g., some laminated marine deposits). Another distinguishing characteristic of lake sediments is that individual lake beds tend to be thin and laterally continuous compared to associated fluvial deposits (although total lake sediments can be very thick). Otherwise, no uniquely diagnostic structures occur in lake sediments. Many sedimentary structures of lacustrine deposits are similar to those of shallow marine sediments.

Owing to high sedimentation rates in lakes and the fact that they are essentially closed systems with respect to sediment transport, all lakes are ephemeral features. Lake basins eventually fill with sediment, and most are converted into fluvial plains as they are overrun by fluvial systems. Therefore, lake filling is commonly regarded as a regressive process. That is, coarser, nearshore sediments are believed to gradually encroach on finer lake basin sediments and to be covered in turn with fluvial sediments. This postulated process of filling theoretically generates shallowing and coarsening upward successions of lake facies. Although the ultimate filling of lakes and their encroachment by prograding fluvial or other coarser grained deposits may generate a gross coarsening-upward pattern of facies, ideal coarsening-upward successions of lake sediments probably rarely occur, except perhaps in some very small lakes (Picard and High, 1981).

Ancient Lake Deposits

Lake sediments are preserved in a variety of tectonic settings, including extensional rift systems, strike-slip basins, foreland basins, and cratonic basins. Ancient lake sediments are known from many parts of the world in sedimentary successions ranging in age from Precambrian to Holocene (e.g., Gierlowski-Kordesch and Kelts, 1994a).

Some of the better-known lacustrine deposits in North America include the Pliocene Glenn Ferry Formation of the Snake River Plain; the Eocene Green River Formation of Utah, Colorado, and Wyoming, known for its oil-shale deposits; much of the Jurassic Morrison Formation of the Colorado Plateau, renowned for its dinosaur remains; parts of the Triassic Chugwater Group of Wyoming; Triassic Supergroup rift basins of eastern North America; the Devonian Escuminac Formation of southern Quebec; and the Carboniferous Strathlorne Formation of Nova Scotia. Some well-known lake deposits from other parts of the world include the Cenozoic rift-basin deposits of East Africa; the Cretaceous rift-basin deposits of Brazil (which are source-rocks for much of Brazil's oil; Abrahão and Warme, 1990); the clastics, evaporites, and carbonates of the Triassic Keuper Marl of south Wales; the Permo-Triassic Beaufort strata of the eastern Karoo Basin, Natal, South Africa; parts of the Lower Permian Rotliegend deposits of southwest and eastern Germany; and the middle Devonian sediments from the Old Red Sandstone of the Orcadian Basin of northeast Scotland. Many of these lake deposits throughout the world include thick successions of organic-rich shales that are important source rocks for petroleum (Katz, 1990). A recent monograph edited by Gierlowski-Kordesch and Kelts (2000), entitled "Lake Basins through Space and Time," summarizes the characteristics of 60 additional ancient lakes ranging in age from Carboniferous to Quaternary.

Because lakes vary widely in size and hydrologic characteristics (e.g., open vs. closed), and lake sediments are correspondingly diverse, it is not possible to select a single example that illustrates a "typical" ancient lake. The Green River Formation (Eocene), which extends over more than 100,000 km^2 in Wyoming, Colorado, and Utah, provides an example of a large, well-studied, lacustrine deposit that formed under fluctuating hydrologic and climatic conditions. The formation exceeds 2 km in thickness and has vast reserves of oil shale and trona, a sodium carbonate. It was deposited in two Eocene lakes, Lake Uinta and Lake Goshiute. Lake Uinta, which extended over the Uinta Basin of Utah and the Piceance Basin of Colorado (e.g., Ryder et al., 1976), was a perennial, moderately deep lake in which oil shales and carbonates were deposited. Lake Goshiute, located in the Green River Basin of Wyoming, was a shallow, ephemeral playa lake that appears to record changes from pluvial (wet) climatic conditions during its early history to arid and then back to pluvial through time.

The principal deposits of the Green River Formation in the Green River Basin (Lake Goshiute) are shown in Figure 8.26, after Roehler (1992). During its early history (early Eocene) when the Luman Tongue and Tipton Shale Member were deposited, Lake Goshiute was a freshwater lake enriched in calcium carbonate and fine-size organic matter. These conditions favored deposition of oil shales and dolomitic mudstones, together with mudstone, sandstone, tuff, and limestone. [As discussed in Chapter 7, oil shales are dark-colored shales that contain significant quantities of kerogen, which can be converted into oil by heating.] During deposition of the Wilkins Peak Member in early to middle Eocene time, arid conditions prevailed and the lake became hypersaline. Beds of evaporites (trona and halite) as much as 10 m thick were deposited along with dolomitic mudstones, oil shales, sandstone, and algal limestone. Evaporite deposits grade laterally to mudflat, sandflat, and alluvial-fan deposits. Wet conditions returned in middle Eocene time, bringing about a change from evaporitic deposition to deposition of freshwater oil shales, mudstones, siltstones, and algal limestones that make up the Laney Member. The lacustrine deposits of the Green River Formation as a whole interfinger laterally with fluvial deposits of other, equivalent-age, formations (Wasatch Formation, Battle Springs Formation, Bridger Formation).

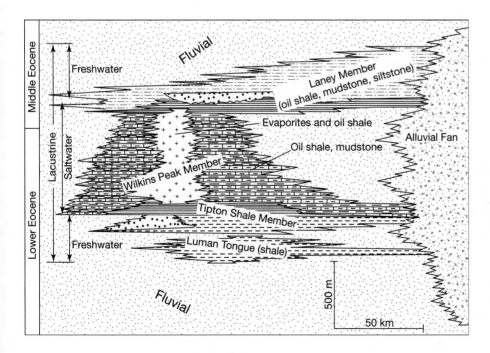

Figure 8.26
Lacustrine deposits of the Green River Formation in the Green River Basin, Wyoming. [Redrawn from Roehler, H. W., 1992, Correlation, composition, areal distribution, and thickness of Eocene stratigraphic units, greater Green River Basin, Wyoming, Utah, and Colorado: U.S. Geological Survey Professional paper 1506-E, Fig. 1, p. E2, reproduced by permission.]

8.5 GLACIAL SYSTEMS

Introduction

I have placed glacial systems last in this discussion of continental environments because the glacial environment, in a broad sense, is a composite environment that includes fluvial, eolian, and lacustrine environments. It may also include parts of the shallow-marine environment. Glacial deposits make up only a relatively minor part of the rock record as a whole, although glaciation was locally important at several times in the geologic past, particularly during the late Precambrian, late Ordovician, Carboniferous/Permian, and Pleistocene (Eyles and Eyles, 1992). Glaciers presently cover about 10 percent of Earth's surface, mainly at high latitudes. They exist primarily as large ice masses on Antarctica (~86 percent of the world's glaciated area) and Greenland (~11 percent of the world's glaciated area) and as smaller masses on Iceland, Baffin Island, and Spitsbergen. Small mountain glaciers occur at high elevations in all latitudes of the world. About 80 percent of the world's fresh water is tied up in glacial ice, of which most is in Antarctica (Hambrey, 1994, p. 31). By contrast to their present distribution, ice sheets covered about 30 percent of Earth during maximum expansion of glaciers in the Pleistocene and extended into much lower latitudes and elevations than those currently affected by continental glaciation.

The glacial environment is confined specifically to those areas where more or less permanent accumulations of snow and ice exist. Such environments are present in high latitudes at all elevations (continental glaciers) and at low latitudes (mountain or valley glaciers) above the snowline—the elevation above which snow does not melt in summer. Mountain glaciers form above the snowline by accumulation of snow. They move downslope below the snowline only if rates of accumulation of snow above the snowline exceed rates of melting of ice below. The factors affecting glacier movement and the mechanisms of ice flow (e.g., Martini et al., 2001; Menzies, 1995) are not of primary interest here. Our concerns are the sediment transport and depositional processes associated with glacial movement and melting and the sediments deposited by glaciers.

Environmental Setting

The glacial environment proper is defined as all those areas in direct contact with glacial ice. It is divided into the following zones: (1) the **basal** or **subglacial zone**, influenced by contact with the bed, (2) the **supraglacial zone**, which is the upper surface of the glacier, (3) the **ice-contact zone** around the margin of the glacier, and (4) the **englacial zone** within the glacier interior. Depositional environments around the margins of the glacier are influenced by melting ice but are not in direct contact with the ice. These environments make up the **proglacial environment**, which includes glaciofluvial, glaciolacustrine, and glaciomarine (where glaciers extend into the ocean) settings (Fig. 8.27). The area extending beyond and overlapping the proglacial environment is the **periglacial environment**.

The basal zone of a glacier is characterized by erosion and plucking of the underlying bed. Debris removed by erosion is incorporated into the bed of the glacier. This debris causes increased friction with the bed as the glacier moves and thus aids in abrasion and erosion of the bed. The supraglacial and ice-contact zones are zones of melting or ablation where englacial debris carried by the glacier accumulates as the glacier melts. The glaciofluvial environment is situated downslope from the glacier front and is characterized by fluctuating meltwater

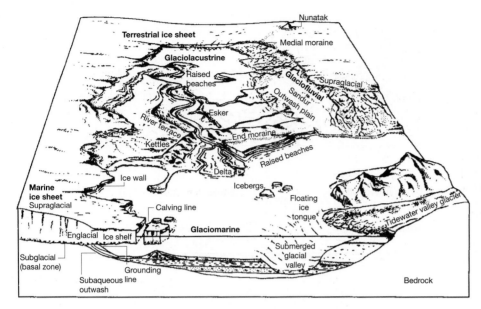

Figure 8.27
Glacial and associated proglacial environments. [After Edwards, M. B., 1986, Glacial environments, *in* Reading, H. G. (ed.), Sedimentary environments and facies, 2nd ed., Fig. 13.2, p. 448, reproduced by permission of Elsevier Science Publishers, Amsterdam.]

flow and abundant coarse englacial debris that is available for fluvial transport. The glaciofluvial environment is one of the characteristic environments in which braided streams develop. Extensive outwash plains or aprons may also be present along the margins of outwash glaciers. Lakes are very common proglacial features, created by ice damming or damming by glacially deposited sediments. Meltwater streams draining into these lakes may create large coarse-grained deltas along the lake edge, while finer sediment is carried outward in the lake by suspension or as a density underflow (Fig. 8.23). Glaciers that extend out to sea create an important environment of glaciomarine sedimentation where sediments are deposited close to shore by melting of the glacier in contact with the ocean or farther out on the shelf or slope by melting of ice blocks, or icebergs.

The glacial environment may range in size from very small to very large. **Valley glaciers** are relatively small ice masses confined within valley walls of a mountain. **Piedmont glaciers** are larger masses or sheets of ice formed at the base of a mountain front where mountain glaciers have debouched from several valleys and coalesced. **Ice sheets**, or continental glaciers, are huge sheets of ice that spread over large continental areas or plateaus.

Transport and Deposition in Glacial Environments

Transport of sediment by ice is a kind of fluid-flow transport, although ice flows very slowly as a high-viscosity, non-Newtonian pseudoplastic. Glaciers can flow at rates as high as 80 m per day during sporadic surges; however, typical flow rates are on the order of centimeters per day (Martini et al., 2001, p. 50). In *A Tramp Abroad,* Mark Twain describes his (fictitious) disappointment, after pitching camp on an alpine glacier in expectation of a free ride down the valley, to find that the view from his camp remained the same day after day. Glaciers advance if the rate of accumulation of snow in the upper reaches (head) of the glacier exceeds the rate of ablation (melting) of ice in the lower reaches (snout). The balance between accumulation and melting is illustrated in Figure 8.28. Ice must flow internally from the head of the glacier to replace that lost by melting at the snout. Flow of ice is laminar, and flow velocity is greatest near the top and center of the glacier. Velocity decreases toward the walls and floor, although not necessarily to zero.

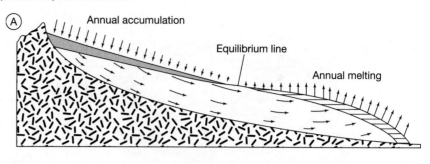

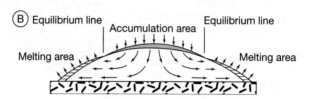

Figure 8.28
Diagrammatic two-dimensional illustration of the balance between glacier accumulation and melting and the movement of ice within a glacier. A. Valley glacier. B. Ice sheet. [(A) after Sharp, R. P., 1988, Living ice—Understanding glaciers and glaciation: Cambridge University Press, Fig. 3.5, p. 58, and Fig. 3.6, p. 59, reproduced by permission; (B) based on Sugden and John, 1976.]

Glaciers retreat if the rate of melting exceeds the rate of accumulation. They reach a state of equilibrium, neither retreating nor advancing, when rates of melting and accumulation are equal, although internal movement of ice continues.

Sediment is entrained by glaciers by quarrying and abrasion by ice as the glacier erodes its bed and by falling or sliding of material from the valley walls. Some of this sediment is transported in contact with the valley walls and floors and is responsible for much of the abrasion. Part of the remaining load is carried on the upper surface of the glacier and part is carried within. The internal load is derived either from the joining of ice streams from two or more valleys or by the washing or falling of material from the surface into crevasses (Fig. 8.29A). Much of the sediment transported by glaciers is carried along the bottom and sides, as illustrated in Figure 8.29B. The entrained sediment load includes large and small blocks of rock as well as extremely fine sediment, called **rock flour**, produced by grinding of the rock-studded glacier base over bedrock. Thus, the glacier sediment load typically consists of an extremely heterogeneous assortment of particles ranging from clay-size grains to meter-size boulders. Glaciers never become overloaded with debris to the point that they become immobilized. As glacial ice melts, however, the sediment load is dropped to form various kinds of glacial **moraines**. See Kirkbride (1995) for details of sediment transport by glaciers.

As glaciers move downslope below the snowline, they eventually reach an elevation where the rate of melting at the front of the glacier equals or exceeds the rate of new snow accumulation above the snowline. If the rate of melting approximately equals the rate of accumulation, the glacier achieves a state of equilibrium in which it neither advances nor retreats. Within such an equilibrium glacier, internal movement of ice continues to carry the rock load along and supply rock debris to the melting snout of the glacier. This process causes a ridge of unsorted sediment, called an **end moraine** or terminal moraine, to accumulate in front of the glacier. **Lateral moraines,** or marginal moraines, can accumulate from concentrations of debris carried along the edges of the glacier where ice is in contact with the valley wall. **Medial moraines** may form where the lateral moraines of two glaciers join (Fig. 8.29). When the rate of melting at the snout of a glacier exceeds the rate of new snow accumulation above the snow line, the glacier retreats back up the valley. If a glacier retreats steadily, it drops its load of rock debris as lateral moraines, medial

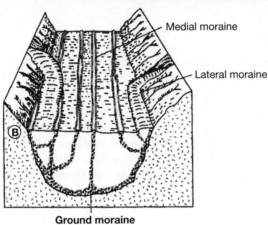

Medial moraine

Lateral moraine

Ground moraine

Figure 8.29
A. Susitna Glacier, eastern Alaska Range, Alaska. The dark stripes are sediments acquired by joining of ice streams from the various valleys. [A. Photograph by Austin Post, American Geographic Society Collection archived at the National Snow and Ice Data Center, University of Colorado, Boulder, and obtained from the Internet at http://-nsidc.colorado.edu, downloaded Dec. 9, 1998]. B. Schematic representation of sediment transport paths within a glacier and the various kinds of glacial moraines; compare with Fig. A. [After Sharp, R. P., 1988, Living ice—Understanding glaciers and glaciation: Cambridge University Press, Fig. 2.5, p. 30, reproduced by permission.]

moraines, and a more or less evenly distributed sheet of **ground moraine**. If the glacier retreats in pulses, it leaves a succession of end moraines, called **recessional moraines**.

As glaciers melt on land, large quantities of water run along the margins, beneath, and out from the front of the glacier to create a meltwater stream. Such streams flow with high but variable discharge in response to seasonal and daily temperature variations. Near the glacier front, the meltwater quickly becomes choked with suspended sediment and loose bedload sand and gravel, leading to formation of branching and anastomosing braided-stream channels. Streams that discharge into glacial lakes tend to build prograding delta systems into the lakes with steeply inclined foresets that grade downward to gently inclined bottomset beds (Edwards, 1986). Very fine sediment discharged into the lake from streams may be dispersed basinward in suspension by wind-driven waves or currents. If a large enough concentration of sediment is present in suspension to create a density difference in the water, a density underflow or turbidity current will result that can carry sediment along the lake bottom into the middle of the basin. Strong winds blowing over a glacier or an ice sheet pick up fine sand from exposed, dry outwash plains and deposit the sand downwind in nearby areas as sand dunes. Fine dust picked up by wind can be kept in suspension and transported long distances before being deposited as loess in the periglacial environment or as pelagic sediment in the ocean.

Where glaciers extend beyond the mouths of river valleys to enter the sea, their sediment load is dumped into the ocean to form **glacial-marine** sediments. Sedimentation under these circumstances may take place in four different ways:

1. Melting beneath the terminus of the glacier allows large quantities of glacial debris to be released onto the seafloor with little reworking (Fig. 8.30).
2. Large blocks of ice calve off from the front of the glacier and float away as icebergs. These icebergs gradually melt, allowing their sediment load to drop onto the seafloor, either on the shelf or in deeper water.
3. Fresh glacial meltwater charged with fine sediment can rise to the surface to form a low-density overflow above denser saline water. Silt and flocculated clays then gradually settle out of suspension from this freshwater plume.
4. Mixing of fresh meltwater and seawater may produce a high-density underflow that can carry sand-size sediment seaward.

Glacial Facies

Because the broad glacial environment encompasses the proglacial and periglacial environments as well as the glacial environment proper, it is necessary, to avoid confusion, to distinguish between glacial facies deposited directly from the glacier and facies transported and reworked by processes operating beyond the margins of glaciers. Furthermore, it is desirable to distinguish between glacial facies deposited on land and those deposited on the seafloor. Table 8.1 illustrates the range of sedimentary facies that are affected in some way by glacial processes. Many of these facies are subtypes of facies deposited in fluvial, lacustrine, eolian, shallow-marine, and deep-marine environments, which are treated elsewhere in this book. Therefore, we focus our discussion here primarily on grounded-ice facies and proximal marine-glacial facies.

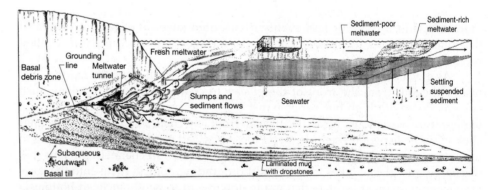

Figure 8.30
Model for glaciomarine sedimentation in front of a wet-based tidewater glacier. Rapid mixing of fresh water and seawater adjacent to tunnel mouths may produce a high-density underflow capable of transporting sand-grade sediment and possibly coarser material. Much of the fresh glacial meltwater rises to the surface of the sea as a low-density overflow layer; as this layer mixes with seawater, silt and flocculated clay gradually settle from suspension. Note also the settling of dropstones from melting ice blocks. [After Edwards, M. B., 1986, Glacial environments, *in* Reading, H. G. (ed.), Sedimentary environments and facies, 2nd ed., Fig. 13.5, p. 453, reproduced by permission of Elsevier Science Publishers, Amsterdam.]

Table 8.1	Facies of glacial environments

Facies of continental glacial environments
  Grounded ice facies
  Glaciofluvial facies
  Glaciolacustrine facies
    Facies of proglacial lakes
    Facies of periglacial lakes
  Cold-climate periglacial facies
Facies of marine glacial environment
  Proximal facies
  Continental shelf facies
  Deep-water facies

Source: Eyles, N., and A. D. Miall, 1984, Glacial facies, *in* R. G. Walker (ed.), *Facies models*, 2nd ed.: Geoscience Canada Reprint Ser. 1, p. 15–38.

Sediment deposited directly from glaciers on land is called **till.** Several kinds of till are recognized, including basal melt-out till, ablation till (supraglacial melt-out till), and lodgment till, deposited under a sliding glacier (Martini et al., 2001). Glacial sediment melted from glaciers in lakes or the ocean is called **waterlaid till.** If direct deposition from a glacier cannot be proved, the term **diamict** is used for poorly sorted, unconsolidated glacial deposits; the term **diamicton** is used for their consolidated equivalents.

Continental Ice Facies

Grounded Ice Facies

Unstratified Diamicts. Till deposited directly from ice on land in various kinds of moraines consists of unstratified, unsorted pebbles, cobbles, and boulders (Fig. 8.31) with an interstitial matrix of sand, silt, and clay. They are thus characterized by a bimodal particle-size distribution in which pebbles predominate in the coarser fraction, with cobbles and boulders scattered throughout (Easterbrook, 1982). Some pebbles are rounded, indicating that they are probably stream pebbles entrained by the ice. Others may be faceted, striated, or polished owing to glacial abrasion. Elongated pebbles and cobbles tend to show some preferred orientation, commonly with their long dimensions parallel to the direction of glacial advance. They may also be crudely imbricated, with long axes dipping upstream. Pebble composition can be highly diverse and may include rock types derived from bedrock located hundreds of kilometers distant. Sands and silts are commonly angular or subangular. Much of the silt in glacial deposits is produced by glacial abrasion and grinding.

Stratified Diamicts. In addition to deposition directly from melting ice, glacial debris can be deposited from meltwaters flowing upon (supraglacial), within (englacial), underneath (subglacial), or marginal to the glacier. The deposits of these meltwaters form on, against, or beneath the ice and thus are commonly known as **ice-contact** sediments. They are reworked to some degree by meltwater and thus exhibit some stratification. They are also better sorted than sediments

Figure 8.31
Thick, poorly stratified, poorly sorted glacial diamict (Quaternary), Alaska. [Photograph courtesy of Gail M. Ashley.]

deposited directly from ice, commonly lack the characteristic bimodal size distribution of direct deposits, and may contain pebbles rounded by meltwater transport. These stratified deposits can accumulate in channels or as mounds or ridges known as kames, kame terraces, or eskers. **Kames** are small mound-shaped accumulations of sand or gravel that form in pockets or crevasses in the ice. **Kame terraces** are similar accumulations deposited as terraces along the margins of valley glaciers. **Eskers** are narrow, sinuous ridges of sediment oriented parallel to the direction of glacial advance. They are the deposits of meltwater streams that probably flowed through tunnels within the glacier. The deposits were then let down onto the subglacial surface after the ice melted. Stratified diamicts are commonly characterized by slump or ice collapse features, including contorted bedding and small gravity faults. Stratified glacial facies can include gravels, sands, and silts, some of which may be extremely well stratified, as illustrated in Figure 8.32.

Facies of Proglacial and Periglacial Environments

As discussed, meltwaters issuing from glaciers transport large quantities of glacial debris downslope and deposit it as **glaciofluvial** sediment in braided streams or as **glaciolacustrine** sediment in glacial lakes, formed by ice damming or moraine damming. These transported and reworked deposits take on the typical characteristics of the environment in which they are deposited; however, they may retain some characteristics that identify them as glacially derived materials. For example, the large daily to seasonal fluctuations in meltwater discharge may be reflected in abrupt changes in particle size of sediments deposited in meltwater streams or lacustrine deltas. Sediments deposited in streams or lakes very close to the glacier front may also display various slump deformation structures caused by melting of supporting ice. As mentioned in the discussion of lakes, one of the most

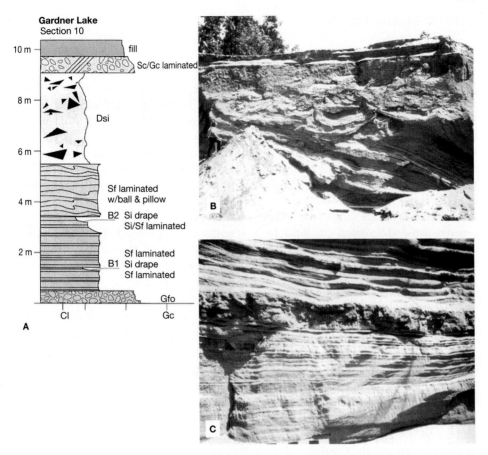

Figure 8.32

A. Vertical stratigraphic profile of glacial sediments deposited in an esker system. The environment is interpreted as ice-tunnel gravel overlain by laminated, marine fan sand and diamictite. Lithofacies code: G, gravel; S, sand; Si, silt; Dsi, silty diamictite; Cl, clay. Grain size: c, coarse; f, fine. Other: B1 and B2, bedding surfaces upon which dip of surface fan measured. B. Photograph showing laminated sand overlain by fine sand deformed into ball and pillow structures, in turn overlain by a massive diamictite (5.5 to 9 m interval in A), interpreted as a debris flow on the fan surface. Person on left gives scale. C. Close-up view of laminated sand within a thin diamictite interbed. Scale increments are 5 cm. [After Ashley, G. M., et al., 1991, Sedimentology of late Pleistocene (Laurentide) deglacial-phase deposits, eastern Maine: An example of a temperate marine grounded ice-sheet margin: Geol. Soc. America Spec. paper 261, Fig. 5, p. 113.]

characteristic properties of glacial lakes is the presence of varves, which form in response to seasonal variations in meltwater flow. Additional details on the characteristics of glaciofluvial and glaciolacustrine sediments are given by Brodzikowski and van Loon (1991), Maizels (1995), and Ashley (1995).

Sand can be transported by wind from outwash plains and deposited as dunes in periglacial areas adjacent to glaciers (e.g., Derbyshire and Owen, 1996); however, the primary wind deposits in many periglacial environments are silts. Deflation of rock flour and other fine sediment from outwash plains and alluvial plains provides enormous quantities of silt-size sediment that is transported by wind and deposited as widespread sheets of fine, well-sorted loess. Because of the even size of its grains, loess typically lacks well-defined stratification. It is composed dominantly of angular grains of quartz but may also contain some clays.

Marine Glacial Facies

Proximal Facies

In environments where marine water is in direct contact with the glacier margin (e.g., Fig. 8.33), substantial quantities of sediment are deposited directly from meltwater conduits or tunnels into subaqueous fans, with additional sediment supplied by melting of rafted ice (Fig. 8.34; Eyles and Miall, 1984; Powell and Domack, 1995). Coarse cobbles and gravels accumulate at the tops of fans, and sands and gravels accumulate within channels owing to sediment gravity underflows. Mud and sand are contributed by ice melting and "rain-out" of suspended sediment. Some reworking of sediment occurs by downslope sediment-gravity flows and episodic traction-current activity. Proximal glacial-marine sediments may thus range from poorly sorted, poorly stratified diamicts that resemble those deposited on land to coarse-grained stratified diamicts, with a muddy sandy matrix, that may display current-produced structures. Melting of buried ice masses can cause surface subsidence and associated deformation and faulting of sediment.

Distal Facies

Away from the proximal environment in which glaciers are in direct contact with marine water, glacial sediment is supplied by floating ice masses, and deposition is dominated by marine processes. Melting of icebergs supplies both fine sediment and coarse debris to the ocean floor by "rain-out," or fallout (Fig. 8.30 and 8.34). Ice-rafted debris deposited on the continental shelves may be reworked to some degree by marine waves and currents (and possibly turbidity currents) and may be affected by iceberg grounding (where icebergs touch bottom). In deeper water on the continental shelf, the debris fallout from floating ice may or may not be retransported by turbidity currents to deeper water. Ice-rafted debris

Figure 8.33
Tidewater glaciers flowing into the ocean. Note the nearshore plume of sediment derived from the melting glaciers and the many small blocks of floating ice. College Fiord Glacier, Chugach Mountains, southeastern Alaska. [U.S. Navy photograph, American Geographic Society Collection archived at the National Snow and Ice Data Center, University of Colorado, Boulder, and obtained from the Internet at http://-nsidc.colorado.edu, downloaded Dec. 9, 1998.]

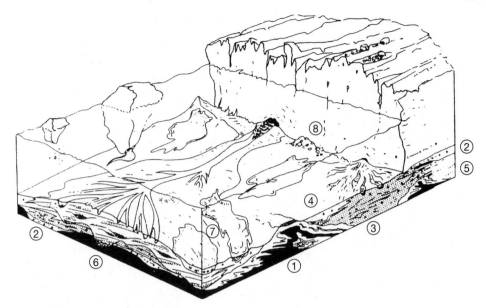

Figure 8.34

Proximal subaqueous sedimentation from glaciers: (1) glacially tectonized marine sediment, (2) lensate lodged diamict units, (3) coarse-grained stratified diamicts, (4) pelagic muds and diamicts, (5) coarse-grained proximal outwash, (6) interchannel cross-stratified sands with channel gravels, (7) resedimented facies (debris flows, slides, turbidites), and (8) supraglacial debris. Sediment deformation results from ice advances, melt of buried ice, and ice-turbation. [From Eyles, N., and A. D. Miall, 1984, Glacial facies, in R. G. Walker (ed.), Facies models: Geoscience Canada Reprint Ser. 1, Fig. 8, p. 20, reprinted by permission of Geological Association of Canada.]

that settles to the deep ocean floor is probably little modified by further depositional processes, except for deposition of a mantle of hemipelagic or pelagic sediment. In general, glacial-marine sediments are distinguished from grounded glacial diamicts by the presence of some stratification and from all on-land glacial diamicts by the presence of marine fossils and dropstones. Dropstones are scattered cobbles or boulders that drop to the seafloor from melting ice blocks or icebergs. Fossil evidence that particularly suggests a glacial-marine origin includes fossils preserved in growth position as whole shells (i.e., fossils entombed by fallout sediment); marine molluscs or barnacles attached to glacially faceted pebbles; preservation of delicate ornamentation on shells; and the presence of forams and diatoms in the matrix material (Easterbrook, 1982).

Vertical Facies Successions

Successive advances and retreats of valley glaciers and ice sheets produce complex vertical successions of facies as ice progressively overrides proglacial environments during glacial advance, and conversely as direct ice deposits and ice-contact deposits are reworked in the proglacial environment as a glacier retreats. These facies are much too varied and complex to attempt description here; however, Figure 8.35 illustrates some typical vertical profiles that might develop in different parts of a glacial environment during a single phase of glacier advance and retreat. See also Brodzikowski and van Loon (1991, Chapter 9) and Edwards (1986, p. 465–466).

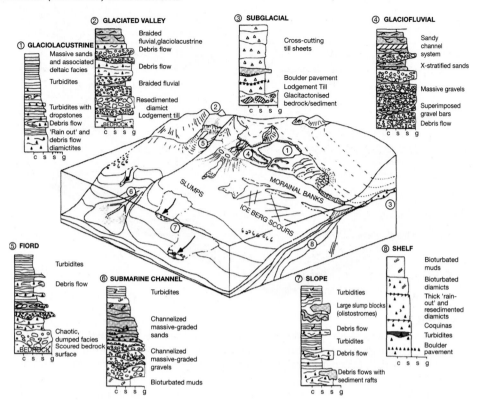

Figure 8.35
Typical vertical profiles of facies deposited during a single phase of glacial advance and retreat in various parts of the glacioterrestrial and glaciomarine setting; c, s, s, g = clay, silt, sand, gravel. [From Eyles, N., and C. H. Eyles, 1992, Glacial depositional systems, *in* R. G. Walker and N. P. James (eds.), Facies models: Geological Association of Canada, Fig. 3, p. 74, reproduced by permission.]

Ancient Glacial Deposits

Glacial deposits range in size from small bodies deposited by valley glaciers to till sheets, deposited by continental glaciers, that cover many thousands of square kilometers. The most characteristic feature of continental tills, or grounded-ice facies, is their extremely poor sorting and lack of stratification. Ancient facies of glaciofluvial, glaciolacustrine, and glaciomarine environments tend to be better stratified and better sorted. Because the characteristics of these proglacial sediments reflect the environment in which they are deposited, it may be very difficult in ancient stratigraphic successions to distinguish proglacial sediments from other types of continental sediments. For example, glaciofluvial sediments may appear much the same as other fluvial sediments. Nevertheless, a few characteristics of these deposits may reveal their relationship to glacial environments. As mentioned, the presence of varves may be diagnostic of glacial lakes, and abrupt changes in sediment size related to variable meltwater discharge may be suggestive of proglacial deposition in general. Ancient glaciomarine sediments are distinguished from other types of glacial deposits by the presence of marine fossils and possibly from other marine deposits by their generally poorer sorting and stratification. They tend to display extreme variation in clast type, reflecting multiple sources. Also, the presence of dropstones, which may deform sedimentary structures such as laminae when they drop into soft sediment, suggests deposition from rafted ice. Anderson and Ashley (1991) further describe several ancient glacial-marine deposits and the paleoclimatic significance of these sediments.

Ancient glacial deposits are best known from sedimentary units of Pleistocene age, which are widespread in many parts of the world. At least four major pulses of continental glaciation occurred during the Pleistocene, plus numerous smaller pulses. Extensive continental glaciation appears to have been important also during the late Precambrian and early Proterozoic, late Ordovician, and late Paleozoic. Carboniferous to Permian glacial deposits are known from South America, southern Africa, Antarctica, India, and Australia. Late Ordovician diamictons have been reported from South America, several parts of Africa, and possibly Ethiopia. Late Precambrian deposits are known on all continents except Antarctica, and early Proterozoic glacial sediments have been reported in North America in a belt extending from Wyoming to Quebec.

FURTHER READING

Fluvial Systems

Best, J. L., and C. S. Bristow (eds.), 1993, Braided rivers: Geol. Soc. London Spec Publ. 75, Bristol, 432 p.

Bridge, J. S., 2003, Rivers and floodplains: Blackwell Science Ltd., Oxford, 491 p.

Collinson, J. D., and J. Lewin (eds.), 1983, Modern and ancient fluvial systems: Int. Assn. Sedimentologists Spec. Publ. 6, Blackwell Scientific Publications, Oxford, 575 p.

Ethridge, F. G., and R. M. Flores (eds.), 1981, Recent and ancient nonmarine depositional environments: Part II: Alluvial fan and fluvial deposits: Soc. Econ. Paleontologists and Mineralogists, Spec. Pub. 31, Tulsa, Okla., p. 49–212.

Ethridge, F. G., R. M. Flores, and M. D. Harvey (eds.), 1987, Recent developments in fluvial sedimentology: Soc. Econ. Paleoatologists and Mineralogists, Spec. Pub. 39, Tulsa, Okla., 389 p.

Fielding, C. R. (ed.), 1993, Current research in fluvial sedimentology: Special Issue of Sedimentary Geology, v. 85, 656 p.

Flores, R. M., F. G. Ethridge, A. D. Miall, W. E. Galloway, and T. D. Fouch, 1985, Recognition of fluvial depositional systems and their resource potential: Soc. Econ. Paleontologists and Mineralogists Short Course Notes 19, Tulsa, Okla., 290 p.

Fraser, G. S., and L. Suttner, 1986, Alluvial fans and fan deltas: International Human Resources Development Corporation, Boston, 199 p.

Marzo, M., and C. Puigdefábregas (eds.), 1993, Alluvial sedimentation, Special Publication No. 17, International Association of Sedimentologists: Blackwell Science, Oxford, 586 p.

Miall, A. D., 1996, The geology of fluvial deposits: Springer-Verlag, New York, 586 p.

Miall, A. D., and N. Tyler (eds.), 1991, Three-dimensional facies architecture of terrigenous clastic sediments and its implications for hydrocarbon discovery and recovery: Concepts in Sedimentology and Paleontology 2, SEPM (Society for Sedimentary Geology), Tulsa, Okla. 309 p.

Nilsen, T. H. (ed.), 1984, Fluvial sedimentation and related tectonic framework, western North America: Sed. Geology, v. 36, 523 p.

North, C. P., and D. J. Prosser (eds.), 1993, Characterization of fluvial and aeolian reservoirs: Geological Society Special Publication 73, Geological Society London, 450 p.

Rachocki, A., 1981, Alluvial fans: John Wiley & Sons, New York, 16l p.

Rachocki, A. H., and M. Church (eds.), 1990, Alluvial fans: John Wiley & Sons, Chichester and New York, 391 p.

Smith, N. D., and J. Rogers, 1999, Fluvial Sedimentology VI, Special Publication No. 28, International Association of Sedimentologists: Blackwell Science, Oxford, 478 p.

Eolian Systems

Barndorff-Nielsen, O. E., and B. B. Willets (eds.), 1991, Aeolian grain transport-1 mechanics: Springer-Verlag, Wien, New York, 181 p.

Fryberger, S. G., L. F. Krystinik, and C. J. Schenk, 1990, Modern and ancient eolian deposits: Petroleum exploration and production: Rocky Mountain Section, SEPM, Denver, various pagination.

Glennie, K. W., 1970, Desert sedimentary environments: Developments in sedimentology 14, Elsevier, Amsterdam, 222 p.

Goudie, A. S., I. Livingstone, and S. Stokes (eds.), 1999, Aeolian environments, sediments and landforms: John Wiley & Sons, Ltd., Chichester, 325 p.

Hesp, P. A. (ed.), 1998, Eolian environments: Special issue of Geomorphology, v. 22. p. 111–204.

Kocurek, G. (ed.), 1988, Late Paleozoic and Mesozoic eolian deposits of the western interior of the United States: Special Issue of Sed. Geology, v. 56, 413 p.

Kocurek, G., 1991, Interpretation of ancient eolian sand dunes: Ann. Rev. Earth and Planetary Sciences, v. 19, p. 43–75.

North, C. P., and D. J. Prosser (eds.), 1993, Characterization of fluvial and aeolian reservoirs: Geological Society Special Publication 73, Geological Society London, 450 p.

Pye, K. (ed.), 1993, The dynamics and environmental context of aeolian sedimentary systems: Geological Society Special Publication 72, Geological Society London, 332 p.

Pye, K., and N. Lancaster (eds.), 1993, Aeolian sediments: Ancient and modern: International Association of Sedimentologists Special Publication 16: Blackwell Scientific Publications, Oxford, 167 p.

Pye, K., and H. Tsoar, 1990, Aeolian sand and sand dunes: Unwin Hyman, London, 396 p.

Lacustrine Systems

Anadón, P., L. I. Cabrera, and K. Kelts (eds.), 1991, Lacustrine facies analysis: Internat. Assoc. Sedimentologists Spec. Pub. 13, Blackwell, Oxford, 318 p.

Gierlowski-Kordesch, E., and K. Kelts (eds.), 1994, Global geological record of lake basins, v. 1: Cambridge University Press, 427 p.

Gierlowski-Kordesch, E. H., and K. R. Kelts (eds.), 2000, Lake basins through space and time: AAPG Studies in Geology No. 46, 648 p.

Lerman, A., D. Imboden, and J. Gat (eds.), 1995, Physics and chemistry of lakes, 2nd ed.: Springer-Verlag, Berlin, 334 p.

Noe-Nygaard, N. (ed.), 1998, Limno-geology—Research and methods in ancient and modern lacustrine basins: Special issue of Palaeogeography, Palaeoclimatology, Palaeoecology, v. 140, p. 1–478.

Renaut, R. W., and W. M. Last, 1994, Sedimentology and geochemistry of modern and ancient saline lakes: SEPM (Society for Sedimentary Geology) Spec. Publ. 50, Tulsa, Okla., 334 p.

Glacial Systems

Anderson, J. B., and G. M. Ashley (eds.), 1991, Glacial marine sedimentation; Paleoclimatic significance: Geol. Soc. America Spec. Paper 261, 232 p.

Brodzikowski, K., and A. J. van Loon, 1991, Glacigenic sediments: Elsevier, Amsterdam, 674 p.

Dowdeswell, J. D., and J. D. Scourse (eds.), 1990, Glaciomarine environments: Processes and sediments: Geol. Soc. London Spec. Pub. 53, 423 p.

French, H. M., 1996, The periglacial environment, 2nd ed.: Addison Wesley Longman Ltd., Edinburgh Gate, Harlow, 341 p.

Hambrey, M., 1994, Glacial environments: UCL Press Ltd., London, 296 p.

Jopling, A. V., and B. C. McDonald, 1975, Glaciofluvial and glaciolacustrine sedimentation: Soc. Econ. Paleontologists and Mineralogist Special Pub. No. 23, 320 p.

Knight, P. G., 1999, Glaciers: Stanley Thornes (Publishers) Ltd., Cheltenham, 261 p.

Martini, I. P., M. E. Brookfield, and S. Sadura, 2001, Principles of glacial geomorphology and geology: Prentice Hall, Upper Saddle River, N. J., 381 p.

Menzies, J. (ed.), 1995, Modern glacial environments: Processes, dynamics and sediments: Butterworth-Heinemann, Oxford, 621 p.

Menzies, J. (ed.), 1996, Past glacial environments: Butterworth-Heinemann, Oxford, 598 p.

Piotrowski, J. A. (ed.), 1997, Subglacial environments: Special issue of Sedimentary Geology, v. 111, 327 p.

9

Marginal-Marine Environments

9.1 INTRODUCTION

The marginal-marine setting lies along the boundary between the continental and the marine depositional realms. It is a narrow zone dominated by river, wave, and tidal processes. Salinities may range in different parts of the system from fresh through brackish to supersaline water, depending upon river discharge and climatic conditions. Intermittent to nearly constant subaerial exposure characterizes some environments of the marginal marine setting. Others are continuously covered by shallow water. Many marginal-marine environments are further characterized by high-energy waves and currents, although some lagoonal and estuarine environments are dominated by quiet-water conditions.

Because of the large quantities of siliciclastic sediment delivered by rivers to the coastal zone throughout geologic time, the volume of marginal marine deposits preserved in the geologic record is significant. The principal depositional settings for marginal-marine sediments are deltas; beaches, strand plains, and barrier bars; estuaries; lagoons; and tidal flats (Fig. 9.1). Estuaries and lagoons are particularly characteristic of transgressive coasts; deltas are features of prograding coasts. A wide variety of sediment types—including conglomerates, sandstones, shales, carbonates, and evaporites—can accumulate in these various marginal-marine environments. We begin study of these environments by examining first deltas, followed, in turn, by beach and barrier-island systems, estuaries, and lagoons.

9.2 DELTAIC SYSTEMS

Introduction

The word **delta** was used by the Greek philosopher Herodotus about 490 B.C. to describe the triangular-shaped alluvial plain formed at the mouth of the Nile River by deposits of the Nile distributaries. Most modern deltas are less triangular and more irregular in shape than the Nile delta (e.g., Fig. 9.2). Nevertheless, the term (alluvial) delta is still applied to any deposit, subaerial or subaqueous, formed by fluvial sediments that build into a standing body of water. Deltas are

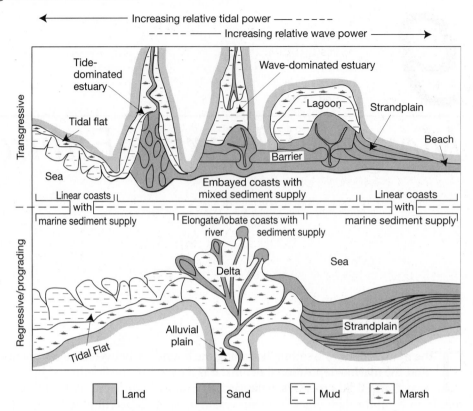

Figure 9.1
The principal coastal environments of the marginal-marine depositional setting. The figure is organized to show the relative influence of tidal power (increasing to the left) and wave power (increasing to the right) on each environment. Note that deltas are features of prograding (regressive) coasts, whereas estuaries and lagoons are particularly characteristic of transgressive coasts. [From Boyd et al., 1992, Classification of clastic coastal depositional environments: Sed. Geology, v. 80, Fig. 2, p. 141, reproduced by permission.]

Figure 9.2
The Misissippi River Delta, Gulf of Mexico. Image courtesy of NASA Landsat Project Science Office and USGS EROS Data Center. Downloaded from the Internet 4/19/04.

"discrete shoreline protuberances formed where rivers enter oceans, semi-enclosed seas, lakes or lagoons and supply sediment more rapidly than it can be redistributed by basinal processes" (Elliott, 1986a). Thus, deltas can form in lakes and inland seas as well as in the ocean, but they are most important in the open ocean. Much of the siliciclastic sediment transported to coastal zones throughout geologic time has been deposited in deltas.

Ancient deltaic deposits have been identified in stratigraphic successions of many ages, and deltaic sediments are known to be important hosts for petroleum and natural gas, coal, and some minerals such as uranium. Although ancient deltaic sediments are common in the rock record, much of what we know about delta systems comes from study of modern deltas. Deltas are particularly common in the modern ocean owing to post-Pleistocene sea-level rise coupled with high sediment loads carried by many rivers. High sea level increases sedimentation rates on deltas because sediment is trapped by the rising water, inhibiting sediment removal by currents. The locations, dimensions, and discharge characteristics of some modern deltas are given in Table 9.1.

Modern deltas occur on all continents, with the possible exception of Antarctica. (Trough-mouth, glacially influenced submarine fans are present on the Weddell Sea continental margin of Antarctica; however, these fans may not be true deltas.) Deltas form where large, active drainage systems with heavy sediment loads exist. These conditions appear to be met particularly well on trailing-edge or passive coasts such as the east coasts of Asia and the Americas where tectonic activity is low. Fewer than 10 percent of major modern deltas occur on collision coasts, where tectonic activity is high and drainage divides are close to the sea (Inman and Nordstrom, 1971; Wright, 1978). Under such conditions, the large drainage systems necessary to supply heavy sediment loads are not developed. Their potential importance as oil and gas reservoirs has generated considerable interest in deltaic deposits since the 1950s. Consequently, the literature on deltas and

Table 9.1 Characteristics of selected modern deltas

Delta	Location	Delta-plain area (km^2)	Annual water discharge (m^3/sec)	Annual sediment discharge (tons × 10^6)
Alta	Norway—Alta Fiord	10	?	?
Burdekin	Queensland, Aust.—Pacific	2112	475	?
Copper	U.S.—Gulf of Alaska	1920	1236	70
Ganges-Bramaputra	India & Pakistan—Bay of Bengal	105,641	30,769	1,670
Irrawaddy	Burma—Gulf of Martaban	20,571	13,562	265
Mackenzie	NW Terr., Canada—Beaufort Sea	13,000	9,100	126
Mahakam	Indonesia—Makassar Strait	5000	?	16
Mississippi	U.S.—Gulf of Mexico	28,568	15,631	349
Niger	Nigeria—Gulf of Guinea	19,135	8,769	40
Ord	Western Australia—Timor Sea	3,896	163	22
Punta Gorda	Belize—Gulf of Honduras	0.4	?	?
São Francisco	Brazil—Atlantic	734	3420	6
Skeidarar Sandur	Iceland—North Atlantic	600	400	?
Yallahs	Jamacia—Caribbean Sea	10.5	17.5	?

Data source: Orton and Reading, 1993

deltaic deposits is extensive. The volumes listed under "Further Reading—Deltaic Systems" at the end of this chapter provide a starting point for further literature research.

Delta Classification and Sedimentation Processes

The distribution and characteristics of deltas are controlled by a complex set of interrelated fluvial and marine/lacustrine processes and environmental conditions. These factors include climate, water and sediment discharge, river-mouth processes, nearshore wave power, tides, nearshore currents, and winds (Coleman, 1981). Other factors that influence the formation of deltas are slope of the shelf, rates of subsidence and other tectonic activity at the depositional site, and geometry of the depositional basin. Among these variables, river (sediment) input, wave-energy flux, and tidal flux are the most important processes that control the geometry, trend, and internal features of the progradational framework sand bodies of deltas (Galloway and Hobday, 1983).

Deltas can be classified in several ways (Nemec, 1990); however, classification on the basis of delta-front regime (Galloway, 1975) appears to be favored by most geologists. Deltas are classified thus as (1) fluvial-dominated, (2) tide-dominated, or (3) wave-dominated (Fig. 9.3). Each of these kinds of deltas can be

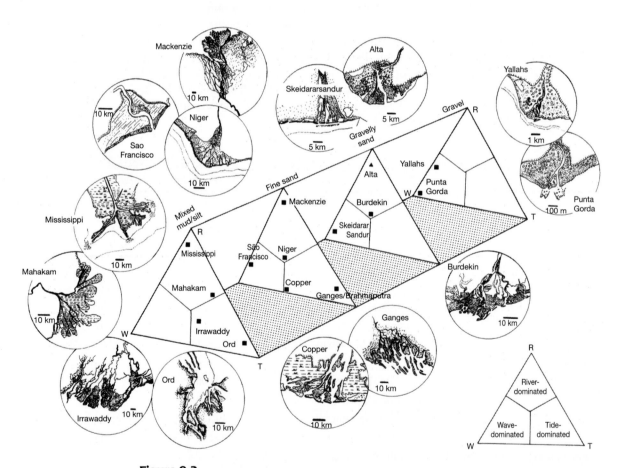

Figure 9.3
Classification of deltas on the basis of dominant process of sediment dispersal at the delta front, and the prevailing grain size of sediment delivered to the front. Dispersal processes: R, river; W, wave; T, tidal. The insets illustrate the shapes of selected modern deltas plotted in the classification diagram; see Table 9.1 for their locations. [After Orton, G. J., and H. G. Reading, 1993, Variability of deltaic processes in terms of sediment supply, with particular emphasis on grain size: Sedimentology, v. 40, Fig. 1, p. 477, reproduced by permission.]

further distinguished on the basis of dominant grain size of sediments (Orton, 1988; Orton and Reading, 1993), that is, mud/silt, fine sand, gravelly sand, or gravel. Figure 9.3 shows selected modern examples of each of these kinds of deltas. The locations, areal extents, and discharge characteristics of these deltas are listed in Table 9.1.

Fluvial-Dominated Deltas

The discharge of river water and sediment into a lake or ocean is referred to as a **jet**. Bates (1953) contrasted the behavior of sediment-laden river water as it enters equally dense, more dense, and less dense basin water. River water entering basin water of almost equal density, referred to by Bates as **homopycnal flow**, leads to rapid, thorough mixing and abrupt deposition of much of the sediment load. This type of jet outflow is particularly common at the mouths of coarse-grained rivers and presumably causes the formation of Gilbert-type deltas that display a topset, foreset, and bottomset arrangement of beds, created as sediment deposition pro-grades basinward (Fig. 9. 4). River water that has higher density than basin water flows beneath the basin water, commonly during floods, generating a vertically oriented, plane-jet flow called **hyperpycnal flow**. This type of jet flow moves along the bottom as a density current that deposits its load along the more gentle slopes of the delta front to form turbidites. If river outflow is less dense than basin water, as in rivers that flow into denser seawater or a saline lake, it flows outward on top of the basin water as a horizontally oriented plane jet called **hypopycnal flow**. Fine sediment may thus be carried in suspension some distance outward from the river mouth before it flocculates and settles from suspension. [Floccula-tion involves aggregation of fine sediment into small lumps owing to the presence of positively charged ions in seawater that neutralize negative charges on clay particles.] Hypopycnal flow tends to generate a large, active delta-front area, typ-ically dipping at 1° or less, as contrasted with the 10°–20° dip of most Gilbert-type deltas (Miall, 1984). Hypopycnal flow is probably the most important type of river outflow in marine basins.

Wright (1977) suggested that the characteristics of delta sediments de-posited by river-mouth processes in coastal areas with low tidal range and low wave energy depend upon the relative dominance of (1) outflow inertia (velocity),

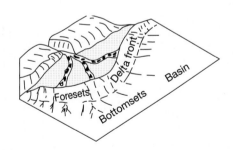

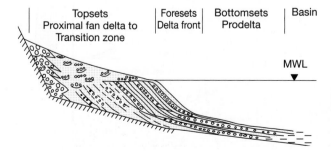

Figure 9.4

Schematic representation of a Gilbert-type delta. [From Reading, H. G., and J. D. Collinson, 1996, Clastic coasts, *in* Reading, H. G. (ed.), Sedimentary environments: Process-es, facies and stratigraphy, 3rd ed., Blackwell Science, Ox-ford, Fig. 6.22, p. 174, Reproduced by permission.]

(2) turbulent bed friction seaward of the river mouth, and (3) outflow buoyancy. For example, outflows dominated by inertial forces produce narrow river-mouth bars of the Gilbert type. Outflows dominated by turbulent friction generate triangular "middle-ground" bars and channel bifurcation, whereas buoyant outflow leads to formation of elongate distributaries with parallel banks, called subaqueous levees; few channel bifurcations; and narrow distributary-mouth bars that grade seaward to fine-grained distal-bar deposits and prodelta clays. Bar sands or **bar-finger** sands are typical components of such deltaic assemblages.

Orton and Reading (1993) extended Wright's work by further suggesting that discharge processes and the nature of the river mouth cannot be considered independently of the sediment load and that mixing behavior at the river mouth differs for suspended load, mixed load, and bedload channel types and for very coarse grained, gravelly bedloads and mass-flow dominated alluvial regimes. Bates' (1953), Wright's (1977), and Orton and Reading's (1993) concepts are combined in Figure 9.5, which graphically depicts the relationship between the kind of sediment load and the type of outflow dominant.

Buoyant dominated river mouths (Fig. 9.5A) form where the inflow extends as a plume (hypopycnal flow) into relatively deep, generally marine, waters. The flow detaches from the bed and is thus unable to move bedload beyond the detachment point (Reading and Collinson, 1996). Turbulent mixing is intense near

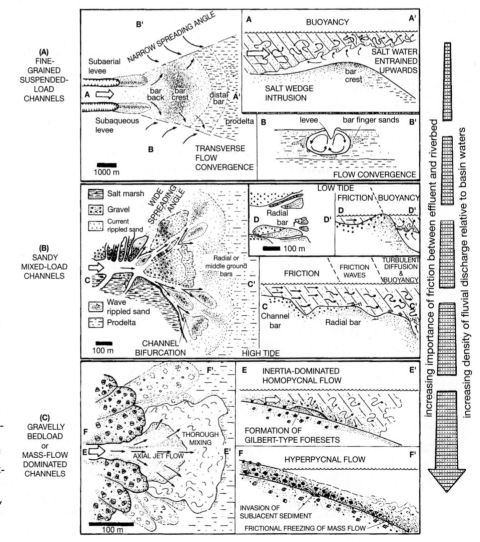

Figure 9.5
Schematic depiction of the complex relationship arising from sediment grain size, river outflow inertia (velocity), outflow friction with the bed, and outflow buoyancy—and its effect on delta formation. [After Orton, G. J., and H. G. Reading, 1993, Variability of deltaic processes in terms of sediment supply, with particular emphasis on grain size: Sedimentology, v. 40, Fig. 4, p. 487, reproduced by permission.]

the river mouth, and much of the coarser suspended and bed load is deposited, whereas finer grained sediment is transported farther into the basin before deposition. Note from Figure 9.5A the position of levees and bars and the presence of bar-finger sands. **Friction-dominated river mouths** (Fig. 9.5B) occur where rivers enter water so shallow that the inflow can expand only in a horizontal direction. So much friction is present between the incoming flow and the sediment surface that the inflow jet spreads laterally, decelerates, and generates a triangular "middle-ground" bar in the river mouth and leads to channel bifurcation. **Inertia-dominated** river mouths (Fig. 9.5C) form where the slope is steep enough to allow expansion of the inflow in both the horizontal and vertical direction. This expansion commonly occurs where high-velocity rivers enter fresh water and/or carry substantial quantities of coarse sediment. Thorough mixing of the axial jet flow (homopycnal flow) may cause rapid deposition of the bed load, leading to formation of Gilbert-type foresets, as mentioned; however, underflows (hyperpycnal flow) may develop lateral to the axial jet that transports sediment as mass flows.

The modern Mississippi River delta is is a classic example of a birdsfoot-type, buoyancy-dominated delta. It is among the largest deltas in the world outside of Asia (Table 9.1). The Mississippi delta consists of seven distinct sedimentary lobes (Fig. 9.2) that have been active during the past 5000–6000 years, indicating that periodic channel or distributary abandonment, to be discussed, is a common process. The generalized characteristics of the Mississippi delta system are shown in Figure 9.6. This figure illustrates the well-developed birdsfoot distributary system, typical of the delta, with bar-finger sands developed at the mouths of the distributaries. Common sediment facies on the Mississippi delta include marsh and natural-levee deposits, delta-front silts and sands, and prodelta clays. Other modern deltas that are largely fluvial-dominated include the Mackenzie Delta (Canada, Beaufort Sea) and the Alta delta (Norway, Alta Fiord), as indicated in Figure 9.3.

Tide-Dominated Deltas

The processes and deposits described above may be significantly modified under conditions of high tidal range or high wave energy. If tidal currents are stronger than river outflow, these bidirectional currents can redistribute river-mouth sediments, producing sand-filled, funnel-shaped distributaries. The distributary mouth

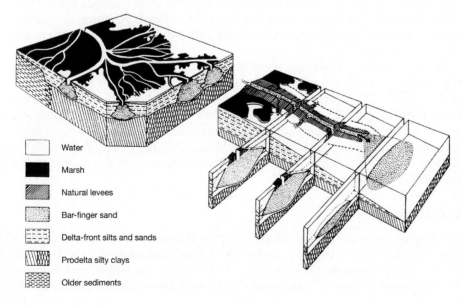

Water

Marsh

Natural levees

Bar-finger sand

Delta-front silts and sands

Prodelta silty clays

Older sediments

Figure 9.6
The fluvial-dominated, Mississippi delta system. [From Reineck, H. E., 1970, Marine sandkörper, rezent und fossil: Geol. Rundschau, v. 60, Fig. 2, p., 305, reprinted by permission. Originally modified from Fisk, H. N., E. Mcfarland, C. R. Kolb, and L. J. Wilbert, 1954, Sedimentary framework of the modern Mississippi delta: Jour. Sed. Petrology, v. 24, Fig. 1, p. 77.]

bar may be reworked into a series of linear tidal ridges that replace the bar and extend from within the channel mouth out onto the subaqueous delta-front platform.

The modern Ganges-Brahmaputra delta is a well-known example (Michels et al., 1998; Goodbred and Kuehl, 2000) of a tide-dominated delta (Fig. 9.7). The areal size of this delta is more than three times that of the Mississippi delta. It has a mean river discharge about twice that of the Mississippi and an exceedingly high discharge during the monsoon season when extreme flooding is common. Mean tidal range is large, about 4 m, and tidal currents can be as strong as 3.8 m/s (Michels et al., 1998); wave energy is relatively low. Sand transport is intense during the monsoon season, leading to deposition of sandy deposits similar to those in braided streams. The delta is characterized by tidal-flat environments, natural levees, and floodbasins in which fine sediment is deposited from suspension. The strong tidal influence is manifested by a network of tidal sand bars and channels oriented roughly parallel to the direction of tidal current flow (Fig. 9.7A, B). A variety of sediment types thus accumulate on the Ganges-Brahmaputra delta, such as tidal-bar or tidal-ridge sands; braided, channel-fill sands; and natural levee, tidal-flat, and floodbasin muds. The Ord Delta (Timor Sea, Australia) provides another example of a modern delta largely dominated by tidal processes (Fig. 9.3).

Wave-Dominated Deltas

Strong waves cause rapid diffusion and deceleration of river outflow and produce constricted or deflected river mouths. Distributary-mouth deposits are reworked by waves and are redistributed along the delta front by longshore currents to form wave-built shoreline features such as beaches, barrier bars, and spits. A smooth delta front, consisting of well-developed, coalescent beach ridges, may eventually be generated.

The Paraiba do Sul delta of Brazil (Fig. 9.8) displays many of the characteristic features of wave-dominated deltas. The Rio Paraiba do Sul coastal plain consists of Pleistocene and Holocene littoral marine sediments and Holocene fluvial and lagoonal sediments (Martin et al., 1987). Tidal range over the delta is moderate (mesotidal); however, wave energy is extremely high. Sediment discharge from the river is high and middle-ground bars are common at the mouth of the river. Sediment is transported at high rates across the mouth, from west to east, leading to development of amalgamated, sandy beach ridges (Bhattacharya and Giosan, 2003). Owing to this extreme wave energy, the Paraiba do Sul delta is dominated by high-energy environments in which sand deposition takes place. Muds accumulate locally in lagoons, but the interdistributary bay mud deposits characteristic of the Mississippi delta are absent. Paraiba do Sul delta deposits are dominated by beach-ridge barrier sands that cover much of the delta surface adjacent to the ocean.

Other examples of modern wave-dominated deltas include the Skeidarar Sandur (Iceland, North Atlantic), the Punta Gorda (Belize, Gulf of Honduras), as shown in Figure 9.3, the Tseng-wen (southwest Taiwan; Liu et al., 2003), and the São Francisco (Brazil; Dominguez, Martin, and Bittencourt, 1987; Dominguez, 1996), often cited as the classic example of a wave-dominated delta. Several additional examples, from the Black Sea, Gulf of Mexico, Mediterranean Sea, South Atlantic, Bay of Bengal, and the Thyrrenian Sea, are discussed by Bhattacharya and Giosan (2003).

Mixed-Process Deltas

The examples discussed above illustrate some differences in characteristics of modern deltas that are shaped by processes that are predominantly fluvial, tidal, or wave-related. Many deltas have characteristics that are transitional between these "end-member" types. The Copper River delta in the Gulf of Alaska (Fig. 9.9)

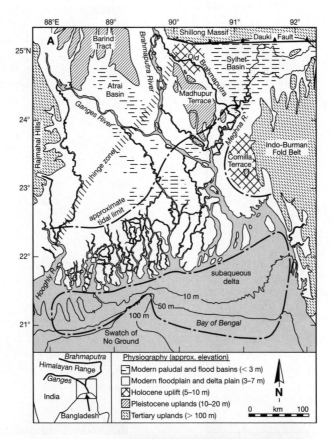

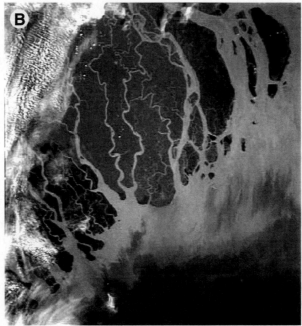

Figure 9.7
The Ganges-Brahmaputra delta, a modern tide-dominated delta. A. Regional map of the
Bengal Basin showing physiography and geology of the delta and surrounding area.
[From Goodbred, S. L., and S. A. Kuehl, 2000, The significance of large sediment supply,
active tectonism, and eustasy on margin sequence development: Late Quaternary stratig-
raphy and evolution of the Ganges-Brahmaputra delta: Sedimentary Geology, v. 133, Fig.
2, p. 229. Reproduced by permission.] B. Space image of the delta, NASA Johnson Space
Center, downloaded from the Internet 4/8/04.

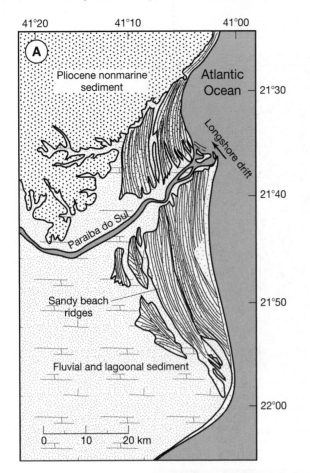

Figure 9.8
The Paraiba do Sul delta, Brazil, which is characterized by sandy beach ridges oriented roughly parallel to the shoreline. A. Map of the Rio Paraiba do Sul coastal plain. [Redrawn from Martin et al., 1987, Quaternary evolution of the central part of the Brazilian coast: The role of relative sea-level variation and shoreline drift, *in* Quaternary coastal geology of West Africa and South America, UNESCO reports in Marine Science, No. 43, Fig. 16, p. 130. [Address: Marine Information Centre, Division of Marine Sciences, UNESCO, Place de Fontenoy, 75700, Paris, France]. B. Space image of the delta and coastal plain, NASA image downloaded from the Internet 5/28/04.

provides an example of a delta that is strongly influenced by tides but also experiences high wave power (e.g., Galloway, 1976). Thus, the Copper River delta falls somewhere between a tide-dominated and a wave-dominated delta (see Fig. 9.3). Tidal range on the Copper River delta may exceed 3 m, and tidal currents of up to 2 m/s have been measured. Strong, wind-driven swells, coupled with the westerly marine current, set up a net westward longshore drift. Thus, deltaic sediments are modified by both tidal and wave-related processes. Major delta environments include (1) the subaerial deltaic plain, consisting of marsh deposits and distributary channel fills; (2) the tidal lagoon, composed of tidal sands and mudflat deposits interlaced with a complex of tidal channel fills (Fig. 9.9); and (3) the wind-influenced shoreface, consisting of marginal island, breaker bar, and middle shoreface sands, lower shoreface sand and mud, and prodelta/shelf mud. As indicated in Figure 9.3, other modern deltas influenced fairly strongly by both tides and wind include the Irrawaddy (Burma), Niger (Nigeria), and Burdekin (Australia).

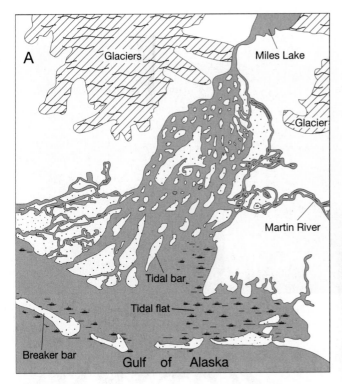

Figure 9.9
The Copper River delta, Gulf of Alaska. A. Physiographic map of the delta, showing tide-influenced channels and bars and wave-influenced breaker bars; based on U.S. Dept. of Agriculture map, "The Copper River," Alaska Region R110-RG-100. B. Tidal channels and tidal sand bars on the delta; photograph by Allan Cline.

Fan Deltas

The concept of fan deltas was introduced by Holmes (1965, p. 554). A fan delta, as defined by Holmes and modified slightly by Nemec and Steel (1988a), is a coastal prism of sediments delivered by an alluvial-fan system and deposited, mainly or entirely subaqueously, at the interface between the active fan and a standing body of water (e.g., Fig. 9.10). Fan deltas were recognized first in modern settings but fan-delta deposits have now been reported in many ancient sedimentary successions (e.g., Nemec and Steel, 1988b; Chough and Orton, 1995).

The relationship between alluvial fans, which constitute the subaerial component of fan deltas, and the subaqueous delta is illustrated in Figure 9.11. The alluvial fans can include any of the fan types discussed in Chapter 8 and may form in settings ranging from glacial to humid to arid. Like other deltas, the subaqueous portion of fan deltas may be fluvial dominated, wave dominated, or tide dominated. Sediments are deposited downslope in the subaqueous part of fan deltas by processes such as slumping and debris avalanching, turbidity-current flow, and inertia (hyperpycnal) flow that takes place particularly during flood stages; the riverborne load achieves sufficient density to overcome buoyancy and frictional effects at the river mouth and can transport even gravel and coarse sand downslope (e.g., Prior and Bornhold, 1990).

Physiographic and Sediment Characteristics of Deltas

Variations in sediment input, outflow velocity, and wave and current energy cause the depositional features of deltas to exhibit a high degree of variability from one

Figure 9.10

Kurobegawa fan, a large fan delta at the mouth of the Kurobe River, Toyama Bay, Japan Sea, off central Japan. [From Fujii, S., and N. Nasu, 1988, Kaiteirin (submerged forest; text in Japanese), Tokyo University Press, Color plate of Kurobegawa fan. Photograph courtesy of S. Fujii; reproduced by permission of University of Tokyo Press.]

Figure 9.11

Schematic diagram of a fan delta, showing the subaerial alluvial fan and subaqueous delta. [After Nemec, W., and R. J. Steel, 1988, What is a delta and how do we recognize it? in Nemec, W., and R. J. Steel (eds.), Fan deltas: Sedimentology and tectonic setting: Blackie, Glasgow, Fig. 1A, p. 6.]

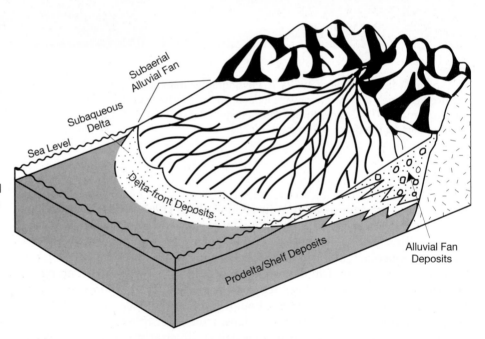

delta to another. Nevertheless, all deltas can be divided into subaerial and subaqueous components, each of which can be further subdivided (Fig. 9.12). The subaerial component of deltas, called the **deltaic plain**, is generally larger than the subaqueous component. It is divided into an upper delta plain, which lies largely above high-tide level, and a lower delta plain, lying between low-tide mark and the upper limit of tidal influence. The upper delta plain is commonly the oldest part of the delta and is dominated by fluvial processes. The lower delta plain is exposed during low tide but is covered by water during high tide. Thus, it is subjected to both fluvial and marine processes.

The upper delta plain is influenced mainly by fluvial processes. Sedimentation is dominated by distributary-channel migration and associated fluvial sedimentation processes such as channel and point-bar deposition, overbank flooding, and crevassing into lake basins (Chapter 8). The principal depositional environments

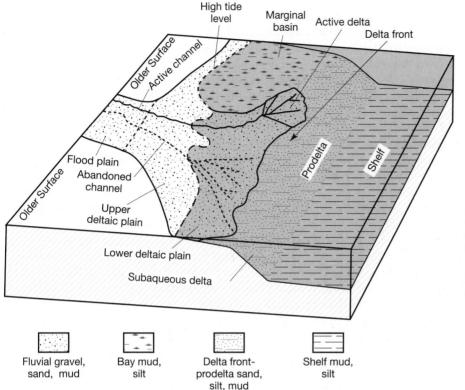

Fluvial gravel,
sand, mud

Bay mud,
silt

Delta front-
prodelta sand,
silt, mud

Shelf mud,
silt

Figure 9.12
Principal components of a fluvial-dominated delta system. The physiographic subdivisions of tide- and wave-dominated deltas are similar; however, sediments on the delta front and the lowermost deltaic plain are reworked by tides and waves. [Based in part on Coleman and Prior, 1982.]

include braided channels, meandering channels, back swamps, and floodplain environments such as swamps, marshes, and freshwater lakes. Therefore, upper delta-plain sediments are predominantly fluvial sands, gravels, and muds that may be closely associated with lake, swamp, and marsh deposits. The width of the lower deltaic plain is greatest on deltas where tidal range is large. This plain includes the active distributary system of the delta as well as abandoned distributary-fill deposits, and it may be flanked by marginal-basin or bay-fill deposits. Distributary channels are numerous, but environments between channels make up the largest percentage of the lower delta plain. These environments include actively migrating tidal channels, natural levees, interdistributary bays, bay fills (crevasse splays), marshes, and swamps (e.g., Coleman and Prior, 1982).

The **subaqueous delta plain** lies seaward of the lower deltaic plain below low-tide water level. The uppermost part of the subaqueous delta, lying at water depths down to 10 m or so, is commonly called the **delta front**. The remaining seaward part of the subaqueous delta is called the **prodelta**, or prodelta slope. The subaqueous delta may extend outward for distances of a few kilometers to tens of kilometers, and the prodelta may extend to water depths as much as 200–300 m. On fluvial-dominated deltas, deposits typically consist in part of sand, and possibly gravel, deposited near river mouths, forming distributary-mouth-bar deposits such as those of the Mississippi delta (Fig. 9.6). On the other hand, the delta front may be dominated by high-energy marine processes, including waves, longshore currents, and tides. On wave- and tide-dominated deltas, sediment is reworked and winnowed by these processes, creating well-sorted delta-front sheet sands that are cross-bedded on a variety of scales. The finest silts and clays are transported still farther seaward and settle on the prodelta on the outermost part of the subaqueous delta. Previously deposited sediments may be reentrained, transported, and redeposited farther downslope on the subaqueous delta by gravity-driven mass-movement processes such as landslides, slumps, turbidity-current flows, and mud flows.

Delta Cycles

During the active, constructional phases of delta outbuilding, deltaic deposits prograde seaward, leading to generation of a coarsening-upward vertical succession of facies as delta-front sands advance seaward over prodelta silts and clays. The sedimentary succession on the progradational lobes of the fluvial-dominated Mississippi River delta provides a good example (Fig. 9.13). Active delta outbuilding may be interrupted by major changes in the hydrologic and sediment regime related to tectonism, climate variation, major diversions of rivers upstream, or changes in sea level. Smaller-scale changes may result from processes such as switching of delta lobes, distributaries, or tidal channels. Major changes, in particular, can result in a relative rise in sea level that may halt progradational delta growth and bring about a transgressive phase of deposition as the shoreline advances in a landward direction. Thus, growth of deltas tends to be cyclic. During active prograding phases, prodelta fine silts and clays are progressively overlain by delta-front silts and sands, distributary-mouth sands, and finally marsh, fluvial, and possibly eolian deposits as the delta builds seaward, producing a coarsening-upward regressive succession. Interruption of progradation by delta-lobe abandonment or marine transgression brings on a destructive phase in which erosion and redistribution of river-mouth deposits predominates. Subsequent distributary shifting or regression may bring on another phase of active progradation. Sediments that make up a complete delta cycle may range in thickness from 50 to 150 m. Smaller scale cycles representing progradation of individual distributaries range from only about 2 to 15 m (Miall, 1984).

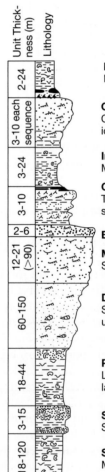

Bay and marsh deposits.
Mud, silt; bioturbated

Crevass splay deposits.
Coarsening-upward successions of muds to sands

Interdistributary bay deposits.
Mud, sand-silt stringers; shells

Overbank splay deposits.
Thin peat & soils. Thin sand, silt, shale stringers; root traces

Beach & dune sands.

Mouth bar & channel deposits.
Sand, silt beds; cross-bedded

Distal bar deposits.
Sand & silt, rippled; coarsening upward; mud laminae; shells

Prodelta deposits.
Laminated mud; silt-sand laminae; shells

Slump-block deposits.
Sand & silt; flow structures

Shelf deposits.
Mud, bioturbated or laminated

Figure 9.13
Idealized vertical succession of facies in a fluvial-dominated (Mississippi) delta. Note the thickness of individual units shown in the column. [From Coleman, J. M., 1981, Deltas: Processes of deposition and models for exploration, 2nd ed., Fig. 4.3, p. 91, reprinted by permission of IHRDC Publications, Boston.]

Ancient Deltaic Systems

Ancient deltaic sediments have been reported in stratigraphic successions of most ages, but they appear to be particularly common in rocks of Carboniferous and Tertiary age. Reported examples of fluvial-dominated deltaic deposits include the Carboniferous of the British Isles (e.g., Martinsen, 1990; Pulham, 1989), the Wilcox Group (Eocene) of Texas (Fisher and McGowan, 1967), and the Dunvegan Formation (Cretaceous) of Canada (Bhattacharya and Walker, 1991). An example is provided by Horne et al. (1978), who describe fluvial deltaic sediments from the Carboniferous of Kentucky (Fig. 9.14). The deposits shown in Figure 9.14 are distributary-mouth-bar sandstones that grade laterally into bay-fill muds. The sand bodies are 1.5–5 km wide and 15–25 m thick. They are widest at the base and have gradational lower and upper contacts. Grain size increases upward in the succession and toward the center of the bars. Fining-upward, graded beds are common on the flanks of the bars, as are oscillation- and current-rippled surfaces. Pebble-lag conglomerates are present at the bases of the channel deposits. Note by comparison with Fig. 9.6 that these ancient river-dominated deltaic sediments are very similar in geometry and sediment characteristics to those of the modern Mississippi delta. Other ancient river-dominated deltas are discussed by Bhattacharya and Walker (1991), Fisher and McGowan (1967), Galloway (1975), and Pulham (1989).

Weise (1980) describes Upper Cretaceous sediments from the San Miguel Formation in the subsurface of Texas that are interpreted from core and well-log information to be wave-dominated deltaic sediments. She constructed thickness (isopach) maps of ten sandy delta lobes having a variety of sand-body shapes ranging from those with the characteristics of river-dominated deposits (reflecting least influence of marine processes) to those with the characteristics of marine wave-dominated deltas (Fig. 9.15). During periods of high sediment input and a low rate of sea-level rise, redistribution of sediment by waves was minimal, and lobate deltas resulted. During periods of low sediment input and high rates of sea-level rise, sediment was extensively reworked by waves to form elongate, strike-aligned sandstone bodies, as illustrated in Figure 9.15. Other ancient wave-dominated deltas are described by Bhattacharya and Giosan (2003), Elliott (1986a), Leckie and Walker (1982), and Liu et al. (2003). Howell and Flint (2003) consider the Cretaceous strata in the Book Cliffs of Utah to be an ancient analogue to the deposits of the Paraiba do Sul delta (Brazil), shown in Figure 9.8.

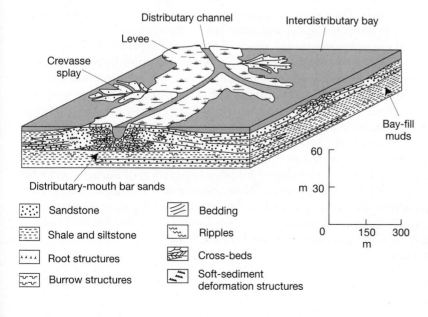

Figure 9.14
Three-dimensional model of fluvial-dominated delta deposits from eastern Kentucky. [After Horne, J. C., et al., Depositional models in coal exploration and mine planning in Appalachian region: Am. Assoc. Petroleum Geologists Bull., v. 62, Fig. 6, p. 2387.]

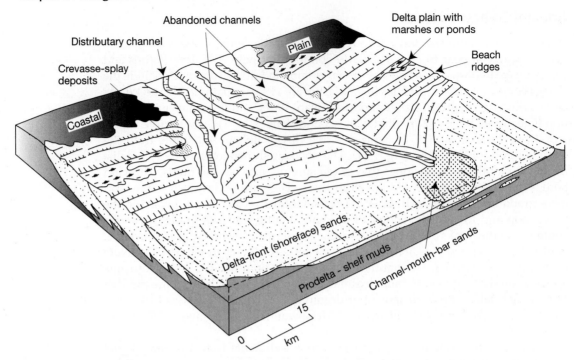

Figure 9.15
Three-dimensional model illustrating the sand-body geometry and facies of a wave-dominated delta system in the San Miguel Formation (Cretaceous), South Texas. [After Weise, B. R., 1980, Wave-dominated delta systems of the Upper Cretaceous San Miguel Formation, Maverick Basin, South Texas: Bureau of Economic Geology, University of Texas at Austin, Report of Investigations 107, Fig. 26, p. 20, reproduced by permission.]

The Eocene Misoa Formation, Maracaibo Basin, Venezuela, is a huge alluvial-deltaic-shallow marine complex extending northeasterly about 250 km from its source that exhibits marked characteristics of a tide-dominated delta. A small part of this complex (64 km^2) was reconstructed by Maguregui and Tyler (1991) as shown in Figure 9.16. A tidal delta plain bisected by narrow, highly sinuous tidal channels formed low-relief highs separating straight estuarine (river mouth) distributary channels. The interdistributary areas have low sand content. Estuarine distributary systems transport and discharge sediments at the end of estuaries, where tidal currents redistribute these sediments to form a tidally modified distributary-mouth-bar complex. Tidal channels within interdistributary areas are not connected to the sediment distributary system but simply drain the tidal plain during low-tide flow. Sands accumulated in the estuarine channels to form estuarine distributary-channel complexes (Fig. 9.16a). Seaward, the estuarine distributary channels diverge outward and merge into tidally modified distributary-mouth-bar areas (Fig. 9.16b) in this zone of maximum tidal-current energy. Tidal sand ridges also formed as coarsening-upward, highly bioturbated sand bodies (Fig. 9.16c, d), which developed by lateral migration of higher energy bar crests (sands) over lower energy muds. Note the orientation of these ridges, from right to left across the block, parallel to the estuaries and perpendicular to the shoreline. Additional examples of ancient tide-dominated deltas are discussed by Dalrymple et al. (2003), Mellere and Steel (1996), Roberts and Sydow (2003), and Willis et al. (1999).

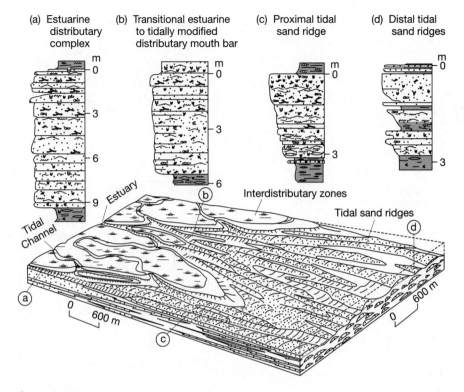

(a) Estuarine distributary complex

(b) Transitional estuarine to tidally modified distributary mouth bar

(c) Proximal tidal sand ridge

(d) Distal tidal sand ridges

Estuary

Tidal Channel

Interdistributary zones

Tidal sand ridges

600 m

600 m

Figure 9.16

Three-dimensional model of the delta front and lower delta plain of the tide-dominated delta in the Misoa Formation (Eocene), Maracaibo Basin, Venezuela. A typical estuarine distributary complex is shown in (a). A transitional lithofacies occurs farther seaward (b), where greater influence of shallow-marine conditions and a high degree of reworking by tidal currents modify the typical estuarine distributary-channel complex. Proximal tidal sand-ridge facies (c) are present in areas of high sediment discharge near the ends of the estuaries (c). Distal sand-ridge facies lie farther from the estuaries, where the sediment supply is limited and tidal currents are weaker (d). Note that both the sand facies (dot pattern) and mud facies (dark shade) are extensively bioturbated. [After Maguregui, J., and N. Tyler, 1991, Evolution of Middle Eocene tide-dominated deltaic sandstones, Lagunillas Field, Maracaibo Basin, western Venezuela, *in* Miall, A. D., and N. Tyler (eds.), Three-dimensional facies architecture of terrigenous clastic sediments and its implications for hydrocarbon discovery and recovery: Concepts in sedimentology and paleontology, 3, SEPM (Society for Sedimentary Geology), Tulsa, Okla., Fig. 16, p. 244, reproduced by permission.]

The Late Carboniferous-Permian Reinodden Formation of western Spitsbergen provides an example of an ancient fan-delta depositional system (Kleinspehn et al., 1984). Several fan-delta successions are present in this area, one of which is shown in Figure 9.17. The stratigraphic succession in Figure 9.17 begins with carbonate and siliciclastic muds that were deposited in a prodelta setting. These fine-grained prodelta deposits are succeeded upward by barrier/spit and distal mouth-bar sands, cross-bedded in part, deposited by turbidity currents and other processes on the delta front. Sandy deposits are overlain by planar to cross-bedded gravels that formed in fluvial channels on the fan-delta plain. A thin unit of wave-reworked gravels caps the succession, representing a minor phase of marine transgression. The small maps in Figure 9.17 show the postulated evolution of the fan with time as it prograded over the delta-front and prodelta environments to generate the coarsening-upward succession of facies shown.

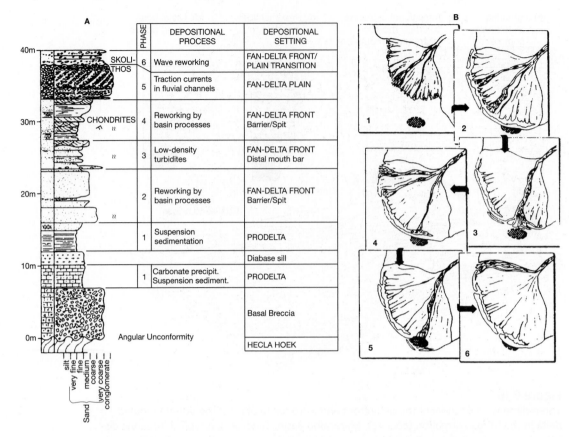

Figure 9.17

Vertical facies succession. A. Developed in Carboniferous-Permian fan-delta deposits of western Spitsbergen in response to various stages of fan progradation. B. Phases 1–6 in the stratigraphic column correspond to phases 1–6 in the plan-view sketches. The depositional site, represented by the striped oval area, remained fixed in space while the fan-delta geometry changed relative to that area. [From Kleinspehn, K. L., et al., 1984, Conglomeratic fan-delta sequences, Late Carboniferous–Early Permian, western Spitsbergen, *in* Koster, E. H., and R. J. Steel (eds.), Sedimentology of gravels and conglomerates: Canadian Soc. Petroleum Geologists Mem. 10, Fig. 8, p. 289, reproduced by permission.]

9.3 BEACH AND BARRIER ISLAND SYSTEMS

Introduction

Mainland beaches are long, narrow accumulations of sand aligned parallel to the shoreline and attached to land (e.g., Fig. 9.18). Bodies of beach sand are typically cut across here and there by headlands and sea cliffs, estuaries, river deltas, tidal inlets, bays, and lagoons. Barrier-island beaches are similar to mainland beaches but are separated from land by a shallow lagoon, estuary, or marsh (Fig. 9.19). They are also commonly dissected by tidal channels or inlets. Beaches may form within delta systems, along a depositional strike from deltas, or in other marine or even lacustrine settings that have no connection with deltas. They are the most dynamic of all depositional environments and are subject to both seasonal and longer range changes that keep them in a state of virtually constant flux. In contrast to deltas, which are influenced by both fluvial and marine processes, beach and barrier-island systems are generated predominantly by marine processes, aided to a minor degree by eolian sand transport.

Figure 9.18
A narrow strip of beach extending south from Cape Sebastian, southern Oregon coast.

Figure 9.19
Barrier-island system at Edingsville Beach, South Carolina. Note the large tidal inlet that cuts across the barrier and the smaller channels in the low-energy lagoon behind (to the left of) the barrier. Photograph courtesy of Michael F. Stephen; reproduced by permission.

Modern and Holocene beaches have perhaps been studied more extensively than any other depositional environment owing to their recreational use; their accessibility; their economic potential as a source of placer gold, platinum, and various minerals; and their importance as an erosion buffer between the sea and the land. Much of the study of modern beaches has been carried out by coastal engineers, geographers, and geomorphologists. Geologists also have a strong scientific interest in beaches owing to the insight they provide into ancient depositional processes and environments. Ancient beach deposits also have been extensively studied. In addition to their significance as indicators of ancient nearshore processes and conditions, ancient beach and barrier-island sediments have considerable economic importance as reservoirs for petroleum and natural gas and as host rocks for uranium. Although most beaches are composed of siliciclastic sediments, some modern beaches on carbonate shelves are made up predominantly of carbonate grains consisting of skeletal fragments, ooids, pellets, and other particles. Carbonate beach deposits are known also from the geologic record (e.g., Inden and Moore, 1983).

Depositional Setting

Beach and barrier-island complexes are best developed on wave-dominated coasts where tidal range is small to moderate. Coasts are classified on the basis of tidal range into three groups: (1) **microtidal** (0–2 m tidal range), (2) **mesotidal** (2–4 m tidal range), and (3) **macrotidal** (>4 m tidal range) (Fig. 9.20). Hayes (1975) has shown that barrier-island and associated environments occur preferentially along microtidal coasts, where they are well developed and nearly continuous. They are less characteristic of mesotidal coasts and, when present, are typically short or stunted, with tidal inlets common. Barriers are generally absent on macrotidal coasts; the extreme tidal range causes wave energy to be dispersed and dissipated over too great a width of shore zone to effectively form barriers.

Beach and barrier islands can occur as one of the following (see Fig. 9.1): (1) a **single beach** attached to the mainland, (2) a broader beach-ridge system that constitutes a **strand plain**, which consists of multiple parallel beach ridges and parallel swales, but which generally lacks well-developed lagoons or marshes; a type of strand plain consisting of sandy ridges elongated along the coast and separated by coastal mudflat deposits is called a **chenier plain**, or (3) a **barrier island** separated wholly or partly from the mainland by a lagoon or marsh.

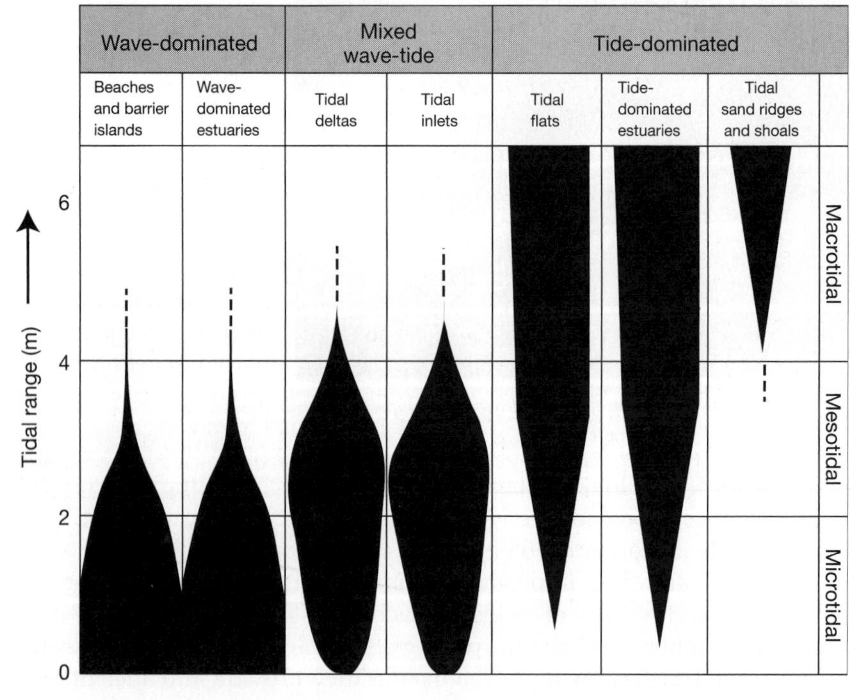

Figure 9.20
Types of coastline with respect to tidal range, grouped into wave-dominated, tide-dominated, and mixed wave-tide types. [After Hayes, M. O.,1979, Barrier island morphology as a function of tidal and wave regime, *in* Leatherman, S. P. (ed.), Barrier islands from the Gulf of St. Lawrence to the Gulf of Mexico: Academic Press, New York, Fig. 2, p. 4, reproduced by permission.]

Beaches

Morphology

The beach environment can be divided into several zones: the **backshore**, which extends landward from the beach berm above high-tide level and commonly includes back-beach dune deposits; the **foreshore**, which mainly encompasses the intertidal (littoral) zone between low-tide and high-tide levels; and the **shoreface**, also called the nearshore, which extends from about low-tide level to the transition zone between beach and shelf sediments (Fig. 9.21), that is to fair-weather wave base at a depth of about 10–15 m. Figure 9.21 illustrates also the approximate zones of shoaling and breaking waves and the position of the surf zone and swash zone, discussed in a succeeding section.

Depositional Processes

Erosion, sediment transport, and depositional processes on beaches have been studied extensively by engineers interested in coastal processes as well as by geologists. The published results of engineering studies tend to be expressed in mathematical terms that may not be of much interest to the average geologist. Perhaps the most detailed and mathematically rigorous descriptions of beach processes written for geologists are those by Komar (1998) and Hardisty (1990). Only a very brief description of these processes is given here. As mentioned, beaches are best developed on wave-dominated coasts where tidal ranges are small. Beaches are constructed primarily by wave-related processes, which include wave swash, storm waves, and nearshore currents (longshore and rip currents). Wind also plays a role in sediment transport on beaches.

Wave Processes.   Water beneath waves moving across deep water is constrained to move in orbital (circular) paths (discussed in Chapter 10). These orbits decrease in diameter downward in the water column. As deep-water orbital waves approach shallow water where depth is about one-half the wave length, the orbital motion of the water is impeded by interaction with the bottom. Orbits become progressively more elliptical and, eventually, near the bottom, develop a nearly horizontal to-and-fro motion that can move sediment back and forth. This to-and-fro movement is important in generating ripple bedforms as well as in producing some net sediment transport. As waves progress farther shoreward into the shallow **shoaling zone** (Fig. 9.21), forward velocity of the wave slows, wave length decreases, and wave height increases. The waves eventually steepen to the point where orbital velocity exceeds wave velocity and the wave breaks, creating the

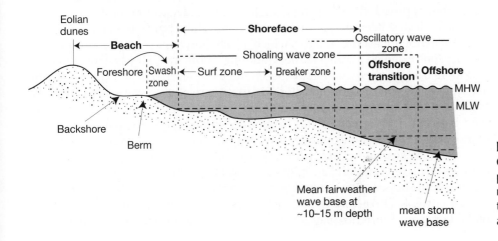

Figure 9.21
Generalized cross-sectional profile of the beach and nearshore zone, showing also the principal zones of wave activity.

breaker zone. Breaking waves generate turbulence that throws sediment into suspension and also brings about a transformation of wave motion to create the **surf zone**. In this zone, a high-velocity translation wave (a wave translated by breaking into a current), or bore, is projected up the upper shoreface, causing landward transport of bedload sediment and generation of a short-duration "suspension cloud" of sediment. At the shoreline, the surf zone gives way to the **swash zone,** in which a rapid, very shallow swash flow moves up the beach, carrying sediment in partial suspension, followed almost immediately by a backwash flow down the beach. The backwash begins at very low velocity but accelerates quickly. (If heavy minerals are present in the suspended sediment, they settle rapidly to generate a thin heavy-mineral lamina.) The width of the surf and swash zones is governed by the steepness of the shoreface and foreshore. Very steep shorefaces may develop no surf zone at all and waves break very close to shore, whereas gentle shorefaces commonly have very wide surf zones.

Sediment transport on beaches is particularly important landward of the shoaling zone. In the high-energy breaker zone, coarse sediments move by saltation in a series of elliptical paths that move sediment parallel to the coast, while finer sediment is thrown into suspension. So-called translation waves, which are actually currents, transport sediment through the surf and swash zone up the beach face. If waves approach the shoreline obliquely (a very common occurrence), sediment is transported alongshore in a zigzag manner owing to the fact that the upswash is directed across the beach at an angle, whereas the backswash flow is perpendicular to the beach face. Thus, normal waves of moderate to low energy tend to produce a net landward and alongshore transport of sediments in a largely constructive sedimentation regime in which the beach builds owing to deposition. Repeated deposition and reentrainment of sediment in the beach regime tends to winnow and remove the finest sediment, producing generally well sorted, positively skewed deposits. High-energy conditions created by storms generate steep, long-period storm waves, which cause considerable erosion of the beach area and a net displacement of sediment in a seaward direction (Davis, 1985). During storms great quantities of sediment are thrown into suspension for transport by surf-zone currents, causing sand bars on the inner beach to be planed off and displaced seaward considerable distances. Thus, it is quite common to observe marked seasonal changes on modern beaches, which often build in a landward direction during low-energy summer conditions but are eroded and reduced in size during winter storm conditions.

Wave-Induced Currents.   As breakers and winds pile water against the beach, not only do they create bidirectional translation waves that move up and down the swash zone, but they also create two different types of unidirectional currents: longshore currents and rip currents. **Longshore currents** are generated when waves that approach the shore at an angle break, and a portion of the translation wave is deflected laterally parallel to the shore. These currents move parallel to shore following longshore troughs, which are shallow troughs in the lower part of the surf zone oriented parallel to the strandline (shoreline). This system of parallel longshore troughs between shallow beach ridges is referred to as a **ridge and runnel** system. The velocity of longshore currents is related to wave height and the angle at which the waves approach shore. As water piles up between shallow sand bars and the shoreline with continued shoreward movement of waves, it cannot go back against incoming waves the way it came. It must find a different way to return seaward. Thus, it moves parallel to shore as a longshore current until it finds a topographic low between sand bars, where it converges with flow moving in the opposite direction (Fig. 9.22) and moves seaward as a narrow, near-surface current. These converging, seaward-moving currents are called **rip currents**. Longshore currents play a very important role in sediment transport and

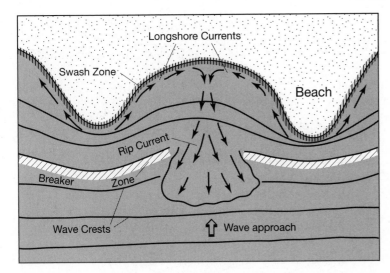

Figure 9.22
Schematic representation of longshore currents that move locally in opposite directions, generated owing to bending (refraction) of wave crests as they move over an irregular seafloor, leading to the formation of rip currents that flow seaward through the breaker zone.

deposition on beaches because they achieve velocities great enough to transport sand. Together with the processes producing transport in the swash zone, they are primary agents of alongshore sand movement. Rip currents are primarily surface phenomena and thus are less important in near-bed sediment transport than longshore currents. Nonetheless, they can entrain considerable quantities of sediment (and perhaps an occasional unwary swimmer) and move it through the breaker zone out into shoal water.

Wind.   In addition to the indirect role that wind plays in generating normal waves, storm waves, and longshore currents, wind also plays a direct role in sediment transport on beaches. The subaerial parts of beaches, above high tide level, are more or less continuously under the influence of wind. Large quantities of sand may be transported, in a largely onshore to alongshore direction, by wind. (Sea breezes tend to move from cool ocean water onto warmer land surfaces.) Wind may also move sand about on the lower shoreface as sands dry out during low-tide phases.

Barrier-Island Systems

The barrier-island setting is not a single environment but a composite of three separate environments (Fig. 9.23): the sandy barrier-island chain itself (the subtidal to subaerial barrier-beach complex); the enclosed lagoon, estuary, or marsh behind it (the back-barrier, subtidal-intertidal region); and the channels that cut through the barrier and connect the back-barrier lagoon to the open sea (the subtidal-intertidal delta and inlet-channel complex). Identification and interpretation of ancient barrier-island complexes requires that this intimate association of lagoonal, estuarine, and tidal-flat facies be recognized. Barrier-island systems are not simply barrier-beach complexes. Sediment transport and depositional processes on barrier beaches are similar to those on mainland beaches; however, additional processes take place in the tidal channels, tidal flats, marshes, and lagoons of barrier complexes. These processes are discussed in succeeding sections of this chapter. Holocene (modern) barrier-island systems are well developed along the shorelines of many parts of the world ocean. For an excellent discussion of these barriers, with emphasis on examples in the United States, see Davis (1994).

Considerable difference of opinion exists about the origin of barrier-island complexes. Proposed mechanisms of origin include (1) shoal and longshore-bar aggradation, that is, upward building and eventual emergence of the offshore bar,

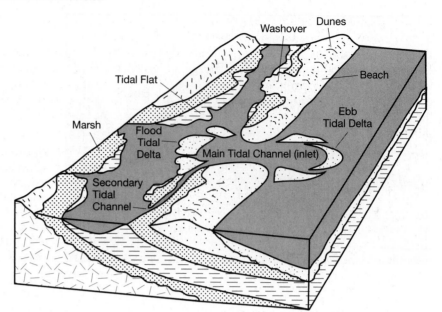

Figure 9.23
Generalized model illustrating the various subenvironments in a transgressing barrier-island system. [From Reinson, G. E., 1992, Transgressive barrier island and estuarine systems, *in* Walker, R. G., and N. P. James (eds.), Facies models, Fig. 3, p. 180, reproduced by permission of Geological Association of Canada.]

(2) spit segmentation by breaching and detachment of spits oriented parallel to the coast, (3) mainland ridge engulfment owing to submergence and drowning of shoreline-attached beaches, (4) welding or veneering of Holocene dune, beach, and foreshore sand into and over pre-Holocene topographic highs, and (5) lateral shifting of coastal sands during transgression to form the barrier islands. Mechanisms (2), (3), and (5) appear most feasible; however, composite modes of origin seem possible. The origin of old barrier systems remains unresolved because most of the evidence pertaining to origin has been destroyed by subsequent modification; however, observations on barriers that have developed in historic time suggest that these barriers formed mainly by mechanism (1) shoaling through wave action (Davis, 1997).

Characteristics of Modern Beach and Barrier-Island Systems

The mainland-beach and barrier-island system as a whole generates a narrow body of sediments elongated parallel to the depositional strike, or strike of the shoreline. This body of sediment is composed predominantly of sand that originates on the beach shoreface, foreshore, and backshore and is commonly tens to hundreds of meters broad, up to hundreds of kilometers long, and 10–20 m thick (Reineck and Singh, 1980). It may be interrupted in many places along its length by deltaic, estuarine, bay, and other deposits where these features cut across the beach (Fig. 9.1). Where barrier islands occur, sands of the barrier beach grade landward into back-barrier sediments that may include washover sands; tidal-delta sands and muds; lagoonal silts and muds; and sandy, muddy tidal-flat and marsh deposits (Fig. 9.23).

Beach deposits form on the beach face or foreshore, which is the intertidal zone extending from mean low-tide level to mean high-tide level, corresponding to the zone of wave swash (Fig. 9.21). Sediments of the foreshore consist predominantly of fine to medium sand but may also include scattered pebbles and gravel lenses or layers. Sedimentary structures are mainly parallel laminae, formed during swash-backwash flow, that dip gently (2°–3°) seaward. Thin, heavy-mineral laminae are commonly present, alternating with layers of quartzose sand. Thin, lenticular sets of low-angle, landward-dipping laminae, possibly formed by antidune migration during backwash, may be present also. Some foreshore sands display high-angle, landward-dipping cross-beds caused by migration of fore-shore ridges. The foreshore is separated from the backshore by a break in slope at the berm crest, which is formed by sand thrown up by storm waves. The **backshore** is inundated only during storm conditions and is thus a zone dominat-ed by intermittent storm-wave deposition and eolian sand transport and deposi-tion. Faint, landward-dipping, nearly horizontal laminae, interrupted locally by crustacean burrows, record deposition by storm waves. These beds may be over-lain by small- to medium-scale eolian trough cross-bed sets, which are commonly disturbed by root growths and burrows of land-dwelling organisms.

Shoreface deposits form in an environment that extends from mean low-tide level on the beach down to the lower limit of fair-weather wave base. Wave base is the depth below which normal waves do not react with the bottom. The depth of wave base on the shoreface is commonly on the order of 10–15 m, but this depth can be lowered significantly during storms. The shoreface can be divided into the upper, middle, and lower shorefaces, which correspond roughly to the surf, breaker, and outer shoaling zones. Each is distinguished by characteristic fa-cies (Fig. 9.24).

Upper-shoreface (surf-zone) deposits form in an environment dominated by strong bidirectional translation waves and longshore currents. Deposits of this zone consist mainly of multidirectional trough cross-bed sets. Trace fossils such as *Skolithos* (Fig. 4.4.1, Box 4.1) are common but not abundant. Middle-shoreface (break-er-zone) deposits form under high-energy conditions owing to breaking waves and associated longshore and rip currents. Sediments are mainly fine- to medium-grained sand, with minor amounts of silt and shell material, that may display both landward- and seaward-dipping trough cross-beds as well as subhorizontal plane laminations. Trace fossils consisting of vertical burrows, such as *Skolithos* and *Ophiomorpha* (Fig. 4.4.1, Box 4.1) are common. Lower shoreface (outer shoaling zone) deposits form under relatively low-energy conditions and grade seaward into open-shelf deposits. They are composed dominantly of fine to very fine sand but may contain thin, intercalated layers of silt and mud. Sedimentary structures can include small-scale cross-stratification; planar, nearly horizontal laminated

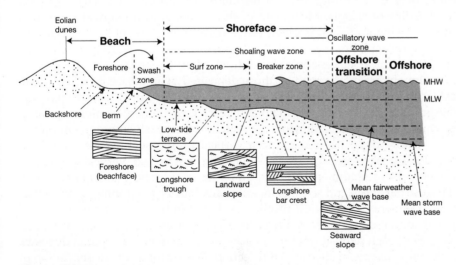

Figure 9.24

Typical sedimentary structures formed in the beach and nearshore zone (same profile as shown in Fig. 9.21). [Sedimentary structures after Davidson-Arnott, R. G. D., and B. Greenwood, 1976, Facies relationships on a barred coast, Kouchibouguac Bay, New Brunswick, Canada, *in* R. A. Davis, Jr., and R. L. Ethington (eds.), Beach and nearshore sed-imentation: Soc. Econ. Paleontologists and Mineralogists Spec. Pub. 24, Fig. 4, p. 154, re-produced by permission of SEPM, Tulsa, Okla.]

bedding; and hummocky cross-stratification (owing to storm events). Trace fossils such as *Thalassinoides* (Fig. 4.4.2, Box 4.1) may be common.

Back-barrier sediments are deposited in several subenvironments in the back-barrier lagoon landward of barrier beaches. **Washover deposits** occur where storm-driven waves cut through and overtop barriers, washing lobes of sandy beach sediment into the back-barrier lagoon (Fig. 9.23). Washover sediment consists dominantly of fine- to medium-grained sand that displays subhorizontal planar laminations and small- to medium-scale landward-dipping foreset bedding. Where tidal channels cut through barriers into the inner lagoon, sediments are deposited in a number of tide-related environments, including tidal channels, tidal deltas, and tidal flats (Fig. 9.23). **Tidal-channel deposits** consist dominantly of sand, and the deposits commonly have an erosional base marked by coarse lag sands and gravels. Sedimentary structures may include bidirectional large- to small-scale planar and trough cross-beds that may display a general fining-upward textural trend. **Tidal-delta deposits** form on both the lagoonal side of the barrier (flood-tidal delta) and the seaward side of the barrier (ebb-tidal delta). They are predominantly sandy deposits to tens of meters thick with a gross parabolic shape or geometry. They are characterized by a highly varied succession of planar and trough cross-bed sets that may dip in either a landward or a seaward direction. **Tidal-flat deposits** form along the margins of the mainland coast and the back of the barrier. They grade from fine- to medium-grained ripple-laminated sands in lower areas of the tidal flats through flaser- and lenticular-bedded fine sand and mud in midtidal flats to layered muds in higher parts of the flats. **Lagoonal** and **marsh deposits** accumulate in the low-energy back-barrier lagoon and grade laterally into higher energy, sandy deposits of tidal channels, deltas, and washover lobes. They consist largely of interbedded and interfingering fine sands, silts, muds, and peat deposits that may be characterized by disseminated plant remains, brackish-water invertebrate fossils such as oysters, and horizontal to subhorizontal layering.

Ancient Beach and Barrier-Island Sediments

Shorelines may shift through time in response to factors such as change in sea level and sediment supply. Movement of the shoreline in a landward direction (e.g., during rising sea level) is called **transgression**; movement in a seaward direction (e.g., during falling sea level) is called **regression**. Transgression or regression creates a vertical succession of facies (rock types), as sediments deposited in one environment are overlapped by those deposited in an adjacent environment. Vertical facies profiles developed during regression differ from those formed during transgression. Regression of a mainland beach environment (shoreline progradation) causes stacking of sediments deposited in the beach environment on top of more offshore deposits, generating a vertical profile such as that shown in Figure 9.25. Transgression produces essentially a reversed vertical profile, in which more offshore sediments are stacked on top of nearshore sediments.

Because barrier-island settings include back-barrier environments as well as beach environments, the vertical profile developed in barrier complexes is more complex than that developed on mainland beaches. Transgression causes deposition of barrier-beach deposits on top of back-barrier lagoonal and marsh deposits, as the shoreline moves in a landward direction. Regression leads to deposition of back-barrier lagoonal and marsh deposits over sandy deposits of the barrier beach-beach complex as the shoreline progrades. The generation of transgressive beach and barrier-island deposits has been suggested to occur by two different mechanisms: landward advance of the shoreline owing to shoreface erosion, as might take place during slow rise of sea level; and relatively sudden upward

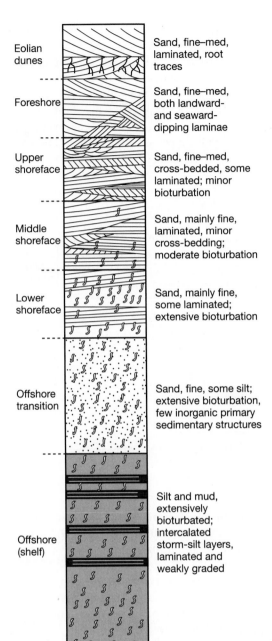

Eolian dunes — Sand, fine–med, laminated, root traces

Foreshore — Sand, fine–med, both landward- and seaward-dipping laminae

Upper shoreface — Sand, fine–med, cross-bedded, some laminated; minor bioturbation

Middle shoreface — Sand, mainly fine, laminated, minor cross-bedding; moderate bioturbation

Lower shoreface — Sand, mainly fine, some laminated; extensive bioturbation

Offshore transition — Sand, fine, some silt; extensive bioturbation, few inorganic primary sedimentary structures

Offshore (shelf) — Silt and mud, extensively bioturbated; intercalated storm-silt layers, laminated and weakly graded

Figure 9.25
Idealized succession of beach sediments on a low-energy, prograding, Holocene beach. [After Reineck, H. E., and I. B. Singh, 1980, Depositional sedimentary environments, 2nd ed., Fig. 534, p. 387, reprinted by permission of Springer-Verlag, Heidelberg.]

"jumps" of the shoreline during rapidly rising sea level. These alternate mechanisms are illustrated in Figure 9.26.

During shoreline retreat (transgression), beach and upper shoreface deposits are presumably eroded and transported to the lower shoreface, or offshore as storm beds, or to the lagoon as washover deposits (Fig. 9.26A). The surface generated by marine reworking and erosion during transgression is referred to as a **ravinement surface**. In-place drowning owing to inundation during a rapid rise in sea level could cause a barrier to be covered by water, resulting in the wave zone moving landward until a new sand barrier forms on the inner side of the lagoon (Fig. 9.26B). Barrier islands can prograde (regression), under conditions of high sediment supply relative to sea-level change, to produce regressive barrier-island facies. Under these conditions, barriers tend to be transformed into strand

A Transgression by shoreface retreat

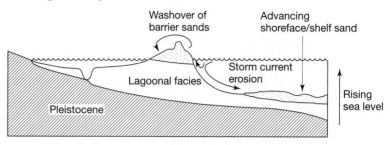

Figure 9.26

Barrier-island facies generated by transgression and regression. A. Transgression owing to shoreface retreat during gradual sea-level rise. B. Effects of rapid sea-level rise, producing in-place drowning (SL = sea level). C. Facies formed as a result of progradation under conditions of high sediment supply relative to sea-level change. [A and B after Rampino, M. R., and J. E. Sanders, 1980, Holocene transgression in south-central Long Island, New York: Jour. Sed. Petrology, v. 50, Fig. 8, p. 1075, reproduced by permission of SEPM, Tulsa, Okla.; Elliott, T., 1986, Siliciclastic shorelines, *in* Reading, H. G. (ed.), Sedimentary environments and facies: Blackwell Scientific Publications, Fig. 7.33, p. 180. Part C after Galloway, W. E., and D. K. Hobday, 1983, Terrigenous clastic depositional systems: Springer-Verlag, New York, Fig. 6.10, p. 126.]

B Transgression by in-place drowning

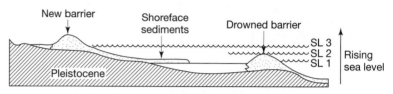

C Regression

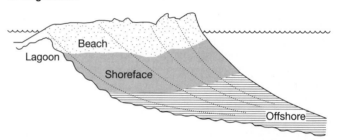

plains, producing dominantly sandy facies in which beach (backshore and fore-shore) deposits overlie foreshore deposits. Galveston Island, Texas (Fig. 9.26C), is an example of such a progradational deposit. A generalized succession of facies deposited in an ancient back-barrier environment is illustrated in Figure 9.27. The cyclic nature of this deposit suggests recurring episodes of transgression and regression. Such a succession might be confused with some deltaic successions, although the absence of delta-front muddy sands and prodelta clays and mud turbidites should help to distinguish it from deltaic deposits.

Many ancient examples of beach and barrier deposits are known. For example, the Cretaceous Gallup Sandstone of northwestern New Mexico is reported to be a progradational (regressive) succession of barrier deposits (McCubbin, 1982), whereas, the Cretaceous Cliff House Sandstone in the San Juan Basin of northwestern New Mexico has been interpreted as a transgressive barrier complex (Donselaar, 1989; McCubbin, 1982). Other siliciclastic sedimentary successions identified as beach and barrier-island complexes are present in rocks of widely differing ages in North America. Such successions have been reported from several Pennsylvanian formations in the Appalachian Basin of Kentucky, Virginia, West Virginia, and Tennessee; the Lower Cretaceous Muddy Sandstone of Wyoming and Montana; the Eocene Wilcox Group of east Texas; and the Quaternary of California (e.g., Davis, 1992). Several ancient carbonate deposits have also been interpreted as beach complexes. The Lower Cretaceous Edwards Formation of west Texas, the Lower Cretaceous Cow Creek Formation of central Texas, the Mississippian Newman Formation in eastern Kentucky, and the Mississippian Mission Canyon Formation in the Williston Basin in the Montana area are examples of ancient stratigraphic units that contain putative carbonate beach deposits.

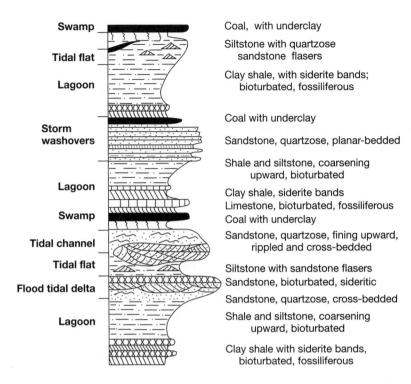

Swamp	Coal, with underclay
Tidal flat	Siltstone with quartzose sandstone flasers
Lagoon	Clay shale, with siderite bands; bioturbated, fossiliferous
Storm washovers	Coal with underclay
	Sandstone, quartzose, planar-bedded
	Shale and siltstone, coarsening upward, bioturbated
Lagoon	Clay shale, siderite bands
	Limestone, bioturbated, fossiliferous
Swamp	Coal with underclay
Tidal channel	Sandstone, quartzose, fining upward, rippled and cross-bedded
Tidal flat	Siltstone with sandstone flasers
Flood tidal delta	Sandstone, bioturbated, sideritic
	Sandstone, quartzose, cross-bedded
Lagoon	Shale and siltstone, coarsening upward, bioturbated
	Clay shale with siderite bands, bioturbated, fossiliferous

Figure 9.27

Generalized succession of facies deposited in a back-barrier environment, Carboniferous of eastern Kentucky and southern West Virginia. Such successions range from 7.5 to 24 m thick. [After Horne, J. C., J. C. Ferm, F. T. Caruccio, and B. P. Baganz, 1978, Depositional models in coal exploration and mine planning in Appalachian region: Am. Assoc. Petroleum Geologists Bull., v. 62, Fig. 4, p. 2385, reprinted by permission.]

9.4 ESTUARINE SYSTEMS

Introduction

Relatively small, partly enclosed coastal embayments are loosely called **coastal bays**. Two broad types of coastal bays are recognized: estuaries and lagoons. **Estuaries** (term derived from the Latin word *aestus*, meaning tide, and from the adjective *aestuarium*, meaning tidal) are considered in a general sense to be the lower courses of rivers open to the sea (e.g., Figure 9.28). Dalrymple, Zaitlin, and Boyd (1992) suggest that the concept of net landward movement of sediment derived from outside the estuary mouth is necessary to distinguish estuaries from deltas. Therefore, they define an estuary as "the seaward portion of a drowned valley system which receives sediment from both fluvial and marine sources and which

Figure 9.28

Wave-dominated estuary of the Klamath River, northern California coast. Note the large, northward-projecting (toward bottom of photograph) spit that partially blocks the mouth of the estuary.

contains facies influenced by tide, wave and fluvial processes. The estuary is considered to extend from the landward limit of tidal facies at its head to the seaward limit of coastal facies at its mouth." According to Dalrymple, Zaitlin, and Boyd, estuaries can form only in the presence of a relative sea-level rise (i.e., transgression). Progradation tends to fill and destroy estuaries, causing them to change into deltas.

Physiographic, Hydrologic, and Sediment Characteristics of Estuaries

Based on the physiographic characteristics of relative relief and degree of channel-mouth blocking, seven basic types of modern estuaries are recognized (Fig. 9.29; Fairbridge, 1980). **Fjords** are high-relief estuaries with a U-shaped valley profile formed by drowning of glacially eroded valleys during Holocene sea-level rise. **Fjards** or firths are related to fjords but have lower relief. Estuaries developed in winding valleys with moderate relief are **rias**. **Coastal-plain estuaries** are low-relief estuaries, funnel-shaped in plan view, that are open to the sea. Low-relief estuaries that are L-shaped in plan view and that have lower courses parallel to the coast are **bar-built estuaries**. Similar estuaries that are seasonally blocked by longshore drift or dune migration are called **blind estuaries**. Bar-built estuaries and blind estuaries are very similar to lagoons and might be considered lagoons by some workers. **Deltaic estuaries** occur on delta fronts as ephemeral distributaries. Flask-shaped, high-relief rias backed by a low-relief plain created by tectonic activity are called compound estuaries or **tectonic estuaries**.

Figure 9.29

Principal types of estuaries based on physiographic characteristics. [From Fairbridge, R. W., The estuary: Its definition and geodynamic cycle, *in* Olausson, E., and I. Cato (eds.), 1980, Chemistry and biochemistry of estuaries, Fig. 2, p. 9, John Wiley and Sons, New York, reprinted by permission.]

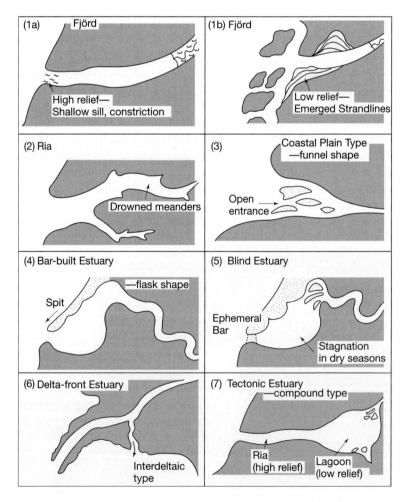

Perillo (1995a) proposes a slightly different classification scheme in which he divides estuaries into the following main types: (1) Estuaries of former river valleys, a) Coastal plain, b) Rias; (2) Estuaries of former glacial valleys, a) Fjord, b) Fjard; (3) River-dominated estuaries, a) Tidal river, b) Delta front; and (4) Structural estuaries. The characteristics of many of these estuaries are described in detail in Perillo (1995b).

Dalrymple, Zaitlin, and Boyd (1992) present a model for estuaries, based on dominant hydrologic characteristics and the kinds of sediment and sediment bodies formed in the estuary. They suggest that estuaries can be classed as wave dominated, tide dominated, or mixed wave and tide dominated.

Wave-Dominated Estuary

In a wave-dominated estuary, tidal influence is small, and the mouth of the estuary experiences high wave energy. Hydrologic conditions may range from partially mixed to well stratified. Sediments tend to move alongshore and onshore into the mouth of the estuary, where a subaerial barrier/spit or submerged bar develops. This barrier prevents most of the wave energy from entering the estuary; thus, only internally generated waves are present behind the barrier. Depending upon tidal range and current velocity, a small number of inlets may be kept open in this barrier. Alternatively, the barrier may close the estuary entirely at times to produce a blind estuary or coastal lake. The relative influence of marine and river processes and the kinds of sediment bodies formed in different parts of a wave-dominated estuary are illustrated in Figure 9.30.

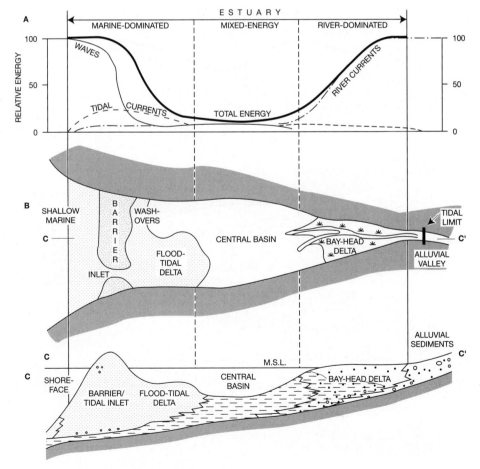

Figure 9.30
Distribution of (A) energy types, (B) morphological components in plan view, and (C) sedimentary facies in longitudinal section within an idealized wave-dominated estuary. The shape of the estuary is schematic. The barrier/sand plug is shown as headland attached; however, on low-gradient coasts, it may be separated from the mainland by a lagoon. The section in Part C represents the onset of estuary filling following a period of transgression. [From Dalrymple, R. W., B. A. Zaitlin, and R. Boyd, 1992, Estuarine facies models: Conceptual basis and stratigraphic implications: Jour. Sed. Petrology, v. 62, Fig. 4, p. 1134, reproduced by permission of SEPM (Society for Sedimentary Geology), Tulsa, Okla.]

Note that marine-derived sands are carried landward only a short distance beyond the barrier as washover deposits or flood-tidal delta sands. Muddy sediments, supplied mainly by the river, accumulate in the central part of the estuary where total energy is lowest. Deposition of mud is enhanced by mixing of river and marine water, which causes flocculation of clay particles owing to positively charged ions in seawater that neutralize the negative charges on clay particles. Muddy sediment may include biogenic debris such as molluscan shells, wood fragments, and organically produced pellets. Horizontal and subhorizontal stratification and bioturbation traces are common structures. In the head of the estuary, coarse, river-derived sediments are deposited in channels; marsh deposits may form adjacent to the channels. Examples of modern wave-dominated estuaries include San Antonio Bay, United States; the Miramichi River, Canada; and Hawksbury Estuary, Australia.

Tide-Dominated Estuaries

Tide-dominated estuaries occur mainly on macrotidal coasts where tidal-current energy exceeds wave energy at the mouth of the estuary, creating energy conditions in the estuary higher than those typical of wave-dominated estuaries. Water in the estuary is commonly well mixed. Elongate sand bars develop parallel to the length of the estuary from sand carried into the estuary from marine sources. These bars tend to dissipate tidal energy. On the other hand, constriction of incoming tidal currents between the tidal bars causes an increase in their velocity for some distance up the estuary. Figure 9.31 shows the hydrologic conditions and sediment bodies developed in tide-dominated estuaries.

Figure 9.31
Distribution of (A) energy types, (B) morphological components in plan view, and (C) sedimentary facies in longitudinal section within an idealized tide-dominated estuary. UFR = upper-flow regime; M.H.T. = mean high tide. The section in Part C is taken along the axis of the channel and does not show the marginal mudflat and salt-marsh facies; it illustrates the onset of progradation following transgression. [From Dalrymple, R. W., B. A. Zaitlin, and R. Boyd, 1992, Estuarine facies models: Conceptual basis and stratigraphic implications: Jour. Sed. Petrology, v. 62, Fig. 7, p. 1136, reproduced by permission of SEPM (Society for Sedimentary Geology), Tulsa, Okla.]

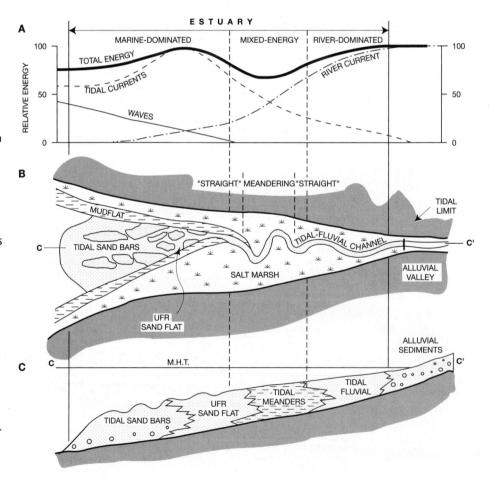

In addition to forming tidal bars in the mouth of the estuary, sand may be transported landward through the estuary in the tidal-fluvial channel. Bedforms ranging in size from ripples to large dunes develop on sandy sediment in the bars and tidal channels, and cross-bedding generated by migration of these bedforms can dip in either a landward or a seaward direction. Flaser bedding may form during slack water owing to deposition of suspended mud over sand ripples. Muddy sediment is deposited also in lower energy parts of the estuary floor and in salt marshes adjacent to the channel along the edges of the estuary. Muddy sediments are characterized by nearly planar alternations of silt, clay, very fine sand, and carbonaceous (plant) debris. Bioturbation by burrowing and feeding organisms may locally mix and homogenize these layers. Estuarine sediments typically contain a brackish-water fauna that may include oysters, mussels, other pelecypods, and gastropods. Examples of tide-dominated estuaries include Cook Inlet, Alaska; Ord River, Australia; Gironde Estuary, France; and the Severn River, United Kingdom.

Mixed Wave- and Tide-Dominated

Many estuaries have characteristics that are intermediate between wave-dominated and tide-dominated types. For example, as tidal energy increases relative to wave energy, the barrier system of wave-dominated estuaries becomes progressively more dissected by tidal inlets and elongate sand bars develop in locations previously occupied by barrier segments and the channel-margin linear bars of ebb-tidal deltas. Marine-derived sand is transported greater distances up the estuary, and the generally muddy central basin is replaced by sandy tidal channels flanked by marshes. Examples of mixed-energy estuaries include the St. Lawrence River, Canada; Willipa Bay, United States; and Oosterschelde Estuary, The Netherlands.

Ancient Estuarine Facies

Estuaries and lagoons are both ephemeral features. Because they tend to fill with sediments in geologically short periods of time, the preservation potential of estuarine and lagoonal sediments is generally high. Nevertheless, relatively few estuarine deposits have been reported from the geologic record, possibly because they have not been widely recognized and distinguished from associated fluvial, deltaic, lagoonal, or shallow marine deposits.

Estuarine deposits tend to have restricted faunal assemblages that include brackish-water species and that may be characterized by trace fossil assemblages reflecting brackish to stressed conditions (Reinson, 1992); however, no unique physical criterion exists for these deposits. Depending upon location within an estuary, estuarine deposits may consist almost entirely of cross-bedded sands, laminated or bioturbated muds, or combinations of sand and mud. Gradation from fluvial channel sands at the base of a vertical section through mixed fluvial-marine muds in the middle of the section to marine (tidal) sands at the top suggests a transgressive estuarine deposit. The exact vertical succession of facies that develops in estuaries depends, however, upon the kind of estuary (wave or tide dominated) and the location within the estuary. Facies dominated by cross-bedded, bioturbated sand are present near the mouths of estuaries and in fluvial-tidal channels, whereas laminated to well-bioturbated muds occupy the nonchannel middle and upper parts of the estuary. Many estuaries are subjected in time to transgression. Transgression brings about a landward shifting of environments, resulting in vertical stacking of estuary-mouth sands on top of middle-estuary muds and/or fluvial-tidal channel sands. By contrast, regression causes filling and destruction of the estuary and seaward progradation, changing it into a delta.

Figure 9.32
Model for a tide-dominated, transgressive estuarine-embayment depositional system based on the Woburn Sands (Lower Cretaceous), southern England. (a) Idealized vertical section showing facies in the more seaward part of the estuary. (b) Block diagram showing sand body characteristics of the inner and outer estuarine embayments. [After Johnson, H. D., and B. K. Levell, 1995, Sedimentology of a transgressive, estuarine sand complex: The Lower Cretaceous Woburn Sands (Lower Greensand), southern England, *in* Plint, A. G. (ed.), Sedimentary facies analysis: A tribute to the research and teaching of Harold G. Reading, International Association of Sedimentologists Spec. Publ. 22, Blackwell Science, Fig. 18, p. 41, reproduced by permission.]

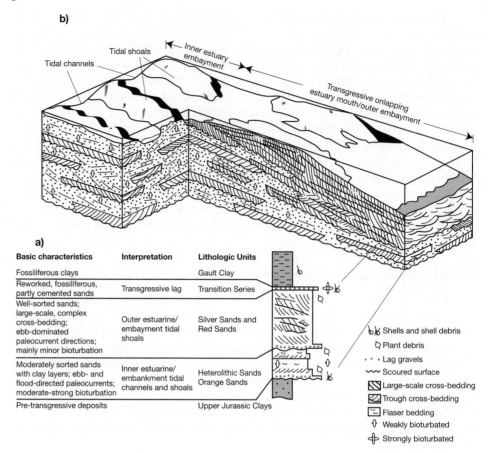

Figure 9.32 shows facies developed in a tide-dominated, transgressive estuarine sand complex in Lower Cretaceous Woburn Sands of southern England (Johnson and Levell, 1995). Johnson and Levell suggest that The Orange and Heterolithic Sands were deposited in ebb and flood tidal channels and intervening tidal shoals in an inner estuarine environment. The Silver and Red sands were deposited in the higher energy outer reaches of an estuary where greater water depth allowed large-scale bedforms to build. Slowly deposited fossiliferous marine beds of the Transition Series and Basal Gault were then laid down on top of the succession as transgression proceeded.

9.5 LAGOONAL SYSTEMS

A coastal **lagoon** is defined as a shallow stretch of seawater—such as a sound, channel, bay, or saltwater lake—near or communicating with the sea and partly or completely separated from it by a low, narrow elongate strip of land, such as a reef, barrier island, sandbank, or spit (Bates and Jackson, 1980), e.g., Figure 9.33. Most modern lagoons are formed behind spits or offshore barriers of some type and thus are elongated bodies lying parallel to the coast with a narrow connection to the open ocean. Lagoons also form behind barrier reefs and atolls. Lagoons commonly extend parallel to the coast, in contrast to estuaries, which are oriented approximately perpendicular to the coast. Many lagoons have no significant freshwater runoff; however, some coastal embayments that otherwise satisfy the general definition of lagoons do receive river discharge. Lagoons may occur in close association with river deltas, barrier islands, and tidal flats.

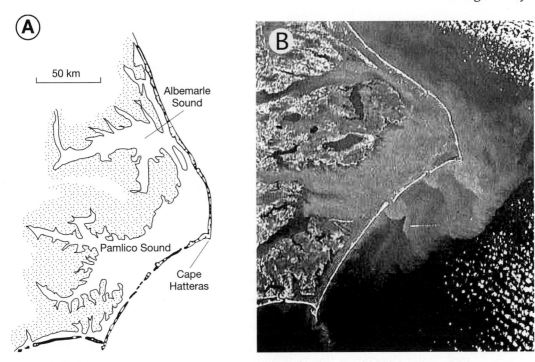

Figure 9.33
Cape Hatteras, South Carolina; a lagoonal system enclosed by a barrier-island chain. A. Diagrammatic sketch of the barrier chain and lagoon. B. Cape Hatteras as seen from Apollo 9; Pamlico Sound is partly obscured by clouds. [A. From Barnes, R. S. K., 1980, Coastal lagoons, Fig. 1.3, p. 5, Cambridge University Press, reprinted by permission; B. NASA photograph, downloaded from the Internet 4/19/2004.]

Many factors affect water flow, water mixing, and sediment transport in lagoons, such as tides, wind waves, freshwater runoff, episodic storms, density gradients, sea-level changes, and changes in climate and temperature. Even so, water circulation patterns in lagoons are much less affected by freshwater inflow than they are in estuaries, and many lagoons receive no freshwater discharge. Also, circulation with the open ocean is restricted by the barrier. Consequently, the principal movement of water within lagoons is in the form of tidal currents (which move in and out through the narrow inlets between barriers) and wind-forced waves.

On the basis of geomorphology and the nature of water exchange with the coastal ocean, Kjerfve and Magill (1989) identify three types of lagoons: choked, restricted, and leaky (Fig. 9.34). **Choked lagoons** occur along coasts with high wave energy and significant alongshore drift (e.g., Coorong Lake, southern Australia). They are characterized by one or more long, narrow entrance channels; long residence times of water within the lagoon; and dominant water movement by wind forcing. Intense solar radiation coupled with inflow events can cause intermittent vertical stratification. **Restricted lagoons** commonly exhibit two or more entrance channels or inlets, have a well-defined tidal circulation, are strongly influenced by winds, and are generally vertically mixed (e.g., Lake Pontchartrain, Louisiana). **Leaky lagoons** typically occur along coasts where tidal currents are a more important factor in sediment transport than are wind waves (e.g., Belize Lagoon, Belize). They may stretch along coasts for more than 100 km but commonly are no more than a few kilometers wide. They are characterized by wide tidal passes, efficient water exchange with the ocean, strong tidal currents, and sharp salinity and turbidity fronts.

Except within tidal channels that extend into the lagoon, lagoons are predominantly areas of low water energy. Tidal deltas commonly develop at the ends

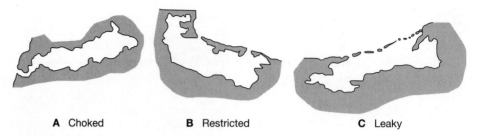

A Choked   B Restricted   C Leaky

Figure 9.34
Principal kinds of coastal lagoons (choked, restricted, leaky) based on the degree of water exchange with the adjacent coastal ocean. [From Kjerfve, B., and K. E. Magil, 1989, Geographic and hydrodynamic characteristics of shallow coastal lagoons: Marine Geology, v. 88, Fig. 2, p. 190, reproduced by permission.]

of these tidal inlets, both within the lagoon and on the ocean sides, and sandy sediment may also be deposited within the higher energy tidal channels inside the lagoon. Otherwise, sedimentation within lagoons is dominated by deposition of silt and mud, although occasional high wave activity during storms can cause washover of sediment from the barrier.

Salinity within lagoons can range from hypersaline to essentially that of fresh water, depending upon the hydrologic conditions and the climate. Lagoons formed in arid or semiarid coastal areas, where little freshwater influx occurs, are commonly hypersaline, with salinities well above that of normal seawater. Lagoons in more humid regions may be characterized by brackish water. Salinity within lagoons may vary in response to seasonal rainfall and evaporation rates. Also, salinity at a particular time may not be uniform throughout a lagoon. Lagoons receiving considerable freshwater inflow commonly display distinct, lateral salinity zones.

The sediments deposited in lagoons can be derived from several sources, which can include (depending upon the nature of the lagoon) rivers, the ocean, shores, and barriers. Sediment can also be derived internally by organic production, chemical precipitation, and erosion of older deposits (Nichols and Boon, 1994). The deposits of lagoons may differ from those of estuaries in several ways. First, because many lagoons do not receive freshwater discharge from rivers, most or all the sediment in such lagoons is from marine sources. Lagoons are typically low-energy environments, although tidal currents move into lagoons through inlets between barriers, winds create some wave action along shorelines, storms provide occasional episodes of high-energy waves that wash over barriers into the lagoon, and prevailing winds may more or less continuously blow small amounts of sediment from barriers into the lagoon. Because of the dominance of low-energy conditions in lagoons, lagoonal deposits consist mainly of fine-grained sediments. Sandy sediments are confined principally to tidal deltas constructed at the mouths of the tidal inlets, to some tidal channels that extend into the lagoon, to washover lobes behind barriers, and to some parts of the lagoonal shoreline (lagoonal beaches). Small amounts of sandy sediment blown from barriers may also be scattered throughout the lagoon. Sandy sediment in tidal channels is characterized by current ripples and internal small-scale cross-bedding that may dip in either a landward or a seaward direction. Most of the lagoonal bottom is covered with silty or muddy sediments, commonly extensively bioturbated, that may contain thin intercalations of sand brought in by storms or blown in by wind. This sand is generally horizontally laminated, but it may display ripple cross-laminations. The faunas that inhabit lagoons are highly variable depending upon the salinity conditions of the lagoon, but they are generally characterized by low diversity. Lagoons with normal salinity show faunas similar to those of the open ocean, whereas brackish-water faunas dominate lagoons in front of river mouths. Hypersaline lagoons commonly contain few organisms because few species are adapted to such high salinities.

In areas where little siliciclastic sediment is available and climatic conditions are favorable, sedimentation in lagoons is dominated by chemical and biochemical deposition. Under very arid conditions, lagoonal sedimentation may be characterized by deposition of evaporites, which are mainly gypsum but may include some halite and minor dolomites (e.g., lagoons in the Persian Gulf). Under less hypersaline conditions, carbonate deposition prevails, particularly in lagoons developed behind barrier reefs (e.g., Australia). Deposits in such lagoons may consist largely of carbonate muds and associated skeletal debris, although ooids may form in more agitated parts of the lagoon. Algal mats, commonly developed in the supratidal and shallow intertidal zone, may trap fine carbonate or siliciclastic mud to form stromatolites. Algal mats in the supratidal zone generally display mudcracks with curled margins.

Ancient Lagoonal Deposits

Lagoonal deposits may form in many settings, including parts of barrier-island complexes. Criteria that can be used to distinguish ancient lagoonal deposits from estuarine and other deposits include evidence for restricted circulation such as the presence of evaporites or anoxic facies (e.g., black shales), lack of strong tidal influence, slow rates of terrigenous sediment influx and dominantly fine grained sediments, low faunal diversity, and extensive bioturbation (Davis, 1983). The St. Mary River Formation of southern Alberta, Canada, contains back-barrier deposits that illustrate some of the characteristics of lagoonal sediments (Fig. 9.35).

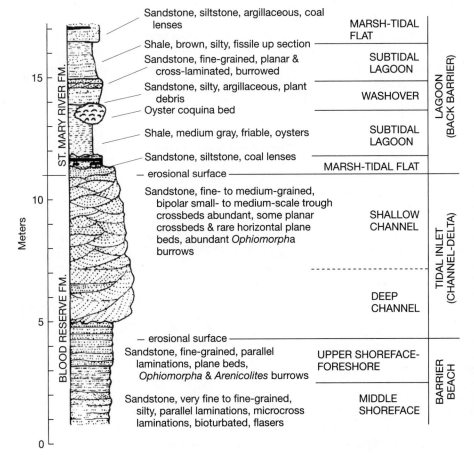

Figure 9.35
Composite stratigraphic section of Cretaceous formations in southern Alberta, Canada. Back-barrier lagoonal deposits of the St. Mary River Formation overlie tidal-inlet and tidal-delta deposits of the Blood Reserve Formation. [From Reinson, G. E., 1984, Barrier island and associated strand-plain systems, in Walker, R. G. (ed.), Facies models, 2nd ed.: Geoscience Canada Reprint Ser. 1, Fig. 16, p. 128, reproduced by permission of Geological Association of Canada.]

This lagoonal succession begins with a basal section of sandstones, siltstones, and coals that represent marsh-tidal-flat deposits (Young and Reinson, 1975). These deposits lie on the eroded surface of tidal-inlet deposits of the underlying Blood Reserve Formation. The basal coal-bearing beds are succeeded upward by gray shales containing oysters, which are brackish-water faunas, and disseminated carbonaceous materials and imprints of plant remains. These deposits are interpreted as subtidal lagoonal sediments. They are overlain by fine-grained, planar and cross-laminated sandstones, which are probably washover deposits from the barrier beach. More subtidal shales lie above these washover sands, and the succession is capped by sandstones and siltstones with coal lenses of probable marsh-tidal-flat origin. Overall, the Blood Reserve and St. Mary River formations appear to represent a progradational barrier-island complex, in which the lagoonal deposits of the St. Mary River Formation form the topmost part of the regressive succession.

9.6 TIDAL-FLAT SYSTEMS

Introduction

Tidal flats form primarily on mesotidal and macrotidal coasts (Fig. 9.36) where strong wave activity is absent. They develop either along open coasts of low relief and relatively low wave energy or behind barriers on high-energy coasts where protection is afforded from waves by barrier islands, spits, reefs, and other structures. Thus, they occur within estuaries, bays, the backshores of barrier-island complexes, and deltas, as well as along open coasts. They are particularly common in the modern ocean along the coasts of Europe, Africa, South Asia, North and East Asia, Australia, New Zealand, large Pacific islands, and western and eastern America (Eisma et al., 1998).

Tidal flats are marshy and muddy to sandy areas partially uncovered by the rise and fall of tides. They constitute almost featureless plains dissected by a network of tidal channels and creeks that are largely exposed during low tide (e.g., Fig. 9.37). As tide level rises, flood-tide waters move into the channels until at high tide the channels are overtopped and water spreads over and inundates the adjacent shallow flats. Ebb tide again exposes the channels and intervening flats. In temperate regions, salt marshes commonly cover the upper parts of tidal flats, and muds and silts accumulate near high-water level. At the same time, mixed mud

Figure 9.36
Global classification of coastlines by tidal range (microtidal 0–2 m; mesotidal 2–4 m; macrotidal >4 m). [From Klein, G. deV., 1985, Intertidal flats and intertidal sand bodies, *in* Davis, R. A. Jr. (ed.), Coastal sedimentary environments, 2nd ed.: Springer-Verlag, New York, Fig. 3.1, p. 189, Redrawn from Davies, J. L., 1964, Zeitschrift für Geomorphologie, v. 8, Fig. 4, p. 136.]

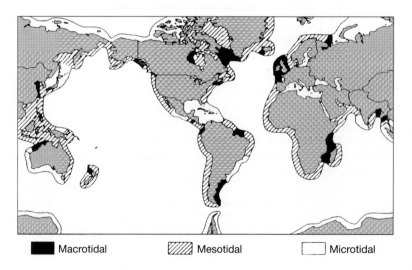

■ Macrotidal   ▨ Mesotidal   □ Microtidal

Figure 9.37
Tidal flat in the Ashe Island area, about 70 km (45 mi) south of Charleston, South Carolina, exposed at low tide. Note tidal channels and areas covered by shallow water (dark patches) on the flats. National Oceanographic and Atmospheric Administration (NOAA) photograph. Downloaded from the Internet 4/23/04.

and sand are deposited in the middle tidal-flat region, and sands accumulate in channels and on the lower parts of the tidal flat. In arid to semiarid regions, tidal flats may become desiccated and marked by mudcracks and by gypsum and halite crystals that form in muds. The surface of tidal flats in subarctic regions may be marked by surficial scars, caused by ice floes and ice-pushed boulders, and ice-rafted pebbles and cobbles. Modern tidal flats are primarily sites of siliciclastic deposition; however, carbonate sediments and, in a few areas, evaporites accumulate on some modern tidal flats such as those in the Bahamas, the Persian Gulf, Florida Bay, and the western coast of Australia (e.g., Hardie and Shinn, 1986).

Much of what is known about ancient tidal-flat sediments comes from research on modern tidal flats. Modern tidal flats have been studied intensively in many parts of the world since the 1950s, particularly in Germany, the North Sea coastline of The Netherlands, England, the Bay of Fundy in Nova Scotia, the Yellow Sea of Korea, and the Gulf of California. Eisma et al. (1998) describes and discusses many of the major tidal flats of the world. Oil and gas deposits have been discovered in both siliciclastic and carbonate tidal facies, and uranium is present in some sandy tidal facies. Therefore, tidal deposits have economic significance as well as general scientific interest.

Depositional Setting

Although tidal currents may operate in the ocean to depths of 2000–2500 m, the tidal-flat environment is confined to the shallow margin of the ocean. The vertical distance between the high- and low-tide line in most modern tidal environments commonly ranges from 1 and 4 m (mesotidal coasts), depending upon the locality, although tidal ranges of 10–15 m or more (macrotidal coasts) occur in some localities, such as the Bay of Fundy. The total width of tidal flats may range from a few kilometers to as much as 25 km. Topographic relief within the tidal-flat environment is generally rather small except for tidal channels, and slopes of the tidal flat are gentle although commonly irregular.

The tidal-flat environment is divided into three zones: subtidal, intertidal, and supratidal (Fig. 9.38). The **subtidal zone** encompasses the part of the tidal flat that normally lies below mean low tide level. It is inundated with water most of the time and is normally subjected to the highest tidal-current velocities. Tidal influence in this part of the environment is particularly important within tidal channels, where bedload transport and deposition are predominant, although this zone is also influenced to some extent by wave processes. The **intertidal zone** lies between mean high and low tide levels. It is subaerially exposed either once or twice each day, depending upon local wind and tide conditions, but it commonly does not support significant vegetation. Both bedload and suspension sedimentation take place in this zone. The **supratidal zone** lies above normal high-tide level but is incised by tidal channels and flooded by extreme tides. This part of the tidal flat is exposed to subaerial conditions most of the time but may be flooded by spring tides twice each month or by storm tides at irregular intervals. Sedimentation is dominantly from suspension. On some tidal flats, the supratidal zone is a salt-marsh environment incised by tidal channels. In arid or semiarid climates, it is commonly an environment of evaporite deposition and is often referred to as a **sabkha**.

Sedimentary Processes and Sediment Characteristics of Tidal-Flats

Physical sedimentation on siliciclastic tidal flats takes place in response to both tidal processes and waves, producing sediments with characteristic grain-size and structural properties in different parts of the tidal flat (Fig. 9.39). Sedimentation in the channels of tidal flats is dominated by tidal currents, but wind-driven waves and the currents generated by these waves also play an important role in deposition on the flats between channels (e.g., Ridderinkhof, 1998). Tidal currents move up the gentle slope of the tidal flat during flood tide and back down during ebb

Figure 9.38
Schematic diagram showing the relationship of subtidal, intertidal, and supratidal zones in the tidal-flat environment. Note that mud is the dominant deposit in the upper part of the intertidal zone, mixed mud and sand predominate in the lower intertidal zone, and sand is deposited in the subtidal zone and in tidal channels. Muddy marsh deposits characterize the supratidal zone.

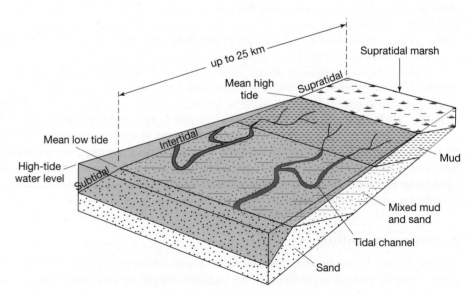

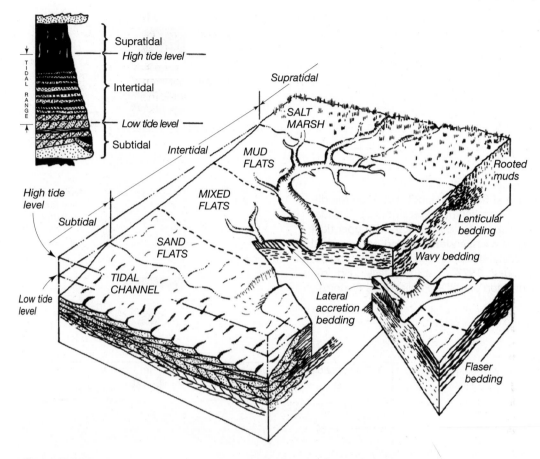

Figure 9.39
Schematic diagram of a typical siliciclastic tidal flat. The tidal flat fines toward the high-tide level, passing gradationally from sandflats, though mixed flats, to mudflats and salt marshes. An example of the upward-fining succession produced by tidal-flat progradation is shown in the upper-left corner. [From Dalrymple, R. W., 1992, Tidal depositional systems, *in* Walker, R. G., and N. P. James (eds.), Facies models: Geol. Assoc. Canada, Fig. 12, p. 201, reproduced by permission.]

tide. The tidal velocities achieved during reversing tides are commonly asymmetrical, and the velocities of flood tides may differ significantly from those of ebb tides. Within the channels, tidal currents can reach velocities of 1.5 m/s or more, and velocities on the flats range from 30 to 50 cm/s (Reineck and Singh, 1980). These velocities are adequate to cause transport of sandy sediment and produce ripple and dune bed forms, cross-bedding, and plane bedding. Thus, sand deposition dominates in the shallow subtidal zone as well as in the lower intertidal zone and the channels.

Channel sands are characterized by ripples and internal cross-bedding that may display bimodal directions of foreset dip, generated by reversing tides. The sands thus display herringbone cross-stratification; that is, cross-laminated sediment deposited during flood tide dip in the opposite direction to those formed almost immediately afterward during ebb tide (see Fig. 9.40). Reversing tides during an asymmetric tidal cycle can also cause erosion of ripple crests during the next tidal cycle, producing reactivation surfaces (Fig. 9.41).

As water fills tidal channels during rising tide, it spills out from the channels and spreads at relatively low velocities across the intervening flats between channels. Both fine sand and mud can be deposited on these low-energy, flat areas.

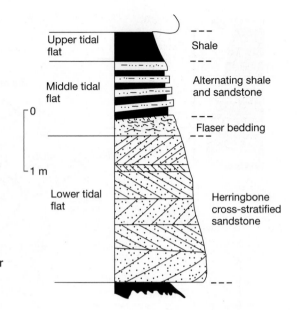

Figure 9.40

Progradational succession of tidal-flat deposits based on Middle Member, Wood Canyon Formation (late Precambrian), Nevada. Suspension sediment transport was dominant in the upper tidal flat, alternating bedload and suspension transport occurred in the middle tidal flat, and bedload transport was dominant in the lower tidal flat. [After Klein, G. deV., 1977, Clastic tidal facies, Fig. 76, p. 85, reproduced by permission of IHRDC Publications, Boston.]

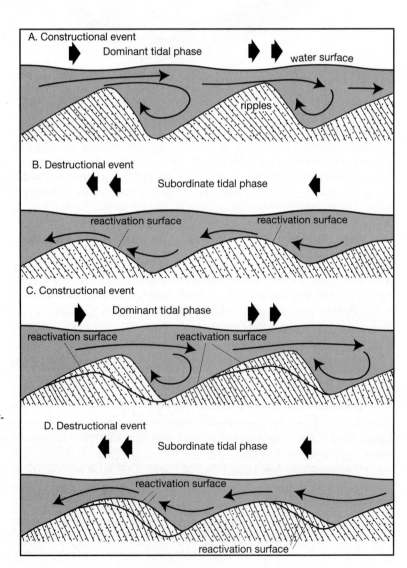

Figure 9.41

Schematic representation of reactivation surface developed owing to alternation of a dominant tidal phase (constructional event) with a subordinate phase (destructional event). [After Klein, G. deV., 1970, Depositional and dispersal dynamics of intertidal sand bars: Jour. Sed. Petrology, v. 40, Fig. 28, p. 1118, reproduced by permission of SEPM (Society for Sedimentary Geology), Tulsa, Okla.]

Some tidal flats are dominated by mud whereas others are dominated by sand. Sandy and muddy sediment deposited on the flats between channels is characterized by small-scale ripple cross-lamination, flaser bedding, wavy bedding, lenticular bedding, and, more rarely finely laminated bedding (Fig. 9.39).

The supratidal zone is only slightly affected by tidal currents and only marginally affected by waves. The supratidal zone is the zone of lowest energy on the tidal flat. The deposits of this zone are mainly mud but may include abundant plant debris in supratidal marshes, which may eventually form peat. Desiccated, cracked muds are characteristic features of the supratidal zone.

Although most tidal flats are sites of siliciclastic deposition, some tidal flats are dominated by deposition of carbonate sediments. Thus, deposition on these tidal flats is characterized by chemical and biological processes as well as physical processes. Lime muds and sand-size skeletal fragments generated within the subtidal zone may be transported into the intertidal and supratidal zones by waves and currents. In arid and semiarid climates, gypsum, anhydrite, and dolomite may be chemically precipitated in the supratidal and upper intertidal zones owing to strong evaporation. Organisms such as pelecypods, crustaceans, polychaete worms, foraminifers, diatoms, and blue-green algae (cyanobacteria) inhabit tidal flats and produce fecal pellets, cause extensive bioturbation of sediment, and generate burrows belonging to the *Skolithos* ichnofacies (Chapter 4). In the supratidal and intertidal zones blue-green algae are particularly important agents that trap and bind fine sediment to produce stromatolites.

Transgression and regression cause deposits of laterally adjacent tidal-flat environments to become superimposed, generating characteristic successions of vertical facies. Progradation produces a generalized fining-upward succession that begins with subtidal and lower intertidal cross-bedded sands, followed upward by mixed sand and mud in the middle intertidal zone and mud and peat in the upper intertidal and supratidal zones. A typical vertical regressive (progradational) succession developed on a siliciclastic tidal flat is illustrated schematically in Figure 9.40; see also Figure 9.39. Transgression presumably generates a coarsening-upward succession that displays the same general facies but in reverse order; however, transgression may rework and destroy intertidal deposits. Similar patterns of subtidal, intertidal, and supratidal carbonate facies could be expected to develop on coasts characterized by carbonate tidal flats (e.g., Hardie and Shinn, 1986). On a local scale, lateral channel migration may also produce small-scale coarsening-upward vertical successions.

Ancient Tidal-Flat Sediments

Tidal-flat deposits have several distinctive characteristics that help to differentiate them from sediments of most other environments; however, their overall characteristics are similar to those of estuarine deposits. The most important criteria for recognition of ancient tidal flat deposits are commonly regarded to include the following: (1) bimodal direction of current-formed cross-bedding resulting from reversing tidal currents, (2) the occurrence of sediments that reflect repeated, small-scale alternations in sediment transport conditions (tidal rhythmites or tidalites; e.g., Klein, 1998) and the joint occurrence of large-scale (channel) and small-scale (sandflat and mudflat) structural units in superposition or juxtaposition, (3) abundant reactivation surfaces and flaser bedding, and (4) a high frequency of erosional contacts and abrupt facies changes. Other supporting criteria include the typical vertical succession of facies discussed above, the high degree of bioturbation of many tidal-flat sediments, the presence of mudcracked stromatolites, and other evidence of subaerial exposure such as raindrop imprints, hail marks, and animal or bird tracks.

Deposits influenced by tidal action occur in several of the environmental settings already discussed (e.g., delta, barrier-island systems, estuaries, lagoons) and

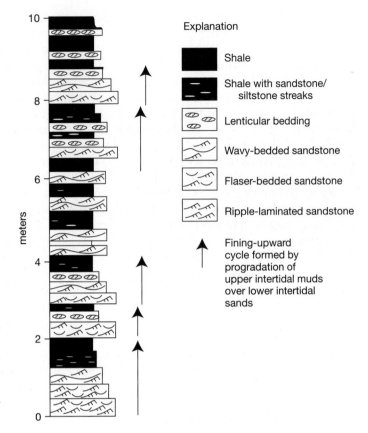

Figure 9.42
Representative lithostratigraphic column of the upper member of the Baraichari Shale Formation (late Miocene-Pliocene), Bengal Basin, Bangladesh, interpreted as a cyclic succession of progradational tidal-flat deposits. [After Alam, M. M., 1995, Tide-dominated sedimentation in the upper Tertiary succession of the Sitapahar anticline, Bangladesh, *in* Flemming, B. W., and A. Bartholomä (eds.), Tidal signatures in modern and ancient sediments, International Association of Sedimentologists Spec. Pub. 24, Blackwell Science, Oxford, Fig. 4, p. 332, reproduced by permission.]

also form on shallow, tide-dominated shelves (to be discussed). Numerous examples of such tidal deposits have been reported from stratigraphic units of virtually all ages from Precambrian to Holocene. Fewer examples of ancient tidal deposits formed specifically in tidal-flat environments have been reported but such deposits are certainly known, for instance, those described in Alexander, Davis, and Henry (1998).

Alam (1995) describes a stratigraphic succession in the upper member of the Baraichari Shale Formation (Late Miocene-Pliocene) from the Bengal Basin, Bangladesh, that is characterized by many of the classic features of tidal-flat deposits (Figure 9.42). This cyclic succession contains repeated fining-upward units of fine-grained sandstone overlain by gray shale that commonly contains thin layers and streaks of sandstone or siltstone. Wavy bedding, flaser bedding, and ripple lamination characterize many of the sandstone beds. The sandstone units are interpreted as lower intertidal deposits and the shales as upper intertidal deposits. Stacking of sand and mud units took place during repeated cycles of progradation (regression), as upper intertidal deposits advanced over lower intertidal deposits.

FURTHER READING

Deltaic Systems

Broussard, M. L. (ed.), 1975, Deltas: Models for exploration: Houston Geol. Society, 555 p.

Chough, S. K., and G. Orton (eds.), 1995, Fan deltas: Depositional styles and controls: Sedimentary Geology, v. 98, 292 p. [Special Issue].

Colella, A., and D. B. Prior (eds.), 1990, Coarse-grained deltas: Internat. Assoc. Sedimentologists Spec. Pub. 10, Blackwell, Oxford, 357 p.

Coleman, J. M., 1981, Deltas: Processes of deposition and models for exploration, 2nd ed.: Burgess Publishing Co., Minneapolis, Minn., 124 p.

Coleman, J. M., and D. B. Prior, 1980, Deltaic sand bodies: Am. Assoc. Petroleum Geologists, Ed. Course Notes 15, Tulsa, Okla., 171 p.

Miall, A. D., and N. Tyler (eds.), 1991, Three-dimensional facies architecture of terrigenous clastic sediments and its implications for hydrocarbon discovery and recovery: Concepts in sedimentology and paleontology, 3: SEPM (Soc. for Sedimentary Geology), Tulsa, Okla., 309 p.

Morgan, J. P. (ed.), 1970, Deltaic sedimentation—Modern and ancient: Soc. Economic Paleontologists and Mineralogists Spec. Pub. 15, Tulsa, Okla., 312 p.

Nemec, W., and R. J. Steel (eds.), 1988, Fan deltas: Sedimentology and tectonic settings: Blackie, Glasgow and London, 444 p.

Roberts, H. K., and J. Sydow, 2003, Late Quaternary stratigraphy and sedimentology of the offshore Mahakam Delta, East Kalimantan (Indonesia), *in* Hasan, S. F., D. Nummedal, P. Imbert, H. Darman, and H. Posamentier (eds.), Tropical deltas of Southeast Asia: sedimentology, stratigraphy, and petroleum geology: Society for Sedimentary Geology (SEPM) Special Publication, v. 76, Tulsa, Okla., p. 125–145.

Shirley, M. L., and J. A. Ragsdale (eds.), 1966, Deltas in their geologic framework: Houston Geol. Society, 251 p.

Sidi, F. H., D. Nummedal, P. Imbert, H. Darman, and H. W. Posamentier (eds.), 2003, Tropical deltas of southeast Asia-Sedimentology, stratigraphy, and petroleum geology: SEPM Spec. Publ. 76, Society for Sedimentary Geology, Tulsa, Okla., 269 p.

Whateley, M. K. G., and K. T. Pickering (eds.), 1989, Deltas—Sites and traps for fossil fuels: Geol. Soc. London Spec. Pub. 41, Blackwell, Oxford, 360 p.

Beach and Barrier-Island Systems

Davis, R. A., Jr. (ed.), 1994, Geology of Holocene barrier island systems: Springer-Verlag, Berlin, 464 p.

Davis, R. A., Jr., 1997, The evolving coast: Scientific American Library, A division of HPHLP, New York, 233 p.

Hardisty, J., 1990, Beaches - Form and process: Unwin Hyman, London, 324 p.

Komar, P. D., 1998, Beach processes and sedimentation, 2nd ed.: Prentice Hall, Upper Saddle River, N.J., 544 p.

Oertel, G. F., and S. P. Leatherman (eds.), 1985, Barrier islands: Marine Geology, v. 63, p. 1–396.

Pilkey, O. H., 1983, The beaches are moving: Duke University Press, Durham, N.C., 336 p.

Estuarine and Lagoonal Systems

Ashley, G. M. (ed.), 1988, The hydrodynamics and sedimentology of a back-barrier lagoon-salt marsh system, Great Sound, New Jersey: Marine Geology, v. 82, p. 1–132.

Kjerfve, B. (ed.), 1994, Coastal lagoon processes: Elsevier, Amsterdam, 577 p.

Mehta, A. J. (ed.), 1993, Nearshore and estuarine cohesive sediment transport: American Geophysical Union, Washington D.C., 581 p.

Nordstrom, K. F., and C. T. Roman (eds.), 1996, Estuarine shores: John Wiley & Sons, Chichester, 486 p.

Perillo, G. M. E. (ed.), 1995, Geomorphology and sedimentology of estuaries, Developments in Sedimentology 53: Elsevier Science B.V., Amsterdam, 471 p.

Prandle, D. (ed.), 1992, Dynamics and exchanges in estuaries and the coastal zone: American Geophysical Union, Washington D.C., 647 p.

Tidal-Flat Systems

Alexander, C. R., R. A. Davis, and V. J. Henry (eds.), 1998, Tidalites: Processes and products: SEPM Special Publication No. 61, 171 p.

Black, K. S., D. M. Paterson, and A. Cramp (eds.), 1998, Sedimentary processes in the intertidal zone: Geological Society Spec. Publ. 139, The Geological Society, London, 409 p.

Eisma, D., P. L. de Boer, G. C. Cadée, K. Dijkema, H. Ridderinkhof, and C. Philippart, 1998, Intertidal deposits: River mouths, tidal flats, and coastal lagoons: CRC Press, Boca Raton, 524 p.

Flemming, B. W., and A. Bartholomä (eds.), 1995, Tidal signatures in modern and ancient sediments, International Association of Sedimentologists Spec. Pub. 24, Blackwell Science, Oxford, 358 p.

Hardie, L. A., and E. A. Shinn, 1986, Carbonate depositional environments modern and ancient—Part 3: Tidal flats: Colorado School of Mines Quarterly, v. 81, No. 1, 74 p.

Open University Course Team, 1989, Waves, tides and shallow-water processes: Pergamon Press, Oxford, 187 p.

Smith, D. G., G. E. Reison, B. A. Zaitlin, and R. A. Rahmani (eds.), 1991, Clastic tidal sedimentology: Canadian Soc. Petroleum Geologists Mem. 16, 307 p.

10
Siliciclastic Marine Environments

10.1 INTRODUCTION

The marine environment is that part of the ocean lying seaward of the zone dominated by shoreline processes. Water depth in the marine realm ranges from a few meters to more than 10,000 m. The salinity of seawater in the open ocean averages about 35 ‰, although higher or lower salinities can occur locally in restricted bodies of the ocean. Marine life forms are characterized by generally high diversity and large populations, and most are low-tolerance organisms adapted to conditions of normal salinity. The energy of the bottom water lying immediately above the ocean floor is generally low, except on the shallow continental shelf, which is affected by a variety of tidal processes and wind- and storm-wave activity, and on some parts of the deeper ocean floor that are swept by bottom currents.

The major subdivisions of the oceanic realm are the **continental margin** and the **ocean basin**. These regions in turn can be further subdivided as shown in Figure 10.1. The **continental shelf** extends seaward from the shoreline at a gentle slope of about 1° to a point where a perceptible increase in rate of slope, the **shelf break**, takes place. The shelf break occurs in the modern ocean at an average distance from shore of about 75 km, although the distance ranges from a few tens of meters to more than 1000 km. Average water depth at the shelf break is about 130 m. The **continental slope** descends from the shelf break to the deep seafloor with a typical slope of about 4°. On passive, or divergent, continental margins, the foot of the continental slope merges with the **continental rise**, which is a gently sloping surface created by coalescing submarine fans at the base of the slope. The continental rise passes gradually into the floor of the ocean basin. Parts of the deep ocean floor consist of nearly flat areas called **abyssal plains,** which are covered by sediment. Other parts of the ocean floor are characterized by volcanic hills that rise above the seafloor to elevations ranging from a few hundred meters to more than 1000 m. The central part of the major ocean basins is occupied by a gigantic **mid-ocean ridge** that may protrude more than 2.5 km above the seafloor. On active, or convergent, margins, the continental slope may descend into a **deep-sea trench**, and the continental rise is absent. On the basis of water depth, we divide the ocean into two major zones: the neritic zone and the oceanic zone. The shallow **neritic zone** extends from the shoreline to the shelf break. The **oceanic zone** extends from shelf break to shelf break and encompasses the deeper part of the ocean.

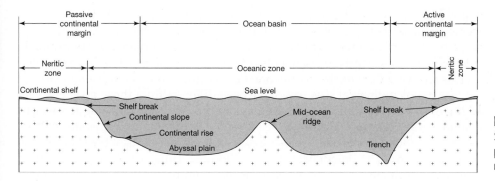

Figure 10.1
Schematic cross-sectional profile of the marine environment. Not to scale.

In this chapter, we consider marine environments characterized by transport and deposition of siliciclastic sediment. Carbonate marine environments are discussed in Chapter 11.

10.2 THE SHELF ENVIRONMENT

The neritic zone encompasses the shallow-water areas of the ocean lying shoreward of the shelf break. Although the shelf break on modern shelves lies at an average depth of about 130 m, as indicated, it may be located on some shelves at depths as shallow as 18 m or as deep as 915 m (Bouma et al., 1982). In the modern ocean, the shallow-marine environment occupies mainly the continental shelf area around the margin of the continents, forming what is referred to as a **pericontinental,** or marginal, sea (Heckel, 1972). At various times in the geologic past, broad, shallow **epicontinental,** or epeiric, seas occupied extensive areas within the continents (Fig. 10.2), somewhat like the present-day Hudson Bay area of the North American Arctic region. The following discussion of the neritic environment is focused primarily on the continental shelf environment because we can draw on the modern continental shelf environment as a model. Readers should keep in mind, however, that many of the shallow-marine deposits preserved in the geologic record may have been deposited in broad epicontinental seas, for which we may have no truly representative modern analogs, although some modern continental shelves are very wide (e.g., continental shelves of the

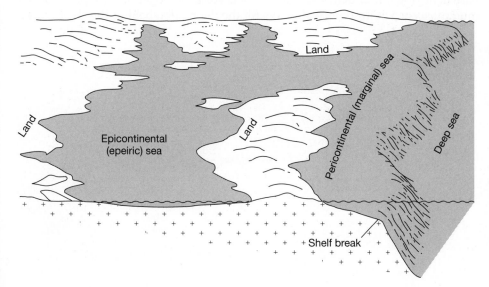

Figure 10.2
Schematic diagram illustrating the difference between pericontinental (continental shelf) and epicontinental shallow-marine environments. [After Heckel, P. H., 1972, Recongintion of ancient shallow marine environments, *in* Rigby, J. K., and W. K. Hamblin (eds.), Recognition of ancient sedimentary environments: SEPM (Society for Sedimentary Geology) Spec. Pub. 16, Fig. 1, p. 227, reprinted by permission.]

North Sea, Yellow Sea, and Timor-Arafura Sea). We may assume that similar sedimentological processes operated on continental shelves and in epicontinental seaways, but, in fact, differences exist between these two environments. For example, epicontinental seas received sediments from nearly all sides, whereas continental shelves receive sediments from only one side. Furthermore, the wave and current regimes in epicontinental seas may have been different from those on shelves. In addition, modern continental shelves may not provide a good analog for ancient marginal seas because rapid rise of sea level following the final episode of Pleistocene glaciation has stranded coarse sediment in deeper parts of the shelves, creating conditions of sediment–water disequilibrium. Thus, sediment grain size on some parts of modern shelves is not consistent with present water depth, energy conditions, and sedimentation processes on the shelves.

Both siliciclastic and carbonate sediments can accumulate in the marine-shelf environment, although most modern continental shelves are covered by siliciclastic sediments. Carbonate sediments (Chapter 11) are restricted to a few shelves, mainly (but not exclusively) in tropical areas.

Physiography and Depositional Setting

The siliciclastic shelf environment is bounded by various coastal environments on the landward side and by the continental slope on the seaward side. It can be divided into the shallow **inner shelf**, which is dominated by tidal, wind-driven, and storm-wave processes; the **middle shelf**; and the deeper-water **outer shelf** (Fig. 10.3). Wright (1995) suggests that the inner shelf extends offshore to depths of about 30 m; however, the boundaries between the inner, middle, and outer shelves are not well defined and their positions fluctuate with changing sea level. In fact, during greatly lowered sea level, the normal inner and middle shelves are subaerially exposed, and the outer shelf may be exposed or covered only by very shallow water.

The width of shelves varies according to their plate-tectonic settings (Shepard, 1973; Eisma, 1988). Shelves along the fore-arc region of convergent continental margins tend to be very narrow. By contrast, broad shelves and platforms occur in the back-arc basins of convergent margins, on divergent or trailing-edge continental margins, and on cratonic downwarps that open to the sea. Because continental margins may evolve from diverse structural origins, the barrier forming the shelf break, which marks the transition from shallow shelf to steep slope, may also have diverse origins as shown in Figure 10.4. It may have formed initially as a basement ridge, fold belt, fault block, reef, volcanic ridge, or a diapir (a salt or mud intrusion). With time, sediments fill the depression behind the shelf-margin

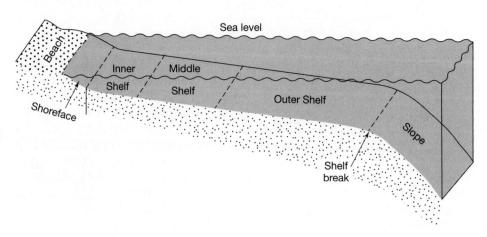

Figure 10.3
Subdivisions of the continental shelf. [Modified from Galloway, W. E., and D. K. Hobday, 1983, Terrigenous clastic depositional systems, Fig. 7.1, p. 144, reprinted by permission of Springer-Verlag, Heidelberg.]

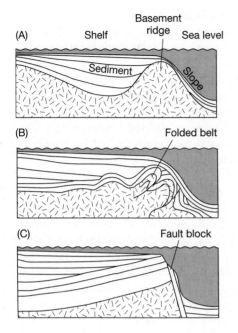

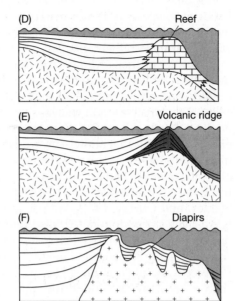

Figure 10.4
Various kinds of structural barriers that form the seaward margins of continental shelves. [After Hedberg, H. D., 1970, Continental margins from viewpoint of the petroleum geologist: American Association Petroleum Geologists Bull., v. 54, Fig. 18, p. 22, reproduced by permission.]

barrier and may drape over the shelf break onto the slope. During this process, shelf sediments build or aggrade upward to wave base, the depth below which wave-generated processes have little effect on sediment movement. Once wave base is reached by an aggrading prism of sediment, the shelf achieves a state of near equilibrium (Swift and Thorne, 1991) by which little further deposition occurs unless additional subsidence takes place or sea level rises (see also Wright, 1995, Chapter 2). That is, under equilibrium conditions, the sediment simply moves across the shelf (bypassing) on its way toward the slope. The mechanisms responsible for movement of sediment across the shelf are considered in the next section. Although shelves are fundamentally low-relief platforms, the shelf surface can vary considerably. It may be relatively smooth or covered by a variety of small- to large-scale bedforms. It may also contain banks, islands, or shoals near its offshore edge (Bouma et al., 1982).

Shelf Sediment Transport and Deposition

Waves are constantly moving from deeper water in the open ocean across the shelf to the shore zone; there, they eventually break and become translated into wave swash and longshore currents. Given that the direction of wave movement across the shelf is dominantly shoreward, how is it that sediment can apparently move across the shelf in a seaward direction? Johnson (1919) suggested the idea of a "graded shelf," which he believed to display progressive decrease in grain size from coarse at the shoreline to very fine at the shelf edge, in response to presumed decrease in water energy seaward. The graded-shelf concept for modern shelves eventually fell into disfavor when it became obvious with additional study in the 1930s–1960s that the grain-size distribution on many modern shelves is patchy or irregular. The concept of **relict** shelf sediments was introduced to explain such irregular distribution patterns as the presence of coarse sands and gravels in deep water. Relict sediments are deposits that are apparently not in equilibrium with present hydrodynamic conditions. They were deposited on the shelf by fluvial or glacial processes during low stands of sea level. Relict sediments were initially believed to remain on the shelf floor without significant reworking as they were inundated by rising sea level (Shepard, 1932; Emery, 1968). It was subsequently recognized, however, that some so-called relict sediments had likely been reworked to some extent during sea-level rise. This reworking caused the sediments to be brought into partial or complete equilibrium with present shelf processes and conditions. Such reworked sediment, having partly relict and partly modern characteristics, was called **palimpsest** (Swift, Stanley, and Curray, 1971).

Studies of shelf sedimentation since the 1960s suggest that purely relict sediments may be less common than originally thought and that Johnson's concept of a graded or equilibrium shelf may still have merit (e.g., Johnson and Baldwin, 1996). Also, many ancient shelves may not have been affected by drastically lowered sea levels. Therefore, it becomes necessary to consider what mechanisms exist on the shelf that allow seaward transport of sediment across shelves during high stands of sea level, as at the present time. We know that modern shelves are affected by waves and storms, tidal currents, major surface ocean currents that intrude onto the edges of some shelves, and possibly density currents (e.g., Swift, Stanley, and Curray, 1971). We commonly divide shelves into two main types: wave- and storm-dominated, also referred to as weather-dominated, and tide-dominated, although most shelves are actually influenced by a mixture of processes. Approximately 80 percent of modern shelves are wave- and storm-dominated and 17 percent are tide-dominated (Swift, Hans, and Vincent, 1986). About 3 percent of shelves are dominated by intruding ocean currents; density currents play only a minor role in shelf transport.

Wave- and Storm-Dominated Shelves

A complex spectrum of transport processes operate on wave- and storm-dominated shelves, including fair-weather waves and swells, storm waves, wind-driven surface currents, river-generated plumes, and density currents. These processes are summarized graphically in Figure 10.5. During fair weather, waves generated locally by wind move across the shelf from deeper water onto the shallow-water inner shelf. As briefly described under beach processes in Chapter 9, these waves have an oscillatory motion that causes water to move in nearly circular orbits. As a wave passes, water moves forward in the crest of the wave, then downward, and finally backward under the trough and upward. Thus, individual water particles do not move forward with the passing wave but simply trace an orbital path (Fig. 10.6a).

Fair-Weather Waves

Orbital motion of water generated by passage of these orbital waves dies out downward at a depth equal to about one-half of the wave length (distance between

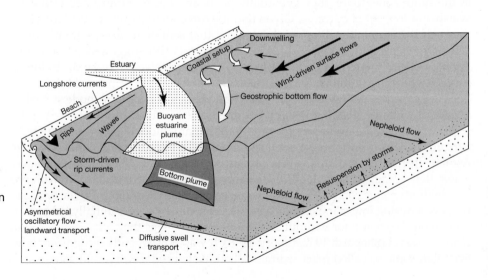

Figure 10.5
Schematic representation of the major physical processes that operate on the shelf to transport sediment. Based on Nittrouer and Wright, 1994; Swift et al., 1986; Swift and Thorne, 1991; and Vincent, 1986.

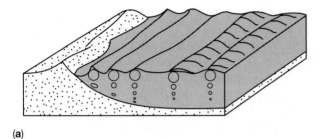

(a)

(b)

Figure 10.6
Behavior of oscillatory waves in shoaling water. (a) Flattening of orbits as waves enter water shallower than about one-half wave length. (b) Time-velocity record of bottom flow during passage of a shoaling wave. The landward stroke as the crest passes has higher velocity and moves more sediment than does the return stroke associated with the passage of the trough. [After Swift, D. J. P., and J. A. Thorne, 1991, Sedimentation on continental margins, I: a general model for shelf sedimentation, *in* Swift, D. J. P., G. F. Oertel, R. W. Tillman, and J. A. Thorne (eds.), Shelf sand and sandstone bodies: Geometry, facies, and sequence stratigraphy: International Association of Sedimentologists Spec. Publ. 14, Blackwell, Oxford, Fig. 5, p. 12, reproduced by permission.]

wave crests). This depth is referred to as the **wave base** (Fig. 9.21). Fair-weather wave base is commonly on the order of 10–15 m. Because orbital motion in deep water is unimpeded by the bottom, orbits are nearly circular. As a wave moves into shallow water, where depth is less than one-half the wave length, the bottom begins to interfere with orbital motion and thus begins to affect the shape of the orbits. By the time that waves reach very shallow water, where depth is less than about 1/20 the wave length, the motion of the particles is strongly affected by interaction with the bottom, and the orbits become much more elliptical (Fig. 10.6a). They become progressively flatter downward below the surface until near bottom they are essentially linear, generating a to-and-fro oscillating motion as waves pass. This motion produces bidirectional flow of water along the seafloor as each wave passes over the surface. The velocity of this bottom flow is referred to as the **orbital velocity** because it varies directly as a function of the magnitude of the orbital diameter and indirectly as a function of the wave period (the time required for passage of one wave length).

Orbital velocity is commonly greater in one direction than the other. This difference in velocity becomes important when the stronger velocity flow exceeds the threshold of movement for grains, resulting in net transport of grains in one direction (Fig. 10.6b). Because the fair-weather wave base is shallow, movement of sediment by fair-weather waves takes place mainly on the innermost shelf and is dominantly in an onshore direction. The eventual breaking of waves in very shallow water generates currents (wave swash), which also move in an onshore direction, as well as longshore currents directed laterally along shore (Chapter 9). This overall shoreward movement of water in the nearshore zone above the fair-weather wave base creates a "littoral energy fence" (Swift and Thorne, 1991) that tends to trap sediment in the nearshore zone. For sediment to move from land onto the shelf, it must escape through the littoral energy fence by a mechanism such as river-mouth bypassing that occurs as a flood-stage jet of sediment-laden water from a river mouth (Fig. 10.5) or rip currents (Fig. 9.22).

Swells, Storm Waves, and Wind-Forced Currents

Swells are low-relief, long-period, long-wave-length waves generated by storms that may originate far out to sea. When swells move onto the shelf they tend to "stir" the bottom to greater depths than do fair-weather waves. Storms (seasonal storms, typhoons, hurricanes) that move across a shelf have an even greater effect on shelf transport and deposition. First, highly energetic storm waves in the shore zone, perhaps accompanied by elevated tides, vigorously erode the beachface and upper shoreface. During such strong storms, beach sediment is flushed seaward, probably mainly by a combination of storm-enhanced rip currents and downwelling currents that evolve from wind-driven or wind-forced currents.

Wind-forced currents are unidirectional currents generated by wind shear stress as wind blows across the water surface, gradually putting into motion deeper and deeper layers of water (Ekman transport). Deeper layers of water are deflected by the Coriolis force, so that their direction of movement diverges from that of surface layers. The Coriolis force is generated by Earth's rotation, causing moving objects to be deflected to the right in the Northern Hemisphere and to the left in the Southern Hemisphere. If the velocity and duration of wind are great enough, water movement may extend to the seabed with enough velocity to transport sediments. Strong winds commonly create wind-forced currents that flow parallel to shore and therefore do not provide much offshore sediment transport. If, however, currents moving along the shoreline are deflected landward owing to the Coriolis force, an onshore pile-up of water takes place (e.g., the Oregon Coast during winter). Piling up of water onshore creates an elevation of the water surface—a **coastal setup** (Fig. 10.5)—of perhaps a meter or two. This setup can apparently be enhanced by very low atmospheric pressure. The different water levels at the coast and offshore result in a hydrostatic pressure difference on the ocean floor that drives a bottom flow seaward (downwelling). As the bottom water flows seaward, it is deflected laterally to form a **geostrophic current.** This current initially moves obliquely offshore but subsequently veers around, owing to the Coriolis force, to assume a direction roughly parallel to the bathymetric contours or isobaths, that is, roughly parallel to the shoreline.

These geostrophic flows can achieve velocities at water depths of 10–20 m of as much as 60 cm/s (Walker and Plint, 1992). Flows of this magnitude may not be capable of transporting much sandy sediment unless they are accompanied by strong wave-driven oscillatory motion at the bed. Oscillatory wave motion can provide the shear stress needed to lift grains off the bottom, and those grains are then transported by the geostrophic currents (Snedden, Nummedal, and Amos, 1988). Unidirectional and oscillatory currents operating together are called combined flows. Tropical storms and hurricanes accompanied by strong winds that blow directly onshore can generate higher coastal setups than those generated by seasonal storms, and thus create much stronger seaward-flowing currents. Current-flow velocities of as much as 2 m/s have been reported on shelves during some tropical storms (Morton, 1988), and currents reaching velocities of 2 m/s have also been recorded flowing down submarine canyons (Hubbard, 1992). Sediment moves obliquely offshore, and some sand moves from the beach shoreface into offshore settings. Note again, however, that such transport by geostrophic currents tends to be mostly parallel to isobaths and not directly seaward. Therefore, sediment may not be transported by such currents to any great distance outward onto the shelf.

Storm waves also affect sediment movement in deeper water on the middle and outer shelf. Because of their longer wave length and period, storm waves moving across a shelf may be able to rake the seafloor to depths of as much as 200 m (Komar et al., 1972). The orbital velocities generated by these storm waves may be nearly equal so that no net seaward transport of sediment occurs; however, the waves resuspend bottom mud and tend to spread or dissipate it around the seafloor. On the other hand, net sediment transport in either a seaward or landward direction can occur.

Sediment Plumes

Sediment discharged at river mouths into the ocean may be carried onto the shelf either as buoyant plumes or underflows (bottom plumes), as suggested in Figure 10.5. Buoyant (hypopycnal) plumes commonly develop where low-salinity water issuing from rivers and estuaries flows out on top of higher-salinity seawater. These plumes generally do not reach farther than the inner or middle shelf before being carried parallel to the coast owing to the Coriolis force (e.g., Nittrouer and Wright, 1994). There the suspended fine sediment (coarser sediment drops out near the river mouth) gradually settles to the shelf floor.

Underflows (hyperpycnal flows) are generated where the mouths of rivers carry unusually high suspended-sediment loads, causing the inflowing river water to be more dense than the ocean water. These flows are, in fact, the steady-flow turbidity currents discussed in Chapter 2. Many of these underflows may be checked rather quickly, resulting in most of the sediment being deposited near the river mouth. Some may be carried greater distances onto the inner shelf, however, and apparently need not be confined by submarine canyons (Seymour, 1990).

Nepheloid Flow

A nepheloid layer is a turbid body of suspended sediment that may reach heights of several hundred meters above the seafloor. These layers were first surveyed and named by Ewing and Thorndike (1965), who discovered them by using an optical nephelometer to measure light scattering at various levels in the water column. A nepheloid layer is more dense than the surrounding ambient water but not dense enough to sink rapidly. Thus, sediment may remain suspended in such layers for a long period of time. Most of the material of nepheloid layers consists of very fine clay particles. Some of this material may have reached the nepheloid layer directly by settling through the water column from buoyant plumes. Most of it is probably fine sediment resuspended from the ocean floor owing to erosion of the seabed by storm waves (Fig. 10.5), or possibly swell conditions, or it is fine material injected into the water column by turbidity currents or other mechanisms. Owing to its low settling velocity, fine sediment may remain in suspension in the nepheloid layer for periods ranging from days to weeks in the lowest 15 m of the water column and from weeks to months in the lowest 100 m (Kennett, 1982). The turbid nepheloid layer is slowly advected (flows laterally) seaward as a kind of density flow. Sedimentation and resuspension would likely occur many times before fine sediment could be moved completely across a shelf by this process.

Sediment Characteristics of Storm-Dominated Shelves

As mentioned, storm-dominated shelves predominate on most of the world's coasts. They are characterized by low tidal current velocities (commonly <25 cm/s), and fair-weather wave base is normally shallow (~10 m). Because of these characteristics, little coarse sediment moves on these shelves except during intense storms. Examples of modern storm-dominated shelves include the Atlantic shelf off the eastern coast of the United States, the Pacific shelf off Oregon and Washington, and the Bering Sea.

Sedimentation patterns on storm-dominated shelves may be quite complex, depending in part upon the extent to which the shelves are mantled by relict sediments or modern sediments. Shelves with abundant relict sediment, such as the Atlantic, are characterized particularly by sand bodies. The most controversial sand bodies, with respect to their origin, are **shelf sand ridges**. Modern shelf sand ridges are elongate, coastal- to shelf-sand bodies that are larger than subaqueous dunes (Chapter 4), with lengths on the order of 10 km and heights that are more than 20 percent of the water depth (Snedden and Dalrymple, 1999). These ridges are also present on tide-dominated shelves and share many similarities. Snedden and Dalrymple suggest that shelf sand ridges pass through three stages of development: an initial irregularity (the ridge nucleus) forms by coastal or shelf processes; storm- or tide-driven nearshore/shelf currents interact with this irregularity, causing upward growth and down-current migration of the incipient ridge; and the ridge evolves as a result of continued current action. Shelf sand ridges are most likely to form during transgressions. Thus, most shelf sand ridges probably overlie transgressive ravinement (erosion) surfaces.

Shelves with a greater component of modern vs. relict sediments, such as the Pacific shelf off Oregon and Washington, are typically characterized by less relief and a greater proportion of finer grained sediments (muds) than are Atlantic-type shelves. Although sediment on these shelves may display a general trend of seaward fining, relict sands or gravels show through extensive "windows" in a discontinuous blanket of muddy modern sediment. Also, mixing of relict sands and modern muds may take place in some areas. Muds are typically thoroughly bioturbated.

Coarse-grained "storm layers" and hummocky cross-stratification are sedimentary structures that appear to be especially characteristic of storm-dominated shelf sediments. "Storm layers" are commonly thin layers consisting of concentrations of coarser grains interlayered or embedded in finer grained muds (Fig. 10.7). The coarser material typically consists of coarse silt, fine sand, shell fragments or, less commonly, gravel. The layers characteristically show vertical size grading. The exact origin of these layers is controversial; however, Cheel and Leckie (1992) suggest that they form by a two-stage process: transport from a beach by offshore-directed, storm-generated combined (geostrophic) flows, followed by reworking and selective sorting of bed material by asymmetrical oscillatory currents generated by shoaling swell waves propagating onshore (Fig. 10.7). Storm layers, also called tempestites, are described in considerable detail by Aigner (1985). They are best developed on the inner shelf, but they have been found as far as 40 km from the coast (Reineck and Singh, 1980).

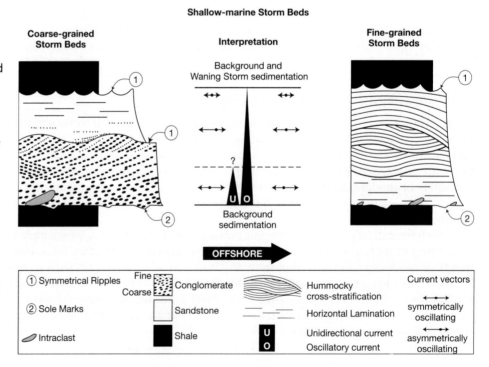

Figure 10.7
Schematic comparison of idealized coarse-grained storm beds and fine-grained hummocky cross-stratified beds on storm-dominated shelves. The lengths of the current vectors are proportional to the strength of the current in a given direction rather than duration. [From Cheel, R. J., and D. A. Leckie, 1992, Coarse-grained storm beds of the Upper Cretaceous Chungo Member (Wapiabi Formation), southern Alberta, Canada: Jour. Sed. Petrology, v. 62, Fig. 14, p. 943, reproduced by permission of Society of Economic Paleontologists and Mineralogists, Tulsa, Okla.]

Hummocky cross-stratification has been identified in few, if any, m
shelf sediments, but it has been described in numerous ancient shelf sedi
ranging in age from Precambrian to Pleistocene. It is thought to be partic
characteristic of shelf sediments, although it has been described also in sho
(beach) and some lake sediments. As discussed in Chapter 4, it consists of cu
gently dipping laminae, both convex-up (hummocks) and concave-up (sw
that intersect at a low angle (Fig. 10.7). It is commonly interbedded with b
bated mudstones. Most workers appear to agree that hummocky cross-stratifi
forms as a result of storm waves acting in some manner below fair-weather
base; however, the exact mechanism of formation remains controversial (s
view in Duke, Arnott, and Cheel, 1991).

Tide-Dominated Shelves

Tidal Processes

As discussed in Chapter 9, tidal processes can affect sedimentation on tidal flats,
in estuaries, and on deltas. They also strongly influence sedimentation on some
continental shelves. Vertical rise and fall of tides is accompanied by horizontal
movements of water (Howarth, 1982; The Open University Team, 1989) that we
refer to as tidal currents. The currents generated on the shelf by tides are bidirectional but asymmetrical with respect to velocity. That is, flood-tide and ebb-tide
velocities are commonly different. Asymmetrical currents may result in net sediment transport in the direction of the stronger current. If both current phases are
able to transport sand, herringbone cross-stratification and reactivation surfaces
may form (see tidal flat systems, Chapter 9). Also, tidal rise and fall in some environments can produce couplets of sand-mud laminae, with mud deposited on top
of sand during a standstill. Stacking of these couplets produces tidal successions
referred to as **tidal rythmites** or **tidalites** (e.g., Dalrymple, Makino, and Zaitlin,
1991; Smith et al., 1991; Alexander, Davis, and Henry, 1998). Such beds may show

Box 10.1 Tides

Tides are generated by the gravitational attraction of the moon and sun for
Earth in conjunction with the rotation of Earth. Tidal influence is manifested at
any given coastal locality by daily rise and fall of the sea over an average range
of about 1–4 m on open coasts, but tidal range may exceed 15 m in some enclosed basins (e.g., the Bay of Fundy, Nova Scotia). Some localities (e.g., the Atlantic coast of the United States) experience **semidiurnal tides**, characterized
by two highs (of approximately equal height) and two lows (of approximately
equal height) each day. Others, owing to configuration of the generation basin
and complexities of the shoreline (e.g., parts of the Gulf of Mexico coast), have
diurnal tides, distinguished by one high and one low each day. Still others
(e.g., the United States, Pacific coast) have **mixed tides**—two highs (of unequal
height) and two lows (of unequal height) each day. Alignment of the sun and
moon (new or full moon) gives rise to **spring tides** that are about 20 percent
higher than normal. When the sun and moon are at right angles in relation to
the Earth, **neap tides** that are about 20 percent lower than normal result. Tidal
currents on continental shelves are propagated as a large wave or tidal bulge
generated in deep ocean basins (Fox, 1983). In major ocean basins, this tidal
bulge rotates around a central point of no tidal movement called an
amphidromic point. The tidal wave follows an elliptical path that is almost circular in the open ocean. In more restricted areas, the ellipse is strongly elongated, forming a narrow, rectilinear pattern.

cyclic changes in thickness that may represent neap-tide and spring-tide variations in tidal current velocity.

Tidal-current velocity decreases with water depth; thus, tidal-current transport is most important in shallow water. Tidal-current velocities ranging up to about 2 m/s have been measured in some enclosed basins—the Bay of Fundy, for example. Tidal currents on some shelves, such as those around the British Isles, have velocities that may exceed 1.5 m/s; however, tidal velocities on most shelves are less than about 1 m/s. Even so, close to the seabed many tidal currents are strong enough to rework and transport significant quantities of sand and possibly gravel. On the other hand, tidal-current velocities on some shelves are so low that they are below the threshold velocities required for sediment entrainment and transport. Much of the movement of sediment by tidal currents occurs when tidal currents are aided by wave action. The orbital motion of waves may be sufficient to lift grains off the seafloor, which are then transported some distance by currents too weak to move the grains unaided (Komar, 1976; The Open University Team, 1989). An outstanding example of a modern shelf dominated by strong tidal currents is under the North Sea, which lies between the United Kingdom and the coasts of Denmark and Norway.

Sediments of Tide-Dominated Shelves

As discussed, tide-dominated shelves are distinguished by the presence of tidal currents with velocities ranging from about 50 to more than 150 cm/s. Modern examples include the North Sea; the Korea Bay of the Yellow Sea; the Gulf of Cambay, India; the shelf around the British Isles; Georges Bank in the outer part of the Gulf of Maine; and the northern Australia shelf. Tide-dominated shelves are characterized particularly by sand bodies of various types and dimensions. Large sand waves (dunes) a few meters to more than 20 m high with wave lengths of tens to hundreds of meters typically occur in fields that may cover areas of 15,000 km^2 or more. Sand waves may have symmetrical cross-sectional shapes if produced by tidal currents with equal ebb and flood peak speeds; however, asymmetrical shapes caused by unequal ebb and flood velocities are more common (Belderson, Johnson, and Kenyon, 1982). Tidal sand ridges similar to the shelf sand ridges present on wave- and storm-dominated shelves are also common. For example, such ridges, covering areas up to 5000 km^2, have been reported from the North Sea shelf (e.g., Swift, 1975). In addition to sand waves and shelf sand ridges, tide-dominated shelves also include sand sheets, sand patches, and gravel sheets, all characterized by small-scale bedforms, and patches of bioturbated muds in areas sheltered from tidal currents and waves (Stride et al., 1982).

Because most of the shelf is constantly covered by water, the characteristics of sand waves, sand ridges, and other bedforms on modern shelves must be studied largely by indirect methods. Small-scale bedforms can be observed and photographed by divers or by remote-controlled cameras. Larger bedforms are investigated by sonar bottom-profiling and side-scan sonar techniques (e.g., Belderson, Johnson, and Kenyon, 1982). Small-scale, internal sedimentary structures can be studied in cores of bottom sediment, and sub-bottom seismic profiling methods, described in Chapter 13, may be used to study some large-scale features such as bedding. None of these methods allows detailed examination of modern shelf structures. The idealized distribution of bedforms along the sediment transport path on tide-dominated shelves is illustrated in Figure 10.8. At high tidal velocities of about 150 cm/s, the seafloor may be eroded, leaving furrows and gravel waves. With progressively diminishing velocity farther down the transport path, eroded sediments are deposited to form flow-parallel sand ribbons, large dunes, small dunes, a rippled sand sheet, and finally sand patches. Sand ridges may form in the dune belt if enough sand is present.

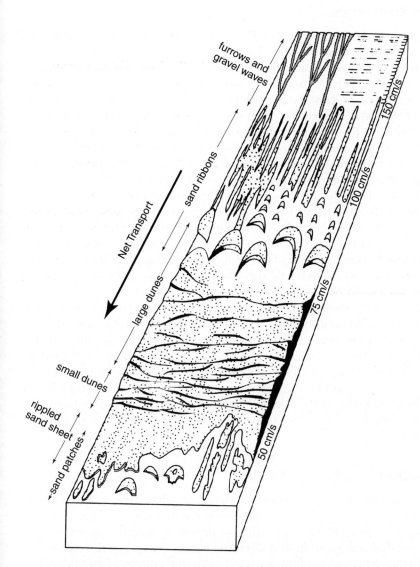

Figure 10.8
Idealized sequence of bedforms developed along a sediment transport path on a tide-dominated shelf. Maximum spring-tide current velocities associated with each bedform type are shown along the edges of the diagram. Sand ridges may form in the dune belt if sufficient sand is present. [After Belderson, R. H., M. A. Johnson, and N. H. Kenyon, 1982, Bedforms, *in* Stride, A. H. (ed.), Offshore tidal sands: Chapman and Hall, London, Fig. 3.1, p. 28. Reproduced by permission.]

Most tidal shelf sands are characterized by cross-bedding. Small-scale cross-bedding and ripple cross-lamination, produced by migration of ripples and small dunes, and large-scale cross-bedding generated by migration of dunes and sand ridges are both common. Foreset dip directions of cross-lamination may be bi-directional or unidirectional depending upon tidal influence. Plane beds also develop under some upper-flow-regime flow conditions. Physical structures thus tend to dominate tidal shelf sands, which typically display fewer bioturbation structures than do muddy shelf sediments (deposited under other conditions), which are commonly highly bioturbated with few physical structures except possibly planar lamination.

Shelves Affected by Intruding Ocean Currents

Major surface ocean currents flow through the world oceans like gigantic rivers, driven by the prevailing wind system around Earth (e.g., Open University Course Team, 1989). These prevailing winds, in conjunction with the configuration of the continents and the Coriolis force arising from Earth's rotation, drive the ocean currents into gigantic circulating cells, or gyres, that rotate clockwise in the Northern Hemisphere (North Pacific and North Atlantic) and counterclockwise in the Southern Hemisphere (South Pacific, South Atlantic, Indian Ocean). The Gulf Stream System off the southeast and east coast of the United States, the Kuroshio

Current off the Asian coast, and the Agulhas Current off the southeast tip of Africa are examples of some prominent ocean currents.

These semipermanent ocean currents intrude onto some shelves with sufficient bottom velocity to transport sandy sediment. About 3 percent of modern shelves are dominated by these ocean currents, which operate most effectively on the outer shelf. Modern examples of such shelves include the northwestern Gulf of Mexico, which is affected by the Gulf Stream System; shelves swept by the Panama and North Equatorial Current off the northeast coast of South America; the Taiwan Strait between Taiwan and mainland China, which is intruded by a branch of the Kuroshio Current flowing north from the Philippines; and the outer shelf of southern Africa, which is crossed by the southward-flowing Agulhas Current of the western Indian Ocean. These currents commonly contribute little if any new sediment to the shelf, but they are capable of transporting significant volumes of fine sediment along the shelf. Some achieve bottom velocities great enough to transport sandy sediment and create sand waves and other bedforms, for example the Kuroshio Current (Boggs, Wang, and Lewis, 1979) and the Agulhas Current (Fleming, 1980).

Sediment on such shelves raked by intruding ocean currents is largely relict, but it is commonly reworked by intruding currents to form sand waves (dunes), sand ribbons, and coarse sand and gravel lag deposits. The best-documented example of this type is the southeastern shelf of South Africa, which is intruded by the Agulhas Current of the Indian Ocean (Fleming, 1980). Sand waves up to 17 m high with wave lengths up to 700 m occur in sand-wave fields as much as 10 km wide and 20 km long (Fig. 10.9). The Taiwan Strait between Taiwan and China is another broad shelf invaded by ocean currents that create extensive sand-wave fields (Boggs, 1974).

Shelf Transport by Density Currents

Density currents are created by density differences within water masses. The buoyant plumes and underflows described above and illustrated in Figure 10.5, as well as nepheloid flows, are all density currents driven by density differences arising from suspended sediment. In arid climates, excessive nearshore evaporation may generate dense brines that flow seaward along the bottom as an underflow. Such underflows could conceivably occur also on shelves where cold surface waters sink and flow seaward. Density currents generated as a result of variations in

Figure 10.9
Sediment transport by the Agulhas Current off the southeastern tip of Africa. Sand in the current-controlled central shelf (B) migrates under the influence of the Agulhas Current; sand-wave fields are up to 20 km long and 10 km wide, and individual sand waves are up to 17 m high. Black streaks indicate sand ribbons. The stippled pattern indicates coarse lag deposits in the sand-depleted outer shelf (C). The nearshore sediment wedge (A) is dominated by wave processes. [From Fleming, B. W., 1980, Sand transport and bedforms on the continental shelf between Durban and Port Elizabeth (southeast Africa continental margin): Sed. Geology, v. 26, Fig. 15, p. 194, reproduced by permission of Elsevier Science Publishers, Amsterdam.]

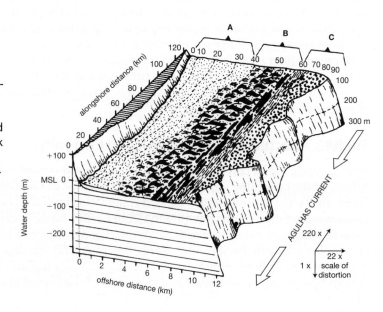

temperature or salinity are not significant agents of shelf transport, and no modern shelf is dominated by such processes.

Effects of Sea-Level Change on Shelf Transport

Because geographic environments shift rapidly and change their form during sea-level fluctuations, sea-level changes constitute an important, and sensitive, depositional variable on shelves. They can affect both erosional and depositional processes and thus the kinds of sediments deposited on shelves. Among other things, sea-level changes are an important factor in establishing the stratigraphic architecture of shelf sediments, a topic that we explore in Chapter 13. See also Walker and James (1992) and Reading and Levell (1996).

Biological Activities on Shelves

Modern continental shelves are among the environments most densely populated by organisms, and the geologic record suggests that ancient epeiric seas were also inhabited by large populations of organisms. The shelf floor is habitat for highly diverse invertebrate organisms such as molluscs, echinoderms, corals, sponges, worms, and arthropods. Both infauna and vagrant and sessile epifauna are represented. Organisms are most abundant in lower-energy areas of the shelf, and the greatest populations occur on the inner shelf just below the wave base.

Organisms on siliciclastic-dominated shelves are particularly important as agents of bioturbation. Both type and abundance of bioturbation structures vary with sediment type and water depth. As discussed in Chapter 4, many burrows in the nearshore high-energy zone are escape structures that tend to be predominantly vertical. The burrow style changes to oblique or horizontal feeding structures with deepening of water across the shelf. In general, muddy sediments of the shelf are more highly bioturbated than are sandy sediments, and physical sedimentary structures in these sediments may be almost completely obliterated by bioturbation. By contrast, only a few species of organisms can survive in the very high energy nearshore shelf and beach zone. Therefore, sandy sediments of the beach–shelf transition zone are dominated by physical structures such as cross-bedding rather than bioturbation structures. Nonetheless, some bioturbation structures may be present if they escape destruction by reworking. Sandy layers deposited in deeper water on the shelf may be bioturbated to some degree in their upper part. In addition to their importance as bioturbation agents, some organisms produce fecal pellets from muddy sediment; these pellets may become hardened and coherent enough to behave as sand grains. Organisms with shells or other fossilizable hard parts also leave remains that may be preserved to become part of the sediment record.

Ancient Siliciclastic Shelf Sediments

Although recognition of ancient shelf sediments is aided by study of modern continental shelves, modern shelves are not necessarily good analogs of ancient shelves. For example, the prevalence of relict sediments on modern shelves may be atypical. Conversely, some structures believed to be diagnostic of ancient shelf sediments, notably storm-generated hummocky cross-stratification, have apparently not been recognized in modern shelf environments. In general, ancient shelf sediments appear to be distinguished by the following: (1) tabular shape, (2) extensive lateral dimensions (thousands of square kilometers) and great thickness (hundreds of meters), (3) moderate compositional maturity of sands with quartz dominating feldspars and rock fragments, (4) generally well-developed, even, laterally extensive bedding, (5) storm beds in some shelf deposits, (6) wide diversity and abundance of normal marine fossil organisms, and (7) diagnostic associations of trace fossils.

More specific characteristics are related to deposition under tide-dominated or storm-dominated conditions. The deposits of ancient tide-dominated shelves are characterized particularly by cross-bedded sandstone. Paleocurrents are mainly unimodal, apparently because of ancient regional net-sediment-transport paths, although bipolar cross-stratification is present locally (Dalrymple, 1992). Reactivation surfaces are abundant. Ancient storm-dominated shelf deposits likely contain a greater proportion of mud than do tide-dominated deposits, and hummocky cross-stratification and storm layers are common (but not in modern deposits!).

Several kinds of vertical successions may thus be generated in shelf sediments, depending upon whether deposition takes place during transgression or regression and depending upon the dominant type of shelf processes operating during deposition. It is difficult to generalize about these successions except to say that transgression tends to produce fining-upward successions that may begin with coarse lag deposits, and regression produces coarsening-upward successions. Some idealized vertical shelf successions produced under different postulated sedimentation conditions are illustrated in Figure 10.10. These successions should be considered only as working models. Actual transgressive and regressive successions may differ markedly in detail from these idealized profiles.

Ancient shelf deposits are known from stratigraphic units of all ages and all continents. They are probably the most extensively preserved rocks in the geologic record. Readers interested in pursuing case histories of specific shelf deposits can consult the extensive list of pertinent references provided by Leeder (1999, p. 463); these references include some of the best accounts of ancient shelf deposits on the basis of study of Cretaceous sediments deposited in the Western Interior Seaway of North America. Additional accounts of these deposits, derived from study of magnificent exposures stretching from Colorado to Alberta, can be found in Bergman and Snedden (1999).

One example of shelf deposits from the Western Interior Seaway is shown in Figure 10.11. Banerjee (1991) describes Cretaceous sandstones of southern Alberta, Canada, which he interprets as tidal sand sheet facies. One of these is the (lowermost) Sunkay Member of the Lower Cretaceous Alberta Group. This sandstone unit, approximately 10 m thick, rests with a sharp erosional base on nonmarine

Figure 10.10

Idealized diagrams illustrating typical fining-upward transgressive shelf successions on (A) a tide-dominated shelf and (B) a storm-dominated shelf, and a coarsening-upward regressive shelf succession (C) on a storm-dominated shelf. [After Galloway, W. E., and D. K. Hobday, 1983, Terrigenous clastic depositional systems. Fig. 7.14, p. 159, Fig. 7.15, p. 160, Fig. 7.17, p. 162, reprinted by permission of Springer-Verlag, Heidelberg.]

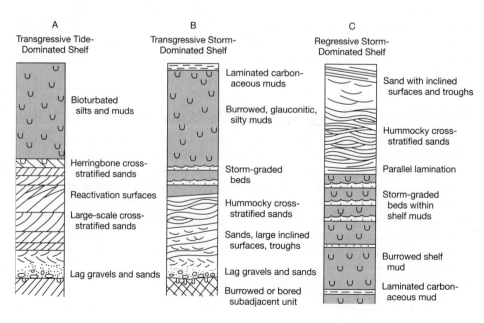

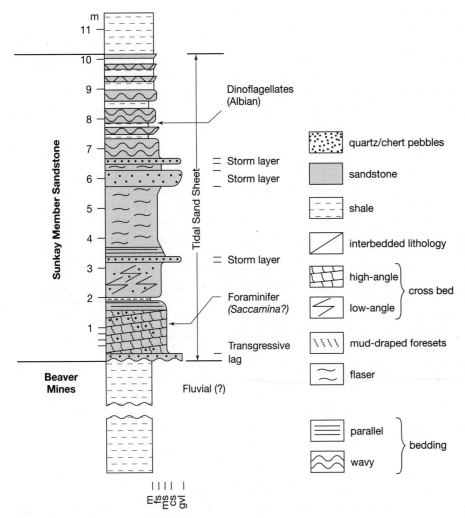

Figure 10.11
Vertical succession of sandy tidal shelf deposits in the Sunkay Sandstone Member of the Lower Cretaceous Alberta Group, southern Alberta, Canada. Symbols in the grain-size scale are gvl = gravel, cs = coarse sand, ms = medium sand, fs = fine sand, and m = mud (silt-clay). [After Banerjee, I., 1991, Tidal sand sheets of the late Albian Joli Fou-Kiowa-Skull Creek marine transgression, Western Interior Seaway of North America in Smith, R. G., et al. (eds.), Clastic tidal sedimentology: Canadian Soc. Petroleum Geologists Mem. 16, Fig. 9, p. 334.]

shales of the Beaver Mines Formation and is overlain by black marine shales. The Sunkay Member consists mainly of sandstone, with a thin lag gravel deposit at the base and a few thin interbedded conglomerate layers, interpreted as storm layers. The succession as a whole fines upward. The basal sandstone units are cross-bedded; the uppermost ones are characterized by flaser and wavy bedding. Foraminifers and dinoflagellates are present in some layers. Bioturbation is not common.

Banerjee interprets these sandstones as tidal sand sheets. Tidal interpretation is based in part on the presence of the following: cross-stratification (>20° dip and sets 30–40 cm thick), with thin shale drapes adhering to foreset surfaces, in the lower part of the section; and flaser, wavy, and lenticular bedding in the upper part. Transgression deepened the water at this site, causing black marine shales to be deposited on top of the tidal sandstones.

10.3 THE OCEANIC (DEEP-WATER) ENVIRONMENT

Introduction

In the discussion of depositional environments to this point, I have focused on the continental, marginal-marine, and shallow-marine environments because much of the preserved sedimentary record was deposited in these environments. In terms of size of the environmental setting, however, these nonmarine and shallow-water marine environments actually cover a much smaller area of Earth's surface than do deep-water environments. By far, the largest portion of Earth's surface lies seaward of the continental shelf in water deeper than about 200 m. Approximately 65 percent of Earth's surface is occupied by the continental slope, the continental rise, deep-sea trenches, and the deep ocean floor. Even so, most textbooks that discuss sedimentary environments typically give only modest coverage to oceanic environments. This bias probably exists because, as a whole, deep-water sediments are much more poorly represented in the exposed rock record than are shallow-water sediments. Deep-water deposits are less abundant than shallow-water deposits in the exposed rock record because sedimentation rates overall are slower in deeper water; thus, the sediment record is thinner. [An exception is the submarine fan environment near the base of the slope where sedimentation rates from turbidity currents can exceed 10 m/1000 yr and turbidite sediments can achieve thicknesses of thousands of meters (e.g., Bouma, Normark, and Barnes, 1985)]. Also, part of the sediment record of the deep seafloor may have been destroyed by subduction in trenches, and those deep-water sediments that have escaped subduction have required extensive faulting and uplift to bring them above sea level where they can be viewed. Deepwater sediments other than turbidites have not been studied as thoroughly as shallow-water sediments—perhaps in part because deep-water sediments have less economic potential for petroleum. Owing to the advent of seafloor spreading and global plate tectonics concepts, however, the deep seafloor has taken on enormous significance for geologists. Consequently, intensive research has been focused on the continental margins and deep seafloor since the early 1960s. Also, the continuing need to add to our fossil fuel reserves is pushing petroleum exploration into deeper and deeper water, and the possibility of mining manganese nodules and metalliferous muds from the seafloor is also causing increased economic interest in the deep ocean.

Deep-sea research has been particularly stimulated by the Deep Sea Drilling Program (DSDP), which began in 1968 and shifted to the Ocean Drilling Program (ODP) in 1984. Since initiation of these programs, several hundred holes have been drilled by DSDP and ODP teams throughout the ocean basins of the world to an average depth below seafloor of about 300 m (~1000 ft) and to maximum depths exceeding 1000 m. In addition to deep coring by DSDP and ODP, many thousands of shallow piston cores have been collected from the seafloor throughout the ocean by marine geologists from major oceanographic institutions of the world. Also, hundreds of thousands of kilometers of seismic profiling lines (Chapter 13) have been run in crisscross patterns across the ocean floor in an attempt to unravel the sub-bottom structure of the ocean. Although much of this research has been aimed at understanding the larger scale features of the ocean basins that illuminate the origin and evolutionary history of the ocean basins along plate tectonics concepts, many data on sedimentary facies and sedimentary environments have also been collected. Much additional new information on ocean circulation and sediment transport systems has also been generated by oceanographers who study ocean-bottom currents and bottom-water masses. Thus, a significant increase in understanding of the ocean basins and the deep ocean floor has come about since the 1950s. We shall concentrate discussion here on the fundamental processes of sediment transport and deposition on continental slopes and the deep ocean floor and the principal types of facies developed in these environments.

Depositional Setting

Continental Slope

The continental slope extends from the shelf break, which occurs at an average depth of about 130 m in the modern ocean, to the deep seafloor (Fig. 10.12). The

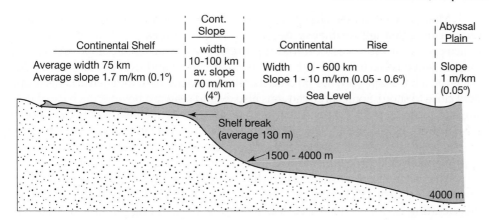

Figure 10.12
Principal elements of the continental margin. [After Drake, C. L., and C. A. Burk, 1974, Geological significance of continental margins, *in* Burk, C. A., and C. L. Drake (eds.), The geology of continental margins, Fig. 9, p. 8, reprinted by permission of Springer-Verlag, Heidelberg.]

lower boundary is typically located at water depths ranging from about 1500 to 4000 m, but locally in deep trenches it may extend to depths exceeding 10,000 m. Continental slopes are comparatively narrow (10–100 km wide), and they dip seaward much more steeply than does the shelf. The average inclination of modern continental slopes is about 4°, although slopes may range from less than 2° off major deltas to more than 45° off some coral islands.

The origin and internal structure of continental slopes are not of primary concern here; however, a brief description of differences in the characteristics of continental slopes on passive (Atlantic-type) and active (Pacific-type) continental margins is pertinent to succeeding discussion. As shown in Figure 10.4, various kinds of barriers form the boundary between the continental shelf and the continental slope. The principal kinds of passive continental margins can include margins where siliciclastic sediments drape over basement ridges, folds, or faults (Fig. 10.4 A, B, C); margins with a carbonate bank or platform (Fig. 10.4D), and margins dominated by salt tectonics (flowage of salt to produce salt domes and diapirs; Fig. 10.4F). More than one of these passive-margin types may be present within a given geographic area. Active continental margins may be characterized by the presence of a fore-arc region only or by both fore-arc and back-arc regions, as, for example, the Japan margin (Fig. 10.13). Sedimentation can take place in both back-arc and fore-arc basins, on back-arc and fore-arc slopes, and in the fore-arc trench.

Continental slopes may have a smooth, slightly convex surface morphology, such as that found on passive siliciclastic margins (e.g., Fig. 10.4A) or they may be irregular on a small to very large scale. Active-margin slopes tend to be particularly irregular. For example, the Pacific slope off Japan, which descends to a depth of about 7000 m into the Japan Trench, is characterized by structural terraces and basins together with anticlinal welts and fault-bounded ridges arranged in an echelon pattern roughly parallel to the Japan coast (Boggs, 1984). These ridges and folds form prominent structural "dams" behind which sediments are ponded. In general, structural barriers on highly irregular slopes can inhibit movement of bottom sediment across the slope and create catchment basins for sediment.

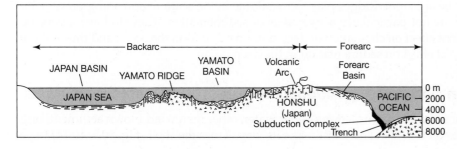

Figure 10.13
Schematic representation of an active continental margin (Japan), showing both the fore-arc and back-arc characteristics of the margin. [From Boggs, S., Jr., 1984, Quaternary sedimentation in the Japan arc-trench system: Geol. Soc. America Bull., v. 95, Fig. 2, p. 670.]

Modern continental slopes are gashed to various degrees by submarine canyons oriented approximately normal to the shelf break (see Fig. 10.16, 10.19), which provide accessways for turbidity currents moving across the slope. Most submarine canyons have their heads near the slope break and do not cross the shelf; however, a few major canyons on modern shelves extend onto the shelf and head up very close to shore. Some large canyons also extend seaward beyond the base of the slope to form deep-sea channels that may meander over the nearly flat ocean floor for hundreds of kilometers. The Toyama Deep Sea Channel in the Japan Sea, for example, winds its way across the seafloor from the mouth of the Toyama Trough for approximately 500 km before emptying onto the Japan Sea abyssal plain.

The origin of submarine canyons has been debated since the early part of the twentieth century (Pickering, Hiscott, and Hein, 1989, p. 134–136). Although downcutting by rivers that extended across the shelf during periods of lowered sea level may have initiated the formation of some canyons on the shelf, turbidity currents are the main agents of canyon cutting on the slope and deeper seafloor. Canyon development may be initiated by local slope failure (slumping), followed by headward growth of erosional scars. Turbidity currents are erosive in their initial stages and thus can deepen and lengthen the incipient canyons over time, aided by further slumping on the upper part of the slope. The locations and shapes of some submarine canyons may have been influenced by the presence of faults and folds (Green, Clarke, and Kennedy, 1991).

Continental Rise and Deep Ocean Basin

The continental rise and deep ocean basin encompass that part of the ocean lying below the base of the continental slope. Together, they make up about 80 percent of the total ocean seafloor. The deeper part of the ocean seaward of the continental slope is divided into two principal physiographic components: the deep **ocean floor,** which is characterized by the presence of abyssal plains, abyssal hills (volcanic hills <1 km high), and seamounts (volcanic peaks >1 km high); and **oceanic ridges**. Off passive continental margins, a **continental rise** (Fig. 10.1) is present at the base of the slope. The continental rise is a gently sloping surface that leads gradually onto the deep ocean floor and is built in part from submarine fans extending seaward from the foot of the slope. It commonly has little relief other than that resulting from incised submarine canyons and protruding seamounts. Continental rises are generally absent on convergent or active margins where subduction is taking place, such as along much of the Pacific margin. On margins of this type, a long, arcuate **deep-sea trench** commonly lies at the foot of the continental slope, and the rise is absent. Trenches in less active subduction zones, such as along the Oregon-Washington coast, may be filled with sediment. Abyssal plains are extensive, nearly flat areas punctuated here and there by seamounts. Some abyssal plains are also cut by deep-sea channels, as mentioned. **Mid-ocean ridges** extend across some 60,000 km of the modern ocean and overall make up about 30–35 percent of the area of the ocean. Mid-ocean ridges are particularly prominent in the Atlantic, where they rise about 2.5 km above the abyssal plains on either side. Rocks on these ridges are predominantly volcanic; the ridges are cut by numerous transverse fracture zones along which significant lateral displacements may be apparent. Ridges play a crucial role in the seafloor spreading process, but they are not particularly active areas of sedimentation. They do have a very important effect on circulation of deep bottom currents in the ocean and thus have an indirect effect on sedimentation in the deep ocean.

Transport and Depositional Processes to and within Deep Water

Most sediment deposited in deeper water, other than wind-blown sediment, originates on the shelf and must make its way across the shelf (Fig. 10.5, 10.14) to get

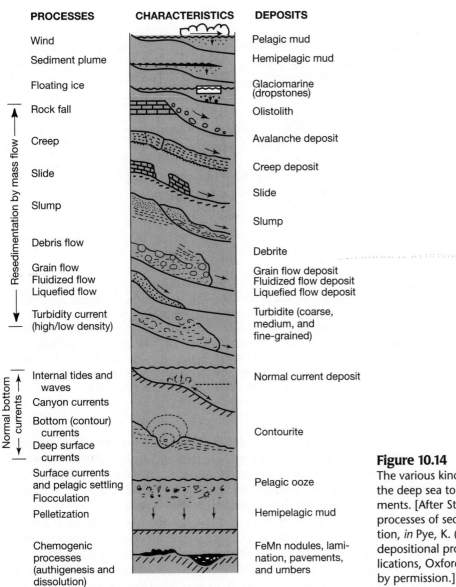

PROCESSES	CHARACTERISTICS	DEPOSITS
Wind		Pelagic mud
Sediment plume		Hemipelagic mud
Floating ice		Glaciomarine (dropstones)
Rock fall		Olistolith
Creep		Avalanche deposit
Slide		Creep deposit
Slump		Slide
		Slump
Debris flow		Debrite
Grain flow / Fluidized flow / Liquefied flow		Grain flow deposit / Fluidized flow deposit / Liquefied flow deposit
Turbidity current (high/low density)		Turbidite (coarse, medium, and fine-grained)
Internal tides and waves		Normal current deposit
Canyon currents		
Bottom (contour) currents / Deep surface currents		Contourite
Surface currents and pelagic settling / Flocculation		Pelagic ooze
Pelletization		Hemipelagic mud
Chemogenic processes (authigenesis and dissolution)		FeMn nodules, lamination, pavements, and umbers

(left margin, top-to-bottom: Resedimentation by mass flow; Normal bottom currents)

Figure 10.14
The various kinds of processes that operate in the deep sea to transport and deposit sediments. [After Stow, D. A. V., 1994, Deep sea processes of sediment transport and deposition, *in* Pye, K. (ed.), Sediment transport and depositional processes, Blackwell Scientific Publications, Oxford, Fig. 8.2, p. 261, reproduced by permission.]

to deeper water environments. Across-shelf sediment movement (discussed under shelf transport in preceding sections) includes transport of coarser sediment by turbidity currents and seaward advection of fine sediment by sediment plumes and nepheloid flows. A variety of processes is capable of transporting and depositing sediment within the deep ocean, such as wind transport from continents, airfall and submarine settling of pyroclastic particles generated by explosive volcanism within and outside ocean basins, sediment plumes, floating ice, mass-flow processes, various kinds of bottom currents, surface currents, and pelagic settling (e.g., Stow, 1994). These processes are summarized in Figure 10.14. The locations within the deep ocean where the various processes operate are summarized more graphically in Figure 10.15.

Sediment Plumes, Wind Transport, Ice Rafting, Nepheloid Transport

Where continental shelves are narrow, freshwater surface plumes carrying fine sediment across the shelf (Fig. 10.5) can move considerable distances into deeper water, possibly as far as 100 km offshore (Reineck and Singh, 1980), before mixing

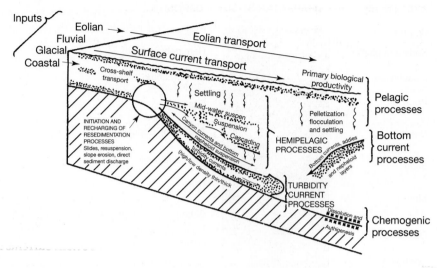

Figure 10.15
Schematic representation of principal processes responsible for transport and deposition of sediments to the deep ocean. Note that most of the processes deposit fine sediment; however, glacial (floating ice), turbidity current, and resedimentation processes can move both coarse and fine sediment. Chemogenic refers to minor processes that are largely chemical in nature. [After Stow, D. A. V., H. G. Reading, and J. D. Collinson, 1996, Deep seas, *in* Reading, H. G. (ed.), Sedimentary environments: Processes, facies and stratigraphy, Blackwell Science Ltd., Oxford, p. 395–453, reproduced by permission.]

and flocculation cause clay particles to settle. Winds blowing over continents, particularly desert areas, can also transport fine suspended dust particles seaward, where they settle out over the ocean hundreds of kilometers from shore. In fact, wind transport may be the primary mechanism by which clay-size siliciclastic sediment is transported to the distal part of the deep ocean.

During glacial episodes of the Pleistocene when sea levels were low and many land areas were covered by ice, rafting of sediment of all sizes into deeper water by icebergs was a particularly important transport process. Ice transport is still going on today on a more limited scale at high latitudes in the Arctic and Antarctic regions (e.g., Anderson and Ashley, 1991; Dowdeswell and Scourse, 1990; Kempema, Reimnitz, and Barnes, 1988). Melting of the floating ice dumps sediments of mixed sizes, commonly referred to as glacial-marine sediment (Chapter 9), onto the shelf and the deep ocean floor. The overall quantitative significance of iceberg transport into deep water through geologic time has probably not been significant, but locally and at certain times it may have been important.

Fine sediment resuspended by storms on the outer shelf can move off the shelf and down the slope in near-bottom nepheloid suspension. Less-dense suspensions may move seaward along density interfaces as midwater suspensions that gradually settle to the bottom (Fig. 10.15). Fine sediment may also be injected into the water column by turbidity flows moving downslope. Nepheloid flows are reported to extend seaward for hundreds of kilometers and to water depths of 6000 m or more. Deep bottom currents (to be discussed) may aid in resuspending sediment into nepheloid layers.

Currents in Canyons

Tidal currents measured in submarine canyons at depths exceeding 1000 m may be capable of transporting silt and fine sand (Shepard, 1979; Pickering, Hiscott, and Hein, 1989, p. 143). Two types of currents have been detected in submarine valleys:

ordinary tidal currents that rarely exceed 50 cm/s and that flow alternately up and down the valley in response to tidal reversal, and occasional surges of strong downcurrent flow with velocities up to 100 cm/s. Shepard (1979) interprets the surges as low-velocity turbidity currents. There are few data as yet to support the quantitative importance of net down-canyon transport owing to tidal currents; however, surge currents of the magnitude measured by Shepard are certainly capable of transporting fine sediment seaward. Together, these currents probably help to winnow canyon deposits and keep them free from fine sediment.

Contour Currents

Density differences in surface ocean water caused by temperature or salinity variations create vertical circulation of water masses in the ocean commonly referred to as **thermohaline circulation**. Circulation is initiated primarily at high latitudes as cold surface waters sink toward the bottom, forming deep-water masses that flow along the ocean floor as bottom currents. The path of these bottom currents is influenced by the position of oceanic ridges and rises and other topographic features such as narrow passages through fracture zones. Owing to density stratification of ocean water, bottom currents adjacent to continental margins tend to flow parallel to depth contours or isobaths and thus are often called **contour currents**. The movement of these currents is also affected by the Coriolis force, which likewise tends to deflect them (left in the Southern Hemisphere and right in the Northern Hemisphere) into paths parallel to depth contours; thus, they are sometimes also called geostrophic contour currents.

In the modern ocean, Antarctic bottom water runs down the continental slope, circulates eastward around the Antarctic continent possibly several times, and then flows northward into the Atlantic, Indian, and Pacific oceans (Stow, 1994). In the North Atlantic, deep bottom water flows south out of the Norwegian-Greenland seas, Labrador Sea, and other parts of the North Atlantic. Interaction of these deep-water masses creates a highly complex ocean circulation system.

Because contour currents are best developed in areas of steep topography where the bottom topography extends through the greatest thickness of stratified water column (Kennett, 1982), they are particularly important on the continental slope and rise. Photographs of the deep seafloor have revealed current ripples in some areas and suspended sediment clouds and seafloor erosional features in others, both of which suggest that some contour currents can achieve velocities on or near the seafloor great enough to erode the seabed and transport sediment. Evidence is now available (e.g., Hollister and Nowell, 1991) which suggests that the speed of these bottom currents may be accelerated in some parts of the ocean to velocities on the order of 40 cm/s, perhaps because of the superimposed influence of large-scale, wind-driven circulation at the ocean's surface. That is, eddy kinetic energy may be transmitted from the surface of the ocean to the deep seafloor. Intensification may also occur where the Coriolis Force causes deep flows to bank up against the continental slope on the western margins of ocean basins, where it is unable to move upslope against gravity and thus becomes restricted and intensified (Stow, 1994). Where bottom currents are intensified, resulting motions near the seafloor are so energetic that they have been referred to as "abyssal storms" or "benthic storms" (Hollister and Nowell, 1991), particularly because huge amounts of fine sediments are stirred up and transported by these energetic pulses. Contour currents are believed to have had a particularly important role in shaping and modifying continental rises, such as those off the eastern coast of North America.

Pelagic Rain

Calcareous- and siliceous-shelled planktonic organisms settle through the ocean water column to the seafloor upon death, a process called pelagic rain or pelagic settling. The geographic distribution of these organisms in surface waters is affected by nutrients and prevailing ocean currents. After the tiny shells settle onto the ocean floor, they may be retransported by turbidity currents or contour currents. These skeletal materials form extensive biogenic deposits, or **oozes,** in some areas of the modern ocean floor. They are also important contributors to ancient deep ocean sediments, particularly in Jurassic and younger rocks, forming thick deposits of chalk, radiolarian chert, and diatomite (Chapters 6 and 7). Fine-size siliciclastic particles (e.g., clay minerals, quartz, feldspars) transported from land by surface plumes, wind, or floating ice may settle along with organic remains to form mixed biogenic and siliciclastic sediment referred to as hemipelagic deposits (Fig. 10.15).

Explosive Volcanism

Volcanism within and along the margins of marine basins may contribute important quantities of sediment to both the shelf and deeper water, particularly near volcanic arcs. Volcanic ash, lapilli, and bombs can be ejected both subaerially and subaqueously. Coarse material ejected subaerially tends to be deposited by air fall close to the eruption column on all sides of the vent. If strong prevailing winds are blowing during eruption, fine ash will be carried downwind for considerable distances before settling. Pyroclastic particles ejected beneath the sea as well as air-fall particles can be dispersed still more widely within the ocean basin by various transport processes. Pumice may even be dispersed to some extent by floating on the ocean surface.

Turbidity Currents and Other Mass-Transport Processes

As indicated in Figure 10.14, a variety of mass-flow processes can take place within the deep ocean. Catastrophic or surge-type, high-velocity turbidity currents generated on the shelf or upper slope are probably the single most important mechanism for transporting sands and gravels to deeper water through submarine channels (e.g., Normark and Piper, 1991). On passive margins and in back-arc basins, the deposits of these flows spread out from the mouths of the canyons onto the deep seafloor to form deep-sea fans (see Fig. 10.16), and they contribute in part to building of the continental rise. In the fore-arc region of active margins, submarine canyons discharge turbidites into fore-arc basins on the slope or into deep-sea trenches where they may spread out along the canyon axis. Down-canyon grain flow of beach sands swept into the heads of nearshore submarine canyons during storms and submarine debris flows may also be locally important. Under some conditions, submarine debris flows may be transformed downslope into turbidity currents (e.g., Piper, Cochonat, and Morrison, 1999). In addition to these processes, other mass-transport processes, such as creep, gliding (sliding), and slumping, appear to be responsible for large-scale en masse retransport of sediments on oversteepened continental slopes and ridge slopes. Some of these slump masses can be of enormous size—up to 300 m thick and 100 km long (e.g., Saxov and Nieuwenhuis, 1982; Bouma, Normark, and Barnes, 1985; Stow, Reading, and Collinson, 1996). Also, some sediment can be retransported off oceanic ridges and seamounts after accumulating there by pelagic rainout or volcanic processes.

Principal Kinds of Modern Deep-Sea Sediments

Little agreement exists regarding the classification of deep-sea sediments. Suggested classifications range from those that are largely genetic (Shepard, 1973; Berger, 1974) to those that are largely descriptive (Dean, Leinen, and Stow, 1985; Pickering, Hiscott, and Hein, 1989). No entirely satisfactory scheme that gives due regard to both genesis and descriptive properties of all kinds of deep-sea sediments

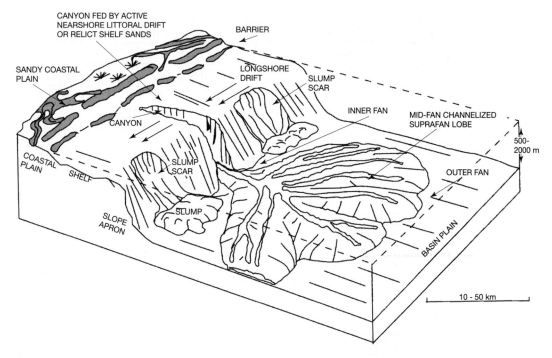

Figure 10.16
Depositional model for a point-source sand-rich submarine fan. [From Reading, H. G., and M. Richards, 1994, Turbidite systems in deep-water basin margins classified by grain size and feeder system: American Association Petroleum Geologists Bull., v. 78, Fig. 7, p. 805, reproduced by permission.]

has yet been devised. Two broad classes of deep-sea sediment, **terrigenous** and **pelagic**, are often mentioned; however, these two terms are difficult to define precisely. Terrigenous deposits include gravel, sand, and mud derived from land and transported within the more proximate parts of the deep ocean by a variety of processes (e.g., turbidity currents, contour currents, ice rafting). Some pelagic deposits (clays) are also derived from land but are deposited by slow settling in the more distal parts of the ocean; others consist of pelagic biogenic remains that rain down from near-surface waters. A third minor category of deeper-water sediments is shallow-water carbonate sediments (Chapter 11) that have been retransported from the shelf into deeper water (**allochthonous deep-water carbonates**). The principal kinds of deep-sea sediments are summarized in Table 10.1; additional details are discussed below.

Terrigenous Sediments

A wide variety of deep-sea siliciclastic sediments are grouped under this heading, including, as shown in Table 10.1, hemipelagic muds, turbidites and other sediment gravity-flow deposits, contourites, glacial-marine sediments, and slumps and slides. These deposits are all composed mainly of siliciclastic materials, which may range in size from gravel to clay. Some deposits exhibit well-developed bedding and may or may not display vertical size grading. Others have a disorganized fabric with poorly developed bedding. For convenience of discussion, these terrigenous deposits are grouped below under mainly genetic subheadings that reflect their principal mode of transport and deposition.

Hemipelagic Muds. Hemipelagic ("half-pelagic") deposits are difficult to define precisely. According to Stow and Piper (1984a), they are muddy deposits that

Table 10.1   Principal kinds of deep-sea sediments

Terrigenous siliciclastic deposits

Hemipelagic mud—mixtures of terrigenous mud and biogenic remains; deposited from
    nepheloid plumes and by suspension settling and pelagic rain-out

Turbidites—graded gravel/sand/mud; deposited by turbidity currents

Contourites—sandy or muddy sediments deposited and/or reworked by contour cur-
    rents

Glacial-marine sediments—Gravel, sand, and mud deposited by ice rafting

Slump and slide deposits—Terrigenous or pelagic deposits emplaced downslope by
    mass-wasting processes

Pelagic deposits

Pelagic clay— >2/3 siliciclastic clay; deposited by suspension settling and authigenic for-
    mation of clay minerals

Oozes— >2/3 planktonic biogenic remains; deposited by pelagic rain-out

    Calcareous—dominantly C_4CO_3 biogenic remains

    Siliceous—dominantly SiO_2 biogenic remains

Allochthonous deep-sea carbonates

    Shallow-water carbonates emplaced downslope by storms or sediment gravity flows

contain more than 5 percent biogenic remains and a terrigenous component of more than 40 percent silt, although some geologists may find this definition too restrictive. They are deposited under very low current velocities (e.g., by suspension settling), and they are probably the principal kind of sediment deposited on continental slopes and in fore-arc basins on slopes. Much of the compendium volume on fine-grained sediments and deep-water processes and facies edited by Stow and Piper (1984b) is devoted to discussion of hemipelagic sediments.

Hemipelagic muds range in color from gray to green and, more rarely, to reddish brown. Textures range from clay to silty, sandy clay. In addition to biogenic remains, they are commonly composed of fine terrigenous quartz, feldspar, micas, and clay minerals and/or volcanogenic sediments such as ash, fine pumice, and palagonite. Volcanic ash or glass may be intermixed with other hemipelagic sediment or concentrated into distinct **tephra** layers that range in thickness from less than 1 cm to more than 25 cm (cf. Boggs, 1984). Granules and pebbles of pumice may occur as isolated fragments, pockets, or distinct layers associated with hemipelagic muds. Hemipelagic muds may contain the remains of siliceous organisms, particularly diatoms, and calcareous organisms such as foraminifers and nannofossils, as well as fine lime muds swept off carbonate platforms into deeper water.

Hemipelagic muds are poorly laminated to massive, and generally are moderately to highly bioturbated. In general, they are deposited closer to shore than are pelagic muds. They are widely distributed on continental slopes of volcanic arcs, such as those of the Western Pacific. They also occur in back-arc basins, on inner trench walls, and on the tops of some rises. Hemipelagic muds are probably deposited mainly from nepheloid layers and plumes. Deposition may be aided in some settings by planktonic organisms that aggregate fine sediment into fecal pellets that settle rapidly.

Turbidites.   The general characteristics of turbidites are described in Chapter 4. Turbidites may occur in the lower reaches of submarine canyons and farther seaward in deep-sea channels, but most are deposited in broad, cone-shaped fans. These turbidite fans, or submarine fans, spread outward on the seafloor from the mouths of canyons. Where submarine canyons are closely spaced along the slope, the fans at the base of the slope may coalesce to build a broad, gently sloping continental rise. On active margins where a trench is present, turbidites commonly occur on the trench floor throughout the length of the trench owing to deposition from turbidity currents flowing longitudinally through the trench. Most turbidites are composed of sands, silty sands, or gravelly sands interbedded with pelagic clays. They are commonly characterized by normal size grading and may or may not display complete Bouma sequences (see Fig. 2.8). Figure 10.17 is an example of a sediment core composed of turbidites. As discussed in Chapter 2, many turbidites lack either the basal (A unit) or upper (D unit) part of the Bouma sequence, or both. Sole markings, such as flute casts, groove casts, and load casts, are common on the base of many turbidite sequences. In addition to sandy turbidites, mud turbidites occur on many parts of the modern ocean floor. These turbidites are composed of normally graded silt and clay that may be either laminated or massive and that commonly lack extensive bioturbation. The depositional processes that generate fine-grained turbidites and the characteristics of these deposits are described in detail in Bouma and Stone (2000). A comparison of the characteristics of mud-rich and sand-rich turbidites is provided by Bouma (2000). Turbidites are widely distributed in the modern ocean on passive margins and in both the back-arc and the fore-arc regions of active margins.

Because the main depositional environments of turbidites are submarine fans, and most modern turbidites occur in such fans, geologists have displayed

Figure 10.17
Graded volcaniclastic turbidite with Bouma divisions marked, from an Ocean Drilling Program (ODP) Leg 127 core of Miocene sediments in the Japan Sea back-arc basin. [Photograph courtesy of Ocean Drilling Program, Texas A & M University.]

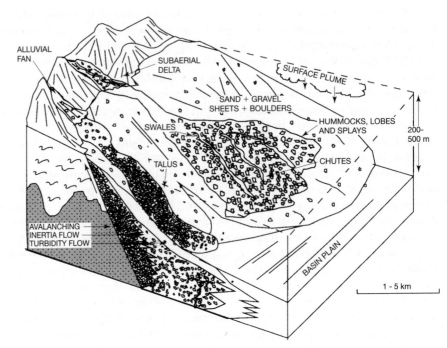

Figure 10.18
Depositional model for a point-source gravel-rich submarine fan. [From Reading, H. G., and M. Richards, 1994, Turbidite systems in deep-water basin margins classified by grain size and feeder system: American Association Petroleum Geologists Bull., v. 78, Fig. 9, p. 807, reproduced by permission.]

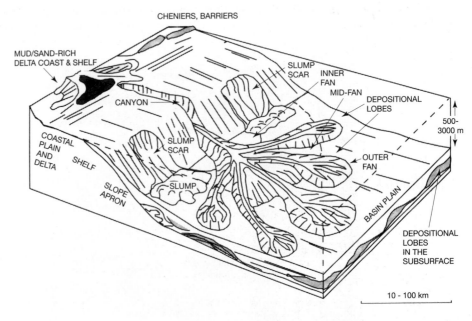

Figure 10.19
Depositional model for a point-source mud/sand-rich submarine fan. [From Reading, H. G., and M. Richards, 1994, Turbidite systems in deep-water basin margins classified by grain size and feeder system: American Association Petroleum Geologists Bull., v. 78, Fig. 5, p. 803, reproduced by permission.]

considerable interest in developing models for submarine fan sedimentation (e.g., Normark, 1978; Walker, 1978, 1992; Mutti, 1985). Reading and Richards (1994) point out that submarine fan systems may originate from point sources or linear sources and can include gravel-rich, sand-rich, mud/sand-rich, and mud-rich systems. They suggest that sand-rich systems are characterized by braided channels and channelized sediment lobes (Fig. 10.16). Gravel-rich systems display chutes but lack distinct sediment lobes (Fig. 10.18). Mud/sand-rich systems are characterized by channel levees, and distinct depositional lobes (Fig. 10.19); mud-rich systems may contain sheetlike deposits in addition to channel levees and lobes. Linear-source submarine fan systems differ from the point-source models shown in Figures 10.16, 10.18, and 10.19 in that they form coalescing lobes along the base of a slope. Miall (2000) and Galloway (1998) further discuss fans and other slope and base-of-slope depositional systems.

Modern submarine fans occur in many parts of the world's ocean. The largest of these fans include the Amazon fan in the South Atlantic (330,000 km^2 surface area; Manley and Flood, 1988); the Rhone fan (70,000 km^2) in the Mediterranean Sea (Droz and Bellaiche, 1985); the Indus fan (1,100,000 km^2) off the coast of Pakistan and India (Kolla and Coumes, 1985); the Laurential fan (300,000 km^2) off the Grand Banks of New Foundland (Piper, Stow, and Normark, 1984); and the Mississippi fan (300,000 km^2) in the Gulf of Mexico (Weimer, 1989). Pickering et al. (1995) discuss additional examples.

Attempts to relate the characteristics of modern fans to ancient turbidite systems are complicated by problems of scale (many features on modern fans are much larger than corresponding features in ancient fan systems) and by the nature of the features themselves (e.g., the abundance of channel-levee systems and the scarcity of depositional lobes in modern fans). For example, the presence of thickening- and thinning-upward successions in ancient turbidites has commonly been attributed respectively to progradation of fan lobes and gradual channel filling and abandonment. This interpretation is difficult to reconcile with the scarcity of lobes in modern fans (Walker, 1992).

Contourites.   Coring of sediments on continental rises has shown that in addition to turbidites, rise deposits also include sediments interpreted to be the deposits of contour currents. Several different countourite facies are recognized, such as muddy, silty, sandy, and gravel-rich (Stow et al., 1998). A composite contourite facies model is illustrated in Figure 10.20. This model shows overall negative grading

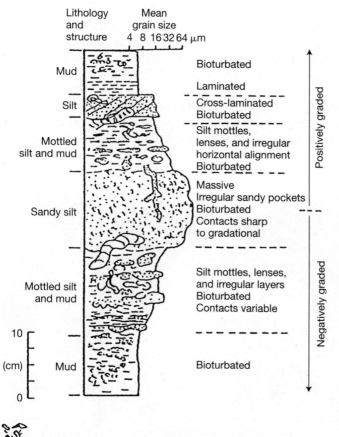

Figure 10.20

Composite contourite facies model showing grain-size variations and sedimentary structures through a mud-silt-sand contourite succession. [From Stow et al., 1998, Fossil contourites: A critical review: Sedimentary Geology, v. 105, Fig. 2, p. 12. Reproduced by permission of Elsevier Science.]

(coarsening upward) from muddy through silty to sandy contourites followed up section by positive grading (fining upward) back through silty to muddy contourite facies. Muddy contourites tend to be homogeneous, poorly bedded, and highly bioturbated. Silty contourites are also bioturbated and commonly display a mottled appearance. Sandy contourites occur as thin, irregular layers within the finer grained facies and are also commonly bioturbated, although some primary horizontal and cross-lamination may be preserved. Contourites may have either sharp or gradational bed contacts. The composition of contourites can include both terrigenous constituents and biogenic components. Pickering, Hiscott, and Hein (1989, p. 219–245); Stow, Reading, and Collinson (1996); and Stow et al. (1998) provide further discussion of contourites.

As discussed in a preceding section ("Contour Currents"), recent work has shown that modern contour currents develop velocities high enough to stir up and transport huge quantities of fine sediments during what is called abyssal or benthic "storms" (Hollister and Nowell, 1991). It thus appears that transport and deposition of sediment by contour currents may be even more important than previously believed. Contourites have been reported in the modern ocean from numerous places, including the continental rise of eastern North America, the continental margin off northwest Britain, and the southern Brazil margin; ancient contourites have been described from many continents (e.g., Stow et al., 1998).

Glacial-Marine Sediments.   Sediments ice-rafted to deep water are typically poorly sorted gravelly sands or gravelly muds that show crude to well-developed stratification. The coarse fraction may include angular, faceted, and striated pebbles. Significant areas of the modern ocean floor at high latitudes are covered by these glacial-marine sediments, particularly the subpolar North Atlantic, the circum-Antarctic, some parts of the Arctic Ocean, the North Pacific, and the Norwegian Sea (Fig. 10.21). Glacial-marine deposits are discussed in further detail under "Glacial Systems" in Chapter 9.

Slump and Slide Deposits.   These deposits consist of previously sedimented pelagic or terrigenous deposits that have been emplaced downslope owing to mass-movement processes. During the transport process, consistency of the slump masses is

Figure 10.21
Distribution and dominant types of deep-sea sediments in the modern ocean. [From Davies, T. A., and D. S. Gorsline, 1976, Oceanic sediments and sedimentary processes, *in* Riley, J. P., and R. Chester (eds.), Chemical oceanography, v. 5, 2nd ed., Fig. 24.7, p. 26, reprinted by permission of Academic Press, Orlando, Fla.]

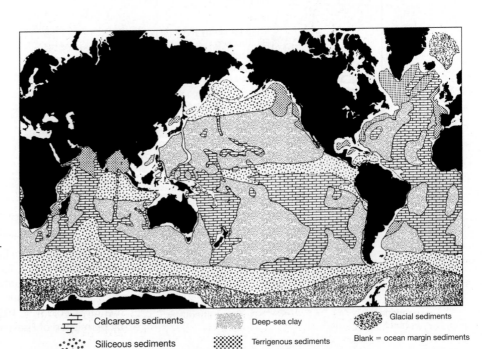

Calcareous sediments   Deep-sea clay   Glacial sediments

Siliceous sediments   Terrigenous sediments   Blank = ocean margin sediments

disturbed, resulting in faulted, contorted, and chaotic bedding and internal structure. Studies of the ocean floor with sidescan sonar and bottom and sub-bottom acoustical and seismic profiling show that slump and slide deposits are particularly common on continental slopes with high rates of deposition, such as off the Mississippi and Rhone deltas, and on slopes with glacial-marine deposits (e.g., Nardin, Edwards, and Govsline, 1979; Schwab, Lee, and Twichell, 1993). These deposits consist of slides, in which failure took place elastically and only minor internal deformation of strata occurred, and of various kinds of failed masses that were emplaced plastically and are characterized by different degrees of internal deformation. Emplacement of sediment along some modern lower continental slopes by mass-transport processes appears to be the dominant sediment transport process.

Pelagic Sediments

The term **pelagic sediment** has been defined in different ways, but it is generally taken to mean sediment deposited far from land influence by slow settling of particles suspended in the water column. Pelagic sediments may be composed dominantly of clay-size particles of terrigenous or volcanogenic origin, or they may contain significant amounts of silt- to sand-size planktonic biogenic remains. **Pelagic clays** are siliciclastic muds that contain clay minerals, zeolites, iron oxides, and windblown dust or ash. They are commonly red to red-brown owing to oxidation by oxygen-bearing deep waters in areas of very slow sedimentation. These clays cover vast areas of the deeper parts of the ocean below about 4500 m (Fig. 10.21). Pelagic sediments that contain significant quantities of biogenic remains are called oozes, as mentioned. Little agreement exists with regard to the amount of biogenic remains required to qualify a sediment as an ooze. In Table 10.1, I suggest that oozes have more than two-thirds biogenic components. Oozes composed predominantly of $CaCO_3$ tests are **calcareous oozes**; those composed mainly of siliceous tests are **siliceous oozes**. Calcareous oozes are dominated by the tests of foraminifers and nannofossils such as coccoliths, but may they include somewhat larger fossils such as petropods, which are planktonic molluscs. Calcareous oozes are widespread in the modern deep ocean at depths shallower than about 4500 m, the calcium carbonate compensation depth (Chapter 6), particularly in the Atlantic Ocean (Fig. 10.21). In deeper ocean basins such as the Pacific, they may occur on the shallow tops of ridges and rises. Lithified equivalents of calcareous oozes are called chalks (limestones).

Siliceous oozes are particularly abundant in the modern ocean at high latitudes in a belt more than 200 km wide stretching across the ocean (Fig. 10.21). They occur also in some equatorial regions of upwelling where nutrients are abundant and productivity of siliceous organisms is high. Siliceous oozes are composed primarily of the remains of diatoms and radiolarians but may include other siliceous organisms such as silicoflagellates and sponge spicules. Diatom oozes occur mainly in high-latitude areas and along some continental margins, whereas radiolarian oozes are more characteristic of equatorial areas. As discussed in Chapter 7, siliceous oozes are modified and transformed during burial into bedded cherts.

Planktonic sediments that settle onto steep slopes, such as those of seamounts and ridges, may be retransported to adjacent basins by turbidity currents or slumping and sliding. Excellent discussions of pelagic environments and sediments are given in Jenkyns (1986); Kennett (1982); Scholle, Arthur, and Ekdale (1983); Stow and Piper (1984b); and Stow, Reading, and Collinson (1996).

Chemical Sediments

Chemical processes in the ocean can operate to modify sediments deposited by other processes, such as dissolution of calcareous oozes; however, chemical processes may also form some new sediment. Clay minerals, zeolites, ferromanganese crusts and nodules, and phosphates are formed in minor amounts by these authigenic chemical processes.

Ancient Deep-Sea Sediments

As mentioned, deep-sea sedimentary deposits other than turbidites are not as abundant in the rock record as shallow-water sediments because the potential for preservation and uplift of these sediments above sea level is much less. Nonetheless, they are known from stratigraphic units of most ages. Typically, deep-sea sedimentary rocks consist predominantly of siliciclastic turbidite sandstones, shales, and conglomerates; pelagic and hemipelagic shales, which may be associated with bedded cherts formed by recrystallization of siliceous oozes; chalks and marls (lithified, clayey pelagic calcareous oozes); limestone breccias (slope deposits); and carbonate turbidites (Chapter 11). Except for turbidites and carbonate breccias, which may be very coarse grained deposits, deep-sea deposits are distinguished in general by their fine grain size. Other than turbidites, most deep-sea deposits do not show vertical facies successions that change upward in any fixed order. Physical sedimentary structures in ancient deep-sea sediments consist predominantly of thin, horizontal laminations, although rippled bedding and graded bedding are common in turbidites, and cross-lamination occurs in some contourites. The bedding of many deep-sea deposits is well developed, even, and laterally persistent.

Colors of deep-water sediments are typically dark gray to black; red pelagic shales are much rarer. Deep-water muds may be well bioturbated or essentially nonbioturbated; they are commonly characterized by distinctive deep-water trace-fossil associations. Fine-grained, deep-water sediments are characterized also by the presence of much greater concentrations of planktonic organisms than occur in shallow-water sediments. These organisms include diatoms, radiolarians, foraminifers, coccoliths, and, in older rocks, graptolites and ammonites. Deep-water sedimentary rocks occur in extensive tabular- or blanket-shaped deposits and may be underlain by ocean crustal rocks such as submarine basalts and ophiolite assemblages consisting of serpentinized peridotite, dunite, gabbros, sheeted dikes, and pillow lavas.

Figure 10.22
Rhythmically bedded turbidites in the Canning Formation (early Tertiary), Arctic National Wildlife Refuge, Alaska. Note the large, low-angle truncation in the middle of the outcrop. [Photograph by D. W. Houseknecht, U.S. Geological Survey Open File Report 98–34, 1999, The oil and gas resource potential of the Arctic National Wildlife Refuge 1002 Area, Alaska.]

Turbidites appear to be, by far, the most abundant kind of ancient deep-sea sedimentary rocks. Turbidites commonly display repetitive, well-bedded successions of thin, graded units that are often referred to as **rhythmites.** Such successions are also called **flysch** facies. Figure 10.22 is a fairly typical example of rhythmically bedded turbidites. This exposure is part of the Canning Formation (early Tertiary), which crops out in the Arctic National Wildlife Refuge, northeastern Alaska. The deposits shown in this outcrop consist of shale and siltstone, with generally thin beds of fine-grained sandstone. Many of the sandstone beds display graded bedding (Bouma sequences) and groove and flute casts are common on the bases of sandstone beds. These turbidites formed at water depth estimated to be about 1200 m (ANWR Assessment Team, 1999). Additional description of both ancient and modern turbidites can be found in Mutti (1992), the excellent compendium volume edited by Pickering et al., (1995), and in Bouma and Stone (2000).

FURTHER READING

Siliciclastic Shelf Systems

Bergman, K. M., and J. W. Snedden (eds.), 1999, Isolated shallow marine sand bodies: Sequence stratigraphic analysis and sedimentologic interpretation: SEPM Special Publication No. 64, p. 13–28.

De Batist, M. D., and P. Jacobs (eds.), 1996, Geology of siliciclastic shelf seas: Geol Society London Spec. Publ. 117, Geol. Soc. Publ. House, Bath, UK, 345 p.

Fleming, B. W., and A. Bartholomä (eds.), 1995, Tidal signatures in modern and ancient sediments: International Association Sedimentologists Spec. Publ. 24, Blackwell Science, Ltd., Oxford, 358 p.

MacDonald, D. I. M. (ed.), 1991, Sedimentation, tectonics and eustasy: Sea-level changes at active margins: Internat. Assoc. Sedimentologists Spec. Pub. 12, Blackwell, Oxford, 518 p.

Pedlosky, J., 1996, Ocean circulation theory: Springer-Verlag, Berlin, 453 p.

Reading, H. G. (ed.), 1996, Sedimentary environments: Processes, facies and stratigraphy, 3rd ed.: Blackwell Science Ltd., Oxford, 688 p.

Smith, G. G., G. E. Reinson, B. A. Zaitlin, and R. A. Rahmani, 1991, Clastic tidal sedimentology: Canadian Soc. Petroleum Geologists Mem. 16, 387 p.

Swift, D. J. P., G. F. Oertel, R. W. Tillman, and J. A. Thorne (eds.), 1991, Shelf sand and sandstone bodies, Geometry, facies and sequence stratigraphy: International Association of Sedimentologists Spec. Publ. 14, Blackwell Scientific Publications, Oxford, 532 p.

Wilgus, C. K., B. S. Hastings, C. G. St. C. Kendall, H. W. Posamentier, C. A. Ross, and J. C. Van Wagoner (eds.), 1988, Sea-level changes: An integrated approach: Soc. Econ. Paleontologists and Mineralogists Spec. Pub. 42, 407 p.

Wright, L. D., 1995, Morphodynamics of inner continental shelves: CRC Press, Boca Raton, Fla., 241 p.

Continental Slope and Deep-Sea Systems

Bouma, A. H., W. R. Normark, and N. E. Barnes (eds.), 1985, Submarine fans and related turbidite systems: Springer-Verlag, New York, 351 p.

Bouma, A. H., and C. G. Stone (eds.), 2000, Fine-grained turbidite systems: AAPG Memoir 72 and SEPM Special Publication No. 68, 342 p.

Hartley, A. J., and D. J. Prosser (eds.), 1995, Characterization of deep marine clastic systems: Geological Society Special Publ. 94, Geological Society, London, 247 p.

Mutti, E., 1992, Turbidite sandstones: Agip, Instituto di Geologia, Universitá di Parma, Milan, 275 p.

Nowell, A. R. M. (ed.), 1991, Deep ocean transport: Marine Geology, v. 89, no. 3/4, p. 275–460.

Osborne, R. H. (ed.), 1991, From shoreline to abyss: Contributions in marine geology in honor of Francis Parker Shepard: SEPM (Society for Sedimentary Geology) Spec. Pub. No. 46, 320 p.

Pedlosky, J., 1996, Ocean circulation theory: Springer-Verlag, Berlin, 453 p.

Pickering, K. T., R. N. Hiscott, N. H. Kenyon, F. Ricci Lucchi, and R. D. A. Smith (eds.), 1995, Atlas of deep water environments: Architectural style in turbidite systems: Chapman and Hall, London, 333 p.

Saxov, S., and J. K. Nieuwenhuis (eds.), 1982, Marine slides and other mass movements: Plenum Press, New York, 353 p.

Siebold, E., and W. H. Berger, 1996, The sea floor: An introduction to marine geology, 3rd ed.: Springer-Verlag, Berlin, 356 p.

Stow, D. A. V. (ed.), 1992, Deep-water turbidite systems: International Association Sedimentologists Reprint Series, V. 3, Blackwell Scientific Publications, Oxford, 473 p.

Stow, D. A. V., and J-C. Faugères (eds.), 1998, Contourites, turbidites and process interaction: Sedimentary Geology, v. 115, 386 p. (Special Issue)

Stow, D. A. V., J-C. Faugères, A. Viana, and E. Gonthier, 1998, Fossil contourites: A critical review: Sedimentary Geology, v. 115, p. 3–31.

Weimer, P., and M. H. Link (eds.), 1991, Seismic facies and sedimentary processes of submarine fans and turbidite systems: Springer-Verlag, New York, 447 p.

Winn, R. D., Jr., and J. M. Armentrout (eds.), 1995, Turbidites and associated deep-water facies: SEPM Core Workshop No. 20, SEPM (Soc. for Sed. Geol.), Tulsa, Okla., 176 p.

11

Carbonate and Evaporite Environments

11.1 INTRODUCTION

Chapters 6 and 7 describe the distinguishing physical, chemical, and biological characteristics of carbonate and evaporite deposits and discuss some of the factors that affect their deposition. In this chapter, we examine the depositional environments in which these deposits form. Carbonate rocks make up roughly one-quarter of the sedimentary rocks in the geologic record. They are an extremely important group of rocks owing to the information they provide about Earth's history and environments and to their economic importance as hosts for petroleum and some metallic elements. Evaporite deposits make up a scant percent or so of the total sedimentary rock record at most, but they are nonetheless very important. Their presence in the rock record affords significant insight into Earth's past climates and they too have considerable economic significance.

Carbonates

Although most modern continental shelves are mantled by siliciclastic sediments, carbonate deposits constitute the dominant sediment cover on a few shelves. Modern carbonate shelves are located primarily at low latitudes in clear, shallow, tropical to subtropical seas (Fig. 11.1) where little terrigenous siliciclastic detritus is introduced. Most of these tropical, carbonate-producing shelves, such as Florida Bay and western Australia, are attached to the mainland. A few smaller shelves surround oceanic islands—the Bahama Platform and the narrow shelves around Pacific atolls, for example (e.g., Vacher and Quinn, 1997). Carbonate sediments also form on some higher latitude (30–60°), cool-water shelves, where they consist predominantly of shell remains (Lees and Buller, 1972; Nelson, 1988; James and Clarke, 1997). Several temperate (cool-water) carbonate environments are present in the modern ocean, including the shelf off southern Australia between ~32 and 40° south latitude, portions of the northwest European shelf, and the Orkney shelf off northeast Scotland (Fig. 11.1).

A few carbonates form in nonmarine environments—in lakes, streams, caves, soils, and dune settings. These carbonates have value as paleoenvironmental indicators, but their volume in the ancient record is quite small. They are not considered further in this volume; however, a brief description of terrestrial (nonmarine) carbonates is given in Boggs (1992, Ch. 10).

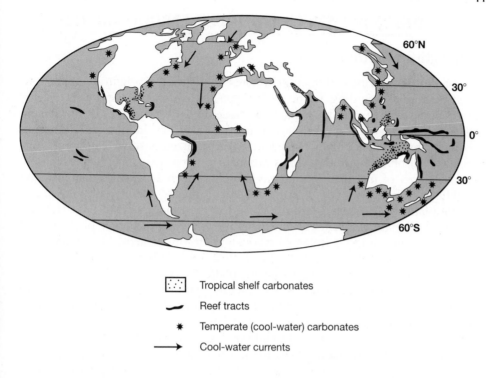

Figure 11.1
Distribution of tropical platform (shelf) carbonates, reefs, and cool-water carbonates in the modern ocean. [Based on Wilson, 1975; Nelson, 1988; Whalen, 1995; and James, 1997.]

The relatively minor importance of modern carbonate deposition is decidedly atypical of many geologic periods of the past when widespread deposition of carbonate sediments characterized sedimentation in broad epeiric seas hundreds to thousands of kilometers wide (see Fig. 10.2). During the middle Paleozoic, for example, carbonate deposition prevailed in shallow inland seas that spread over much of the continental interior of North America. In spite of the small areal extent of modern shelf carbonate environments, carbonate-dominated shelves nonetheless provide outstanding natural laboratories for studying the mechanisms of carbonate sedimentation. Much of what we now understand about carbonate textures and the basic processes of carbonate deposition has come from study of modern carbonate environments. On the other hand, we must turn to the ancient rock record itself for insight into the environmental conditions that typified carbonate-dominated epeiric seas.

Evaporites

Evaporites form in both nonmarine and marine environments; however, marine evaporites are commonly of greatest geologic interest. Marine evaporite deposits, like carbonate deposits, cover relatively small areas of the modern world ocean. Marine evaporites form where rates of evaporation exceed water input, mainly in warm areas of the world. Today, marine evaporite deposits are confined to coastal supratidal settings and sites where marine waters seep into low-lying pools and small basins (Kendall and Harwood, 1996). Such occurrences include coastal salinas (salt ponds, lakes) around the edges of the Mediterranean, the Black Sea, the Red Sea, and the southern and western coasts of Australia, as well as sabkhas (marine to continental salt flats), which are particularly common in the Persian Gulf. Some small-scale ancient evaporites formed in similar environments; however, many ancient evaporite deposits are of giant proportions compared to modern deposits. These giants have no modern analogs and appear to have formed under rather different conditions compared to modern evaporites.

11.2 CARBONATE SHELF (NONREEF) ENVIRONMENTS

Depositional Setting

As mentioned, marine carbonate sediments are deposited primarily on shallow-shelf platforms, including, in the geologic past, broad epeiric platforms covered by shallow water (e.g., Simo, Scott, and Masse, 1993; Tucker et al., 1990). Carbonate platforms can occur on the margins of cratonic blocks, in intracratonic basins, across the tops of major offshore banks, and on localized positive features on wide shelves (Wilson and Jordan, 1983). Carbonate environments may be present also in some parts of marginal-marine environments such as beaches, lagoons, and tidal flats. With respect to the nature of the platform edge, four basic types of carbonate platforms or shelves are recognized in the modern ocean (Fig. 11.2): (1) rimmed carbonate platforms, (2) unrimmed (open shelf) carbonate platforms, (3) carbonate ramps, and (4) isolated carbonate platforms (Harris, Moore, and Wilson, 1985; James and Kendall, 1992; Read, 1982, 1985). When we consider ancient environments, we must add broad, epeiric platforms (Fig. 11.2) to this list.

Rimmed carbonate shelves are shallow platforms marked at their outer edges (margins) by a pronounced break in slope into deeper water. They have a nearly continuous rim or barrier along the platform edge. This barrier consists of either a reef buildup or a skeletal/ooid sand shoal that absorbs wave action and may restrict water circulation, creating a low-energy shelf environment, sometimes called a "lagoon," landward of the shelf-edge barrier. The lagoon commonly grades landward into a low-energy tidal-flat environment rather than a high-energy beach zone. An **unrimmed platform** has no pronounced marginal barrier. Unrimmed platforms occur today on the leeward side of large tropical banks and in all cool-water carbonate settings (James and Kendall, 1992). A **ramp** is a gently sloping ($<{\sim}1°$) unrimmed platform on which shallow-water deposits pass downslope with only a slight break in slope into deeper water facies. The break in slope on a ramp is not marked by a pronounced reef trend, but discontinuous sand shoals may be present along the shelf edge where water energy is high. Water circulation across an unrimmed platform may be adequate to allow a moderately high-energy beach zone to develop alongshore and skeletal or ooid-pellet sand shoals to form along the shelf edge. Thus, unrimmed carbonate platforms are affected by much the same physical processes as siliciclastic shelves. **Isolated platforms** (Bahama type) are shallow-water platforms tens to hundreds of kilometers wide, commonly located offshore of shallow continental shelves, surrounded by deep water that may range from several hundreds of meters to a few kilometers deep. The platforms may have gently sloping, ramplike margins or more steeply sloping margins resembling those of rimmed shelves. Such isolated platforms are essentially free of clastic sediments. Although the Bahamas are probably the best studied example of a modern isolated carbonate platform, numerous other "carbonate islands" are present in the modern ocean, such as Bermuda, Barbados, and the Cook Islands, where carbonate sediments are presently accumulating or were deposited during Pleistocene time (see Vacher and Quinn, 1997).

No modern examples of carbonate **epeiric platforms** exist; however, such platforms (Figure 11.2) were common in the past, particularly during the Paleozoic and parts of the Mesozoic. Some platforms were hundreds or thousands of kilometers across and covered millions of square kilometers (Wright and Burchette, 1996). We can only guess at the hydrologic processes that operated on such broad shelves. They were likely connected to the open sea, and storms and winds may have strongly affected water circulation. Tidal activity may also have been important. During times of major carbonate deposition on epeiric platforms, influx of clastic detritus must have been at a minimum.

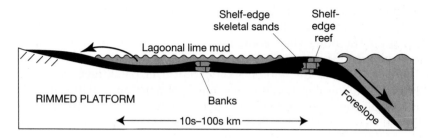

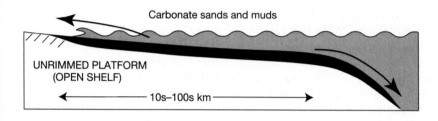

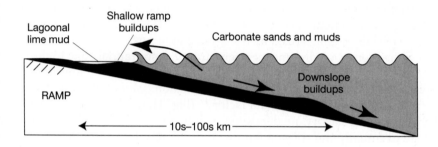

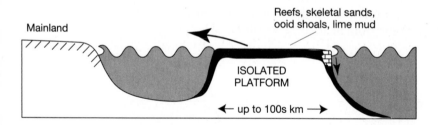

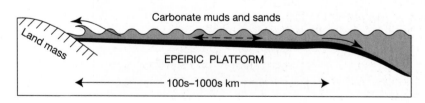

Figure 11.2
Schematic representation of principal kinds of carbonate platforms, shown in cross section. Arrows indicate directions of sediment movement. [Based on James and Kendall (1992) and Wright and Burchette (1996).]

In contrast to most siliciclastic shelves, many modern carbonate platforms, particularly rimmed platforms, are characterized by some kind of topographic buildup at the shelf margin of the outer shelf. This buildup may be caused by organic reefs or banks, lime sand shoals, or small islands that create a barrier to incoming waves. This outer barrier is commonly dissected by a network of tidal channels that allow high-velocity tidal currents to flow through onto the shelf.

Water may be only a few meters deep over this buildup, but depth increases over the middle shelf to perhaps several tens of meters (e.g., the rimmed platform illustrated in Fig. 11.2). The outer shelf is the highest energy zone of such shelves. Much of the middle shelf is commonly below fair-weather wave base. Water energy is thus low over most of the middle shelf except over patch reefs, localized banks, or shoals and along the shoreline of some carbonate ramp platforms. The elevation and lateral continuity of the shelf-edge carbonate barrier control water circulation over the entire shelf. The effect of this barrier on water circulation, coupled with the width of the shelf, strongly influences the type and distribution of carbonate facies that develop on the shelf. If a well-developed barrier is present, or if the shelf is very wide, water circulation on the shelf may be restricted to some degree because water energy is expended in friction with the bottom, leading to poor water circulation. On the other hand, the geologic record does not necessarily indicate that water circulation on wide epeiric shelves was strongly restricted. Restricted water circulation leads to development of salinity conditions that deviate from normal (~35 ‰). Salinities may rise well above normal in arid or semi-arid climates where evaporation rates are high, or they may fall below normal in areas that receive considerable freshwater runoff. Variations in salinity affect the diversity and numbers of organisms living on shelves; the organisms, in turn, strongly affect carbonate deposition owing to the extremely important role they play in carbonate sedimentation processes (Chapter 6). The innermost part of the shelf may be especially characterized by restricted conditions.

Although carbonate environments extend from the supratidal zone to deeper basins off the shelf, the shallow platform basin that constitutes the middle and outer shelves is the primary site of carbonate production. James (1984a) refers to this platform as the "subtidal carbonate factory" (Fig. 11.3). The sediments produced in this carbonate factory are deposited mainly on the shelf; however, some sediments are eventually transported landward onto tidal flats and beaches and into subtidal settings. Others are transported seaward off the shelf onto the slope and into the deeper basin. Little carbonate sediment is generated in the deeper water basin environment off the shelf except for fallout of calcium-carbonate-secreting plankton from near-surface waters.

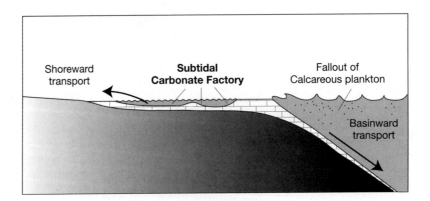

Figure 11.3
The main areas of marine carbonate production. Most carbonates accumulate in water less than about 30 m deep—the "subtidal carbonate factory." The example shown depicts carbonate production on a rimmed platform. Similar production takes place on the other platforms illustrated in Fig. 11.2. [After James, N. P., 1984, Introduction to carbonate facies models, *in* Walker, R. G. (ed.), Facies models, 2nd ed.: Geoscience Canada Reprint Ser. 1, Fig. 2, p. 210, reprinted by permission of Geological Association of Canada.]

Sedimentation Processes

Chemical and Biochemical Processes

The principal chemical and biological/biochemical controls on carbonate deposition are discussed in Chapter 6; they are reviewed only briefly here. The solubility of calcium carbonate is controlled by pH, temperature, and carbon dioxide content of seawater. Loss of carbon dioxide owing to increased temperature, decreased pressure, or plant photosynthesis exerts a major control on inorganic precipitation of $CaCO_3$. Nonetheless, the relative importance of chemical (inorganic) precipitation of calcium carbonate in the modern and ancient oceans, as compared to organic production of $CaCO_3$, is not definitely known (e.g., Shinn et al., 1989). Carbonate deposition brought about by organisms capable of extracting calcium carbonate from the seawater to build their shells or skeletal structures may be a more important process in the modern ocean than are purely inorganic processes. Such biogenic processes have likely been important throughout post-Precambrian time, and some may have played a role in carbonate production during the Precambrian. Organisms also contribute to the formation of carbonate sediment through their feeding and bioturbation activities, which cause breakdown of skeletal fragments and other carbonate materials and generate various kinds of trace fossils.

The organisms primarily responsible for carbonate production in the modern ocean are not necessarily the same as those that were major carbonate formers in the past. Figure 11.4 shows the relative importance of some major groups of organisms as carbonate formers during Phanerozoic (post-Precambrian) time. Note that the principal carbonate formers have changed somewhat with time. For example, crinoids, byrozoans, and brachiopods were more important during the Paleozoic than during the Cenozoic, whereas coccoliths, planktonic foraminifers, coralline algae, and green algae were particularly important carbonate formers during the Cenozoic. The list of carbonate-sediment formers shown in Figure 11.4 is not complete. Other groups, such as sponges and stromatoporoids, were also important sediment formers at times.

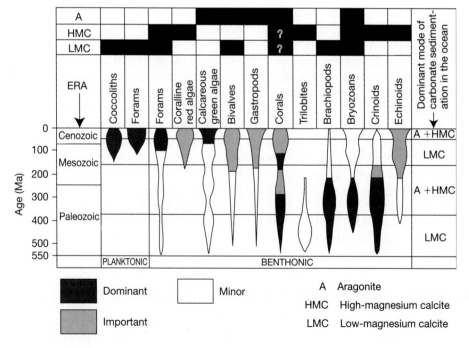

Figure 11.4
The relative importance through time of various calcareous marine organisms as sediment producers. This diagram also shows the skeletal mineralogy of the organisms, which, in some cases, changed as the sea chemistry changed from a "calcite sea" to an "aragonite sea" through time. Calcite seas favored precipitation of low-magnesian skeletal structures, and argonite seas favored precipitation of aragonite and high-magnesian calcite, although not all organisms responded to such changes in sea chemistry. [Based on Wilkinson, 1979; Jones and Deroschers, 1992; James, 1997; Stanley and Hardie, 1999.]

Note also from Figure 11.4 that different groups of organisms secrete different carbonate minerals to build their skeletal structures. As an example, coccoliths and forams are composed of low-magnesian calcite, crinoids and echinoids are composed of high-magnesian calcite, and calcareous green algae and gastropods are composed of aragonite. Figure 11.4 also shows the times when the Phanerozoic ocean precipitated predominantly low-magnesian calcite (calcite seas) and the times when aragonite and high-magnesian calcite were the favored carbonate precipitates (aragonite seas), as discussed in Chapter 6 (see Fig. 6.9). Some groups of organisms, such as corals, secreted different minerals at different times in their history in response to changes in sea chemistry, primarily changes in the ratio of magnesium to calcium (e.g., Stanley and Hardie, 1998, 1999). Corals formed calcite (low- or high-Mg?) skeletons during much of the Paleozoic when a "calcite sea" characterized the world ocean (Fig. 6.9), but more recent corals, especially those living during the late Cenozoic, secreted aragonite skeletons. Although the skeletal mineralogy of many groups of organisms appears to parallel that of inorganic precipitates formed during times of calcite or aragonite seas, some other organisms, such as echinoids, crinoids, and brachiopods, secreted the same skeletal minerals throughout their history in spite of changing sea chemistry. Skeletal structures composed of aragonite are chemically less stable than calcite structures and are thus more susceptible to dissolution and destruction during diagenesis.

Physical Processes

Physical processes are important primarily in the reworking and transport of carbonate materials on the shelf, but they also aid in the production of carbonate sediments. Circulation of water onto the shelf brings fresh nutrients, necessary for organic growth, from deeper water. Breaking waves against reef barriers on the outer shelf increases oxygen content in the water by interacting with the atmosphere and decreases CO_2 because of decreased water pressure. Thus, modern reefs are best developed in wave-agitated zones, and biogenic production of carbonate sediment in general is stimulated by strong water movement. On the other hand, strong waves crashing on the reef front break down reef rock, producing sand- and gravel-size bioclasts that subsequently undergo transport both seaward and landward from the reef.

Agitated water is important to the formation of ooids, and currents help to generate and preserve grapestones and hardened fecal pellets by submarine accretion and cementation. Waves and currents also winnow fine carbonate mud from coarser sediment and transport this mud off the shelf platform or into sheltered or protected areas of the shelf. Depending upon water energy, the coarser sediment itself may either remain as a winnowed lag deposit, forming sand- or gravel-covered flats, or be transported and deposited to create wave-formed bars and shoals, beaches, spits, or tidal deltas and bars. Wave- and current-transported and -winnowed carbonate sand deposits are particularly common along the outer edge of the shelf platform, where water energy is highest. In resuspension and transport of sediment, storms are as important on carbonate shelves as they are on siliciclastic shelves. For example, storms transport most sediment from the subtidal shelf into the intertidal (tidal flat) environment. Absence of wave and current activity on the shelf leads to stagnant circulation, consequent deviations from normal salinity, and possibly anoxic conditions. Such restricted environments constitute unfavorable habitats for many normal marine organisms.

Skeletal and Sediment Characteristics of Carbonate Deposits

The deposition of carbonate sediments is favored in moderately shallow, warm water that receives little terrigenous siliciclastic sediment. Although carbonates form predominantly in warm water settings, they can accumulate also in some cool-water, higher latitude environments, as mentioned. In these cool-water environments, the carbonate sediment is composed almost entirely of the skeletal remains of organisms. Cool-water assemblages of organic remains have commonly been referred to as **foramol** assemblages (Lees and Buller, 1972; Jones and Desrochers, 1992), named for the dominance of foraminifers and molluscs. They are composed of benthic (bottom-dwelling) foraminifers, molluscs, barnacles, bryozoans, and calcareous red algae. By contrast, warm-water ($>\sim 20°C$) assemblages of organisms, called **chlorozoan** assemblages (named from chlorophyta plus zoantharia corals), are dominated by hermatypic corals (corals that live primarily in the photic zone) and calcareous green algae in addition to foramol components. James (1997) suggests that **heterozoan association** (named for organisms that feed through heterotrophic means) is a more appropriate term than foramol assemblage. He proposes to replace the term chlorozoan assemblage by **photozoan association**, to emphasize the light-dependent nature of the major biotic constituents.

In any case, cool-water carbonates make important contributions to the deposits of some modern shelves (e.g., Farrow, Allen, and Akpan, 1984; Nelson, 1988; James and Clarke, 1997). These cool-water shelves range from those located in middle to low latitude settings where cool-water currents intrude (Fig. 11.1) to those located at high latitudes such as Spitsbergen Bank in the Barents Sea. Cool-water carbonate shelf deposits have also been reported in ancient rocks ranging in age from Tertiary to Paleozoic on several continents, including North America, Australia, and Europe (e.g., James and Clarke, 1997; Anastas et al., 1998).

Warm-water carbonates may contain, in addition to skeletal remains, substantial amounts of ooids, aggregate grains, peloids, and lime mud. Table 11.1 provides a more complete list of modern warm-water and cool-water organisms and their ancient counterparts. This table also suggests the manner in which these organisms contribute to the makeup of carbonate sediment. Notice the extremely important roles that organisms play in the formation of carbonate sediment. Modern warm-water carbonates, especially reef carbonates, accumulate at a much faster rate than do cool-water carbonates. On the other hand, modern cool-water carbonates appear to accumulate at about the same rate as did most ancient carbonates (James, 1997). Reasons for the slow accumulation rates of many ancient carbonates are poorly understood.

As stated, carbonate sediment accumulates primarily in shallow-water settings (Fig. 11.3). The outer shelf is commonly the highest energy environment of the shelf. It is characterized by the development of lime sand or gravel sheets and shoals. The middle shelf is a zone of generally low water energy, particularly on rimmed shelves. Sediments on the middle shelf are typically poorly winnowed, with a high ratio of micrite to skeletal fragments and other carbonate grains. The inner shelf in most carbonate environments is also typically a low-energy, tidal-flat environment in which predominantly fine grained, tidal-flat sediments accumulate. On some ramp platforms, however, a higher-energy nearshore zone may be present where carbonate beaches or lime shoals develop that are composed of skeletal fragments, ooids, pellets, and possibly intraclasts. In many cases, carbonate beach sands are retransported and reworked by wind to form so-called **eolianites**. Numerous examples of Quaternary eolianites have been reported (e.g., Abegg, Harris, and Loope, 2001); however, the pre-Quaternary record of carbonate eolinites is meager.

Some carbonate sediments are deposited in deeper water beyond the shelf edge. Most carbonate sediment deposited in deeper water results from the fallout of calcareous plankton (Fig. 11.3)—foraminifers, green algae (coccoliths), and tiny gastropods. These pelagic calcareous organisms evolved mainly in Jurassic and post-Jurassic time; therefore, deeper water pelagic carbonates are not important in

Table 11.1   Modern warm- and cool-water marine organisms and their counterparts in the fossil record

Modern, warm-water	Modern, cool-water	Ancient counterpart	Sedimentary aspect
Corals	Absent	Corals, stromatoporoids, stromatolites, coralline sponges, rudist bivalves	Large components of reefs and biogenic mounds
Bivalves, red algae, echinoderms	Bivalves, red algae, brachiopods, echinoderms, barnacles	Red algae, brachiopods, cephalopods, trilobites	Remain whole or break apart into several pieces to form sand- and gravel-size particles
Gastropods, benthic foraminifera	Gastropods, benthic foraminifera	Gastropods, benthic foraminifera	Whole skeletons that form sand- and gravel-size particles
Green (codiacean) and red algae	Red algae, bryozoans	Phylloid algae, crinoids and other echinoderms, bryozoans	Spontaneously disintegrate upon death to form many sand-size particles
Ooids, peloids	Absent	Ooids, peloids	Concentrically laminated or micritic sand-size particles
Planktonic foraminifera, coccoliths, pteropods	Planktonic foraminifera, coccoliths, pteropods	Planktonic foraminifera, coccoliths (post-Jurassic), styliolinoids	Medium sand-size and smaller particles in basinal deposits
Encrusting foraminifera, red algae, bryozoans	Encrusting foraminifera, red algae, bryozoans, serpulid worms	Red algae, renalcids, encrusting foraminifera, bryozoans	Encrust on or inside hard substrates; build up thick deposits or fall off upon death to form sand grains
Dasyclad green algae	Absent	Dasyclad green algae	Spontaneously disintegrate upon death to form lime mud
Cyanobacteria and other calcimicrobes	Cyanobacteria and other calcimicrobes	Cyanobacteria and other calcimicrobes (especially pre-Ordovician)	Trap, bind, and precipitate fine-grained sediments to form mats and stromatolites or thrombolites

Source: James, N. P., and A. C. Kendall, 1992, Introduction to carbonate and evaporite facies models, *in* Walker, R. G., and N. P. James (eds.), Facies models—Response to sea level change: Geol. Assoc. Canada. Table 2, p. 269.

older rocks. In addition to pelagic carbonate, some shallow-water carbonate sediment may be swept off carbonate platforms into deeper water by storm waves or be transported by sediment gravity-flow processes (e.g., turbidity currents).

Examples of Modern Carbonate Platforms

Modern carbonate shelves include ramps, unrimmed platforms (open shelves), rimmed platforms, and isolated platforms. Examples of tropical unrimmed shelves or carbonate ramps include the eastern Gulf of Mexico off the Florida coast; the Yucatan Shelf, Mexico, in the southern part of the Gulf of Mexico; and the Trucial Coast of the Persian Gulf. As mentioned, most cool-water carbonates accumulate on unrimmed (open) shelves (see Fig. 11.1). Examples of rimmed shelves include Florida Bay, the Bahama Platform (an isolated platform), the Belize Shelf in the western Caribbean off Guatemala, and the Great Barrier Reef area of Australia. Other important deposits of carbonate sediments in Australian waters lie along the western coast. The characteristics of several of these modern

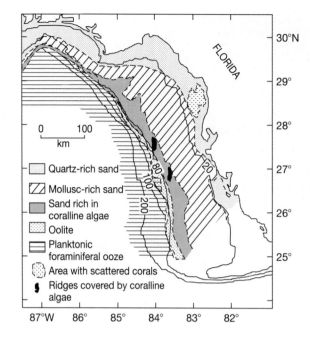

Figure 11.5

Example of an open shelf or carbonate ramp, the West Florida Shelf in the eastern Gulf of Mexico. [From Sellwood, B. W., 1978, Shallow-water carbonate environments, *in* Reading, H.G. (ed.), Sedimentary environments and facies, Fig. 10.17, p. 276, reprinted by permission of Elsevier Science Publishers, Amsterdam. Originally after Ginsburg, R. N., and N. P. James, 1974, Holocene carbonate sediments of continental shelves, *in* Burk, C. A., and C. L. Drake (eds.), The geology of continental margins, Fig. 6, p. 140, Springer-Verlag, New York.]

platforms are summarized by Jones and Desrochers (1992), Sellwood (1986), Wilson and Jordan (1983), and Wright and Burchette (1996).

Sediment facies maps of three well-known modern carbonate shelves are presented here to show some of the facies-distribution patterns on these types of shelves. Figure 11.5 illustrates an open shelf or carbonate ramp, the West Florida Shelf. Mollusc-rich siliciclastic (quartz) sands dominate the inner ramp down to about 60 m depth, grading downslope into coralline algal carbonate sands. Relict oolitic sands with pelagic and benthic foraminifers dominate between 80 and 100 m, and planktonic foraminiferal oozes are common in deeper water. South Florida Bay (Fig. 11.6) is a good example of a rimmed shelf. The inner shelf margin is marked by the Florida Keys, an emergent Pleistocene reef-ooid-shoal complex. Florida Bay lies inboard of the Keys, bordered on its northern margin by coastal swamps of the Everglades. Florida Bay is filled with carbonate mud and muddy carbonate sands enriched in molluscs and foraminifers. The belt of muddy carbonate sands lying between the Keys and the outer reef tract is composed mainly of calcareous algae (*Halimeda*) and molluscs. Carbonate sands within the outer reef tract are composed of *Halimeda* and coralline algae; corals are also present.

The best-studied example of a modern isolated platform is the Bahama Platform, which is also rimmed (Fig. 11.7). Coralgal sands and ooid facies are present along the platform margin in zones affected by wave turbulence and tidal currents, and discontinuous coral reefs are present along the windward (east) margin. Ooid facies are best developed in water less than 3 m deep and occur as sandwaves and subaqueous dune fields up to 50 km long (e.g., Fig. 11.8). Grapestone facies, which cover large areas of the platform interior down to depths of 9–10 m, contain little carbonate mud. They are stabilized by cyanobacteria mats, calcareous algae, and sea grasses. Pellet mud and mud facies accumulate in the lowest energy parts of the platform interior, commonly at water depths less than 4 m. The sediment consists of highly bioturbated aragonite mud, which is rich in fecal pellets in some areas.

Modern carbonate sediments on the Bahama Platform are underlain by Pliocene and Pleistocene carbonates, which were recently investigated by coring (Ginsburg, 2001). The cores reveal seaward progradation of the leeward margin of the bank with overall shallowing. The Pliocene-Pleistocene sediments grade from

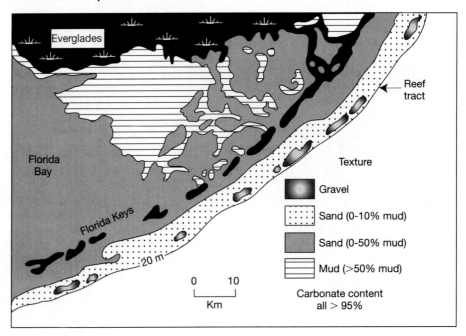

Figure 11.6
Sediment map of South Florida Bay area, an example of a modern rimmed carbonate shelf, showing the distribution of carbonate sediment by grain size on the shelf platform. [After Sellwood, B. W., 1978, Shallow-water carbonate environments, *in* Reading, H. G. (ed.), Sedimentary environments and facies, Fig. 10.21A, p. 281, reprinted by permission of Elsevier Science Publishers, Amsterdam. Originally after Ginsburg, R. N., and N. P. James, 1974, Holocene carbonate sediments of continental shelves, *in* Burk, C. A., and C. L. Drake (eds.), The geology of continental marings, Fig. 23, p. 150, Springer-Verlag, New York.]

skeletal grainstones and packstones at the base upward to reefal and coral-bearing deposits that in turn are capped by nonskeletal grainstones similar to modern sediments in the interior of the Bahama Banks (Manfrino and Ginsburg, 2001).

Note the general progression of facies on the rimmed shelves (Fig. 11.6, 11.7) from reef buildups and shelf-edge sands on the higher energy outer shelf to carbonate muds and muddy carbonate sands on the lower energy middle and inner shelves. By contrast, most of the open-shelf, carbonate ramp in Figure 11.5 is covered by carbonate sand deposits at depths less than about 100 m, mixed, on this shelf, with some terrigenous quartz sands.

Examples of Ancient Carbonate Shelf Successions

Isolated Platforms

Examples of carbonate sediments that may have formed in ancient settings similar to all of the platform types illustrated in Figure 11.2 have been reported in the published literature. For example, early to middle Triassic carbonates in the Dolomite Alps of northern Italy are thought to represent deposition on isolated platforms much like the modern-day Bahama Banks (Bosellini, 1991). These ancient carbonates consist of flat-lying successions up to 800 m thick composed of meter-scale, cyclic, peritidal carbonates with teepee structures. These deposits probably formed within the interiors of isolated carbonate platforms under moderately low energy, open to restricted, shallow, subtidal conditions. Platform-interior sediments grade in a seaward direction to peloidal skeletal grainstones/packstones and algal boundstones with sponges, which were deposited on the higher energy platform margin.

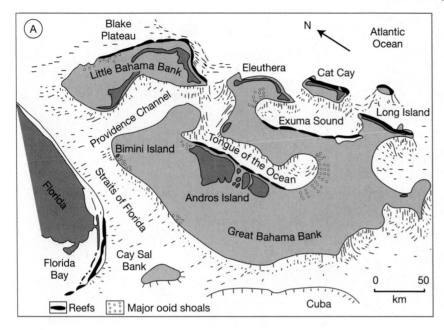

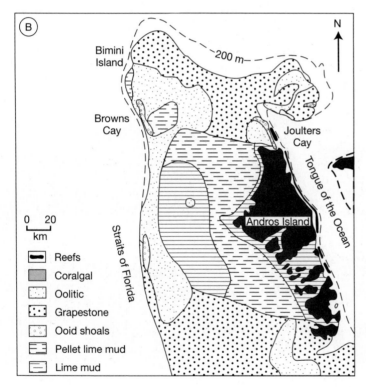

Figure 11.7
Carbonate sediment distribution on an isolated carbonate platform, the Great Bahama Banks. Figure A shows the positions of the major banks and channels in the Bahama area. [After Geblein, C. D., 1974, Guidebook for modern Bahaman plaform environments: Geological Society of America Annual Meeting, 1974, Fig. 18, p. 22.] Figure B shows sediment distribution on a portion of Great Bahama Bank surrounding Andros Island. [After Sellwood, B. W., 1978, Shallow-water carbonate environments, in Reading, H. G. (ed.), Sedimentary environments and facies. Fig. 10.21B, p. 281, reprinted by permission of Elsevier Science Publishers, Amsterdam. Originally after E. G. Purdy, 1963, Recent calcium carbonate facies of the Great Bahama Banks: Jour. Geology, v. 71, Fig. 1, p. 473.]

Rimmed Shelves

The Permian (Guadalupian) carbonate deposits of the Delaware Basin, western Texas and southern New Mexico, have often been cited as an outstanding example of an ancient rimmed shelf deposit (e.g., Ward, Kendall, and Harris, 1986; Saller et al., 1999). These deposits extend throughout an area of about 100,000 km² and are completely encircled by a reef margin, the Capitan and Goat Seep reefs (Fig. 11.9). Outer-shelf deposits make up a belt about 10 to 15 km across, dominated by shallowing-upward cycles of lagoonal mudstone and wackestone, capped by laminated anhydrite. The outer-shelf belt grades abruptly to an inner lagoonal to

Figure 11.8
Ooid shoal in shallow water of the Great Bahama Bank. Photograph courtesy of the National Oceanic and Atmospheric Administration (NOAA). Downloaded from the Internet 4/30/04.

Figure 11.9
Bold cliffs of massive Capitan Limestone (reef) that formed the rim of a huge Permian carbonate platform. The platform (shelf) carbonates extend toward the back of the photograph; deeper-water, basinal sediments are present in the foreground. Photograph courtesy of Gregory Retallack.

marginal-coastal playa facies about 40 km wide, composed of intercalated dolomite mudstones and siltstones, and evaporitic units up to 7 m thick. Between the reef and shelf-lagoonal deposits is a belt of carbonate sand bodies, 800 m to 3 km wide, consisting of a coarsening-upward succession of cross-laminated to structureless fine-grained carbonate grainstones.

Ramps

Carboniferous limestones of southwestern Britain provide an example of an ancient carbonate ramp deposit (Burchette, Wright, and Faulkner, 1990). The ramp succession forms a wedge that begins with outer-ramp carbonate muds and bio-clastic limestones containing abundant crinoid and brachiopod remains. Reef mounds with relief up to 200 m also formed in the outer-ramp setting. Outer-ramp deposits give way in a shoreward direction to bioclastic sand bodies of the mid-ramp and finally to ooid sand bodies laid down as shoals or beaches in the nearshore area.

Epeiric Platforms

The great bulk of carbonate sediments formed throughout geologic time have probably been deposited on epeiric platforms (Fig. 11.2). Such platforms were widespread at various times, such as the late Precambrian (e.g., China), Cambro-Ordovician (e.g., North America; Middle East), Mississippian, Triassic-Jurassic (e.g., western Europe), Permian, and Tertiary (e.g., Middle East) (Wright and Burchette, 1996; Tucker and Wright, 1990). Storms and hurricanes were likely dominant processes on these broad shelves; however, tidal processes may also have been important (e.g., Pratt and James, 1986). Shelf carbonate facies are characterized by distinctive suites of largely normal marine organisms and carbonate textures that are generally muddy, although lithofacies types range from lime mudstones, wackestones, grainstones, and packstones to stromatolitic bound-stones and patch-reef boundstones. Bedding of shelf carbonates is variable and lens- or wedge-shaped layers are common, although some shelf carbonate beds may be laterally extensive. Carbonates are commonly interbedded with thin shale beds. Sedimentary structures include cross-bedding in lime-sand units, extensive bioturbation structures and burrows, and flaser and nodular bedding.

Many epeiric deposits appear to be dominated by shallowing-upward cyclic successions that range from a few tens of meters to hundreds of meters thick. Many successions begin with a high-energy carbonate sand or conglomerate unit followed upward progressively in the depositional succession by sediments de-posited in the lower energy, subtidal, open-marine shelf; intertidal zone; supratidal zone; and possible nonmarine environment. Some depositional cycles appear to have ended with deposition of evaporites. Such a succession is basically regressive (progradational); however, because rates of carbonate sedimentation commonly exceed rates of basin subsidence or sea-level rise, sediments also build upward toward sea level. Sediment is thus deposited in progressively shallower water as the sediment surface accretes toward sea level, generating the shallowing-upward successions (e.g., James, 1984b). Intraplatform basins, with water depth commonly less than 100–200 m, can form within epeiric platforms. During sea-level highstands, water within these basins may be stratified—oxygenated at the top but suboxic to anoxic at the bottom. Sea-level lowstands may lead to isolation of the basin and onset of evaporitic conditions (Wright and Burchette, 1996).

Repetition of large-scale shallowing-upward successions may be largely the result of repeated episodes of rapid sea-level rise, flooding the carbonate platform, followed by periods of standstill during which shallowing-upward successions develop, e.g., Wilkinson (1982). Osleger and Read (1991) suggest that meter-scale cyclicity is the result of Milankovich-forced sea-level oscillations (see discussion of stratigraphic cycles in Chapter 12), with a cyclicity on the order of 20,000–40,000 years. Several additional examples of ancient carbonate platform deposits are discussed in Alsharhan and Scott (2000) and Zempolich and Cook (2002).

11.3 SLOPE/BASIN CARBONATES

Although we tend to think of carbonate sediments as strictly shallow-water deposits, as mentioned, deeper-water carbonates have been identified in several areas of the modern ocean, such as the slope and adjacent basin floor around the Bahama

Platform. They have also been reported from many Phanerozoic-age stratigraphic successions. As shown in Figure 11.3, carbonate sediments are generated primarily on the shelf. No important source of carbonate sediments exists within deep water except that provided by the rain of calcareous pelagic organisms. Therefore, with the exception of calcareous oozes, carbonate sediments in deep water are derived from the shelf by transport processes that include storm waves, turbidity currents, debris and grain flows, slumping, sliding, and rock falls. Carbonate sediment deposited on the slope and basin by these processes generally consists of bioclastic debris and limestone blocks derived from the talus slopes off reef fronts. Also, sediments may be transported downslope from carbonate sand shoals or lime-mud deposits on the platform margin. Modern examples of carbonate slopes have been reported from the northern Bahamas-Florida region as well as from Belize, Jamaica, Grand Cayman, the northeast Australian coast, and several atolls in the Pacific and Indian Oceans (e.g., Coniglio and Dix, 1992). Modern slope carbonates consist mainly of pure carbonates, in contrast to many ancient slope deposits that include a large percentage of terrigenous clastic rocks.

Three types of carbonate slopes are recognized (Fig. 11.10): **erosional** (steep slope angle), **by-pass** (moderate slope angle), and **accretionary** (low slope angle). Only accretionary slopes are sites of significant carbonate deposition, although minor amounts of sediments may be deposited on by-pass slopes. Erosional and by-pass slopes mainly serve as conduits by which carbonate sediment moves from shallow to deeper water. Several kinds of carbonate sediments can be deposited on slopes. **Periplatform oozes** consist of fine sediment swept off the shelf mixed

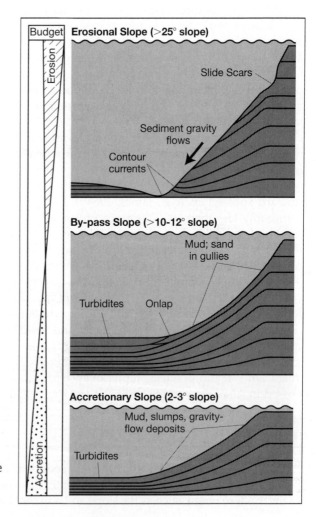

Figure 11.10
Principal kinds of carbonate slopes, based on examples from the Bahama Platform. [After Schalger, W., and R. N. Ginsburg, 1981, Bahama carbonate platforms—The deep and the past: Marine Geology, v. 44, Fig. 10, p. 15. Reproduced by permission.]

with foraminifers and coccoliths that settle from the water column. Talus deposits and **debrites** are carbonate breccias and conglomerates derived from shallow water. **Turbidites** are the carbonate equivalents of siliciclastic turbidites.

Two basic models for accretionary (depositional) carbonate slopes have been proposed: slope aprons and submarine fans (Fig. 11.11). Carbonate aprons are distinguished by having a line source or multiple sources that feed sediment seaward through closely spaced gullies, generating wedge-shaped aprons of sediment (e.g., Mullins and Cook, 1986). Aprons may develop on the slopes of both rimmed and unrimmed platforms. **Slope aprons**, which extend without break from the basin up to the shallow-water margin, putatively develop along rimmed platforms where slopes are less than ~4°. Such aprons have not been described from the modern ocean; however, ancient examples have been reported. Open-platform slopes are similar in form to slope aprons but develop on unrimmed platforms. Base-of-slope aprons (Fig. 11.11A) form along rimmed platforms downslope from the platform-slope break on steeper slopes ranging between 4 and 15°. Sediment bypasses the upper slope and is transferred to its foot by way of numerous gullies. Proximal sediments in the uppermost part of the apron consist of talus, debrites, thick turbidites, and periplatform oozes. Sediments in the distal apron consist mainly of finer grained turbidites.

Carbonate **submarine fans** (Fig. 11.11B) are similar in form to the siliciclastic submarine fans described in Chapter 10. Sediment is supplied to fans from a single point source through a major channel that may bifurcate in its lower reaches to

A. Base-of-Slope Apron

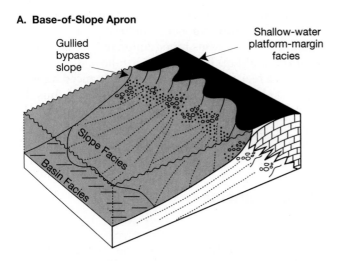

B. Submarine Fan

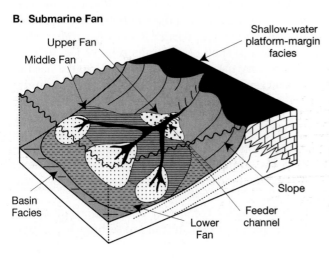

Figure 11.11
Schematic block diagrams illustrating models for two fundamental kinds of carbonate slopes: slope aprons and carbonate submarine fans. A. Shows the kind of apron that develops along rimmed platforms where slopes range between ~4 and 15°. B. Shows an idealized carbonate submarine fan. [A. after Mullins, H. T., and H. E. Cook, 1986, Carbonate apron models: Alternatives to the submarine fan model for paleoenvironmental analysis and hydrocarbon exploration: Sedimentary Geology, v. 48, Fig. 24, p.66. B. After Coniglio, M., and G. R. Dix, 1992, Carbonate slopes, *in* Walker, R. G., and N. P. James (eds.), Facies models: Response to sea level change: Geological Association of Canada, Fig. 22a,b, p. 367. Reproduced by permission.]

generate different sediment lobes. Commonly, coarsest sediment is deposited in the upper fan and sediment becomes finer in more distal parts of the fan. No modern carbonate submarine fans are known; however, a few examples of ancient fans have been reported (see reviews in Tucker and Wright, 1990, and Coniglio and Dix, 1992).

11.4 ORGANIC REEF ENVIRONMENTS

As mentioned, the outer shelf of many rimmed platforms is characterized by nearly continuous carbonate reefs that constitute an effective barrier to wave movement across the shelf. Reefs may also be developed as fringing masses along the shoreline or as isolated patches within the inner shelf. Reefs constitute a unique depositional environment that differs greatly from environments in other parts of the shelf. They have been studied intensively for years; however, discussion of reefs has long been plagued by confusion over the precise meaning of the term reef. Carbonate workers have been unable to agree on whether to restrict use of the term **reef** to carbonate buildups or bioherms that have a rigid organic framework or core, built of colonial organisms, or to extend the definition to include carbonate buildups of other types that do not have a rigid-framework core. The word **bioherm** is a nonspecific term used for lenslike bodies of organic origin that are enclosed in rocks of different lithology or character; a bioherm may or may not have a rigid internal organic framework. The term carries no connotation of the internal structure or composition of the lens. By contrast, a **biostrome** is a tabular body of carbonate rock such as typically forms in nonreef platform environments. Wilson (1975) uses the term **carbonate buildup** for a body of locally formed, laterally restricted, carbonate sediment that possesses topographic relief, without regard to the internal makeup of the buildup. In this book, I follow the usage of Longman (1981, p. 10), who defines a reef as "any biologically influenced buildup of carbonate sediment which affected deposition in adjacent areas (and thus differed to some degree from surrounding sediments), and stood topographically higher than surrounding sediments during deposition." Most reefs, defined in this way, are built by larger organisms that are capable of thriving in energetic environments.

Modern Reefs and Reef Environments

Depositional Setting

Most modern reefs (e.g., Fig. 11.12) form in shallow water. The most striking are the linear reefs located along platform margins, commonly called **barrier reefs**. These reefs are more or less laterally continuous, and the reef trend may extend for hundreds of kilometers—for example, the Great Barrier Reef of Australia, which runs for some 1900 km along the eastern shelf of Australia (Fig. 11.13). In a few modern localities where shelves are very narrow, linear reefs are located hard up against the shoreline, with no intervening lagoon, and thus are called **fringing reefs**. Isolated, doughnut-shaped reefs called **atolls** occur around the tops of some Pacific seamounts that rise out of deeper water. These reefs form an outer wave-resistant barrier that encloses a shallow lagoon. **Faro reefs** are ringlike (atoll-like) structures that form within lagoons or on atoll margins. Small isolated reef masses commonly referred to as **patch reefs**, **pinnacle reefs**, or **table reefs** occur along some shelf margins or scattered on the middle shelf. Flat-topped table reefs may also form in deeper water (Fig. 11.14).

Mounds are structures built by smaller, commonly delicate and/or solitary organisms, possibly aided by inorganic processes, in tranquil settings (e.g., James

Figure 11.12
Modern coral reef. Photo-
graph courtesy of National
Oceanic and Atmospheric Administration (NOAA). Downloaded from the Internet
4/30/04.

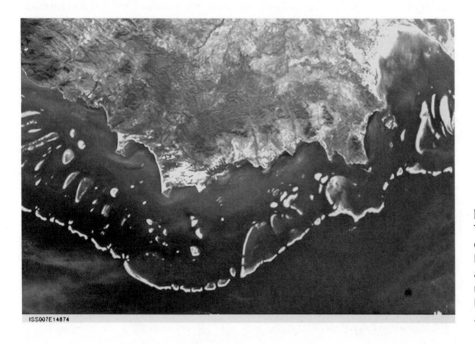

ISS007E14874

Figure 11.13
The Great Barrier Reef off the
coast of Queensland, Australia.
Photograph courtesy of National Aeronautics and Space Administration (NASA).
Downloaded from the Internet
5/4/04.

and Bourque, 1992; Monty et al., 1995) in either shallow or deeper water. **Microbial mounds** are built by stromatolites/thrombolites and calcimicrobes (microbes capable of mediating carbonate precipitation). **Skeletal mounds** are composed of reef-building organisms (see below) plus calcareous algae, broyzoans, sponges, ahermatypic hexacorals, and some kinds of brachiopods and bibalves. **Mud mounds** are formed by lime mud (inorganic?) accumulations with various amounts of fossils. Mounds range in size from small structures (1–5 m high) to gigantic edifices that may reach 100 m high (e.g., Wendt et al., 1997).

Figure 11.14
Schematic representation of the principal kinds of reefs. Based on Tucker, M. E., and V. P. Wright, 1990, Carbonate sedimentology: Blackwell Scientific Publications, Fig. 4.86, p. 192.

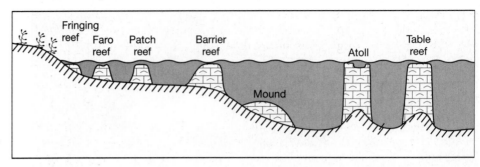

Reef Organisms

We tend to think of all reefs as coral reefs (e.g., Fig. 11.12); however, many organisms in addition to corals can contribute to the formation of reefs. These organisms include blue-green algae (cyanobacteria), coralline red algae, green algae, encrusting foraminifera, encrusting bryozoa, sponges, and molluscs (e.g., Fig. 11.15). In the geologic past, reef-building organisms also included some now-extinct groups such as the archaeocyathids, stromatoporoids, fenestellid bryozoans, and rudistid clams. Nonetheless, corals are certainly dominant constituents of modern reefs, and two types of corals are recognized. The principal corals in shallow-water reefs are hermatypic (zoanthellae) hexacorals. Hermatypic corals carry out a symbiotic relationship with several kinds of unicellular organisms, mainly algae, referred to collectively as zooxanthellae. These algae live in or between the living cells of the corals and aid them in gaining energy by producing photosynthetic products (Cowen, 1988). They may also facilitate the process of secreting calcium carbonate by removing CO_2 from the tissues during photosynthesis. Because the zooxanthellae require sunlit waters, hermatypic corals are restricted to living in very shallow water. Ahermatypic (azooxanthellae) corals lack the symbiotic relationship (or do not require it) and are not restricted to shallow water (e.g., Martin Willison

Figure 11.15
Some common organisms that act as frame-builders, sediment contributors, bafflers, binders, and precipitators in reefs and mounds. The thickness of the horizontal bars indicates relative importance. [After Tucker, M. E., and V. P. Wright, 1990, Carbonate sedimentology: Blackwell Scientific Publications, Fig. 4.88, p. 194. Reproduced by permission.]

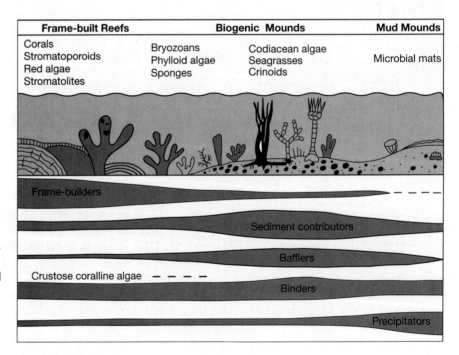

et al., 2001). Some coral species can apparently have life strategies ranging from zooxanthellate-hermatypic to azooxanthellate-ahermatypic (Best, 2001). They are one of the principal organisms today that form carbonate buildups in deeper water. Their distribution ranges from shallow water to water depths exceeding 2000 m (Stanley and Cairns, 1988). Differences in the attributes of shallow- and deep-water corals are explored by Hatcher (2001).

Some reef-building organisms such as corals and stromatoporoids are important **frame-builders** (Fig. 11.15), which construct wave-resistant cores of reefs. Others, such as crinoids and codiacian algae (e.g., *Halimeda*), whose skeletal elements may disintegrate into smaller fragments, are important **sediment contributors**. **Bafflers** are organisms such as seagrass that provide a protective baffle against currents and thus generate a localized, low-energy environment in which fine sediment can accumulate. **Binders** such as cyanobacteria (which form stromatolites) trap and bind sediment, and **precipitators** are primarily microbes, such as cyanobacteria, that help mediate precipitation of carbonate muds. The relative importance of some common organisms as frame-builders, sediment contributors, bafflers, binders, and precipitators is illustrated in Figure 11.15.

The growth forms of reef-building organisms may range from delicate branching forms to globular or massive structures (Fig. 11.16A). The form of the organisms is closely related to the water energy over the reef and, thus, varies over different parts of the reef (Fig. 11.16B). Organisms that live in low-energy parts of the reef tend to have delicate branching or platelike forms. Those living in higher energy zones of the reef develop hemispherical, encrusting, or tabular forms that are better able to withstand strong wave action.

A Growth Forms		Environment	
		Wave Energy	**Sedimentation**
	Delicate, branching	low	high
	Thin, delicate, plate-like	low	low
	Globular, bulbous, columnar	moderate	high
	Robust, dendroid, branching	moderate-high	moderate
	Hemispherical, domal, irregular, massive	moderate-high	low
	Encrusting	intense	low
	Tabular	moderate	low

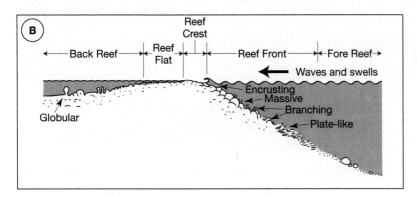

Figure 11.16

Growth forms of reef-building organisms. A. Principal kinds of growth forms and their relationship to water energy. B. Distribution of growth forms across a typical reef. [A. after James, N. P., 1983, Reef environment, *in* Scholle, P. A., D. G. Bebout, and C. H. Moore (eds.), Carbonate depositional environments: Am. Assoc. Petroleum Geologists Mem. 33, Fig. 59, p. 374. B. after James, N. P., 1984, Reefs, *in* Walker, R. G. (ed.), Facies models, 2nd ed.: Geoscience Canada Reprint Ser. 1. Fig. 9, p. 233, reproduced by permission of Geological Association of Canada.]

Reef Deposits

We cannot discuss all types of modern reefs and reef facies here; however, we will examine the zoning and facies development of high-energy, platform-margin reefs as a general model for high-energy reef environments. Figure 11.17 illustrates schematically the principal facies subdivisions of a platform margin reef. Note that the reef consists of a central core, the **reef framework**, which grades seaward into the **reef slope,** and a loose accumulation of reef debris called the **fore-reef talus.** The nearly flat, uppermost, shallowest part of the reef is called the **reef flat,** which grades landward into **back-reef skeletal (coralgal) sands** and **subtidal lagoonal deposits.** Reefs are divided physiographically into **fore-reef, reef-front, reef-crest, reef-flat,** and **back-reef** zones (Fig. 11.16B).

Figure 11.17 also shows the types of carbonate materials typically formed in different zones of the reef. The words rudstone, bafflestone, bindstone, and framestone in Figure 11.17 are terms used by Embry and Klovan (1971) as modifications of Dunham's (1962) limestone classification (see Table 6.3). Floatstone and rudstone are unbound carbonate grains, more than 10 percent of which are more than 2 mm in size; floatstones are mud supported; and rudstones are grain supported. Bafflestones are carbonate components bound together at the time of deposition by stalked organisms that trapped sediment by acting as baffles. Bindstones were bound during deposition by encrusting and binding organisms such as encrusting foraminifers and bryozoans, and framestones were bound by organisms such as corals, which build a rigid framework structure.

These different carbonate facies represent variations in water energy, dominant sedimentation processes, and types of organisms in each zone of the reef. Water energy is highest on the reef crest, which also contains the highest percentage of framework constituents (framestones). As water energy decreases toward both the fore reef and the back reef, the percentage of framework constituents also decreases. Note that overall the framework component of reefs is commonly much smaller than the volume of nonframework constituents. Longman (1981) compares the structure of reefs to that of an apple, which has a central core, or framework, surrounded by the much larger edible fruit. The nonframework fraction of reefs consists of organisms such as echinoderms, green algae, and molluscs, which do not build framework structures, bioclasts broken from the reef by wave activity, and, in lower-energy zones of the reef, some lime mud. The fore-reef talus slope and back-reef coralgal sand zone are made up entirely

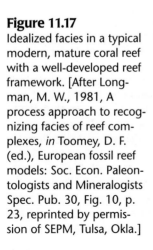

Figure 11.17
Idealized facies in a typical modern, mature coral reef with a well-developed reef framework. [After Longman, M. W., 1981, A process approach to recognizing facies of reef complexes, *in* Toomey, D. F. (ed.), European fossil reef models: Soc. Econ. Paleontologists and Mineralogists Spec. Pub. 30, Fig. 10, p. 23, reprinted by permission of SEPM, Tulsa, Okla.]

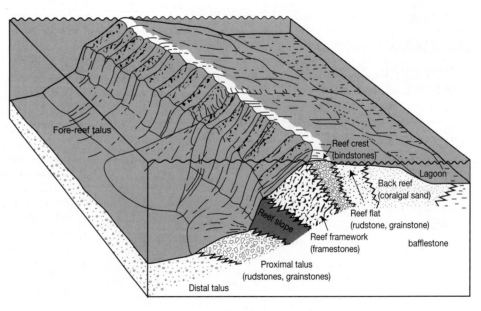

of nonframework constituents that consist principally of reef-derived bioclasts. Relatively few organisms live in these zones.

Low-Energy Reef Facies

The facies of modern, high-energy, platform margin-type reefs thus consist fundamentally of a central framework core composed largely of corals and coralline algae; the core grades seaward through a zone of rubbly fore-reef talus to deeper water lime muds or shales and landward through back-reef coralgal sands to finer grained lagoonal deposits. This model serves reasonably well for high-energy reefs developed in most settings; however, some reefs form under much lower energy conditions. Low-energy reefs do not develop the characteristic zoning of high-energy reefs and tend to be circular to elliptical in plan view. Organisms growing on such reefs are dominated by the more delicate, branching forms (Fig. 11.16A). Some low-energy reefs do not contain the typical reef structure described above but are constructed simply of carbonate sands and muds built by organisms that are very similar to reef-type organisms in composition (e.g., James, 1984c). Other low-energy buildups are composed largely of nonreef-type organisms. They consist of mound-shaped piles of skeletal fragments and/or bioclastic lime muds rich in skeletal organisms and minor amounts of organic boundstone. As mentioned, these structures are called reef mounds or simply mounds (e.g., James and Bourque, 1992).

Ancient Reefs

Reef Deposits

Reefs as they appear in the fossil record differ in some important respects from modern reefs. First, we commonly see ancient reefs only in vertical exposures. We observe a two-dimensional limestone body composed of different components formed at different times (James, 1983). Thus, we may not be able to detect all of the facies zones displayed by modern reefs. The nature of the flank and inter-reef facies exposed in outcrop obviously may differ depending upon whether the vertical exposure cuts through the reef to expose a cross section from fore reef to back reef or a longitudinal section running parallel to the reef crest (e.g., Fig. 11.9). In many cases, only the massive reef core is clearly exposed, as in Figure 11.18. The reef structure and nature of reef facies also depend upon the kind of reef (e.g., bar-

Figure 11.18
The Resolution Reef (Pennsylvanian) exposed near Minturn, Colorado. A. Distal view showing the biohermal shape of the reef core, enclosed in finer sediment. B. Close-up view of the front of the massive reef core.

rier reef, fringing reef, patch reef). An additional factor that can further complicate recognition of ancient reef facies is diagenesis, which may cause selective dolomitization or solution that can obliterate or destroy parts of the reef complex.

Depending upon their age, ancient reefs and mounds may also differ markedly from modern reefs and mounds in terms of the dominant organisms that make up these structures. The hermatypic corals that dominate modern coral reefs first appeared in the Mesozoic and thus are not components of older reefs. Older reefs and mounds are dominated by other kinds of organisms, such as tubiphytes (problematic microstructures of possible algal origin), sponges, bryozoans, and microbes (e.g., cyanobacteria) (Fig. 11.19). See Fagerstrom (1987), Stanley and Fagerstrom (1988), and Kiessling, Flügel, and Golonka (1999, 2002) for additional information about reef organisms and the evolution of reef ecosystems through time.

Because many kinds of ancient reefs, with different kinds of reef-building organisms, are present in the sedimentary record, it is difficult to generalize about the structure and makeup of ancient reefs. Still, some effort in that direction is required. Figure 11.20 is a highly generalized sketch of an idealized ancient reef enclosed in nonreef deposits. The core framework consists of skeletal bindstones and rudstones. The specific kinds of skeletal components depend upon the age of the reef. These core sediments grade seaward into the forereef facies, consisting in large part of rudstones and floatstones broken from the framework by wave action. The reef core grades landward into a high-energy, skeletal sand apron consisting also largely of material broken off the reef core. In turn, skeletal sands grade into a back-reef facies deposited under quieter water conditions. This facies may contain skeletal sands and muds, pelleted muds, and micrites, and stromatolites may be common. Evaporites may also be present in some back-reef facies, reflecting restricted water circulation in the back-reef environment. Reef growth eventually terminates because of drowning of the reef or possibly, in some cases, burial by siliciclastic muds. The topography of the ancient reef is commonly visible beneath this covering of younger sediment, as illustrated in Figure 11.20.

Occurrence of Ancient Reefs

Reefs of some type are present in Phanerozoic (post-Precambrian) carbonate rocks of most ages (e.g., see the compendium volume edited by Kiessling, Flügel, and Golonka, 2002). Although carbonate-secreting organisms were not present during the Precambrian, Precambrian carbonate buildups composed of stromatolites have been reported from various localities in North America, Europe, Africa, and Australia. Phanerozoic rocks contain reefs composed of calcium carbonate-secreting organisms. Many of these reefs were built by framework-constructing or encrusting organisms. Reef development was not uniform throughout geologic time, however, and reefs are much more abundant in some parts of the rock record than in others. Figure 11.19 graphically depicts the distribution of reefs and mounds through time and also shows the major kinds of organisms that were responsible for reef building at different times, as previously mentioned. Figure 11.19 is based on a database of 2470 reefs compiled by Kiessling, Flügel, and Golonka (1999). These authors also show the geographic distribution of reefs of various ages and discuss reef type, dimensions, and environmental setting. For an update of these data, see Kiessling (2002). These variations in reef distribution and abundance through time reflect times when carbonate production on reefs flourished and times when reefs were in crisis and carbonate production declined (e.g., Flügel and Kiessling, 2002).

11.5 MIXED CARBONATE-SILICICLASTIC SYSTEMS

To avoid possible confusion, the carbonate depositional environments described in the preceding sections are discussed as if only carbonate sediments are deposited

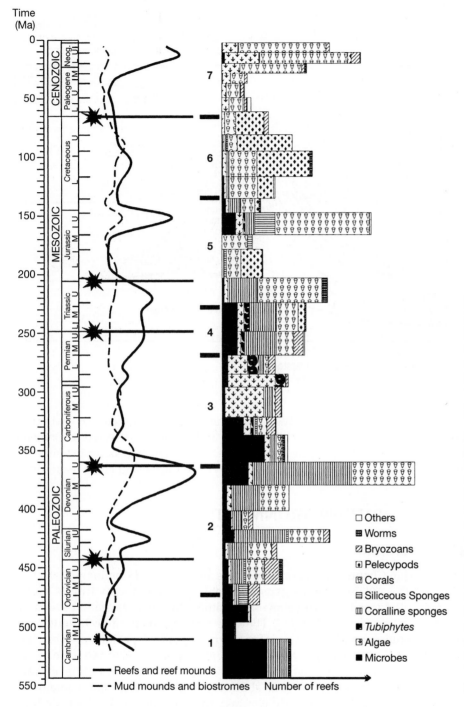

Figure 11.19
Distribution of reefs through Phanerozoic time showing both reefs and mud mounds as well as the dominant organisms that make up the reefs and mounds. "Others" refers to brachiopods, pelmatozoans, and foraminifers. The numbers 1–7 refer to seven major cycles of Phanerozoic reef building. Starred lines indicate times of mass extinctions. Curves to the left show the cumulative number of reefs and mounds in each time interval, and horizontal bars on the right depict the cumulative number of reefs in which a particular fossil group is dominant. [From Kiessling, W., E. Flügel, and J. Golonka, 1999, Paleoreef maps: Evaluation of a comprehensive database on Phanerozoic reefs: American Association Petroleum Geologists Bulletin, v. 83, Fig. 13, p. 1576. Reproduced by permission.]

in these systems. In fact, mixed carbonate and siliciclastic sediments are present in many stratigraphic successions. Such deposits are referred to as mixed carbonate-siliciclastic successions or carbonate-clastic transitional successions. Carbonate and siliciclastic sediments can mix owing to lateral facies mixing (spatial variations in environments) that produces lateral interfingering of carbonate and clastic sediment. Mixtures may also result from sea-level change and/or variations in sediment supply, which cause vertical variations in the stratigraphic succession of facies, called temporal variability (Budd and Harris, 1990). Thus, siliciclastic facies may occur in a lateral interfingering relationship (see Chapter 12) with carbonate facies or as distinct interbeds within carbonate successions. Carbonate-siliciclastic

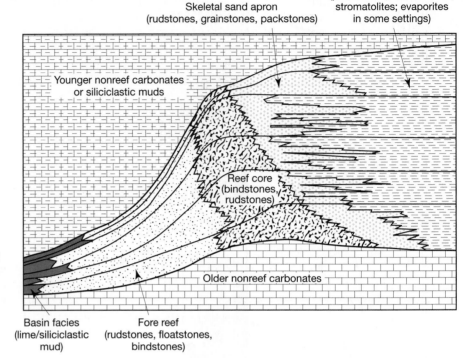

Skeletal sand apron
(rudstones, grainstones, packstones)

Back reef
(packstones, wackestones,
stromatolites; evaporites
in some settings)

Younger nonreef carbonates
or siliciclastic muds

Reef core
(bindstones,
rudstones)

Older nonreef carbonates

Basin facies
(lime/siliciclastic
mud)

Fore reef
(rudstones, floatstones,
bindstones)

Figure 11.20
Idealized, schematic representation of an ancient barrier reef deposit illustrating typical reef, fore-reef, and back-reef facies. Based in part on the Upper Devonian Swan Hills Reef Buildup, central Alberta, Canada (Viau, 1983).

transitions are known in a variety of environments, including coastal and inner-shelf, middle- and outer-shelf (including reef), and the slope-to-basin environments (Doyle and Roberts, 1988; Lomando and Harris, 1991). They may occur in temperate as well as tropical shelf environments (e.g., Haywick, Carter, and Henderson, 1992).

11.6 EVAPORITE ENVIRONMENTS

General Statement

As discussed in Chapter 7, evaporite deposits are composed dominantly of halite (NaCl) and the sulfate minerals gypsum ($CaSO_4 2H_2 0$) and anhydrite ($CaSO_4$). Evaporites are deposited under climatic conditions where evaporation losses exceed precipitation (rain and snow). They are forming today in both nonmarine and marine environments. The sites of modern evaporite deposition are very limited in size, however, compared to many ancient evaporite basins that were gigantic. To illustrate, one of the larger modern evaporite deposits occurs in Lake Macleod, western Australia, which is about 100×50 km in size and contains up to 9 m of Holocene evaporites. By contrast, the Messinian (Miocene) Mediterranean basins extended over an area of about 2400×600 km and have evaporite fills up to 2 km thick (Kendall and Harwood, 1996). Although no modern analogs for these giant evaporite basins exist, we can nonetheless gain important insight into ancient evaporite environments by studying modern environments.

Modern Evaporite Environments

Nonmarine Environment

Figure 7.5 (Chapter 7) illustrates the principal environments in which evaporites are forming today. They are present in a variety of small-scale continental settings such as springs, desert dunefields, and soils (Smoot and Lowenstein, 1991). They typically form and are most important, however, in closed basins characterized by playas (continental sabkhas) that contain an ephemeral lake or salt pan, commonly surrounded by saline mudflats (Kendall, 1992). Evaporite minerals are deposited in salt pans when ephemeral water evaporates and also displacively and as cements within adjacent mudflat sediment. In a few continental settings where intermontaine basins are fed by perennial streams, perennial saline lakes may exist (e.g., Great Salt Lake, Utah). Nonmarine evaporite deposits may contain minerals such as borax, epsomite, and trona (see Chapter 7) that are not common in marine deposits; however, gypsum, anhydrite, and halite tend to dominate nonmarine deposits as they do in marine deposits.

Shallow Marine Environment

Modern marine evaporites are forming in two principal kinds of settings: coastal sabkhas and salinas (e.g., Handford, 1991; Warren, 1991). Marine **sabkhas** are coastal supratidal mudflats. Evaporite minerals do not precipitate from standing water but instead form displacively within sabkha sediments, consisting of carbonates and/or siliciclastic deposits, in a capillary zone above a saline water table. Water lost by evaporation is replaced by downward seepage of storm-driven seawater or by interflow of groundwater from continental sources. As illustrated by the sabkhas of the Arabian Gulf (Fig. 11.21), evaporite minerals that crystallize within the sabkha sediment are mainly gypsum, anhydrite, and carbonates. Algal mats (stromatolites) commonly form in associated intertidal to supratidal sediments.

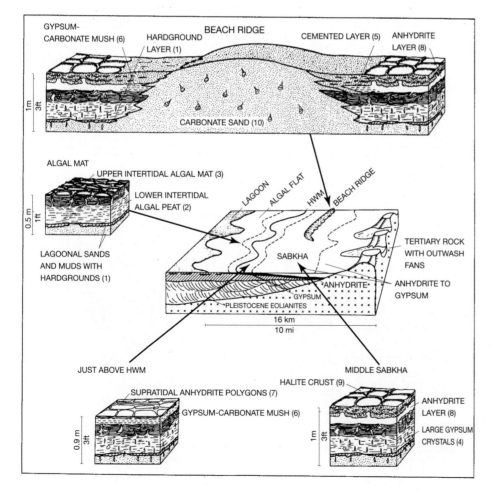

Figure 11.21
Schematic representation of vertical and lateral facies relationships in sabkhas of the Arabian Gulf.
HWM = high-water mark. [From Warren, J. K., and G. St. C. Kendall, 1985, Comparison of sequences formed in marine sabkha (subaerial) and salina (subaqueous) settings—Modern and ancient: Am. Assoc. Petroleum Geologists Bull., v. 69, Fig. 2, p. 1015, reprinted by permission of AAPG, Tulsa, Okla.]

Other modern shallow-marine evaporites are forming in a few marine-fed coastal lagoons and salt pans called **salinas**. Salinas occur in depressions on sabkhas, between coastal dunes, on deltas, or in tectonic downwarps behind coastal barriers. Modern salinas are particularly common in southern and western Australia (e.g., Fig. 11.22), but they are present also around the margins of the Mediterranean, Black, and Red Seas. Salinas differ from sabkhas particularly in that evaporite deposits precipitate primarily from surface brine rather than within sediment. Gypsum is the most common mineral in salina deposits; however, some salinas such as Lake Macleod, Australia, contain abundant halite.

Deep-Water Environment

Water depth in most modern salinas is quite shallow, a few meters at most. The Dead Sea, lying between Israel and Jordan, provides a single modern example of a deeper water evaporite environment. The Dead Sea, about 15–20 km wide and 80 km long, is fed by the Jordan River. The southern basin of the Dead Sea is shallow, but the northern basin has a brine depth of more than 200 m. Clayey silt is the predominant sediment deposited in the basin; however, evaporitic minerals such as gypsum, halite, and aragonite are also forming (Garber, Levy, and Friedman, 1987). Although called a sea, the Dead Sea is actually a large, closed, perennial saline lake that lies in a rift valley more than 400 m below sea level, and the composition of its brine differs considerably from that of sea water (Kendall and Harwood, 1996). Thus, although a deeper water evaporite basin, the Dead Sea is not a

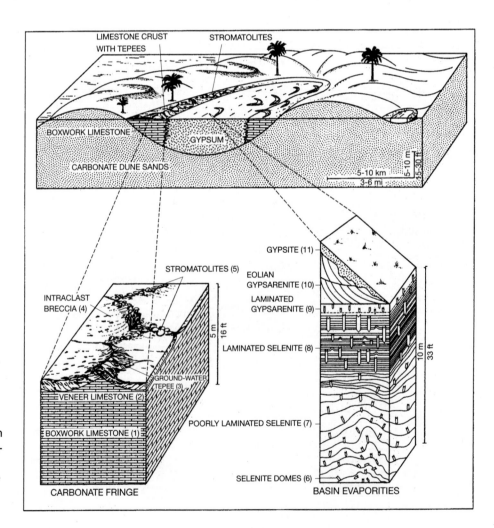

Figure 11.22
Environmental setting and typical evaporite facies in southern Australia salinas. [From Warren, J. K., and G. St. C. Kendall, 1985, Comparison of sequences formed in marine sabkha (subaerial) and salina (subaqueous) settings—Modern and ancient: Am. Assoc. Petroleum Geologists Bull., v. 69, Fig. 4, p. 1017, reprinted by permission of AAPG, Tulsa, Okla.]

modern analog for large central-basin, marine evaporite deposits (discussed below) that formed in the geologic past.

Ancient Evaporite Environments

Nonmarine Environment

Because the sites of modern continental evaporite deposition are rather small on the whole, geologists have commonly assumed that ancient nonmarine deposits are also quite small. Hardie (1984) takes exception to this general perception. He maintains that nonmarine evaporite deposits can be as large or as thick as marine evaporites, and cites several examples of Tertiary nonmarine salt deposits in California and Nevada, that are 200–600 m thick, to support his view. Still, few large ancient nonmarine evaporite deposits are known, and relatively few nonmarine deposits of any size have been reported compared to the number of putative ancient marine evaporite deposits.

The nonmarine versus marine origin of ancient evaporites is determined mainly on the basis of the kinds of fossils in and the sedimentology of associated nonevaporite facies. For example, nonmarine evaporites may be associated with continental redbeds, alluvial fan sediments, or desert dune deposits. An example of this kind of association is illustrated by the Wilkins Peak Member of the Green River Formation (Eocene), Wyoming, shown in Figure 8.26 (Chapter 8). Mineralogy may also provide clues to nonmarine origin because certain evaporite minerals appear to form only under nonmarine conditions (Chapter 7). Other putative ancient nonmarine evaporite deposits include the sediments of the Tajo Basin (Miocene), Spain, filled with more than 1500 m of siliciclastic sediments and associated evaporites; evaporites associated with fluvial and lacustrine siliciclastic sediments in the Newark Basin (Triassic-Jurassic) of the northeastern United States; and parts of the Permian Rotliegendes Formation of Europe (Smoot and Lowenstein, 1991).

Marine Environment

As mentioned, the great bulk of ancient evaporites and the truly large evaporite deposits appear to be of marine origin. Some ancient marine evaporites were probably deposited in settings similar to modern coastal sabkhas and salinas. On the other hand, many ancient evaporites appear to have formed in broad shelf environments or within marine basins for which there are no good modern analogs. Thus, the nature of these ancient environments must be deduced from the rock record itself. Consequently, considerable controversy has arisen with respect to the nature of these environments, and the question of deep- vs. shallow-water origin is particularly troubling. In any case, it now appears that ancient marine evaporites likely accumulated in three kinds of settings: small coastal settings similar to modern sabkhas and salinas; wide, basin-margin shelves or platforms (Fig. 11.23A); and extensive basin-central environments (Fig. 11.23B).

Basin-margin evaporites likely formed in vast expanses of evaporitic lagoons and mudflats, possibly extending tens to thousands of kilometers, throughout which brine depths did not exceed a few meters (Kendall and Harwood, 1996). The evaporites consist of anhydrite (originally gypsum) and/or halite, which may be interbedded with carbonates and siliciclastic sediment. Cyanobacterial mats may have been present in less saline parts of the setting. For example, the San Andreas Formation (Permian) of the Palo Duro Basin, Texas, contains more than twenty cycles consisting of anhydritic mudstone overlain by a regressive succession of carbonate, anhydrite, and bedded halite. These cycles range from 1 to 100 m thick and are traceable over more than 10,000 km^2, with only minor changes in thickness and facies (Hovorka, 1987).

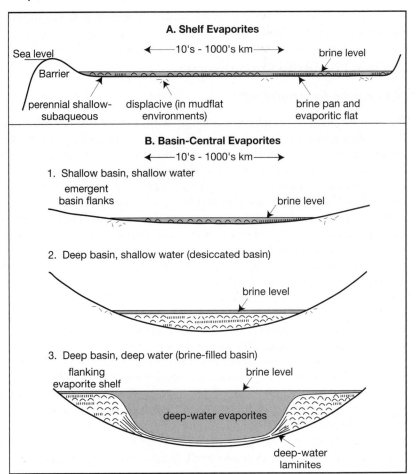

Figure 11.23
Principal kinds of ancient marine evaporite environments (excluding small-scale coastal sabkhas and salinas). [After Kendall, A. C., 1992, Evaporites, *in* Walker, R. G., and N. P. James (eds.), Facies models—Response to sea-level change: Geological Association of Canada, Fig. 5, p. 380, reproduced by permission.]

Basin-central evaporites accumulated in subsiding basins that may also have ranged from tens to thousands of kilometers across (Fig. 11.23B). Some basins had very little relief, and evaporites accumulated from shallow bodies of brine or within saline mudflats (**shallow-basin, shallow-water model;** Fig. 11.23B1). Similar shallow-water evaporites could form from shallow bodies of brine that existed within basins having much greater relief, which were subjected to substantial evaporative drawdown (**deep-basin, shallow-water model;** Fig. 11.23B2). Finally, deeper water evaporites formed in basins having substantial relief (tens to hundreds of meters) and which were largely filled with brine (**deep-basin, deep-water model;** Fig. 11.23B3). Deep-water evaporites appear to be especially characterized by fine, even, widely distributed laminations (see Fig. 7.3). They may also include gravity-displaced deposits. The Castille Formation (Permian), Texas and New Mexico, is a classic example of a laminated, deep-water evaporite in which individual lamina about 2 mm thick can be traced for a distance of 113 km (Dean and Anderson, 1978).

As noted in Chapter 7, some kind of barrier must be present between evaporative basins and the open ocean to provide partial restriction of ocean-water circulation into the basins (see Fig. 7.6). Brine levels in these basins can fluctuate in response to changes in rates of sea-water inflow, outflow, and evaporation. Thus, deep basins can alternate between deep-water evaporite deposition and shallow-water/mudflat evaporite deposition. Also, deep basins can eventually fill up with evaporites and be converted into shallow basins.

Two of the most impressive ancient marine evaporite deposits are the Permian Zechstein evaporites and the Miocene Messinian evaporites. The **Zechstein** evaporites of the North Sea area were deposited in a huge intercratonic basin that extended from the British Isles to eastern Poland and Lithuania. More than 2 km of carbonates, evaporites, and siliciclastics were deposited during a period of five million years (Menning, Katzung, and Lutzner, 1988). Evaporites include carbonates, anhydrite, halite, and potash salts, and deposition took place in settings that ranged, at various times, from water as deep as 200 m to shallow-water brine flats.

The Miocene **Messinian** evaporites of the Mediterranean region cover an area of about 2400 × 600 km in Sicily and the eastern Apennines. The evaporite consists of carbonates, gypsum/anhydrite, halite, and potash salts that range in thickness from a few meters at the basin margin to more than 2 km in the basin center. The evaporite succession is underlain and overlain by deep-water marine sediments, suggesting desiccation of the entire Mediterranean Basin (Kendall and Harwood, 1996). A few of the evaporites appear to be sabkha deposits; however, most apparently formed in shallow subaqueous environments, whereas a few contain evenly laminated beds that suggest deeper water.

FURTHER READING

Carbonates

Abegg, F. E., P. M. Harris, and D. B. Loope (eds.), 2001, Modern and ancient carbonate eolianites: Sedimentology, sequence stratigraphy, and diagnesis: SEPM Special Publication No. 71, 207 p.

Alsharhan, A. S., and R. W. Scott (eds.), 2000, Middle East models of Jurassic/Cretaceous carbonate Systems: SEPM Special Publication No. 69, 364 p.

Camoin, G. F., and P. J. Davies, 1998, Reefs and carbonate platforms in the Pacific and Indian Oceans: International Association of Sedimentologists, Spec. Publ. No. 25, Blackwell Science Ltd., Oxford, 328 p.

Franseen, E. K., M. Esteban, W. C. Ward, and J-M. Rouchy, 1996, Models for carbonate stratigraphy from Miocene reef complexes of Mediterranean regions: Concepts in sedimentology and paleontology, v. 5, Society for Sedimentary Geology, Tulsa, Okla., 391 p.

James, N. P., and J. A. D. Clarke, 1997, Cool-water carbonates: SEPM Special Publication No. 56, Society for Sedimentary Geology, Tulsa, Okla., 440 p.

Kiessling, W., E. Flügel, and J. Golonka (eds.), 2002, Phanerozoic reef patterns: SEPM Special Publication No. 72, Society for Sedimentary Geology, Tulsa, Okla., 775 p.

Lomando, A. J., and P. M. Harris (eds.), 1991, Mixed carbonate-siliciclastic sequences: Soc. Econ. Paleontologists and Mineralogists Core Workshop No. 15, 568 p.

Monty, C. L. V., D. W. J. Bosence, P. H. Bridges, and B. R. Pratt (eds.), 1995, Carbonate mud-mounds—Their origin and evolution: International Association of Sedimentologists, Spec. Pub. No. 32, Blackwell Science Ltd., Oxford, 537 p.

Simo, J. A. T., R. W. Scott, and J. P. Masse (eds.), 1993, Cretaceous carbonate platforms: AAPG Memoir 56, American Association Petroleum Geologists, Tulsa, Okla., 479 p.

Stanley, G. D., Jr., and J. A. Fagerstrom (eds.), 1988, Ancient reef ecosystems: Palaios, v. 3, p. 111–254.

Tucker, M. E., J. L. Wilson, P. D. Crevello, J. R. Sarg, and J. F. Read, 1990, Carbonate platforms—Facies, successions and evolution: Internat. Assoc. Sedimentologists Spec. Pub. No. 9, Blackwell, Oxford, 328 p.

Tucker, M. E., and V. P. Wright, 1990, Carbonate sedimentology: Blackwell, Oxford, 482 p.

Vacher, H. L., and T. M. Quinn (eds.), 1997, Geology and hydrology of carbonate islands: Elsevier Science B. V., Amsterdam, 948 p.

Zempolich, W. C., and H. E. Cook (eds.), 2002, Paleozoic carbonates of the Commonwealth of Independent States (CIS): Subsurface reservoirs and outcrop analogs: SEPM Special Publication No. 74, Society for Sedimentary Geology, Tulsa, Okla., 245 p.

Evaporites

Busson, G., and Schreiber, B. C. (eds.), 1997, Sedimentary deposition in rift and foreland basins in France and Spain: Columbia University Press, New York, 479 p. (A volume devoted entirely to evaporite deposits.)

Melvin, J. L. (ed.), 1991, Evaporites, petroleum, and mineral resources: Elsevier, Amsterdam, 556 p.

Schreiber, B. C. (ed.), 1988, Evaporites and hydrocarbons: Columbia University Press, New York, 475 p.

Warren, J. K., 1989, Evaporite sedimentology: Prentice Hall, Englewood Cliffs, N. J., 285 p.

PART V

Stratigraphy and Basin Analysis

Well-stratified Paleozoic sedimentary rocks exposed in the North Rim of the Grand Canyon, Arizona.

397

Emphasis in preceding parts of this book has been on sedimentary processes, the environments in which these processes take place, and the properties of sedimentary rocks generated in these environments. In this final part, we focus on a different aspect of sedimentary rocks. Our concern here is not so much sedimentary processes and detailed rock properties but rather the larger scale vertical and lateral relationships between units of sedimentary rock that are defined on the basis of lithologic or physical properties, paleontological characteristics, geophysical properties, age relationships, and geographic position and distribution. It is the study of these characteristics of layered rocks that encompasses the discipline of **stratigraphy**. Understanding the principles and terminology of stratigraphy is essential to geologic study of sedimentary rocks because stratigraphy provides the framework within which sediments can be studied systematically. It allows the geologist to bring together the details of sediment composition, texture, structure, and other features into an environmental and temporal synthesis from which we can interpret the broader aspects of Earth history.

Prior to the 1960s, the discipline of stratigraphy was concerned particularly with stratigraphic nomenclature; the more classical concepts of lithostratigraphic, biostratigraphic, and chronostratigraphic successions in given areas; and correlation of these successions between areas. **Lithostratigraphy** deals with the lithology or physical properties of strata and their organization into units on the basis of lithologic character. **Biostratigraphy** is the study of rock units on the basis of the fossils they contain. **Chronostratigraphy** (chrono = time) deals with the ages of strata and their time relations. These established principles are still the backbone of stratigraphy; however, today's students must go beyond these basic principles. They must also acquire a thorough understanding of depositional systems and be able to apply stratigraphic and sedimentological principles to interpretation of strata within the context of global plate tectonics. This means, among other things, becoming familiar with comparatively new branches of stratigraphy that have developed since the early 1960s as new concepts and methods of studying sedimentary rocks and other rocks by remote-sensing techniques have unfolded. For example, the concept of depositional sequences, which are packages of strata bounded by unconformities, has gained particular prominence since the late 1970s. This concept has now become so important that we refer to study of sequences as **sequence stratigraphy**. Two new offshoots of stratigraphy that have made particularly important contributions to our understanding of the physical stratigraphic relationships, ages, and environmental significance of subsurface strata and oceanic sediments are **seismic stratigraphy**, which is the study of stratigraphic and depositional facies as interpreted from seismic data, and **magnetostratigraphy**, which deals with stratigraphic relationships on the basis of magnetic properties of sedimentary rocks and layered volcanic rocks. We also recognize sub-branches of stratigraphy such as **event stratigraphy** (correlation of sedimentary units on the basis of marker beds or event horizons), **cyclostratigraphy** (study of short-period, high-frequency, sedimentary cycles in the stratigraphic record, particularly by use of oxygen-isotope data), and **chemostratigraphy** (correlation on the basis of stable isotopes such as oxygen, carbon, and strontium). Application of these new concepts and techniques makes possible subdivision of stratigraphic successions into relatively small units, e.g., <1000–3000 years for Quaternary strata and 225,000–2,000,000 years for Triassic strata (Kidd and Hailwood, 1993). Such fine-scale stratigraphic resolution is now commonly referred to as **high-resolution stratigraphy**.

We explore all of these stratigraphic concepts in the next few chapters. In the final chapter of the book, we take a look at basin analysis. **Basin analysis** is a kind of umbrella under which we integrate and apply all of the sedimentologic and stratigraphic principles presented in this book. Together with fundamental tectonic concepts, basin analysis allows us to develop an understanding of the rocks that fill sedimentary basins in order to interpret their geologic history and evaluate their economic significance.

12

Lithostratigraphy

12.1 INTRODUCTION

Perhaps the most fundamental kind of stratigraphic study is recognition, subdivision, and correlation (establishing equivalency) of sedimentary rocks on the basis of their lithology, that is, lithostratigraphy. The term **lithology** is used by geologists in two different but related ways. Strictly speaking, it refers to study and description of the physical character of rocks, particularly in hand specimens and outcrops (Bates and Jackson, 1980). It is used also to refer to these physical characteristics: rock type, color, mineral composition, and grain size are all lithologic characteristics. For example, we may refer to the lithology of a particular stratigraphic unit as sandstone, shale, limestone, and so forth. Thus, lithostratigraphic units are rock units defined or delineated on the basis of their physical properties, and lithostratigraphy deals with the study of the stratigraphic relationships among strata that can be identified on the basis of lithology.

In this chapter, we begin study of stratigraphic principles by briefly discussing the nature of lithostratigraphic units, followed by exploration of the various types of contacts that separate these units. We then consider the important concepts of sedimentary facies and depositional sequences. The essentials of stratigraphic nomenclature and classification as they apply to lithostratigraphic units are discussed next, including examination of the North American Code of Stratigraphic Nomenclature. Finally, correlation of lithostratigraphic units is explained and the various methods of correlation described.

12.2 TYPES OF LITHOSTRATIGRAPHIC UNITS

Lithostratigraphic units are bodies of sedimentary, extrusive igneous, metasedimentary, or metavolcanic rock distinguished on the basis of lithologic characteristics. A lithostratigraphic unit generally conforms to the **law of superposition**, which states that in any succession of strata, not disturbed or overturned since deposition, younger rocks lie above older rocks. Lithostratigraphic units are also commonly stratified and tabular in form. They are recognized and defined on the basis of observable rock characteristics. Boundaries between different units may be placed at clearly identifiable or distinguished contacts or may be drawn arbitrarily within a zone of gradation. Definition of lithostratigraphic units is based on a **stratotype** (a designated type unit), or type section, consisting of readily accessible rocks, where possible, in natural outcrops, excavations, mines, or bore holes. Lithostratigraphic units are defined strictly on the basis of lithic criteria as determined by descriptions of actual rock materials. They carry no connotation of age. They cannot be defined on the basis of paleontologic criteria, and they are independent of time concepts. They may be established in subsurface sections as well as in rock units exposed at the surface, but they must be established on the basis of lithic characteristics and not on geophysical properties or other criteria. Geophysical criteria, described in Chapter 13, may be used to aid in fixing boundaries of subsurface lithostratigraphic units, but the units cannot be defined exclusively on the basis of remotely sensed physical properties.

Wheeler and Mallory (1956) introduced the term **lithosome** to refer to masses of rock of essentially uniform character and having intertonguing relationships with adjacent masses of different lithology. Thus, we speak of shale lithosomes, limestone lithosomes, sand-shale lithosomes, and so forth. Krumbein and Sloss (1963) clarify the meaning of lithosome by asking readers to imagine the body of rock that would emerge if it was possible to preserve a single rock type, such as sandstone, and dissolve away all other rock types. The resulting sandstone body would thus appear as a roughly tabular mass with intricately shaped boundaries. These irregular boundaries would represent its surfaces of contact with erosion surfaces and with other rock masses of differing constitution above, below, and to the sides. Lithosomes have no specified size limits and may range in gross shape from thin, sheetlike or blanketlike units to thick prisms, or narrow, elongated shoestrings.

Of course, stratigraphic units of a single lithology rarely exist as isolated bodies. They are commonly in contact with other rock bodies of different lithology. An important part of lithostratigraphy is identifying and understanding the nature of contacts between vertically superposed or laterally adjacent bodies. Another important aspect is the identification of single lithosomes, groups of lithosomes, or subdivisions of lithosomes that are so distinctive that they form lithostratigraphic units that can be distinguished from other units that may lie above, below, or adjacent.

The fundamental lithostratigraphic unit of this type is the formation. A **formation** is a lithologically distinctive stratigraphic unit that is large enough in scale to be mappable at the surface or traceable in the subsurface. It may encompass a single lithosome, or part of an intertonguing lithosome, and thus consist of a single lithology. Alternatively, a formation can be composed of two or more lithosomes and thus may include rocks of different lithology. Some formations may be divided into smaller stratigraphic units called **members**, which, in turn, may be divided into smaller distinctive units called **beds**. Beds are the smallest formal lithostratigraphic units. Formations having some kind of stratigraphic unity can be combined to form **groups**, and groups can be combined to form **supergroups**. All formal lithostratigraphic units are given names that are derived from some geographic feature in the area where they are studied.

Subdivision of thick units of strata into smaller lithostratigraphic units such as formations is essential for tracing and correlating strata both in outcrop and in the subsurface. We will come back to discussion of these formal stratigraphic units near the end of this chapter. First, however, we examine the nature of contacts between stratigraphic units and the lateral and vertical facies relationships that characterize strata.

12.3 STRATIGRAPHIC RELATIONS

Different lithologic units are separated from each other by **contacts**, which are planar or irregular surfaces between different types of rocks. Vertically superposed strata are said to be either conformable or unconformable depending upon continuity of deposition. Conformable strata are characterized by unbroken depositional assemblages, generally deposited in parallel order, in which layers are formed one above the other by more or less uninterrupted deposition. The surface that separates conformable strata is a **conformity**, that is, a surface that separates younger strata from older rocks but along which there is no physical evidence of nondeposition. A conformable contact indicates that no significant break or hiatus in deposition has occurred. A **hiatus** is a break or interruption in the continuity of the geologic record. It represents periods of geologic time (short or long) for which there are no sediments or strata.

Contacts between strata that do not succeed underlying rocks in immediate order of age, or that do not fit together with them as part of a continuous whole, are called unconformities. Thus, an **unconformity** is a surface of erosion or nondeposition, separating younger strata from older rocks, that represents a significant hiatus. Unconformities indicate a lack of continuity in deposition and correspond to periods of nondeposition, weathering, or erosion, either subaerial or subaqueous, prior to deposition of younger beds. Unconformities thus represent a substantial break in the geologic record that may correspond to periods of erosion or nondeposition lasting millions or even hundreds of millions of years.

Contacts are also present between laterally adjacent lithostratigraphic units. These contacts are formed between rock units of equivalent age that developed different lithologies owing to different conditions in the depositional environment. Excluded from discussion here are contacts between laterally adjacent bodies that arise from postdepositional faulting. Contacts between laterally adjacent bodies may be gradational, where one rock type changes gradually into another, or they may be intertonguing, that is, pinching or wedging out within another formation (see Figures 12.1–12.4).

Contacts between Conformable Strata

Contacts between conformable strata may be either abrupt or gradational. **Abrupt contacts** directly separate beds of distinctly different lithology (Figs. 12.1, 12.2). Most abrupt contacts coincide with primary depositional bedding planes that formed as a result of changes in local depositional conditions, as discussed in Chapter 4; thus, the contacts are commonly very sharp. In general, bedding planes represent minor interruptions in depositional conditions. Such minor depositional breaks, involving only short hiatuses in sedimentation with little or no erosion before deposition is resumed, are called **diastems**. Abrupt contacts may be caused also by postdepositional chemical alteration of beds, producing changes in color owing to oxidation or reduction of iron-bearing minerals, changes in grain size owing to recrystallization or dolomitization, or changes in resistance to weathering owing to cementation by silica or carbonate minerals.

Conformable contacts are said to be **gradational** if the change from one lithology to another is less marked than abrupt contacts, reflecting gradual change in depositional conditions with time (Fig. 12.1). Gradational contacts may be of either the progressive gradual type or the intercalated type. **Progressive gradual contacts** occur where one lithology grades into another by progressive, more or less uniform changes in grain size, mineral composition, or other physical characteristics. Examples include sandstone units that become progressively finer grained upward until they change to mudstones, or quartz-rich sandstones that become progressively enriched upward in lithic fragments until they change to lithic arenites. **Intercalated contacts** are gradational contacts that occur because of an increasing number of thin interbeds of another lithology that appear upward in the section (Figs. 12.1, 12.3).

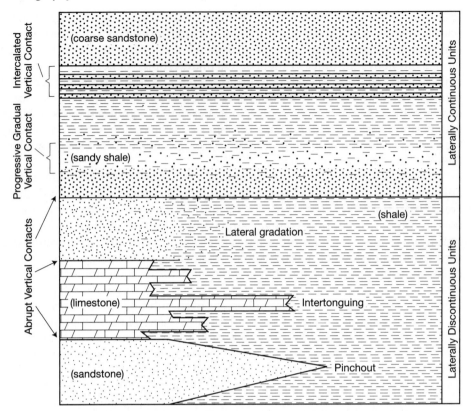

Figure 12.1
Schematic representation of the principal kinds of vertical and lateral contacts between lithologic units. Vertical contacts include abrupt, progressive gradual, and intercalated. Lithologic units may be laterally continuous or they may change laterally by pinchout, intertonguing, or lateral gradation. See Figures 12.2–12.4 for examples of actual contacts.

Figure 12.2
Abrupt contact (arrow) between massive-bedded sandstone below and fine-grained conglomerate above. Miocene deposits near Blacklock Point, Southwest Oregon Coast.

Contacts between Laterally Adjacent Lithosomes

In addition to vertical boundaries delineated by contacts, stratigraphic units also have finite lateral boundaries. They do not extend indefinitely laterally but must eventually terminate, either abruptly as a result of erosion or more gradually by change to a different lithology. Some sedimentary units are laterally discontinuous in the sense that lateral changes in lithology may occur within single outcrops or at least within a local area. Many nonmarine deposits, such as alluvial-fan deposits, exhibit such lateral discontinuity. Lateral changes may be accompanied by progressive thinning of units to extinction—**pinch-outs** (Figs. 12.1, 12.4); lateral splitting of a lithologic unit into many thin units that pinch-out independently—**intertonguing**;

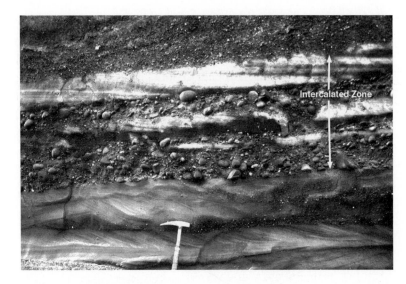

Figure 12.3
Gradation from sandstone below through a zone of intercalated thin conglomerates and sandstones (light) to conglomerate at the top of the section. Miocene deposits near Blacklock Point, Southwest Oregon Coast.

Figure 12.4
Pinch-out. Note how the sandstone bed (light) pinches out abruptly to the right and disappears into the conglomerate. Cretaceous (?) deposits, near Ashland, southern Oregon.

or **progressive lateral gradation**, similar to progressive vertical gradation. We speak of sedimentary units that do not terminate within single outcrops or local areas as being laterally continuous (Fig. 12.1). If traced far enough, of course such units must also eventually terminate. Many marine strata such as shelf sandstones, limestones, and central-basin evaporites exhibit lateral continuity.

Unconformable Contacts

As mentioned, contacts between strata that do not succeed underlying rocks vertically in immediate order of age are called unconformities. Four types of unconformable contacts (unconformities) are recognized: (1) angular unconformity, (2) disconformity, (3) paraconformity, and (4) nonconformity (Fig. 12.5). Unconformities are recognized by an angular relationship between strata (angular unconformities), the presence of a marked erosional surface separating these strata (e.g., disconformities), marked disparity in age of rocks above and below the unconformity (e.g., paraconformities), and the nature of the rocks underlying the surface of

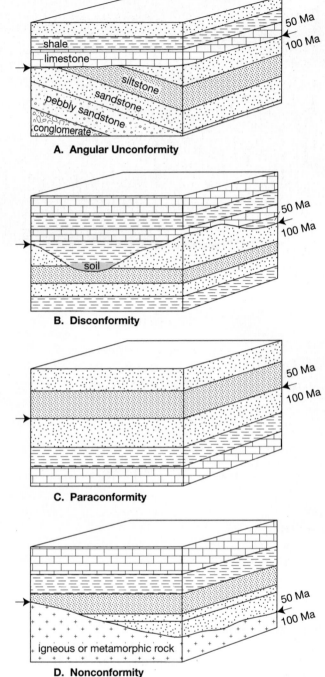

Figure 12.5
Schematic representation of four basic kinds of unconformities. Arrows indicate the unconformity surface. For the purpose of illustration, the youngest strata below the unconformity surface in each diagram is shown to have a (hypothetical) age of 100 million years and the oldest strata above the unconformity surface an age of 50 million years, indicating a hiatus in each case of 50 million years. [Modified from Dunbar, C. O., and J. Rodgers, 1957, Principles of stratigraphy: John Wiley & Sons, New York, Fig. 57, p. 117, reprinted by permission.]

unconformity. The first three types of unconformities occur between bodies of sedimentary rock. Nonconformities occur between sedimentary rock and metamorphic or igneous rock.

Angular Unconformity

An angular unconformity is a type of unconformity in which younger sediments rest upon the eroded surface of tilted or folded older rocks; that is, the older rocks dip at a different, commonly steeper, angle than do the younger rocks (Fig. 12.5A). The unconformity surface may be essentially planar or markedly irregular. Angular

Figure 12.6
Major angular unconformity (arrow) between tilted sedimentary rocks of the Grand Canyon Group (Precambrian) and overlying, nearly horizontal, Tapeats Sandstone (Cambrian). Younger Paleozoic strata are exposed above the Tapeats. View from Desert Overlook, Southeast Rim, Grand Canyon, Arizona.

unconformities may be confined to limited geographic areas (local unconformities) or may extend for tens or even hundreds of kilometers (regional unconformities). Some angular unconformities are visible in a single outcrop (Fig. 12.6). By contrast, regional unconformities between stratigraphic units of very low dip may not be apparent in a single outcrop and may require detailed mapping over a large area before they can be identified.

Disconformity

An unconformity surface above and below which the bedding planes are essentially parallel and in which the contact between younger and older beds is marked by a visible, irregular or uneven erosional surface is a disconformity (Fig. 12.5B). Disconformities are most easily recognized by this erosional surface, which may be channeled and which may have relief ranging to tens of meters. Disconformity surfaces, as well as angular unconformity surfaces, may be marked also by "fossil" soil zones (paleosols) or may include lag-gravel deposits that lie immediately above the unconformable surface and that contain pebbles of the same lithology as the lithology of the underlying unit. Disconformities are presumed to form as a result of a significant period of erosion throughout which older rocks remained essentially horizontal during nearly vertical uplift and subsequent downwarping.

Paraconformity

A paraconformity is an obscure unconformity characterized by beds above and below the unconformity contact that are parallel and in which no erosional surface or other physical evidence of unconformity is discernible. The unconformity contact may even appear to be a simple bedding plane (Fig. 12.5C). Paraconformities are not easily recognized and must be identified on the basis of a gap in the rock record (because of nondeposition or erosion) as determined from paleontologic evidence such as absence of faunal zones or abrupt faunal changes. In other words, rocks of a particular age are missing, as determined by fossils or other evidence.

Nonconformity

An unconformity developed between sedimentary rock and older igneous or massive metamorphic rock that has been exposed to erosion prior to being covered by sediments is a **nonconformity** (Fig. 12.5D). Nonconformity surfaces probably represent an extended period of erosion.

Significance of Unconformities

The presence of unconformities has considerable significance in sedimentological studies. Many stratigraphic successions are bounded by unconformities, indicating that these successions are incomplete records of past sedimentation. Not only do unconformities show that some part of the stratigraphic record is missing, but they also indicate that an important geologic event took place during the time period (hiatus) represented by the unconformity—an episode of uplift and erosion or, less likely, an extended period of nondeposition.

12.4 VERTICAL AND LATERAL SUCCESSIONS OF STRATA

Nature of Vertical Successions

As discussed, conformities and unconformities divide sedimentary rocks into vertical successions of beds, each characterized by a particular lithologic aspect. Different types of beds can succeed each other vertically in a great variety of ways, and distinctions can be drawn between rock units characterized by lithologic uniformity, lithologic heterogeneity, and cyclic successions. Rock units that have complete lithologic uniformity are rare, although many beds may display a high degree of uniformity in color, grain size, composition, or resistance to weathering. Beds that are most likely to be uniform are fine-grained sediments deposited slowly under essentially uniform conditions in deeper water, or coarser sediments that have been deposited rapidly by some type of mass sediment transport mechanism such as grain flow. By contrast, heterogeneous bodies of sedimentary strata are characterized by internal variations or irregularities in properties. Heterogeneous units may include strata such as extremely poorly sorted debris-flow deposits, as well as thick units broken internally by thinner beds characterized by differences in grain size or bedding features.

Cyclic Successions

Many stratigraphic successions display repetitions of strata that reflect a succession of related depositional processes and conditions that are repeated in the same order. Such repetitious events are referred to as **cyclic sedimentation** or rhythmic sedimentation. Cyclic sedimentation leads to the formation of vertical successions of sedimentary strata that display repetitive orderly arrangement of different kinds of sediments. The term cyclic sediment has been used for a wide variety of repetitious strata including such small-scale features as presumed annually deposited varves in glacial lakes as well as large-scale sediment cycles caused by long-period, recurring migration of depositional environments. Other common examples of repetitive deposits include rhythmically bedded turbidites, laminated evaporite deposits, limestone-shale rhythmic successions, coal cyclothems (repeated cycles involving coal deposition), black shale deposits, and chert deposits. Cyclic successions occur on all continents in essentially every stratigraphic system. They are produced by processes that range in geographic scope and duration from very local, short-term events—such as seasonal climatic changes that generate varves—to global changes in sea level that may involve entire geologic periods.

a **AUTOCYCLIC MECHANISMS**

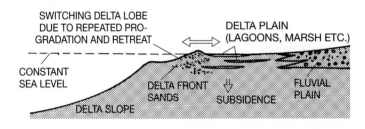

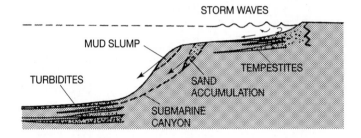

b **ALLOCYCLIC MECHANISMS**

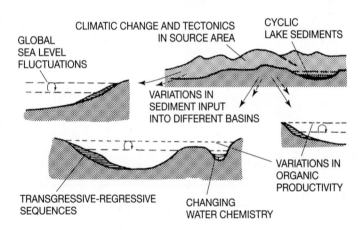

Figure 12.7
Autocyclic (a) and allocyclic (b) mechanisms that generate rhythmic and cyclic successions. [From Einsele, G., W. Ricken, and A. Seilacher, 1991, Cycles and events in stratigraphy—Basic concepts and terms, *in* Einsele, G., W. Ricken, and A. Seilacher (eds.), Cycles and events in stratigraphy. Fig. 4, p. 9, reproduced by permission of Springer-Verlag, Berlin.]

On the basis of the mechanisms that form cyclic deposits, two kinds of cyclic successions are recognized: autocyclic and allocyclic (Fig. 12.7). **Autocyclic successions** are controlled by processes that take place within the basin itself, and their beds show only limited stratigraphic continuity. Examples include nonperiodic storm beds and turbidites. **Allocyclic successions** are caused mainly by variations external to the depositional basin. Fundamentally, these variations are caused by changes in climate and tectonic movements. For example, climate can influence sea level, waxing and waning of continental glaciers, and sedimentary processes such as deposition of evaporites. Tectonic movements affect sea level and water depth. Allocyclic successions may extend over long distances and perhaps even from one basin to another (Einsele, Ricken, and Seilacher, 1991a).

Several scales of allocyclic successions have been postulated, all of which are related in one way or another to changes in climate and sea level and tectonism (Fig. 12.8). Very long term eustatic cycles (rise and fall of global sea level) that appear to take 200–500 million years (Ma) were referred to by Vail, Mitchum, and Thompson (1977b) as **first-order cycles** (Table 12.1). [Note: Miall (1997, p. 50) recommends against using the hierarchal classification of cycles as first-order, second-order, etc.; however, this practice is now thoroughly entrenched in the literature and is widely followed.] These cycles are too large to see in normal outcrops and must be deduced from study of large sets of field exposures or from subsurface information

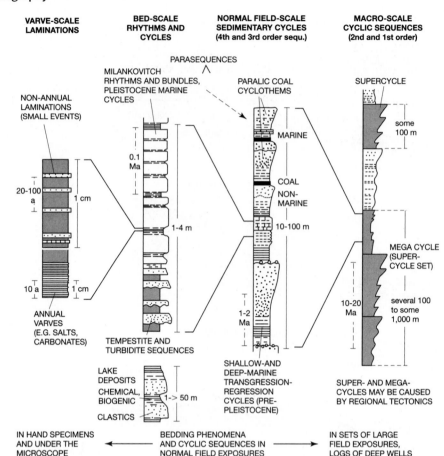

Figure 12.8
Schematic representation of the scale of cyclic sedimentation in the stratigraphic record. Ma = million years, a = years; parasequences are discussed in Chapter 13. [After Einsele, G., W. Ricken, and A. Seilacher, 1991, Cycles and events in stratigraphy—Basic concepts and terms, *in* Einsele, G., W. Ricken, and A. Seilacher (eds.), Cycles and events in stratigraphy, Springer-Verlag, Berlin, Fig. 2, p. 3, reproduced by permission.]

Table 12.1   Stratigraphic cycles and their postulated causes

Type	Other terms	Duration, m.y.	Probable cause
First-order	—	200–400	Major eustatic cycles caused by formation and breakup of supercontinents
Second-order	Super cycle (Vail, Mitchum, and Thompson, 1977b); sequence (Sloss, 1963)	10–100	Eustatic cycles induced by volume changes in global mid-ocean spreading ridge system
Third-order	Mesothem (Ramsbottom, 1979); mega-cyclothem (Heckel, 1986)	1–10	Possibly produced by ridge changes and continental ice growth and decay
Fourth-order	Cyclothem (Wanless and Weller, 1932); major cycle (Heckel, 1986)	0.2–0.5	Milankovitch glacioeustatic cycles, astronomical forcing
Fifth-order	Minor cycle (Heckel, 1986)	0.01–0.2	Milankovitch glacioeustatic cycles, astronomical forcing

Source: Vail, P. R., R. M. Mitchum, Jr., and S. Thompson, III (1977b); Miall (1990, p. 447).

derived from deep wells or seismic data (Chapter 13). They are not identified directly from lithologic data but rather by application of the techniques of sea-level analysis described in Chapter 13. They are postulated to develop by assembly of supercontinents (e.g., Pangea) by seafloor spreading and their subsequent breakup and dispersal. Sea level falls during times when supercontinents are assembled and rises during times of rapid seafloor spreading when continents are rifting apart. Two first-order, or supercontinent, cycles appear to have taken place during Phanerozoic time (Fig. 12.9).

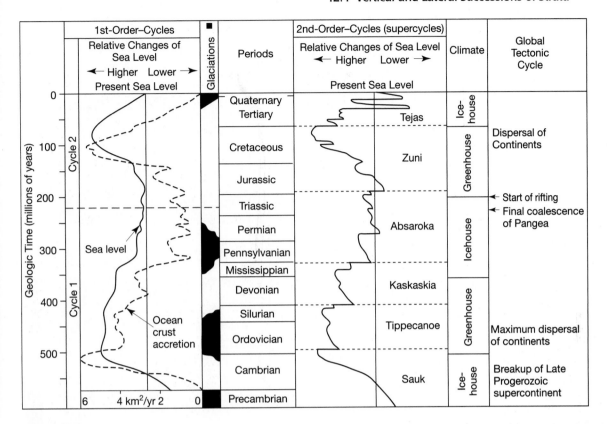

Figure 12.9
Illustration of first-order and second-order global sea-level cycles. First-order cycles reflect variations in production of new oceanic crust (in km²/yr) related to formation and breakup of continents, which cause major, long-term changes in sea level. Second-order cycles are related to changes in volume of oceanic spreading centers. [Modified from Plint et al., 1992, Control of sea-level changes, *in* Walker, R. G., and N. P. James (eds.), Facies models—Response to sea level change: Geol. Assoc. Canada, Fig. 3, p. 18, reproduced by permission. Climate states are after Fischer (1981); the second-order cycles (Tejas, Zuni, etc. are named for Sloss' (1963) sequences.]

Fischer (1984) refers to climate states on Earth during cool periods as **icehouse states** and those during warm periods, when greenhouse gases such as CO_2 were abundant, as **greenhouse states**. Fischer (1984) and Frakes, Francis, and Syktus (1992) suggest that three icehouse states and two greenhouse states existed on Earth during late Precambrian and Phanerozoic time (Fig. 12.10). Note from Figure 12.10 that fluctuations in ancient sea levels tend to track variations in ancient temperatures. Temperatures also affect moisture conditions. The wettest periods seem to have been during cool periods of the Carboniferous to early Permian and the early Tertiary; the driest conditions apparently occurred during the Triassic-Jurassic and Devonian warm periods. Evaluating the exact causes of shifts in Earth's climate from warm to cool periods and back again is fraught with difficulty; however, these changes are related in a complex way to changes in Earth's orbital parameters, changes in sea levels, increase in carbon dioxide levels in the atmosphere owing to volcanic emissions, and removal of carbon dioxide from the atmosphere by weathering of silicate rocks. For example, during times of rapid seafloor spreading and accretion of crustal rocks (Chapter 13), sea level rises and volcanic emissions of CO_2 increases dramatically (Fig. 12.9). As mentioned, an increase in CO_2 levels brings on a greenhouse state. Thus, there appears to be a reasonably good, but not perfect, correlation between higher sea levels and warmer climates (Fig. 12.10).

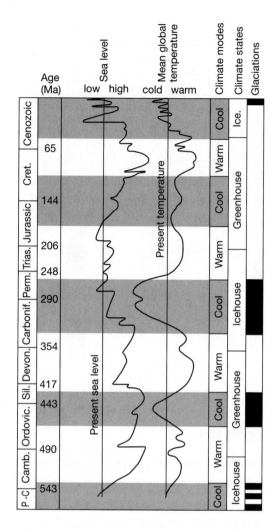

Figure 12.10
Estimated mean global temperature curve for Phanerozoic time and corresponding climate modes (Frakes et al., 1992, p.194), sea-level curve (Vail et al., 1977b), greenhouse-icehouse climate states (Fischer, 1984), and times of major glaciation (Eyles, 1993). Ages from GSA 1999 Geologic Time Scale (see Fig 15.3).

Second-order cycles have durations on the order of tens to hundreds of million years. Like first-order cycles, these macro-scale cycles are too large to study in normal field exposures. The major cause of these cycles is attributed to volume changes in oceanic spreading centers; that is, the volume of spreading ridges increases (owing to increased heating) and sea level rises during rapid spreading, and the volume decreases, with concomitant sea-level fall, during slow spreading. Broad, regional cycles of basement elevation (crustal flexing) that occur in response to convergent, divergent, and transcurrent plate motions may also contribute to second-order cycles (Miall, 1997, p. 53). Sloss (1963) defined six major stratigraphic cycles on the North American continent which he called **sequences** and to which he gave Indian names. These sequences, ranging in age from Tertiary to late Precambrian, are second-order cycles that can be recognized and correlated with similar cycles on other continents (e.g., Sloss, 1979). These cycles, generated as a result of global sea-level change, represent shorter-duration pulses superimposed on longer-duration first-order cycles, also caused by global sea-level change.

Third-order cycles have episodicities on the order of one to ten million years. These cycles can be recognized in normal field exposures, as well as in subsurface well records and seismic reflection profiles. They have been attributed to fluctuations in eustatic sea level (see Chapter 13) owing to changes in spreading ridges and/or continental ice growth and decay; however, their origin has not been fully established and is still controversial. Also, it has not yet been definitely proven that these cycles can be correlated on a global basis. If they are strictly regional rather than global in scope, their origin might be related more to tectonic mechanisms than to eustatic sea-level changes.

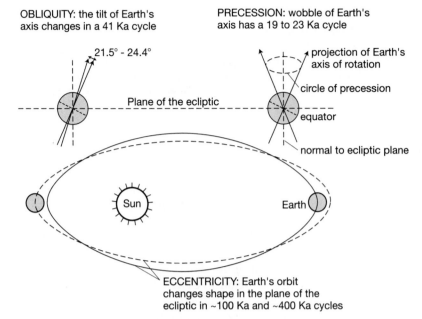

OBLIQUITY: the tilt of Earth's axis changes in a 41 Ka cycle

PRECESSION: wobble of Earth's axis has a 19 to 23 Ka cycle

21.5° - 24.4°

Plane of the ecliptic

projection of Earth's axis of rotation

circle of precession

equator

normal to ecliptic plane

Sun

Earth

ECCENTRICITY: Earth's orbit changes shape in the plane of the ecliptic in ~100 Ka and ~400 Ka cycles

Figure 12.11
Diagram of the Earth-Moon-Sun system, illustrating the causes of oscillations that produce changes in the amount of solar radiation reaching Earth. These oscillations may, in turn, lead to orbitally forced changes in Earth's climate and thus the sedimentary record (e.g., cycles). [Modified from House, M. R., 1995, Orbital forcing timescales: An introduction, *in* House, M. R., and A. S. Gale (eds.), Orbital forcing timescales and cyclostratigraphy: Geological Society Special Publication 85, Fig. 9, p. 10, reproduced by permission.]

Many cycles of smaller scale than third-order cycles have now been identified. These cycles, often referred to informally as bed-scale or meter-scale cycles, have durations less than one million years. Cycles with durations ranging from 0.2 to 0.5 million years are called **fourth-order cycles**, and those with durations from 0.01 to 0.2 million years are called **fifth-order cycles**. It now appears that most of these cycles are related to changes in Earth's orbital parameters (Fig. 12.11). Earth's axis of rotation precesses (the position of the rotational pole wobbles) in two predominant periods averaging 19,000 and 23,000 years. The axis also changes its inclination, called obliquity, from 21.5° to 24.4° in a cycle of about 41,000 years. In addition, its orbit changes from almost circular to almost elliptical (eccentricity) in two main cycles, one with a cycle of 106,000 years, the other with a period of 410,000 years. These orbital variations produce cyclic variations in the intensity and seasonal distribution of incoming solar radiation. Because of such variations, incoming solar radiation may at times be reduced sufficiently to prevent complete summer melt of winter snowpack, leading eventually to snowpack buildup and subsequent development of continental glaciers with resulting removal of large amounts of water from the ocean (lowered sea level).

These variations in Earth's orbital behavior produce periodic changes of climate, called **Milankovitch cycles**, which, in turn, influence sea level and depositional patterns and facies (e.g., de Boer and Smith, 1994; Gale, 1998; Schwarzacher, 1993). Milankovitch was a Serbian mathematician who calculated orbital variations accurately for the first time and showed how these variations affected the amount of solar radiation reaching Earth. He suggested that these orbital cycles caused climatic changes that led to the ice ages, thus affecting sea levels. This postulated link between orbital cycles, climate, and sea level is sometimes referred to as **orbital forcing** (e.g., de Boer, 1991; House and Gale, 1995). Cycles of Milankovitch frequency are particularly well developed in Quaternary strata, and a high-resolution orbital time scale graduated in precession units of 21,000 years has been constructed for strata extending back to the base of the Miocene. Milankovitch cycles may also be features of older sedimentary succession; however, their frequencies are more difficult to identify (Gale, 1998). Milankovitch cycles have been recognized in a variety of rock types including limestone-marl (clayey carbonate) successions, limestone-shale successions, limestone-shale-coal successions (cyclothems), chert-shale successions, evaporite deposits, and muds and shales consisting of alternating light and dark (organic-rich) layers. The study of short-period, high-frequency cycles (e.g., Milankovich cycles) is commonly referred to as **cyclostratigraphy** (e.g., Gale, 1998; House and Gale, 1995).

Note from Figure 12.8 that annual varves constitute cycles of smaller scale than Milankovitch cycles. Stratigraphic cycles are discussed further under "Sea-Level Analysis" in Chapter 13. See also Anderson and Goodwin (1990); de Boer and Smith (1994); Einsele, Ricken, and Seilacher (1991b); Gale (1998); and House and Gale (1995).

Sedimentary Facies

Discussion in preceding chapters dealing with depositional environments referred to many examples of sediments of one type that grade laterally into sediments of a different type, deposited in laterally contiguous parts of a given depositional setting. For example, sandy sediments of the beach shoreface may grade seaward to muddy sediments of the shallow inner-shelf; delta-front sands and silts commonly grade seaward to prodelta muds; and shelf-edge skeletal or oolitic carbonate sands grade toward the open shelf to pelleted carbonate muds. I have already referred to such laterally equivalent bodies of sediment with distinctive characteristics as **facies**. Thus, a deposit may be characterized by shale facies, sandstone facies, limestone facies, and so forth. The concept of facies is so important in stratigraphy that a more detailed explanation of the meaning and significance of facies is necessary at this point.

The term facies was introduced into the geological literature by Nicolas Steno in 1669 (Teichert, 1958); however, modern scientific use of the term is credited to the Swiss geologist Amanz Gressly, who used the term in 1838 in his description of Upper Jurassic strata in the region of Solothurn in the Jura Mountains to record marked changes in lithology and paleontology of these strata. [See Cross and Homewood (1997) for a discussion of Gressly's contributions to the science of stratigraphy.] Krumbein and Sloss (1963) maintain that Gressly intended to confine usage of the term to lateral changes within a stratigraphic unit, such as those illustrated in Figure 12.12. Other workers have interpreted Gressly's usage to include vertical changes in the character of rock units as well (Teichert, 1958). Subsequently, the term has been used with numerous meanings, many of which bear little resemblance to Gressly's original meaning. These various meanings have been summarized and discussed by Moore (1949), Teichert (1958), Weller (1958), Markevich (1960), Walker (1992), Pirrie (1998), and many others. The extended meanings of facies have included referring to all strata of a particular type as a certain facies, such

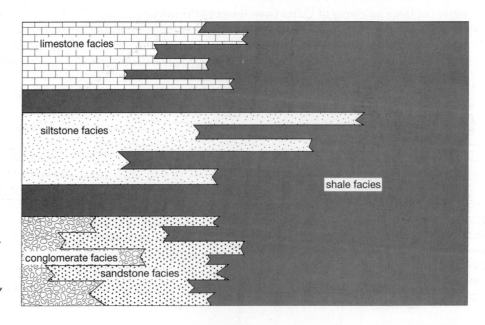

Figure 12.12
Simplified, schematic representation of lithofacies. Note that one facies may change into another laterally or vertically; see also Figure 12.1. Such relationships are rarely seen in a single outcrop (e.g., Fig. 12.4).

limestone facies

siltstone facies

shale facies

conglomerate facies

sandstone facies

as referring to all redbeds as the "redbed facies," and even such nonstratigraphical usage as "metamorphic facies," "igneous facies," and "tectonic facies." Because of these rather loose and inconsistent usages of the term, the meaning of facies has become considerably clouded.

Moore (1949) described facies as "any areally restricted part of a designated stratigraphical unit which exhibits characters significantly different from those of other parts of the unit." Facies comprise "one or any two or more different sorts of deposits which are partly or wholly equivalent in age and which occur side by side or in somewhat close neighborhood." According to Moore's definition, facies are restricted in areal extent, but the same facies could be found at different levels within the same stratigraphic unit. A different usage of facies that is closer to that of Gressly usage—and to that of many European geologists—is to consider facies simply as stratigraphic units distinguished by lithological, structural, and organic aspects detectable in the field. The areal distribution of facies thus designated may not be well known (Blatt, Middleton, and Murray, 1980), in contrast to the restricted areal distribution required by Moore's definition. See also the discussion by Walker (1992). Regardless of the exact definition followed in defining facies, it is now common practice to designate facies identified on the basis of lithologic characteristics as **lithofacies,** and facies distinguished by paleontologic characteristics (fossil content) without regard to lithologic character as **biofacies**. A facies may be divided into **subfacies**. For example, a thick cross-bedded sandstone facies might be divisible into trough cross-bedded and tabular cross-bedded subfacies. Very small-scale facies that can be recognized within microscope thin sections or polished sections of rock are referred to as **microfacies** (Flügel, 1982).

An important objective of facies studies is to ultimately make environmental interpretations from the facies. Thus, some geologists designate facies on the basis of assumed depositional environment and speak of "continental facies," "fluvial facies," "delta facies," and so on. Such generic usage involves subjective judgments that may not always be justified. It is better to make the usage of facies purely descriptive and objective and then make subjective interpretations of environment on the basis of these descriptive facies (Hallam, 1981).

Walther's Law of Succession of Facies

Relationship of Lateral and Vertical Facies

It is implicit in the concept of facies that different facies represent different depositional conditions and environments. As laterally contiguous environments in a given region shift with time in response to shifting shorelines or other geologic conditions, facies boundaries also shift so that eventually the deposits of one environment may lie above those of another environment. This deceptively simple idea embodies one of the single most important concepts in stratigraphy—the concept that a direct environmental relationship exists between lateral facies and vertically stacked or superimposed successions of strata. This concept was first formally stated by Johannes Walther in 1894 and is now called the **law of the correlation (or succession) of facies**, or simply **Walther's Law**. This law has often been misstated as "the same facies sequences are seen laterally as vertically." The correct statement of the law as translated by Middleton (1973, p. 979) is

> The various deposits of the same facies-area and similarly the sum of the rocks of different facies-areas are formed beside each other in space, though in a cross-section we see them lying on top of each other . . . it is a basic statement of far-reaching significance that only those facies and facies-areas can be superimposed primarily which can be observed beside each other at the present time. (Walther, 1894).

Walther suggests that comparison with recent environments could always provide the essential clues for interpretation of ancient facies. Middleton (1973) is careful to point out that the law states not that vertical successions always reproduce the horizontal succession of environments, but merely that only those facies can be superimposed that can now be seen side by side. For example, the beach and barrier-island environmental setting discussed in Chapter 9 may include several laterally adjacent environments such as beach, back-barrier lagoon, marsh, tidal flat, tidal channel, and tidal delta. Depending upon the manner in which these lateral environments shift with time, the vertical successions produced by deposition in a particular barrier-island setting might consist only of beach sands overlain by lagoonal muds and capped by marsh peats. The entire lateral succession of deposits formed in the contiguous environments may not be preserved.

Transgressions and Regressions

The principles embodied in Walther's Law are well illustrated by considering transgressive and regressive sedimentary successions. As discussed in Chapters 9 and 10, **transgression** refers to movement of a shoreline in a landward direction, also called retrogradation. A seaward movement of a shoreline is called **regression**, or progradation. Consider the growth of a delta as illustrated in Figure 12.13. As the delta builds seaward, coarser grained deltaic sediments are deposited on top of finer grained prodelta muds. The result is a coarsening upward vertical succession of facies, generated by the progradation of laterally adjacent depositional

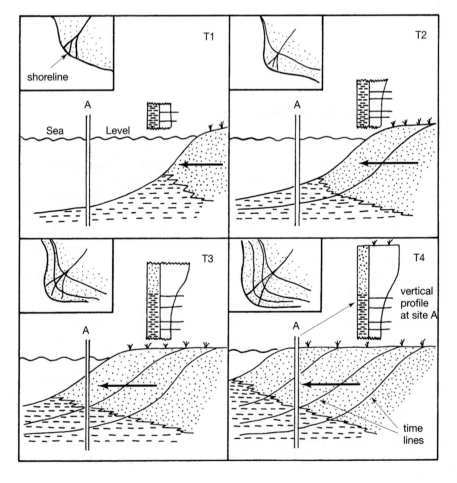

Figure 12.13
Walther's Law illustrated by the growth of a delta through time. Note the successive outbuilding of the delta at four different time periods (T1–T4). With time, the shoreline progrades from right to left, so that at a single location depicting a vertical succession (A), a gradual transition from prodelta mud to coarser grained delta deposits takes place, generating a coarsening-upward succession. [After Pirrie, D., 1998, Interpreting the record: Facies analysis, *in* Doyle, P., and M. R. Bennett (eds.), Unlocking the stratigraphical record: Advances in modern stratigraphy, John Wiley and Sons, Ltd., Chichester, reproduced by permission.]

environments. Note that the depositional surface (sediment-water interface) represents a time line, indicating that at any given time prodelta mud was deposited at the same time as coarser grained delta sediments. It is the seaward migration of the deltaic environment (regression) that brings deltaic sediments on top of prodelta muds to create the coarsening-upward vertical succession.

Transgressions occur during a relative rise in sea level when influx of terrigenous sediments from land sources is low enough to allow deeper water marine sediments to encroach landward over nearshore deposits (coastal encroachment). Transgression will not occur during rising sea level if the influx of terrigenous sediments is so high that outbuilding of the shoreline takes place; instead, regression occurs. That is, regression may occur during a relative rise in sea level or during static sea level if the influx of terrigenous clastics is high. Regression may also occur during a relative fall in sea level. To summarize, transgression occurs only during rising sea level. Regression can occur either during rising sea level, if influx of terrigenous detritus is high, or during falling sea level.

Transgression followed by regression tends to produce a wedge of sediments in which deeper water sediments are deposited on top of shallower water sediments in the basal part of the wedge, and shallower water sediments are deposited on top of deeper water sediments in the top part of the wedge (Fig. 12.14). Note the marked coastal onlap illustrated in Figure 12.14. The initial depositional surface at the base of a transgressive succession is commonly an unconformity. The bounding surface at the base of a regressive succession can also be an unconformity if a relative fall in sea level is accompanied by erosion.

Effects of Climate and Sea Level on Sedimentation Patterns

The preceding discussion shows that both the rate of influx of terrigenous clastic sediments and the change in relative sea level exert control on sedimentation

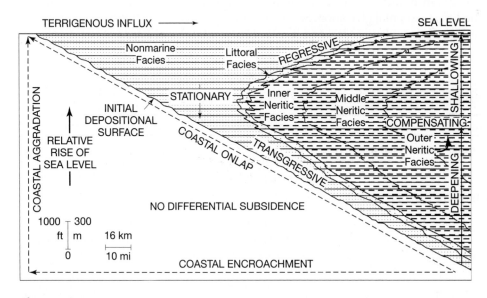

Figure 12.14
Coastal onlap owing to marine transgression and regression. During relative rise in sea level, littoral facies may be transgressive, stationary, or regressive. Neritic (shallow shelf) facies may be deepening, shallowing, or compensating (maintaining a given depth). Note the wedge of sediment formed during a cycle of transgression-regression. [From Vail, P. R., R. M. Mitchum, Jr., and S. Thompson, III, 1977, Seismic stratigraphy and global change of sea level. Part 3: Relative changes of sea level from coastal onlap, in Payton, C. E. (ed.), Seismic stratigraphy—Application to hydrocarbon exploration: Am. Assoc. Petroleum Geologists Mem. 26, Fig. 4, p. 67, reprinted by permission of AAPG, Tulsa, Okla.]

patterns in coastal areas and on the continental shelf. In turn, terrigenous influx is itself influenced by tectonism and climatic conditions (e.g., Church and Coe, 2003). Tectonism produces changes in elevation of sediment source areas and thus affects rates of erosion, which generally increase with increase in land elevation. Also, source areas at higher elevations and with steeper slopes tend to shed coarser sediment than do those at lower elevations. Climate regulates sediment influx by controlling rates of weathering and erosion, sediment transport conditions, and sedimentation mechanisms. For example, in a given geographic area, significantly greater terrigenous influx will occur during periods of heavy rain (e.g., during the winter rainy season), when erosion rates are accelerated and stream transport is increased, than during dry periods. On a shorter time scale, more sediment, and coarser sediment, may be eroded and transported during a single unusually large, high-velocity flood that occurs only once every hundred years than during all the smaller floods that may have occurred during the preceding hundred years. [See Clifton (1988) and Ager (1993a) for discussion of the sedimentologic consequences of large, rare, convulsive geologic events.] Thus, the rates of sediment influx and the grain sizes of sediments delivered to coastal areas from continents have varied throughout geologic time in response to these variables of tectonism and climate.

Changes in sea level also affect sedimentation patterns in coastal areas. Changes in sea level that are worldwide and that affect sea level on all continents essentially simultaneously are called **eustatic sea-level changes,** as mentioned. Changes of sea level that affect only local areas are referred to as **relative sea-level changes.** Relative sea level changes may involve some global eustatic change but are also affected by local tectonic uplift or downwarping of the basin floor and sediment aggradation (buildup). Local tectonics and rates of sedimentation have little or no effect on worldwide sea levels.

Eustatic sea level changes have been attributed to a variety of causes, all of which can be lumped under changes in volume of water and changes in volume of the ocean basins (Table 12.2). The most important changes in water volume are tied to continental glaciation. Sea level drops during glacial stages when seawater is locked up on land as ice, and it rises during interglacial stages as continental ice

Table 12.2   Postulated mechanisms of sea-level change

Mechanisms	Time scale (yr)	Order of magnitude
1. Ocean steric (thermohaline) volume changes		
Shallow (0–500 m)	0.1–100	0–1 m
Deep (500–4000 m)	10–10,000	0.01–10 m
2. Glacial accretion and wastage		
Mountain glaciers	10–100	0.1–1 m
Greenland Ice Sheet	100–100,000	0.1–10 m
East Antarctic Ice Sheet	1,000–100,000	10–100 m
West Antarctic Ice Sheet	100–10,000	1–10 m
3. Liquid water on land		
Groundwater aquifers	100–100,000	0.1–10 m
Lakes and reservoirs	100–100,000	0.01–0.1 m
4. Crustal deformation		
Lithosphere formation and subduction	$100,000–10^8$	1–100 m
Glacial isostatic rebound	100–10,000	0.1–10 m
Continental collision	$100,000–10^8$	10–100 m
Seafloor and continental epirogeny	$100,000–10^8$	10–100 m
Sedimentation	$10,000–10^8$	1–100 m

Source: Revelle, 1990.

sheets melt. Water may also be tied up on land in lakes, reservoirs, and ground-water aquifers. Finally, fluctuations in ocean temperature (Table 12.2) may produce small variations in sea level. Changes in volume of an ocean basin may be brought about by a variety of causes. Sediment infill of an ocean basin, for example, would cause sea level to rise. Changes in the volume of the mid-ocean ridge system may be another cause. Changes in volume of mid-ocean ridges occur as a result of variations in rates of seafloor spreading. An increase in rates of seafloor spreading causes an increase in volume of mid-ocean ridges and a consequent rise in sea level, and a decrease in spreading rates generates a decrease in ridge volume and a corresponding fall in worldwide sea level. Pitman (1978) suggests, for example, that change in the rate of seafloor spreading from 2 cm/yr to 6 cm/yr in the modern ocean could produce a rise in sea level of more than 100 m during a period of seventy million years. Correspondingly, a decrease in spreading rate back to 2 cm/yr for the next seventy million years would cause sea level to drop by more than 100 m. Other possible ways of changing the volume of an ocean basin include glacial isostatic rebound, upwarping or downwarping of the seafloor, and continental collision (Table 12.2). Isostatic rebound after glaciation produces gradual rise in a land surface, which has been depressed owing to weight of the ice, after weight of the ice has been removed by melting.

Changes in sea level and the methods that stratigraphers use to determine the magnitude of sea level changes from the stratigraphic record are further discussed under "Sequence Stratigraphy" in Chapter 13. The effects of sea-level change on the stratigraphic characteristics of sedimentary rocks are examined in detail in that section.

12.5 NOMENCLATURE AND CLASSIFICATION OF LITHOSTRATIGRAPHIC UNITS

To bring order to strata and to understand to the fullest extent the geologic history recorded in these strata, it is necessary to have a formal system for defining, classifying, and naming geologic units. Such a stratigraphic procedure promotes systematic study of the physical properties and successional relationships of sedimentary strata and is essential for interpretation of depositional environments and other aspects of Earth history. The need for systematic organization of strata was recognized as early as the latter half of the 18th century by European scientists such as Johann Gottlob Lehman, Giovanni Arduino, and Georg Christian Füchsel, who made early attempts to organize strata on the basis of relative age (Krumbein and Sloss, 1963). The gradual evolution of these efforts to organize and classify strata continued through the 18th and 19th centuries and eventually culminated in formulation of the internationally used Geologic Time Scale and the Geologic (Stratigraphic) Column (Chapters 14 and 15). This evolution is one of the more fascinating chapters in the history of stratigraphic study. Succinct summaries of these early efforts at stratigraphic classification are given by Weller (1960), Krumbein and Sloss (1963), and Dunbar and Rogers (1957).

Development of the Stratigraphic Code

Local study of rock strata requires subdivision of the Stratigraphic Column into smaller units that are systematically arranged on the basis of inherent properties and attributes. The purpose of stratigraphic classification is thus to promote understanding of the geometry and successions of rock bodies. To ensure uniform usage of stratigraphic nomenclature and classification, attempts have been underway for several decades to adopt a code of stratigraphic nomenclature that formulates views on stratigraphic principles and practices designed to promote standardized classification and formal nomenclature of rock materials. In the United States, such codes have been drafted by the Committee on Stratigraphic Nomenclature, 1933, and its successors, the American Commission on Stratigraphic Nomenclature, 1961, and the North American Commission on Stratigraphic Nomenclature, 1983. The Code of Stratigraphic Nomenclature published by the American Commission on Stratigraphic Nomenclature in 1961, and revised slightly in 1970, standardized terminology and practices used in stratigraphy in the United States at that time and was widely accepted by North American geologists. New concepts and techniques, particularly the concept of global plate tectonics, have developed in the past few decades. These developments revolutionized the earth sciences and necessitated revision of the 1961 Code. In order to incorporate new concepts and techniques, the North American Commission on Stratigraphic Nomenclature published a new North American Stratigraphic Code in May 1983. For the convenience of readers, this Code is reproduced in full in Appendix C.

Since publication of the 1983 Code, the North American Commission on Stratigraphic Nomenclature (NACSN) has become aware of several places in the Code where inconsistencies are present, clarification is needed, or updating and revision are required. Such revision and updating are currently in progress. A revision of the 1983 Code will likely be published in 2005 (Randall Orndorff, U.S. Geological Survey, personal communication, 2004). Unfortunately, this new Code will not be available in time to incorporate into the fourth edition of this book. It should, however, be available on the Internet prior to publication as an open-file report at http://www.agiweb.org/nacsn/. Some proposed changes to the Code have already been published as North American Commission on Stratigraphic Nomenclature Notes 63 and 64 (see Ferrusquia-Villafranca et al., 2001, and Lenz et al., 2001).

The International Stratigraphic Guide, published by the International Subcommission on Stratigraphic Classification in 1976 and 1994 (Hedberg, 1976; Salvador, 1994), provides a comprehensive treatment of stratigraphic classification, terminology, and procedures from an international point of view. In this book I accept and use the terminology of the North American Stratigraphic Code. Readers should be aware, however, that some departures from the Code may appear in the International Stratigraphic Guide. See also Whittaker et al., 1991, *A Guide to Stratigraphical Procedure* (in the U.K.).

Major Types of Stratigraphic Units

The various categories of stratigraphic units recognized by the Code are summarized in Table 12.3. Note that some stratigraphic units (e.g., lithostratigraphic units, biostratigraphic units) are based on observable characteristics of rocks. Such units are identified in the field on the basis of physical or biological properties that can be measured (e.g., grain size), sensed by instruments (e.g., magnetic polarity), or described (e.g., sedimentary structures, kinds of fossils). Others are related to geologic ages of rocks. Stratigraphic units having time significance may be actual units of rock (e.g. chronostratigraphic units) that formed during particular time intervals or they may simply be divisions of time (e.g., geochronologic units) and not actual rock units.

Categories and ranks of all stratigraphic units as defined in the 1983 Code, and modified slightly in the proposed 2005 Code (Ferrusquia-Villafranca et al., 2001), are shown in Table 12.4. Procedures and requirements for defining formal stratigraphic units are set forth in detail in the Code (Appendix C). These procedures include requirements for picking a name, designating a stratotype or type section, describing the units, specifying the boundaries between units, and publishing appropriate descriptions of the units in a recognized scientific medium.

Table 12.3 Categories of stratigraphic units defined by the 1983 North American Stratigraphic Code

---

Material categories based on content or physical limits (composition, texture, fabric, structure, color, fossil content)

  Lithostratigraphic units—conform to the law of superposition and are distinguished on the basis of lithic characteristics and lithostratigraphic position

  Lithodemic units—consist of predominantly intrusive, highly metamorphosed, or intensely deformed rock that generally does not conform to the law of superposition

  Magnetopolarity units—bodies of rock identified by remnant magnetic polarity

  Biostratigraphic units—bodies of rock defined and characterized by their fossil content

  Pedostratigraphic units—consist of one or more pedologic (soil) horizons developed in one or more lithic units now buried by a formally defined lithostratigraphic or allostratigraphic unit or units

  Allostratigraphic units—mappable stratiform (in the form of a layer) bodies defined and identified on the basis of bounding discontinuities

Categories expressing or related to geologic age

  Material categories to define temporal spans (stratigraphic units that serve as standards for recognizing and isolating materials of a particular age)

    Chronostratigraphic units—bodies of rock established to serve as the material reference for all rocks formed during the same spans of time

    Polarity-chronostratigraphic units—divisions of geologic time distinguished on the basis of the record of magnetopolarity as embodied in polarity-chronostratigraphic units

  Temporal (nonmaterial) categories—(not material units but conceptual units, i.e., divisions of time)

    Geochronologic units—divisions of time distinguished on the basis of the rock record as expressed by chronostratigraphic units

    Polarity-chronologic units—divisions of geologic time distinguished on the basis of the record of magnetopolarity as embodied in polarity-chronostratigraphic units

    Diachronic units—comprise the unequal spans of time represented by one or more specific diachronous rock bodies, which are bodies with one or two bounding surfaces that are not time synchronous and thus "transgress" time

    Geochronometric units—isochronous units (units having equal time duration) that are direct divisions of geologic time expressed in years

---

Source: North American Commission on Stratigraphic Nomenclature, 1983, North American Stratigraphic Code: *Am. Assoc. Petroleum Geologists Bull.*, v. 67.

Our immediate concern here is with subdivision and nomenclature of lithostratigraphic units. Other types of stratigraphic units are described in subsequent parts of the text. [Note: Only minor changes in treatment of lithostratigraphic units, as described in the 1983 Code, are expected in the revised (2005) version of the Code (Randall Orndorff, U.S. Geological Survey personal communication, 2004). Some of the proposed minor changes can be viewed in North American Commission on Stratigraphic Nomenclature Note 63; see Ferrusquia-Villafranca et al., 2001, or http://www.agiweb.org/nacsn/.]

Formal Lithostratigraphic Units

The concept of formations and other formal lithostratigraphic units is briefly introduced in Section 12.2. In terms of size, the hierarchy of lithostratigraphic units in descending order is supergroup, group, formation, member, and bed (Table 12.5). Although a **formation** is not the largest lithostratigraphic unit, it is nonetheless the fundamental unit of lithostratigraphic classification. *All other lithostratigraphic units are defined as either assemblages or subdivisions of formations.* Note from Table 12.5 that a formation is defined strictly on the basis of lithology. Formations may be defined on the basis of a single lithic type, repetitions of two or more lithic types, or extreme lithic heterogeneity where such heterogeneity constitutes a form of unity when the rock unit is compared with adjacent units. For example, a formation might be composed entirely of shale, entirely of sandstone, or of an

Table 12.4 Categories and ranks of stratigraphic units as defined in North American Commission on Stratigraphic Nomenclature Note 63

I. Material categories based on content or physical limits

Lithostratigraphic	Lithodemic	Magnetopolarity	Biostratigraphic	Pedostratigraphic	Allostratigraphic
Supergroup	Supersuite				
Group	Suite	Polarity Superzone			Allogroup
Formation	*Lithodeme* (Complex)	*Polarity Zone*	*Biozone* (Interval, Assemblage or Abundance)	*Geosol*	*Alloformation*
Member (or Lens, or Tongue)		Polarity Subzone	Subbiozone		Allomember
Bed(s) or Flow(s)					

IIA. Material categories used to define temporal spans

IIB. Nonmaterial categories related to geologic age

Chrono-stratigraphic	Polarity Chrono-stratigraphic	Geochronologic	Polarity Chronologic	Diachronic	Geochronometric
Eonothem	Polarity Superchronozone	Eon	Polarity Superchron		Eon
Erathem (Supersystem)		Era (Superperiod)			Era (Superperiod)
System (Subsystem)	*Polarity Chronozone*	*Period* (Subperiod)	*Polarity Chron*	*Episode* (Diachron)	*Period* (Subperiod)
Series		Epoch		Phase (Diachron)	Epoch
Stage (Substage)	Polarity Subchronozone	Age (Subage)	Polarity Subchron	Span (Diachron)	Age (Subage)
Chronozone		Chron		Cline	Chron

*Fundamental units are italicized.

Source: Ferrusquia-Villafranca et al., 2001.

intimate mixture of sandstone and shale beds that is distinctive because of the mixed lithology. Boundaries of formations, as with all lithostratigraphic units, are placed at the position of lithic change. Boundaries between different formations may, therefore, occur both vertically and laterally. That is, a formation may be located above or below another formation or be positioned laterally adjacent to another formation where lateral facies changes occur. Illustrations of different types of formation boundaries are given in Figure 2, Appendix C. A formation must be of sufficient areal extent and thickness to be mappable at the scale of mapping commonly used in the region where it occurs.

Formal lithostratigraphic units are assigned names that consist of a geographic name combined with the appropriate rank (formation, member, etc.) or an appropriate lithic term, such as limestone, or both. Formation names thus consist of a geographic name followed by either the word *Formation* or a lithic designation. For example, a particular formation might be called the Otter Point Formation (geographic name only) or the Eureka Quartzite (geographic name plus lithic designation). The names of members include a geographic name and the word

Table 12.5	Hierarchy of lithostratigraphic units

Supergroup—a formal assemblage of related or superposed groups or of groups and formations.

  Group—consists of assemblages of formations, but groups need not be composed entirely of named formations.

    Formation—a body of rock, identified by lithic characteristics and stratigraphic position, that is prevailingly but not necessarily tabular and is mappable at Earth's surface and traceable in the subsurface. Must be of sufficient areal extent to be mappable at the scale of mapping commonly used in the region where it occurs. The fundamental lithostratigraphic unit—formations are grouped to form higher-rank lithostratigraphic units and are divided to form lower-rank

      Member—the formal lithostratigraphic unit next in rank below a formation and always part of some formation. A formation need not be divided entirely into members. A member may extend laterally from one formation to another.

      Lens (or lentil)—a geographically restricted member that terminates on all sides within a formation.

      Tongue—a wedge-shaped member that extends beyond the main boundary of a formation or that wedges or pinches out within another formation.

      Bed—distinctive subdivisions of a member; the smallest formal lithostratigraphic unit of sedimentary rock. Members commonly are not divided entirely into beds.

      Flow—the smallest formal lithostratigraphic unit of volcanic rock.

Source: North American Commission on Stratigraphic Nomenclature, 1983, North American Stratigraphic Code: *Am. Assoc. Petroleum Geologists Bull.*, v. 67.

Member, or the name may have an intervening lithic designation such as Eau Claire Sandstone Member. A group name combines a geographic name with the word *Group*, as in Arbuckle Group. The first letters of all words used in formal names of lithostratigraphic units are capitalized.

The North American Stratigraphic Code of 1983 recognizes that some lithostratigraphic bodies are bounded top and bottom by discontinuities (unconformities or diastems). The code introduces the name **allostratigraphic unit** for such mappable stratiform bodies of sedimentary rock that are defined on the basis of bounding, laterally traceable discontinuities rather than on the basis of lithologic change. The International Stratigraphic Guide (Salvador, 1994) refers to an unconformity-bounded unit as a **synthem**.

Informal names may be used for lithostratigraphic units when there is insufficient need, insufficient information, or an inappropriate basis to justify designation as a formal unit (Hedberg, 1976). Informal names may be applied to such units as oil sands, coal beds, mineralized zones, quarry beds, and key or marker beds. Informal names are not capitalized. Examples of informally designated names are "shaley zone," "coal-bearing zone," "pebbly beds," and "siliceous-shale member."

12.6 CORRELATION OF LITHOSTRATIGRAPHIC UNITS

Introduction

In the simplest sense, stratigraphic correlation is the **demonstration of equivalency of stratigraphic units**. Correlation is a fundamental part of stratigraphy, and much of the effort by stratigraphers that has gone into creating formal stratigraphic units has been aimed at finding practical and reliable methods of correlating these units from one area to another. Without correlation, treatment of stratigraphy on anything but a purely local level would be impossible.

The concept of correlation goes back to the very roots of stratigraphy. The fundamental principles of correlation have been presented in numerous early textbooks on geology and stratigraphy; especially interesting reviews of these general principles are given in Dunbar and Rodgers (1957), Weller (1960), and Krumbein and Sloss (1963). The continued strong interest in correlation is demonstrated by more recent publication of several books and articles dealing with correlation, particularly statistical methods of correlation (e.g., Agterberg, 1990; Cubitt and Reyment, 1982; Mann, 1981; Merriam, 1981).

The fundamental concepts of stratigraphic correlation were already firmly established by the 1950s and 1960s. These basic principles are still important today; however, the emergence of new concepts and more advanced analytical tools have changed our perception of correlation to some degree, as well as adding new methods for correlation. The development of the field of magnetostratigraphy since the late 1950s (Chapter 13), for example, has provided an extremely important new tool for global time-stratigraphic correlation on the basis of magnetic polarity events. Also, rapid advances in computer technology and availability and the application of computer-assisted statistical methods to stratigraphic problems have added a new quantitative dimension to the field of stratigraphic correlation. This section will attempt to bring out some of these new developments, along with discussion of the more "classical" concepts of stratigraphic correlation.

Definition of Correlation

In spite of the fact that the concept of correlation goes back to the early history of stratigraphy, disagreement has persisted over the exact meaning of the term. Historically, two points of view have prevailed. One view rigidly restricts the meaning of correlation to demonstration of time equivalency, that is, to demonstration that two bodies of rock were deposited during the same period of time (Dunbar and Rodgers, 1957; Rodgers, 1959). From this point of view, establishing the equivalence of two lithostratigraphic units on the basis of lithologic similarity does not constitute correlation. A broader interpretation of correlation allows that equivalency may be expressed in lithologic, paleontologic, or chronologic terms (Krumbein and Sloss, 1963). In other words, two bodies of rock can be correlated as belonging to the same lithostratigraphic or biostratigraphic unit even though these units may be of different ages. It is clear, from a pragmatic point of view, that most geologists today accept the broader view of correlation. Petroleum geologists, for example, routinely correlate subsurface formations on the basis of lithology of the formations, the specific "signatures" recorded within the formations by instrumental well logs, or the reflection characteristics on seismic records. The 1983 North American Stratigraphic Code (Appendix C) recognizes three principal kinds of correlation:

1. **Lithocorrelation**, which links units of similar lithology and stratigraphic position
2. **Biocorrelation**, which expresses similarity of fossil content and biostratigraphic position
3. **Chronocorrelation**, which expresses correspondence in age and chronostratigraphic position

Even though our concern in this chapter is correlation on the basis of lithology, it is important to clarify the relationship between chronocorrelation and lithocorrelation. Chronocorrelation can be established by any method that allows matching of strata by age equivalence. Correlation of units defined by lithology may also yield chronostratigraphic correlation on a local scale, but when traced regionally many lithostratigraphic units transgress time boundaries. Stratigraphic units deposited during major transgressions and regressions are notably time-transgressive. Perhaps the most famous North American example of a time-transgressive formation is the Cambrian Tapeats Sandstone in the Grand Canyon region. This sandstone is all early Cambrian in age at the west end of the Canyon and all middle Cambrian in age at the eastern end (Fig. 12.15). Thus, the Tapeats Sandstone, which can be traced continuously through the Canyon region, correlates from one end of the Canyon to the other as a lithostratigraphic unit but not as a chronostratigraphic unit. The important point stressed here is that the boundaries defined by criteria used to establish time correlation of stratigraphic units may not be the same as those defined by criteria used to establish lithologic correlation. Because of this fact, different methods of correlation (lithocorrelation, biocorrelation, chronocorrelation) may yield different results when applied to the same stratigraphic succession.

Another point that requires some clarification is the difference between matching of stratigraphic units and correlation of these units. Matching has been defined simply as correspondence of serial data without regard to stratigraphic units (Schwarzacher, 1975; Shaw, 1982). For example, two rock units identified in stratigraphic sections at different localities as having essentially identical lithology (e.g., two black shales) can be matched on the basis of lithology; however, these units may have neither time equivalence nor lithostratigraphic equivalence. Physical tracing of the units between the localities may show that one unit lies stratigraphically above the other. Matching by lithologic characteristics in this particular case does not constitute demonstration of equivalence. Shaw (1982) states that the process of correlation is the demonstration of geometric relationships between rocks, fossils, or successions of geologic data for interpretation and inclusion in facies models, paleontologic reconstructions, or structural models. The object of correlation is to establish equivalency of stratigraphic units between geographically separated parts of a geologic unit. Implicit in this definition is the concept that correlation is made between stratigraphic units, that is, lithostratigraphic units, biostratigraphic units, or chronostratigraphic units. The difference between correlation and matching is illustrated in Figure 12.16. Figure 12.16A

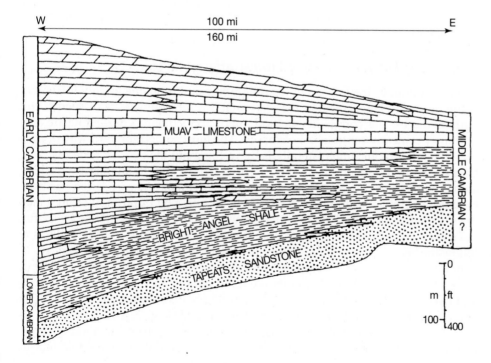

Figure 12.15
Change in age of the basal Cambrian Tapeats Sandstone across the Grand Canyon region. [From Clark, T. H., and C. W. Stern, 1968, Geological evolution of North America, 2nd ed., Fig. 7.10, p. 126, reprinted by permission of John Wiley & Sons, Inc. Originally from E. D. McKee, 1954, Cambrian history of the Grand Canyon region, Part 1. Stratigraphy and ecology of the Grand Canyon Cambrian: Carnegie Inst. Washington Pub. 563, Washington, D.C.]

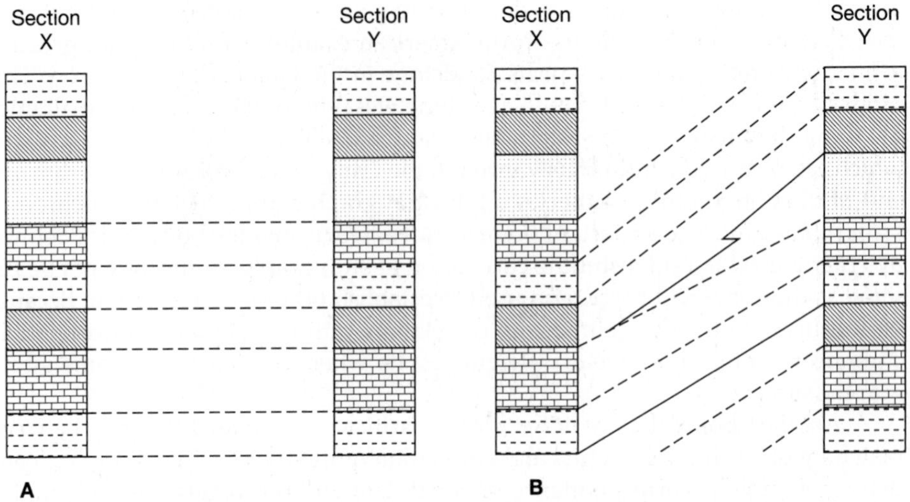

Section X   Section Y   Section X   Section Y

Figure 12.16
Illustration of the difference between matching and correlation. A. Apparent correlation achieved by matching of similar-appearing strata. B. Actual lithocorrelation. [After Shaw, A. B., 1964, Time in stratigraphy, Fig. 30.1, p. 214, McGraw Hill, New York.]

A   B

shows two stratigraphic sections that appear to be perfectly matched. The actual lithocorrelation is shown in Figure 12.16B. The tie lines in Figure 12.16A do not constitute correlation because they do not encompass equivalent lithostratigraphic units.

Correlation can be regarded as either direct (formal) or indirect (informal) (Shaw, 1982). **Direct correlation** can be established physically and unequivocally. Physical tracing of continuous stratigraphic units is the only unequivocal method of showing correspondence of a lithic unit in one locality to that in another. **Indirect correlation** can be established by numerous methods, such as visual comparison of instrumental well logs, polarity reversal records, or fossil assemblages; however, such comparisons have different degrees of reliability and can never be totally unequivocal.

Lithocorrelation

We turn now to the methods used for correlating strata on the basis of lithology. Methods of biocorrelation and chronocorrelation are discussed in appropriate sections of Chapters 14 and 15.

Continuous Lateral Tracing of Lithostratigraphic Units

Direct, continuous tracing of a lithostratigraphic unit from one locality to another is the only correlation method that can establish the equivalence of such a unit without doubt. This correlation method can be applied only where strata are continuously or nearly continuously exposed. The most straightforward way of tracing lithostratigraphic units laterally is by walking out the beds. A geologist who traces a stratigraphic unit continuously from one locality to another by walking along the top of a particular bed can be quite confident that correlation has been established. Thus, the application of field boots and a bit of physical effort yields the satisfaction of achieving a virtually unequivocal correlation. Another useful, but somewhat more equivocal, method of tracing stratigraphic units laterally is to follow the beds on aerial photographs. In areas where surface exposures are abundant and visibility is little hampered by soil or vegetation cover, lateral tracing of thick, distinctive stratigraphic units on aerial photographs can be done rapidly and effectively. This method is limited to tracing of distinctive beds that are thick enough to show up on photographs of a suitable scale (e.g., Fig. 12.17).

Although physical tracing of beds is the only unequivocal method of correlation, it is not without limitations. The most serious of these is the fact that in

Figure 12.17
Excellent exposure of nearly flat-lying Permian-Pennsylvanian strata, allowing continuous tracing of beds for considerable distances, Goosenecks of the San Juan River, Utah.

most areas, geologists cannot trace beds continuously for more than a very short distance before encountering areas covered by soil or vegetation, structural complications (faults), or erosional terminations, as across a large valley. In fact, it is often impossible to trace a given stratigraphic unit more than a few hundred meters before the unit is lost for one of these reasons. An additional problem may arise if the beds being traced pinch out or merge with others laterally, a very common occurrence in nonmarine strata. In such a case, tracing of an individual bed or bedding plane will be impossible. Therefore, in practice, geologists commonly trace a gross lithostratigraphic unit (e.g., a member or a formation) consisting of beds of like character, rather than trying to trace individual beds.

Lithologic Similarity and Stratigraphic Position

Lithologic Similarity. Geologists working in areas where direct lateral tracing of beds is not possible must depend for correlation of lithostratigraphic units upon methods that match strata from one area to another on the basis of lithologic similarity and stratigraphic position. Because matching of strata does not necessarily indicate correlation, correlation by lithologic similarity has varying degrees of reliability. The success of such correlation depends upon the distinctiveness of the lithologic attributes used for correlation, the nature of the stratigraphic succession, and the presence or absence of lithologic changes from one area to another. Facies changes that take place in lithostratigraphic units between two areas under study obviously complicate the problem of lithologic correlation.

Lithologic similarity can be established on the basis of a variety of rock properties. These properties include gross lithology (e.g., sandstone, shale, or limestone), color, heavy mineral assemblages or other distinctive mineral assemblages, primary sedimentary structures such as bedding and cross-lamination, and even thickness and weathering characteristics. The greater the number of properties that can be used to establish a match between strata, the stronger the likelihood of a reliable match. A single property such as color or thickness may change laterally within a given stratigraphic unit, but a suite of distinctive lithologic properties is less likely to change. I caution again that matching of strata on the basis of lithology is not a guarantee that correlation has been established. Strata with very similar lithologic characteristics can form in similar depositional environments widely separated in time or space. It may be quite possible, for example, to obtain an excellent lithologic match between a clean, well-sorted, cross-bedded, eolian sandstone unit of Triassic age and a lithologically virtually identical sandstone of Jurassic age, yet these sandstones do not correlate as either lithostratigraphic or chronostratigraphic units. Correlation on the basis of lithologic identity is particularly difficult between cyclic successions, such as the Pennsylvanian cyclothems of the United States midcontinent region. Very similar successions of units can be repeated over and over in the stratigraphic section owing to the fact that similar environmental conditions can reappear in a region time after time during repeated transgressive-regressive cycles of deposition.

The most reliable lithologic correlations are made when it is possible to match not just one or two distinctive beds or rock types but a succession of several distinctive units. For example, the Triassic and Jurassic formations of the Colorado Plateau in the western United States consist of a highly distinctive succession of largely nonmarine red to green siltstone and mudstone units (the Moenkopi, Chinle, Kayenta, Summerville, and Morrison Formations) interstratified with red to white, cross-bedded eolian (?) sandstones (the Wingate, Navajo, and Entrada Formations). Some of these formations are shown in Figure 12.18. This succession of formations is so distinctive that it can be recognized and correlated lithologically with considerable confidence over wide areas on the Colorado Plateau (Figure 12.19). In some cases, it may be possible to improve the reliability of correlation by applying statistical and computer-assisted techniques. These quantitative methods can provide a probability measure of whether a proposed correlation is valid or invalid (Agterberg, 1990).

Stratigraphic Position in a Succession. The preceding illustration points out the importance of position in a stratigraphic succession when correlating units by lithologic identity. Several of the Colorado Plateau formations are lithologically similar, but because they occur in a succession of strata distinctive enough to be correlated from one area to another, individual formations can be correlated also by their position in this succession. Another way in which position in a stratigraphic succession is important has to do with establishing correlation of strata by relation to some highly distinctive and easily correlated unit or units. Such distinctive beds serve as control units for correlation of other strata above and below. For example, a thin, ash-fall unit or bentonite bed may be present and easily recognized throughout a particular region. If it is the only such bed in the stratigraphic succession in the region, and thus cannot be confused with any other bed, it can serve as a **key bed,** or **marker bed,** to which other strata are related. Strata immediately above or below this control unit can be correlated with a reasonable

Figure 12.18
Well-exposed, distinctive Triassic Formations (Wingate, Chinle, Moenkopi) that can be correlated over wide areas in Utah and Colorado.

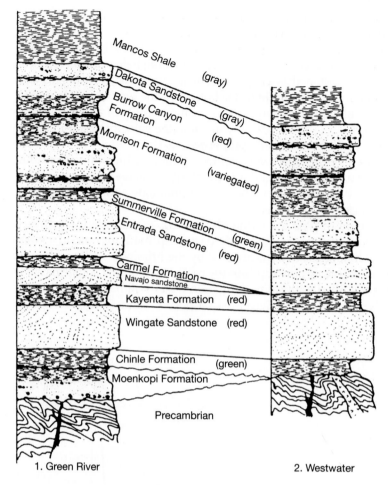

1. Green River   2. Westwater

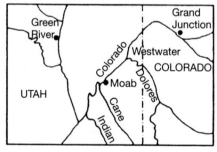

Figure 12.19
Correlation of strata between two localities on the Colorado Plateau on the basis of similar lithology of distinctive stratigraphic units. [From Mintz, L. W., 1981, Historical geology, The science of a dynamic Earth, 3rd ed., Fig. 10.1, p. 241, reprinted by permission of Charles H. Merrill Pub. Co., Columbus, Ohio.]

degree of confidence with strata that are in a similar stratigraphic position with respect to the control unit in other areas. If two or more marker beds are present in a succession, they give even greater reliability to correlation of units that lie between the marker beds. Obviously, correlation becomes more equivocal with increasing stratigraphic separation above or below the control units.

Correlation by Instrumental Well Logs

Well logs are simply curves sketched on paper charts that are produced from data obtained from measurements in well bores. These traces record variations in such rock properties as electrical resistivity, transmissibility of sound waves, or adsorption and emission of nuclear radiation in the rocks surrounding a borehole. These variations are a reflection of changes in features such as gross lithology, mineralogy, fluid content, and porosity in the subsurface formations. Thus, correlation by use of well logs is not based totally on lithology. Nonetheless, most of the rock properties measured by well logs are closely related to lithology.

Well logs are obtained by the following procedure. After an exploratory well is drilled by a petroleum company, the well is logged before being completed as an oil or gas producer or abandoned as a dry hole. The logging procedure begins with lowering an instrument called a **sonde** to the bottom of the well bore (Fig. 12.20). The sonde may be designed to measure the electrical resistivity of a rock unit, natural or induced gamma radiation emitted by the unit, the velocity of sound waves passing through the rock, or other rock properties. As the sonde is slowly withdrawn from the bore hole through a succession of stratigraphic units, it continuously measures the particular property of the rock that it is designed to analyze, and it electrically transmits this information to a digital tape and display unit located in a logging truck at the surface.

One common type of well log is the **electric log**, or resistivity log, which records resistivity of rock units as the sonde passes up the bore hole in contact with the wall of the hole. Resistivity is affected by the lithology of the rock units and the amount and nature of pore fluids in the rock. For example, a marine shale whose pore spaces are filled with saline formation water will have a much lower electrical resistivity (higher conductivity) than a porous sandstone or limestone

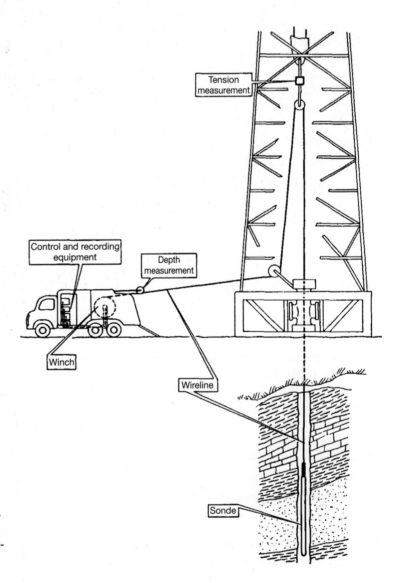

Figure 12.20
Schematic diagram illustrating how instrumental well logs are obtained in a well bore. [From Desbrandes, R., 1985, Encyclopedia of well logging. Editions Technip, Paris, Fig. 2.1, p. 95.]

filled with oil or gas. With experience in a given geological province, petroleum geologists can recognize the particular signatures represented by the traces on the log and can relate these signatures to particular types of lithostratigraphic units or to a specific formation. Lithology cannot be read directly from such logs, but the characteristics of the log traces are a reflection of lithology (and fluid content). Commonly, the lithology of cores and cuttings obtained from the well bore during drilling are identified and a "lithologic log" is prepared from this information. This lithologic log can then be matched against the instrumental log tracing to aid in interpreting lithology from the instrumental log.

A variety of other types of logging tools are in common use by petroleum companies. **Gamma ray logs** measure the natural gamma radiation in rock units. **Sonic logs** measure the velocity with which a sound signal passes through rock units. In addition to their usefulness in correlation, sonic logs can also be used to determine the porosity of subsurface formations because of the fact that sound waves are slowed in their passage through rocks by the presence of fluid-filled pores. **Formation density logs** provide information about porosity and lithology. Very specialized logs include geochemical, formation microscanner, and magnetic susceptibility logs. All logs share the common characteristic that they consist of electrically produced signatures or traces that represent some particular property of subsurface lithostratigraphic unit that is related in some way to lithology, fluid content, bed thickness, or other properties.

One common type of log data display consists of two different types of traces that are arranged on either side of a central column that represents the well bore. This central column is calibrated in feet (or meters) to show depth below the surface. Figure 12.21 illustrates a section of a well log showing a sonic curve opposite a gamma ray curve. The curve shapes generated by a particular lithostratigraphic unit are not unique, but a trained, experienced well-log analyst can learn to recognize the signature of a particular formation or succession of formations and can match up the signatures in logs from one area to those from nearby wells. See Asquith (1982), Desbrandes (1985), Rider (1986), and Whittaker (1998) for additional descriptions of well logs and log interpretations.

Characteristically, the well-log curves of adjacent wells are very similar, but the degree of similarity decreases in more distant wells. By working with a series of closely spaced wells, however, a geologist can carry a correlation across an entire sedimentary basin, even when pinch outs or facies changes occur. In fact, one of the reasons why petroleum geologists find correlation of well logs so useful in petroleum exploration is that correlation permits recognition of pinch-outs and facies changes that may be potential traps for oil and gas. Figure 12.22 is an example of correlation by gamma-ray and sonic logs across a portion of the East Irish Sea Basin, lying amid Ireland, Wales, Scotland, and England. These wells penetrate a stratigraphic section consisting mainly of halite and mudstone with minor anhydrite, dolomite, and sandstone. Geologists often add lithologic information obtained from drill cores or cuttings to the well logs; however, I stress again that correlation by well logs is not necessarily correlation based entirely upon lithologic identity because the shapes of the curves can represent a variety of rock properties, such as porosity and fluid content. Note, however, that there is generally good agreement between the shapes of the well-log curves and the lithology. Correlation by well logs is actually based more upon the position of each unit in a succession of units represented on the logs rather than on the character of any individual unit reflected in the curves. Correlation by well logs is thus the approximate subsurface equivalent of correlation of surface sections by position in the succession.

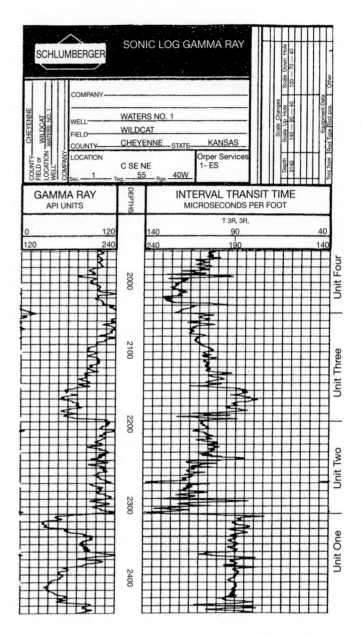

Figure 12.21
A section of a sonic-gamma ray log. The gamma ray curve is shown on the left. The sonic log (interval transit time log) is on the right. Depths (in feet) below the surface are shown in the central column. Four distinct correlatable units are indicated.

Correlation by instrumental well logs can be a laborious process involving large numbers of logs; it is also subject to considerable subjectivity owing to the similarity of the log curves or traces in different parts of a logged stratigraphic section. Differences between stratigraphic units may be manifested only by very subtle differences in the digital plots and can be difficult to discern visually. The availability of computers and sophisticated statistical techniques now makes it possible to apply automated approaches to stratigraphic correlation of well logs, removing some of the subjectivity in correlation. These approaches involve using digital tapes to segment the logs for use in computational systems. These systems then provide a statistical match for correlation purposes. Details of automated well-log correlation are given by Shaw and Cubitt (1978), Griffiths (1982), and Olea (1988).

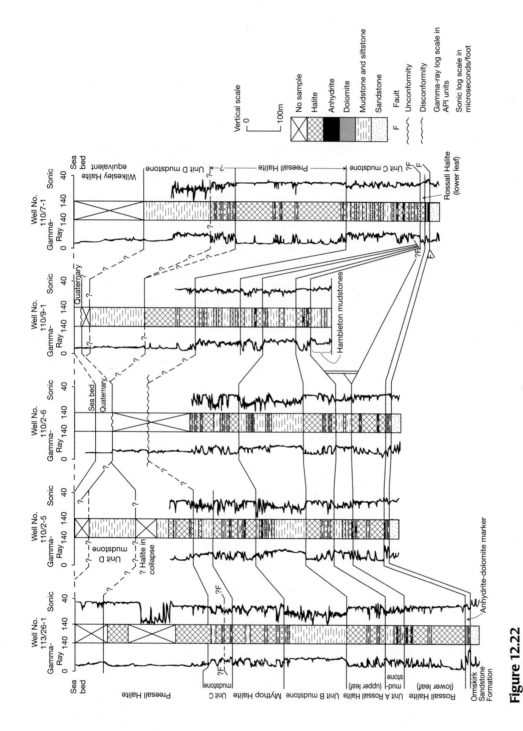

Figure 12.22

An illustration of correlation by use of well logs. The logs for each well contain a gamma ray and a sonic log. Lithologic information has been added to the logs to aid in correlation; however, correlation is made primarily on the basis of the "signatures" of the well-log curves. Selected boreholes from the Mercia Mudstone Group (Triassic), East Irish Sea Basin. [After Jackson, D. I., et al., 1995, United Kingdom offshore regional report: The geology of the Irish Sea, HMSO for the British Geological Survey, London, Fig. 58, p. 75.]

FURTHER READING

Ager, D. V., 1993, The new catastrophism: The importance of the rare event in geologic history: Cambridge University Press, Cambridge, 231 p.

Ager, D. V., 1993, The nature of the stratigraphical record, 3rd ed.: John Wiley & Sons, Chichester, 151 p.

Agterberg, F. P., 1990, Automated stratigraphic correlation: Elsevier, Amsterdam, 424 p.

Coe, A. L. (ed.), 2003, The sedimentary record of sea-level change: The Open University, Cambridge University Press, Cambridge, 287 p.

de Boer, P. L., and D. G. Smith (eds.), 1994, Orbital forcing and cyclic sequences: International Assoc. Sediamentologists Spec. Pub. 19, Blackwell Scientific Publications, Oxford, 559 p.

Doyle, P., and M. R. Bennett (eds.), 1998, Unlocking the stratigraphical record: Advances in modern stratigraphy: John Wiley and Sons, Chichester, 532 p.

Doyle, P., M. R. Bennett, and A. N. Baxter, 1994, The key to Earth history: An introduction to stratigraphy: John Wiley & Sons, Chichester, 231 p.

Einsele, G., and A. Seilacher (eds.), 1982, Cyclic and event stratification: Springer-Verlag, Berlin, 536 p.

Einsele, G., W. Ricken, and A. Seilacher (eds.), 1991, Cycles and events in stratigraphy: Springer-Verlag, Berlin, 955 p.

Emery, D., and K. J. Meyers (eds.), 1996, Sequence stratigraphy: Blackwell Science Ltd., Oxford, 297 p.

Hailwood, E. A., and R. B. Kidd (eds.), 1993, High resolution stratigraphy: Geological Society London Spec. Pub. 70, 357 p.

Hallam, A., 1981, Facies interpretation and the stratigraphic record: W. H. Freeman, San Francisco, 287 p.

Haq, B. U. (ed.), 1995, Sequence stratigraphy and depositional response to eustatic, tectonic and climatic forcing: Kluwer Academic Publishers, Dordrecht, 381 p.

House, M. R., and A. S. Gale (eds.), 1995, Orbital forcing timescales and cyclostratigraphy: Geological Society Special Publication 85, 210 p.

Miall, A. D., 1997, The geology of stratigraphic sequences: Springer-Verlag, Berlin, 433 p.

North American Commission on Stratigraphic Nomenclature, 1983, North American Stratigraphic Code: Am. Assoc. Petroleum Geologists Bull., v. 67, p. 841–875.

Posamentier, H. W., C. P. Summerhayaes, B. U. Haq, and G. P. Allen (eds.), 1993, Sequence stratigraphy and facies associations: Blackwell Scientific Publications, Oxford, 644 p.

Salvador, A. (ed.), 1994, International stratigraphic guide: A guide to stratigraphic classification, terminology, and procedure: International Union of Geological Sciences and Geological Society of America, Inc., Trondheim, Norway, 214 p.

Schwarzacher, W., 1993, Cyclostratigraphy and the Milankovitch theory: Elsevier, Amsterdam, 225 p.

Weimer, P., and H. W. Posamentier (eds.), 1993, Siliciclastic sequence stratigraphy: Recent developments and applications: AAPG Memoir 58, American Association Petroleum Geologists, Tulsa, Okla., 492 p.

Wilgus, C. K., B. S. Hastings, C. G. St. C. Kendall, H. W. Posamentier, C. A. Ross, and J. C. Van Wagoner (eds.), 1988, Sea-level changes: An integrated approach: Soc. Econ. Paleontologists and Mineralogists Spec. Pub. 42, 407 p.

13

Seismic, Sequence, and Magnetic Stratigraphy

13.1 INTRODUCTION

Seismology is the study of earthquakes and the structure of Earth on the basis of the characteristics of seismic waves. Although the broad subject of seismology lies outside the scope of this book, some aspects of seismology have a very important application to stratigraphy. The emphasis of this chapter is on what is commonly referred to as exploration seismology and, more specifically, on application of the techniques of exploration seismology to stratigraphic study. Exploration seismology deals with the use of artificially generated seismic waves to obtain information about the geologic structure, stratigraphic characteristics, and distributions of rock types. The techniques of exploration seismology were developed initially to locate structural traps for petroleum deposits, and they are still used extensively for that purpose; however, seismic methods can also be applied to stratigraphic problems.

Seismic stratigraphy is thus the study of seismic data for the purpose of extracting stratigraphic information. Seismic stratigraphy is a relatively new science, born in the 1960s. Because of its wide applicability to subsurface study both on land and at sea, where other types of stratigraphic data are few, it has already achieved an important position alongside the more traditional branches of stratigraphy.

Sequence stratigraphy is an outgrowth of seismic stratigraphy, although the practice of sequence stratigraphy is not limited to study of seismic records. Sequence stratigraphic concepts can be applied also to outcrop and well data. A sedimentary sequence is a stratigraphic unit composed of a relatively conformable succession of genetically related strata that is bounded at its top and base by unconformities or their correlative conformities (Mitchum, Vail, and Sangree, 1977). A sequence represents one cycle of deposition bounded by nonmarine erosion, deposited during one significant cycle of rise and fall of base level. Because base level in marine basins is controlled by sea level, a sequence is thus the product of a cycle of rise and fall of sea level. The sequence concept has assumed such an important role in stratigraphic thinking that we now refer to application of sequence concepts as sequence stratigraphy.

Magnetostratigraphy is also a relatively new branch of stratigraphy. It is based on the principle that iron-bearing minerals such as magnetite become magnetized at the time they crystallize in a magma or lava flow. The magnetized

433

minerals acquire magnetic polarity (north and south pole alignments) in keeping with Earth's magnetic polarity, a property called remanent magnetism. Geologists discovered in the 1960s that the remanent magnetism of some volcanic rocks displays reversed polarity, indicating that Earth's magnetic field has reversed at times in the geologic past. Sedimentary rocks can also become magnetically polarized because small iron-bearing minerals are mechanically aligned with earth's magnetic field as they settle in water. Patterns of magnetic reversals in ancient sedimentary rocks constitute a powerful tool for stratigraphic subdivision of these rocks, which can be applied to a variety of geologic problems such as correlation, geochronology (study of time in relation to Earth history), and paleoclimatology.

13.2 SEISMIC STRATIGRAPHY

Early Development of Seismic Methods

The use of seismic methods for obtaining information about subsurface rocks and structures involves the natural or artificial propagation of seismic (elastic) waves. These waves pass downward into Earth until they encounter a discontinuity and are reflected back to the surface, where they can be picked up by detectors. Seismic waves travel at velocities ranging from less than 2 km/s (some sediments) to more than 8 km/s (some ultramafic rocks), depending upon the kinds of rocks through which they pass and their depth below Earth's surface. If we know the velocity with which seismic waves travel through a particular kind of rock and if we can time their passage downward to a reflector and back to the surface, then we can calculate the depth to the reflecting horizon. This principle forms the basis for application of seismic methods to geologic study.

Much of the theory of elasticity and propagation of seismic waves through rock materials that constitutes the theoretical basis for seismology was developed in the early part of the 19th century. An English seismologist, Robert Mallet, was the first scientist to measure the velocity of seismic waves in subsurface materials. He initiated experimental seismology in 1848 by measuring the speed at which seismic waves passed through comparatively near-surface materials. He used black powder as an energy source to create a disturbance in the rocks and the surface of a bowl of mercury as the detector for the arriving seismic waves. The possibility of using seismic techniques to define the characteristics of subsurface rocks was apparently first put forward by a scientist named Milne in 1898. Two other interesting applications of the principles of seismology were experimented with in the early part of the 20th century. These were a method for detecting icebergs (after the sinking of the *Titanic* by an iceberg in 1912) and the use of mechanical seismographs to detect the position of large enemy guns during World War I.

Principles of Reflection Seismic Methods

On-Land Surveying

Practical application of seismology to the detection of rock structures began immediately after the end of World War I in both the United States and Europe, especially in Germany and England. The first applications were in petroleum exploration in Germany and in the Gulf Coast region of the United States, particularly in exploration for petroleum traps associated with salt domes. These early exploration efforts used the refraction seismic method for determining the structure of subsurface formations. This method is based on the principle that artificially generated seismic waves (produced in the early years of exploration by explosives) are refracted or bent at discontinuity surfaces as they travel downward below the surface. The waves then travel along these discontinuities before being refracted back to the surface, where their arrival is picked up by detectors placed at various distances away from the explosion (shot) point. The time that elapses during passage of the seismic waves downward to the discontinuity and back to the surface is used to compute the depth to the discontinuity.

Although several shallow petroleum deposits in salt domes were discovered during the early years of seismic exploration by the refraction method, this method did not work well for deeper structures because of the excessive distances required between shot points and detectors. Therefore, it was soon largely supplanted in petroleum exploration by the **reflection** seismic method. In the reflection method, waves created by an explosion are reflected back to the surface directly from subsurface rock interfaces without being refracted and traveling laterally along discontinuity surfaces. Therefore, detectors can be located at relatively short distances from the shot points, and reflection seismic techniques can be used for delineating very deep structures. After introduction of the reflection method about 1930, it quickly became the primary tool in the petroleum industry for locating buried anticlines and other structural traps for oil.

A very brief summary of basic principles of reflection seismology is given here. Additional details of the physical principles upon which reflection methods are based can be found in standard seismology textbooks, such as Lay and Wallace (1995) and Stein and Wysession (2003). As mentioned, the reflection seismic method for delineating the structure of subsurface rock units is based on the principle that elastic or seismic waves travel at known velocities through rock materials. These velocities vary with the type of rock (typical average velocities: shale = 3.6 km/s; sandstone = 4.2 km/s; limestone = 5.0 km/s; Christie-Blick, Mountaiun, and Miller, 1990). Where the subsurface lithology is known relatively well from drill hole information, it is possible to make accurate calculations of the time required for a seismic signal to travel from the surface to a given depth and then be reflected back to the surface.

The reflection technique involves first generating elastic waves at the surface at a point source, originally called a shot point because explosives were first used to create the seismic waves. Nonexplosive energy sources located on the surface are now also in common use. These nonexplosive energy sources include vibratory devices that produce continuous vibrations at the surface or devices that drop heavy weights onto a metal plate placed on the ground surface. Seismic detectors, called geophones, are laid out in arrays extending outward from the shot point. Seismic waves reflected back from subsurface discontinuities are picked up by these detectors and fed electronically to a recording device. The principal discontinuities that reflect seismic waves are bedding planes and unconformities. By multiplying the travel velocity by one-half of the travel time elapsed from initiation of the elastic waves at the point source to their arrival at the detector, geophysicists can accurately calculate depths to the discontinuities. This procedure thus allows the subsurface position of the discontinuities to be determined. The data obtained in this manner can then be displayed as seismic sections or profiles that depict the structure of the major rock units as they appear in cross section. Alternatively, the data may be used to prepare structure contour maps (see Chapter 16) on the tops of particular reflecting horizons.

The general principles of on-land reflection seismic "shooting" are illustrated in the interesting old diagram (Fig. 13.1) from Nettleton's 1940 *Geophysical Prospecting for Oil*. The equipment and techniques for surveying locations, shooting, and recording and processing seismic data have changed and significantly improved since the 1940s; however, the basic principles illustrated in this figure still apply. As seismic waves pass downward and outward from the point energy source through the subsurface formations, they are reflected from successively deeper formations back to the surface where they are picked up by the electronic

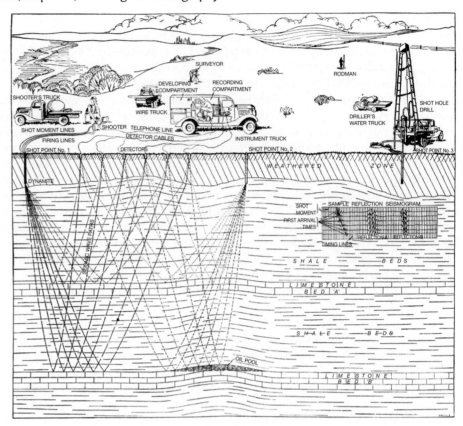

Figure 13.1
Diagram illustrating the equipment and procedures used in seismic exploration in the 1940s. [From Nettleton, L. L., 1940, Geophysical prospecting for oil, Fig. 155, following p. 332, McGraw-Hill Book Co.]

detectors. The signal from the detectors is then amplified, filtered to remove excess "noise," digitized, and fed to a recording truck to be recorded on magnetic tape or disk.

The data recorded on the magnetic tape or disk must then be presented in visual form for monitoring and interpretation. Prior to the use of magnetic tapes or disks for recording, the visual seismic records, or seismograms, were mechanically produced, wiggly-trace records such as that shown in the sample reflection seismogram in Figure 13.1. Photographic or dry-paper recording methods are now used for visual display, and several modes of displaying the amplitude of arriving seismic waves against arrival time are in use. A common type of display, called a **variable-density mode** display, is generated by a technique by which light intensity is varied to display differences in wave amplitude (one-half the height of a wave above the adjacent trough) by producing alternating light and dark areas on film or paper, thereby accentuating the amplitude of waves from a particular reflecting surface. For example, all wave traces having an amplitude greater than a given value are shaded black; traces with lower amplitudes are unshaded. Thus, a strong reflection event will show up as a black line on the record, as illustrated in Figure 13.2.

Marine Seismic Surveying

Early seismic surveys were carried out on land; however, reflection seismic methods can be used also in the ocean or in lakes. Some marine operations in very shallow water began in the late 1920s and 1930s, but extensive marine seismic surveys did not get underway until about the middle 1940s. Marine seismic operations employ the same principles as those used on land, but they differ in the speed at which they take place and in the specific details of the shooting and detection processes. Sound sources and detectors are towed behind the survey ship, which can operate at a speed of 6 knots or more on a continuous 24-h/day basis. In the

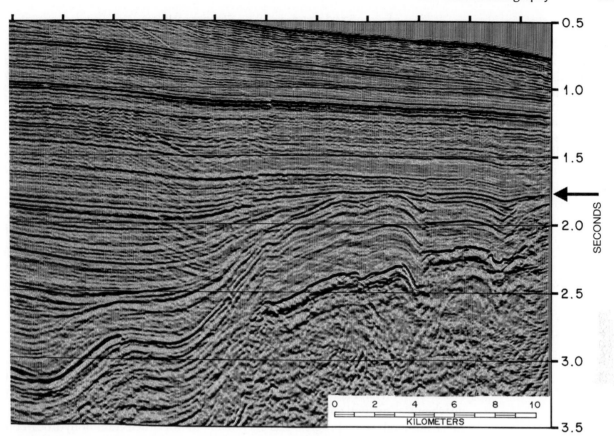

Figure 13.2
Example of a seismic record displayed using the variable-density method of printing. The vertical scale is given in two-way seismic-wave travel time (in seconds) rather than depth. Note the presence of an unconformity (arrow), at a depth of about 1.7 seconds, separating folded strata below from nearly horizontal strata above. [From Brown et al., 1995, Sequence stratigraphy in offshore South African divergent basins: AAPG Studies in Geology 41, Fig. 43, p. 54, reproduced by permission.]

early years of marine operations, a half-pound block of TNT was tossed over the ship's side every 3 minutes to provide a continuous seismic record. This method was potentially dangerous and damaging to fish and other ocean life. It has now been largely replaced by techniques that use acoustic sources such as airguns, which produce sound energy by releasing highly compressed air.

Early marine operations were severely hampered by problems of accurately locating the shot point and detector positions and the operations had to be carried out within sight of land so that locations could be determined by land-based surveying methods. The development about 1949 or 1950 of radio navigation methods made possible operations in the ocean away from land. Subsequent development of satellite navigation methods (so-called GPS systems) that "home in" on orbiting satellites to fix the position of ships at sea (or positions on land) now allows the position of the survey ships in the open ocean to be accurately and continuously determined. Another significant development that came about as a result of new advancements during World War II was the invention of the floating streamer cable, which allowed detectors (called hydrophones) to be towed in a floating cable behind the ship. Streamers may be up to several kilometers in length. These technical advances made possible rapid progress in marine seismic surveying methods in the years following. The general principles of marine seismic profiling are illustrated in Figure 13.3.

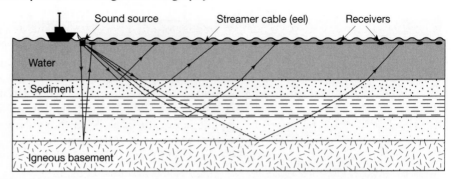

Figure 13.3
Diagram illustrating the principles of marine seismic surveying. Only a few of the possible paths for travel of seismic waves are shown. [Based on Kennett, 1982, Fig. 2.13, p. 40.]

Subsequent progress in both marine and land-based seismic surveying techniques have improved energy sources and devised sources that do not require explosives; developed new detection equipment and procedures; and improved treatment and analysis of seismic data. In particular, computer analysis of seismic data has allowed a quantum leap forward in filtering and enhancing seismic signals and in methods of displaying and interpreting seismic data.

Application of Reflection Seismic Methods to Stratigraphic Analysis

The science of seismic stratigraphy was developed largely by petroleum companies out of pragmatic necessity to locate petroleum deposits in deep, unexplored basins both on land and offshore. Geologists have not yet discovered a successful geochemical method for directly detecting oil or gas in deep subsurface formations, although some progress has been made in direct detection of hydrocarbon deposits by seismic methods. Therefore, the successful search for petroleum still requires that explorationists locate and drill petroleum traps such as anticlines and salt domes. Because most shallow petroleum traps were located and tested long ago during the earlier phases of petroleum exploration, petroleum companies have been forced to extend exploration efforts to deeper formations and frontier basins, which are undrilled or sparsely drilled basins, on land and offshore. Inasmuch as successful oil finding depends upon knowledge of stratigraphic relationships as well as structural anomalies, and because poorly explored basins lack sufficient well control for stratigraphic analysis, new techniques had to be developed that would allow stratigraphic information to be extracted from seismic data. Thus, seismic stratigraphy was born in the 1960s as a tool that made possible the integration of stratigraphic concepts with geophysical data—that is, a geologic approach to stratigraphic interpretation of seismic data (Payton, 1977; Berg and Wolverton, 1985; Vail, 1987; Cross and Lessenger, 1988; Whittaker, 1998).

Seismic reflections are generated by physical surfaces in subsurface rocks. In the conventional structural application of seismic data, seismic reflections are used to identify and map the structural attitudes of subsurface sedimentary layers. By contrast, seismic stratigraphy uses seismic reflection patterns to identify depositional sequences, to predict the lithology of seismic facies by interpreting depositional processes and environmental settings, and to analyze relative changes in sea level as recorded in the stratigraphic record of coastal regions. Seismic stratigraphy thus makes possible many types of stratigraphic interpretations, such as geologic time correlations, definition of genetic depositional units, and thickness and depositional environment of genetic units.

Parameters Used in Seismic Stratigraphic Interpretation

To accomplish the objective of interpreting stratigraphy and depositional facies from seismic data, geologists must identify characteristic features of seismic reflection records (seismograms) and relate these features to the geologic factors responsible for the reflections. An understanding of the factors that generate seismic reflections is therefore critical to the entire concept of seismic stratigraphy. Fundamentally, primary seismic reflections occur in response to the presence of significant density-velocity changes at either unconformity or bedding surfaces. Reflections are generated at unconformities because unconformities separate rocks having different structural attitudes or physical properties, particularly different lithologies. The density-velocity contrast along unconformities may be further enhanced if rocks below the unconformity have been altered by weathering. Reflections are generated at bedding surfaces because, owing to lithologic or textural differences, a velocity-density contrast exists between some sedimentary beds; however, not every bedding surface will generate a seismic reflection. Also, a given reflection event identified on a seismic record may not be caused by reflection from a single surface, but it may represent the sum or average of reflections phased together from several bedding surfaces, particularly if beds are thin.

The seismic records produced as a result of primary reflections from unconformities or bedding surfaces have distinctive characteristics that can be related to depositional features such as lithology, bed thickness and spacing, and continuity. Two particularly important seismic parameters that have stratigraphic significance are reflection configuration and reflection continuity (Table 13.1).

Reflection Configuration.   Reflection configuration refers to the gross stratification patterns identified on seismic records. As illustrated in Figure 13.4, **parallel**, **divergent**, and **prograding** patterns can be interpreted in terms of primary

Table 13.1   Seismic reflection parameters commonly used in seismic stratigraphy, and the geologic significance of these parameters

Seismic facies parameters	Geologic interpretation
Reflection configuration	Bedding patterns
	Depositional processes
	Erosion and paleotopography
	Fluid contacts
Reflection continuity	Bedding continuity
	Depositional processes
Reflection amplitude	Velocity-density contrast
	Bed spacing
	Fluid content
Reflection frequency	Bed thickness
	Fluid content
Interval velocity	Estimation of lithology
	Estimation of porosity
	Fluid content
External form and areal association of seismic facies units	Gross depositional environment
	Sediment source
	Geologic setting

Source: Mitchum, R. M., Jr., P. R. Vail, and J. B. Sangree, 1977, Seismic stratigraphy and global change of sea level, Part 6: Stratigraphic interpretation of seismic reflection patterns in depositional sequences, *in* C. E. Payton (ed.), Seismic stratigraphy—Applications to hydrocarbon exploration: *Am. Assoc. Petroleum Geologists Mem. 26*, Table 2, p. 122, reprinted by permission of AAPG, Tulsa, Okla.

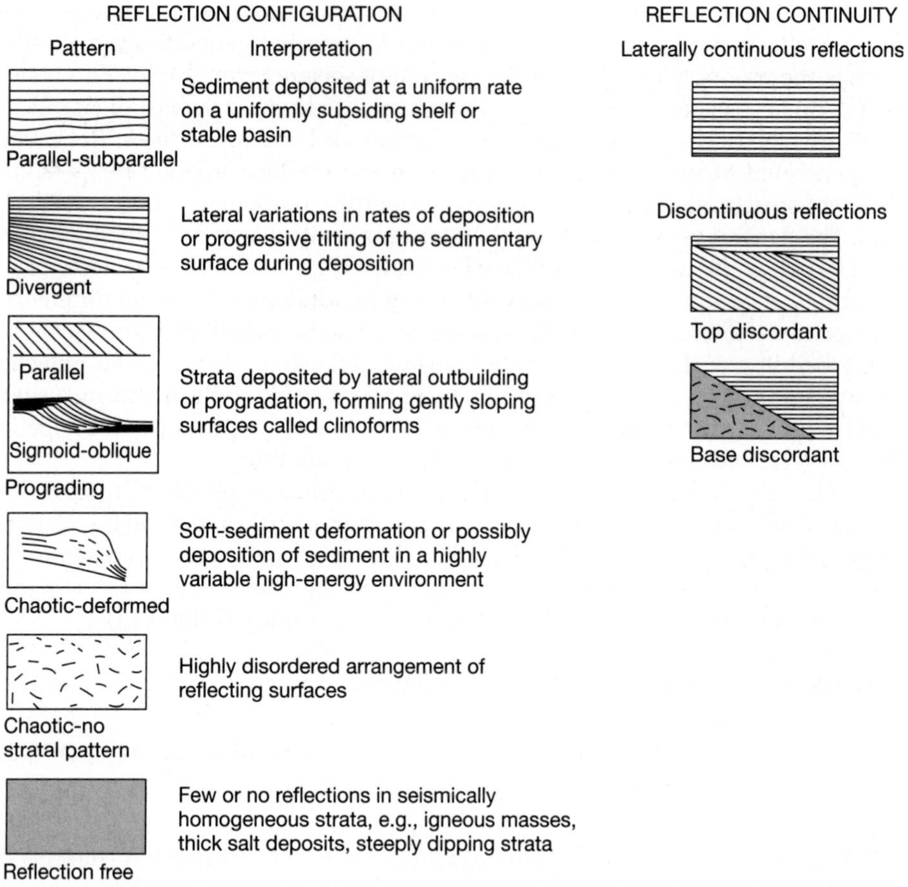

REFLECTION CONFIGURATION

Pattern	Interpretation

Parallel-subparallel — Sediment deposited at a uniform rate on a uniformly subsiding shelf or stable basin

Divergent — Lateral variations in rates of deposition or progressive tilting of the sedimentary surface during deposition

Parallel / Sigmoid-oblique / Prograding — Strata deposited by lateral outbuilding or progradation, forming gently sloping surfaces called clinoforms

Chaotic-deformed — Soft-sediment deformation or possibly deposition of sediment in a highly variable high-energy environment

Chaotic-no stratal pattern — Highly disordered arrangement of reflecting surfaces

Reflection free — Few or no reflections in seismically homogeneous strata, e.g., igneous masses, thick salt deposits, steeply dipping strata

REFLECTION CONTINUITY

Laterally continuous reflections

Discontinuous reflections

Top discordant

Base discordant

Figure 13.4
Schematic illustration of seismic reflection configuration and reflection continuity. [Based on Mitchum, R. M., Jr., P. R. Vail, and J. B. Sangree, 1977, Stratigraphic interpretation of seismic reflection patterns in depositional sequences, *in* Payton, C. E. (ed.), Seismic stratigraphy—Applications to hydrocarbon exploration: Am. Assoc. Petroleum Geologists Mem. 26, Fig. 4, p. 123; Fig. 5, p. 124; Fig. 2, p. 119; Fig. 3, p. 120, reproduced by permission of AAPG, Tulsa, Okla.]

depositional conditions or characteristics (the upper part of Fig. 13.2 further illustrates parallel reflectors). **Chaotic** patterns are the result of soft-sediment deformation or other kinds of deformation that produce a disordered arrangement of reflecting surfaces. So-called **reflection-free** patterns display no identifiable stratal reflections; reflections appear simply as random "noise."

Note that prograding patterns display sloping surfaces called clinoforms. The terms undaform, clinoform, and fondaform were introduced by Rich (1951) to describe depositional environments in relation to wave base (Fig. 13.5A). The undaform is the more or less flat topographic surface that exists in an aqueous environment above wave base where bottom sediments are moved or stirred by waves and currents, particularly during storms. The **clinoform** is the sloping surface extending from wave base down to the generally flat floor, called the fondaform, of the water body. Bedding characterized by initial dips owing to deposition in the clinoform zone is referred to as clinoform bedding. The sloping strata deposited on delta fronts constitute an example of clinoform bedding, as illustrated in the seismic record in Figure 13.5B.

Reflection Continuity. Reflection continuity depends upon the continuity of the density-velocity contrast along bedding surfaces or unconformities. It is closely associated with continuity of strata, and it provides information about depositional process and environment. Continuous reflections, such as some of those in the upper part of Figure 13.2 that extend for more than 20 kilometers, characteristically indicate stratified deposits that are continuous over large areas. In contrast to continuous reflections, reflection patterns showing reflection terminations (e.g., some reflections in the lower-left part of Fig. 13.2) indicate stratigraphic relationships

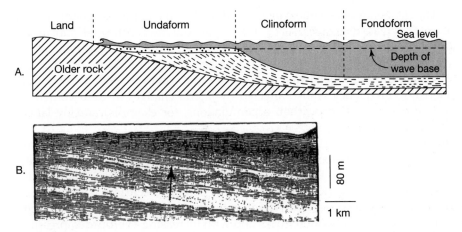

Figure 13.5
A. Sketch illustrating the meaning of the terms undaform, clinoform, and fondoform as used by Rich (1951); after Rich, J. L., 1951, Geol Soc. America Bull, v. 62, Fig. 1, p. 3. B. Example of clinoform (prograding) bedding in a seismic record from the Rhone Delta, Southeast France. After Posamentier and Allen, 1993, Variability of the sequence stratigraphic model: Effects of local basin factors: Sedimentary Geology, v. 86, Fig. 8, p. 98; reproduced by permission.

such as onlap, downlap, and toplap (discussed subsequently) that occur in coastal regions in response to transgressions and regressions.

Other Reflection Parameters. Other reflection parameters that have some significance in seismic stratigraphic interpretation are listed in Table 13.1. The **amplitude** of seismic waves displayed on seismograms is, among other things, an indication of bed thickness and spacing. (As mentioned, amplitude is equal to one-half the height of the wave above the adjacent trough.) If bed thickness is less than the wavelength of the seismic wave—for example, one-fourth of a wavelength—the reflections from the top and base of the bed can be phased together to give exceptionally large amplitudes. On the other hand, when beds are very thick (commonly >50 m), the reflections from the top and base of the beds are completely separate and wave amplitude may be small (Sheriff, 1980). Analysis of wave amplitudes allows geophysicists to calculate the thickness of beds, if a nearby contact with a layer thicker than 1 wavelength is present for calibration (Christie-Blick, Mountain, and Miller, 1990).

The amplitude of reflected seismic waves can be affected also by fluid content of sedimentary beds or by accumulations of gas in the beds. The presence of hydrocarbons in beds can produce a marked increase in amplitude of waves that shows up on seismic records as so-called "bright spots." These bright spots actually appear blacker than surrounding events, so the meaning of the name is not clear; perhaps it simply means that they stand out clearly on seismic records. Bright-spot analysis was introduced in the petroleum industry in the early 1970s and is now used as a method for direct detection of hydrocarbon deposits.

Reflection frequency refers to the number of vibrations or oscillations of seismic waves per second. It is numerically equal to wave velocity divided by wave length. The frequency of a seismic wave is commonly expressed in hertz (Hz) or kilohertz (kHz). A hertz is a unit of frequency equal to one cycle per second; a kilohertz is 1000 hertz. The frequency of seismic waves affects both the depth of penetration of the waves into the subsurface and the resolution of the seismic records, that is, the sharpness with which details of the seismograms can be distinguished. Lower frequencies give greater depth of penetration but less resolving power. The frequency of seismic waves is induced by the particular energy sound source used to create the waves. As the waves pass downward through subsurface formations and are reflected back to the surface, the initial induced frequency is attenuated by bed thickness, which controls the spacing of reflectors. Thus, attenuations of the initial induced frequency of seismic waves is related to bedding characteristics. Frequency is also affected by lateral changes in fluid content of beds (the presence of hydrocarbon accumulations, for example) and by lateral thickness changes in beds.

Interval velocity refers to the average velocity of seismic waves between reflectors. Seismic wave velocity is affected by several factors, especially porosity, density, external pressure, and pore (fluid) pressure. Porosity has a particularly significant effect on velocity, which increases as porosity decreases. Thus, because porosity commonly decreases with depth, velocity increases with depth. Velocity also increases with density of the rocks and with increasing overburden pressure. For example, the velocity for a typical sandstone increases from about 4 km/s at the surface to more than 5 km/s at a depth of 5000 m. Velocity decreases with increasing interstitial fluid pressure; the presence of gas at low saturations in the pore spaces of the rocks also causes a decrease in velocity. Seismic velocity is of particular interest because of the possibility that different rock types, which are characterized by different densities, porosities, pore fluid pressures, and other characteristics, can be differentiated on the basis of seismic velocity.

The **external form** or geometry of stratigraphic bodies that generate seismic reflections can be interpreted from seismic data (e.g., Fig. 13.13). Thus, these data can be used to identify "seismic facies," which may be interpreted in terms of depositional environments of the lithologic analogs of these seismic facies. This procedure of interpreting the external form of stratigraphic packages from seismic data is part of the process of seismic facies analysis, which also provides information on sediment source and geologic setting, including major facies changes. Seismic facies analysis is an extremely important aspect of seismic stratigraphy and is discussed in greater detail in the following paragraphs.

Procedures in Seismic Stratigraphic Analysis

The significance of the seismic stratigraphic approach to the study of subsurface sedimentary rocks lies in the fact that it permits geologists and geophysicists to interpret stratigraphic relationships and depositional processes as well as to use seismic data for conventional structural mapping. Interpretation is a subjective process, but when seismic stratigraphic analysis is pursued in a logical manner and interpretation is based upon analogy with established stratigraphic and depositional models that have been generated by other types of studies, seismic stratigraphic analysis becomes an extremely valuable tool. Seismic stratigraphy can thus provide insight into such stratigraphic and depositional factors as lithofacies changes, relief and topography of unconformities, paleobathymetry (depth relationships and topography of ancient oceans), geologic time correlations, depositional history, and subsidence and tilting history (burial history). The procedures for interpreting stratigraphy from seismic data involve three principal stages: seismic sequence analysis, seismic facies analysis, and interpretation of depositional environments and lithofacies (Vail, 1987). Seismic stratigraphic analysis is applied also to interpretation of ancient sea-level changes.

Seismic Sequence Analysis

The term sequence is often used informally by geologists to refer to any grouping or succession of strata. Sequence is also used in a more restricted sense to identify distinctive stratigraphic units that are commonly bounded by unconformities (i.e., similar to allostratigraphic units, described in Appendix C). Sloss (1963) considered sequences to be major rock-stratigraphic units of interregional scope that are separated and delimited by interregional unconformities. He recognized and named (using Indian names) six major sequences on the North American craton (see Fig. 12.9), each separated by demonstrable regional unconformities that can be traced from the Cordilleran region of western North America to the Appalachian Basin in the east. Each succession or sequence represents a major cycle of transgression and regression, that is, advance and retreat of shorelines. Recognition of the sequences is based on physical relationships among rock units, although Sloss indicates that the sequences also have time-stratigraphic significance.

The sequence concept was subsequently extended and redefined by Mitchum, Vail, and Thompson (1977). These authors define a **depositional sequence** as "a stratigraphic unit composed of a relatively conformable succession of genetically related strata and bounded at its top and base by unconformities or their correlative conformities." Sequences as thus defined differ from Sloss's sequences in that they may be much smaller rock units (a few tens of meters to as much as a thousand meters; Wilson, 1992). Also, because they are bounded by interregional unconformities and their equivalent conformities, they may be traceable over major areas of ocean basins as well as continents. Distinct, related groups of depositional sequences superposed one on another are designated by Mitchum, Vail, and Thompson (1977) as supersequences. These supersequences are of the same general order of magnitude as Sloss's original sequences. The basic concept of depositional sequences is illustrated in Figure 13.6. Because sequences are defined on the basis of physical relationships of the strata, that is, bounded at the top and base by unconformities or their correlative conformities, they are not primarily dependent for recognition upon determination of rock types, fossils, or depositional processes.

Internal Relationships.   The strata that make up a depositional sequence may be either concordant, that is, essentially parallel to the sequence boundary, or discordant, lacking parallelism with respect to the sequence boundaries. **Concordant**

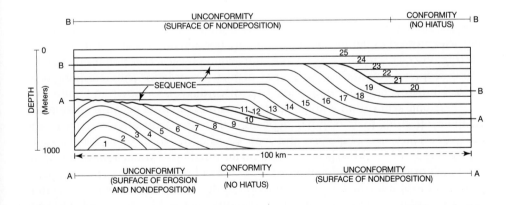

Figure 13.6

Illustration of the concept of depositional sequences. A depositional sequence is composed of relatively conformable, genetically related strata bounded at its base (A) and top (B) by unconformities that pass laterally to correlative conformities. Individual units of strata 1 through 25 are traced by following stratification surfaces; they are assumed to be conformable where successive strata (no missing strata) are present. Where units of strata are missing, hiatuses are present. [From Mitchum, R. M., Jr., P. R. Vail, and S. Thompson, III, 1977, Seismic stratigraphy and global change of sea level. Part 2: The depositional sequence as a basic unit for stratigraphic analysis, in Payton, C. E. (ed.), Seismic stratigraphy—Applications to hydrocarbon exploration: Am. Assoc. Petroleum Geologists Mem. 26, Fig. 1, p. 54, reproduced by permission of AAPG, Tulsa, Okla.]

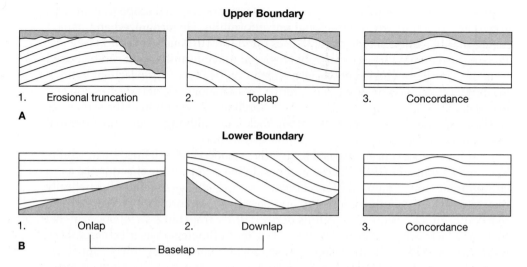

Figure 13.7
Relations of strata to the (A) upper boundary and (B) lower boundary of a depositional sequence. A1. **Erosional truncation:** Strata terminate against the upper boundary mainly
owing to erosion. A2. **Toplap:** Initially inclined strata terminate against an upper boundary created mainly as a result of nondeposition (e.g., foreset strata terminating against an
overlying surface where no erosion or deposition took place). A3. **Top concordance:**
Strata at the top of a sequence do not terminate against an upper boundary. B1. **Onlap:**
Strata terminate updip against an inclined surface. B2. **Downlap**: Initially inclined strata
terminate downdip progressively against initially horizontal or inclined surfaces. B3. **Base-
concordance:** Strata at base of a sequence do not terminate against lower boundary.
[From Mitchum, R. M., Jr., P. R. Vail, and S. Thompson, III, 1977, Seismic stratigraphy and
global change of sea level. Part 2: The depositional sequence as a basic unit for stratigraphic analysis, *in* Payton, C. E. (ed.), Seismic stratigraphy—Applications to hydrocarbon
exploration: Am. Assoc. Petroleum Geologists Mem. 26, Fig. 2, p. 58, reproduced by permission of AAPG, Tulsa, Okla.]

relations can occur at either the upper or the lower boundary of a sequence and
may be expressed as parallelism to an initially horizontal, inclined, or uneven surface (Fig. 13.7). **Discordance** is the most important physical criterion used in determining sequence boundaries. Strata may display discordance with overlying beds
(top discordance) or underlying beds (base discordance), as shown in Figure 13.8.

Erosional **truncation** is the lateral termination of strata because erosion has
cut them off from their original depositional limits. Truncation occurs at the
upper boundary of a sequence and may be of either local or regional extent.
Toplap (Fig 13.7A.2, 13.8B) is lapout (lateral termination of strata against a boundary at their original depositional limit) at the upper boundary of a depositional sequence, for example, the lateral termination updip of the foreset beds of a deltaic
complex. Toplap is evidence of a nondepositional hiatus. Mitchum, Vail, and Sangree (1977) suggest that toplap results from a depositional base level, such as sea
level, being too low to permit the strata to extend farther updip, thus allowing
sedimentary bypassing and possibly minor erosion to occur above base level
while prograding strata are deposited below base level.

Baselap occurs at the lower boundary of a depositional sequence and may itself be of two types. **Onlap** (Fig. 13.7B.1, 13.8C) is baselap in which an initially horizontal or inclined stratum terminates against a surface of greater inclination.
Downlap (Fig. 13.7B.2, 13.8D) is baselap in which an initially inclined stratum terminates downdip against an initially horizontal or inclined surface. Onlap and
downlap indicate nondepositional hiatuses and not erosional breaks in deposition.

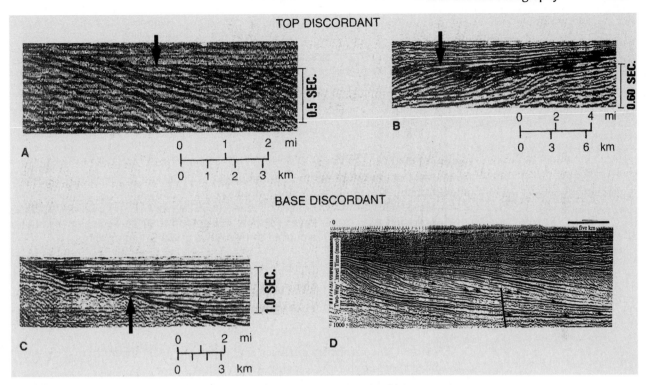

TOP DISCORDANT

BASE DISCORDANT

Figure 13.8
Top-discordant seismic reflection patterns: A. Erosional truncation (arrow). B. Toplap (arrow). **Base-discordant** seismic reflection patterns: C. Onlap (arrow). D. Downlap; downlap surfaces shown by small, black arrows. [A, B, C after Mitchum, R. M., Jr., P. R. Vail, and J. B. Sangree, 1977, Stratigraphic interpretation of seismic reflection patterns in depositional sequences, *in* Payton, C. E. (ed.), Seismic stratigraphy—Applications to hydrocarbon exploration: AAPG Mem. 26, Fig. 2, p. 119; Fig. 3, p. 120, reproduced by permission; D after Posamentier and Allen, 1999, Siliciclastic sequence stratigraphy—Concepts and applications: SEPM Concepts in Sedimentology and Paleontology 7, Fig. 3.55, p. 94, reproduced by permission.]

Toplap and baselap relations of the strata in a depositional sequence should not be confused with the foreset bedding of cross-laminated units, which form parts of beds rather than sequences. Figure 13.9 diagrammatically illustrates in a regional setting the relationships of strata in depositional sequences to sequence boundaries.

Identification of Depositional Sequences.   Depositional sequences can be identified both in outcrop sections and in subsurface sections by searching for unconformities (erosional surfaces, truncation of strata, missing strata). Subsurface identification depends largely upon the use of instrumental well logs, such as electric logs that measure resistivity of rock units (Chapter 12) and seismic data to locate and trace unconformities, truncations, and lapout relationships (see Van Wagoner et al., 1990). In fact, the sequence concept, as conceived by Vail and his co-workers at Exxon, was developed initially from seismic data.

Because sequences are defined as stratigraphic units separated by unconformities or their correlative conformities (Mitchum, Vail, and Thompson, 1977), the mapping of unconformities is thus the key to seismic sequence analysis. Seismic sequence analysis aims to identify major reflection "packages" that can be delineated by recognizing surfaces of discontinuity. We commonly think of discontinuities as unconformities, which are surfaces of erosion or nondeposition that represent

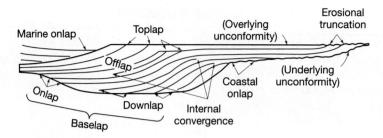

Figure 13.9

Terminology for relations that define unconformable boundaries of a depositional sequence. [After Mitchum, R. M., Jr., P. R. Vail, and J. B. Sangree, 1977, Stratigraphic interpretation of seismic reflection patterns in depositional sequences, *in* Payton, C. E. (ed.), Seismic stratigraphy—Applications to hydrocarbon exploration: Am. Assoc. Petroleum Geologists Mem. 26, Fig. 1, p. 118; Fig. 3, p. 120, reproduced by permission of AAPG, Tulsa, Okla.]

major hiatuses, and we identify four different kinds of unconformities (see Fig. 12.5). In seismic stratigraphic analysis, however, two kinds of discontinuities are recognized:

1. **Erosional unconformity surfaces** that represent a significant hiatus owing to subaerial or subaqueous erosional truncation
2. Unconformable surfaces called **downlap surfaces**, which are marine surfaces representing a hiatus but without evidence of erosion.

Discontinuities are generally good reflectors and also commonly separate rock units having different dips, at least on a regional scale. Discontinuities may thus be recognized by interpreting systematic patterns of reflection terminations along the discontinuity surfaces. Two patterns, onlap and downlap, occur above discontinuities. Three patterns, truncation, toplap, and apparent truncation (Fig. 13.10), occur below discontinuities (Vail, 1987).

Seismic resolution is generally not adequate to delineate minor sedimentary sequences because the practical vertical resolution is on the order of 10 to 50 m. On

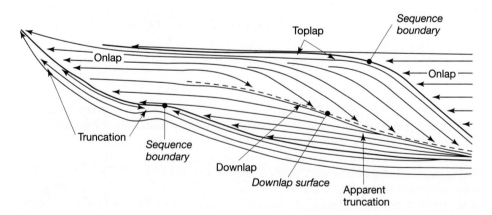

Figure 13.10

Diagram illustrating sequence boundaries (unconformities), downlap (maximum flooding) surfaces, and various kinds of reflection terminations. Apparent truncation refers to termination by depositional thinning. [From Vail, P. R., 1987, Seismic stratigraphic interpretation using sequence stratigraphy, *in* Bally, A. W. (ed.), Seismic stratigraphy: Am. Assoc. Petroleum Geologists Studies in Geology 27, Fig. 1, p. 2, reproduced by permission of AAPG, Tulsa, Okla.]

the other hand, major depositional units or systems such as progradational delta-slope systems, carbonate shelf-margin systems, or marine offlap-onlap systems can be identified. Once depositional sequences have been correlated basinwide, these sequences provide a first-order stratigraphic framework within which seismic facies can be studied in more detail. Figure 13.11, a seismic section from offshore Newfoundland, provides an example of sequences delineated by unconformities that separate different depositional units.

The procedure for carrying out seismic sequence analysis thus involves the following steps:

1. Picking unconformities in a given area by recognizing reflection terminations along their surfaces.
2. Extending or extrapolating these boundaries throughout the complete section, including areas where the reflectors are conformable, to define the sequences completely.
3. Repeating the process of delineating sequence boundaries on seismic records from other parts of a basin or region and correlating the sequences throughout the seismic grid to produce a three-dimensional framework of successive stratified seismic sequences separated by unconformities or correlative conformities.
4. Mapping sequence units on the basis of thickness, geometry, orientation, or other features to see how each sequence relates to neighboring sequences.

Details of seismic sequence mapping and its application to geologic problems are given by Hubbard, Pape, and Roberts (1985a, 1985b), Vail (1987), and Vail et al. (1991). The sequence concept has become so important in stratigraphic studies that it has taken on the role of a specialized branch of stratigraphy called **sequence stratigraphy**. Sequence stratigraphy is further discussed in Section 13.3.

Seismic Facies Analysis

Seismic facies analysis takes the interpretation process one step beyond seismic sequence analysis by examining within sequences smaller reflection units that may be the seismic response to lithofacies. In seismic facies analysis, the most common reflection characteristics used to distinguish one seismic facies from another are the geometry of reflections or reflection terminations with respect to the two unconformity surfaces bounding the sequence, the external geometry of the facies, and the internal configuration and character of the reflections. A **seismic facies unit** is a mappable, areally definable, three-dimensional unit composed of seismic reflections whose characteristic reflection elements differ from those of adjacent units. It is considered to represent or express the gross lithologic aspect and stratification characteristics of the depositional unit that generates the reflections. Some of the more important seismic reflection patterns that constitute seismic facies are offlap, onlap, submarine mounds, channel/overbank complexes, slope-front fill, toplap, and drape (Figs. 13.10, 13.12).

Procedures for Interpreting Seismic Facies.   The objective of seismic facies analysis is regional interpretation of lithology, depositional environments, and geologic history. The interpretation process proceeds in several distinct steps (Mitchum and Vail, 1977; Vail, 1987).

The first step is recognizing and delineating seismic facies units within each sequence on all of the seismic sections in the region being mapped. The most useful seismic parameters in seismic facies analysis are the following:

1. The geometry of reflections (Fig. 13.12) and reflection terminations
2. Reflection configuration (e.g., parallel, divergent)
3. Three-dimensional form (Fig. 13.13)

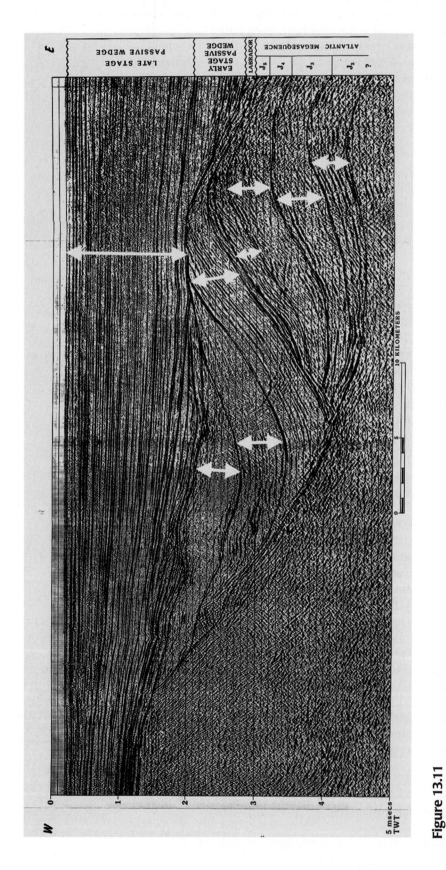

Figure 13.11

Depositional sequences as defined from seismic records. In this example from offshore Newfoundland, seismic sequence boundaries are shown by solid black lines. The vertical scale on this record is given in seismic wave two-way travel time (TWT) rather than depth. [From Hubbard, R. H., J. Pape, and D. G. Roberts, 1985, Depositional sequence mapping to illustrate the evolution of a passive continental margin, *in* Berg, O. R., and R. G. Wolverton (eds.), Seismic stratigraphy II—An integrated approach: Am. Assoc. Petroleum Geologists Mem. 39, p. 104, Fig. 8, reprinted by permission of AAPG, Tulsa, Okla.]

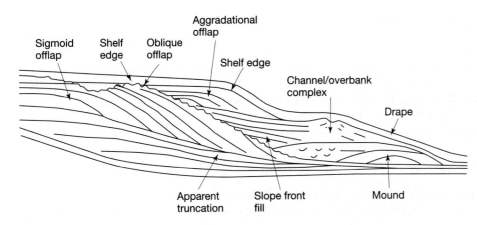

Sigmoid offlap
Shelf edge
Oblique offlap
Aggradational offlap
Shelf edge
Channel/overbank complex
Drape
Apparent truncation
Slope front fill
Mound

Figure 13.12
A simulated seismic section illustrating some common seismic facies patterns that can be identified from seismic records. [From Vail, P. R., 1987, Seismic stratigraphic interpretation using sequence stratigraphy, *in* Bally, A. W. (ed.), Seismic stratigraphy: Am. Assoc. Petroleum Geologists Studies in Geology 27, Fig. 6, p. 6, reproduced by permission of AAPG, Tulsa, Okla.]

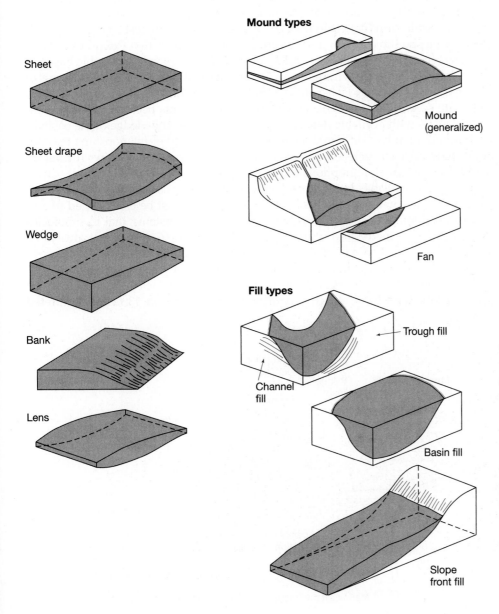

Sheet
Sheet drape
Wedge
Bank
Lens

Mound types
Mound (generalized)
Fan

Fill types
Trough fill
Channel fill
Basin fill
Slope front fill

Figure 13.13
External form of some stratigraphic bodies as interpreted from seismic facies units. [After Mitchum, R. M., Jr., P. R. Vail, and J. B. Sangree, 1977, Stratigraphic interpretation of seismic reflection patterns in depositional sequences, *in* Payton, C. E. (ed.), Seismic stratigraphy—Applications to hydrocarbon exploration: Am. Assoc. Petroleum Geologists Mem. 26, Fig. 12, p. 131, reproduced by permission of AAPG, Tulsa, Okla.]

Reflection terminations and configurations can be analyzed visually from two-dimensional seismic profiles. External three-dimensional geometry must be determined by mapping based on many different seismic profiles. Each seismic facies unit is distinguished from adjacent units on the basis of these parameters.

After the distribution (geometry) and thickness of the reflection packages have been mapped, the next step in seismic facies analysis is to combine this information with any other distinctive seismic information, such as interval velocity, and any available nonseismic data, such as well and outcrop data, that shed light on the regional geology. Vail et al. (1991) provide an in-depth overview of these procedures.

Interpretation of Lithofacies and Depositional Environments. Once the objective aspects of delineating seismic sequences and facies have been completed, the final objective is to interpret the facies in terms of lithofacies, depositional environments, and paleobathymetry. For example, seismic facies that show prograding reflection characteristics commonly indicate deltaic deposits. Reflection patterns showing laterally adjacent undaform (above wave base), clinoform (seaward sloping), and fondaform (flat basin floor) beds (see Fig. 13.5) suggest changes in water depth from shelf to slope to deep basin. Parallel reflectors that extend over large areas suggest shelf deposits or possibly deeper water deposits in a stable basin.

To a large extent, environmental interpretation of seismic facies is a process of elimination (Brown and Fisher, 1980). The interpreter may be able to immediately eliminate certain lithofacies or depositional environments because of obvious inconsistencies with available data or because of personal knowledge of the basin or region under study. Additional analysis involving study of relationships to other units, reflections characteristics, and other properties commonly allows further reduction of the remaining options until only one or two depositional or lithofacies models fit the available data. Lateral facies equivalents must be given special attention in the interpretation process and the interpreter must be experienced in and have a good knowledge of depositional processes and systems, that is, lithofacies composition, geometry, and spatial relationships. Even so, it may not always be possible to arrive at a final, unique interpretation, and the interpreter may have to settle for the best conclusion that can be made consistent with available data.

Figure 13.14 demonstrates the process of seismic facies interpretation by showing how the seismic reflection patterns depicted in Figure 13.12 are interpreted

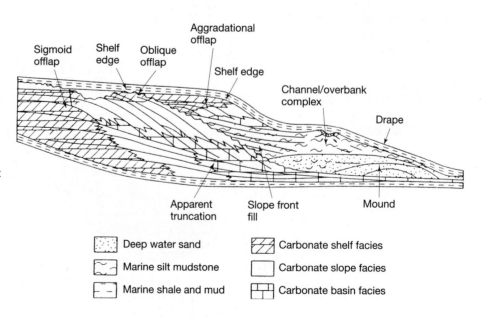

Figure 13.14
Schematic illustration of lithologic and environmental interpretation of the simulated seismic facies patterns shown in Figure 13.12. [From Vail, P. R., 1987, Seismic stratigraphic interpretation using sequence stratigraphy, *in* Bally, A. W. (ed.), Seismic stratigraphy: Am. Assoc. Petroleum Geologists Studies in Geology 27, Fig. 9, p. 10, reproduced by permission of AAPG, Tulsa, Okla.]

in terms of lithofacies and environmental setting. The first step in making such an interpretation is to learn as much as possible about the regional geology from well and outcrop control. Such nonseismic data can commonly show whether the sedimentary section consists of carbonates (and/or evaporites), siliciclastics, or mixed carbonates and siliciclastics. The seismic facies patterns can then be interpreted. For example, the depositional unit lying directly above the sequence boundary (i.e., the unconformity shown by the wiggly line) in Figure 13.12 is interpreted as mainly siliciclastic; the channel/overbank complex consists of marine silt and clay (mudstone); and the mound is composed of deep-water sand. The depositional unit lying immediately below the sequence boundary consists of carbonates, which are shelf, slope, or basin carbonates depending upon their relation to the off-lap pattern (Vail, 1987). For additional examples of seismic sequence analysis, see Light et al., 1993.

13.3 SEQUENCE STRATIGRAPHY

Fundamental Principles

Sequence stratigraphy is based on the premise that sedimentary successions can be divided into unconformity-bounded units (sequences) that form during a single, major cycle of sea-level change, as mentioned. Sequences can be split into smaller units, which are genetically linked and which form during different stages of a single sea-level cycle (Table 13.2). The entire three-dimensional assemblage of lithofacies enclosed within sequence boundaries is referred to as a **depositional system.** Depositional systems, in turn, are made up of smaller stratigraphic units: **system tracts** and **parasequences** (which are arranged into parasequence sets). Sequence stratigraphy sets out to place these stratal units into a predictable, chronostratigraphic framework by demonstrating how their generation is related to accommodation space (Vincent, Macdonald, and Gutteridge, 1998).

Accommodation space is the space available at any point in time in which sediments can accumulate (Jervey, 1988). Fundamentally, it is the space between a conceptual equilibrium surface that separates erosion from deposition in which potential sediment can accumulate. In the marine realm, this equilibrium surface, called base level, is sea level. Accommodation in marine settings is governed by eustasy (sea-level rise and fall) and tectonism (subsidence/uplift) (Fig. 13.15). Relative rise in sea level or subsidence of the seafloor creates more sedimentation space. Fall in sea level or elevation of the seafloor decreases accommodation. These accommodation variables and rate of sediment supply govern water depth, and thus transgression and regression, in several ways (Emery and Meyers, 1996). If sediment is added while accommodation remains constant (e.g., static sea level and no subsidence) or the sediment supply is greater than the rate of relative sea level rise, water depth will decrease (regression of facies belts). If sediment supply is less than the rate of relative sea-level rise, water depth increases (transgression of facies belts). If sediment supply is so great that it fills the accommodation to sea level, a drop in relative sea level owing either to eustatic fall or to uplift of the seafloor can cause sediment to be eroded.

Fundamental Units of Sequence Stratigraphy

Parasequences and Parasequence Sets

Parasequences are the smallest facies units of depositional sequences and range in thickness from about 10 to 100 m. A parasequence is a small-scale succession of beds or bed sets bounded by **flooding surfaces**, which are surfaces that separate younger strata from older strata, across which there is evidence of an abrupt increase in

Table 13.2   Hierarchy of sequence-stratigraphic units

Depositional Sequence—genetically related strata bounded by surfaces of erosion or nondeposition or their correlative conformities.

Stratal units within sequences include:

Depositional System—a three-dimensional assemblage of lithofacies, genetically linked by active (modern) or inferred (ancient) processes and environments (e.g., fluvial, deltaic, barrier-island).

System tract—a subdivision of a depositional system. Four main kinds are recognized: **highstand** (sediment deposited during high sea level), **falling-stage** (sediment deposited as sea falls from high to low), **lowstand** (sediment deposited during low sea level and early rising sea level), and **transgressive** (sediment deposited during rising sea level).

Parasequence Set—a succession of genetically related parasequences that form a distinctive stacking pattern that is bounded, in many cases, by major marine-flooding surfaces and their correlative surfaces.

Parasequence—a relatively conformable succession of genetically related beds or bedsets (within a parasequence set) bounded by marine flooding surfaces or their correlative surfaces.

Marine flooding surface—a surface that separates younger from older strata, across which there is evidence of an abrupt increase in water depth.

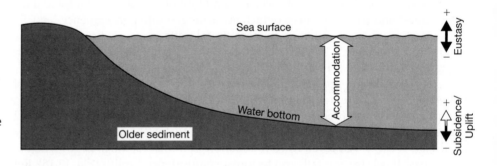

Figure 13.15
Diagram illustrating accommodation, the space available for sediment accumulation, in the marine realm. [Based on Posamentier, Jervey, and Vail, 1988].

water depth. They commonly form in coastal environments where the rate of increase of accomodation space is less than the rate of sediment supply. Under these conditions, sediment entering the sea will fill nearshore areas first, then move seaward (prograde) into more distal areas. The result is a shallowing-upward, and commonly coarsening-upward, vertical succession of facies. If this stage is followed by an increase in accomodation space (owing to eustatic sea-level rise or decrease in rate of sediment supply), transgression of the sea over the succession will take place, effectively terminating shallowing-upward deposition.

Subsequent decrease in accommodation space, owing to eustatic sea-level fall or increase in rate of sediment supply, can initiate deposition of a new parasequence. Short-term changes in rate of sediment supply can cause each succeeding parasequence to start prograding at a different point. Vertical stacking of parasequences owing to these variations in relative sea level thus generates a **parasequence set** (Fig. 13.16). In the case of the parasequences described here, each succeeding parasequence will advance in a seaward direction, forming a **progradational parasequence** set. Under different conditions of relative sea-level rise and fall, parasequences may advance in a landward direction to form **retrogradational** parasequence sets or build vertically to generate **aggradational**

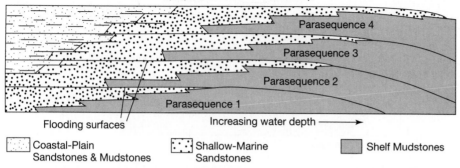

Progradational Parasequence Set
(Deposition > accommodation)

Parasequence 4

Parasequence 3

Parasequence 2

Parasequence 1

Flooding surfaces   Increasing water depth ⟶

☐ Coastal-Plain   ☐ Shallow-Marine   ☐ Shelf Mudstones
  Sandstones & Mudstones   Sandstones

Figure 13.16
Illustration of parasequences and parasequence sets formed under prograding conditions (rate of deposition exceeds the rate at which accommodation space is being created). Flooding surfaces are surfaces that separate younger strata from older, across which there is evidence of an abrupt increase in water depth. [After Van Wagoner et al., 1990, Siliciclastic sequence stratigraphy in well logs, cores, and outcrops: AAPG Methods in exploration series, No. 7, Fig. 4, p. 12, reproduced by permission.]

parasequence sets (e.g., Coe and Church, 2003). For example, carbonate parasequences are commonly aggradational and also shallow upward.

System Tracts

As shown in Table 13.2, the strata that make up a depositional sequence are referred to collectively as a depositional system. A depositional system constitutes the sediments deposited during a complete cycle of sea-level change, from high sea level to low sea level and back to high sea level. The sediments deposited during particular parts of this cycle are referred to as system tracts, which, in turn, are composed of parasequences. Thus, system tracts may lie immediately above or below a sequence boundary or lie in the middle part of a sequence. Four kinds of system tracts are recognized, depending upon the sea-level conditions under which they formed.

Highstand system tracts lie immediately below a sequence boundary. They form during the late part of a sea-level rise, during a sea-level standstill, or during the early part of a sea-level fall (Fig. 13.17A). They form under conditions that allow progradation to aggradation. Alluvial and coastal plain sediments characterize the later parts of highstand system tracts, which grade seaward into shallow-marine (that may include deltaic sediments) and offshore-marine deposits. Highstand system tracts are terminated by the unconformity produced by the next eustatic sea-level fall.

Falling-stage system tracts (Fig. 13.17B), referred to by some authors as early lowstand system tracts, form as sea level falls from a highstand position. Falling sea level, due either to the rate of eustatic sea-level fall exceeding the rate of tectonic subsidence or the rate of eustatic sea-level rise being less than the rate of tectonic uplift, brings on a condition referred to as **forced regression**. During forced regression, accommodation space is *reduced* as the shoreline moves in a seaward direction (progrades) and also moves lower down the depositional profile. As a result, the coastal plain is not a site of deposition but rather a zone of sedimentary bypass, forming an unconformity surface. That is, sediment eroded onshore by fluvial incision will bypass the coastal plain and be deposited in a more seaward position. Falling-stage system tract deposits include shallow-marine sediments, offshore-marine sediments, and submarine-fan sediments.

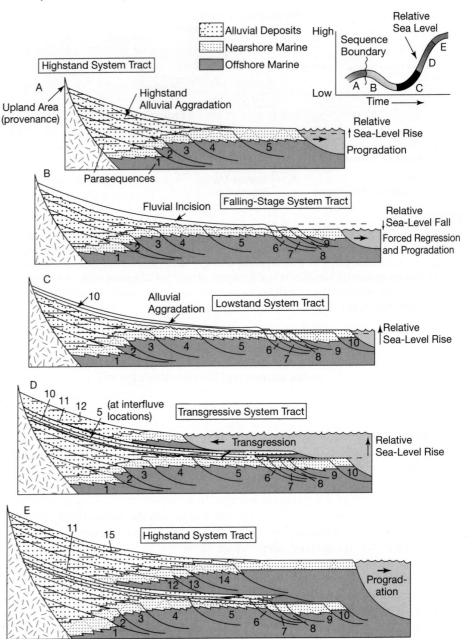

Figure 13.17
Schematic illustration of system tracts formed at different stages in a cycle of eustatic sea-level change from high sea level to low sea level and back. Note that each system tract comprises several small-scale regressive-transgressive cycles (i.e., parasequences). At each phase of normal shoreline regression within a parasequence, fluvial accommodation is created and deposition occurs, except during falling-stage conditions, when erosion occurs. [After Posamentier and Allen, 1999, Siliciclastic sequence stratigraphy—Concepts and applications: SEPM Concepts in sedimentology and paleontology No.7, Fig. 2.41, p. 41, reproduced by permission.]

Lowstand system tracts (Fig. 13.17C) begin to form after relative sea level has fallen to its minimum and begun to rise, creating a small amount of accomodation space. As sea level continues to rise, marine sediments are deposited and the fluvial system ceases to incise. Thus, a lowstand system tract is the package of sediments deposited between minimum relative sea level (reduced accommodation space) and subsequent pronounced increase in accommodation space. The sediment consists of progradational to aggradational parasequence sets that can contain alluvial and coastal-plain sediments, shallow-marine sediments (including deltaic sediments that can form and fill previously incised river valleys), offshore-marine sediments, and submarine-fan sediments (Coe and Church, 2003).

Continued rise in sea level creates conditions whereby the rate at which accommodation space is created is greater than the rate of sediment supply, which brings on transgression. The site of deposition shifts in a landward direction, generating retrogradational parasequences or parasequence sets. Transgressive surfaces may be marked by marine sediments overlying nonmarine sediments. The sediment deposited under these conditions forms a **transgressive system tract** (Fig. 13.17D). Transgressive system tract sediments may contain alluvial and coastal plain sediments, shallow-marine sediments, and offshore marine sediments, but they generally do not include submarine-fan sediments.

When sea level approaches its maximum, the rate of sedimentation eventually exceeds the rate of sea-level rise and aggradation to strong progradation generates a new **highstand system tract** (Fig. 13.17E). During this stage, coastal-plain and deltaic sedimentation predominates and these facies may prograde out over underlying lowstand deposits.

The above discussion presents only the very basic elements of sequence stratigraphy. For additional details, interested readers may wish to consult Coe and Church (2003), Posamentier and Allen (1999), or the excellent Sequence Stratigraphy Web Site maintained by the University of South Carolina (http://strata.geol.sc.edu/).

Methods and Applications of Sequence Stratigraphy

Methods

The preceding discussion indicates that the concepts involved in sequence stratigraphy are fairly complex. Students can be reasonably expected to ask at this point exactly what advantage sequence stratigraphy has over other stratigraphic methods that makes it worthwhile to develop an understanding of these concepts. The fundamental aim of sequence stratigraphy is to provide a high-resolution chronostratigraphic (time-stratigraphic) framework for carrying out facies analysis. Vertical facies analysis must be done within conformable packages of stratal units to accurately correlate coeval (equivalent age), lateral facies relationships along a single depositional surface. Other researchers have accomplished this end for many years by using transgressive and regressive cycles of strata for regional correlation of time and facies. The proponents of sequence stratigraphy maintain that sequence stratigraphy is a much better way of doing this (e.g., Van Wagoner et al., 1990, p. 6). A fundamental aspect of sequence stratigraphy is the recognition that sedimentary rocks are composed of a hierarchy of stratal units including laminae, beds, bedsets (see Chapter 4), parasequences, parasequence sets, and system tracts. With the exception of laminae, each of these units is a genetically related succession of strata bounded by chronostratigraphically significant surfaces. Facies above these surfaces or boundaries have no physical or temporal (time) relationship to the facies below. Therefore, correlation of these surfaces provides the high-resolution chronostratigraphic framework necessary for facies analysis.

The actual practice of sequence stratigraphy requires that stratigraphers be able to recognize the stratal expression of parasequences, parasequence sets, system tracts, and sequences. Early work relied mainly on seismic sequence and seismic facies analysis, as discussed in the preceding section—recognition of sequence boundaries on the basis of stratal terminations, for example. Seismic stratigraphy alone does not offer the necessary precision to recognize and analyze smaller scale sedimentary units; therefore, well logs, cores, and outcrops are also used to analyze sequences. Identification of shallowing upward units allows recognition of parasequences. Groups of parasequences can be observed to stack into retrogradational, progradational, and aggradational patterns to form parasequence sets, which correspond roughly to a system tract (Van Wagoner et al., 1990, p. 3). System tracks are identified by distinct associations of facies and position within a sequence. Thus, using an appropriate combination of seismic data, well logs, cores, and outcrop information, it is possible to generate a high-resolution chronostratigraphic framework of sequences and parasequence boundaries, defined solely by the relationship of the strata.

Environmental Applications

Sequence stratigraphic concepts were originally applied primarily to analysis of siliciclastic sediments deposited along continental margins, because these silici-clastic environments are particularly affected by cycles of relative sea-level change. As sea level swings from highstand to lowstand, a succession of system tracts are laid down, as documented in Figure 13.17. Subsequently, attempts have been made to extend the concepts of sequence stratigraphy to carbonate and evaporite environments, deep-sea environments, epicontinental (cratonic) marine environments, and even fluvial systems (see Emery and Meyers, 1996; Witzke, Ludvigson, and Day, 1996; Vincent, Macdonald, and Gutteridge, 1998; Coe, 2003). Although extending sequence-stratigraphy techniques to these environmental settings is apparently possible, important differences exist between sedimentation patterns in these environments and the siliciclastic marine shelf-slope environment.

For example, the rate of production of carbonate sediment in carbonate environments is typically much higher than the rate of accumulation of siliciclastic sediment in siliciclastic settings. Consequently, the rate of carbonate production generally exceeds the rate at which accommodation is created, causing the basins to fill to sea level and generating a shallowing-upward succession of facies (Chapter 11). Therefore, the pattern of system tracts in carbonate sediments may not be quite the same as that in siliciclastic sequences. Also, sequence boundaries are commonly more difficult to distinguish in carbonate successions than in siliciclastic deposits. Furthermore, the effects of subaerial exposure on carbonate platforms depends on climate. Humid climates will cause widespread dissolution and reprecipitation of carbonate; arid climates will cause less carbonate diagenesis but tend to promote precipitation of evaporites.

The deep-marine environment is affected far less by changes in relative sea level of a few hundred meters than is the shelf environment. Nonetheless, sea-level changes do affect deposition in deep-ocean basins, particularly deposition of turbidites in submarine fan systems. Although submarine fans can develop during sea-level highstands, turbidity currents appear more likely to move sediment from shelf environments to the deep ocean basin during lowstands of sea level than during highstands (Fig. 13.17). Analysis of deep-marine turbidite systems appears to be the main application of sequence-stratigraphy methods to the deep-sea environment (e.g., Emery and Meyers, 1996, p. 178).

Although sequence-stratigraphic concepts have been applied to marine epicontinental (cratonic) environments (e.g., Witzke, Ludvigson, and Day, 1996), problems arise with such applications because of the low rates of subsidence of cratonic areas. Sloss (1996) points out that vast areas of cratonic platforms appear to have subsided at rates of 5 m/m.y., or less, for epoch and period-length spans of time. Under such conditions, the bathometric relief required for clinoforms, downlap surfaces, lowstand tracts, and other characteristics of basins with a shelf break are rarely attainable except under special circumstances. Sloss suggests that meaningful progress is inhibited by forcing cratonic stratigraphy to conform to principles, definitions, and practices developed for a different set of conditions.

Application of sequence-stratigraphic techniques to fluvial systems presents particular problems because the base level for fluvial sedimentation, and thus accommodation, is more difficult to define than that for marine systems. The conceptual equilibrium surface that defines the upper limit of accommodation space (Fig. 13.15) in fluvial systems is commonly taken as the graded profile, or profile of equilibrium, of a stream. (A graded profile is the longitudinal profile of a graded stream or of a stream whose gradient at every point is just sufficient to enable the stream to transport the load of sediment made available to it.) The level to which a stream can ultimately grade is called the **bayline**, which is effectively sea level for streams that drain into the ocean. Changes in a stream's graded profile can either create or remove accommodation space. Such changes can include changes in discharge, sediment supply, channel form, and uplift, as well as the position of the bayline (sea level). Application of sequence-stratigraphy concepts to nonmarine systems is being actively researched but is still controversial. The general view is that the lower reaches (100–150 km) of fluvial systems are most likely to be greatly affected by base-level changes and that it is this portion of fluvial systems that is most likely to be preserved in the stratigraphic record (see Shanley and McCabe, 1994, and Vincent, Macdonald, and Gutteridge, 1998).

Global Sea-Level Analysis

General Principles.   One of the most controversial applications of sequence-stratigraphy concepts is the analysis of ancient sea levels. As discussed throughout this book, changes in sea level have an important bearing on sedimentation patterns. Studies of sea-level changes have special relevance with respect to analysis of cyclic successions in the stratigraphic record. Sea-level changes through time have been studied particularly intensively by P. R. Vail and his associates at the Exxon research laboratory in Houston (e.g., Vail, Mitchum, and Thompson, 1977a, 1977b; Haq, Hardenbol, and Vail, 1988). These authors used seismic data and surface outcrop data to integrate occurances of coastal onlap, marine (deep-water) onlap, baselap, and toplap into a model that involves asymmetric cycle oscillations of relative sea level.

Vail and his group inferred changes in relative sea level by reference to coastal onlap charts. These charts were constructed by estimating from seismic profiles the magnitude of sea-level rise, as measured by coastal aggradation (the thickness of coastal sediments deposited during sea-level rise). The amount of sea-level drop is determined by measuring the magnitude of downward shifts in coastal onlap, that is, the elevation (vertical) difference between the point of maximum coastal onlap reached at maximum sea level and the point of maximum sea-level fall, which is determined from the seismic records by the position where the next (younger) onlap unit lies above the unconformable surface produced during the sea-level fall (Vail, Mitchum, and Thompson, 1977a; Vail, Hardenbol, and Todd, 1984). The procedures used in constructing a relative coastal onlap chart from coastal and marine sequences are illustrated in Figure 13.18.

The first step involves analysis of sequences such as those shown as units A through E of Figure 13.18A. Sequence boundaries, areal distributions, and the presence or absence of coastal onlap and toplap are determined by tracing reflections on seismic profiles. Available age controls from well data are used to establish the geologic-time range of each sequence. An environmental analysis is also made from seismic and other available data to distinguish coastal facies from marine facies. The second step is to construct a chronostratigraphic (time-stratigraphic) chart of the sequences, a procedure first described by Wheeler (1958). Both stratal surfaces and unconformities give time-stratigraphic information. Because they are depositional surfaces, the seismic response to strata surfaces are assumed to be chronostratigraphic reflectors. In addition, because seismic reflectors are isochronous (have the same age everywhere), they can cross lithologic boundaries. That is, the seismic reflections from a given surface may extend laterally through a variety of lithofacies. Seismic reflectors may be traced continuously, for example, through a shelf system, over the shelf edge, and downward through an equivalent slope system. Unconformities are not isochronous surfaces; however, strata below an unconformity are older than strata above it. Therefore, strata between unconformities constitute time-stratigraphic units. After determining the ages of depositional sequences, such as those shown on the stratigraphic cross section in Figure 13.18A, from well-control or other information, workers plot the stratigraphic

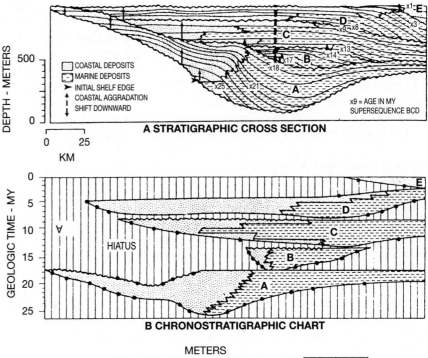

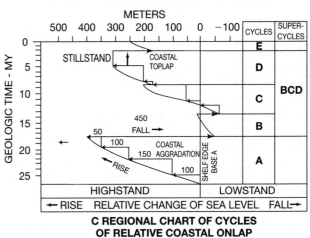

Figure 13.18

Diagram illustrating the procedures used by Exxon geologists for constructing a regional cycle of relative coastal onlap. [After Vail, P. R., R. M. Mitchum, Jr., and S. Thompson, III, 1977, Seismic stratigraphy and global change of sea level, Part 3: Relative changes of sea level from coastal onlap, *in* Payton, C. E. (ed.), Seismic stratigaphy—Applications to hydrocarbon exploration: Am. Assoc. Petroleum Geologists Mem. 26, Fig. 13, p. 78, reprinted by permission of AAPG, Tulsa, Okla.]

information against geologic time to construct a chronostratigraphic correlation chart (Fig. 13.18B). Such charts are referred to as **Wheeler diagrams**. The final step in the procedure is to identify cycles of relative coastal onlap in each seismic sequence, measure the magnitude of aggradation and seaward downshifts in coastal onlap that result from relative rise and fall of sea level, and plot these changes and sea-level standstills against geologic time as shown in Figure 13.18C. Thus, the magnitude of coastal aggradation is a measure of a relative rise in sea level; a relative standstill is indicated by coastal toplap; and seaward shifts in coastal onlap indicate a relative sea-level fall. The plots of relative coastal onlap are repeated for each sequence (cycles A though E of Fig. 13.18C) to complete the relative coastal onlap chart. Finally, the changes in relative coastal onlap are used as the basis of inferring changes in relative sea level. For simplicity, it is assumed that there has been no subsidence of the margin and that the bed-thinning effects of compaction under deep burial have been considered (original thickness restored).

The time interval occupied by a relative rise and fall of sea level, as interpreted from a coastal onlap chart such as that shown in Figure 13.18C, constitutes a cycle of relative sea-level change in a region. Vail, Mitchum, and Thompson

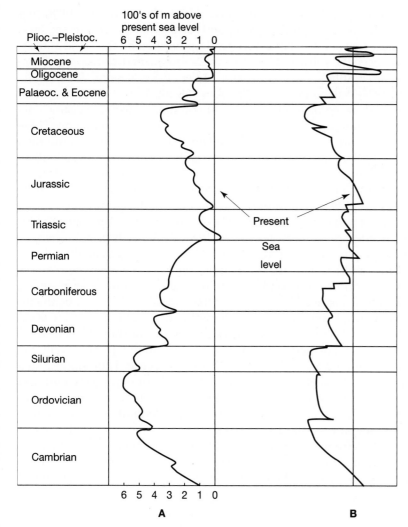

100's of m above
present sea level

Plioc.–Pleistoc.   6 5 4 3 2 1 0

Miocene
Oligocene
Palaeoc. & Eocene
Cretaceous
Jurassic
Triassic
Permian
Carboniferous
Devonian
Silurian
Ordovician
Cambrian

Present

Sea

level

6 5 4 3 2 1 0
A   **B**

Figure 13.19
Eustatic sea-level curves for Phanerozoic
time. (A) Hallam, 1984; (B) Vail, Mitchum,
and Thompson,1977b. [From Hallam, A.,
1984, Pre-Quaternary sea-level changes:
Ann. Rev. Earth and Planetary Sciences, v.
12, Fig. 5, p. 220, reprinted by permission.]

(1977b) correlated regional cycles and used this information to construct composite charts of global cycles of relative sea-level change. They interpreted these cycles to be worldwide and apparently to be controlled by absolute or eustatic
sea-level changes. Figure 13.19 shows the chart published by Vail, Mitchum, and
Thompson (1977b); a somewhat different chart constructed by Hallam (1984) from
compilations of continental flooding is shown also for comparison. Upon publication, the Exxon coastal onlap charts, and the relative sea-level curves inferred from
these charts, immediately generated lively interest, discussion, and controversy
among geologists. They continue to be a focus of controversy.

Reliability of Sea-level Analysis from Sequence-Stratigraphic Data. Following publication of the coastal onlap curves of Vail, Mitchum, and Thompson (1977b), several
workers (e.g., Brown and Fisher, 1980; Kerr, 1984; Miall, 1986) challenged the basic
premise of the Vail group that these curves primarily reflect eustatic changes in
sea level, and they pointed out that tectonism (basin subsidence) and rates of sediment supply also affect these curves. Nonetheless, interest by the geologic community in sea-level analysis continued at a high level. The publication in 1988 of
the SEPM Special Publication *Sea-Level Changes: An Integrated Approach*, edited by
Wilgus et al., is an indication of this interest. This volume contains several articles
on analysis of sea-level changes as well as articles dealing with sea-level changes

and sequence stratigraphy. In this volume, Haq, Hardenbol, and Vail (1988) present a new generation of coastal onlap curves and corresponding eustatic sea-level curves for the Mesozoic and Cenozoic (originally published by the same authors in 1987) that have been the subject of considerable subsequent discussion and controversy (e.g., Christie-Blick, Mountain, and Miller, 1990; Hallam, 1998; Miall, 1991, 1992, 1994, 1997; Walker, 1990). Haq, Hardenbol, and Vail (1988) indicate that these new curves are based on well-log and outcrop data as well as seismic data and have greater resolution than that obtainable from seismic data alone. These highly detailed curves are not reproduced in full here; however, a much simplified portion of the late Tertiary and Quaternary curve is shown in Figure 13.20 to illustrate the coastal onlap chart and show the nature of the second-order and third-order sea-level cycles interpreted from this chart. Figure 13.21 shows a more complete, but still simplified, part of the cycle chart for the Tertiary.

Christie-Blick, Mountain, and Miller (1990) provide a comprehensive critique of seismic stratigraphy and its usefulness in sea-level analysis and interpretation of the stratigraphic record. With respect to the Haq, Hardenbol, and Vail (1988) curves, Christie-Blick et al. are particularly concerned by what they call the uncritical interpretation of all second- and third-order boundaries as resulting from eustatic sea-level changes. They maintain that amplitudes of eustatic fluctuations cannot be inferred from seismic stratigraphic data alone because coastal aggradation (the vertical component of onlap) is primarily a result of basin subsidence, not sea-level rise. Furthermore, downward shifts in onlap reflect only the rate of sea-level fall in relation to the rate of basin subsidence. They suggest that the large component of basin subsidence cannot be easily or objectively removed to derive the smaller eustatic signal. Christie-Blick et al. point out further that another major limitation of the global onlap chart is the uncertainties about the calibration of many boundaries to the geologic time scale. Another potential problem in interpretation arises from the operation of autocyclic mechanisms (Chapter 12), such as delta switching, that can generate small-scale cycles. Miall (1991, 1992, 1994) also concludes that the implied precision of the Mesozoic-Cenozoic global

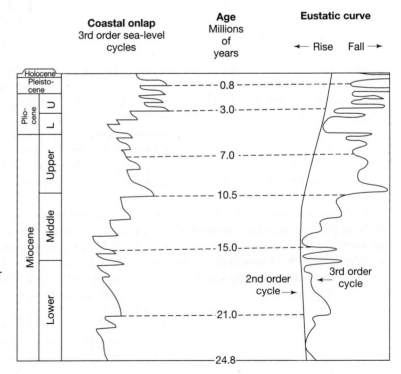

Figure 13.20
Coastal onlap and long-term (second-order) and short-term (third-order) sea-level curves for the late Tertiary and Quaternary, as determined by Haq et al., 1988. [Redrawn from Haq, B. U., J. Hardenbol, and P. R. Vail, 1988, Mesozoic and Cenozoic chronostratigraphy and cycles of sea-level change, *in* Wilgus, C. K., et al. (eds.), Sea-level changes: An integrated approach: Soc. Econ. Paleontologists and Mineralogists Spec. Pub. 42, Fig. 14, p. 94, reproduced by permission of SEPM, Tulsa, Okla.]

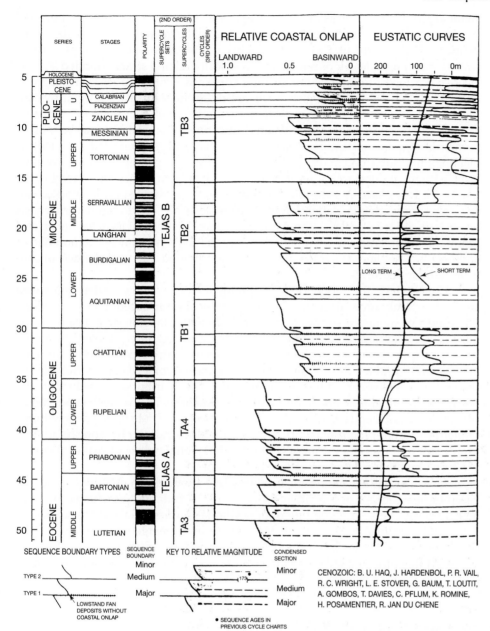

Figure 13.21
Simplified global sequence chart for part of the Tertiary and Quaternary. The polarity column shows the geomagnetic time scale (explained in Section 13.4); series and stages are discussed in Chapter 15. [After Haq, B. U., J. Hardenbol, and P. R. Vail, 1987, Chronology of fluctuating sea levels since the Triassic: Science, v. 235, Fig. 2, p. 1158. Copyright 1987 by the AAAS. As redrawn by Vail et al., 1991.]

cycle chart of Haq, Hardenbol, and Vail (1988) is not justified. He maintains that the chart is not an independently tried and tested global standard and that the implied precision is unsupportable because it is greater than that of the best available chronostratigraphic techniques, such as those used to construct the global standard time scale (Chapter 15).

In spite of these criticisms, interest in sea-level analysis and sequence stratigraphy in general remains high. To quote Plint et al. (1992), "The Exxon sea-level curve is here to stay, although it is likely to undergo progressive evolution and refinement as more data are gathered and better chronostratigraphic control becomes available." Over the next several years, geologists all over the world will surely be watching the outcome of the sea-level controversy. Hallam (1998) points out that relative sea levels can be studied by using an alternate approach that employs facies analysis. Fundamentally, this approach involves estimating relative sea-level changes by studying shallowing-upward successions in carbonate and siliciclastic rocks.

Summary Remarks.   The sequence concept produced a revolution in stratigraphy during the 1980s and early 1990s and gave rise to the term sequence stratigraphy, defined as the study of rock relationships within a chronostratigraphic (time stratigraphic) framework of repetitive, genetically related strata bounded by surfaces of erosion or nondeposition or their correlative conformities (Van Wagoner et al., 1988). Numerous geologists have embraced the sequence stratigraphy concept, and it is currently widely applied, particularly within the petroleum industry, to a variety of stratigraphic successions. Since the initial concept was proposed, numerous papers on sequence stratigraphy have been presented at scientific meetings and published in the geological literature. The concept has not, however, been without its detractors, as mentioned. It has been criticized, among other things, for being largely theoretical and introduced without specific, worked-out examples; for inadequate attention to problems of scale; and for an overemphasis on eustatic sea-level cycles without due regard to the effects of local tectonism and sediment supply. Also, it is difficult to apply the sequence concept to nonmarine and deep-sea sediments, which may not be subdivided into well-defined unconformity-bounded units.

Nonetheless, the sequence-stratigraphic concept is now widely embraced and has certainly revived interest in stratigraphy and produced many new ideas. The aspect of sequence stratigraphy that deals with global eustasy has been the most severely criticized part of the theory. As discussed, many geologists simply do not believe that reliable global sea-level charts can be constructed from sequence-stratigraphic data. On the other hand, most geologists agree that relative sea-level changes do affect the geometry and facies distribution of sediments deposited on continental margins and within sedimentary basins. Such facies can be effectively studied and interpreted by using outcrop and subsurface data within a sequence-stratigraphic framework. Thus, sequence stratigraphy appears to be a useful tool for stratigraphic analysis even if the global nature of sea-level curves produced by the Exxon group cannot be proven.

Galloway (1989) proposed an alternative method for defining sequences based on the concept of repetitive episodes of progradation punctuated by periods of transgression and flooding of depositional surfaces. Galloway's sequences are bounded by features formed during maximum submergence, in contrast to the Exxon sequences, which are bounded by features formed during maximum emergence. At this time, the Exxon scheme for defining sequences appears to be more widely understood and used.

13.4 MAGNETOSTRATIGRAPHY

General Principles

Magnetic stratigraphy, or **magnetostratigraphy**, is a relatively new branch of stratigraphy developed largely since about the middle 1960s. The principles of magnetic stratigraphy were initially applied to the study of volcanic rocks and sediments younger than about 5 million years. Magnetic stratigraphic techniques have now been applied to much older rocks, and a detailed magnetic polarity time scale has been extended to the Jurassic and parts of the older record. Magnetic stratigraphy came about through the discovery that magnetic iron-rich minerals in igneous and sedimentary rocks can preserve the orientation or field direction of Earth's magnetic field at the time the rocks were formed. During the cooling of molten rock, iron-bearing minerals become magnetized in alignment with Earth's magnetic field as they cool through a critical temperature of about 500°C–600°C (for magnetite), the **Curie point**. As they approach this temperature, the influence of the magnetic field exerts itself, and the atomic-scale magnetic fields within the crystal lattices of the minerals begin to line up parallel to one another and to the direction of the magnetic lines of force around Earth. With further cooling, these atoms become locked into this orientation, and each mineral in essence becomes a small magnet, having polarity consistent with Earth's magnetic field. These magnetic properties are retained for geologically long periods of time unless the rocks are again heated to near the Curie point. Therefore, this residual magnetism is called remanent magnetism, or **thermal remanent magnetism (TRM)**.

During deposition of sediments, small magnetic mineral grains are able to rotate in the loose, unconsolidated sediment on the depositional surface and thus align themselves mechanically with Earth's magnetic field. This (statistically) preferred orientation of magnetic minerals in sedimentary rocks imparts bulk magnetic properties to the rocks, referred to as depositional remanent magnetism or **detrital remanent magnetism**. Because sediment grains can be disturbed by bioturbating organisms or by physical and chemical processes during burial and diagenesis, the magnetization of sedimentary rocks is less stable, as well as weaker, than that of volcanic lavas. The study of remanent magnetism in rocks of various ages to determine the intensity and direction of Earth's magnetic field in the geologic past is called **paleomagnetism**.

Remanent magnetism is measured by instruments called magnetometers. Early magnetometers were capable of making paleomagnetic measurements only in igneous rocks and highly magnetized iron-bearing red sediments. Modern superconducting magnetometers can measure the magnetism in much more weakly magnetic sediments, including carbonates. Remanent magnetism is complex and can include secondary magnetism caused by prolonged effects of Earth's present magnetic field or by chemical changes owing to alteration of one magnetic mineral to another. Demagnetization techniques are available for destroying this secondary magnetic effect in the laboratory so that the primary magnetization can be measured. It is this primary magnetic component, which records Earth's geomagnetic field at the time volcanic or sedimentary rocks formed, that is of interest in stratigraphic studies.

The significance of primary remanent magnetism for stratigraphic studies stems from the fact that Earth's magnetic field has not remained constant throughout geologic history but has frequently reversed. The geomagnetic field is generated in some poorly understood way by the motion of highly conducting nickel-iron fluids in the outer part of Earth's core; this motion is assumed to be controlled by thermal convection and by the Coriolis force generated by Earth's rotation, constituting what is called a **self-exciting dynamo** (e.g., Butler, 1992; Merrill, McElhinny, and McFadden, 1996). Studies of the remanent magnetism in igneous and sedimentary rocks show that the dipole (main) component of Earth's magnetic field has reversed its polarity at irregular intervals from Precambrian time onward, apparently because of instabilities in outer-core convection (Merrill, McElhinny, and McFadden, 1996). When Earth's magnetic field has the present orientation, it is said to have **normal** polarity. When this orientation changes 180°, it has **reversed** polarity.

Figure 13.22 illustrates diagrammatically the magnetic lines of force around Earth during normal and reversed polarity epochs. Note that during a time of normal polarity the magnetic lines of force appear to "flow" out of Earth at the geographic South Pole, bend around Earth, and reenter at the geographic North Pole. Thus, a compass needle hinged vertically (swings up and down) would point down (into Earth) at the North Pole and up (out of Earth) at the South Pole. The flow direction of the magnetic lines of force, and the direction of a compass needle, would reverse during an episode of reversed polarity.

Reversals of Earth's magnetic field are recorded in sediments and igneous rocks by patterns of normal and reversed remanent magnetism. The direction of magnetization of a rock is defined by its **north-seeking magnetization**. If the

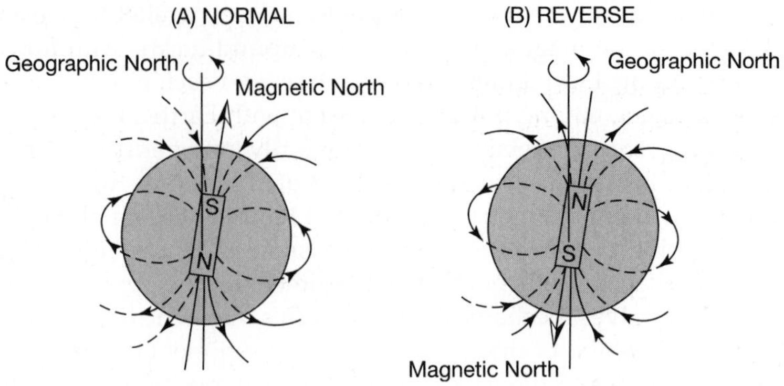

Figure 13.22
Schematic representation of Earth's magnetic field during episodes of (A) normal and (B) reversed polarity. [After Wylie, P. J., 1976, The way the Earth works: John Wiley and Sons, Inc., Fig. 9.1, p. 120, reprinted by permission.]

north-seeking magnetization of rocks points toward Earth's present magnetic north pole, the rock is said to have **normal-polarity magnetization**. If the north-seeking magnetization points toward the present-day south magnetic pole, the rock has **reversed-polarity magnetization,** or reversed polarity. Thus, sedimentary and igneous rocks that display bulk remanent magnetic properties of the same magnetic polarity as the present magnetic field of Earth have **normal** polarity, whereas those that have the opposite magnetic orientation have **reverse** polarity.

These geomagnetic reversals are contemporaneous worldwide phenomena. Thus, they provide unique stratigraphic markers in igneous and sedimentary rocks. The reversal process is thought to take place over a period of 1000–10,000 years (e.g., Clement, Kent, and Opdyke, 1982). The intensity of the magnetic field decreases by 60–80 percent over a period of about 10,000 years preceding reversal. The actual reversal requires about 1000–2000 years, followed by a buildup of intensity for the next 10,000 years (Cox, 1969). The last unquestioned reversal of the magnetic field took place approximately 700,000 years ago, although a brief reversal or excursion of the field may have occurred about 20,000 years ago.

[Note: The intensity of Earth's magnetic field has been declining over the past few centuries. It has decreased by 10 percent in the past 300 years, and the rate of decline is increasing. It is possible that the magnetic field will fall to nearly zero in the next millennium, leading scientists to speculate that another magnetic reversal is in the making (Nova, 2003). Think what that would do to compass directions!]

In the early years of paleomagnetic study, intervals of reversed or normal polarity lasting 100,000 years or more were called epochs and those having a duration of about 10,000–100,000 years were called events. Geomagnetic polarity reversals are now known to occur on a much broader spectrum of time scales ranging from less than 10,000 to greater than 10 million years. Magnetic stratigraphy is based on these changes in polarity recorded in sediments or volcanic rocks that produce recognizable patterns of alternating-polarity stratigraphic units that can be used for chronological and correlation purposes.

Sampling, Measuring, and Displaying Remanent Magnetism

To determine remanent magnetism in sedimentary rocks, geologists commonly remove samples from the field for subsequent laboratory analysis (see McElhinny and McFadden, 2000, and Tauxe, 2002, for details), although techniques have been developed to measure remanent magnetism in well bores (e.g., Bouisset and Augustin, 1993). Three kinds of samples may be taken:

1. Samples cored with a portable drill. Cores taken by this method are commonly 2.5 cm in diameter and 6–12 cm long. Before cores are broken out of the rock on the outcrop, the orientation of the cores must be determined and

marked on the sample. The inclination (dip) of the core axis is determined and the azimuth of the core axis (deviation from geographic north) is measured by use of a magnetic and/or sun compass.

2. Oriented block (hand) samples. These samples, which are broken from the outcrop with a hammer, are easier to obtain than core samples but present more difficulties with later orientation in the laboratory instruments.

3. Cores of lake- or ocean-bottom samples. Sediment cores, commonly obtained by piston coring apparatus, are assumed to penetrate the sediment vertically but are azimuthly unoriented.

In the laboratory, the remanent magnetism of a rock specimen is measured by means of a magnetometer. Several types of magnetometers (balanced fluxgate, cyrogenic, astatic, spinner) are used (Hailwood, 1989; Butler, 1992). Three orthogonal components of magnetism are commonly measured; these three components are then combined to give the direction and intensity of the magnetic vector of the specimen. Because most rocks carry different components of magnetization that have been acquired at different times, the signature of secondary magnetism must be removed to reveal the primary remanent magnetism. Secondary magnetization may be removed by progressive demagnetization methods such as alternating field demagnetization, thermal demagnetization, and chemical demagnetization.

Once primary remanent magnetism has been measured, vector directions in paleomagnetism are described in terms of the following:

1. **Inclination** (with respect to component in the original bed at the collecting station)
2. **Declination** (with respect to geographic north) (Fig. 13.23)

Inclination is called positive if downward directed and negative if upward directed. Positive inclination in the Northern Hemisphere indicates normal polarity; negative inclination means reversed polarity. Inclination directions are opposite relative to the polarity in the Southern Hemisphere. Inclination and declination together define the geomagnetic field vector (F in Fig. 13.23). Inclination is a function of the latitude at which the rock specimens formed, and declination shows the deviation of the ancient paleomagnetic pole from the geographic pole. The polarity data are commonly displayed graphically in polarity-reversal stratigraphic columns by plotting intervals of normal polarity in black and intervals of reversed

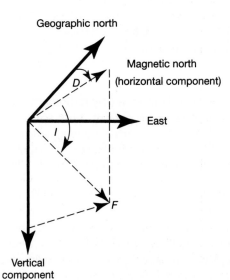

Figure 13.23
Description of the direction of the geomagnetic field. The total magnetic field vector F can be divided into a vertical component and a horizontal component. The angle D is the declination, the azimuthal angle between magnetic north and geographic north. The angle I is the inclination (dip), the angle between the horizontal and F.

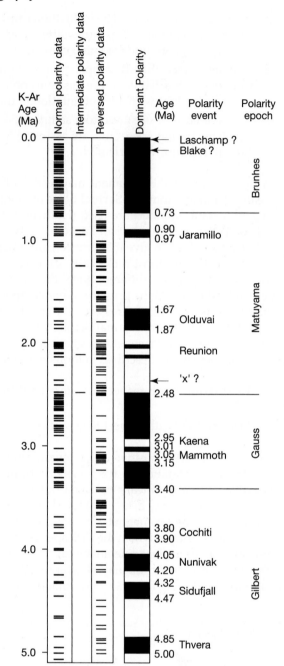

Figure 13.24
Late Cenozoic (Pliocene-Pleistocene) geomagnetic polarity time scale. Each horizontal line in the columns labeled normal polarity, intermediate polarity, or reversed polarity represents an igneous rock for which both K-Ar age and geomagnetic polarity have been determined. Auxiliary information from marine magnetic anomaly profiles and deep-sea core paleomagnetism have also been used. Black pattern in the dominant-polarity column indicates normal polarity; white indicates reversed polarity. Arrows indicate disputed short polarity intervals or geomagnetic "excursions"; numbers to the right of the polarity column indicate interpreted ages of polarity boundaries. [After Maniken, E. A., and G. B. Dalrymple, 1979, Revised geomagnetic polarity time scale for the interval 0–5 m.y. B. P.: Jour. Geophysical Research, v. 84, Fig. 3, p. 624.]

polarity in white (e.g., Fig. 13.24). Dating of the polarity intervals by radiometric and biochronologic methods forms the basis of the magnetic polarity time scale, described in the next paragraphs. Data may be plotted also as virtual geomagnetic pole (VGP) latitudes. The VGP represents the position of the effective north geomagnetic pole calculated from both inclination and declination information (Hailwood, 1989).

Magnetic Polarity Time Scales

The concept of remanent magnetism is well known to today's students of geology; however, only a few studies of rock magnetism had been made prior to the 1960s. The basic principles of magnetostratigraphy were developed in the early and middle 1960s in the remarkably short time of about five years by two groups of scientists working independently and competitively—one group in northern California and one in Australia. The initial development of a magnetic polarity sequence by these groups of scientists is summarized by Cox (1973), Glen (1982), McDougall (1977), and Watkins (1972). These scientists quickly realized the geologic potential of magnetic reversals. If the absolute age of reversals could be established, a quantitative time scale for reversals could be set up that would be extremely useful for stratigraphic correlation and other purposes.

The first such polarity scales were achieved by measuring the ages and magnetic polarities of young volcanic rocks on land by using potassium-argon (K-Ar) techniques to estimate the ages of the rocks younger than about 5 million years (Cox, Doell, and Dalrymple, 1963). During the 1960s, the geomagnetic time scale evolved through numerous versions as new data points were added and age estimates were refined (McDougall, 1979). Figure 13.24 shows the version of the time scale that had emerged by 1979. Note from Figure 13.24 that this polarity time scale is subdivided into polarity "epochs," each named for a distinguished scientist who contributed to development of the field of geomagnetism; shorter "events" were named for localities where definitive paleomagnetic characteristics of specific groups of rocks has been studied, commonly localities where the rocks were first sampled. We now understand that there is no fundamental distinction between polarity epochs and polarity events and that polarity intervals of a wide spectrum of durations are possible (e.g., Butler, 1992, p. 210).

In addition to study of the polarity of volcanic rocks on land, a second very important source of information about magnetic reversal sequences is provided by the linear anomaly patterns discovered in volcanic rocks of the ocean floor, particularly along mid-ocean ridges, and first interpreted by Vine and Matthews (1963). (Magnetic anomalies are significant deviations from Earth's magnetic background on either a local or regional scale.) These linear "stripes" of normal- and reversed-polarity magnetic rocks (e.g., Fig. 13.25) are roughly parallel to ridge crests and are typically 5–50 km wide and hundreds of kilometers long. These anomalies were produced owing to reversals in Earth's magnetic field as successive

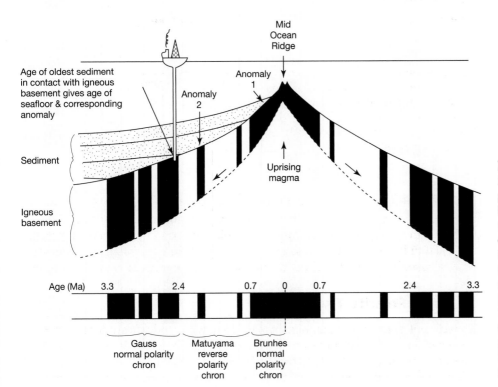

Figure 13.25
Schematic illustration of the principles by which the biostratigraphic age of sediments (determined by deep-sea drilling) overlying igneous basement can be used to date a particular marine magnetic anomaly. [From Hailwood, E. A., 1989, Magnetostratigraphy: Geological Soc. Spec. Report 19, Blackwell Scientific Publications. Fig. 17, p. 29, reproduced by permission.]

flows of lava erupted along ridge crests and cooled below the Curie point. Previously magnetized volcanic rock was pushed or pulled aside from the ridges as new volcanic rock formed and became magnetized. Vine and Matthews (1963) hypothesized that the linear magnetic anomaly patterns on the ocean floor correlate with normal and reversed polarity intervals in the geomagnetic scale established on land, allowing the ages of the anomalies to be estimated. The fact that the magnetic anomalies are roughly symmetrical about spreading ridges was a critically important piece of evidence used in developing the concept of seafloor spreading.

The principal problem with the oceanic record is that it is very difficult to date directly. Paleontologic ages on the oldest sediments overlying marine magnetic anomalies are available where basement has been reached by Deep Sea Drilling Program (DSDP) or Ocean Drilling Program (ODP) drill holes, but large uncertainties are often associated with these age determinations. Figure 13.25 illustrates the method of dating seafloor anomalies by use of paleontologic data. By making use of data from the ocean floor magnetic record, a detailed paleomagnetic time scale dating back to the Jurassic has now been established (Fig. 13.26).

Figure 13.26
Magnetic polarity time scale for the Cenozoic-late Mesozoic. Intervals of normal polarity are shown in black, reversed polarity in white. Intervals lacking magnetostratigraphic studies or having uncertain validity are cross-hachured. [Compiled by Ogg, J. G., 1995, Magnetic polarity time scale of the Phanerozoic, in Ahrens, T. J. (ed.), Global Earth physics—A handbook of physical constants: AGU Reference Shelf 1, American Geophysical Union, Washington, D.C., Fig. 2, p. 252, reproduced by permission.]

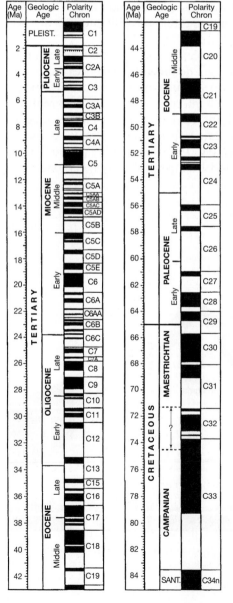
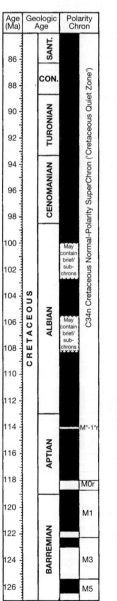
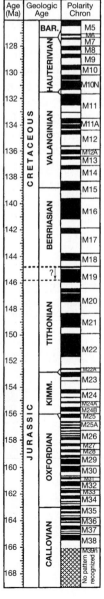

A detailed magnetic-polarity time scale for rocks older than the Jurassic has been more difficult to establish because the continuous oceanic geomagnetic time scale cannot be extrapolated beyond the age of the oldest oceanic crust (about 160–170 m.y; older oceanic crust has been destroyed in subduction zones); however, magnetic reversals are known in terrestrial sections in rocks at least as old as 1.5 billion years (Conde, 1982). Although study of on-land stratigraphic sections from various parts of the world has now provided many data on the paleomagnetic characteristics of early Mesozoic and older rocks on land, our knowledge of the polarity time scale for these rocks is much less refined than that for younger rocks. Figures 13.27 and 13.28 show generalized polarity reversal patterns for the late Mesozoic and Paleozoic; not all of these reversal patterns have been definitely verified (Ogg, 1995). See also Gradstein et al. (1995).

Terminology in Magnetostratigraphy

Because magnetostratigraphic polarity units are of primary interest in stratigraphy, these units are formally subdivided into polarity superzones, polarity zones, and polarity subzones depending upon their duration (Table 13.3). The equivalent geochronologic time units are referred to as **chrons** or superchrons. This terminology is recommended by the IUGS International Subcommission on Stratigraphic Nomenclature and the IUGS/IAGA Subcommission on a Magnetic Polarity Time Scale (see Salvador, 1994). The **polarity zone** is the fundamental polarity unit for subdivision of stratigraphic sections. A polarity zone may consist of strata with a single direction of polarization throughout, may be composed of an intricate alternation of normal and reversed units, or may be dominantly either normal or reversed but with minor subdivisions of the opposite polarity. A **polarity superzone** consists of two or more polarity zones, and a **polarity subzone** is a subdivision of a polarity zone. The North American Stratigraphic Commission follows approximately the same scheme of nomenclature as that proposed by the IUGS (see the North American Stratigraphic code in Appendix C). The principal polarity zone names now in use are the well-established names such as Brunhes, Matuyama, Gauss, and Gilbert, which are used for the most recent 5 million years of Earth's history. Historically, these units have been called epochs (Fig. 13.24); however, it is now recommended that these "epochs" be called the Brunhes, Matuyama, Gauss, and Gilbert polarity zones, or the Brunhes, Matuyama, Gauss, and Gilbert polarity chrons, if referring to time. Similarly, the so-called "events," such as the Jaramillo, Olduvai, and Reunion, should now be referred to as the Jaramillo, Olduvai, and Reunion polarity subzones (in rocks) or subchrons (for time).

Applications of Magnetostratigraphy and Paleomagnetism

Correlation

The primary application of magnetostratigraphy lies in its use as a tool for global correlation of marine strata. Magnetostratigraphic correlation is particularly important where paleontologic or lithologic correlation is difficult. It has special significance for international correlation because geomagnetic reversals are contemporaneous, synchronous, worldwide phenomena. They have worldwide scope owing to the fact that reversals of Earth's magnetic field affect the magnetic field everywhere on Earth at the same time. Because the polarity time scale can be calibrated radiometrically or paleontologically, polarity events thus provide a precise tool for chronostratigraphic (time) correlation. The first significant application of magnetostratigraphic techniques to correlation and age determinations of rocks was correlation of linear ocean-floor magnetic anomalies to on-land sections of volcanic strata whose ages had been determined by radiometric methods. These correlation techniques were subsequently extended to cores of oceanic sediments.

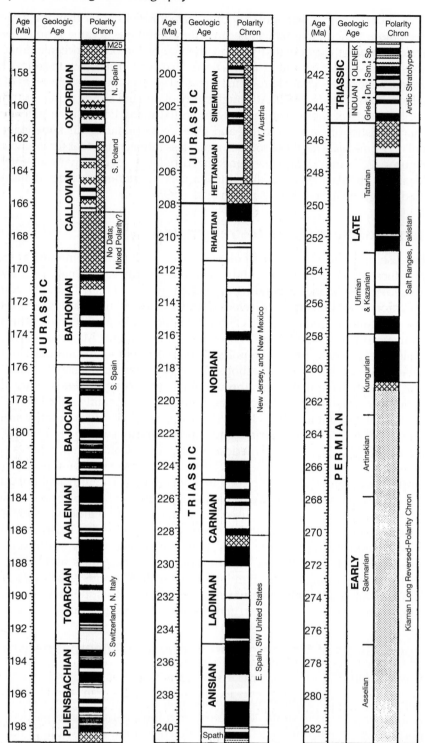

Figure 13.27
Magnetic polarity time scale for Jurassic to Permian. Intervals of normal polarity are shown in black, reversed polarity in white. Intervals lacking magnetostratigraphic studies or having uncertain validity are cross-hachured. [Compiled by Ogg, J. G., 1995, Magnetic polarity time scale of the Phanerozoic, *in* Ahrens, T. J. (ed.), Global Earth physics—A handbook of physical constants: AGU Reference Shelf 1, American Geophysical Union, Washington, D.C., Fig. 3, p. 260, reproduced by permission.]

Until very recently, correlation of sediment cores by use of magnetic polarity events had its greatest application in the study of marine sediments younger than about 6 to 7 million years. Correlation was previously restricted to very young rocks because the magnetic time scale had not been developed beyond about 7 Ma, and because most gravity and piston cores of ocean-floor sediment did not penetrate deep enough to sample older sediments. As mentioned, the detailed geomagnetic time scale has subsequently been extended to about 160–170 Ma.

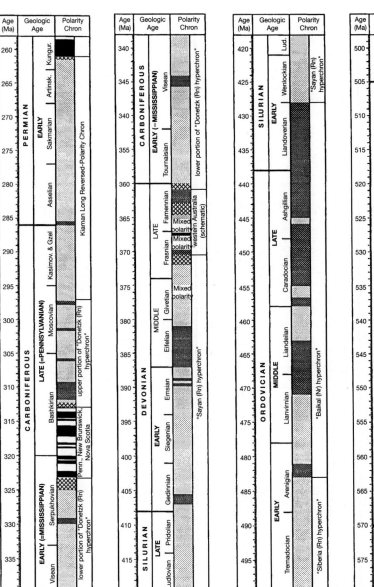
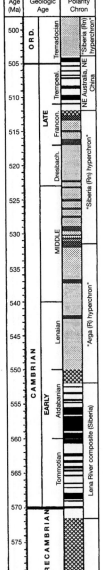

Figure 13.28
Magnetic polarity time scale for the Paleozoic. Intervals of normal polarity are shown in black, reversed polarity in white. Intervals lacking magnetostratigraphic studies or having uncertain validity are cross-hachured. Unverified Paleozoic polarity intervals from the Soviet scale (Khramov and Rodionov, 1981) are shown as shades of gray. [Compiled by Ogg, J.G., 1995, Magnetic polarity time scale of the Phanerozoic, *in* Ahrens, T. J. (ed.), Global Earth physics—A handbook of physical constants: AGU Reference Shelf 1, American Geophysical Union, Washington, D.C., Fig. 4, p. 265, reproduced by permission.]

Furthermore, deeper coring by use of hydraulic piston cores now makes it possible to obtain undisturbed cores of sediments as old as about middle Miocene. Longer cores obtained during the Deep Sea Drilling Program and Ocean Drilling Program by rotary coring methods have recovered rocks as old as about middle Jurassic; however, these rotary cores are commonly too badly disturbed to provide unambiguous paleomagnetic data. Because paleomagnetic methods have now been extended to correlation of on-land sections, this development opens up the possibility of even more extensive future use of paleomagnetic methods for correlating on-land stratigraphic sections. Magnetostratigraphy thus becomes an important tool for international correlation of older, on-land strata as the magnetic polarity time scale is extended farther back into geologic time.

Figure 13.29 provides an easily visualized example of paleomagnetic correlation in cores of young oceanic sediment. Beginning with the Brunhes Normal Epoch (polarity chron) at the top of the cores, the correlation can be carried downward on the basis of the patterns of reversed and normal polarity. With longer

Table 13.3 Nomenclature of magnetostratigraphic polarity units

Magnetostratigraphic polarity units	Description	Geochronologic (time) equivalent	Chronostratigraphic equivalent	Example	
				New name	Old name
Polarity superzone (10^6–10^7 years duration)	Magnetostratigraphic unit composed of two or more polarity zones	Chron (or superchron)	Chronozone (or superchronozone)		
Polarity zone (10^5–10^6 years duration)	Magnetostratigraphic unit distinguished by a single direction of magnetic polarization or by a distinctive alternation of normal and revised polarities	Chron	Chronozone	Brunhes Normal Polarity Zone or Brunhes Normal Polarity Chron	Brunhes Normal Epoch
Polarity subzone (10^4–10^5 years duration)	Subdivision of a polarity zone	Chron (or subchron)	Chronozone (or subchronozone)	Jaramillo Polarity subzone or Jaramillo Polarity subchron	Jaramillo Event

Note: See Chapter 15 for an explanation of geochronologic and chronostratigraphic units.

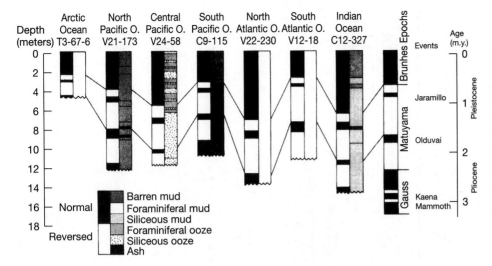

Figure 13.29
Paleomagnetic correlations of cores from the Arctic, Pacific, Indian, and Atlantic oceans. Cores have different lithologies and fossil assemblages. [From Opdyke, N., 1972, Paleomagnetism of deep-sea cores: Reviews Geophysics and Space Physics, v. 10, Fig. 20, p. 244, American Geophysical Union, Washington D.C., reproduced by permission.]

cores and older sediment, correlation becomes more difficult because the magnetostratigraphic record consists of many sets of reversals (Figs. 13.24 and 13.26) that may look very much alike. Correlation of these reversal patterns may require independent radiometric or paleontologic age evidence to first establish stratigraphic position. Paleomagnetic reversal patterns are particularly useful for correlating long distances across biogeographic boundaries where correlation by fossils, even planktonic fossils, may be difficult owing to the fact that different biogeographic provinces are marked by different fossil assemblages. Figure 13.29 shows that paleomagnetic correlation can be carried across the Arctic, Pacific, Indian, and Atlantic ocean basins, each of which is characterized by sediments composed of different lithologies with different fossil assemblages. In a similar manner, Figure 13.30 illustrates how on-land stratigraphic sections can be correlated on the basis of magnetic polarity reversal stratigraphy. Additional examples of correlation of on-land sections by means of magnetostratigraphy are given by Butler (1992), Aissaoui, McNeill, and Hurley (1993), and Belkaaloul et al. (1997).

Geochronology

Although magnetostratigraphic sequence in itself does not normally provide unequivocal ages for geologic events preserved in strata, correlation of magnetic polarity zones or anomalies from areas where the ages of magnetic events have been established by radiometric methods or paleontologic data to areas where the ages of the strata are unknown, or poorly known, provides a means of estimating the ages of events in the new areas. Magnetostratigraphic geochronology may be particularly useful in determining ages within stratigraphic successions that are so nonfossiliferous that little biostratigraphic age control exists. For example, Heller and Tungsheng (1984) used magnetostratigraphic methods to work out the chronology of nonfossiliferous Chinese loess (eolian) deposits.

Magnetostratigraphic chronometry can also provide absolute ages for sediments that have been zoned by fossils and whose ages have been estimated from fossil data, as described by Hailwood (1989, p. 51). For example, McFadden and Hunt (1998) correlated magnetopolarity zones within the Oligocene-Miocene Arikaree Group of northwestern Nebraska to the magnetic polarity time scale to establish absolute ages of important land-mammal zones (Fig. 13.31). The ability to correlate a local magnetic polarity zonation to the magnetic polarity time scale requires either a unique pattern of magnetic reversals with one or more distinctive polarity intervals, excellent biochronological data (Chapter 14), and/or high-resolution radioisotope age determinations from interbedded datable rocks such

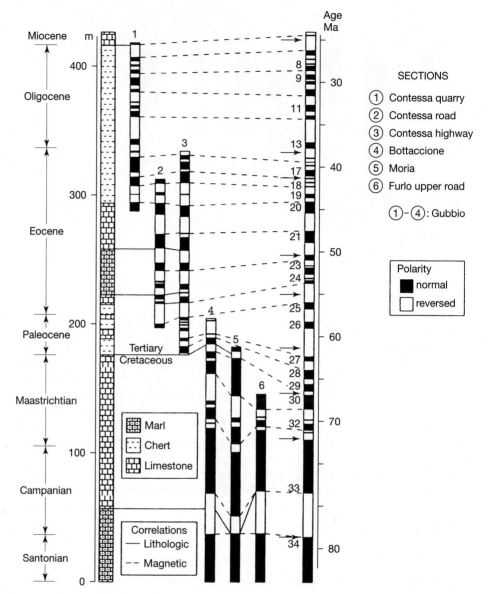

Figure 13.30
Correlation of Late Cretaceous through Cenozoic stratigraphic sections in the Umbrian Apennines on the basis of magnetostratigraphy. The sections are also correlated with the seafloor magnetic anomaly sequence (right column). Age from biostratigraphic (foraminiferal) zonation (Miocene-Santonian) is given at the left of the column showing dominant lithology. Magnetic anomaly numbers and paleontologic calibration points (shown by arrows) are noted at the left side of the seafloor polarity column. [Redrawn and modified from Lowrie, W., and W. Alvarez, 1981, One hundred million years of geomagnetic history: Geology, v. 9, Fig. 1, p. 393.]

as volcanic rocks (Chapter 15). In the case of the Arikee Group rocks, which are between about 18 and 30 million years old, correlation is solid because Earth's geomagnetic history during the past 30 million years was relatively "busy," with numerous polarity transitions. Thus, the correlation is comparatively easy and unambiguous.

Another example is the use of magnetostratigraphy of sediment cores or on-land sections as a tool for estimating the ages of volcanic eruptions that took place either on land or in the ocean. This is done by determining the ages of erupted ash that fell or were washed into on-land depositional sites or the ocean and preserved as ash beds in sediments (e.g., Shane et al., 1996). By establishing the magnetic chronology from the reversal patterns in cores or outcrop sections, geologists can determine the ages of ash layers in the sediments by reference to the paleomagnetic time scale, a technique called **tephrochronology.**

A related application is in determination of rates of sedimentation for deep-sea sediments. Paleomagnetic correlation of deep-sea cores with rocks on land whose ages have been determined radiometrically allows absolute ages to be assigned to the boundaries between different geomagnetic events in the cores. The thickness of sediments between horizons within the cores whose ages are thus

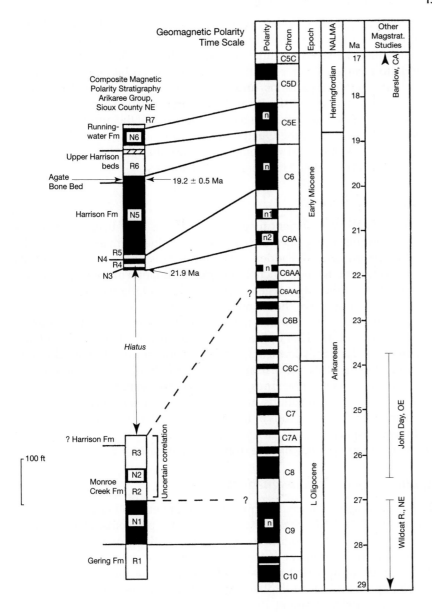

Figure 13.31
Correlation of the Arikaree Group and lower Runningwater Formation, northwest Nebraska, to the magnetic polarity time scale of Berggren et al. (1995). [From MacFadden, B. J., and R. M. Hunt, Jr., 1998, Magnetic polarity stratigraphy and correlation of the Arikaree Group, Arikareean (late Oligocene-early Miocene) of northwestern Nebraska, *in* Terry, D. O., H. E. LaGarry, and R. M. Hunt, Jr. (eds.), Depositional environments, lithostratigraphy and biostratigraphy of the White River and Arikaree Groups (late Eocene to early Miocene, North America), GSA Spec. Paper 325, Fig. 16, p. 162.]

determined can then be used to calculate the sedimentation rate. For example, if we assume that 10 m of sediment were deposited in a given area of the ocean during the time represented in a core by the Matuyama Reversed Polarity chron extending from 2.4 to 0.7 Ma, a time interval of 1.7 million years, the sedimentation rate for this area of the ocean can be calculated as 10 m/1.7 million yr = 5.8 m/m.y.

Paleoclimatology

Ages of sediment cores determined by paleomagnetic methods have also been used to study paleoclimate oscillations during the Quaternary and late Pliocene. For example, the magnetostratigraphy of deep-sea sediments provided a means of estimating the ages of ice-rafted debris in piston cores. It also furnished a method of studying the timing of siliceous ooze deposition, which reflects increased biologic productivity owing to increased oceanic upwelling and resulting increase in nutrients during cooler periods. Quantitative determinations of variations in microfossil assemblages, particularly planktonic foraminiferal assemblages, have also been studied in relation to climatic cycles. Some microfossil species are much more abundant during cooling trends, whereas others are more abundant during warming trends. Also, fluctuations in oxygen-isotope ratios in carbonate shells in deep-sea sediments are related to climate changes (Chapter 15). Use of paleomagnetic methods to estimate the ages of these climate-related biologic oscillations, fluctuations in oxygen isotope ratios, and variations in distribution of ice-rafted material in the oceans has added significantly to our knowledge of climatic fluctuations on land. It has also radically changed our ideas about the number of climatic cycles that occurred during the Quaternary. We now know, for example, that many more cycles of cooling and warming took place than the four major glacial advances and retreats postulated from land-based studies.

Study of Displaced Terranes

The magnetic inclination of ancient magnetized rocks is now being used extensively as a tool to examine presumed movements of continental masses and smaller blocks (see McElhinny and McFadden, 2000, for details and case histories). By measuring the remanent magnetic inclination and declination in ancient rocks, geologists can reconstruct the original geographic position of these rocks at the time they formed. These studies have shown not only that major continents have shifted their positions with time, but also that many smaller blocks of rock have moved from their original locations. That is, these blocks are now located in different latitudes from those in which they formed. Quite commonly the blocks have different lithologies and structural attitudes from those of adjoining areas. These exotic blocks are often called **suspect terranes**, referring to the probability that they are not now in the geographic positions in which they originally formed. There is growing evidence that large portions of many continental margins are made up of a collage of these suspect terranes, assembled by seafloor spreading and subduction processes over long periods of time from different parts of Earth.

Other Applications of Paleomagnetism

The study of paleomagnetism can be applied to a number of other geologic problems, not all of which, strictly speaking, are stratigraphic problems. These applications include dating of archaeological materials; tracing the source or provenance of these materials; study of magnetic fabrics in sedimentary rocks (e.g., paleocurrent analysis); study of apparent polar-wandering paths of Earth through time; and paleogeographic and tectonic plate reconstruction (Aissaoui, McNeill, and Hurley, 1993; Butler, 1992; Khramov, 1987; McElhinny and McFadden, 2000; Piper, 1987; Tarling, 1983; Tauxe, 2002, Ch. 6; Van der Voo, Scotese, and Bonhommet, 1984; Van der Voo, 1993).

FURTHER READING

Seismic Stratigraphy

Bally, A. W. (ed.), 1987, 1988, 1989, Atlas of seismic stratigraphy, Am. Assoc. Petroleum Geologists Studies in Geology 27, 3 volumes, Tulsa, Okla.

Berg, O. R., and D. G. Woolverton (eds.), 1985, Seismic stratigraphy II: An integrated approach to hydrocarbon exploration: AAPG Memoir 39, Am. Assoc. Petroleum Geol., Tulsa, Okla., 276 p.

Hardage, B. A. (ed.), 1987, Seismic stratigraphy: Handbook of geophysical exploration, v. 9: Geophysical Press, London, 432 p.

Sheriff, R. E., 1980, Seismic stratigraphy: International Resources Development Corporation, Boston, 227 p.

Vail, P. R., 1987, Seismic stratigraphy—A 46-minute video coproduced by the Open University and the American Association of Petroleum Geologists. Available from the American Association of Petroleum Geologists, Tulsa, Okla.

Sequence Stratigraphy

Brink, G. J., N. J. S. van Wyk, and L. F. Brown, Jr. (eds.), 1995, Sequence stratigraphy in offshore South African divergent basins: AAPG Studies in Geology 41, Tulsa, Okla., 184 p. [Many excellent seismic profiles in this volume.]

Carter, R. M., T. R. Nash, M. Ito, and B. J. Pillans (eds.), 1998, Sequence stratigraphy in the Plio-Pleistocene: An evaluation: Sedimentary Geology, v. 122, 288 p. [Special Issue].

Coe, A. L. (ed.), 2003, The sedimentary record of sea-level change: Cambridge University Press, Cambridge, 287 p.

Emery, D., and K. J. Meyers (eds.), 1996, Sequence stratigraphy: Blackwell Science Ltd., Oxford, 297 p.

Haq, B. U. (ed.), 1995, Sequence stratigraphy and depositional response to eustatic, tectonic and climatic forcing: Kluwer Academic Publishers, Dordrecht, 381 p.

Miall, A. D., 1997, The geology of stratigraphic sequences: Springer-Verlag, Berlin, 433 p.

Posamentier, H. W., and G. P. Allen, 1999, Siliciclastic sequence stratigraphy—Concepts and Applications: SEPM Concepts in Sedimentology and Paleontology, Tulsa, Okla., 210 p.

Posamentier, H. W., C. P. Summerhayes, B. U. Haq, and G. P. Allen (eds.), 1993, Sequence stratigraphy and facies associations: International Association Sedimentologists Spec. Pub. 18, Blackwell Scientific Publications, Oxford, 644 p.

Van Wagoner, J. C., R. M. Mitchum, K. M. Campion, and V. D. Rahmanian, 1990, Siliciclastic sequence stratigraphy in well logs, cores, and outcrops: Concepts for high-resolution correlation of time and facies: AAPG Methods in Exploration Series No. 7, Am. Assoc. Petroleum Geologists, Tulsa, Okla., 55 p.

Weimer, P., and H. Posamentier (eds.), 1993, Siliciclastic sequence stratigraphy: Recent developments and applications: Am. Assoc. Petroleum Geologists Memoir 58, Tulsa, Okla., 492 p.

Wilgus, C. K., B. S. Hastings, C. G. St. C. Kendall, H. W. Posamentier, C. A. Ross, and J. C. Van Wagoner (eds.), 1988, Sea-level changes: An integrated approach: Society of Economic Paleontologists and Mineralogists Spec. Pub. 42, Tulsa, Okla., 407 p.

Williams, G. D., and A. Dobb (eds.), 1993, Tectonics and sequence stratigraphy: The Geological Society, London, 226 p.

Magnetostratigraphy

Aissaoui, D. M., D. F. McNeill, and N. F. Hurley (eds.), 1993, Applications of paleomagnetism to sedimentary geology: SEPM Spec. Publ. 49, Tulsa, Okla., 216 p.

Berggren, W. A., D. V. Kent, M-P. Aubry, and J. Hardenbol, 1995, Geochronology, time scales and global stratigraphic correlation: SEPM Spec. Pub. 54, Tulsa, Okla., 386 p.

Butler, R. F., 1992, Paleomagnetism: Magnetic domains to geologic terranes: Blackwell, Boston, 319 p.

Hailwood, E. A., 1989, Magnetostratigraphy: Geol. Soc. Spec. Report 19, Blackwell, Oxford, 84 p.

McElhinny, M. W., and P. L. McFadden, 2000, Paleomagnetism: Continents and oceans: Academic Press, San Diego, 386 p.

Opdyke, N. D., and J. E. T. Channell, 1996, Magnetic stratigraphy: Academic Press, San Diego, 346 p.

Salvador, A. (ed.), 1994, International stratigraphic guide: A guide to stratigraphic classification, terminology, and procedure: International Union of Geological Sciences and Geological Society of America, Inc., Trondheim, Norway, 214 p.

Tarling, D. H., and P. Turner (eds.), 1999, Paleomagnetism and diagenesis in sediments: Geological Society Special Publication No. 151, London, 214 p.

Tauxe, L., 2002, Paleomagnetic principles and practice: Kluwer Academic Publishers, Dordrecht, 299 p.

Van der Voo, R., 1993, Paleomagnetism of the Atlantic, Tethys, and Iapetus Oceans: Cambridge University Press, New York, 411 p.

14

Biostratigraphy

14.1 INTRODUCTION

The preceding two chapters focus on stratigraphic relations in sedimentary successions that exist owing to the physical characteristic of sedimentary rocks, either lithology or physical properties that can be remotely sensed by seismic or magnetic instrumentation. We turn now in the present chapter to examine the exceedingly important role that fossil organisms play in stratigraphy. First, fossils provide an additional and highly useful method for subdividing sedimentary rocks into identifiable stratigraphic (biostratigraphic) units. In addition, they make possible the ordering and relative-age dating of strata and their correlation on both a continental and (in some cases) a global scale. The characterization and correlation of rock units on the basis of their fossil content is called **biostratigraphy**. Stratigraphy based on the paleontologic characteristics of sedimentary rocks is also referred to as **stratigraphic paleontology**, the study of fossils and their distributions in various geologic formations.

Separation of rock units on the basis of fossil content may or may not yield stratigraphic units whose boundaries coincide with the boundaries of lithic stratigraphic units. In fact, lithostratigraphic units such as formations commonly can be subdivided by distinctive fossil assemblages into several smaller biostratigraphic units. Indeed, one of the primary objectives of biostratigraphy is to make possible differentiation of strata into small-scale subunits or zones that can be dated and correlated throughout wide geographic areas, allowing interpretation of Earth history within a precise framework of geologic time. On the other hand, it is quite common for some biologically defined stratigraphic units to span the boundaries of formally defined lithostratigraphic units. Some biostratigraphic units may thus include parts of two members or formations or even encompass two or more entire members or formations. Note in Figure 14.1, for example, that the Naheola Formation contains two biostratigraphic units, the *Pr. pusilla pusilla I.Z.* and *M. angulata I.Z.* planktonic foraminiferal zones, whereas the *Pr. pseudomenardii R.Z.* foraminiferal zone spans the Nanafalia Formation and part of the Tuscahoma Formation and includes several different rock types (sandstone, shale, marl).

The concept of biostratigraphy is based on the principle that organisms have undergone successive changes throughout geologic time. Thus, any unit of strata can be dated and characterized by its fossil content. That is, on the basis of its contained fossils, any stratigraphic unit can be differentiated from stratigraphically

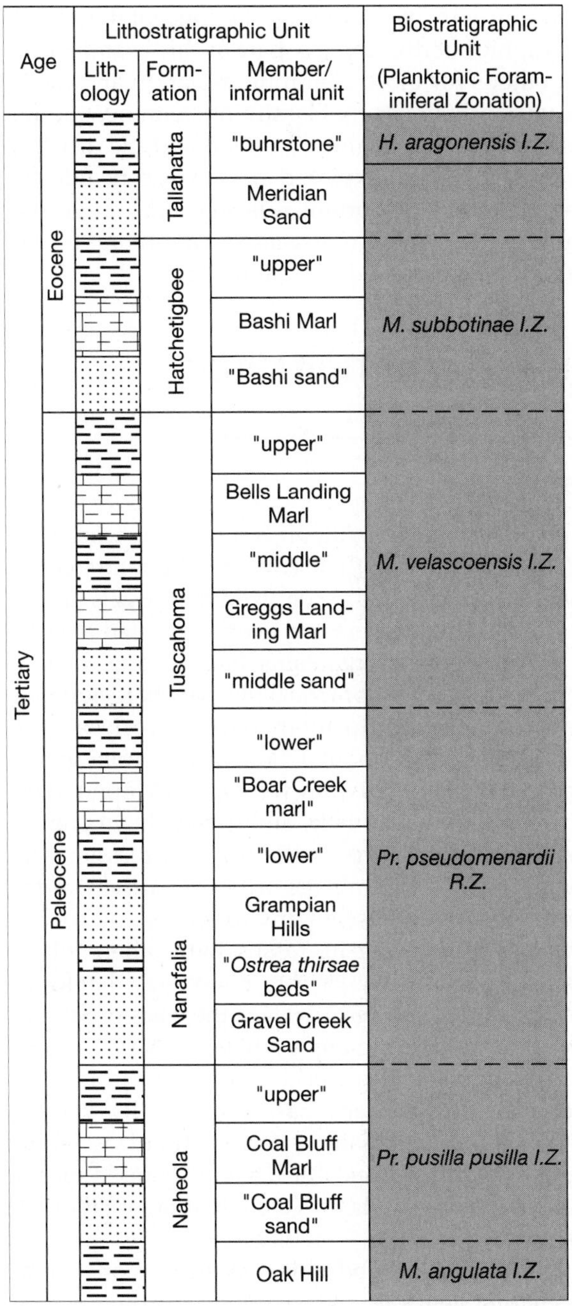

Age	Lithostratigraphic Unit			Biostratigraphic Unit (Planktonic Foraminiferal Zonation)
	Lithology	Formation	Member/informal unit	
Tertiary / Eocene		Tallahatta	"buhrstone"	*H. aragonensis* I.Z.
			Meridian Sand	
		Hatchetigbee	"upper"	*M. subbotinae* I.Z.
			Bashi Marl	
			"Bashi sand"	
Tertiary / Paleocene		Tuscahoma	"upper"	*M. velascoensis* I.Z.
			Bells Landing Marl	
			"middle"	
			Greggs Landing Marl	
			"middle sand"	
			"lower"	*Pr. pseudomenardii* R.Z.
			"Boar Creek marl"	
			"lower"	
		Nanafalia	Grampian Hills	
			"*Ostrea thirsae* beds"	
			Gravel Creek Sand	
		Naheola	"upper"	*Pr. pusilla pusilla* I.Z.
			Coal Bluff Marl	
			"Coal Bluff sand"	
			Oak Hill	*M. angulata* I.Z.

Figure 14.1
Illustration of the relationship between biostratigraphic units and lithostratigraphic units, Tertiary sedimentary rocks of the eastern Gulf Coastal Plain, U.S.A. [Based on Mancini, E. A., and Tew, B. H., 1995, Geochronology, biostratigraphy and sequence stratigraphy of a marginal marine shelf stratigraphic succession: Upper Paleocene and lower Eocene, Wilcox Group, eastern Gulf Coastal Plain, U.S.A., *in* Berggren, W. A., D. V. Kent, M-P. Aubry, and J. Hardenbol (eds.), Geochronology and global stratigraphic correlation, Fig. 1, p. 282, SEPM Spec. Publ. 54, Tulsa, Okla.]

younger and older units. Biostratigraphy is obviously closely allied to paleontology, and a skilled biostratigrapher must also be a well-trained paleontologist. In fact, the application of biostratigraphy is for specialists who have intimate knowledge of large groups of organisms and their temporal and spatial distribution. Because stratigraphic paleontology is such a complex field, comprehensive treatment of this subject is beyond the scope of this book. The aim of this chapter is to introduce some very basic concepts and principles of biostratigraphy. Readers who wish more in-depth treatment of biostratigraphy should consult standard reference works on paleontology as well as more specialized, biostratigraphically oriented monographs such as those listed under "Further Reading" at the end of this chapter.

We begin discussion of biostratigraphy by examining the concept that fossils constitute a valid basis for stratigraphic subdivision. As part of this examination, the origin and development of methods for biostratigraphic zonation are traced, and the stratigraphic procedures currently in use for classifying, naming, and describing biostratigraphic units are discussed. Organic evolution and the distribution of organisms in both time and space are explored next. We conclude the chapter with a discussion of the extremely important role that biostratigraphy plays in correlation of stratigraphic units. The use of fossils for calibrating the geologic time scale, called biochronology, is discussed in Chapter 15.

14.2 FOSSILS AS A BASIS FOR STRATIGRAPHIC SUBDIVISION

Principle of Faunal Succession

An English surveyor and civil engineer named William Smith, who worked in England and Wales in the late 1700s, is credited with discovering the fundamental principle of biostratigraphy. Previous workers had recognized that fossils are the remains of once-living organisms, and some workers had even suggested the possibility that certain species of marine-shelled organisms had become extinct. Smith was evidently the first to utilize fossils as a practical tool for characterizing, subdividing, and correlating strata from one area to another. In his work as a surveyor and canal builder, he had discovered by about 1796 that the strata in and around Bath in Somerset and for some distance outward were always found in the same order of superposition—the order in which rocks are placed above one another. Furthermore, he noted that each layer in the stratigraphic succession was characterized by the same distinctive fossil assemblage wherever it was found throughout the region. Soon, Smith was able to assign any fossil-bearing rock to its proper superpositional interval by comparing its fossils with others whose stratigraphic position he knew from previous study. He thus discovered that fossil-bearing strata occur in a definite and determinable order. On the basis of Smith's discovery, we now know that rocks formed during any particular interval of geologic time can be recognized and distinguished by their fossil content from rocks formed during other time intervals. This concept has consequently become known as the **principle (law) of faunal succession**. Even without assigning names to fossils, Smith successfully used them to establish a stratigraphic succession and to subdivide the rocks into mappable units by a combination of lithologic characteristics and fossil assemblages.

It is important to stress that Smith did not subdivide rock successions on the basis of fossils alone. His strata were first delineated and named according to their lithology. Then, their characteristic fossils were collected and studied. The use of fossils alone to subdivide thick, essentially lithologically homogeneous formations did not come about for another fifteen years. The French scientist George Cuvier, a contemporary of Smith's, recognized the desirability of using fossils to subdivide rocks but did not attempt this process himself. Subdivision of rock successions on the basis of fossils was first carried out on sediments of Tertiary age in the early 1830s. Deshayes in France (1830), Bronn in Germany (1831), and Lyell in England (1833) all proposed subdivisions of Tertiary strata based on fossils (reported in Hancock, 1977). Lyell's subdivisions are historically noteworthy. He split the Tertiary strata into four units on the basis of the proportions of living to extinct species in the rocks (Table 14.1). Thus, we see here for apparently the first time the use of fossils as an essential part of the definition of units of geologic time and the possibility of biostratigraphy freed from lithologic control.

Table 14.1 Lyell's subdivisions of the tertiary

Name of subdivision	Extant species in the rocks (%)
Pliocene (more recent)	
Newer Pliocene	90
Older Pliocene	33–50
Miocene (less recent)	18
Eocene (dawn of recent)	3.5

Concept of Stage

Although Smith's principle of faunal succession was to be the cornerstone for all subsequent biostratigraphy, his own work on biostratigraphic successions led to only vaguely defined time units. Lyell's subdivisions were similarly vague and, in any case, were confined to the Tertiary. A closer look at fossil successions was needed to refine their use in dating and correlation. This important step came with introduction of the concept of stage, credited to the French paleontologist Alcide d'Orbigny. About 1842, d'Orbigny came up with the idea of erecting major subdivisions of strata, each systematically following the other and each bearing a unique assemblage of fossils. Like Smith, d'Orbigny recognized that similarity of fossil assemblages was the key to correlating rock units, but he went a step further to propose that strata characterized by distinctive and unique fossil assemblages might include many formations (lithostratigraphic units) in one place or only a single formation or part of a formation in another place. He defined as **stages** groups of strata containing the same major fossil assemblages. He named these stages after geographic localities with particularly good sections of rock that bear the characteristic fossils on which the stages are based. Using the stage concept, he was able to divide the rocks of the Jurassic system into ten stages and the Cretaceous rocks into seven stages, each characterized strictly by its fossil fauna.

The boundaries of d'Orbigny's stages were defined at intervals marked by the last appearance, or disappearance, of distinctive assemblages of life forms and their replacement in the rock record by other assemblages. He conceived these stages as having worldwide extent and to be the result of repeated catastrophic destruction of life on Earth followed by new creations. His ideas on catastrophic destruction and special new creation, like those of Cuvier who preceded him, failed to gain lasting acceptance among geologists, and subsequent study has shown that his stages and their characteristic faunas are local rather than worldwide. Nonetheless, d'Orbigny's concept of stages as major bodies of strata characterized by large assemblages of fossils unique to that part of the total stratigraphic column was a significant and lasting contribution to the growing discipline of biostratigraphy. Somewhat different interpretations of the meaning of stage have been used by subsequent workers, but d'Orbigny's basic concept is still valid.

Concept of Zone

The stage concept of d'Orbigny permitted subdivision of strata into major successions on the basis of fossils. What they did not provide was a method by which fossiliferous strata could be divided into small-magnitude, clearly delimited units. Friedrich Quenstedt in Germany was particularly critical of d'Orbigny's stages because, according to him, d'Orbigny's method "centered around the acceptance, as the diagnostic faunal aggregate, of species of many strata in many localities, lumped together without enough regard for their precise stratigraphic ranges" (Berry, 1987, p. 125). Quenstedt maintained that only by extremely detailed study of strata on essentially a centimeter-by-centimeter basis could full understanding of the succession of faunas be developed. Quenstedt's own work did not bring this notion of detailed biostratigraphic subdivision into full fruition. It remained for his student, Albert Oppel, to expand, synthesize, and meld Quenstedt's ideas into the concept of the **zone**.

Oppel introduced the concept of zone in 1856 and thereby altered for all time the practice of biostratigraphy. Working with Jurassic rocks in various parts of Germany, he conceived the idea of small-scale units defined by the **stratigraphic ranges** of fossil species irrespective of lithology of the fossil-bearing beds. Oppel noted that the vertical ranges of some species were very short; that is, the species existed for only a very short time geologically. Others were quite long, but most were of some intermediate length. Oppel noted also that the assemblages of fossils that characterized the strata were made up of **overlapping** ranges of fossils. He defined his zones by exploring the vertical range of each separate species. Each zone was characterized by the joint occurrence of species not found together above or below this zone. Thus, the range of some species began at the base of a zone (the first appearance of a species), others ended at the top of a zone (the last appearance of a species), whereas still others ranged throughout the zone or even extended beyond it. Using species ranges, Oppel discovered that he could delineate the boundaries between small-scale rock units and distinguish a succession of unique fossil assemblages. Each of these assemblages was bounded at its base by the appearance of distinctive new species and at its top, that is, the base of the succeeding section, by the appearance of other new species. It is, however, the overlapping stratigraphic ranges of the species that make up the fossil assemblage that typifies a zone (Fig. 14.2). Because a zone represents the time between the appearance of species chosen as the base of the zone and the appearance of other species chosen as the base of the next succeeding zone, recognition of zones thus permits delineation of clear-cut, small-scale **time units**. Each of Oppel's zones was named after a particular distinctive fossil species, called an **index fossil,** or **index species**, which is but one fossil species in the assemblage of species that characterize the zone.

The concept of zone thus allowed subdivision of stages into two or more smaller, distinctive biostratigraphic units that could be recognized and correlated

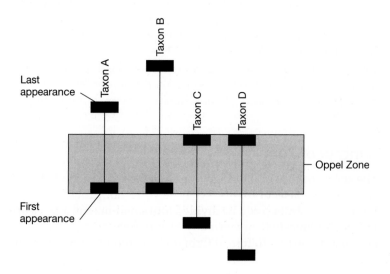

Figure 14.2
Diagrammatic illustration of an Oppel zone defined by overlapping ranges of two or more taxa.

over long distances. Oppel was able, for example, to subdivide the Jurassic rocks of western Europe into 33 zones. It should be noted that Oppel did not start with d'Orbigny's stages and subdivide them into zones. Instead, he delineated zones on the basis of fossil ranges and then combined the zones into stages, all of which did not necessarily fit into one of d'Orbigny's stages as then defined (Hancock, 1977).

Zones were slow to be adopted into stratigraphic practice, especially in the United States, but with minor modifications of Oppel's method they have now become the common denominator of biostratigraphic study. Zones have been extended to all parts of the fossil record, not just the Jurassic, and to all areas of the world. It is now recognized, however, that there are definite geographic limits beyond which most zones cannot be traced. The area within which a zone can be recognized is a **biogeographic province**. Because zones constitute the basic unit of biostratigraphic classification, considerable work has gone into efforts to standardize their usage. The current usage of zone and the different kinds of zones now recognized are described in the following section.

14.3 BIOSTRATIGRAPHIC UNITS

As the preceding discussion makes clear, a biostratigraphic unit is a body of rock strata characterized by its fossil content that distinguishes and differentiates it from adjacent strata. Furthermore, the zone, or **biozone,** is the fundamental biostratigraphic unit. Biozones do not have any prescribed thickness or geographic extent. They may range in thickness from thin beds a few meters thick to units thousands of meters thick and in geographic extent from local units to those with nearly worldwide distribution. Recent attempts to standardize nomenclature and usage of biozones have been made by the International Subcommission on Stratigraphic Classification (Hedberg, 1976; Salvador, 1994) in the International Stratigraphic Guide and by the North American Commission on Stratigraphic Nomenclature in the 1983 North American Stratigraphic Code and pending revisions. The usage in this book follows that recommended in the North American Stratigraphic Code, but reference is made to usage in the International Stratigraphic Guide where appropriate.

Principal Categories of Zones

Revisions to the 1983 North American Stratigraphic Code (Appendix C) contained in North American Stratigraphic Commission Note 64 (Lenz et al., 2001; see also http://www.agiweb.org/nacsn/) subdivides biostratigraphic units into five principal kinds of biozones: range biozones (two kinds), interval biozones (two or more kinds), lineage biozones, assemblage zones, and abundance biozones (Fig. 14.3). Each of these zones is distinguished by different criteria, as explained below.

A **taxon-range biozone** (Fig. 14.3A) is a body of rock representing the known stratigraphic and geographic range of occurrence of a single taxon. A **concurrent-range biozone** (Fig. 14.3B) is a body of rock that includes the concurrent, coincident, or overlapping part of the ranges of two specified taxa.

An **interval biozone**, or subzone, is the body of strata between two specific, biostratigraphic surfaces (biohorizons of the International Stratigraphic Guide, p. 56). The features on which biohorizons are commonly based include lowest occurrences (Fig. 14.3C) and highest occurrences (Fig. 14.3D); however, they may also include distinctive occurrences and changes in the character of individual taxa, such as changes in the direction of coiling in foraminifers or in number of septa in corals.

A **lineage biozone** (Fig. 14.3E) is a body of rock containing species representing a specific segment of an evolutionary lineage.

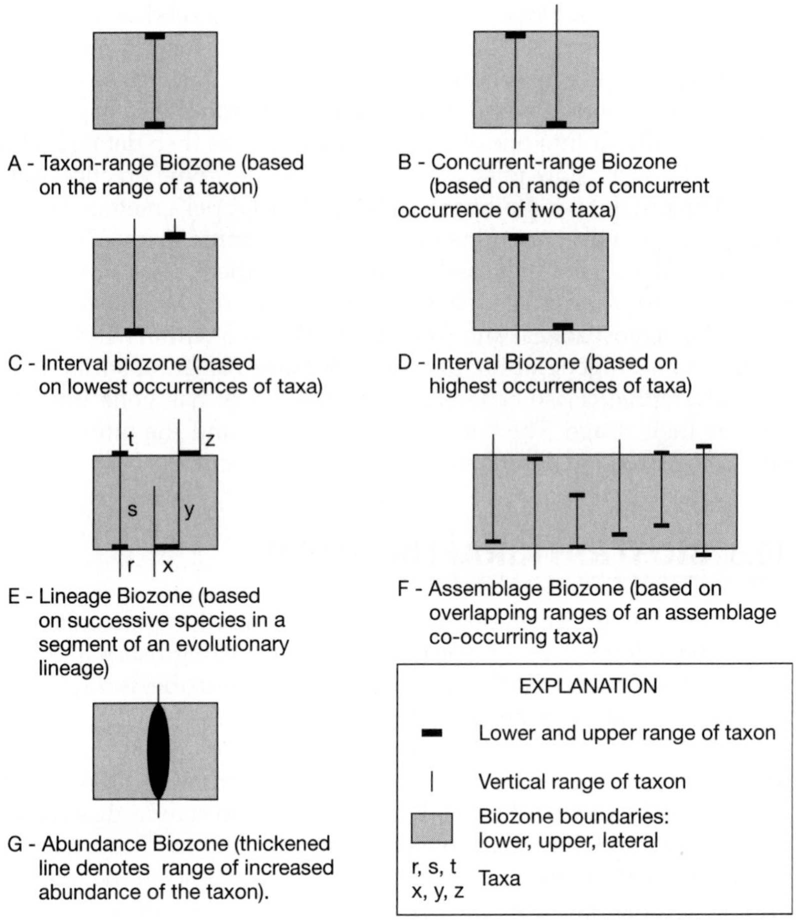

A - Taxon-range Biozone (based on the range of a taxon)

B - Concurrent-range Biozone (based on range of concurrent occurrence of two taxa)

C - Interval biozone (based on lowest occurrences of taxa)

D - Interval Biozone (based on highest occurrences of taxa)

E - Lineage Biozone (based on successive species in a segment of an evolutionary lineage)

F - Assemblage Biozone (based on overlapping ranges of an assemblage co-occurring taxa)

G - Abundance Biozone (thickened line denotes range of increased abundance of the taxon).

EXPLANATION	
▬	Lower and upper range of taxon
│	Vertical range of taxon
▨	Biozone boundaries: lower, upper, lateral
r, s, t x, y, z	Taxa

Figure 14.3
Diagram illustrating the principal kinds of biozones as defined in North American Stratigraphic Commission Note 64, a revision to the 1983 North American Stratigraphic Code. [After Lenz et al., 2001, Note 64—Application for revision of articles 48–54, biostratigraphic units, of the North American Stratigraphic Code: AAPG Bull., v. 85, Figures 4 and 5, p. 373.] (see also http://www.agiweb.org/nacsn/).

An **assemblage biozone** (Fig 14.3F) is a body of rock characterized by a unique association of three or more taxa, the association of which distinguishes it in biogeographic character from adjacent strata. An assemblage biozone may be based on a single taxonomic group, for example, trilobites, or on more than one group, such as acritarchs and chitinozoans.

An **abundance biozone** (Fig. 14.3G) is a body of rock in which the abundance of a particular taxon or specified group of taxa is significantly greater than in adjacent parts of the section. Figure 14.4 further illustrates subdivision of strata on the basis of abundance zones. Note that gaps may exist between zones defined on the basis of taxon abundance.

Rank of Biostratigraphic Units

The biozone is the fundamental unit of biostratigraphic classification. Other biostratigraphic units are formed by either grouping or subdividing biozones. The International Stratigraphic Guide (Salvador, 1994) suggests that some kinds of biozones may be subdivided into subbiozones (subzones) and/or grouped into superbiozones (superzones). The North American Stratigraphic Code provides that a biozone may be completely or partly divided into subbiozones.

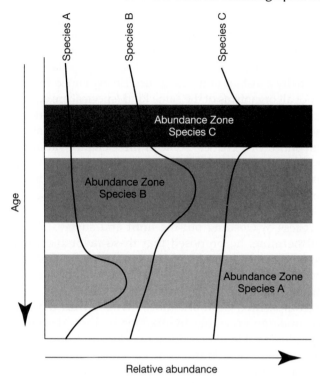

Figure 14.4
Schematic illustration of the abundance zones of three hypothetical fossil species. Note that each species reaches peak abundance (total number of individuals) at a particular time and then declines in abundance. Ages of the strata increase downward, and relative abundance increases to the right.

Naming Biostratigraphic Units

The name of a biozone consists of the name of one or more distinctive taxa found in the biozone, followed by the word Biozone (e.g., *Turborotalia cerrozaulensis* Biozone). The name of the species whose lowest occurrence defines the base of the zone is the most common choice for the biozone name.

14.4 THE BASIS FOR BIOSTRATIGRAPHIC ZONATION: CHANGES IN ORGANISMS THROUGH TIME

Evolution

The practicality of biostratigraphy as a tool for characterizing and correlating strata had been clearly established by Smith, d'Orbigny, and Oppel by the middle of the 19th century. Probably, none of these workers fully understood why fossil assemblages changed from one stratigraphic layer to another, although d'Orbigny apparently gave considerable thought to the origin of his stage boundaries. Charles Darwin provided the answer to this puzzle a few years after Oppel conceived the zone concept. In his monumental work on the origin of species published in 1859, Darwin demonstrated the existence of organic evolution and thereby greatly changed subsequent geological and philosophical thought, although his ideas were by no means accepted by all of his contemporaries (nor are they accepted by some people outside the field of science today).

Darwin was not the first to conceive the general idea of evolution, but previous workers had marshalled little supporting evidence for their ideas. By contrast, Darwin drew together data on the fossil records of extinct organisms, the results of selective breeding of domestic animals, observations on ecological adaptations and variations among living organisms, and details on comparative anatomy. He then wove these data into a powerful argument for organic evolution. Darwin pointed out that all organisms have high reproductive rates, yet populations of these organisms remain essentially constant in the long run. He explained this observation by suggesting that not all organisms of the same kind (species) are equally well equipped to survive, and therefore many individuals die before reproducing. Each individual of a species differs from other individuals as a result of variations that arise within an organism entirely by chance. Some of these chance variations may be an advantage to the organism in coping with its environment in its struggle for existence. Others may be a disadvantage. Successful variations help organisms survive and extend their environment and range. Unsuccessful variations result in extinction.

Darwin termed this process of weeding out the unfit and survival of the fittest **natural selection**. Furthermore, he proposed that these favorable variations are inheritable and can be transmitted from one generation to the next. Darwin's fundamental contribution to understanding evolution was thus recognizing that natural selection was the process by which new species arise. New species appear because the composition of populations changes with time because those individuals that undergo favorable adaptations will stand a better chance of surviving and reproducing. He likely did not understand how variations arose or how these traits were passed on from one generation of organisms to the next. The concept of spontaneous changes in genes that we now call **mutations** was not well known at the time Darwin published *The Origin of Species,* although the basic laws of genetics had been set forth in a paper by the Austrian monk Gregor Mendel in 1865.

Taxonomic Classification and Importance of Species

Organisms can be classified in a variety of ways, including habitat (planktonic, nektonic, benthonic) and environmental distribution (littoral, neritic, bathyal, etc.); however, **taxonomic classification** based on morphological and developmental similarities and presumed genetic relationships is most pertinent to recognizing evolution and biostratigraphic zonation. The basic system of taxonomic classification now in use was introduced in 1735 by the Swedish naturalist Linnaeus, who grouped organisms into a hierarchy of different categories based on the number of distinctive characteristics shared in common. Organisms in the lowest, or least inclusive, category have the greatest number of common characteristics; those in the next highest category have fewer common characteristics; and so on, until the highest, or most inclusive, category is reached. In the last category, organisms share only a very few common characteristics or traits. Linnaeus's system of classification, as modified by some later additions, is illustrated schematically in Figure 14.5.

The Linnaean system of taxonomic classification brought to light the fact that degrees of similarities among organisms differ at different levels of classification. Differences among groups of organisms are greatest at the kingdom level and least at the species level. Species have thus become the fundamental entity of biostratigraphy. Biologists define species as a breeding community that preserves its genetic identity by its ability to exchange genes with other breeding communities. In other words, all members of a given species have the ability to interbreed, but they do not normally breed with members of a different species. Thus, a species constitutes a group of interbreeding organisms that are reproductively isolated from other such groups. The criteria for identifying a biologic species are difficult to apply to fossil organisms. Therefore, fossil species are commonly

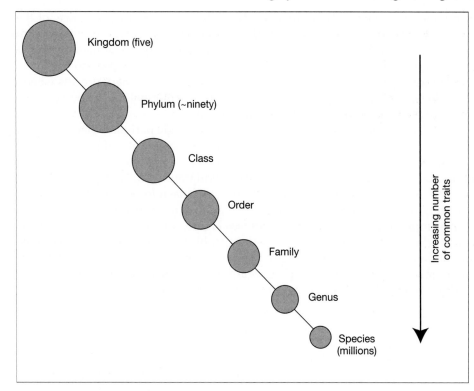

Figure 14.5
Schematic representation of the hierarchical Linnaean system for classifying organisms. All organisms are currently grouped into five kingdoms, about ninety phyla; numerous classes, orders, families, and genera; and millions of species. Note that organisms at the species level share many common characteristics whereas those at higher levels share fewer characteristics.

characterized mainly on the basis of shell, or skeletal, morphology. Because the skeletal morphology of different members of the same species can be quite variable, determination of fossil species must be made by taxonomic specialists. Such determination may require quantitative measurements of shell parameters and computer analysis of measurement data to provide statistical rigor to fossil species identification.

Changes in Species Through Time

The importance of species in biostratigraphic study lies in the fact that species do not remain immutable for all time. If environmental conditions remained absolutely constant through time, perhaps species would change very little. The fact is that environments do change and, as they change, species also change, although environments do not directly cause species to change. Both gene mutation, or gene pool combinations, and shifting environmental conditions are essential to the evolution of species. Most species are well adjusted to their normal environments, but if an appropriate variation appears in a species just at the time when it is becoming inadaptive to a changing environment, the force of natural selection may preserve this novel variant (e.g., Shaw, 1964). Thus, species have evolved through time as a result of natural selection of those random, chance mutations that brought the species into better adjustment with changing environmental conditions.

All indications from the geologic record suggest that species variations are one-directional and nonreversible. Once a species has become extinct, it does not reappear in the fossil record. As members of a new species increase in numbers, they may eventually become abundant and widespread enough to show up in the geologic record as the **first appearance** of the species. When the species is no longer able to adjust to shifting environmental conditions, its members decrease in number and eventually disappear—the extinction, or **last appearance**, of the species. **Extinction** refers to the disappearance by death of every individual member of a species or higher taxonomic group so that the lineage no longer exists. Paleontologists recognize also that a species may experience pseudoextinction. **Pseudoextinction**, or phyletic extinction, refers to an evolutionary process whereby a species evolves into a different species. Thus, the original species becomes extinct, but the lineage continues in the daughter species.

Some species exist for only a fraction of a geologic period. Others may persist for longer periods of time. Organisms that were abundant and geographically widespread and had relatively short ranges have the greatest time-stratigraphic utility, that is, the greatest usefulness for biostratigraphic study (Fig. 14.6). Figure 14.6 shows some of the more important groups of macrofossils that are useful for biostratigraphic zonation. Many other fossil groups, including microfossil groups such as foraminifers, are also important (see Figure 11.4. for example). Particularly important and useful fossils for biostratigraphic purposes are called **guide fossils** or **index fossils**. Ideally, index fossils should be (1) independent of their environment, (2) fast evolving, (3) geographically widespread, (4) abundant, (5) readily preserved, and (6) easily recognizable (e.g., Doyle, Bennett, and Baxter, 1994, p. 37).

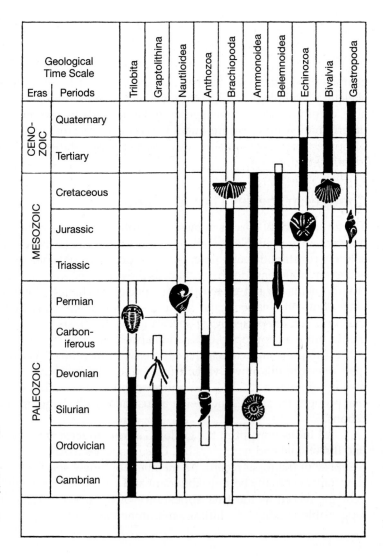

Figure 14.6
Some of the more important macrofossil groups of marine invertebrate organisms for biostratigraphic zonation. The white columns show the time span of distribution, the black columns the time span in which the organisms are important as index fossils. [From Thenius, E., 1973, Fossils and the life of the past, Fig. 50, p. 79, reprinted by permission of Springer-Verlag.]

Box 14.1 Models and Rates of Evolution

There is currently considerable controversy among paleontologists concerning the mode of change in organic evolution. Two principal points of view prevail. One view states that evolution proceeds mainly as a gradual change by slow, steady transformation of well-established lineages—**phyletic evolution,** or **gradualism**. The gradualist concept has been the traditional view of species evolution. The second view holds that many species arise very rapidly from small populations of organisms that have become isolated from the parental range and then subsequently change very little after their successful origin. This latter view represents evolution by speciation or branching of lineages, the so-called **punctuated equilibria** model of Eldredge and Gould (1972). Thus, according to this theory, fossil populations are in stable equilibrium for long periods of time and change very little (called **stasis**), punctuated by sudden introduction of new species. Differences in these two postulated modes of evolution are illustrated graphically in Figure 14.1.1. In the punctuational model, **speciation**, or branching of species, is viewed as a very rapid process,

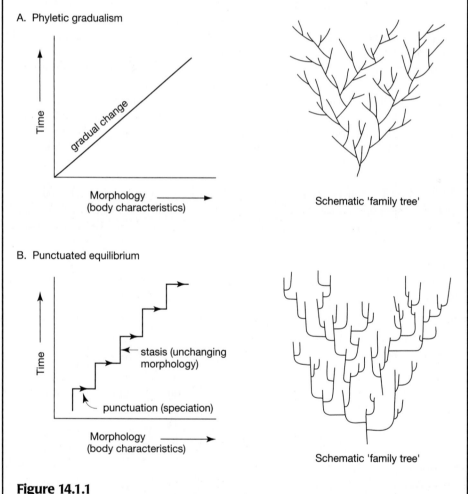

A. Phyletic gradualism

Time

gradual change

Morphology
(body characteristics)

Schematic 'family tree'

B. Punctuated equilibrium

Time

stasis (unchanging morphology)

punctuation (speciation)

Morphology
(body characteristics)

Schematic 'family tree'

Figure 14.1.1
Diagrammatic representation of gradualistic and punctuated models of evolution. [Schematic family trees after Stanley, W. H., 1979, Macroevolution, patterns and processes, W. H. Freeman and Company.]

requiring only tens of thousands of years or possibly as little as a few hundred years (e.g., Stanley, 1979) after a population becomes reproductively isolated from the parent population. Although the duration of species from first appearance to extinction may be measured in millions of years (Table 14.1.1), species are believed by the punctuationalists to change morphologically very little and only very slowly after initial speciation. This concept is stated very succinctly by Eldredge and Gould (1977), who emphasize the importance of speciation (splitting) and claim "that most morphological differences between two species appear in conjunction with the speciation process itself, whereas most of a species' history involves little further change, at least of a progressive nature."

Whether species evolution takes place mainly by gradual evolutionary change, mainly by punctuated speciation, or by both, is still a much discussed and debated issue (see, for example, Gould and Eldredge, 1993, and Sheldon, 1996). Both sides of the controversy continue to be aired in the paleontological literature. Some workers propose that certain groups of organisms, such as mammals, tend to evolve by gradual transformation whereas others, such as many marine invertebrates, tend to evolve by punctuated equilibrium. Sheldon (1996) suggests that organisms on land in the tropics and those in the deep sea may tend to undergo continuous, gradualistic evolution, whereas organisms in temperate zones and shallow water tend more toward stasis and occasional punctuations. Because the majority of the fossil record comes from dynamic shallow-marine environments, many fossil lineages thus show approximate stasis and occasional punctuations. Different groups of species are known to evolve at greatly different rates. Stanley (1985) indicates, for example, that marine bivalve groups evolve at a rate that yields only three or four species in 20 million years. By contrast, mammalian families evolve at a rate that yields roughly 80 species in 20 million years. From a practical point of view, the task of delineating

Table 14.1.1 Estimated mean species duration (in millions of years) for a variety of biological groups

Biological group	Estimated mean species duration (Ma)
Marine diatoms	25
Benthic foraminifers	20–30
Planktonic foraminifers	>20
Bryophytes	>20
Marine bivalves	11–14
Marine gastropods	10–14
Higher plants	8–>20
Ammonites	~5 (but with a mode in the 1–2 Ma range)
Freshwater fish	3
Graptolites	2–3
Beetles	>2
Snakes	>2
Mammals	~1–2
Trilobites	>1

Source: Stanley (1985).

the boundaries of species is more difficult if evolution occurs by phyletic gradualism because a chain of intermediate species is present in the geologic record and the boundaries between successive species are arbitrary. Thus, it is difficult to pick points in the evolutionary sequence at which distinct species boundaries are recognizable. If, on the other hand, evolution occurs by speciation (punctuated equilibrium), most morphological change presumably occurs at branch intersections in the evolutionary line (Fig. 14.1.1B), which represent discrete points in time. Thus, the task of picking species boundaries should theoretically be easier and less error prone if evolution occurs by speciation rather than by phyletic gradualism. On the other hand, the initial appearance of a new species in different provinces may show a time lag owing to lags in migration, which makes identification of the first-appearance species boundary more difficult.

The practicality of identifying species boundaries, and of establishing the boundaries of biostratigraphic zones, is further complicated by problems involving the following: (1) sampling intervals (How small must they be to ensure that species boundaries are detected?), (2) changes in the fossil record induced by burial and the vagaries of preservation, (3) constancy and rates of sedimentation (smaller sampling intervals are required for sediments that accumulated very slowly vs. those that accumulated very rapidly), and (4) intermittent or punctuated patterns of sedimentation and erosion that yield an incomplete stratigraphic record, thus giving the appearance of punctuated speciation. [For some comparatively recent views on punctuated equilibrium, see Gould (2001) and Kemp (1999, Chapter 7).]

DETERMINISTIC VS. PROBABILISTIC EVOLUTION

An interesting side issue to the problem of evolutionary controls relates to the question of whether or not such evolutionary events as adaptive radiation and periods of mass extinction are deterministic or probabilistic. That is to say, are evolutionary events explainable only in terms of causal factors, or are there statistical laws or generalizations that can explain these events on the basis of random variations or processes? Every human, for example, is destined from the instant of his or her birth to age and eventually die. Is every species likewise destined from the time of its birth (initial speciation) to eventually age and become extinct? Raup (1991, p. 6) states that there is absolutely no basis for equating the life span of species with those of humans and that there is no evidence of aging in species or any known reason why a species could not live forever. Nonetheless, in a subsequent chapter of his book entitled "Gambler's Ruin and Other Problems," Raup discusses the probability of extinction of a genus with a limited number of species (e.g., ten). If the chance of extinction is identical to that of speciation (fifty-fifty), the number of species will fluctuate up and down as in a random walk but will finally reach zero. Therefore, the laws of probability suggest that eventual extinction of the genus is inevitable (Raup, 1991, p. 49), although the greater the number of species in the genus the longer it will take for extinction to occur.

Probabilistic evolutionary models are called **stochastic models**. Van Valen (1973, p. 1) asserted, for example, that "all groups for which data exist go extinct at a rate that is constant for a given group." Such statements should not be taken to mean that extinctions occur without cause. Extinction of a species may be the result of any number of specific causes and it may, therefore, be invalid to attribute the death of individuals to chance. If frequency of death is considered at population levels, however, it may be mathematically valid to describe the frequency as being governed by random stochastic processes. In other words, individuals in a given population of organisms may die owing to various specific causes; however, the population as a whole will become extinct at a constant rate, depending upon its size, regardless of the specific causes of death of the individuals. Thus, in the stochastic approach, the pattern of evolution as a whole is perceived to be a random process, although individual fluctuations in this pattern can be explained by cause and effect.

Stochastic models may serve to separate those features of the evolutionary record that are amenable to deterministic explanations from those that do not warrant a search for a specific cause. For example, the fairly rapid demise of the dinosaurs at the end of Cretaceous time can probably be explained by some specific environmental or catastrophic event such as dramatic climatic change resulting from meteorite impact; however, the gradual decline of conodonts and their eventual extinction at the end of the Triassic appears to be more difficult to attribute to a specific cause or causes. Probabilistic extinction can be thought of as a kind of "background" extinction that has acted throughout the fossil record, almost but not quite keeping pace with the rise of new species (global diversity of organisms has increased with time). From time to time, however, major extinction events, called mass extinctions, have occurred and demand a specific causal explanation.

MASS EXTINCTIONS

The fossil record shows that the diversity of both marine and continental life has increased exponentially since the end of the Precambrian (e.g., Benton, 1995; Miller, 2000); however, the record also shows that many groups of organisms became extinct or suffered dramatic reductions in numbers and diversity at particular times. During the past two decades, these episodes of mass extinction have assumed increasing importance to paleontologists and other geologists, judging by the rapid appearance of new articles and books dealing with mass extinctions. Five extinction events have become so important and far reaching that they are now commonly referred to as the **big five**. These major extinction episodes took place near the end of the Ordovician, Devonian, Permian, Triassic, and Cretaceous (Fig. 14.1.2), and the later extinctions affected both terrestrial and marine forms.

As shown in Table 14.1.2, 47–82 percent of extant marine animal genera became extinct during these five major mass extinction episodes and 16–51 percent of the marine animal families became extinct. Such dramatic extinctions demand an explanation linked to some specific cause or causes. Particularly important groups of organisms that became extinct include trilobites and Fusulinid foraminifers (Late Permian), conodonts (Late Triassic), and ammonites and dinosaurs (Late Cretaceous). Many other groups also became extinct or were greatly reduced in numbers. These dramatic extinctions have taxed the imagination of paleontologists and other geologists to provide acceptable causal explanations. The Late Permian extinction phase has received particular attention because of the number of major groups affected and the sharpness of the change with which these groups disappeared from the geologic record at the end of the Permian.

Mass extinctions are of enormous interest to geologists because of the questions they raise, among other things, about possible recurring catastrophic events in Earth's history. The past few decades saw explosive growth in research into the patterns, rates, causes, and consequences of extinction but little overall agreement about the causes of extinction. Theories about extinction fall into three groups: catastrophic extinction, gradual extinction, and stepwise

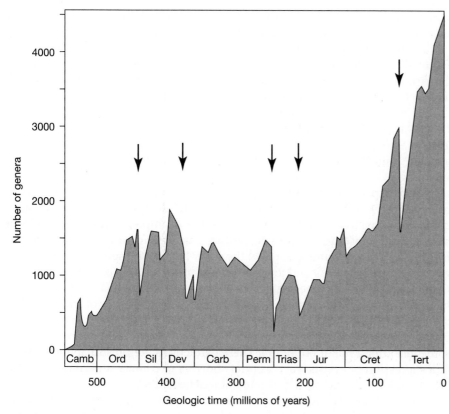

Figure 14.1.2

Diversity of marine animal genera (number of genera) during Phanerozoic time. Arrows point to the "big five," the five great mass extinctions. [After Sepkoski, J. J., Jr., 1995, Patterns of Phanerozoic extinction: a perspective from global data bases, *in* Walliser, O.H. (ed.), Global events and event stratigraphy: Springer-Verlag, Berlin, Fig. 1, p. 38.]

Table 14.1.2 Percentage of decline in marine animal diversity associated with the five great mass extinctions shown in Figure 14.1.2

	Percent decline in diversity	
Mass-Extinction Event	Genera	Families
End-Ordovician	60	26
Late Devonian	57	22
End-Permian	82	51
End-Triassic	53	22
End-Cretaceous	47	16

Source: Sepkoski (1995).

extinction (extinction that occurs in a series of discrete steps in the vicinity of major stratigraphic boundaries, such as the Permian/Triassic boundary). These various theories have been expounded in numerous research papers and a number of recent books, some of which are listed at the end of this chapter.

Causes proposed for the main mass-extinction events are summarized in Figure 14.1.3. Proponents of the catastrophic theory, especially for the sharp Cretaceous/Tertiary boundary event, suggest that the impact of extraterrestrial objects called bolides (meteorites and comets) created major climatic change (global winter) by throwing up huge clouds of dust and/or generated acid rain, tsunamis, and wildfires that caused extinction of some taxonomic groups. Alternatively, intense explosive volcanic activity may have adversely affected climates through discharge of excessive gas clouds (greenhouse warming). Other geologists suggest that such extraterrestrial causes are not needed to explain most extinction events. Gradual, progressive changes in climate (either warming or cooling) together with changes in sea level are adequate, they say, to account for extinctions. For example, lowering of sea level during major episodes of regression reduces habitats for shallow-water organisms and increases competition. On the other hand, widespread marine transgressions appear to be linked to development of anoxic (low oxygen) conditions that adversely affect some organisms and cause extinction (Hallam and Wignall, 1997, p. 251). The actual causes of anoxia during phases of transgression are poorly understood. Still other geologists suggest that some extinctions occur in a stepwise fashion by a series of pulses—some before, some at, and some just after a major boundary. These extinctions are presumably the result of a succession of events, such as brief showers of comets superimposed on a background of progressive environmental deterioration.

In any event, worldwide extinctions of major groups of organisms, while extremely interesting and significant, play only a limited role in biostratigraphy because these major extinctions provide only a few correlation horizons. Changing local environmental conditions are probably a more significant cause of extinction of individual species, which form the most important basis for biostratigraphy.

Extinction Event	Proposed Cause of Extinction					
	Bolide impact	Volcanism	Cooling	Warming	Regression	Transgression/ Anoxia
Late Precambrian						●
Late Early Cambrian					●	●
Late Cambrian			○			●
Late Ordovician			●	●	●	●
Late Devonian			○		○	●
Devonian-Carboniferous						●
Late Permian					●	
End-Permian		●		●	○	●
End-Triassic					●	○
Early Jurassic						●
Late Cretaceous			○			●
End-Cretaceous	●	●	●		●	○
End-Paleocene			●	●		●
Late Eocene				●		

● strong link   ○ possible link

Figure 14.1.3
Proposed causes for the main Phanerozoic extinction events. [After Hallam, A., and P. B. Wignall, 1997, Mass extinctions and their aftermath: Oxford University Press, Table 11.1, p. 248.]

14.5 DISTRIBUTION OF ORGANISMS IN SPACE: PALEOBIOGEOGRAPHY

When d'Orbigny introduced the concept of stage, he believed that the fossil assemblages upon which his stages were based had worldwide distribution. We now know that the fossil species and assemblages that characterize biostratigraphic units are not necessarily present everywhere that rocks of the appropriate ages occur. Few species are distributed throughout the entire world. Most, in fact, are restricted in their geographic range, although some fossil groups ranged widely throughout whole ecological realms at times in the geologic past. The region within which a particular group or groups of plants or animals is distributed is called a **biogeographic province**. Biogeographic provinces are separated by physical or climatic barriers. Land areas are barriers to marine organisms; open marine water is a barrier to land animals and plants; deep water is a barrier to shallow-water, shelf-dwelling organisms; cold water is a barrier to warm-water organisms; fresh water is a barrier to organisms adapted to saline marine conditions; and so forth. A particular type of barrier may be impenetrable by one species of organism but not by another. For example, benthonic organisms that do not have a long-lived, juvenile planktonic larval stage find deep water a barrier to dispersal. By contrast, planktonic organisms, which live in near-surface waters in the ocean, are distributed widely throughout the oceans in both shallow and deep water.

Box 14.2 Dispersal of Organisms

Paleontologists regard species as the fundamental biologic units in nature. They are the basic unit that undergoes evolution; the species niche is the basic functional unit in ecological interactions; and species are the fundamental units of biogeography, biostratigraphic zonation, and correlation. Because of their central importance in biostratigraphy, it is essential that we understand the factors that control the dispersal and distribution of species. Obviously, factors affecting the dispersal of land organisms and plants are different from those that control the dispersal of marine organisms. Also, the distribution of invertebrate marine organisms is controlled by different factors than those that control the distribution of vertebrate marine groups. Because of the overriding importance of marine invertebrates in biostratigraphic studies, we shall confine our discussion of dispersal here to invertebrate organisms in the marine setting, which consists of the pelagic realm (the water column) and the benthic realm (the seafloor, or bottom environment) (Fig. 14.2.1).

Marine invertebrate organisms can be divided into three fundamental types on the basis of habitat: plankton, nekton, and benthos (Table 14.2.1). **Plankton** are mainly microscopic-size organisms that live suspended at shallow depths within the water column (pelagic realm) and have very weak or limited ability to direct their own movements. They are distributed more or less passively by currents and wave action and may be dispersed widely into all types of open-ocean environments. Planktonic organisms are exceptionally useful fossils for biostratigraphic zonation and correlation because of their widespread distribution. They reflect the habitat of the pelagic realm and not the bottom environment into which they fall upon death; therefore, their presence in ancient marine sedimentary rocks is of limited value in environmental interpretation, although they are useful in some paleoceanographic applications (e.g., interpretation of water paleotemperature from oxygen isotopes). A few plankton such as graptolites do have some value as indicators of bottom

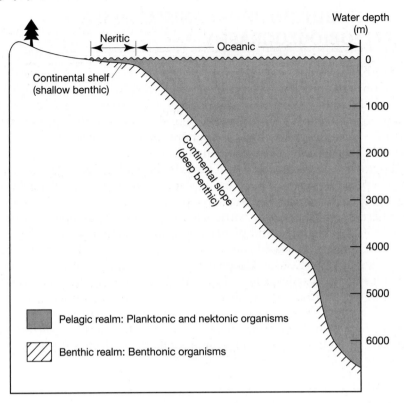

Figure 14.2.1
Subdivision of the marine environment into the pelagic (water column) and benthic (bottom) realms. The pelagic realm is inhabited by planktonic and nektonic organisms; benthonic organisms occupy bottom environments of the benthic realm.

environments. Graptolites were too fragile to survive in high-energy, shallow-water environments and are hence preserved mainly in the facies of quiet-water environments. Thus, they constitute "facies fossils."

Nekton also inhabit the pelagic realm and include all animals that are able to swim freely. Modern nekton are distributed in the ocean at depths ranging from the surface to thousands of meters and encompass many advanced groups of animals such as fish, whales, and mammals. Nekton are less abundant in the fossil record than planktonic and benthonic organisms (below) and thus overall appear to have somewhat less value in biostratigraphic studies. Nektonic fossils include fish remains in some deep-sea clays, belemnites and other mobile cephalopods, and probably conodonts. Conodonts are an interesting type of fossil, with considerable biostratigraphic significance, that occurs in rocks ranging in age from Cambrian to Triassic. They are tiny, toothlike phosphate fossils whose origin remained an enigma until 1982. A complete specimen of the conodont-bearing animal has been found in Lower Carboniferous rocks of the Edinburgh district, Scotland (Briggs, Clarkson, and Aldridge, 1983). The specimen is an elongate, soft-bodied animal 40 mm long and 1.8 mm wide. The conodont apparatus occurs in the head or anterior region of the body and may have served as teeth or possibly some type of internal support.

Benthos are bottom-dwelling organisms that live either on or below the ocean floor (benthic realm). Benthos (benthonic organisms) with preservable hard parts are particularly important for environmental interpretation because

Table 14.2.1   Classification of marine organisms by habitat or "life style"

Classification	Description	Example
Planktonic	Organisms that live suspended in the upper water column and which have only a very weak or limited ability to direct their own movements	
Phytoplankton	Have the ability to carry on photosynthesis; primary food producers, or autotrophs	Diatoms, dinoflagellates coccolithophoridae
Zooplankton	Do not carry on photosynthesis and thus cannot produce their own food (heterotrophs); feed on phytoplankton	Foraminifers, radio-larians, graptolites
Meroplankton	Spend only their juvenile stage as plankton; later become free-swimming or bottom-dwelling organisms	Larvae of most benthonic organisms such as molluscs
Pseudoplankton	Organisms distributed by waves and currents as a result of attachment to floating seaweed, driftwood, etc.	Mussels, barnacles
Benthonic	Bottom-dwelling organisms that live either on or below the ocean floor	
Sessile benthos	Benthos that attach themselves to the substrate (epifauna)	Crinoids, oysters, brachiopods
Vagrant benthos	Benthos that either creep or swim over the bottom (epifauna) or burrow into the bottom (infauna)	Starfish, echinoids, crabs, clams, worms
Nektonic	Organisms able to swim freely and thus move about largely independently of waves and currents	Mobile cephalopods, fish, sharks

their remains are commonly preserved in the same environment in which they lived. Because most benthos live in shallow water and have limited ability to move long distances along the bottom, they tend to be more provincial, and of somewhat less biostratigraphic significance, than plankton. Nonetheless, benthos can be dispersed outside their local environment because many benthonic species have a planktonic, juvenile larval stage during which they can be dispersed by currents. Some workers originally questioned the importance of larval transport as a mechanism for dispersal of shallow-water benthos over long distances, such as across ocean basins. It was initially believed that the duration of the larval stage was so short that the larva would change to the adult phase while the organisms were still over deep water, causing them to perish when they settled to the bottom. Subsequent work (e.g., Scheltema, 1977) has demonstrated, however, that there are many shoal-water or continental-shelf benthic invertebrates whose larva have a pelagic stage lasting from six months to more than one year. Scheltema refers to such larva as **teleplanic,** or "far wandering." Such long-lived larval species could cross the modern Atlantic Ocean, for example, by way of the main surface currents. Many workers interpret this ability to become dispersed by pelagic mechanisms as being favorable for interpopulation migration and, therefore, gene flow. Valentine (1977b, p. 145) summarizes the importance of these long-lived larval forms as follows:

> Species with the more long-lived and hardy pelagic larva have the greater chance to be widely dispersed after reproduction. . . . Therefore, species with such attributes would commonly be able to colonize habitats that lie at some distance from their parental ranges, and would usually be able to maintain gene flow to such outlying populations. Species with shorter planktonic development periods, smaller broods, or more restricted larval requirements would tend to colonize only localities that are fairly close to their parental regions. If a population became established at any considerable distance from others, gene exchange might be sporadic or lacking altogether, leading to divergence between the colonists and the parental population, and a reduction in their usefulness in correlation.
>
> For a species in a given locality, then, a geographic range exists for which colonization is essentially obligatory, as the region lies within the normal migratory range of the population; thus, by some standard time, occupation is virtually assured. This can be called the **local range**.

BARRIERS TO DISPERSAL

Each species thus has a potential geographic range that is determined by its habitat requirements. Few species actually occur throughout their potential range. Their distribution is restricted owing either to barriers of some type that prevent their expansion into all areas of suitable habitat or because the species may not have had time to spread to all suitable areas, especially if barriers are present. At any given time, there are many regions in the world that could be colonized by species if they could reach them in appropriate numbers, but they are barred from reaching them by intervening inhospitable areas. Many species eventually find ways to broach narrow barriers and perhaps in time even to cross wider barriers. Once barriers are crossed, or barriers disappear, the migrant species may find itself in competition for environmental niches with similar species or similarly adapted species in the new province. In the face of this competition, either the indigenous species or the migrant species may become extinct. Alternatively, the less well adapted species could evolve and become adapted to a different environmental niche. Once a barrier is surmounted, the colonizers typically expand their range at the new location until it is circumscribed by other barriers, filling out their new local range. The intruding species may subsequently broach still other barriers, hopping from one habitable region to another across barriers of varying difficulty of penetration and episodically expanding their total species range (Valentine, 1977a).

The broaching of barriers thus leads to expansion of the total range of a species, although in some cases it may lead to extinction of the species in the new region or to its evolution to a more adaptable species. On the other hand, if the opposite situation prevails and a barrier "suddenly" appears and divides a once-continuous area of suitable habitat, the result is the segregation of the species into different populations separated by the barrier. The separated populations would gradually evolve into different species, each with a more restricted geographic range than the parent species (Dodd and Stanton, 1981).

Numerous ecological factors can act as barriers to dispersal of organisms, all of which can be grouped under two major categories: habitat failure, as when shelf habitats give way to deep-sea conditions or to land, and temperature.

Hallam (1981) suggests that the major controls on faunal provinciality are climate and plate movements. Dodd and Stanton (1981) indicate that the principal factors controlling geographic distribution of species are depth and elevation (that is, water depth and land elevation) and temperatures. These possible controls on provinciality are discussed in further detail in the following paragraphs.

Temperature

Temperature is a major barrier to migration of species, and it commonly affects larvae more than adult organisms. Because the distribution of worldwide temperatures is latitudinally controlled, temperature barriers are most important latitudinally, although seasonal and even diurnal temperature changes are also important. The boundaries of all modern biotic provinces are in part temperature controlled, and ancient biotic provinces were undoubtedly similarly controlled. Warm-water taxa are restricted primarily to the equatorial zone of the ocean because no other large parts of the ocean, either at the surface or at depth, are warm enough to sustain these tropical species. Cold-water taxa, on the other hand, can extend their range closer to the equatorial region by migrating down the bathymetric gradient into deeper and colder water, the phenomenon of **submergence**, if they are capable of adapting to greater depths. Also, if a polar species can manage to find a way of breaking through the temperature barrier and crossing the equatorial region, it can find suitable cold-water habitats at or near the surface in the higher latitudes of the other hemisphere.

Some species of organisms are adapted to a wide range of temperatures and may thus be distributed through a much wider range of temperature zones than less tolerant species. Nonetheless, even tolerant species are sensitive to temperature variations and do not occur throughout all temperature zones. It must be recognized also that marine temperature zones have changed throughout geologic time as world climatic zones have shifted in response to plate movements and episodes of glaciation. A given geographic region of the world may thus record a succession of colder water or warmer water faunas through time in response to these shifting climatic conditions.

Geographic Barriers

The terms habitat failure, plate movements, and depth-elevation used above are all different ways of expressing the concept of geographic barriers. These geographic barriers arise out of the distribution pattern of landmasses and oceans and variations in water depths of the oceans. All organisms have limited water depths at which they can survive. Thus, water that is either too deep or too shallow can constitute a barrier to a particular species of organism. Landmasses constitute barriers to the dispersal of marine organisms, and the open ocean is a barrier to migration of land animals and plants from one continent to another. The most important factors influencing geographic barriers appear to be changes in sea level and changes in the nature and geographic distribution of landmasses and the ocean floor brought about by plate movements (discussed in a following section).

Sea-Level Changes

Causes of major cycles of sea-level change are discussed in Chapter 12. Fluctuations in sea level cause significant interruptions in biogeographic provinces because of variations in water depths on the continental shelves. During a major drop in sea level, water is withdrawn from the continental shelves, exposing much of the inner shelf. The habitable area of shallow water is greatly reduced, leading to crowding and increased competition among shallow-water species that cannot move seaward into deeper water, and to probable extinction of less adaptable groups. During major rises in sea level, water depths on the outer continental shelf are increased, and the total area of shallow water along continental margins is also vastly increased owing to spread of the seas over the edges of the continents. The available environmental niches for shallow-water organisms are correspondingly increased, resulting in less competition among species for available space and food. These conditions lead to expansion of the local ranges of species as they move into favorable habitats, and also probably to rapid emergence of new species (speciation) as a result of adaptive radiation of groups that survived the preceding episode of lowered sea level. On the other hand, as discussed in the preceding section, transgressions can be accompanied by the onset of low-oxygen conditions (anoxia) in some environments. Anoxia may have lead to extinction of some invertebrate animal groups at various times in the past (Fig. 14.1.3).

An inverse relationship appears to exist between the area of continents covered by sea and phenomenon of endemism. **Endemism** is the tendency of species or other taxa to have a very limited geographic range, as contrasted with **pandemism**, which is the tendency of species to have worldwide distribution. At times of low sea level and restriction of seas, faunal migrations between continental shelf areas is rendered more difficult, resulting in less gene flow. Thus, more local speciation may occur among the dispersible organisms that occupy shallower water habitats (Hallam, 1981).

Plate Movements

Tectonism is the major factor controlling the distribution of landmasses and ocean basins. Major changes in the environmental framework of the marine realm occur as the geographic positions, configurations, and sizes of continents and ocean basins are changed by global plate tectonic processes. Plate movements can greatly affect topographic barriers by producing changes in oceanic widths and depths. As previously discussed, changes in rates of seafloor spreading may have a major effect on sea level. Plate movements can also alter latitudinal temperature gradients by shifting the geographic position of continents, and they can even affect the distribution patterns of major ocean currents. The creation or destruction of migration barriers may thus be tied closely to plate tectonics events.

Other Barriers

Other barriers, less important than temperature and geographic barriers, may also help to define the boundaries of biogeographic provinces. Salinity differences constitute an important boundary between freshwater and marine provinces; however, salinity is a relatively unimportant barrier within the marine realm itself. Salinity can markedly increase in some small, restricted arms of the ocean where evaporation rates are high. Conversely, lower than normal salinities may ensue in some coastal areas where freshwater runoff is high. These salinity variations can control local communities of organisms but not the distribution of organisms on a provincial level. In the open ocean, salinity tends to be highest in the equatorial region, where evaporation rates are at a maximum, and lowest in the middle latitudes, where some dilution occurs as a result of freshwater runoff from the continents. Even so, the salinity in these regions varies only a few parts per thousand from the average ocean salinity (35 ‰), a variation not adequate to seriously affect the dispersal of organisms in the open ocean.

Currents aid in the dispersal of planktonic species and the larva of benthonic species, but they help also, in some parts of the ocean, to maintain the
temperature gradients that create barriers to dispersal. Thus, currents may act
as either a barrier or an aid to dispersal. The long-term pattern of currents is itself affected by plate movements.

14.6 COMBINED EFFECTS OF THE DISTRIBUTION OF ORGANISMS IN TIME AND SPACE

Both the environmental and the temporal (variations with time) records are important for interpretation of geologic history. If organisms throughout geologic time
had been spread over the world and not confined to specific biogeographic
provinces and environments, worldwide correlation of strata on the basis of fossils
would be greatly facilitated, assuming that evolutionary changes have been simultaneous and worldwide. Under these conditions, however, fossils would provide
little or no help in working out ancient depositional environments because more or
less the same organisms would have lived in all environments. Conversely, if organisms were distributed in biogeographic provinces as they are today, but organic
evolution never occurred, we would be able to interpret local ancient environments
with great confidence because ancient sedimentary rocks would contain the same
species as modern environments. By the same token, these species would be of no
value in correlation and the unraveling of local chronologies because the same
species would have existed throughout geologic time.

The real fossil record reflects the fact that both segregation into biogeographic provinces and organic evolution took place. Owing to organic evolution, we are
able to correlate strata of a given age from one area to another and to work out the
relative chronology of strata in a given area. Because many organisms were confined to biogeographic provinces in the past, however, we cannot always correlate
time-equivalent strata from different environments because the organisms that existed in different biogeographic provinces during the same period of time were
different. Thus, correlation between biogeographic provinces is difficult, and it is
commonly not possible to make worldwide correlations. On the other hand, because different groups of organisms were confined to different provinces and different environments, the provinciality of ancient organisms provides an invaluable
tool for interpreting ancient sedimentary environments.

The provinciality of organisms creates special problems from the standpoint of determining the total vertical stratigraphic range of a species. A species
may exist in one province for long periods of time before broaching a particular
barrier and spreading into a nearby province. After migration into the new
province, the species may die out in the old province while continuing to thrive
for some time in the new region. Therefore, the **local vertical range** of a species
in a given province, sometimes called the teil zone, may be much shorter than
the **total range** of the species on a global scale. Paleontologists must be extremely careful about recognizing this possibility when using fossils for time correlations. This problem is demonstrated in Figure 14.7, which illustrates some of the
factors that can affect the range of a species. This diagram shows that the range
of a species is affected both by evolutionary changes and by the presence of barriers that can regulate the times of migration into and first appearance in nearby
provinces.

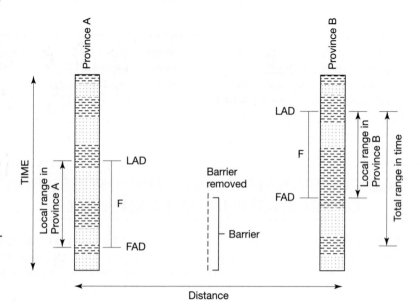

Figure 14.7
Diagram illustrating the difference in local range and total range of a hypothetical species (F). Species F first appears in Province A and is restricted to Province A by a barrier. Later removal of the barrier allows migration to Province B, where the species persists for a time after it has died out in Province A. FAD = first appearance datum; LAD = last appearance datum.

14.7 BIOCORRELATION

Biostratigraphic units are observable, objective stratigraphic units identified on the basis of their fossil content. As such, they can be traced and matched from one locality to another just as lithostratigraphic units are traced. Biostratigraphic units may or may not have time significance. For example, assemblage biozones and abundance biozones may cross time lines (be diachronous) when traced laterally. On the other hand, taxon-range biozones and interval biozones, particularly those defined by first appearances of taxa, yield correlation lines that coincide in general with time lines. Biostratigraphic units may be correlated, irrespective of their time significance, using much the same principles employed in correlation of lithostratigraphic units—matching by identity and position in the stratigraphic sequence, for example. In this section, we will first examine correlation by assemblage biozones and abundance biozones, which can be correlated as biostratigraphic units even though they may not have time-stratigraphic significance. We will then discuss biocorrelation methods based on interval zones and other zones that yield time-stratigraphic correlations.

The following discussion is aimed primarily at biocorrelation on the basis of marine invertebrate organisms. Readers should keep in mind, however, that strata can be zoned on the basis of terrestrial animal and plant remains and that correlation can in some cases be made on the basis of such data (e.g., Flynn and Swisher, 1995).

Correlation by Assemblage Biozones

Assemblage biozones are based on distinctive groupings of three or more taxa without regard to their range limits (Fig. 14.3). They are defined by different successions of faunas or floras, and they succeed each other in a stratigraphic section without gaps or overlaps. Assemblage zones have particular significance as an indicator of environment, which may vary greatly regionally. Therefore, they tend to be of greatest value in local correlations. Nonetheless, some assemblage zones based on marine planktonic assemblages may be used for correlation over much wider areas. The principle of correlation by assemblage zones is illustrated graphically in the very simple example shown in Figure 14.8.

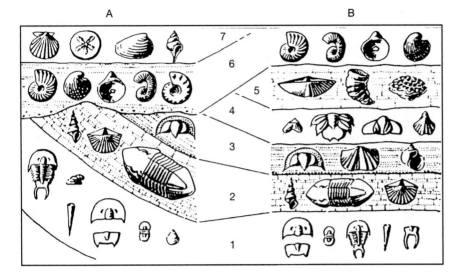

Figure 14.8
Diagram illustrating graphically the principle of correlation by fossil assemblages. [From Moore, R. C., C. G. Lalicker, and A. G. Fischer, 1952, Invertebrate fossils: McGraw-Hill Book Co., Fig. 1.3, p. 8, reproduced by permission.]

Shaw (1964) points out, however, that the boundary between assemblage zones is inherently fuzzy because above and below the limits of this zone will be transition zones in which part of the characteristic fossil assemblage will be missing because it has not yet appeared or has already vanished. Therefore, there are practical limits to the accuracy that can be achieved by assemblage zone correlations. Part of the problem in correlation by assemblage zones stems from the fact that the number of fossil taxa that a biostratigrapher must work with is so large that it is difficult to visually assimilate the data and draw meaningful zone boundaries (e.g., Fig. 14.9). To overcome this problem, earlier workers tended to reduce the number of taxa whose distributions would be studied, or they tried to make composites of the samples. A more recent solution to this problem is to apply the techniques of multivariate statistical analysis to recognition and delineation of assemblage zones. These quantitative techniques provide a rational statistical basis for delineating zones based on large numbers of taxa without taking the decision making out of the hands of the biostratigrapher (e.g., Gradstein et al., 1985).

Correlation by Abundance Biozones

As mentioned, abundance biozones are defined by the quantitatively distinctive maxima of relative abundance of one or more species, genus, or other taxon rather than by the range of the taxon. They represent a time or times when a particular taxon was at the peak of its development with respect to numbers of individuals. Some biostratigraphers previously used abundance zones for time-stratigraphic correlation under the assumption that there is a time in the history of every taxon when it reaches its maximum abundance and that this abundance peak occurs everywhere at the same time. The current prevailing opinion among biostratigraphers is that most abundance zones are unreliable and unsatisfactory for time-stratigraphic correlation. This opinion is based on the apparent fact that not all species achieve a maximum abundance, or that if they do this peak is not necessarily recorded by layers of abundant specimens. Furthermore, peak abundances that are recorded in the stratigraphic record may be related to favorable local ecological conditions that can occur at different times in different areas and that may persist in one area much longer than in another. Maximum abundance may thus represent local, sporadically favorable environments, suddenly unfavorable environments that caused mass mortality, or mechanical concentrations of the shells of organisms after death. Some of the problems of correlating by abundance zones

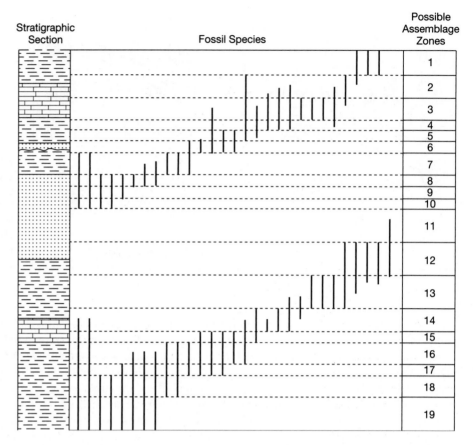

Stratigraphic Section   Fossil Species   Possible Assemblage Zones

Figure 14.9
Hypothetical stratigraphic section illustrating the large number of fossil taxa that may be involved in correlation by assemblage biozones. Vertical black lines represent the composite ranges of the species found at various local sections. The column at the right shows one interpretation that could be drawn from these fossil data. [After Hazel, J. E., 1977, Use of certain multivariate and other techniques in assemblage zonal biostratigraphy: Examples utilizing Cambrian, Cretaceous, and Tertiary benthic invertebrates, *in* Kauffman, G. G., and J. E. Hazel (eds.), Concepts and methods of biostratigraphy: Van Nostrand Reinhold, Fig. 1, p. 289, reproduced by permission.]

are illustrated in Figure 14.10. In short, abundance zones may be used for biostratigraphic correlation but they do not provide a reliable means of time-stratigraphic correlation. Although they are sometimes used locally for correlation within provinces, biostratigraphers usually prefer correlations based on assemblage biozones or taxon-range or interval biozones.

Chronocorrelation by Fossils

Chronostratigraphic correlation is the matching up of stratigraphic units on the basis of time equivalence. Establishing the time equivalence of strata is the backbone of global stratigraphy and is considered by most stratigraphers to be the most important type of correlation. Methods for establishing time-stratigraphic correlation fall into two broad general categories: biological and physical/chemical. As mentioned, time-stratigraphic correlation by biological methods is based mainly on use of concurrent range zones and other interval zones. Biological correlation methods also include statistical treatment of range-zone data and correlation by biogeographical abundance zones, which are biological events related to climate fluctuations. A variety of physical and chemical methods available for chronostratigraphic correlation are discussed in the following chapter. Logically, this discussion of chronostratigraphic correlation by fossils also belongs in the next chapter; however, I am including it here to keep all material relating to correlation by fossils in a single unit. The discussion of biocorrelation that follows represents a very general introduction to this subject. For more rigorous treatment of biocorrelation, see Gradstine et al. (1985) and Guex (1991).

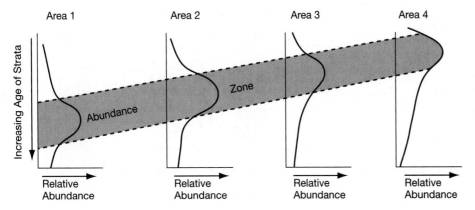

Figure 14.10
Schematic diagram illustrating why correlation by abundance biozones may not yield a true time correlation. The same species may achieve its maximum abundance at different times in different localities. Relative abundance increases to the right.

Correlation by Taxon-Range and Interval Biozones

Taxon-range and interval zones are biozones that constitute the strata that fall between the highest and/or lowest occurrence of taxa. Figure 14.3 illustrates several ways that the first and last appearances of taxa may be used to define biozones. These different biozones have varying degrees of usefulness in time-stratigraphic correlation, as described below.

Taxon-Range Biozones

Taxon-range biozones may be very useful for time correlation if the taxa upon which they are based have very short stratigraphic ranges. They are of little value if the taxa range through an entire geologic period or several periods. Correlation by taxon range zone is often referred to as correlation by **index fossils**. As mentioned, index fossils are considered to be those taxa that have very short stratigraphic ranges, were geographically widespread, were abundant enough to show up in the stratigraphic record, and are easily identifiable. Unfortunately, the term index fossil has also been used in other ways and can have other connotations. Therefore, it is less confusing when speaking of correlation based on the entire range of a taxon to refer to it simply as correlation by taxon-range biozone. Correlation by taxon-range biozone is illustrated diagrammatically in Zone 1 of Figure 14.11.

Interval Biozones

When individual taxon-range biozones are very long, and correlation by taxon-range biozone is thus not suitable, much finer scale correlation is possible by using other types of interval biozones. Interval biozones defined by the first (stratigraphically lowest) appearance of two taxa, for example, are particularly useful in time-stratigraphic correlation because they are based on evolutionary changes, along phyletic lineages, that tend to occur very rapidly. Thus, the interval between the first documented appearance of two taxa may represent a very short span of time, and the age of the strata in this interval may be nearly synchronous throughout their extent. Interval biozones defined on the last (stratigraphically highest) appearances of taxa are commonly considered to have less time significance than those based on first appearances because extinctions of taxa commonly do not occur with the same suddenness that new species appear through phyletic evolution.

Figure 14.11 illustrates some of the various methods that can be used for correlating between two stratigraphic sections on the basis of taxon-range or interval biozones. Note from this illustration that interval biozones can be identified which represent much shorter spans of time than that represented by the taxon-range

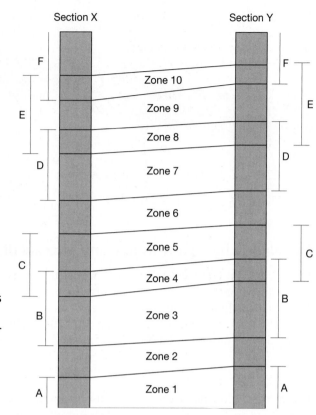

Figure 14.11
Correlation between two hypothetical sections on the basis of taxon-range and interval biozones. Note that several types of interval zones are used here for correlation. For example, Zone 1 is defined by the total vertical range of Species A (taxon-range zone); Zone 2 is an interval zone defined by the last appearance of Species A and the first appearance of Species B; Zone 4 is formed by the overlapping ranges of Species B and C; and so forth.

biozones of most individual taxa. Correlation can be made also between stratigraphic sections simply on the basis of first or last appearances of specific taxa, without correlating entire zones. In other words, a correlation line can be drawn from the stratigraphic position represented by the first appearance of a particular taxon, called the first appearance datum or FAD, to the FAD of the same taxon in another stratigraphic section. Similarly, correlation can be made between the last appearance datums, LADs, of a given taxon in different stratigraphic sections. FADs and LADs are further discussed in the Chapter 15.

Graphic Method for Correlating by Taxon-Range Biozone

Although interval biozones can be used to define units of strata deposited during relatively short periods of time, they do not necessarily yield precise time-stratigraphic correlations. Organisms may migrate laterally and appear in other areas at somewhat later times than their true first appearance (Fig. 14.7), or they may migrate out of a local area before their final extinction elsewhere. These variables of behavior make the boundaries between interval zones inherently "fuzzy." The exact boundary between biozones can never be known because such boundaries are determined empirically. Additional collecting in a new area always holds the possibility of extending the known range of previously defined species or taxa, because they may have appeared earlier or persisted longer in the new area than in the originally defined areas. One way to minimize the problem of fuzzy zonal boundaries is to treat range data statistically, utilizing the first and last appearances of all the species present in a stratigraphic section rather than the ranges of just one or two species. A. B. Shaw in 1964 was the first to propose a graphical method for establishing time equivalence of strata in two stratigraphic sections by plotting first and last appearances of all the species in one section against the first and last appearances of the same species in another section. This method is now widely used by stratigraphers for detailed time-stratigraphic correlation between stratigraphic sections, particularly local sections.

Shaw's method, as further elaborated by Miller (1977), involves first selecting a single stratigraphic section as a reference section to which other sections can be compared and correlated. This reference section should be the thickest section available, should be free of faulting or other structural complications, and should contain a large and varied fossil content. The reference section is measured and sampled as completely as possible, and the first and last appearances of all species are documented in terms of their positions in the stratigraphic section above an arbitrarily chosen reference point, that is, number of meters above the base of the measured section (Fig. 14.12). The species ranges recorded by the first and last appearances in this local reference section may not be the true (total) ranges for all of the species; however, this fact does not preclude using them to help establish correlation, as we shall see. A second stratigraphic section is then chosen to be compared with the reference section, and the first and last appearances of the same species, and any other species, are determined in this section.

From two such stratigraphic sections (Fig. 14.12), a graph is constructed in which distance above the base of the reference section, say section A, is indicated on the vertical axis and distance above the base of the second measured section, section B, is plotted on the horizontal axis (Fig. 14.13). The first and last appearances of each species in the reference section can then be plotted against the first and last appearances of the same species in the second measured section. In Figure 14.12, for example, Species 3 first appears in reference Section A at 42 m above the base of the section and in measured Section B at about 21 m above the base. A single point can be plotted on the graph to represent these values. Similarly, additional points are plotted to represent the first and last appearances of all the species in the two sections. The dashed lines in Figure 14.13 illustrate how the points are plotted. This procedure yields a series of points that tend to cluster around a straight line, the line of correlation in Figure 14.13. This line can be drawn visually to yield a "best-fit" line, or it can be drawn by use of statistical regression

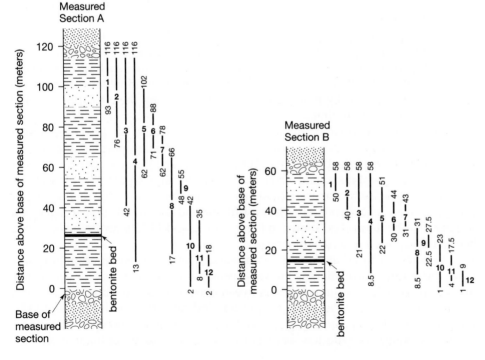

Figure 14.12
Two stratigraphic sections with ranges of fossil species (Species 1 through 12) graphed in meters above the base of the section. Sections A and B contain identical fossils with identical time spans; however, Section B represents only half the rate of sediment accumulation. Use of these fossil ranges in A. B. Shaw's (1964) graphic correlation method is illustrated in Figure 14.18. [After Eicher, D. L., Geologic time, 2nd ed., Prentice-Hall, Fig. 5.8, p. 112; data from Ericson and Wollin, 1968.]

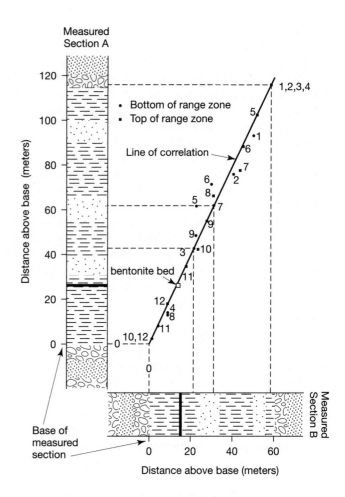

Figure 14.13
Illustration of the graphic correlation method using the data shown in Figure 14.12. The dashed lines illustrate how the base or top of a range zone in one section is plotted against an equivalent base or top in the other section. Once the line of correlation is drawn, any part of Section A can be correlated to the equivalent part of Section B.

methods. The x and y coordinates of any point on this line provide a precise time-stratigraphic correlation between the two sections. In Figure 14.13, for example, the bed at 60 m in Section A correlates with the bed at 30 m in Section B, and the bed at 100 m in Section A correlates with the bed at about 49 m in Section B.

First and last appearances of species represented by points that plot well off the best-fit line in Figure 14.12 indicate species that appear in or disappear from Section A at distinctly different times than in Section B. Either such species are environmentally controlled (facies dependent), or their migration between Sections A and B was impeded by biogeographic barriers causing them to appear in the two sections at different times.

This graphic correlation method can take advantage of physical events such as ash falls or stable isotopic events that have time-stratigraphic significance, to verify the position of the best-fit line. For example, ash falls occur over wide geographic areas almost instantaneously. Their presence in two stratigraphic sections constitutes a precise time marker (that can be dated by radiometric methods) that provides a very reliable point for the best-fit line and should fall almost exactly on this line. The bentonite bed shown in Figures 14.12 and 14.13 is an ash bed that has been partially altered to clay minerals.

In addition to its usefulness in correlating between two stratigraphic sections, the graphic correlation method also provides a powerful tool for evaluating differences in rates of sedimentation between two sections or the presence of a hiatus in a section. The slope of the best-fit line indicates the relative rates of sedimentation between the areas (Fig. 14.14). If an abrupt change occurs in this slope, this change suggests a sudden relative increase or decrease in sedimentation rates

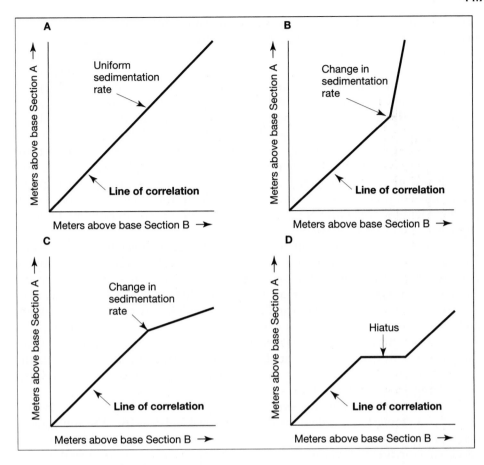

Figure 14.14
Effect of change in sedimentation rate on the shape of the line of correlation. (A) Uniform sedimentation rate in both Section A and Section B. (B) "Dogleg" in the line of correlation indicates a relative decrease in rate of sedimentation in Section B. (C) "Dogleg" indicates a relative decrease in rate of sedimentation in Section A. (D) A hiatus in deposition in Section A (caused by nondeposition, an unconformity, or possibly faulting) shows up as a horizontal segment in the line of correlation.

in the sections. The change in slope in Figure 14.14B, for example, indicates a decrease in the rate of sedimentation in Section B compared to that in Section A; Figure 14.14C illustrates a decrease in the rate of sedimentation in Section A. The presence of a hiatus in deposition in one section shows up as a horizontal line segment in the line of correlation (Fig. 14.14D). In this example, sedimentation ceased for a time at Section A while it continued at Section B; then, sedimentation at Section A resumed. Alternatively, the hiatus might be due to an unconformity or faulting.

Not only can the graphic correlation method be used for correlating between any two local sections, but it also can be expanded by correlating one section after another to compile a composite section or **composite standard** (Shaw, 1964; Carney and Pierce, 1995). Once a database of composited fossil ranges has been compiled, this database can be scaled in chronostratigraphic (time) units (Chapter 15). It is then possible to correlate a particular stratigraphic section against this composite standard in the same manner that one stratigraphic section is correlated against another (Carney and Pierce, 1995). By correlating to a composite standard scaled in time units, the age of any part of the stratigraphic section can be determined. Graphic correlation is now being used in a variety of applications, including sequence stratigraphic studies. See Mann and Lane (1995) for detailed discussion of these various applications.

Correlation by Biogeographical Abundance Biozones

Under the heading of "Biocorrelation," I discussed correlation by fossil abundance biozones and pointed out that abundance zones are unreliable for time-stratigraphic correlation because they are affected by environmental conditions and other factors that can cause them to be diachronous (occur at different times in different areas). A different approach to the use of abundance zones yields correlations that have time-stratigraphic significance; this approach is correlation based on the maximum abundance of a taxon that results from geographical shifts of an environmentally sensitive fossil assemblage (Haq and Worsley, 1982). Because of latitudinally related temperature differences in the ocean, some species or other taxa are restricted to biogeographic provinces that are defined by latitude. Thus, low-latitude taxa are ecologically excluded from high latitudes, and vice versa; however, changes in climate can allow these taxa to shift into a different biogeographic province. During major glacial stages, for example, high-latitude taxa can expand into lower latitudes, and during warming trends between major glacial stages, low-latitude taxa can expand into higher latitudes. From a geochronological point of view, the spreading out of certain planktonic species in response to major climatic fluctuations is essentially isochronous.

Climate-related shifts in planktonic taxa at specific times thus provide biogeographical abundance events that can be correlated from one area to another. For each core or outcrop section studied, climatic curves are constructed on the basis of percentages of warm-climate to cool-climate taxa or relative abundance of a particular taxon. These curves can then be used to identify episodes of warming and cooling that can be correlated from one section to another. Figure 14.15, constructed from this type of information, illustrates how climatically controlled latitudinal shifts in calcareous nannoplankton assemblages in the North Atlantic during Miocene time has been used for chronostratigraphic correlation in Deep Sea Drilling Program (DSDP) cores.

A related approach is time-stratigraphic correlation based on the coiling ratios of planktonic foraminifera, as described by Eicher (1976). The multichambered shells of some foraminifera are known to coil in one direction when the species lives in areas of warm water, and in the opposite direction when it lives in areas of cold water. The foraminifer *Globorotalia truncatulinoides*, for example, has dominantly right-handed coils in warm water and left-handed coils in cold water.

Figure 14.15
Use of biogeographical abundance zones as a means of time correlation. Cycles of latitudinal shifts of calcareous nannoplankton assemblages in the North Atlantic Ocean during the Miocene are interpreted as responses to major fluctuations in climate. The major shifts of relatively warmer, midlatitude assemblages into higher latitudes can be used for the refinement of the biochronological scale in the higher latitudes from which marker, low-latitude taxa are normally excluded. [From Haq, B. V., and T. R. Worsley, 1982, Biochronology—Biologic events in time resolution, their potential and limitations, *in* Odin, G. S. (ed.), Numerical dating in stratigraphy: John Wiley & Sons, Ltd., Fig. 4, p. 27. Reprinted by permission.]

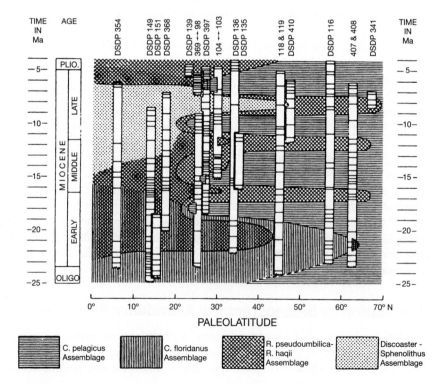

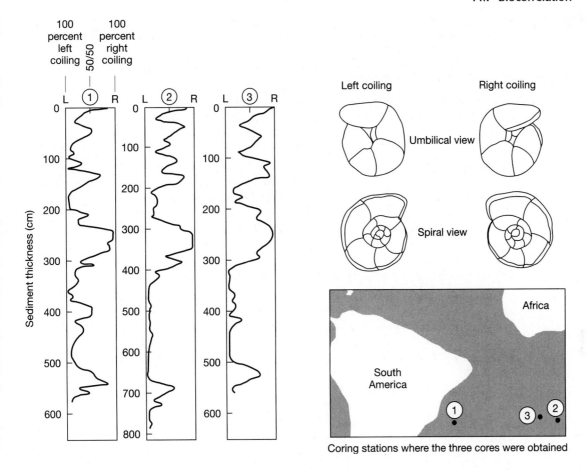

Figure 14.16
Biogeographical abundance zone correlation on the basis of coiling ratios of foraminifers. Correlation is based on coiling ratios of *Globorotalia truncatulinoides* in three South Atlantic Ocean cores. The depositional time represented by the cores is about 1.5 million years. [After Eicher, D. L., Geologic time, 2nd ed., 1976, Prentice-Hall, Fig. 5.12, p. 115.]

Figure 14.16 shows that during times of glacial cooling of the ocean in the Pleistocene, predominantly right-coiled populations of *Globorotalia truncatulinoides* were replaced in middle and low latitudes by dominantly left-coiled populations. These changes in coiling ratios of foraminiferal species provide a means of correlating short-term fluctuations of climatic change in the Pleistocene that are essentially synchronous throughout at least a part of an ocean basin.

The major drawback to these correlation methods based on biologic response to climate fluctuations is that their use is restricted mainly to correlating sediments deposited during the Quaternary and late Tertiary, when several episodes of cooling and warming in the world ocean took place. Nonetheless, they provide a useful supplement to correlation methods based on oxygen isotopes (Chapter 15), which also involve climate fluctuations in the late Tertiary and Quaternary.

FURTHER READING

Berggren, W. A., D. V. Kent, M-P. Aubry, and J. Hardenbol (eds.), 1995, Geochronology, time scales and global stratigraphic correlation: SEPM Spec. Publ. 54, 386 p.

Erwin, D. H., and S. L. Wing (eds.), 2000, Deep time: Paleobiology's perspective: The Paleontologtical Society, Lawrence, Kans., 373 p.

Glen, W., 1994, Mass-extinction debates: How science works in a crisis: Stanford University Press, 370 p.

Hallam, A., and P. B. Wignall, 1997, Mass extinctions and their aftermath: Oxford University Press, Oxford, 320 p.

Jackson, J. B. C., S. Lidgard, and F. K. McKinney (eds.), 2001, Evolutionary patterns: Growth, form, and tempo in the fossil record, 399 p.

Kemp, T. S., 1999, Fossils and evolution: Oxford University Press, Oxford, 284 p.

Lawton, J. H., and R. M. May, 1995, Extinction rates: Oxford University Press, Oxford, 233 p.

Mann, K. O., and H. R. Lane (eds.), 1995, Graphic correlation: SEPM Special Publication 53, 259 p.

Sharpton, V. L., and P. D. Ward (eds.), 1990, Global catastrophes in Earth history: Geol. Soc. America, Spec. Paper 247, 631 p.

Sheldon, P. R., 1996, Plus ça change—A model for stasis and evolution in different environments: Palaeogeography, Palaeoclimatology, Palaeoecology, v. 127, p. 209–227.

Walliser, O. H. (ed.), 1995, Global events and event stratigraphy in the Phanerozoic: Springer-Verlag, Berlin, 333 p.

15

Chronostratigraphy and Geologic Time

15.1 INTRODUCTION

The stratigraphic units described in the preceding chapters are rock units distinguished by lithology, magnetic characteristics, seismic reflection characteristics, or fossil content. As such, they are observable or measurable material reference units that depict the descriptive stratigraphic features of a region. Definition of these units allows the vertical and lateral relationships between rock units to be recognized and provides a means of correlating the units from one area to another. As Krumbein and Sloss (1963) point out, however, descriptive stratigraphic units do not lend themselves to interpretation of the local stratigraphic column in terms of Earth history. To interpret Earth history requires that stratigraphic units be related to geologic time; that is, the ages of rock units must be known. Establishing the time relationship among rock units is called **chronostratigraphy**, and stratigraphic units defined and delineated on the basis of time are **geologic time units**. The relationship between chronostratigraphy and other branches of stratigraphy is illustrated in Figure 15.1.

In this chapter, we examine the concept of geologic time units and explore the relationship of time units to other types of stratigraphic units. We will also see how geologic time units are used to create the Geologic Time Scale and we will discuss methods of calibrating the time scale. Finally, we will examine methods for chronocorrelation—correlation of rock units on the basis of their ages.

15.2 GEOLOGIC TIME UNITS

Geologic time units are conceptual units rather than actual rock units, although most geologic time units are based on rock units. In fact, we recognize two distinct types of formal stratigraphic units that can be distinguished by geologic age: units, called **stratotypes**, based on actual rock sections, and units independent of reference rock sections (see Appendix C)). Ideally, the reference rock bodies for geologic time units are **isochronous units**. That is, they are rock units formed during the same span of time and everywhere bounded by synchronous surfaces, which are surfaces on which every point has the same age.

Figure 15.1
Diagram illustrating the procedures and processes involved in chronostratigraphy and the relationship of geologic time units to other kinds of stratigraphic units. Golden spike refers to internationally agreed upon points or boundaries in stratotype stratigraphic sections selected to serve as reference sections for chronostratigraphic units. [After Holland, C. H., 1998, Chronostratigraphy (global standard stratigraphy): A personal perspective, *in* Doyle, P., and M. R. Bennett (eds.), Unlocking the stratigraphical record: Advances in modern stratigraphy, John Wiley & Sons, Chichester, Fig. 13.1, p. 384.]

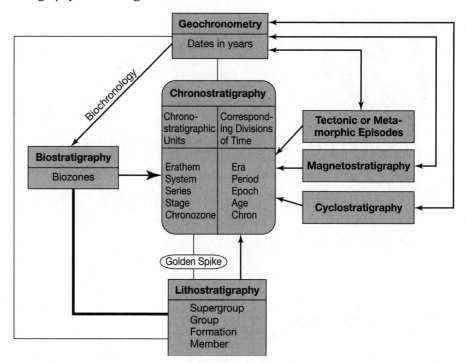

The North American Stratigraphic Code and The International Stratigraphic Guide (Salvador, 1994) recognize two fundamental types of isochronous geologic time units: chronostratigraphic units and geochronologic units. **Chronostratigraphic units** (Table 15.1) are tangible bodies of rock that are selected by geologists to serve as reference sections, or material referents, for all rocks formed during the same interval of time. In other words, a particular section of sedimentary rock having a known age span is selected to represent that particular interval of geologic time. For example, the interval of time from about 275–250 million years ago is called the Permian Period and is represented by rocks of the Permian System located in the Province of Perm, Russia (see Table 15.2). **Geochronologic units**, by contrast, are divisions of time distinguished on the basis of the rock record as expressed by chronostratigraphic units. They are not in themselves stratigraphic units. If the distinction between these two types of units seems somewhat confusing, the following illustration may help to clarify the difference. Chronostratigraphic units have been likened to the sand that flows through an hourglass during a certain period of time. By contrast, corresponding geochronologic units can be compared to the interval of time during which the sand flows (Hedberg, 1976). The duration of the flow measures a certain interval of time, such as an hour, but the sand itself cannot be said to be an hour.

Traditional internationally accepted chronostratigraphic units were previously based primarily on the time spans of lithostratigraphic or biostratigraphic units. We now also formally recognize (as chronostratigraphic units) polarity chronostratigraphic units (Appendix C), which are geologic time units based on the remanent magnetic fields in rocks (Chapter 13).

The characteristics and hierarchical rankings of geologic time units are briefly described in Table 15.1. Chronostratigraphic units are discussed first in this table because they are the reference stratigraphic sections upon which time (geochronologic) units are based. As mentioned, chronostratigraphic units are themselves based upon designated biostratigraphic, lithostratigraphic, or magnetopolarity units (Table 15.1). The fundamental chronostratigraphic unit is the

Table 15.1   Geologic time units

Chronostratigraphic Unit—an isochronous body of rock that serves as the material reference for all rocks formed during the same spans of time; it is always based on a material reference unit, or stratotype, which is a biostratigraphic, lithostratigraphic, or magnetopolarity unit

  Eonothem—the highest ranking chronostratigraphic unit; three recognized: **Phanerozoic**, encompassing the Paleozoic, Mesozoic, and Cenozoic erathems, and the **Proterozoic** and **Archean**, which together make up the Precambrian.

    Erathems—subdivisions of an eonothem; none in the Precambrian; the Phanerozoic erathems, names originally chosen to reflect major changes in the development of life on Earth, are the Paleozoic ("old life"), Mesozoic ("intermediate life"), and Cenozoic ("recent life")

      System—the primary chronostratigraphic unit of worldwide major rank (e.g., Permian System, Jurassic System); can be subdivided into subsystems or grouped into supersystems but most commonly are divided completely into units of the next lower rank (series)

        Series—a subdivision of a system; systems are divided into two to six series (commonly three); generally take their name from the system by adding the appropriate adjective "Lower," "Middle," or "Upper" to the system name (e.g., Lower Jurassic Series, Middle Jurassic Series, Upper Jurassic Series); useful for chronostratigraphic correlation within provinces; many can be recognized worldwide

          Stage—smaller scope and rank than series; very useful for intraregional and intracontinental classification and correlation; many stages also recognized worldwide; may be subdivided into substages

            Chronozone—the smallest chronostratigraphic unit; its boundaries may be independent of those of ranked stratigraphic units

Geochronologic Unit—a division of time distinguished on the basis of the rock record as expressed by chronostratigraphic units; not an actual rock unit, but corresponds to the interval of time during which an established chronostratigraphic unit was deposited or formed; thus, the beginning of a geochronologic unit corresponds to the time of deposition of the bottom of the chronostratigraphic unit upon which it is based and the ending corresponds to the time of deposition of the top of the reference unit; the hierarchy of geochronologic units and their corresponding geochronostratigraphic units are:

Geochronologic Unit	Corresponding Geochronostratigraphic Unit
Eon	Eonothem
Era	Erathem
Period	System
Epoch	Series
Age	Stage
Chron	Chronozone

Geochronometric Units—direct divisions of geologic time with arbitrarily chosen age boundaries; they are not based on the time span of designated chronostratigraphic stratotypes; a geochronometric time scale is commonly used for Precambrian rocks, which cannot be subdivided into globally recognized chronostratigraphic units; ages generally expressed in millions of years before the present (Ma) but may be expressed also in thousands of years (Ka) or billions of years (Ga)

Source: *North American Stratigraphic Code and International Stratigraphic Guide* (Salvador, 1994).

system; higher-ranking units are groupings of systems, and lower ranking units are subdivisions of systems. The ranks of chronostratigraphic units do not correspond directly to those of lithostratigraphic or other stratigraphic units. A system of rocks, for example, does not necessarily correspond to a lithostratigraphic group or a series to a formation. A series might include several formations. In principle, chronostratigraphic units have worldwide extent and can be recognized throughout the world. In practice, worldwide use of chronostratigraphic units depends upon the extent to which the time-diagnostic features that characterize these units can be recognized worldwide.

Geochronologic units represent the interval of time during which a correspondingly ranked chronostratigraphic unit was deposited. The fundamental geochronologic unit is thus the **period**—the time equivalent of a system. Names

Table 15.2   The internationally accepted geologic systems of the Phanerozoic and their Type localities

System name	Type locality	Name proposed by	Date proposed	Remarks
Quaternary	France	Jules Desnoyers	1829	Defined by lithology, including some unconsolidated sediment
Tertiary	Italy	Giovanni Arduino	1760	Originally defined by lithology; redefined with type section in France on the basis of distinctive fossils
Cretaceous	Paris Basin	Omalius d'Halloy	1822	Defined initially on the basis of strata composed of distinctive chalk beds
Jurassic	Jura Mountains, northern Switzerland	Alexander von Humbolt	1795	Defined originally on the basis of lithology
Triassic	Southern Germany	Frederick von Alberti	1843	Defined lithologically on the basis of a distinctive threefold division of strata; also defined by fossils
Permian	Province of Perm, Russia	Roderick I. Murchison	1841	Identified by distinctive fossils
Pennsylvanian	Pennsylvania, United States	Henry S. Williams	1891	Not used outside the United States
Mississippian	Mississippi Valley, United States	Alexander Winchell	1870	Not used outside the United States
Carboniferous	Central England	William Conybeare and William Phillips	1822	Named for lithologically distinctive, coal-bearing strata but recognizable by distinctive fossils
Devonian	Devonshire, southern England	Roger I. Murchinson and Adam Sedgwick	1840	Boundaries based mainly on fossils
Silurian	Western Wales	Roger I. Murchinson	1835	Defined by lithology and fossils
Ordovician	Western Wales	Charles Lapworth	1879	Set up as an intermediate unit between the Cambrian and Silurian to resolve boundary dispute; boundary defined by fossils
Cambrian	Western Wales	Adam Sedgwick	1835	Defined mainly by lithology

Note: The Precambrian has not yet been divided into internationally accepted systems.

for periods and lower ranked geochronologic units are identical with those for their corresponding chronostratigraphic units. For example, the Jurassic Period is the time during which the Jurassic System of rocks was deposited. Periods are divided into epochs. Epochs represent the time during which a series was deposited. They take their name from the period by adding the adjective Early, Middle, and Late (e.g., Early Jurassic Epoch, Middle Jurassic Epoch, Late Jurassic Epoch). Note from Table 15.1 the different usage of Lower, Middle, and Upper for subdivision of series, because series are rock units, not units of time. Most names for eons and eras are the same as the names of the corresponding eonothems and erathems.

Geochronometric units are pure time units. They are not based on the time spans of designated chronostratigraphic stratotypes but are simply time divisions of an appropriate magnitude or scale, with arbitrarily chosen boundaries. At this time, a geochronometric time scale is used to express the ages of Precambrian rocks (see Fig. 15.2) because no globally recognized and accepted chronostratigraphic scale has been developed for these rocks. Precambrian rocks have not yet proven generally susceptible to analysis and subdivision by superposition or by application of other lithologic or biologic principles that we commonly use in subdividing the Phanerozoic rocks; however, a chronostratigraphic scale for Precambrian rocks may be developed in the future. Subdivision of Precambrian rocks is further discussed in the next section.

Eonothem	Erathem	System and Subsystem		Series	Numerical Age (Ma)
PHANEROZOIC	Cenozoic	Quaternary		Holocene	0.1
				Pleistocene	1.8
		Tertiary	Neogene	Pliocene	
				Miocene	23.8
			Paleogene	Oligocene	
				Eocene	
				Paleocene	65.0
	Mesozoic	Cretaceous		Upper	
				Lower	144.2
		Jurassic		Upper	
				Middle	
				Lower	206
		Triassic		Upper	
				Middle	
				Lower	248
	Paleozoic	Permian		Upper	
				Lower	290
		Carbon-iferous	Pennsylvanian	Upper	323
			Mississippian	Lower	354
		Devonian		Upper	
				Middle	
				Lower	417
		Silurian		Upper	
				Lower	443
		Ordovician		Upper	
				Middle	
				Lower	490
		Cambrian		Upper	
				Lower	543
PRECAMBRIAN	PROTEROZOIC	Not formally subdivided			2500
	ARCHEAN	Not formally subdivided			

Source of ages: Geological Society of America 1999 Geologic Time Scale.

Figure 15.2
Nomenclature of Phanerozoic chronostratigraphic units commonly used throughout the world. Precambrian rocks are divided into the Archean and Proterozoic; however, no scheme for further subdivision of the Precambrian is globally accepted.

15.3 THE GEOLOGIC TIME SCALE

Purpose and Scope

Classifying rocks on the basis of time involves systematic organization of strata into named units, each corresponding to specific intervals of geologic time. These units provide a basis for time correlation and a reference system for recording and systematizing specific events in the geologic history of Earth. Thus, the ultimate aim of creating a standardized geologic time scale is to establish a hierarchy of chronostratigraphic units of international scope that can serve as a standard reference to which the ages of rocks everywhere in the world can be related. Establishing the relative ordering of events in Earth's history is the main contribution that geology makes to our understanding of time.

A standard geologic time scale should express any age in any place, and it should be understandable, clear, and unambiguous. It should also be independent of opinion and therefore have some objective reference that is accessible. Finally, it should be stable, that is, not subject to frequent change, and it should be agreed to and used internationally in all languages (e.g., Harland, 1978).

Development of the Geologic Time Scale

Chronostratigraphic Scale

Geologists have been working for more than 200 years to develop a systematic scheme for a global time-stratigraphic classification of rock units. This slow process has evolved through two fundamental stages of development:

1. Determining time-stratigraphic relationships from local stratigraphic sections by applying the principle of superposition, supplemented by fossil control and, more recently, radiometric ages.
2. Using these local stratigraphic sections as a basis for establishing a composite international chronostratigraphic scale, which serves as the material reference for constructing a standardized international geologic time scale.

The international chronographic scale has evolved gradually over the past two centuries into its present form (Table 15.1, Fig. 15.2). Figure 15.2 shows the hierarchy of major chronostratigraphic units in general use throughout most of the world. A more detailed chronostratigraphic scale that also shows subseries and stages is given in Appendix D; this chronostratigraphic and geochronometric scale was compiled by Salvador (1985) as part of the COSUNA (Correlation of Stratigraphic Units of North America) project. [Note: The ages of the boundaries between chronostratigraphic units shown in Appendix D may not agree with the more recent age determinations given in the new geologic time scale shown in Figure 15.3.] Some of the provincial stage names commonly used in North America are also shown in Appendix D; however, there now appears to be a general movement among North American stratigraphers to abandon these provincial stage names and adopt the European (global) stages as standards for North America. Stratigraphers in Europe and many other parts of the world have for many years subdivided the Tertiary into two subsystems, the **Paleogene** and the **Neogene**, with the top of the Oligocene Series as the dividing boundary between the two (Fig. 15.2). Geologists in North America have now also adopted this practice. They have likewise adopted the European usage of the **Carboniferous** as a system name, but with subdivision in North America into the **Mississippian** and **Pennsylvanian** subsystems. Other versions of the chronostratigraphic scale exist (e.g., Cowie and Bassett, 1989; Harland et al., 1990) that differ somewhat from

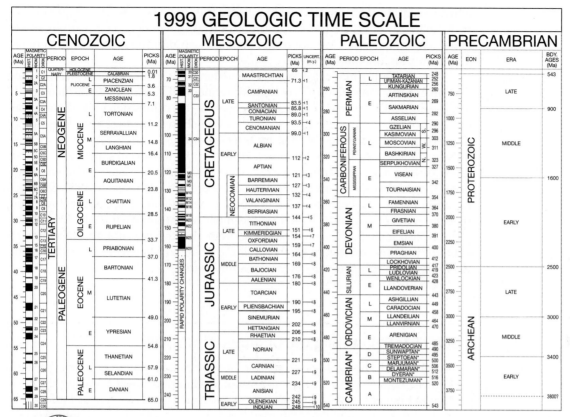

GEOLOGICAL SOCIETY
OF AMERICA

*International ages have not been established. These are regional (Laurentian) only.

Sources for nomenclature and ages: Primarily from Gradstein, F., and Ogg, J., 1996, *Episodes*, v. 19, nos. 1 & 2; Gradstein, F., et al., 1995, SEPM Special Pub. 54, p. 95-128; Berggren, W. A., et al., 1995, SEPM Special Pub. 54, p. 129-212; Cambrian and basal Ordovician ages adapted from Landing, E., 1998, *Canadian Journal of Earth Sciences*, v. 35, p. 329-338; and Davidek, K., et al., 1998, *Geological Magazine*, v. 135, p. 305-309. Cambrian age names from Palmer, A. R., 1998, *Canadian Journal of Earth Sciences*, v. 35, p. 323-328.

Figure 15.3
1999 geologic time scale published by Geological Society of America.

this scale, particularly in naming of series and stages and partial subdivision of the Precambrian into erathems and systems. The geologic community has not yet achieved the ideal of a truly international chronostratigraphic scale that is accepted and used by all geologists worldwide.

Geochronologic (Time) Scale

Figure 15.2 is a chronostratigraphic scale with units and boundaries based on physical divisions of the rock record, but it is not in itself a time scale. To function as a geologic time scale for expressing the age of a rock unit or a geologic event, the chronostratigraphic scale must be converted to a geochronologic scale consisting of units that represent intervals of time rather than bodies of rock that formed during a specified time interval. The geologic time scale is derived from the chronostratigraphic scale by substituting for chronostratigraphic units the corresponding geochronologic units (Table 15.1). Thus, the geologic time scale is expressed in eras, periods, epochs, ages, and chrons rather than erathems, systems, series, stages, and chronozones. The subdivision boundaries of the geologic time scale are calibrated in absolute ages; however, the geologic time scale differs from a true geochronometric scale, which is based purely on time without regard to the rock record. By contrast, the subdivisions of the Phanerozoic time scale are of unequal length, because they are based on chronostratigraphic units that were deposited during unequal intervals of time.

The geologic time scale has been in existence for several decades, and during that time it has continued to evolve, with refinements being made particularly in subdivision of the epochs and ages and absolute-age calibration of the boundaries between periods, epochs, and ages. Figure 15.3 shows the most recent version of the geologic time scale published by the Geological Society of America in 1999. This time scale is subdivided into ages based on the European stages, and boundaries between ages are calibrated in absolute time. Absolute ages are given in millions of years before the present (Ma), where the present refers to 1950. Methods for absolute age calibration of the geologic time scale are discussed below. Note that the magnetic polarity scale for the most recent approximately 160 million years is also included in the time scale. Note also the use of a geochronometric scale for the Precambrian, with the dividing boundary between the Archean and the Proterozoic set arbitrarily at 2500 million years.

Box 15.1 Calibrating the Geologic Time Scale

As mentioned, the geologic time scale has evolved slowly over a long period of time. To develop the scale to its present level of usefulness for fixing the position in time of a particular rock unit or geologic event, two types of information had to be available to stratigraphers: (1) some method of arranging rocks in an orderly succession on the basis of their relative position in time, or relative ages, and (2) a method of determining the ages of the boundaries between rock units on the basis of their absolute position in time with respect to some fixed time horizon, for example, the present.

Placing strata in stratigraphic order in terms of their relative ages has been the guiding principle used by stratigraphers in constructing the geologic time scale. Relative ordering was determined by applying the principle of superposition, aided by use of fossils. The principle of superposition means simply that in a normal succession of strata which have not been tectonically overturned since deposition, the youngest strata are on top and the ages of the strata increase with depth. Most of the divisions in the current global chronostratigraphic scale are based on fossils, and early efforts to create an international chronostratigraphic scale before methods of absolute-age determinations were developed would have been impossible without the use of fossils.

Fortunately, methods are now available not only for determining the relative ages of strata but also for fixing within reasonable limits of uncertainty the absolute ages of some strata. Development of these methods of absolute-age estimation have made it possible to place approximate absolute ages on boundaries of the chronostratigraphic scale initially established by relative-age determination methods. Absolute age data can also be used for determining ages of poorly fossiliferous Precambrian rocks that cannot be placed in stratigraphic order by relative-age determination methods. The principal method for determining the absolute ages of rocks is based on decay of radioactive isotopes of elements in minerals. Other methods of determining the absolute passage of geologic time include counting; lake-sediment varves, which are presumed to represent annual sediment accumulations; growth increments in the shells of some invertebrate organisms; growth rings in trees; and Milankovitch climate cycles in sediments. These alternative methods are useful only for marking the passage of short periods of time in local and regional areas and are not of importance in calibrating the geologic time scale, except possibly some parts of the Pleistocene and Pliocene.

Thus, the major tools for finding ages of sediments to calibrate the geologic time scale are relative-age determinations by use of fossils—biochronology—and absolute age estimates based on isotopic decay—radiochronology. These tools may be used both for calibrating the chronostratigraphic scale directly and for calibrating the succession of reversals of Earth's magnetic field; this succession constitutes the magnetostratigraphic time scale discussed in Chapter 13. We shall now discuss each of these dating methods, beginning with biochronology.

CALIBRATING THE GEOLOGIC TIME SCALE BY USE OF FOSSILS: BIOCHRONOLOGY

Biochronology is the organization of geologic time according to the irreversible process of evolution in the organic continuum (Chapter 14). Useful fossil horizons are more widespread and abundant in Phanerozoic rocks than are horizons whose ages can be estimated by radiochronology. Furthermore, biologic events can commonly be correlated in time more precisely than can radiometric data in all but Cenozoic rocks. Because of these factors, fossils have conventionally provided the most readily available tool for dating and long-distance correlations of Phanerozoic rocks. It is necessary, however, to make a clear distinction between biochronology and biostratigraphy. Biostratigraphy (Chapter 14) aims simply at recognizing the distinctive fossils that characterize a known stratigraphic level in a sedimentary section without regard to the inherent time significance of the fossils. For example, William Smith was able to use fossils very effectively for identifying and correlating strata, even though he had little or no idea of the time relationships or time significance of the fossils. Biochronology, on the other hand, is concerned with the recognition of fossils as having ages that fall at known points in the span of evolutionary time, as measured by fossils of a reference biostratigraphic section. Therefore, by establishing identifiable horizons in reference sections based on fossils, biochronology provides a tool both for international correlation and for worldwide age determination.

The aim of biochronology is to make possible correlation and dating of the geologic record beyond the limits of local stratigraphic sections. To do this most effectively, stratigraphers use features or events in the paleontologic record that are widespread and easily identifiable and that occurred during short periods of geologic time. These events are considered to be biochronologic **datum events** because they mark a particular short period of time in the geologic past. The datum events most commonly used are the immigrations (first appearances) and extinctions (last appearances) of a fossil species or taxon. The first appearance of a species as a result of immigration from another area commonly occurs very rapidly after its initial appearance, owing to evolution from its ancestral morphotype. The first appearance is so rapid, in fact, that geologically speaking we consider speciation and immigration as essentially synchronous events. Extinction of a taxon may also occur very rapidly, although commonly not as rapidly as speciation.

Stratigraphers speak of the first and last appearances of a taxon as the **first appearance datum** (FAD) and the **last appearance datum** (LAD). These FADs and LADs are not totally synchronous owing to the fact that even though immigrations and extinctions can take place quite rapidly, as mentioned, they are not actually instantaneous events. Some planktonic species have been reported to spread worldwide in 100 to 1000 years; however, bioturbation of sediment after deposition can mix fossils through a zone several centimeters thick, and accidents in preservation as well as bias in collection and analytical methods can combine to create uncertainties in the age of the FADs and LADs that can amount to thousands of years. Nevertheless, the duration of the FADs of many planktonic species may be as little as 10,000 years; that is, the ages of the first appearance datum of a species will not vary by more than 10,000 years in different parts of the world (Berggren and Van Couvering, 1978). The error caused by an age discrepancy of this magnitude becomes insignificant when applied to estimation of the ages of rocks that are millions to hundreds of millions of years old. Thus, the FADs and LADs of many fossil species can be considered essentially synchronous for the utilitarian purposes of biochronology.

FADs and LADs are the most easily utilized and communicated types of fossil information upon which to base biochronology, and they can be used across great distances within the range of the defining taxa. Therefore, they have come to dominate global biochronological subdivision. The procedure for establishing the biochronology of any fossil group based on FADs and LADs involves the following steps (described by Haq and Worsley, 1982) and is illustrated graphically in Figure 15.1.1:

1. Identify and locate in local biostratigraphic units the FADs and LADs of distinctive fossil taxa that have wide geographic distribution.

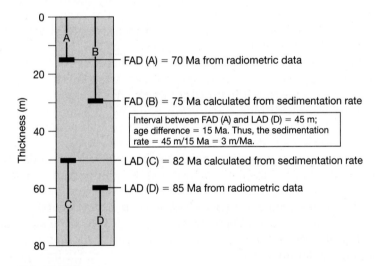

Figure 15.1.1
Schematic illustration of the application of biochronology to age calibration of a local stratigraphic section. The ages of the FAD for Species A and the LAD for Species D are established by radiometric dating of some closely associated physical feature (e.g., an ash bed). The FAD for Species B and the LAD for Species C cannot be dated radiometrically; however, the ages can be calculated from the sedimentation rate determined between FAD (A) and LAD (D). This rate (3 m/Ma) can then be used to determine the age difference between FAD (A) and FAD (B) (3 m/Ma × 15 m = 5 Ma) and between LAD (D) and LAD (C) (3 m/Ma × 10 m = 3 Ma).

2. If possible, assign ages to these events by direct or indirect calibration through radiochronology or magnetostratigraphy. If ages can be assigned to any two events, the sedimentation rates for strata between these events can be calculated by dividing the age difference between the two by the thickness of sediment separating them. The sedimentation rates can then be used to calculate the approximate age of each event enclosed within the dated succession (Fig. 15.1.1).

3. If radiometric or magnetostratigraphic calibration of FADs or LADs in the local section cannot be accomplished, then the ages of the datum levels must be found in a different way. Under these conditions, ages of the FADs and LADs are estimated on the basis of their stratigraphic position with respect to calibrated datum levels of other fossil groups that also occur in the sedimentary succession and whose ages have been found by study of one or more successions elsewhere.

An example of biochronologic calibration is illustrated in Figure 15.1.2, which shows the use of calcareous nannoplankton to establish a biochronology for the Pleistocene by direct correlation with magnetostratigraphic units.

CALIBRATING BY ABSOLUTE AGES: RADIOCHRONOLOGY

General Principles

Radiochronology is based on the principle that radiogenic minerals such as uranium-235 and potassium-40 decay spontaneously at a fixed rate to a "daughter" product. Thus, the age of a radiogenic mineral can be calculated from the measured ratio of parent radionuclide to daughter product in the mineral, by use of the known decay rate of the parent material. The decay rate is commonly expressed as the half-life of the radioactive isotope (i.e., the time required for one-half of the parent material to decay to the daughter product). The number of atoms of the parent radioactive material and the inert daughter product are measured by a **mass spectrometer**, which is an instrument that separates and counts atoms of different masses or charges in a radiogenic mineral such as zircon (Table 15.1.1).

The equation for calculating radiometric age is

$$t = \frac{1}{\lambda} \ln \left[\frac{D - D_0}{N} + 1 \right] \tag{15.1.1}$$

where N is the number of parent atoms of an element (e.g., uranium) present in any given amount of the element, ln is log base e, D is the total number of daughter atoms (e.g., lead), D_0 is the number of original daughter atoms, and λ is the decay constant, which is calculated from the relationship

$$\lambda = \frac{0.693}{T_{1/2}} \tag{15.1.2}$$

where $T_{1/2}$ is the half-life of the radioactive element (Faure, 1986, Chapter 4). N and D are measurable; D_0 is a constant whose value is either assumed or calculated from data for cognetic samples of the same age.

Radiometric Methods

Principal Methods. Some of the most useful radionuclides for estimating absolute ages and the minerals, rocks, and organic materials most suitable for

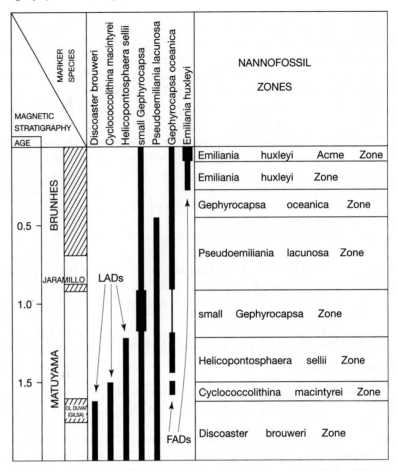

Figure 15.1.2
An example of biochronological dating by use of nannofossil datum
events correlated with magnetic polarity events. [After Gartner, S., 1977,
Calcareous nannofossil biostratigraphy and revised zonation of the Pleistocene: Marine Micropaleontology, v. 2, Fig. 5, reprinted by permission
of Elsevier Science Publishers.]

age determination are shown in Table 15.1.1. The **carbon-14** method is applied
to direct dating of very young sediments. The **protactinium-231** and **thorium-
230** methods are also applied to direct dating of sediments ranging in age to
about 250,000 years. The usefulness and limitations of these methods for direct
dating of sediments are discussed further in succeeding sections.

Most radiometric dating methods cannot be applied to direct dating of
sedimentary rocks. They are used to determine the ages of igneous and metamorphic rocks, which indirectly provide ages for associated sedimentary rocks
(to be discussed). The **potassium/argon method** is widely used because it can
be applied to a number of minerals that are common in igneous and metamorphic rocks, and it gives generally reliable results. It can be used for dating plutonic, igneous, volcanic, and metamorphic rocks (metamorphism resets the
radioactive clock), and even some sedimentary minerals (e.g., glauconite). The
principal problem with the potassium/argon methods is that the decay product, argon-40, is a gas that can leak out of a crystal.

The **argon-40/argon-39** method is a related technique in which potassi-
um-39 is converted to argon-39 by irradiation with fast neutrons in a nuclear

Table 15.1.1 Decay schemes for principal methods of radiometric age determination

Parent nuclide	Daughter nuclide	Half-life (years)	Approximate useful dating range (years B.P.)	Materials commonly used for dating
Carbon-14	*Nitrogen-14	5730	**$<\sim$40,000	Wood, charcoal, $CaCO^3$ shells
Protactinium -231 (daughter nuclide of uranium-235)	*Actinium-227	32,480	$<\sim$150,000	Deep-sea sediment, aragonite corals
Thorium—230 (daughter nuclide of uranium 238/234)	*Radium-226	75,200	$<\sim$250,000	Deep-sea sediment, aragonite corals
Uranium-238	Lead-206	4500 million	10−>4500 million	Zircon, monazite, sphene, uranium/thorium minerals
Uranium-238	Spontaneous fission tracks	—	**$<\sim$65 million	Volcanic glass, zircon, apatite, sphene, garnet
Uranium-235	Lead-207	710 million	10−>4500 million	Zircon, monazite, sphene, uranium/thorium minerals
Potassium-40	Argon-40	1250 million	1−>4500 million	Muscovite, biotite, feldspars, glauconite, whole volcanic rock
Rubidium-87	Strontium-87	48 billion	10−>4500 million	Micas, K-feldspar, whole metamorphic rock, glauconite
Samarium-147	Neodymium -143	106 billion	>200 million	Pyroxene, plagioclase, garnet, apatite, sphene
Lutetium-176	Hafnium-176	35 billion	>200 million	Pyroxene, plagioclase, garnet, apatite, sphene

Half-life data from Bowen (1998).

*Not used in calculating radiometric ages.

**Can be used for dating older rocks under favorable circumstances.

reactor. The ratio of potassium-39 to potassium-40 is known, so argon-39 can serve as a proxy for potassium-40. This relationship permits the potassium determination for a potassium-argon age to be made as part of the argon isotope analysis. In other words, measurement of the amount of argon-39 (which proxies for potassium-40) renders it unnecessary to separate potassium from a mineral and measure the amount of potassium-40. Both argon-39 and argon-40 are measured during the argon analysis. An age can be determined from the argon-40/argon-39 ratio once the conversion rate of argon-40 to argon-39 has been determined by irradiating a standard of known age along with the sample (e.g., Bowen, 1998). The method is so sensitive that very small samples can be used, and it has the further advantage that it allows correction for loss of argon by leakage. Because of these advantages, it is being increasingly used.

Like the potassium/argon method, the **rubidium/strontium method** can also be applied to a number of common minerals; however, it is less commonly used. Rubidium is so rare that a long decay period is required to generate a measurable amount of strontium. The **uranium/lead methods** make use of minerals such as zircon, sphene, and monazite as well as some less common uranium/thorium minerals. These methods give generally reliable ages for older rocks and can be used for dating some rocks as young as about 10 million years. **Fission-track** dating is a technique that relies on counting fission tracks in minerals such as zircon (e.g., Wagner and Van den Haute, 1992). Emission of charged particles from decaying nuclei causes disruption of crystal lattices, creating the tracks, which can be seen and counted under a microscope. The older the mineral the more tracks are present. The **samarium/neodymium** and **lutetium/hafnium** methods are less commonly used dating techniques that may be applied to some rocks that are less amenable to dating by conventional methods. Samarium and lutetium are rare earth elements with long half-lives, making them useful for dating very old (Precambrian) rocks.

Additional, specialized dating methods (e.g., amino-acid racemization method, obsidian hydration method) are available also (Faure, 1986; Geyh and Schleicher, 1990). Details of radiochronologic methods and discussions of errors and uncertainties in radiometric age determinations are available in several published volumes (e.g., Bowen, 1988, 1998; Dickin, 1995; Easterbrook, 1988; Faure, 1986; Geyh and Schleicher, 1990; Mahoney, 1984; McDougall and Harrison, 1988; Odin, 1982; Parrish and Roddick, 1985; Williams, Lerche, and Full, 1988).

Application to Dating Sedimentary Rocks. Although radiochronologic methods can be applied to a variety of rock materials and organic substances (Table 15.1.1), they have limited application to the direct estimation of ages of sedimentary rocks. Most of the potentially usable minerals in sedimentary rocks are terrigenous minerals that when analyzed yield the age of the parent source rock (see Appendix B), not the time of deposition of the sedimentary rock, although a few marine minerals such as glauconite can be used for direct dating of sedimentary rocks. Therefore, much of the geologic time scale has been calibrated by indirect methods of estimating ages of sedimentary rocks on the basis of their relationship to igneous or metamorphic rocks whose ages can be determined by radiochronology. The types of rocks that are most useful for isotopic calibration of the geologic time scale are described in Table 15.1.2. We will now examine in greater detail the most common methods used to find ages of the sedimentary rocks of the international chronostratigraphic scale. These methods are not, of course, restricted to determining the ages of sedimentary rocks that make up the international chronostratigraphic scale. They can be applied to determining the ages of sedimentary rocks in general.

Finding Ages of Sedimentary Rocks by Analysis of Interbedded "Contemporaneous" Volcanic Rocks. Lava flows and pyroclastic deposits such as ash falls can be incorporated very quickly into an accumulating sedimentary succession without significantly interrupting the sedimentation process. Volcanic materials may be erupted onto "soft" unconsolidated sediment and then buried during subsequent, continued sedimentation, leading to a succession of interbedded sedimentary rocks and volcanic rocks that are essentially contemporaneous

Table 15.1.2 Categories of rocks most useful for geochronologic calibration of the geologic time table

Type of rock	Stratigraphic relationship	Reliability of age data
Volcanic rock (lava flows and ash falls)	Interbedded with "contemporaneous" sedimentary rocks	Give actual ages of sedimentary rocks in close stratigraphic proximity above and below volcanic layers
Plutonic igneous rocks	Intrude (cut across) sedimentary rocks	Give minimum ages for the rocks they intrude
	Lie unconformably beneath sedimentary rocks	Give maximum ages for overlying sedimentary rocks
Metamorphosed sedimentary rocks	Constitute the rocks whose ages are being determined	Give minimum ages for metamorphosed sedimentary rocks
	Lie unconformably beneath non-metamorphosed sedimentary rocks	Give maximum ages for the overlying non-metamorphosed sedimentary rocks
Sedimentary rocks containing contemporary organic remains (fossils, wood)		Give actual ages of sedimentary rocks
Sedimentary rocks containing authigenic minerals such as glauconite		Give minimum ages for sedimentary rocks

in age. Thus, estimates of the ages of such associated volcanic rocks also establish the ages of contemporaneous sedimentary rocks.

Ages of whole volcanic rock can be estimated relatively easily by the potassium-argon method, and ages of minerals in these rocks can be determined by the potassium-argon or other methods. Volcanic rocks that occur in association with nearly contemporaneous sedimentary rocks whose ages can also be determined by fossils provide extremely useful reference points for calibration. In fact, establishing the absolute ages of fossiliferous sedimentary rocks by association with contemporaneous volcanic flows whose ages can be radiometrically estimated has probably been the single most important method of calibrating the geologic time scale.

For this method to work, the contemporaneity of the interbedded volcanic and sedimentary rocks must first be established. If a pyroclastic flow such as an ash fall or a lava flow erupts over an older, exposed sedimentary rock surface where erosion is taking place or sedimentation is inactive, the flow is not contemporaneous with the underlying sedimentary rock. The age calculated for such a flow indicates only that the rock below the flow is older and the rock above younger than the flow. A geologist can establish contemporaneity by determining if fossils in sedimentary layers above and below the flow belong to the same biostratigraphic zone or by looking, along the basal contact of the flow unit, for physical evidence that may show that the underlying sediment was still soft at the time of the volcanic eruption. For example, ash fall material may be mixed by bioturbation into underlying sediment, soft

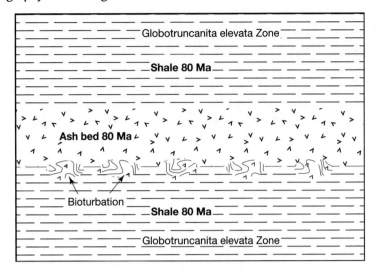

Figure 15.1.3
Diagram illustrating how the contemporaneity of sedimentary rocks
to an associated, datable volcanic layer can be established. The
shale beds below and above the volcanic ash bed belong to the
same Foraminiferal biozone and the base of the ash bed has been
bioturbated, indicating that the underlying sediment was still soft
at the time of the ash fall. Therefore, the shale beds are approximately the same age as the ash bed (80 Ma).

sediment may be mixed into the base of a submarine lava flow, or other such
relationships may exist (Fig. 15.1.3).

Bracketed Ages from Associated Igneous or Metamorphic Rocks. The radiometric ages of igneous rock that are not contemporaneous with associated
sedimentary rocks can be used to estimate the ages of associated sedimentary
rocks if two or more igneous bodies "bracket" the sedimentary unit. In this case,
the age of the sedimentary unit can be established only as lying between those
of the bracketing igneous bodies. The sedimentary unit will be older than an
igneous body that intrudes it, but younger than an igneous body upon which
it rests unconformably (Fig. 15.1.4A). For example, a sedimentary succession
deposited on the eroded, weathered surface of a granite batholith may subsequently be intruded by a dike or a sill. The sedimentary unit is obviously
younger than the batholith but older than the dike or the sill. Unfortunately,
there is no way to determine how much younger or older unless other evidence is available. Because erosional and depositional processes are relatively
slow, the time represented by a bracketed age may be so long as to be of relatively little use in calibrating the geologic time scale. Only a few points on the
time scale have been calibrated by this method.

Metamorphic minerals that develop in sedimentary rocks owing to regional or contact metamorphism can be studied also to provide a method of
bracketing the ages of sedimentary rocks (Fig. 15.1.4B). The radiometric age of
metamorphic minerals gives a minimum age for the metamorphosed sediment; that is, the metamorphosed sedimentary rocks are older than the time of
metamorphism. If a succession of metamorphic rocks is overlain unconformably by nonmetamorphosed sedimentary rocks, the nonmetamorphosed
rocks are obviously younger than the age of metamorphism.

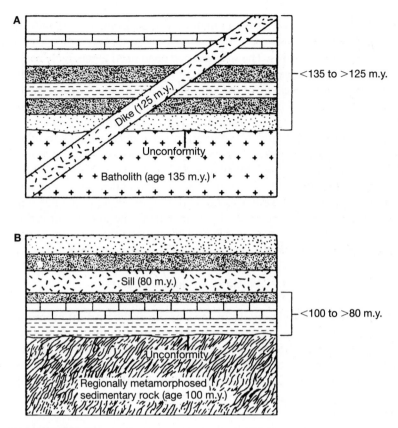

Figure 15.1.4
Determining the ages of sedimentary rocks indirectly by (A) bracketing
between two igneous bodies and (B) bracketing between regionally
metamorphosed sedimentary rocks and an intrusive igneous body.

Direct Radiochronology of Sedimentary Rocks

The calibration methods discussed above allow the estimation of ages of sedimentary rocks only through their association in some manner with igneous or metamorphic rocks whose ages can be determined by radiometric methods. Clearly, the uncertainties involved in finding ages of sedimentary rocks by these indirect methods could be avoided if ages could be estimated directly. As mentioned, terrigenous minerals in sedimentary rocks are not useful for radiochronology because they yield ages for the parent rocks, not the time of deposition of the sediment. The only materials in sedimentary rocks that can be used for direct radiochronology are organic remains (wood, calcium carbonate fossils, and other such remains) that were deposited with the sediment and authigenic minerals that formed in the sediment while still on the seafloor or shortly after burial. The principal methods that have been used for direct radiochronology of sedimentary rocks are (1) the carbon-14 technique for organic materials, (2) the potassium-argon and rubidium-strontium techniques for glauconites, (3) the thorium-230 technique for ocean floor sediments, and (4) the thorium-230/protactinium-231 technique for fossils and sediment.

A short discussion of the advantages and disadvantages of each of these methods follows. For a description of other possible direct dating methods, such as amino-acid racemization and other methods based on radioactive disequilibrium of uranium, thorium, and protactinium, see Geyh and Schleicher (1990).

Carbon-14 Method. The carbon-14 method can be applied to the radiochronology of materials such as wood, peat, charcoal, bone, leaves, and the $CaCO_3$ shells of marine organisms. The method has been used extensively for estimating ages of archaeological materials, but its application in geology is limited to Quaternary geology because of the very short useful age range of the method. Carbon-14 is produced in the atmosphere owing to the impact of cosmic-ray neutrons on ordinary nitrogen-14 atoms. The nitrogen atoms lose a proton and are thus converted to carbon-14, which, in turn, decays backs to nitrogen-14 with a half-life of 5730 years. Carbon-14 is incorporated into carbon dioxide (CO_2), which is assimilated by plants and animals during their life cycles. When organisms die, their tissue no longer assimilates new radioactive carbon; thus, the amount of radiocarbon in the organisms diminishes with time. The age of a sample is determined by measuring the amount of radiocarbon per gram of total carbon in a sample and comparing this amount with the initial amount at the time the organism died. The age equation is

$$t = 19.035 \times 10^3 \log\left(\frac{A_0}{A}\right) years \qquad (15.1.3)$$

where A is the measured activity of the sample at the present moment in disintegrations per minute per gram of carbon (dpm/g) and A_0 is the initial activity (e.g., Bowen, 1998). Burning of trees and fossil fuels in the past few centuries has produced a relative decrease in radioactive carbon in the atmosphere whereas detonation of thermonuclear bombs has caused a slight increase. Corrections must be made for these changes to obtain correct radiocarbon ages.

Because of the short half-life of radioactive carbon, the carbon-14 method is commonly used only for materials less than about 40,000 years old; older materials contain too little carbon-14 to be determined by standard analytical methods. Special techniques that make use of mass spectrometers that allow analysis of smaller amounts of carbon-14, or special proportional counters with high counting efficiencies (e.g., Bowen, 1988), make it possible to extend the usable ages in some cases to as much as 60,000–80,000 years. These special methods are very expensive and have not been widely used in the past. Also, they are exceptionally subject to systematic error because of contamination of samples with young carbon.

The carbon-14 method has been used successfully for such applications as estimating ages of very young sediment in cores of deep-sea sediment and unraveling recent glacial history by analysis of wood in glacial deposits. Its extremely short range renders the method of little value in calibrating the geologic time scale except for very recent Quaternary events.

Radiochronology of Glauconites by Use of Potassium-40/Argon-40 and Rubidium-87/Strontium-87. Radioactive potassium-40 (^{40}K) is incorporated into glauconite grains (green clay minerals composed of complex potassium-aluminum-iron silicates) as they evolve by alteration processes on the seafloor. When the glauconite grains are fully formed, they theoretically become closed systems with respect to gain or loss of potassium or argon; that is, no additional radioactive potassium is taken into the grains and the ^{40}Ar that forms by gradual decay of potassium remains trapped within the glauconite grains (e.g., Odin and Dodson, 1982). Measurement of the ^{40}K/^{40}Ar ratio in the glauconite grains thus allows the age of the grains to be estimated. The half-life of potassium-40 is 1250 million years; therefore, it is theoretically possible to apply the K-Ar method to radiochronology of rocks ranging in age from about one million years (less in some cases) to the age of Earth.

Several workers have reported that glauconite ages tend to be 10–20 percent too young owing to some argon loss. On the other hand, calculated glauconite ages may be too old in some cases owing to the presence of inherited radiogenic argon that was already in sediment at the time the glauconite grains formed. Also, the formation of glauconite grains and their closure to loss of argon do not occur simultaneously with deposition of the enclosing sediment. Glauconite grains, therefore, must yield a slightly younger age than the sediment in which they occur, even if uncertainties about inherited or lost argon are not a problem. Odin and Dodson (1982) suggest that the time required for glauconites to evolve and become closed systems may range up to 25,000 years or more. Thus, in relation to biostratigraphic zonation, the glauconite K-Ar ages are closer to those of fossils in the horizon immediately above the glauconites than to the fossils deposited with the glauconites.

The ages of glauconites can also be estimated by the rubidium-strontium method (Table 15.1.1). Radioactive rubidium (^{87}Rb) is incorporated into glauconites as they form, along with potassium-40. The long half-life of rubidium-87 limits the use of the rubidium-strontium method to radiochronology of rocks older than about 10 million years. Details of the Rb-Sr method as applied to the radiochronology of sedimentary rocks are given by Clauer (1982).

Estimating Ages of Sedimentary Rocks by Use of Other Authigenic Minerals. In addition to glauconite, several other authigenic minerals have been used in direct radiochronology of sedimentary rocks by the K-Ar and Rb-Sr methods. These minerals include clay minerals such as illite, montmorillonite, and chlorite; zeolites; carbonate minerals; and siliceous minerals such as chert and opal. Because of uncertainties about their origin—that is, authigenic or detrital—and time of closure to seawater interactions, none of the clay minerals except glauconite have so far proven to yield reliable ages. Zeolites, carbonate minerals, and siliceous minerals have been used for direct radiochronology of sedimentary rocks with some success, but the overall usefulness and reliability of methods based on these minerals have not yet been adequately investigated.

Thorium-230 and Thorium-230/Protactinium-231 Methods for Estimating Ages of Recent Sediments. Uranium-238 decays through several intermediate daughter products, including uranium-234, to thorium-230. Uranium-238 is fairly soluble in seawater and is present in detectable amounts in seawater. By contrast, the thorium-230 daughter product precipitates quickly from seawater by adsorption onto sediment or inclusion in certain authigenic minerals and becomes incorporated into accumulating sediment on the seafloor. Thorium-230 is an unstable isotope and itself decays with a half-life of 75,000 years to still another unstable daughter product, radium-226. Owing to this fairly rapid decay of ^{230}Th, cores of sediment taken from the ocean floor exhibit a measurable decrease in ^{230}Th content with increasing depth in the cores. If we assume that sedimentation rates and the rates of precipitation of ^{230}Th have remained fairly constant through time, the concentration of ^{230}Th should decrease exponentially with depth. The ages of the sediments at various depths in a core can be calculated by comparing the amount of remaining ^{230}Th at any depth to the amount in the top layer of the core (surface sediment). This method can be applied to the dating of sediments younger than about 250,000 years old, which makes it useful for bridging the gap between maximum carbon-14 ages and minimum K-Ar ages.

Protactinium-231 is the unstable daughter product of uranium-235 and itself decays with a half-life of about 34,000 years to actinium-227. Protactinium-231, like thorium-230, precipitates quickly from seawater and becomes incorporated into sediment along with thorium-230. Because protactinium-231 decays about twice as rapidly as thorium-230, the ^{231}Pa/^{230}Th ratio in the sediments changes with time. Thus, in a sediment core, this ratio is largest in the surface layer of the core and decreases progressively with depth in the core. The age of the sediment at any depth in the core is determined by comparing the ^{231}Pa/^{230}Th ratio at that depth to the ratio in the surface sediment. The reliability of the ages determined by this method rests on the assumption that protactinium-231 and thorium-230 are produced everywhere in the ocean at a constant rate and that the starting ratio of these two isotopes in surface sediment is constant throughout the ocean. (See Faure, 1986, and Bowen, 1998, for details.)

An alternative method for calculating ages of sediment based on protactinium-231 and thorium-230 involves measuring the ratio of these daughter products to their parent isotopes in the skeletons of marine invertebrates such as corals. Dissolved uranium-238 and uranium-235 in seawater are incorporated into corals as they grow, whereas seawater contains no appreciable protactinium-231 and thorium-231, because of the rapid precipitation of these daughter products. Therefore, any protactinium-231 or thorium-230 present in corals results from decay of the parent uranium isotopes within the corals. The ratio of parent isotope to daughter product decreases systematically with time, providing a method for dating the corals. These ratios approach an equilibrium value with increasing passage of time, owing to the fact that the daughter products themselves continue to decay. Thorium-230 reaches a steady state after about 250,000 years and protactinium-231 after about 150,000 years. Thus, these methods can be used only for radiochronology of rocks younger than these ages. Because corals and other skeletal materials tend to recrystallize with burial and diagenesis, the ^{231}Pa/^{230}Th method has severe limitations. Recrystallization may open the initially closed system and allow escape of the daughter or parent isotopes. Therefore, this method cannot be applied to estimating ages of skeletal materials that have undergone recrystallization.

SUMMARY

Radiochronology of sedimentary rocks whose relative positions in the stratigraphic column are already established can be accomplished by several methods. The choice of method depends upon the age of the rocks and the types of materials present in them. In general, calibration of the time scale by estimating ages of volcanic rocks associated with essentially contemporaneous sedimentary rocks that can be easily correlated by marine fossils is the most useful and reliable approach. Radiochronology of sedimentary glauconites or bracketing the ages of sedimentary rocks from associated plutonic intrusive rocks may also yield usable ages—the only ages available in some cases. Therefore, different methods may have to be applied to estimating ages of rocks in each geologic system. Details of the methods used for estimating ages of boundaries between and within the different systems are given in Odin (1982), Snelling (1985a), and Harland et al. (1990).

Figure 15.3 shows the calibration of the Geological Society of America 1999 Geologic Time Scale on the basis of absolute ages obtained from a number of different sources. Readers should be aware, however, that other published geologic time scales have slightly different values for some of these boundaries (e.g., Odin, 1982; Curry et al., 1982; Snelling, 1985b; Harland et al., 1990), indicating differences in opinion about the ages of the boundaries. Calibration of the geologic time table has changed steadily through the years as radiochronologic methods have improved and more absolute ages have become available. Although the ages now used to calibrate the major boundaries of the geologic time scale are unlikely to undergo major revision in the future, it is safe to assume that refinements in these ages will continue for some time.

15.4 CHRONOCORRELATION

Chronostratigraphic units are extremely important in stratigraphy because they form the basis for provincial to global correlation of strata on the basis of age equivalence. We have already established that chronostratigraphic correlation is correlation that expresses correspondence in age and chronostratigraphic position of stratigraphic units. To many geologists, correlation on the basis of age equivalence is by far the most important type of correlation and, in fact, it is commonly the only type of correlation possible on a truly global basis. Methods of establishing the age equivalence of strata by magnetostratigraphic and biologic techniques have already been discussed (Chapters 13, 14). Several other methods of time-stratigraphic correlation are also in common use, including correlation by short-term depositional events, correlation based on transgressive-regressive events, correlation by stable isotope events, and correlation by absolute ages. These methods are discussed below.

Event Correlation and Event Stratigraphy

Event correlation constitutes part of what has come to be known as **event stratigraphy**. Event stratigraphy focuses on the specific events that generate a stratigraphic unit or succession rather than on the physical or biological characteristics of the unit. For example, a eustatic rise in sea level can affect sedimentation patterns worldwide. As a result of this event, sedimentary facies are generated in a variety of environments in various parts of the world. These facies may not be equivalent in terms of their physical characteristics; however, they are equivalent in the sense that they were produced as a result of the same event. Thus, they are chronological equivalents.

Events can be considered to have different scales depending upon their duration (Fig. 15.4), intensity, and geologic effect. Some convulsive events are extraordinarily energetic, occur quickly, and have regional influence (e.g., explosive volcanic eruptions, impact of large extraterrestrial bodies (bolides), great earthquakes, catastrophic floods, large violent storms, large tsunamis). These events may produce widespread effects, including mass extinctions. Because of their magnitude, the deposits of such events may form important parts of the geologic record; in fact, the stratigraphic record tends to overemphasize extraordinary perturbations (Schleicher, 1992). On the other hand, the products of a particular event may not be well enough preserved in the geologic record to be recognized as an event marker (Clifton, 1988), and synchroneity of event deposits from one region to another may not be easily recognized. Other events occur more slowly and produce important stratigraphic successions that may be well preserved and recognized over large areas, such as the rise and fall of sea level that generates a transgressive-regressive stratigraphic succession.

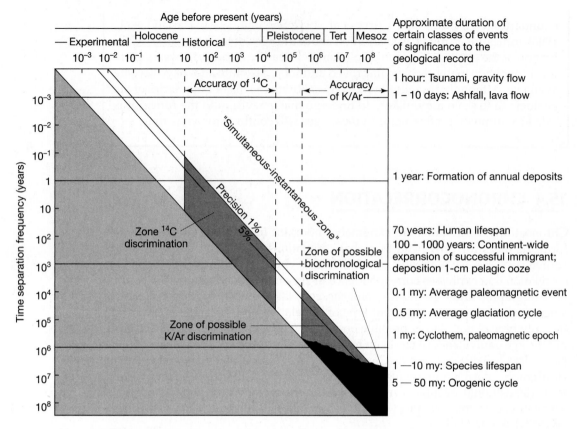

Figure 15.4
Resolving power of geochronologic systems in the Cenozoic on the basis of ^{14}C and K-Ar absolute age discrimination and biochronological discrimination. The vertical axis shows the duration of events ranging from hours to hundreds of millions of years, and the horizontal axis shows age before the present ranging from hours to hundreds of millions of years. Note that ^{14}C dating can resolve events that range in age from tens of years to less than 100,000 years and that are years to tens of thousands of years apart. K-Ar dating can resolve events that are older than 100,000 years and that are separated by at least 10,000 years. Biochronology is most effective in resolving events that are older than about one million years and that are spaced at least one million years apart. [After Berggren, W. A., and J. A. Van Couvering, 1978, Biochronology, *in* G. V. Cohee, M. F. Glaessner, and H. D. Hedberg (eds.), The geologic time scale: Am. Assoc. Petroleum Geologists Studies in Geology 6, Fig. 1, p. 43, reprinted by permission of AAPG, Tulsa, Okla.]

To be useful in chronocorrelation, events should be relatively sudden, thus producing abrupt changes in lithology, chemistry, biology, and/or paleomagnetism that can be recognized. Physical events that meet this criterion include tsunamis, storms, floods, sediment gravity flows, volcanic eruptions, meteorite and comet impacts, rapid sea-level changes, and abrupt reversals of Earth's magnetic field (e.g., Einsele, 1998; Shiki, Chough, and Einsele, 1996). Chemical events, which may be related to physical events, include sudden changes in stable isotope (e.g., oxygen, carbon) content of the ocean and development of anoxic (low-oxygen) conditions in the ocean. Biologic changes such as sudden appearance of new species or sudden extinction of species are also useful events (e.g., Walliser, 1996). Biological events may be related to relatively sudden environmental changes such as major changes in current patterns or to other physical events (e.g., meteorite impacts).

These various types of events are summarized in Figure 15.5. As mentioned, physical, chemical, and biological events generate corresponding event deposits (e.g., a volcanic eruption produces an ash bed). Combining several kinds of event

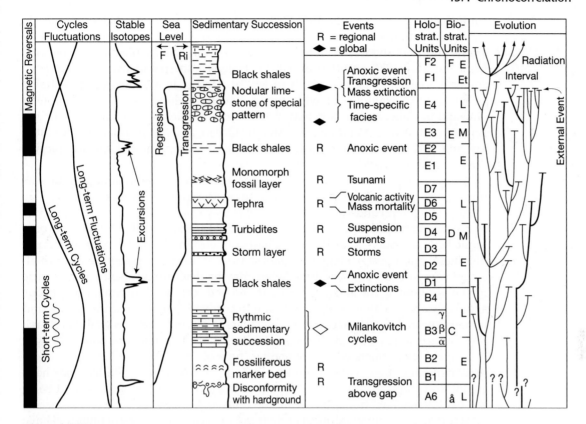

Figure 15.5
Schematic illustration of events and event deposits that are useful in chronostratigraphic correlation. Note that most events are restricted to regions; however, a few, such as anoxic events that cause mass extinctions or mass mortality, may be of global scope. Combining various kinds of events leads to identification of high-resolution stratigraphic units (holostratigraphic units) and biostratigraphic units that have chronostratigraphic significance. Column sea level: F = fall, Ri = rise. Column biostratigraphic units: Et = earliest, E = early, M = middle, L = late. Column evolution: thick lines of species ranges = index species. [After Barnes et al., 1995, Global event stratigraphy, *in* Walliser, O. H. (ed.), Global events and event stratigraphy: Springer-Verlag, Fig. 1, p. 320.]

markers to identify correlatable horizons has come to be known as **high-resolution event stratigraphy** (Kauffman, 1988). Many events are of local or provincial scope; however, some produce event deposits that are globally traceable, holding out the possibility of **global event stratigraphy** (Barnes et al., 1996; Wallister, 1996). Other than paleomagnetic correlation, most worldwide correlation is based on biostratigraphy.

Correlation by Short-Term Depositional Events

Some events produce key beds, or marker beds, that can be traced in outcrop or subsurface sections for long distances. These marker beds are useful for time-stratigraphic correlation, as well as for lithostratigraphic correlation, if they were deposited as a result of a geologic event that took place essentially "instantaneously." The most striking short-term depositional event is ash fall from volcanic eruptions, which can take place in 1 to 10 days (Fig. 15.4). Beds formed from ash falls are called ash layers, tephra layers, bentonite beds (if the ash alters to bentonite clays), or tuff layers. The ash fall from a single eruption may produce ash layers several centimeters thick that can cover thousands to hundreds of thousands of square kilometers. For example, ash from the eruption of Mt. Mazama in southeastern Oregon about 6500–7000 years ago, an eruption that subsequently led to the formation of the Crater Lake caldera, was carried northeastward by winds and deposited as far away as Saskatchewan and Manitoba, Canada. Ash from the May 1980 eruption of Mt. St. Helens also spread over thousands of square kilometers east and north of Mt. St. Helens in Washington and Idaho. Other historic examples of widespread ashfalls include the 1932 eruption of Quizapú in Chile, an eruption that distributed volcanic ash eastward for 1500 km across South America and into the Atlantic Ocean, and the eruption of Perbuatan Volcano at Krakatoa Island, Indonesia, in 1883, an eruption that spread volcanic dust around the world.

Tephra layers make extremely useful reference points in stratigraphic sections. They provide a means for reliable time-stratigraphic correlation if they are of sufficient lateral and vertical extent and if they can be identified as the product of a particular volcanic eruption. Identification of individual ash layers or bentonite beds can often be made on the basis of petrographic characteristics—types of mineral grains, rock fragments, glass shards, or other components—or trace-element composition. Ages of these layers may be determined also by radiometric methods, allowing the layers to be identified and correlated by contemporaneous age. Tephra layers are particularly useful in correlating across marine basins, and it may even be possible to correlate ash layers in marine basins to well-dated lava flows or ash layers on land, thereby extending marine correlations onto land.

Turbidity currents constitute another type of "instantaneous" geologic event that can produce thin, widespread deposits (e.g., Einsele, 1998). Turbidites may have chronostratigraphic significance if a particular turbidite bed, or succession of beds, can be differentiated from other turbidite units and traced laterally. Unfortunately, most turbidites commonly consist of rhythmic or cyclic successions of units that have very similar appearance and are very difficult to differentiate. Thus, in practice, the usefulness of turbidites in time-stratigraphic correlation is rather limited.

Other types of "catastrophic" short-term geologic events include dust storms that produce fine-grained loess deposits on land or silt-sand layers in marine basins. Storms at sea can stir up and transport sediment on the continent shelf to produce thin "storm layers" of sand or silt, as discussed in a preceding chapter.

Slower, noncatastrophic depositional conditions also may generate thin, distinctive, widespread stratigraphic marker beds under some depositional conditions. Deposition of these beds does not necessarily take place "instantaneously." Nevertheless, they can be used for time-stratigraphic correlation if they formed as a result of deposition that took place over a large part of a basin during a relatively short period of time under essentially uniform depositional conditions. For example, changes in ocean circulation patterns may bring about anoxic conditions (Fig. 15.5), leading to widespread deposition of organic-rich black shales. A thin, widespread limestone bed within a dominantly shale or silt succession implies deposition of the limestone under conditions that were in effect essentially simultaneously throughout a geologic province. Such a thin limestone bed within a succession of nonmarine clastic units may represent a brief incursion of marine conditions into a nonmarine environment or the temporary ponding of fresh water to form a large, shallow lake. Thin limestone units in a thick succession of marine clastic deposits may indicate shelf carbonate deposition during brief periods when clastic detritus was temporarily trapped in estuaries or deltaic environments and thus prevented from escaping onto the shelf. By contrast, thin interbeds of sand, clay, or silt in a thick carbonate or evaporite succession may represent temporary incursions of clastic detritus into a carbonate or evaporite basin. Such incursions may be due to a sudden increase in the supply of detritus as a result of tectonic events, periodic flooding on land, or deposition by windstorms or turbidity currents.

Note that chronocorrelation on the basis of physical event stratigraphy re-quires that event beds can be recognized and traced laterally in outcrop or that they can be correlated on the basis of distinctive lithology (lithocorrelation). Be-cause they were produced as a result of an event that took place rapidly, lithocor-relation also results in chronocorrelation. Of course, the actual ages of event beds must be established by the radiometric dating techniques discussed in the preced-ing section.

Biologic events include episodes of punctuated evolution, mass extinctions, mass mortalities (caused, for example, by major ash fall into a basin), and rapid immigration and emigration (Kauffman, 1988; Wallister, 1996). Some of the tech-niques and problems of chronocorrelation by biologic events are discussed in the preceding chapter on biostratigraphy.

Event Correlation Based on Transgressive-Regressive Events

A different approach to event correlation is represented by local correlation based on position within a transgressive-regressive succession or cycle (Ager, 1993b). According to Ager, event correlation in this case is based on the correlation of cor-responding peaks of symmetric sedimentary cycles that are presumed to be syn-chronous. The events represented in this type of correlation are the result of transgressions and regressions that may represent either worldwide, simultane-ous, eustatic changes in sea level or more local changes owing to uplift, subsi-dence, or fluctuation in sediment supply.

The principle of correlation based on transgressive-regressive events is illus-trated in Figure 15.6. The deposits formed during any transgressive-regressive cycle contain one particular time plane that represents the time of maximum in-undation by the sea, that is, the time at which water depth was greatest at any par-ticular locality. Rocks lying stratigraphically below this time plane were deposited during transgression and those above during regression. This time plane can be identified by use of fossil data to determine depth zonation and maximum water depth at various localities, as illustrated in Figure 15.6. The position of the time plane can be established also from lithologic evidence by determining in the verti-cal stratigraphic section at each locality the position within the section where the rocks are symmetrically distributed with respect to the most basinward facies pre-sent. A surface connecting the most basinward rocks in each of the vertical sec-tions defines the approximate position of the time plane and thus the time-stratigraphic correlation between the sections. Figure 15.7 further illustrates the method. Note from this illustration how time-equivalent points on the cycle

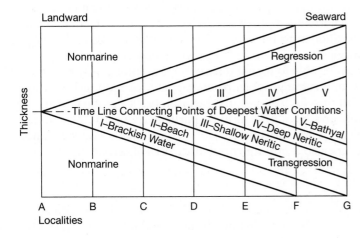

Figure 15.6

Time correlation by position in a transgressive-regressive cycle. The line connecting points of deepest-water condition is a time line. [After Israelski, M. C., 1949, Oscillation chart: Am. Assoc. Petroleum Geologists Bull., v. 33, Fig. 3, p. 98.]

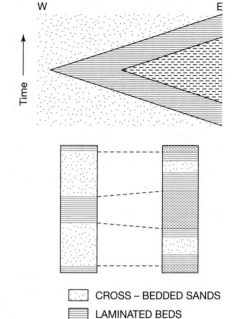

Figure 15.7
Transgressive-regressive cycle sedimentation and event correlation in the Eocene of the Isle of Wight in southern England. [From Ager, D. V., 1993, The nature of the stratigraphical record, 3rd ed., Fig. 7.2, p. 100. Reprinted by permission of John Wiley & Sons Ltd.

are related, resulting in a correlation in which glauconitic clays at the east end of the succession are equated to laminated beds at the west end. Correlation is expressed, as Ager (1993b, p. 101) puts it, in terms of degrees of "marineness." Correlation in this manner can be considered to be a part of sequence stratigraphy (Chapter 13).

Correlation by Stable Isotope Events

Variations in the relative abundance of certain stable, nonradioactive isotopes in marine sediments and fossils, referred to as stable isotope geochemistry (e.g., Valley and Cole, 2001), can be used as a tool for chronostratigraphic correlation of marine sediments. Geochemical evidence shows that the isotopic composition of oxygen, carbon, sulfur, and strontium in the ocean has undergone large fluctuations, or "excursions," in the geologic past—fluctuations that have been recorded in marine sediments. Because the mixing time in the oceans is about 1000 years or less, marine isotopic excursions are considered to be essentially isochronous throughout the world. Variations in isotopic composition of sediments or fossils allow geochemists to construct **isotopic composition curves** that can be used as stratigraphic markers for correlation purposes. To be useful for correlation, fluctuations in isotopic composition must be recognizable on a global scale and must be of sufficiently short duration to show up as a shift on isotopic composition curves. Also, stratigraphers must be able to fix the relative stratigraphic position of these fluctuations in relation to biostratigraphic, paleomagnetic, or radiometric scales. Of the various potentially useful isotopes, oxygen isotopes seem most nearly to meet these requirements and have proven to be particularly useful for chronostratigraphic correlation of Quaternary and late Tertiary sediments. Carbon, sulfur, and strontium isotopes are also useful for correlating rocks of certain ages.

Oxygen Isotopes

The natural isotopes of oxygen are listed in Table 15.3. Most of the oxygen in the oceans occurs as oxygen-16. Oxygen-18 is much rarer (about 0.2 percent of total

Table 15.3 Relative abundances of natural stable
isotopes of oxygen, carbon, sulfur, and strontium

Element	Atomic number	Isotope	Relative Isotopic abundances (%)
O	8	16	99.789
		17	0.037
		18	0.204
C	6	12	98.89
		13	1.11
S	16	32	95.0
		33	0.76
		34	4.22
		36	0.014
Sr	38	84	0.56
		86	9.86
		87	7.02
		88	82.56

Source: *CRC Handbook of Chemistry and Physics.*

oxygen), but it is present in measurable amounts. The ratio of $^{18}O/^{16}O$ in the ocean at any given time in the past is built into contemporaneous marine carbonate minerals and the calcium carbonate shells of marine organisms as a permanent record of the isotopic composition of the ocean at those times. Fluctuations in oxygen isotope ratios in the ocean with time thus show up in the geologic record as fluctuations in the isotopic ratios of these marine carbonates and fossils. Classification of deep-sea sediments on the basis of oxygen isotope ratios in the shells of calcareous marine organisms, particularly foraminifers, has given rise to a new stratigraphy for Quaternary sediments. This stratigraphic method is commonly referred to as **oxygen isotope stratigraphy**. It was first used by Emiliani (1955), who studied the isotopic composition of foraminifers in deep-sea cores and used oxygen isotope ratios to subdivide the core sediments. Oxygen isotope stratigraphy has now developed into a major tool for correlating Quaternary and late Tertiary marine successions, as explained below.

The $^{18}O/^{16}O$ ratio in biogenic marine carbonates reflects both the temperature and the $^{18}O/^{16}O$ ratio of the water in which these carbonates formed. The relationships of ocean paleotemperature (T) to oxygen isotopic composition has been shown by Shackleton (1967) to be

$$T(°C) = 16.9 - 4.38(d_C - d_W) + 0.10(d_C - d_W)^2 \qquad (15.1)$$

where d_C = the equilibrium oxygen isotope composition of calcite and d_W = oxygen isotope composition of the water from which the calcite was precipitated. The d_C and d_W notations refer not to the actual oxygen isotopic abundances in calcite and water but to the per mil (parts per thousand) deviation of the $^{18}O/^{16}O$ ratio in calcite and water from that of an arbitrary standard. A commonly used standard for oxygen isotopes in the past was the University of Chicago PDB standard, where PDB refers to a particular fossil belemnite from the Pee Dee Formation of South Carolina. More commonly now, the isotope composition of ocean water (Standard

Mean Ocean Water, or SMOW) is used as a standard (e.g., Coplen, Kendall, and Hopple, 1983). The per mil deviation from the standard, referred to as $\delta^{18}O$, is expressed by the relationship

$$\delta^{18}O = \frac{[(^{18}O/^{16}O) \text{ sample} - (^{18}O/^{16}O) \text{ standard}]}{(^{18}O/^{16}O) \text{ standard}} \times 1000 \qquad (15.2)$$

Oxygen isotope stratigraphy is based on the fact that $\delta^{18}O$ values in biogenic marine carbonates reflect both the temperature and the isotopic composition of the water from which the calcite precipitates. These factors are both, in turn, a function of the climate. When water evaporates at the surface of the ocean, the lighter ^{16}O isotopes are preferentially removed in the water vapor, leaving the heavier ^{18}O in the ocean. This isotopic fractionation process thus causes water vapor to be depleted of ^{18}O with respect to the seawater from which it evaporates. When water vapor condenses to form rain or snow, the water containing heavy oxygen will tend to precipitate first, leaving the remaining vapor depleted in ^{18}O compared to the initial vapor. Thus, the precipitation that falls near the coast and runs back quickly to the ocean will contain heavier oxygen than that which falls in the interior of continents or in polar regions, where it returns more slowly to the ocean. The $^{18}O/^{16}O$ ratio of precipitates also correlates with the air temperature. The colder the air, the lighter the rain or snow (Odin, Renard, and Grazzini, 1982). For example, the overall average oxygen isotope composition of seawater is –0.28‰ (per mil); however, the precipitation that falls in the crests of the Greenland Ice Sheet is about –35‰ and in relatively inaccessible parts of the Antarctic Ice Sheet it is as negative as –58‰.

The ^{18}O-depleted moisture that falls in polar regions is locked up as ice on land and is thus prevented from quickly returning to the ocean. Because of this retention of light-oxygen water in the ice caps, the ocean becomes progressively enriched in ^{18}O as ^{18}O-depleted ice caps build up during a glacial stage. Marine carbonates that precipitate in the ocean during a glacial stage, particularly biogenic carbonates such as foraminifers, will be enriched in ^{18}O relative to those that precipitate during times when the climate is warmer and ice caps are absent, or are much smaller, on land. Changes in the $\delta^{18}O$ content of biogenic marine calcite thus reflect changes in the volumes of ice on land and concomitant changes in sea level. That is, sea level drops as ice masses build up on land during glacial episodes and rises when continental ice masses melt during interglacial stages. The $\delta^{18}O$ values of seawater track these changes, becoming higher (more positive values) during glacial stages when heavy oxygen is concentrated in the ocean and lower (more negative) during interglacial stages as melting of polar ice caps returns light-oxygen water to the oceans. These principles are illustrated in Figure 15.8.

Decrease in temperature of the seawater in which biogenic calcite precipitates also causes an increase in the $\delta^{18}O$ values that are built into the calcite. Thus, during glacial periods both decrease in temperature of ocean water and changes in isotopic composition of ocean water owing to buildup of ice caps on the continents combine to increase the $\delta^{18}O$ content of biogenic calcites. Conversely, melting of polar ice caps, with consequent return of light-oxygen water to the oceans, and increase in ocean temperature will be reflected in a decrease in $\delta^{18}O$ values in marine biogenic carbonates.

Different kinds of marine organisms tend to incorporate somewhat different ratios of oxygen isotopes into their shells (fractionate oxygen isotopes to different degrees), as indicated in Figure 15.9. Therefore, to evaluate changes in oxygen isotopes in the ocean as a function of time requires that we analyze the same kind of fossil organism in rocks of different ages. Planktonic foraminifers are the most common fossil used in oxygen isotope studies of this kind.

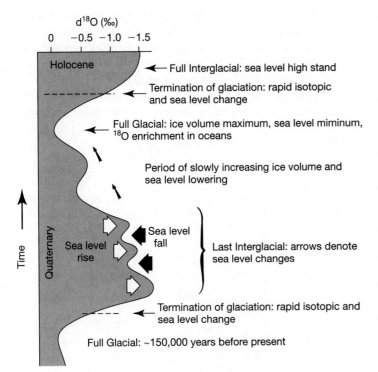

Figure 15.8
Schematic illustration of the relationship between continental glaciation and the $\delta^{18}O$ content of ocean water during the last 150,000 years. Note that $\delta^{18}O$ values become less negative (more heavy oxygen) during stages of continental glaciation and lowered sea level. [After Williams, D. F., I. Lerche, and W. Full, 1988, Isotope chronostratigraphy—Theory and methods: Academic Press, Fig. 30, p. 58. Reproduced by permission.]

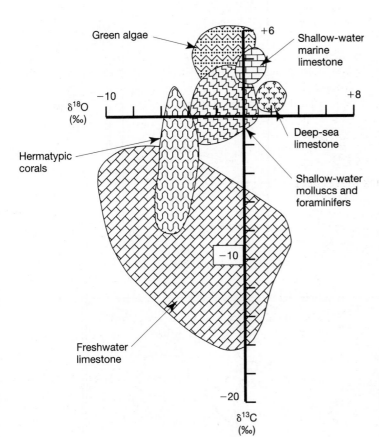

Figure 15.9
Distribution of $\delta^{18}O$ and $\delta^{13}C$ values in various types of marine carbonates. [After Milliman, J. D., 1974, Marine carbonates. Fig. 19, p. 33. Reprinted by permission of Springer-Verlag.]

Oxygen isotopes in sediment cores of late Tertiary to Quaternary age show numerous $\delta^{18}O$ maxima and minima. Figure 15.10 shows an example of oxygen isotope values in three Pleistocene cores from widespread localities in the Pacific and Atlantic oceans (Wei, 1993). The numbers on the isotope curves are oxygen isotope stage numbers (e.g., Kennett, 1982; Ruddiman et al., 1989; Raymo et al., 1989). Where available, the magnetostratigraphic chrons are shown also. Once the isotope stages in a core have been identified and numbered, they can be correlated

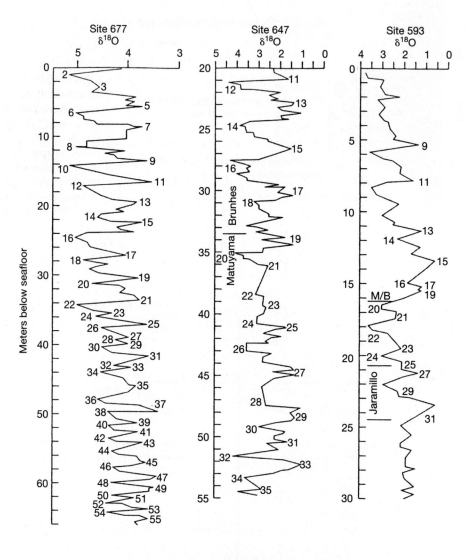

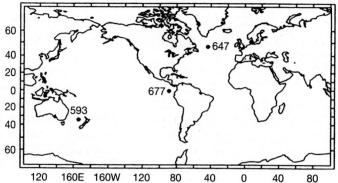

Figure 15.10
Oxygen-isotope stratigraphy of Pleistocene cores from Deep Sea Drilling Project (DSDP) sites in the Pacific and Atlantic oceans. Note the numbered isotope stages, which can be correlated from one core to another. M/B refers to the Matuyama/Brunhes polarity chrons. [After Wei, W., 1993, calibration of upper Pliocene–lower Pleistocene nannofossil events with oxygen isotope stratigraphy: Paleoceanography, v. 8., Fig. 4, p. 91, published by the American Geophysical Union.]

to the same isotope stages in other cores across the world ocean. These isotopic events appear to be related to the Milankovich orbital-climate cycles discussed in Chapter 12. Thus, they provide a record of fourth- and fifth-order cycles that are presumably driven by changes in climate related to changes in Earth's orbital parameters. These orbital changes cause fluctuations in the intensity of solar radiation reaching Earth at different latitudes, which, in turn, result in alternate accumulation and melting of continental ice sheets, producing rise and fall of sea level.

Carbon Isotopes

Carbon-12 and carbon-13 are the nonradioactive isotopes of carbon. Carbon-12 is much more abundant than carbon-13 and makes up most of the carbon in seawater (Table 15.3). The isotopic $^{13}C/^{12}C$ ratio can be expressed in terms of per mil deviation ($\delta^{13}C$) from the PDB standard, just as oxygen isotope ratios are expressed. The $\delta^{13}C$ values in marine carbonates reflect the $^{13}C/^{12}C$ ratio of CO_2 dissolved in deep ocean water; this ratio, in turn, reflects the source of carbon in CO_2. Carbon dioxide dissolves in the ocean by interchange with the atmosphere, and it is generated also by the decay of organic matter that originates both in the ocean and on land. Organisms preferentially incorporate light carbon (^{12}C); therefore, carbon dioxide derived from decaying organic matter is sharply depleted of ^{13}C compared to that derived from the atmosphere. Thus, water runoff from the continents (where soil organic matter is abundant) brings organic-rich waters with low $^{13}C/^{12}C$ ratios into the ocean, significantly lowering the $\delta^{13}C$ content of surface ocean waters near the continents. (Note from Fig. 15.9 the low $\delta^{13}C$ values in freshwater carbonates deposited in lakes.)

Another factor that influences the $\delta^{13}C$ content of ocean water, and thus the $\delta^{13}C$ content in the shells of marine organisms that live in these waters, is the residence time of deep-water masses in the ocean. Carbon-13 is depleted in deep-water masses that have long residence times near the ocean bottom, owing to oxidation of low-$\delta^{13}C$ marine organic matter that sinks from the surface. Oxidation of this low-$\delta^{13}C$ organic matter leads to production of low-$\delta^{13}C$ dissolved bicarbonate (HCO_3^-). Respiration by bottom-dwelling organisms also apparently causes a decrease in $\delta^{13}C$ of deep bottom waters (Kennett, 1982). If low-$\delta^{13}C$ bottom water is later circulated to the surface in some manner, carbonate-secreting organisms will build this low-$\delta^{13}C$ isotope ratio into their shells.

The $\delta^{13}C$ content of ocean water is also related to the primary productivity of photosynthesizing marine organisms (e.g., diatoms). During times of high productivity (large numbers of organisms present), the rate of removal of light carbon (^{12}C) in CO_2 by these organisms is high. Selective removal of the light carbon causes surface ocean waters to be relatively enriched in ^{13}C, resulting in positive values of $\delta^{13}C$. The light carbon is delivered to the seafloor when organisms die, where it is reoxidized (Holser and Magaritz, 1992). During times of low primary productivity, this trend is reversed and surface waters tend to have negative $\delta^{13}C$ values. It has been suggested, for example, that the dramatic decrease in $\delta^{13}C$ values across the Cretaceous-Tertiary boundary (see Fig. 15.11) is the result of marked decrease in marine photosynthesizing organisms owing to a catastrophic meteorite impact (e.g., Hsu et al., 1982).

Because the $\delta^{13}C$ in the calcareous shells of marine organisms is a function of the $\delta^{13}C$ content of the waters in which they live, changes in the $\delta^{13}C$ content of fossil marine organisms indicate changes in ocean water masses. Abrupt decreases in the $\delta^{13}C$ in fossil marine calcareous organisms may reflect changes in primary marine productivity, as suggested, or changes in deep ocean paleocirculation and upwelling patterns that caused low-$\delta^{13}C$ deep waters to spread upward and outward into other parts of the ocean. Or such decreases may reflect changes in

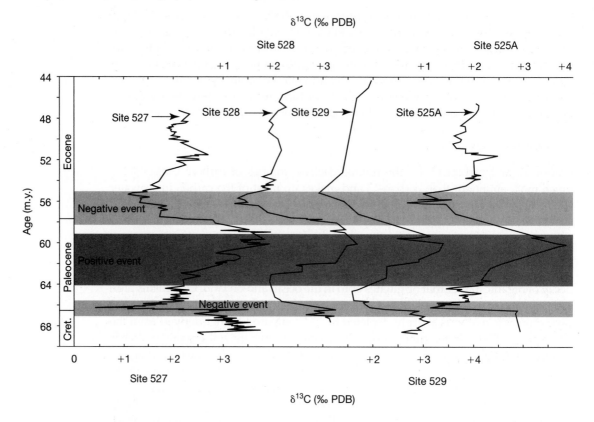

Figure 15.11
Carbon isotope stratigraphy of bulk carbonate sediments from Deep Sea Drilling Project (DSDP) Sites 525A, 527, 528, and 529 in the Atlantic Ocean about 800 km off the coast of Africa. The sites are about 40 to 70 km apart. [After Shackleton, N. J., and M. A. Hall, 1984, Carbon isotope data from Leg 74 sediments, *in* Moore, T. C., Jr., and P. D. Rabinowitz et al., Initial Reports DSDP 74, Washington, U.S. Gov. Printing Office., Fig. 1, p. 614.]

surface circulation patterns that brought low-δ^{13}C surface ocean waters from continental margins into deeper basins. Significant increases in the total biomass produced on the continents during any particular geologic time interval could cause an increase in runoff of low-δ^{13}C to the oceans, an increase that may be reflected also as an episode of low-δ^{13}C surface water. Increased rates of erosion of organic-rich sediments, such as dark shales and limestones, on land could produce much the same effect of increasing runoff of low-δ^{13}C in organic matter to the ocean. These abrupt changes in circulation patterns or organic carbon runoff from the continents may have affected the area of a single ocean basin, such as the Pacific, or in some cases the entire ocean. On the other hand, increased rates of sediment burial in the ocean may have the opposite effect of removing fine organic matter containing low-δ^{13}C from interaction with seawater. This removal would have the effect of increasing δ^{13}C in ocean water. Several increases in δ^{13}C that took place in the middle Miocene between about 12–18 Ma may be associated with exceptionally rapid burial of organic carbon and lowered levels of atmospheric CO_2 (e.g., Woodruff and Savin, 1991).

Inasmuch as changes in the δ^{13}C content of the ocean are reflected in the δ^{13}C content of marine calcareous organisms, these isotopic events or "excursions" can be used for correlation, regardless of their exact causes. For example, several major changes in the Tertiary δ^{13}C record include (1) a pronounced negative δ^{13}C event across the Cretaceous-Tertiary boundary, (2) a positive event from the early

Paleocene into the late Paleocene, and (3) a major $\delta^{13}C$ decrease across the Paleocene-Eocene boundary (Fig. 15.11). There is also a gradual but significant $\delta^{13}C$ deterioration from the mid-Miocene to the Pleistocene (e.g., Williams, Lerche, and Full, 1988, p. 54). Many smaller-scale fluctuations in the Tertiary $\delta^{13}C$ record are also present. These carbon isotopic excursions are essentially synchronous events that can be correlated over wide areas of the ocean in DSDP and ODP cores. Major carbon isotope events are also present in the older sedimentary record. For example, large changes in carbon isotope content of carbonates have been measured at the Precambrian/Cambrian boundary and the Permian/Triassic boundary (Magaritz, 1991).

Sulfur Isotopes

Sulfur has four stable isotopes (Table 15.3); sulfur-32 is the most abundant, followed by sulfur-34. The $^{34}S/^{32}S$ ratio is used in most stratigraphic studies involving sulfur isotopes and is expressed in terms of $\delta^{34}S$, which is per mil deviation of the $^{34}S/^{32}S$ ratio relative to a meteorite standard, troilite (a FeS mineral) from the Canyon Diablo meteorite. Figure 15.12 shows the $\delta^{34}S$ values in various materials relative to the Canyon Diablo standard (CDT).

The major means of sulfur isotope fractionation in the oceans is bacterial reduction of sulfate (SO_4^{2-}) in seawater to sulfides (H_2S, HS^-, HSO_4^-). Bacterial reduction of dissolved seawater sulfate at the sediment-seawater interface causes isotopic fractionation of the sulfate, thus enriching the remaining seawater sulfate in $\delta^{34}S$ by about +20‰ and depletion in the reduced sulfide by about –9‰ (e.g., Schopf, 1980; Canfield, 2001). Precipitation of evaporites from dissolved marine sulfates introduces an additional fractionation ($\sim$+1.65‰), causing the $\delta^{34}S$ of evaporites to be higher than that of dissolved marine sulfates. Other minor factors that can influence the $\delta^{34}S$ content of seawater include oxidation of bacterial H_2S, which produces sulfates depleted of ^{34}S relative to original sulfates, and local emanations of sulfate or sulfide through volcanic activity.

Marine sulfates in the present ocean have a mean $\delta^{34}S$ of about +21‰, however, the $\delta^{34}S$ of ancient marine evaporites ranges from about +10 to +30 (Fig. 15.13). On the basis of sulfur isotope ratios in ancient evaporite deposits, it appears that the sulfur isotope ratios in the surface waters of the world ocean have undergone major changes, or excursions, at various times. These major excursions are characterized by sharp rises in $\delta^{34}S$ in the surface waters of the world ocean followed by significant drops. The early Paleozoic exhibits high $\delta^{34}S$ levels, indicative

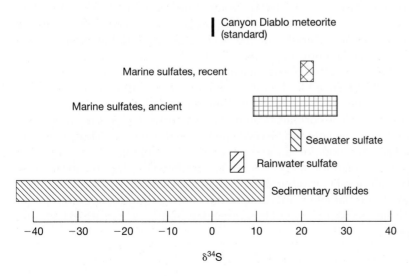

Figure 15.12

$\delta^{34}S$ values of marine sulfates shown relative to that of the Canyon Diablo meteorite. Values for seawater sulfate, rainwater sulfate, and sedimentary sulfides are shown for comparison. [Data from Degens, 1965.]

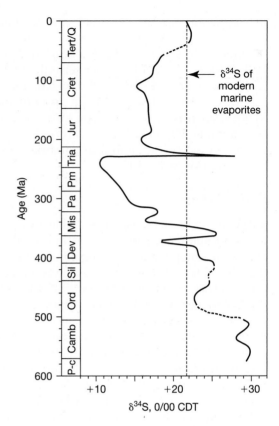

Figure 15.13

Sulfur isotope age curve for Phanerozoic marine evaporites. Dashed portions of the curve indicate lack of data. The dashed vertical line indicates the δ^{34}S composition of marine evaporites in the modern ocean. [After Holser, W. T., M. Magaritz, and J. Wright, 1986, Chemical and isotopic variations in the world ocean during Phanerozoic time, *in* Walliser, O. (ed.), Global bio-events, Lecture Notes in Earth Sciences, v. 8, Fig. 4, p. 69, Springer-Verlag.]

of a high net flux of sulfide from ocean to sediments; δ^{34}S drops to a minimum in the Permian, and it returns to intermediate levels in the Mesozoic (Holser, Magaritz, and Wright, 1986). Many of these sulfur isotope excursions appear to have affected the ocean worldwide.

Chemical events characterized by sharply increased δ^{34}S may be caused by catastrophic mixing of deep ^{34}S-rich brines with surface waters. Brines generated by evaporite deposition are stored in deep basins. Underneath the brines, bacterial reduction of sulfates to form pyrite builds up a store of brine heavy in ^{34}S sulfate. Catastrophic mixing of these ^{34}S-rich brines with surface waters, owing to destruction of the storage basin by tectonism, causes a sharp rise in the δ^{34}S of surface ocean waters and consequently in the evaporite deposits formed from these surface waters (e.g., Holser, 1977; 1984). Gradual decrease in the δ^{34}S of surface ocean waters with time after a catastrophic event is attributed to on-land erosion of predominantly sulfide materials into the ocean in an amount that exceeds evaporite deposition (e.g., Claypool et al., 1980).

In any case, these sulfur isotope excursions constitute a sulfur isotope age curve because each catastrophic chemical event occurred within a very short interval of geologic time. Each major event thus represents a synchronous stratigraphic marker that can be correlated in marine evaporite deposits from one area to another. Some of the events can be correlated on a global basis. Thus, they provide an important method for international chronostratigraphic correlation of evaporite deposits, which commonly cannot be correlated by other means because they do not contain fossils or other datable materials. On the other hand, only a few well-defined, correlatable points are present on the sulfur-isotope curve; thus, sulfur isotopes are less useful overall for correlation purposes than are oxygen and carbon isotopes.

Strontium Isotopes

Strontium has four principal isotopes (Table 15.3), of which ^{87}Sr and ^{86}Sr are of concern here. The relative abundances shown in Table 15.3 are somewhat variable because ^{87}Sr is the daughter product of the radiogenic isotope ^{87}Rb (Table 15.3), as previously discussed. The present-day quantity of ^{87}Sr is thus a fraction of the initial amount of ^{87}Sr in Earth plus the amount of radiogenic ^{87}Sr generated from decay of ^{87}Rb through time (Veizer, 1989).

The amount of strontium in the ocean is derived by three mechanisms and from three sources: (1) leaching of strontium from basaltic rocks owing to hydrothermal circulation at mid-ocean ridges (see discussion in Chapter 1), (2) weathering of continental rocks and delivery to the ocean in river water, and (3) diffusion of strontium from carbonate sediments caused by recrystallization during burial diagenesis (McArthur, 1998). Because the ^{87}Sr/^{86}Sr ratio in each of these sources is different, different ratios of ^{87}Sr/^{86}Sr are thus furnished to the ocean. Strontium in river runoff has the highest ^{87}Sr/^{86}Sr ratio; strontium derived by hydrothermal circulation has the lowest. Weathering and river-runoff furnish the greatest quantity of strontium; hydrothermal circulation at ridge crests is second in importance.

The strontium isotope ratio of ocean water is constant throughout the ocean at any given time but has varied through time owing to variations in strontium contributed by these three processes. Variations in mid-ocean ridge volcanism and the effects of changing world climates on weathering and river flow have exerted the most important controls. Strontium is removed from the oceans by coprecipitation with calcium in calcium carbonate minerals. Therefore, analysis of marine carbonates of various ages allows the strontium isotope composition of the ocean through time to be determined.

Figure 15.14 shows variations of the ^{87}Sr/^{86}Sr ratio of Phanerozoic-age marine carbonates, reflecting variations in these isotope ratios in ocean water since Precambrian time owing to changing proportions of strontium contributed to the ocean from different sources. Note a general decrease in ^{87}Sr/^{86}Sr ratios of the ocean from Precambrian time until the Jurassic, with several pronounced excursions of lowered ratios (e.g., Ordovician, Devonian, Mississippian, Permian/Triassic). A general increase in the ratio since Jurassic time is evident from Figure 15.14.

Strontium isotope data can be reported in terms of deviation from a standard, as in the case of oxygen, carbon, and sulfur isotopes (McArthur, 1998; Veizer, 1989); however, many research results appear to be reported simply as ^{87}Sr/^{86}Sr ratios. Isotope ratios can be used to establish correlation by finding horizons within two different stratigraphic sections that have identical ^{87}Sr/^{86}Sr ratios (Fig. 15.15). Because more than one point on the strontium-isotope curve (Fig. 15.14) can have the same ^{87}Sr/^{86}Sr value, independent evidence must be available (e.g., fossils, magnetostratigraphic data, regional geologic age relationships) to show that the two horizons are approximately of equivalent age. Once correlation has been established, it is possible to determine the numerical age (in millions of years) of the correlation points from the isotopic age curve shown in Figure 15.14 (Faure, 1982; McArthur, 1994, 1998).

Correlation by strontium isotopes has been applied with particularly good results to some Tertiary formations. For example, Depaolo and Finger (1988) use strontium isotopes to correlate marine sediments of the Miocene Monterey Formation of California at resolutions comparable to those of biostratigraphic methods. By correlating the ^{87}Sr/^{86}Sr ratios in these sediments with the strontium isotope vs. time curve for seawater derived from Deep Sea Drilling Program (DSDP) coreholes in the southwestern and central Pacific, these authors determined ages with resolutions ranging from <0.1 to 2.5 Ma. See Bralower et al.

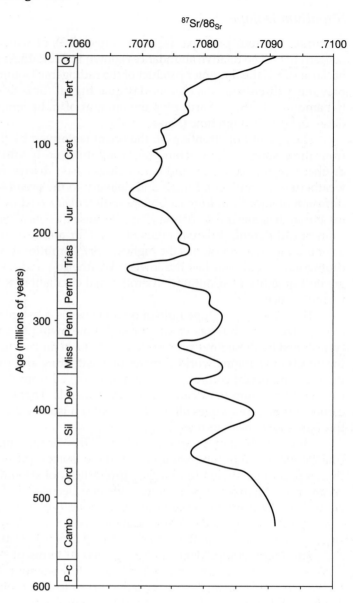

Figure 15.14
Variations of $^{87}Sr/^{86}Sr$ of seawater during Phanerozoic time. [After Veizer, J., 1989, Strontium isotopes in seawater through time: Annual Review Earth and Planetary Sciences, v. 17, Fig. 9., p. 157. Reproduced by permission.]

(1997) for an application to Cretaceous deep-sea sediments recovered during DSDP and Ocean Drilling Program (ODP) coring.

Problems with Isotopic Chronocorrelation

The field of stable isotope geochemistry is very complex; many questions remain with regard to the validity of observed oxygen, carbon, sulfur, and strontium isotope excursions in the geologic record and use of these excursions in chronocorrelation. A discussion of these problems and the details of correlation by stable isotopes is outside the scope of this book. Additional information on this subject is available in several of the publications listed at the end of this chapter.

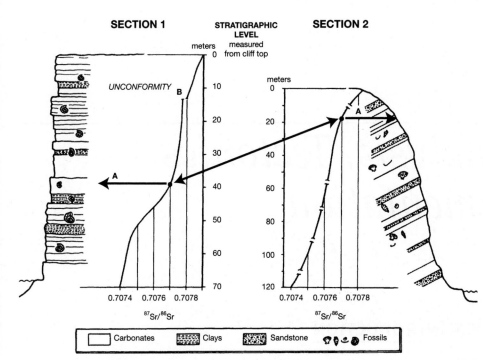

SECTION 1   STRATIGRAPHIC   SECTION 2
                        LEVEL

Figure 15.15
Schematic illustration of correlation by $^{87}Sr/^{86}Sr$ ratios between two widely separated hypothetical stratigraphic sections. If the sections can be established to have approximately equivalent ages on the basis of fossils or other age data, levels within the stratigraphic sections that have identical $^{87}Sr/^{86}Sr$ ratios formed at the same time and therefore correlate. [After McArthur, J. M., 1998, Strontium isotope stratigraphy, *in* Doyle, P., and M. R. Bennett (eds.), Unlocking the stratigraphical record: Advances in modern stratigraphy: John Wiley & Sons Ltd., Fig. 8.3, p. 228.]

FURTHER READING

Berggren, W. A., D. V. Kent, M-P. Aubry, and J. Hardenbol (eds.), 1995, Geochronology, time scales and global stratigraphic correlation: SEPM Spec. Pub. 54, Society for Sedimentary Geology, Tulsa, Okla; 386 p.

Clauer, N., and S. Chaudhuri (eds.), 1992, Isotopic signatures and sedimentary records: Lecture Notes in Earth Sciences 43, Springer-Verlag, Berlin, 529 p.

Clifton, H. E. (ed.), 1988, Sedimentological consequences of convulsive geologic events: Geol. Soc. America Spec. Paper 229, 157 p.

Dickin, A. P., 1995, Radiogenic isotope geology: Cambridge University Press, Cambridge, 452 p.

Faure, G., 1986, Principles of isotope geology, 2nd ed.: John Wiley & Sons, New York, 589 p.

Geyh, M. A., and H. Schleicher, 1990, Absolute age determination: Springer-Verlag, Berlin, 503 p.

Harland, W. B., R. L. Armstrong, A. V. Cox, L. E. Craig, A. G. Smith, and D. G. Smith, 1990, A geologic time scale 1989, 2nd ed.: Cambridge University Press, Cambridge, 263 p.

Hoefs, J., 1997, Stable isotope geochemistry, 4th ed.: Springer-Verlag, Berlin, 201 p.

Odin, G. S. (ed.), 1982, Numerical dating in stratigraphy: John Wiley & Sons, New York, Parts I and II, 1040 p.

Shiki, T., S. K. Chough, and G. Einsele, 1996, Marine sedimentary events and their records: Sedimentary Geology, v. 104, 255 p. [Special Issue]

Snelling, N. J. (ed.), 1985, The chronology of the geological record: Geol. Soc. Mem. 10, Blackwell, Oxford, 343 p.

Valley, J. W., and D. R. Cole (eds.), 2001, Stable isotope geochemistry: Reviews in mineralogy and geochemistry, v. 43, Mineralogical Society of America, 662 p.

Vértes, A., S. Nagy, and K. Süvegh (eds.), 1998, Nuclear methods in mineralogy and geology: Techniques and applications: Plenum Press, New York, 555 p.

Walliser, O. H. (ed.), 1996, Global events and global event stratigraphy in the Phanerozoic: Springer-Verlag, Berlin, 333 p.

Williams, D. F., I. Lerche, and W. E. Full, 1988, Isotope chronostratigraphy: Theory and methods: Academic Press, San Diego, 345 p.

16

Basin Analysis, Tectonics, and Sedimentation

16.1 INTRODUCTION

Preceding chapters of this book introduce the fundamental principles of sedimentology and stratigraphy. I have reviewed in those chapters the basic processes that generate sedimentary rocks; discussed the physical, chemical, and biological properties of these rocks; and described the various environments in which they accumulate. Further, I have emphasized those properties of sedimentary rocks that can be used to subdivide strata into meaningful stratigraphic units on the basis of lithology, seismic reflection characteristics, magnetic polarity, fossil content, and age.

Geologists study sedimentary rocks to develop a critical understanding of their geologic history or to evaluate their economic potential. Effective study requires utilization of all the sedimentological and stratigraphic principles discussed in this book. Because most sedimentary rocks were deposited in basins, it is common to refer to such detailed, integrated study as **basin analysis**. The concept of basin analysis, which goes back to the 1940s, is attributed to one of geology's most famous sedimentologists, Francis Pettijohn (Kleinspehn and Paola, 1988). Basin analysis has been defined in somewhat different ways (e.g., Potter and Pettijohn, 1977, p. 1; Allen and Allen, 1990; Miall, 2000, p. 3; Klein, 1987); however, the common theme that emerges from all of these points of view is that basin analysis is an integrated program of study that involves application of sedimentologic, stratigraphic, and tectonic principles to develop a full understanding of the rocks that fill sedimentary basins for the purpose of interpreting their geologic history and evaluating their economic importance.

During the earlier years of geologic study from about 1860–1960, most geologists regarded marine basins as mainly linear troughs, called **geosynclines**, in which great thicknesses of predominantly shallow-marine deposits accumulated owing to continued subsidence of the geosynclines (see review by Dott, 1974). With the emergence of the plate tectonics concept in the late 1950s and early 1960s, geological thinking shifted away from geosynclines. Today, geologists recognize that there are many kinds of basins and many different mechanisms by which basins form. Under the general rubric of basin analysis, geologists are concerned about the global tectonic controls that create basins and the geologic controls (e.g., sea-level changes, sediment supply, basin subsidence) that govern basin filling. Some of these factors are discussed in preceding chapters.

In this final chapter, attention is focused particularly on the different kinds of basins that we now recognize, the processes that form these basins, the processes that bring about filling of basins, and the nature of the fills. Brief discussions of the methods used to carry out basin analysis and the applications of basin analysis are also included. This chapter rounds out the text coverage by providing an overall perspective regarding the integrated nature of sedimentologic, environmental, and stratigraphic synthesis.

16.2 MECHANISMS OF BASIN FORMATION (SUBSIDENCE)

A sedimentary basin is a depression of some kind capable of trapping sediment. Subsidence of the upper surface of the crust must take place to form such a depression. Mechanisms that can generate sufficient subsidence to create basins include crustal thinning, mantle-lithosphere thickening, sedimentary and volcanic loading, tectonic loading, subcrustal loading, asthenopheric flow, and crustal densification (Dickinson, 1993). A brief explanation of each of these mechanisms is given in Table 16.1.

Note in Table 16.1 that **isostatic compensation** is an important aspect of sedimentary and volcanic loading. The concept of isostasy assumes that local compensation of the crust occurs as if Earth consists of a series of free-floating pistons. Adjacent blocks of crust of different thickness and/or density structure will have different relative relief (Angevine, Heller, and Paola, 1990). Thus, adding a load to the crust (e.g., filling a basin with sediment) causes subsidence; removing a load (e.g., erosion of the crust) causes uplift. It follows from this premise that a basin originally filled with water will be deepened by the sediment load as the basin gradually accumulates sediment. In addition to the effects of loading, flexing of the crust also occurs, to various degrees depending upon the rigidity of the underlying lithosphere, as a result of tectonic forces: overthrusting, underpulling, underthrusting of dense lithosphere. Finally, thermal effects (e.g., cooling of lithosphere, increase in crustal density caused by changing temperature/pressure conditions) may also be important in basin formation.

Table 16.1   Possible mechanisms of crustal subsidence

Crustal thinning:	Extensional stretching, erosion during uplift, and magmatic withdrawal
Mantle-lithospheric thickening:	Cooling of lithosphere following either cessation of stretching or heating due to adiabatic melting or rise of asthenospheric melts
Sedimentary and volcanic loading:	Local isostatic compensation of crust and regional lithospheric flexure, dependent on flexural rigidity of lithosphere, during sedimentation and volcanism
Tectonic loading:	Local isostatic compensation of crust and regional lithospheric flexure, dependent on flexural rigidity of underlying lithosphere, during overthrusting and/or underpulling
Subcrustal loading:	Lithospheric flexure during underthrusting of dense lithosphere
Asthenospheric flow:	Dynamic effects of asthenospheric flow, commonly due to descent or delamination of subducted lithosphere
Crustal densification:	Increased density of crust due to changing pressure/ temperature conditions and/or emplacement of higher-density melts into lower-density crust

Source: Dickinson, 1993; Ingersoll and Busby, 1995.

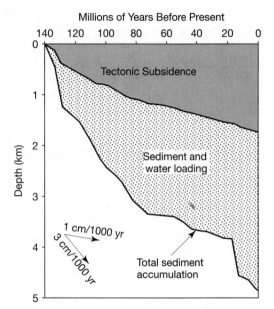

Figure 16.1
Example of a geohistory diagram showing, in this case, burial history of a stratigraphic horizon in COST B-2 Well, Atlantic passive margin of the eastern United States. Note that subsidence owing to tectonism is distinguished from subsidence caused by sediment and water loading. [After Watts, A. B., 1981, The U.S. Atlantic continental margin: Subsidence history, crustal structure and thermal evolution, *in* Bally, A. W., et al., Geology of passive continental margins: AAPG Education Course Note Series 19, Fig. 14, p. 2–33.]

Geologists, particularly petroleum geologists searching for oil and gas, are commonly interested in developing an understanding of the history of basin subsidence to determine the relative importance of the various basin-subsiding mechanisms. Subsidence history is often studied by means of **backstripping**, which attempts to remove the effects of sedimentation in order to analyze the manner in which a basin would have subsided if sediments had not been deposited (e.g., Slingerland, Harbaugh, and Furlong, 1994, p. 12; Watts, 1981). The results of backstripping are generally displayed in **geohistory diagrams**, such as that shown in Figure 16.1. These diagrams allow the effects of sediment loading to be distinguished from those owing to tectonic subsidence. Data required to carry out geohistory analysis include a stratigraphic column showing present-day thickness of stratigraphic units, types of lithologies, ages of stratigraphic horizons, estimated water depths at the sedimentation site at the time of deposition (paleodepth), and sediment porosity (Angevine, Heller, and Paola, 1990). In constructing geohistory diagrams, a correction must be made for sediment compaction. The relative importance of the basin subsidence mechanisms listed in Table 16.1 to the formation of different kinds of sedimentary basins (discussed in the succeeding section) is indicated schematically in Figure 16.2.

16.3 PLATE TECTONICS AND BASINS

I have made numerous references throughout this book regarding the effects of tectonics on sedimentation patterns and stratigraphic characteristics. Plate tectonics provides a first-order control on sedimentation through its influence on the sediment source area. The kinds of basins in which sediments accumulate are also directly related to tectonic processes (e.g., Frostick and Steel, 1993; Ingersoll and Busby, 1995; Miall, 2000, Chapter 7). For example, some basins form as a result of tectonic processes that produce faulting; others are sag basins created by crustal cooling and subsidence or other tectonic processes. In any case, tectonic forces control the size, shape, and location of the basins. Tectonic processes, together with sediment loading, further determine the rate of basin subsidence and thus the space available (accommodation) for sediment accumulation.

Plate-tectonic processes bring about major changes in continental masses and ocean basins through time. Continents break up and drift apart to create

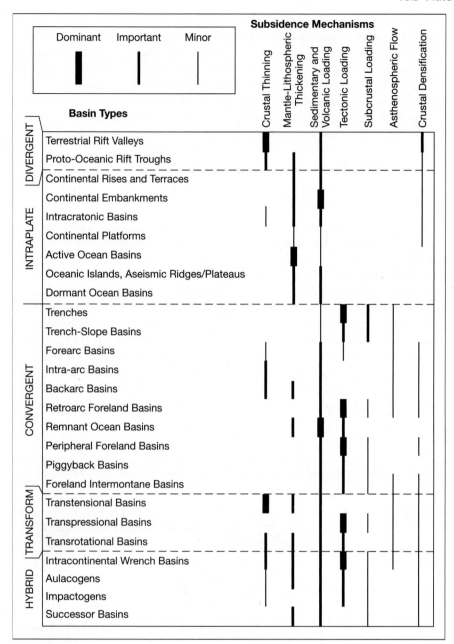

Figure 16.2
Suggested subsidence mechanisms for all kinds of sedimentary basins. [From Ingersoll, R. V., and C. J. Busby, 1995, Tectonics of sedimentary basins, *in* Busby, C. J., and R. V. Ingersoll (eds.), Tectonics of sedimentary basins: Blackwell Science, Cambridge, Mass., Fig. 1.1, p. 8.]

ocean basins as much as 500 km wide, which can subsequently close again as ocean-floor crust is subducted in trenches. The opening and closing of an ocean basin is referred to as a **Wilson cycle** (after Wilson, 1966). Wilson cycles begin with the formation of rift basins (floored by continental crust), which subsequently evolve into proto-oceanic troughs (partially floored by oceanic crust), and eventually into ocean basins, floored by oceanic crust and bordered by passive continental margins. After tens of millions of years or more, subduction zones develop around the ocean margins and the ocean begins to close. Closure culminates with continental collision and the formation of an orogenic belt. The entire process of basin formation and destruction requires perhaps 50 to 150 million years. The geologic record suggests that there have been many Wilson cycles in the history of each continent. Thus, few sedimentary basins remain unchanged with time, or in fixed positions, except perhaps some basins located on cratons well within continental margins.

During the opening phases of a Wilson cycle, tectonic plates are moving apart (by rifting) to form **divergent** (passive) **continental margins**. The closing stages of a Wilson cycle are characterized by plates moving toward each other, as oceanic crust is subducted (consumed) in trenches. Continental margins formed during a stage of ocean closing are called **convergent** (active) **margins**. During opening or closing of an ocean basin, some parts of plates may slide past each other without either diverging or converging. Such a setting is referred to as a **transform margin**. During a Wilson cycle, various kinds of sedimentary basins form in divergent, convergent, and transform settings, as well as in intraplate settings.

16.4 KINDS OF SEDIMENTARY BASINS

We now recognize that the origin of sedimentary basins is related in some way to crustal movements and plate-tectonics processes. Several tectonic classifications for basins have been proposed (e.g., Dickinson, 1974; Bally and Snelson, 1980; Kingston, Dishroon, and Williams, 1983; Mitchell and Reading, 1986; Klein, 1987; Ingersoll, 1988; Ingersoll and Busby, 1995). Ingersoll and Busby (1995) point out that sedimentary basins can form in the four types of tectonic settings discussed above (divergent, intraplate, convergent, transform) as well as in some hybrid settings (Table 16.2). Different kinds of basins are recognized within these various settings on the basis of (1) the type of crust on which the basins rest, (2) the position of the basins with respect to plate margins, and (3) for basins lying close to a plate margin, the type of plate interactions occurring during sedimentation (e.g., Dickinson, 1974; Miall, 2000, p. 468). Detailed discussion of all of these kinds of

Table 16.2   Major kinds of sedimentary basins and their tectonic settings

Divergent Settings		
	Terrestrial rift valleys:	Rifts within continental crust commonly associated with bimodal volcanism. **Modern example: Rio Grand Rift (New Mexico).**
	Proto-oceanic rift troughs:	Incipient oceanic basins floored by new oceanic crust and flanked by young rifted continental margins. **Modern example: Red Sea.**
Intraplate Settings		
	Continental rises and terraces:	Mature rifted continental margins in intraplate settings at continental-oceanic interfaces. **Modern example: East coast of USA.**
	Continental embankments:	Progradational sediment wedges constructed off edges of rifted continental margins. **Modern example: Mississippi Gulf Coast.**
	Intracratonic basins:	Broad cratonic basins floored by fossil rifts in axial zones. **Modern example: Chad Basin (Africa).**
	Continental platforms:	Stable cratons covered with thin and laterally extensive sedimentary strata. **Modern example: Barents Sea (Asia).**
	Active ocean basins:	Basins floored by oceanic crust formed at divergent plate boundaries unrelated to arc-trench systems (spreading still active). **Modern example: Pacific Ocean.**
	Oceanic islands, aseismic ridges and plateaus:	Sedimentary aprons and platforms formed in intraoceanic settings other than magmatic arcs. **Modern example: Emperor-Hawaii seamounts.**

Table 16.2   *(Continued)*

	Dormant ocean basins:	Basins floored by oceanic crust, which is neither spreading nor subducting (no active plate boundaries within or adjoining basin). **Modern example: Gulf of Mexico.**
Convergent Settings		
	Trenches:	Deep troughs formed by subduction of oceanic lithosphere. **Modern example: Chile Trench.**
	Trench-slope basins:	Local structural depressions developed on subduction complexes. **Modern example: Central America Trench.**
	Fore-arc basins:	Basins within arc-trench gaps. **Modern example: Sumatra.**
	Intra-arc basins	Basins along arc platform, which includes superposed and overlapping volcanoes. **Modern example: Lago de Nicaragua.**
	Back-arc basins:	Oceanic basins behind intraoceanic magmatic arcs (including interarc basins between active and remnant arcs), and continental basins behind continental-margin magmatic arcs without foreland fold-thrust belts. **Modern example: Marianas.**
	Retro-arc foreland basins:	Foreland basins on continental sides of continental-margin arc-trench systems (formed by subduction-generated compression and/or collision). **Modern example: Andes foothills.**
	Remnant ocean basins:	Shrinking ocean basins caught between colliding continental margins and/or arc-trench systems, and ultimately subducted or deformed within suture belts. **Modern example: Bay of Bengal.**
	Peripheral foreland basins:	Foreland basins above rifted continental margins that have been pulled into subduction zones during crustal collisions (primary type of collision-related forelands). **Modern example: Persian Gulf.**
	Piggyback basins:	Basins formed and carried atop moving thrust sheets. **Modern example: Peshawar Basin (Pakistan).**
	Foreland intermontane basins (broken forelands):	Basins formed among basement-cored uplifts in foreland settings. **Modern example: Sierras Pampeanas basins (Argentina).**
Transform Settings		
	Transtensional basins:	Basins formed by extension along strike-slip fault systems. **Modern example: Salton Sea (California).**
	Transpressional basins:	Basins formed by compression along strike-slip fault systems. **Modern example: Santa Barbara Basin (California) (foreland).**
	Transrotational basins:	Basins formed by rotation of crustal blocks about vertical axes within strike-slip fault systems. **Modern example: Western Aleutian fore-arc (?).**
Hybrid Settings		
	Intracontinental wrench basins:	Diverse basins formed within and on continental crust owing to distant collisional processes. **Modern example: Quaidam Basin (China).**
	Aulacogens:	Former failed rifts at high angles to continental margins, which have been reactivated during convergent tectonics, so that they are at high angles to orogenic belts. **Modern example: Mississippi Embayment.**

Table 16.2	*(Continued)*	
	Impactogens:	Rifts formed at high angles to orogenic belts, without preorogenic history (in contrast with aulacogens). **Modern example: Baikal Rift (Siberia) (distal).**
	Successor basins:	Basins formed in intermontane settings following cessation of local orogenic or taphrogenic activity. **Modern example: Southern Basin and Range (Arizona).**

Basin classification—modified after Dickinson, 1974, 1976, and Ingersoll, 1988. Source: Ingersoll, R. V., and C. J. Busby, 1995 Tectonics of sedimentary basins, *in* Busby, C. J., and R. V. Ingersoll (eds.), *Tectonics of sedimentary basins*: Blackwell Science, Table 1.1, p. 3, Table 1.2, p. 5. Reproduced by permission.

basins is beyond the scope of this book; however, a few of the more interesting and important types of basins are illustrated schematically in Figure 16.3 and further discussed below. Interested readers should consult Busby and Ingersoll (1995), Einsele (2000), and Miall (2000) for additional insight.

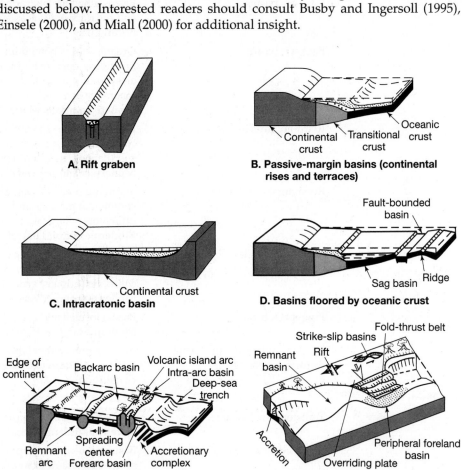

A. Rift graben

B. Passive-margin basins (continental rises and terraces)

C. Intracratonic basin

D. Basins floored by oceanic crust

E. Subduction-related basin

F. Collision-related basins

Figure 16.3
Schematic representation of selected kinds of tectonically formed basins. [After Dickinson and Yarborough, 1976; Kingston, Dishroon, and Williams, 1983; Mitchell and Reading, 1986; Einsele, 1992; Ingersoll and Busby, 1995.]

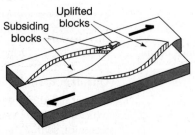

G. Transform-fault-related basin

H. Aulacogen

Basins in Divergent Settings

Divergent tectonic settings are regions of Earth where tectonic plates are separating. These regions are characterized by extensional (stretching) features. Examples of extension include seafloor spreading along mid-ocean ridges and the stretching and downfaulting of continental crust to form grabens. Basins form in divergent settings owing to crustal thinning as well as to sedimentary and volcanic loading and crustal densification (Fig. 16.2).

The early stages of rifting are characterized by breaking of the crust and down dropping of blocks to form fault grabens called **terrestrial rift valleys**. Rifts (Fig. 16.3A) are narrow, fault-bounded valleys that range in size from grabens a few kilometers wide to gigantic rifts such as the East African Rift System, which is nearly 3000 km long and 30–40 km wide. Rifts result from a thermal event of some kind that causes extension or spreading within continental crust. The East African Rift System (Fig. 16. 4A) is an example of a young rift zone. Different stages in the development of the rift are illustrated in Figure 16.4B. The East African Rift is filled mainly with volcanic rocks; however, a great variety of sedimentary environments can exist within rifts, ranging from nonmarine (fluvial, lacustrine, desert) to marginal marine (delta, estuarine, tidal flat) and marine (shelf, submarine fan). Thus, the deposits of rift basins can include conglomerates, sandstones, shales, turbidites, coals, evaporites, and carbonates. Many ancient rift systems are known from Asia, Europe, Africa, Arabia, Australia, North America, and South America (e.g., Sengör, 1995; Leeder, 1995; Ravnås and Steel, 1998). They occur in many tectonic settings (Sengör, 1995) but are particularly common in divergent settings.

As opening of an ocean proceeds, continued extension within continental crust leads to additional thinning of the crust and eventually rupturing, allowing basaltic magma to rise into the axis of the rift and begin the process of forming

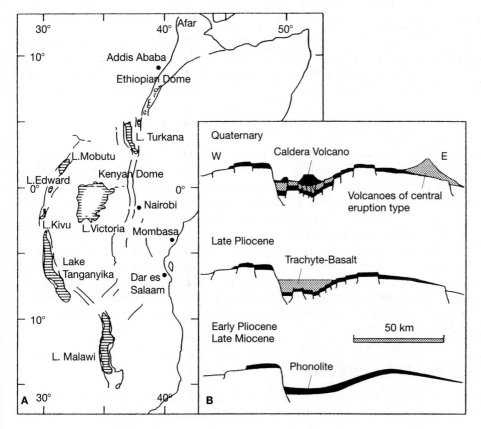

Figure 16.4
Map (A) showing the surface configuration of the East African Rift Zone and cross sections (B) illustrating stages in evolution of the rift from late Miocene through the Quaternary. The rift is floored by volcanic rocks and volcaniclastic detritus. [From Einsele, G., 1992, Sedimentary basins, Fig. 12.4, p. 434. Reprinted by permission of Springer-Verlag, Berlin.]

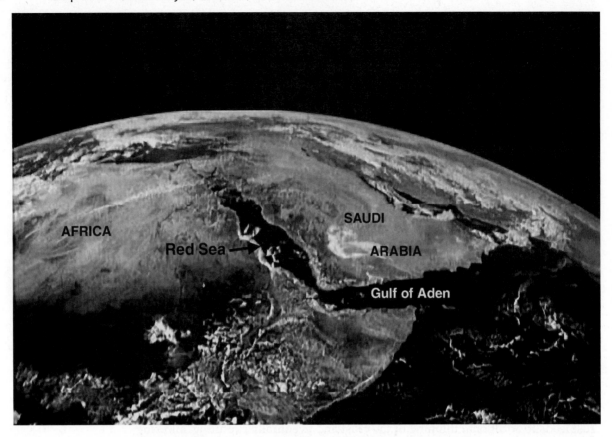

Figure 16.5
View of the Red Sea, lying between northeastern Africa and Saudi Arabia, from Apollo 17 aircraft. NASA photograph, downloaded from the Internet 6/26/04.

new oceanic crust. Thus, terrestrial rift valleys gradually evolve into **proto-oceanic rift troughs**. Proto-oceanic rifts are floored (at least in part) by oceanic crust and flanked by young rifted continental margins. The Red Sea (Fig. 16.5) is the best modern analog of a proto-oceanic rift. The Red Sea, which lies between northeastern Africa and Saudi Arabia, is 2000 km long, is more than 200 km wide in places, and has an axial zone about 50 km wide with some axial deeps that extend to more than 3 km.

The axial region is floored by young (<5 m.y.) oceanic crust in the southern third of the Red Sea. The shelves of the sea are underlain by stretched continental crust in central areas but by an abrupt ocean-continental crust transition in the north (Leeder, 1999, p. 511). To the south, the Red Sea intersects the slow-spreading Gulf of Aden rift. The extension that formed the Red Sea started in middle Tertiary. Early sedimentation that accompanied rifting was characterized by development of marginal alluvial fans and fan deltas, as well as nearshore sedimentation in both siliciclastic and carbonate environments. During Miocene time, significant thicknesses of evaporites were deposited as a result of periodic isolation of the trough. Conditions returned to normal salinities in Pliocene times. Holocene sedimentation was particularly characterized by deposition of foram-pteropod (calcareous) oozes.

Basins in Intraplate Settings

Once an ocean basin has opened fully during a Wilson cycle, sedimentary basins may be present in a variety of settings along the margin and within both continental and oceanic plates. Continental margins formed during opening of an ocean are called **passive margins** (lacking significant seismic activity). Continental crust is commonly thinned on passive margins and a zone of transitional crust is present between fully continental crust and fully oceanic crust (Fig. 16.3B). Thus, sediments may be deposited in settings floored by wholly continental crust, transitional crust, or wholly oceanic crust.

Intraplate Basins Formed on Continental - Transitional Crust

Continental platforms are stable cratons covered by thin, laterally extensive sedimentary strata. Basins developed on these stable platforms are referred to as cratonic basins. They are commonly bowl shaped (ovate), and they are generally filled with Paleozoic and Mesozoic sediments that formed under shallow-water conditions. Sediments can include shallow marine sandstones, limestones, and shales, as well as deltaic and fluvial sediments. The sediments commonly thicken toward the basin centers where they may attain thicknesses of 1000 m or more. The North American craton is an example of a major continental platform marked by numerous cratonic basins (Fig. 16.6). These basins filled with sediment ranging in age from Paleozoic to Mesozoic (Sloss, 1982).

Several different kinds of basins may form in cratonic settings. **Intracratonic basins** (Fig. 16.3C) are broad basins floored by fossil rifts in axial zones. They are relatively large, commonly ovate downwarps that occur within continental interiors away from plate margins. Subsidence in intracratonic basins may be due largely to mantle-lithosphere thickening and sedimentary or volcanic loading (Fig. 16.2), however, several other causes have also been proposed, such as crustal thinning (e.g., Klein, 1995). The presence of fossil rifts beneath intracratonic basins, such as the Michigan Basin, suggests some crustal thinning and possibly crustal densification. Some intercratonic basins are filled with marine siliciclastic, carbonate, or evaporite sediment deposited from epicontinental seas; others contain nonmarine sediments. Ancient North American intracratonic basins include the Hudson Bay Basin (Canada), Michigan Basin, Illinois Basin, and Williston Basin (Fig. 16.6, 16.7). Ancient intracratonic basins on other continents include the Amadeus Basin of Australia; the Parana Basin in southern Brazil, Paraguay, northeast Argentina, and Uruguay; the Paris Basin in France; and the Carpentaria Basin in Australia

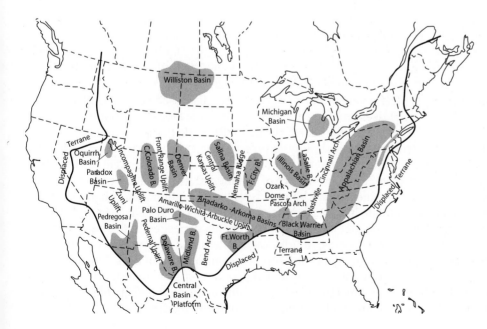

Figure 16.6

Late Mississippian–Early Jurassic cratonic basins on the North American (mainly USA) craton. The Michigan, Illinois, and Williston Basins are intractatonic basins. [After Sloss, 1982, The Midcontinent provinces: United States, *in* Palmer, A. R. (ed.), Perspectives in regional geological synthesis: D-NAG Special Publ. 1, Geological Society of America, Fig. 3, p. 36. Reproduced by permission.]

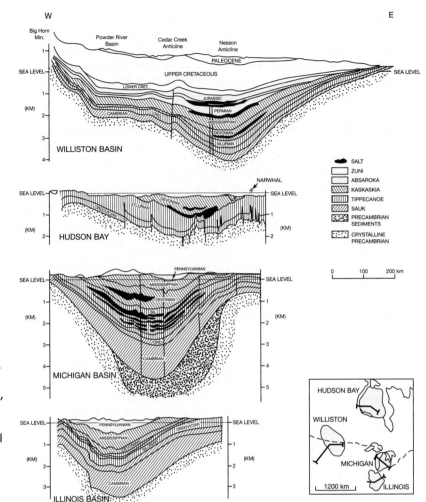

Figure 16.7
Schematic cross section of four intracratonic basins in North America. The locations of the cross sections are shown in the inset map. The terms Zuni, Absaroka, Kaskaskia, Tippecanoe, and Sauk refer to depositional sequences identified on the craton (Sloss, 1982). [From Leighton and Kolata, 1990, Selected interior cratonic basins and their place in the scheme of global tectonics, *in* Leighton, M. W., et al. (eds.), Interior cratonic basins: AAPG Memoir 51, Fig 35.9, p. 743. Reproduced by permission.]

(Klein, 1995; Leighton et al., 1990). The Chad Basin in Africa is an example of a modern intracratonic basin.

Not all basins that form on cratons are intracratonic basins, as defined in Table 16.2. Some of the North American basins shown in Figure 16.6, for example, formed by mechanisms other than rifting. The Paradox basin formed by strike-slip (compressional) processes. Several other basins (e.g., Oquirrh, Denver, Appalachian) are foreland basins, whose origins are related to collisional (compressional) events (G. D. Klein, personal communication, 2004). The Anadarko Basin may be an aulacogen. [These various kinds of basins are discussed subsequently.]

Continental rises and terraces are features characterized by enormous wedges of sediment bounded on the seaward side by the gently sloping continental slope and rise. A structural discontinuity is present beneath the terrace-rise system between normal continental crust and modified or transitional crust (Fig. 16.3B). These rises and terraces are the consequence of continental rifting within passive margins initiated along divergent plate boundaries (Bond, Kominz, and Sheridan, 1995). Sediments accumulate in several parts of the terrace-rise system—shelf, slope, and continental rise at the foot of the slope. Sediments deposited in this setting can include shallow neritic sands, muds, carbonates, and evaporites on the shelf; hemipelagic muds on the slope; and turbidites on continental rises. Thick prisms of sediment may accumulate owing to long-continued subsidence, which may be caused by deep crustal metamorphism (causing increase in density of lower crustal rocks), crustal stretching and thinning, and sediment loading.

Sedimentation on continental terraces and rises occurs after continental rifting is completed and a new basin has begun to form by seafloor spreading (Bond, Kominz, and Sheridan, 1995). The basin is locked into a relatively stable interplate position at the edge of the rifted continent. Good examples of such basins are the basins located off the eastern United States and the southeastern Canada coast (Blake Plateau Basin, Carolina Trough, Baltimore Canyon Trough, Georges Bank Basin, Nova Scotian Basin; Figure 16.8) which were created in late Triassic to early Jurassic time by rifting accompanying the breakup of Pangaea (Manspeizer, 1988). Some of these basins were isolated from the sea and accumulated thick deposits of arkosic clastic sediments and lacustrine deposits, intercalated with basic volcanic rocks. Others, with some connections to the sea, accumulated deposits ranging from evaporites to deltaic sediments, turbidites, and black shales. Figure 16.9 shows some of the sediments in the Baltimore Canyon Trough. Other examples of terrace-slope basins include the Campos Basin, Brazil; the northwest shelf of Australia; and the sedimentary basins of Gabon on the west coast of Africa (Edwards and Santogrossi, 1990). Some terrace-slope basins are prolific producers of petroleum and natural gas.

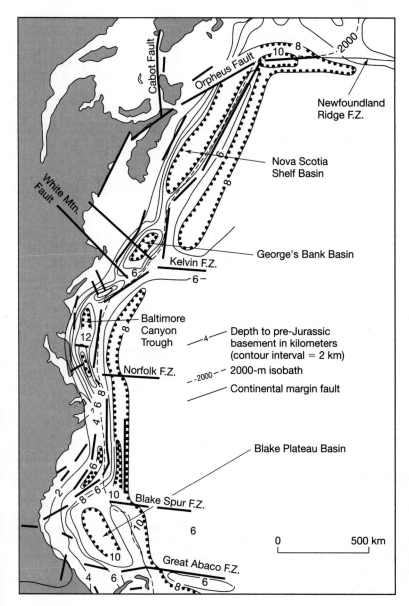

Figure 16.8
Sedimentary basins on the passive continental margin of eastern North America. F.Z. = fracture zone. [After Sheridan, R. E., 1974, Atlantic continental margin of North America, *in* Burk, C. A., and C. L. Drake (eds.), The geology of continental margins, Fig. 12, p. 403. Reprinted by permission of Springer-Verlag, New York.]

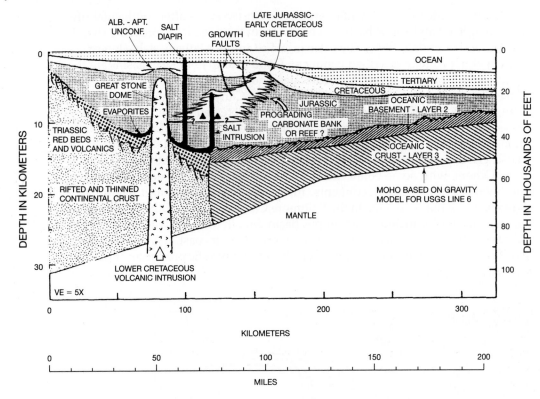

Figure 16.9
Interpretative stratigraphic section across the Atlantic continental margin of North America in the vicinity of Baltimore Canyon Trough. Based on geophysical data and drill-hole information. [After Grow, J. A., 1981, Structure of the Atlantic margin of the United States, *in* Geology of passive continental margins: Am. Assoc. Petroleum Geologists Ed. Course Note Ser. No. 19, Fig. 13, p. 3–20. Reprinted by permission of AAPG, Tulsa, Okla.]

Intraplate Basins Floored by Oceanic Crust

These oceanic basins (e.g., Fig. 16.3D) occur in various parts of the deep ocean floor. They are created by rifting and subsidence accompanying opening of an ocean owing to continental rifting. Oceanic basins may include ocean-floor sag basins as well as fault-bounded basins associated with ridge systems. Sediments that accumulate in these basins are mainly pelagic clays, biogenic oozes, and turbidites. Sediments deposited in oceanic basins adjacent to active margins may eventually be subducted into a trench and consumed during an episode of ocean closing. Alternatively, they may be offscraped in trenches during subduction to become part of a subduction (accretionary) complex (Fig. 16.3E). The Pacific Ocean is a modern example of a major active ocean basin. The Gulf of Mexico is a modern example of a dormant ocean basin, which is floored by oceanic crust that is neither spreading nor subducting.

Basins in Convergent Settings

Subduction-Related Basins

Subduction-related settings (Fig. 16.3E) are features of seismically active continental margins, such as the modern Pacific Ocean margin. These settings are characterized by a deep-sea trench, an active volcanic arc, and an arc-trench gap separating the two. The most important depositional sites in subduction-related settings are deep-sea **trenches, fore-arc basins** that lie within the arc-trench gap

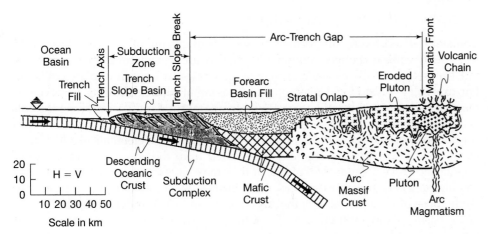

Figure 16.10
Schematic representation of basin structure in the trench and fore-arc zone of a subduction setting. [After Dickinson, W. R., 1995, Forearc basins, *in* Busby, C., and R. V. Ingersoll (eds.), Tectonics of sedimentary basins: Blackwell Science, Cambridge, Mass., Fig. 6.1, p. 221.]

(Fig. 16.10), and **back-arc**, or marginal, basins that lie behind the volcanic arc in some arc-trench systems (Underwood and Moore, 1995; Dickinson, 1995; Marsaglia, 1995). Subduction-related settings may occur also along a continental-margin arc (not shown in Fig. 16.3) rather than an oceanic arc. In these continental-margin settings, so-called **retro-arc** basins (intermontane basins within an arc orogen) may lie on continental crust behind fold-thrust belts (e.g., Jordan, 1995). Sediments deposited in subduction-related basins are mainly siliciclastic deposits derived largely from volcanic sources in the volcanic arc. These deposits include sands and muds deposited on the shelf and muds and turbidites deposited in deeper water in slope, basin, and trench settings. Sediments in the trench may include terrigenous deposits transported by turbidity currents from land, together with sediments scraped from a subducting oceanic plate, which together form an accretionary complex (Fig. 16.3E; Fig. 16.10). The most characteristic kind of rock in an accretionary wedge is **melange**, a chaotically mixed assemblage of rock consisting of brecciated blocks in a highly sheared matrix.

Examples of modern trench and fore-arc sedimentation sites include the Sundra, Japan (Fig. 16.11), Aleutian, Middle America, and Peru-Chili arc-trench systems (Leggett, 1982; Dickinson, 1995; Underwood and Moore, 1995). Examples of ancient fore-arc basins include the Great Valley fore-arc basin, California; Oregon Coast Range; Tamworth Trough, Australia; Midland Valley, Great Britain; and Coastal Range, Taiwan (Dickinson, 1995; Ingersoll, 1982). The Japan Sea is a good modern example of a back-arc basin (Fig. 16.11); the Late Jurassic–Early Cretaceous basin formed behind the Andean arc in southernmost Chile is a well-studied example of an ancient back-arc basin (Marsaglia, 1995). The Taranaka Basin, New Zealand, and the Magdalena Basin, Columbia, both petroleum producers, are additional examples of active-margin basins (Biddle, 1991). For further insight into subduction-related settings, see Busby and Ingersoll (1995).

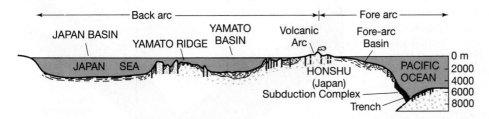

Figure 16.11
Schematic representation of an active continental margin (Japan), showing both the fore-arc and back-arc characteristics of the margin. [From Boggs, S., Jr., 1984, Quaternary sedimentation in the Japan arc-trench system: Geol. Soc. America Bull., v. 95, Fig. 2, p. 670.]

Basins in Collision-Related Settings

Collision-related basins are formed as a result of closing of an ocean basin and consequent collision between continents or active arc systems, or both. Figure 16.3F illustrates some of the basins that may be generated as a result of plate collision. For example, collision can generate compressional forces, resulting in development of fold-thrust belts and associated **peripheral foreland basins** along the collision suture belt where rifted continental margins have been pulled into subduction zones. Figure 16.12 illustrates the fundamental elements of a foreland-basin system. Foreland basins may be isolated from the ocean and receive only nonmarine gravels, sands, and muds, or they may have an oceanic connection and contain carbonates, evaporites, and/or turbidites. Examples of foreland basins include those of western Taiwan, the Alpennines, and eastern Pyrenees; the Magallanes Basin at the southern tip of South America; basins of the northwestern Himalayas; and various basins in the Appalachians, Rocky Mountains, and western Canada (Allen and Homewood, 1986; Macqueen and Leckie, 1992; Dorobek and Ross, 1995).

Because of the irregular shapes of continents and island arcs, and the fact that landmasses tend to approach each other obliquely during collision, portions of an old ocean basin may remain unclosed after collision occurs. These surviving embayments are called **remnant basins**. Modern remnant basins include the Mediterranean Sea, Gulf of Oman, and northeast South China Sea. The Marathon Basin, Texas, provides an example of Pennsylvanian-age sedimentation in an ancient remnant basin adjacent to a fore-arc basin (Fig. 16.13). Structural weaknesses developed in this region in the late Precambrian/early Cambrian and were reactivated in the late Paleozoic as reverse faults in response to compressional stresses (Wuellner, Lehtonen, and James, 1986). An early phase of sedimentation filled part of the fore-arc basin with volcaniclastic detritus. Subsequently, sediments of the Tesnus Formation accumulated in the fore-arc and remnant basin. Later deposition of the Dimple Limestone and Haymond Formation (not shown in Fig. 16.13) generated a total of more than 3400 m of Pennsylvanian sediment in the basin. Sediments include sandstones, shales, and limestones deposited in environments ranging from shelf/platform to submarine fan (turbidite) settings. Other ancient examples of remnant basins include the southern Uplands of Scotland (Silurian-Devonian); Nevadan orogenic belt, California (Jurassic); western Iran (Cretaceous-Paleogene);

Figure 16.12
Schematic illustration of the fundamental elements of an orogen-foreland-basin system: a compressional orogen and thrust belt and the foreland basin in which erosion, sediment transportation, and deposition take place. The basin may be filled to different degrees along the strike zone depending upon the relative rates of mass flux into the orogen, denudation and sedimentation by surface processes, isostatic compensation, and eustatic changes in sea level. [After Johnson, D. D., and C. Beaumont, 1995, Preliminary results from a planform kinematic model of orogen evolution, surface processes and the development of clastic foreland basin stratigraphy, *in* Dorobek, S. L., and G. M. Ross (eds.), Stratigraphic evolution of foreland basins: SEPM Spec. Publ. 52, Fig. 1, p. 4. Reproduced by permission.]

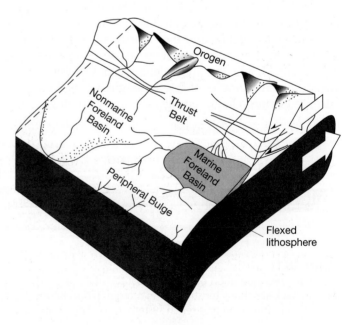

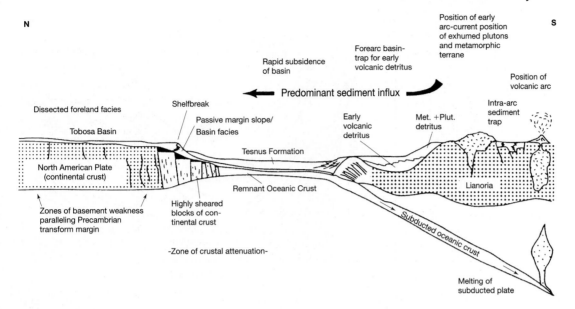

N S

Position of early
arc-current position
of exhumed plutons
and metamorphic
terrane

Forearc basin-
trap for early
volcanic detritus

Rapid subsidence
of basin

Position of
volcanic arc

← Predominant sediment influx →

Dissected foreland facies

Shelfbreak

Intra-arc
sediment
trap

Tobosa Basin

Passive margin slope/
Basin facies

Early
volcanic
detritus

Met. +Plut.
detritus

Tesnus Formation

North American Plate
(continental crust)

Remnant Oceanic Crust

Lianoria

Zones of basement weakness
paralleling Precambrian
transform margin

Highly sheared
blocks of con-
tinental crust

Subducted oceanic crust

-Zone of crustal attenuation-

Melting of
subducted plate

Figure 16.13

Cross-sectional diagram showing the remnant basin and associated basins that existed in the Marathon Basin, Texas, during deposition of Carboniferous sedimentary rocks. [From Wuellner, E. E., L. R. Lehtonen, and W. C. James, 1986, Sedimentary tectonic development of the Marathon and Val Verde basins, West Texas, U.S.A.: A Permo-Carboniferous migrating foredeep, *in* Allen, P. A., and P. Homewood (eds.), Foreland basins: Internat. Assoc. Sedimentologists Spec. Pub. 8, Fig. 5, p. 354. Reproduced by permission of Blackwell Scientific Publications, Oxford.]

and the northeast Caribbean (Tertiary). These basins, and other examples, are discussed by Ingersoll, Graham, and Dickinson (1995).

Basins in Strike-Slip/Transform-Fault-Related Settings

Strike-slip-related basins occur along ocean spreading ridges, along the transform boundaries between some major crustal plates, on continental margins, and within continents on continental crust. Movement along strike-slip faults can produce a variety of pull-apart basins, only one kind of which is illustrated in Figure 16.3G. Faults that define strike-slip basins may be either "transform faults" that define plate boundaries and penetrate the crust or "transcurrent faults," which are restricted to intraplate settings and penetrate only the upper crust (Sylvester, 1988). Most basins formed by strike-slip faulting are small, a few tens of kilometers across, although some may be as wide as 50 km (Nilsen and Sylvester, 1995). They may show evidence of significant local syndepositional relief, such as the presence of fault-flank conglomerate wedges. Because strike-slip basins occur in a variety of settings, they may be filled with either marine or nonmarine sediments, depending upon the setting. Sediments in many of these basins tend to be quite thick, because of high sedimentation rates that result from rapid stripping of adjacent elevated highlands, and may be marked by numerous localized facies changes.

As shown in Table 16.2, basins in transform settings are referred to as **transtensional, transpressional,** or **transrotational** depending upon whether the basins formed by extension, compression, or rotation of crustal blocks along a strike-slip fault system. The Ridge Basin, California (Fig. 16.14), is a good example of an ancient transpressional basin. Strike-slip movement on the San Gabriel fault in Pliocene/Miocene time created a lake basin about 15 km by 40 km in which up

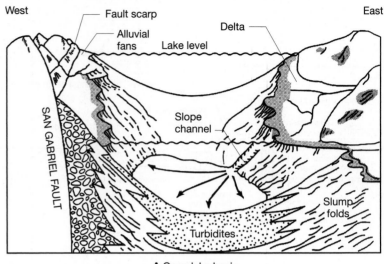

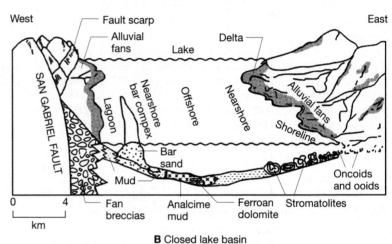

Figure 16.14
Paleoenvironmental reconstruction of the Pliocene/Miocene Ridge Basin, California, during (A) the open, deep-water lacustrine and/or marine phase and (B) the closed, shallow-water lacustrine phase. [After Link, M. H., and R. H. Osborne, 1978, Lacustrine facies in the Pliocene Ridge Basin Group, Ridge Basin, California, *in* Matter, A., and M. E. Tucker (eds.), Modern and ancient lake sediments: Blackwell Scientific Publications, Oxford, Fig. 14, p. 185, and Fig. 15, p. 186, reproduced by permission.]

to 9000 m of sediment eventually accumulated (Link and Osborne, 1978). Initially, the lake basin was open (Fig. 16.14A), allowing deltaic sediments and turbidites to form. As a result of subsequent strike-slip displacement on the fault, external drainage was blocked to the south and the lake basin became a closed system. During the closed phase, alluvial fan, fluvial, deltaic, and barrier-bar sediments accumulated along the margins of the lake, and siliciclastic mud, zeolite mud, dolomite, and stromatolites formed in the central part of the basin (Fig. 16.14B). For additional examples of strike-slip basins, see Nilsen and Sylvester, 1995.

Basins in Hybrid Settings

Aulacogens are special kinds of rifts situated at high angles to continental margins, which are commonly presumed to be rifts that failed but were reactivated during convergent tectonics (Fig. 16.3H). Other suggested origins for alaucogens include doming and rifting, strike-slip-related extension, and continental rotation (Sengör, 1995). The long, narrow troughs that make up the arms of aulacogens extend into continental cratons at a high angle from fold belts. Deposition of thick sequences of sediment can take place in these arms during long periods of time. These deposits may include nonmarine (e.g., alluvial-fan) sediments, marine shelf deposits, and deeper water facies such as turbidites. Examples of aulacogens include

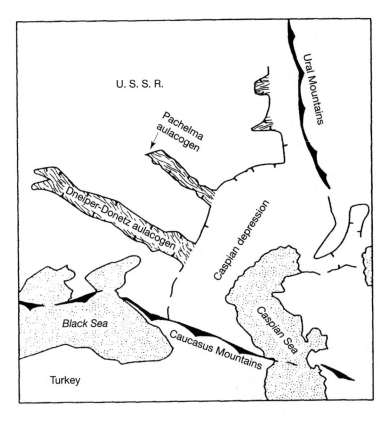

Figure 16.15
Aulacogen north of the Black and Caspian Seas on the Russian platform. [After Burke, K., 1977, Aulacogens and continental breakup: Annual Review of Earth and Planetary Sciences, v. 5. Reproduced by permission of Annual Reviews, Inc.]

the Reelfoot Rift of late Paleozoic age in which the Mississippi River flows, the Amazon Rift in which the Amazon River flows, the Benue Trough of Cretaceous age in which the Niger River is located, the aulacogen north of the Black and Caspian Seas on the Russian platform (Fig. 16.15), and the Anadarko Basin in Oklahoma (Fig. 16.6). **Impactogens** are structures similar to aulacogens in that they formed at high angles to orogenic belts; however, they do not have a preorogenic history.

   Intracontinental wrench basins are hybrid basins that formed within continental crust owing to distant collisional processes (e.g., the Quaidam Basin of China). **Successor basins** are basins that formed in intermountain settings following cessation of local orogenic activity (e.g., the southern Basin and Range, Arizona). See Ingersoll and Busby (1995) and Sengör (1995) for details.

16.5 SEDIMENTARY BASIN FILL

The preceding discussion focuses on the structural characteristics of sedimentary basins and the tectonic processes that create these basins. The particular concern of basin analysis is, however, the sediments that fill the basins. This concern encompasses the processes that produce the filling, the characteristics of the resulting sediments and sedimentary rocks, and the genetic and economic significance of these rocks. The fundamental processes that generate sediments (weathering/erosion) and bring about their transport and deposition; the physical, chemical, and biological properties of these rocks; the depositional environments in which they form; and their stratigraphic significance are discussed in preceding chapters of this book. The factors that control or affect these depositional processes and sediment characteristics, discussed also in appropriate parts of the book, include

1. The lithology of the parent rocks (e.g., granite, metamorphic rocks) present in the sediment source area, which controls the composition of sediment derived from these rocks

2. The relief, slope, and climate of the source area, which control the rate of sediment denudation and thus the rate at which sediment is delivered to depositional basins

3. The rate of basin subsidence together with rates of sea-level rise or fall

4. The size and shape of the basins

The processes that may cause basin subsidence are discussed briefly in Section 16.2 (see Figure 16.2). The rate of basin subsidence coupled with sea level fluctuations controls the available space in which sediments can accumulate (**accommodation**; see Fig. 13.15) at any given time, as well as affecting sediment transport and deposition. Thus, owing to continued subsidence, thousands of kilometers of sediments may accumulate even in shallow-water basins.

The purpose of basin analysis is to interpret basin fills to better understand sediment provenance (source), paleogeography, and depositional environments in order to unravel geologic history and to evaluate the economic potential of basin sediments. Basin analysis incorporates the interpretive basis of sedimentology (sedimentary processes); stratigraphy (spatial and temporal relations of sedimentary rock bodies); facies and depositional systems (organized response of sedimentary products and processes into sequences and rock bodies of a contemporaneous or time-transgressive nature); paleooceanography; paleogeography, and paleoclimatology; sea-level analysis; and petrographic mineralogy as a means of interpreting sediment source (Klein, 1987; 1991). Further, biostratigraphy provides a means of establishing a temporal framework for correlating time-equivalent facies and systems and to constrain timing of specific events, and radiochronology allows, in addition, the dating of specific sedimentological events and stratigraphic boundaries. Recent research in sedimentary geology and basin analysis has focused particularly on analysis of sedimentary facies, cyclic subsidence events, changes in sea level, ocean circulation patterns, paleoclimates, and life history.

Depositional models are being increasingly used to better understand the processes of basin filling and the effects of varying basin-filling parameters such as sediment supply and sediment flux into basins (e.g., Jones and Frostick, 2002), grain size, basin subsidence rates, and sea-level changes (e.g., Tetzlaff and Harbaugh, 1989; Angevine, Heller, and Paola, 1990; Cross, 1990; Slingerland, Harbaugh, and Furlong, 1994; Miall, 2000, Chapters 7, 9). Models may be either geometric or dynamic. Geometric models begin by specifying the geometry of the depositional system rather than calculating it as part of the model. Dynamic models begin with consideration of the transport of sediment in the basin and use some form of approximation to the basic laws that govern sediment transport and deposition.

16.6 TECHNIQUES OF BASIN ANALYSIS

Analyzing the characteristics of sediments and sedimentary rocks that fill basins, and interpreting these characteristics in terms of sediment and basin history, demands a variety of sedimentological and stratigraphic techniques. These techniques require the acquisition of data through outcrop studies and subsurface methods that can include deep drilling, magnetic polarity studies, and geophysical exploration. These data are then commonly displayed for study in the form of maps and stratigraphic cross sections, possibly using computer-assisted techniques. In this section, we look briefly at the more common techniques of basin analysis.

Measuring Stratigraphic Sections

To interpret Earth history through study of sedimentary rocks requires that we have detailed, accurate information about the thicknesses and lithology of the stratigraphic successions with which we deal. To obtain this information, appropriate stratigraphic successions must be measured and described in outcrop and/or from subsurface drill cores and cuttings. We refer to this process of studying outcrops as "measuring stratigraphic sections"; however, the process also involves describing the lithology, bedding characteristics, and other pertinent features of the rocks. Samples for mineralogic or paleontologic analysis may also be collected and keyed to their proper position within the measured sections. Thus, measuring and describing stratigraphic sections is commonly the starting point for many geologic studies, and the measured-section data become an indispensable part of the studies. Such stratigraphic sections are referred to throughout this book; see, for example, Figure 14.12.

A variety of techniques are available for measuring stratigraphic sections, depending upon the nature of the stratigraphic succession and the purpose of the study. The useful little book by Kottlowski (1965) describes these various methods, and the equipment needed to carry out measurements, in considerable detail. One of the most common methods involves use of a **Jacob staff**. A Jacob staff is a lightweight metal or wood pole (rod) marked in graduations of feet or meters. It is commonly cut to about eye-height and is used in conjunction with a Brunton compass placed at or near the top of the staff. The technique is illustrated in Figure 16.16. The clinometer of the Brunton compass is set at the dip angle of the beds, allowing the staff to be inclined normal to the bedding to determine the true stratigraphic thickness of the bed or beds. By stepping uphill and making a succession of measurements, comparatively thick sections of strata can be measured. After measuring several meters of section, the geologist commonly stops measuring for a time to describe the lithology and other pertinent features of the section before resuming measurement. A lithologic column, together with appropriate descriptive notes, is recorded in a field notebook.

Preparing Stratigraphic Maps and Cross Sections

Stratigraphic Cross Sections

Once stratigraphic sections have been measured and described, they can be used to prepare stratigraphic cross sections. Stratigraphic cross sections are used extensively for correlation purposes and structural interpretation, as well as for study

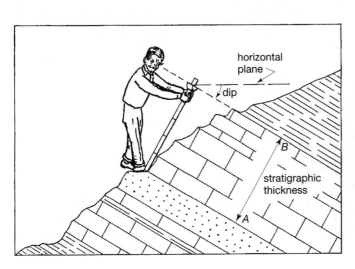

Figure 16.16
Schematic illustration of the Jacob-staff technique for measuring the thickness of stratigraphic units. By setting the clinometer of a Brunton compass attached to the Jacob staff to the dip angle of the beds, the staff can be held normal to the bedding planes, yielding the true stratigraphic thickness (AB) of the measured unit. [From Kottlowski, F. E., 1965, Measuring stratigraphic sections: Holt, Rinehart, and Winston, New York, Fig. 3.2, p. 63.]

of the details of facies changes that may have environmental or economic significance. Cross sections may be drawn to illustrate local features of a basin, often in conjunction with preparation of lithofacies maps (described below), or they may depict major stratigraphic successions across an entire basin. In addition to measured outcrop sections, the information needed to prepare stratigraphic cross sections may be obtained from subsurface lithologic logs (which are prepared by study of drill cores and cuttings) and/or mechanical well logs (petrophysical logs). Most stratigraphic cross sections depict in two dimensions the lithologic and/or structural characteristics of a particular stratigraphic unit, or units, across a given geographic region. Several examples of such cross sections are given in preceding chapters of this book. See, for example, Figure 12.15. Stratigraphic information may be presented also as a **fence diagram**. These diagrams attempt to give a three-dimensional view of the stratigraphy of an area or region (Fig. 16.17). Thus, they have the advantage of giving the reader a better regional perspective on the stratigraphic relationships. On the other hand, they are more difficult to

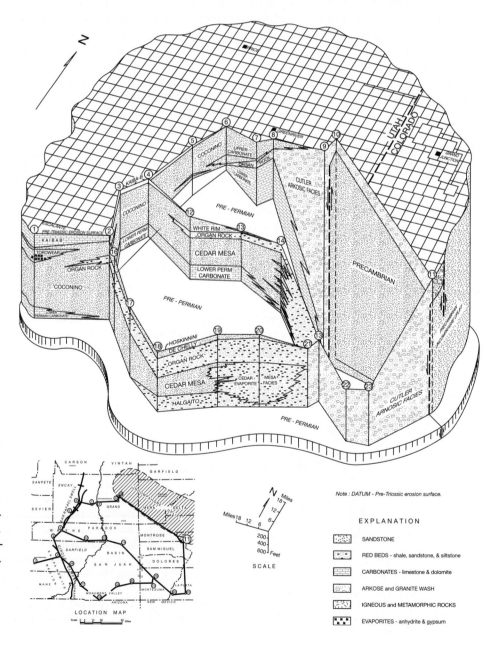

Figure 16.17
Schematic illustration of a fence diagram showing intertonguing facies relationships in Permian strata across the Paradox Basin, Utah and Colorado. [From Kunkle, R. P., 1958, Permian stratigraphy of the Paradox Basin, *in* Sanborn, A. F. (ed.), Guidebook to the geology of the Paradox Basin, Intermountain Association of Petroleum Geologists, Ninth Annual Field Conference, Fig 1, p. 165.]

construct than conventional two-dimensional diagrams, and parts of the section are hidden by the fences in front.

Structure-Contour Maps

It is often desirable in basin studies to determine the regional structural attitude of the rocks as well as the presence of local structural features such as anticlines and faults. Structure-contour maps are prepared for this purpose. These maps provide information about a basin's shape and orientation and the basin-fill geometry. Structure-contour maps are prepared by drawing lines on a map through points of equal elevation above or below some datum, commonly mean sea level. Elevations are typically determined on the top of a particular formation or key bed at a number of control points. Elevation data may be obtained through outcrop study and/or subsurface interpretation of mechanical or lithologic well logs. After control points are plotted on a base map, a suitable contour interval is selected, and structure contour lines are drawn by hand or by use of computers. Figure 16.18 shows an example of a structure-contour map.

Structure contours may also be prepared on the top of prominent subsurface reflectors from seismic data (Chapter 13). Depth to a particular reflector may be plotted initially as two-way travel time. Thus, the initial map shows contour lines of equal travel time. If the seismic-wave velocity can be determined from well information, the travel times can be converted to actual depths, allowing maps to be redrawn in terms of actual elevations on the reflecting horizon.

Structure-contour maps can reveal the locations of subbasins or depositional centers within a major basin as well as axes of uplift (anticlines, domes). Structural features may be related also to syndepositional topography. Thus, analysis of these maps can provide clues to local paleogeography and facies patterns. Structural maps are useful also in economic assessment (e.g., petroleum exploration) of basins.

Isopach Maps

Isopachs are contour lines of equal thickness. An isopach map is a map that shows by means of contour lines the thickness of a given formation or rock unit. A map that shows the areal variation in a specific rock type (e.g., sandstone) is an **isolith map**. The thickness of sediment in a basin is determined by the rate of supply of sediment and the accommodation space in the basin, which in turn is a function of

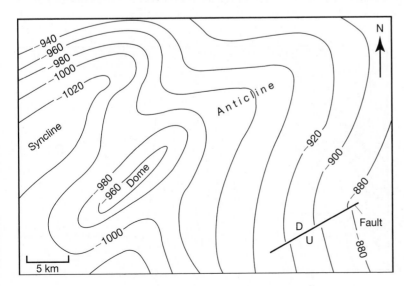

Figure 16.18

Schematic illustration of a structure-contour map drawn on the top of a formation. The contour interval is 20 m. The negative contour values indicate that this formation is located below sea level and is thus a buried (subsurface) formation. Note the presence of a syncline, dome, anticline, and fault.

basin geometry and rate of basin subsidence. Abnormally thick parts of a stratigraphic unit suggest the presence of major depositional centers in a basin (basin lows), whereas abnormally thin parts of the unit suggest predepositional highs or possibly areas of postdepositional erosion. Isopach maps thus provide information about the geometry of the basin immediately prior to and during sedimentation. Furthermore, analysis of a succession of isopach maps in a basin can provide information about changes in the structure of a basin through time.

To construct an isopach map, the thickness of a formation or other stratigraphic unit must be determined from outcrop measurements and/or subsurface well-log data at numerous control points. The thickness of the unit at each of these control points is entered on a base map, and the map is contoured in the same way that a structure-contour map is prepared. Figure 16.19 is an example.

Paleogeologic Maps

Paleogeologic maps are maps that display the areal geology either below or above a given stratigraphic unit. Imagine, for example, that we could strip off the rocks that make up a particular formation (and all the rocks above that formation) to reveal the rocks beneath—the rocks on which the formation was deposited. We could then construct a geologic map on top of these underlying formations. Such a map has been referred to as a **subcrop map** (Krumbein and Sloss, 1963). In a similar manner, the rocks above a formation or rock body may also be mapped. This kind of map, looked at as though from below, is called a **worm's eye view map** or supercrop map. Subcrop and worm's eye maps are commonly constructed at unconformity boundaries; however, they can be constructed at the top and base of any distinctive rock unit, whether or not an unconformity is present. The main purpose of such maps is to illustrate paleodrainage patterns, pattern of basin fill, shifting shorelines, or gradual burial of a preexisting erosional topography (Miall, 2000, p. 250). To construct paleogeologic maps requires identification of the stratigraphic units that lie immediately below (or above) a given formation or other stratigraphic unit at numerous control points. Such data are gathered from outcrops or from subsurface well logs.

Lithofacies Maps

Facies maps depict variation in lithologic or biologic characteristics of a stratigraphic unit. The most common kinds of facies maps, and the only kind discussed here, are lithofacies maps, which depict some aspect of composition or texture. Maps based on faunal characteristics are biofacies maps. Many kinds of lithofacies

Figure 16.19

Example of an isopach map of a hypothetical formation drawn at a contour interval of 40 m. Note that the formation thickens to more than 240 m in the basin low (depositional center), thins over a basin high, and thins to zero along the northwest and north sides of the map.

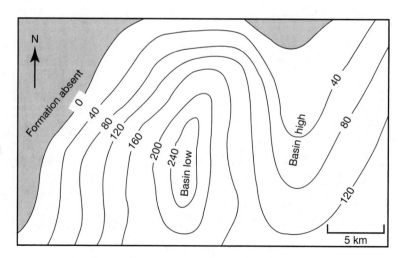

maps are in use. Some are plotted as ratios of specific lithologic units (e.g., the ratio of siliciclastic to nonsiliciclastic components) or as isopachs of such units [e.g., sandstone isopach (or isolith) maps, limestone isopach (or isolith) maps]. Others examine the relative abundance and distribution of three end-member components (e.g., sandstone, shale, limestone). Two kinds of lithofacies maps are discussed here to illustrate the method: clastic-ratio maps and three-component lithofacies maps. Additional examples are given in Krumbein and Sloss (1963) and Miall (2000).

Clastic-ratio maps are maps that show contours of equal clastic ratio, which is defined as the ratio of total cumulative thickness of siliciclastic deposits to the thickness of nonsiliciclastic deposits, for example,

$$\frac{(\text{conglomerate} + \text{sandstone} + \text{shale})}{(\text{limestone} + \text{dolomite} + \text{evaporite})}$$

Values are computed for a number of control points, from outcrop or subsurface data, and plotted on a map. The map is then contoured in the same manner as that described for isopach maps. An example is shown in Figure 16.20. Clastic-ratio maps are useful for showing the relationship of lithologic units along the margin of a basin in which both siliciclastic and nonsiliciclastic deposits accumulated. Such maps provide limited information also about the location of the siliciclastic sediment source.

Three-component (triangle) lithofacies maps show by means of patterns or colors the relative abundance, within a formation or other stratigraphic unit, of three principal lithofacies components. Figure 16.21 is an example of such a map based on the relative thickness of sandstone, shale, and limestone. A ternary diagram, called the standard diagram (inset in Fig. 16.21), is drawn using the three lithofacies components as end members. The triangle is subdivided into fields, each of which is indicated by a suitable pattern or color. The thickness of each end-member component is measured at as many control points in outcrop or in the subsurface as practical. The relative values (normalized to 100 percent if necessary) at each control point are plotted on the ternary diagram. The clastic ratio and sand-shale ratio at each data point are calculated and used to draw contour lines of clastic ratio (CR) and sand/shale ratio (SSR) on the map. These contour lines allow the map to be divided into selected pattern areas corresponding to the patterns in the standard triangle. In this example, a progressive change in facies from dominantly clastic material in the northwest part of the map to a section composed dominantly of limestone in the southern part is evident.

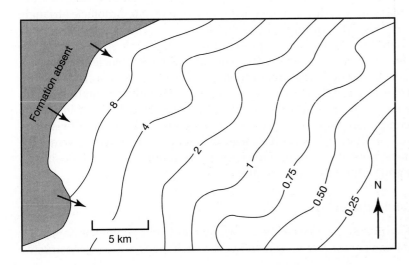

Figure 16.20

Example of a clastic-ratio (clastic/nonclastic) map. The progressive increase in the ratio from southeast to northwest across the map indicates a progressively increasing percentage of siliciclastic components in the stratigraphic section toward the northwest. Thus, the source of the sediments must have been located somewhere to the northwest. The small arrows show the probable direction of sediment transport.

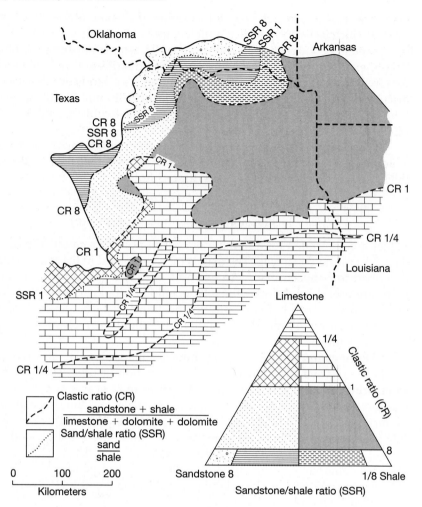

Figure 16.21
Triangle lithofacies map of the Cretaceous Trinity Group, southern U.S.A. Note that the stratigraphic section changes from predominantly siliciclastic in the northwest part of the map to predominantly limestone in the southern part. Not all lithofacies shown in the lithofacies triangle (e.g., shale) are actually present in the mapped area. [Redrawn from Krumbein and Sloss, 1963, Fig. 12.11, p. 464, as modified from Forgotson, 1960.]

The degree of "mixing" of three rock components in a stratigraphic section can be calculated mathematically by applying an entropy-like function. The function is set up so that equal parts of (for example) sandstone, shale, and limestone have an entropy value of 100. As the proportion of one end member or another increases, the entropy value becomes smaller, approaching zero as the composition approaches that of a single end member. An **entropy map** can be prepared from these data by using an overlay of the entropy function on the lithofacies triangle (Krumbein and Sloss, 1963, p. 467; Forgotson, 1960).

If more than three lithofacies are present in a stratigraphic section, the additional lithofacies must be omitted (and the remaining three lithofacies normalized to 100 percent) or combined with other lithofacies to yield a total of three lithofacies. For example, if a conglomerate facies is present, it could be combined with the sandstone facies, or an evaporite facies might be combined with a limestone facies. Three-component lithofacies maps provide a convenient means of visualizing the relative importance of each lithofacies throughout a geographic area. Like clastic-ratio maps, however, they provide only a very rough guide to depositional environments and sediment-source locations.

Computer-Generated Maps

All of the maps discussed above can be drawn by hand, and geologists have been drawing them this way for many years; however, such maps are now constructed more and more by computer. Computer application is particularly prevalent in the petroleum industry, where basin analysis is a commonplace procedure. Computers are able to handle large quantities of data, such as stratigraphic and structural data obtained from well records, and they allow these data to be easily manipulated for a variety of statistical and mapping purposes. Appropriate base maps are stored in the computer and the locations of outcrop sections and subsurface wells can be easily plotted on these maps. Lithologic, structural, and stratigraphic data obtained from study of outcrop and subsurface sections are likewise stored in the computer. Selected data (e.g., thickness of lithologic units, structural elevations) can be retrieved as needed and added to the base maps, which can then be contoured by the computer by using appropriate software and special printers to draw the maps. Thus, any of the maps described above can be generated by computer, as well as other kinds of maps such as **trend-surface maps**. Trend-surface analysis allows separation of map data into two components: regional trends and local fluctuations. The regional trend is mathematically subtracted by the computer, leaving residuals, which correspond to local variations. For example, the regional structural trend might be extracted from a structure-contour map to more clearly reveal the nature of local structural anomalies. Computer-generated maps are not necessarily better or more accurate than hand-drawn maps. Their principal advantage is in the ease and rapidity with which they can be drawn. See, for example, Robinson (1982) and Jones, Hamilton, and Johnson (1986).

Paleocurrent Analysis and Paleocurrent Maps

Paleocurrent analysis is a technique used to determine the flow direction of ancient currents that transported sediment into and within a depositional basin, which reflects the local or regional paleoslope (see Chapter 4, Section 4.6). By inference, paleocurrent analysis also reveals the direction in which the sediment source area, or areas, lay. Further, it aids in understanding the geometry and trend of lithologic units and in interpretation of depositional environments. Paleocurrent analysis is accomplished by measuring the orientation of directional features such as sedimentary structures (e.g., flute casts, ripple marks, cross-beds) or the long-axis orientation of pebbles. Numerous orientation measurements must be made within a given stratigraphic unit to obtain a statistically reliable paleocurrent trend. Grain-size trends, lithologic characteristics, and sediment thickness may also have directional significance when mapped, as previously discussed. An example of a paleocurrent map, constructed mainly on the basis of cross-bedding orientation, is shown in Figure 16.22. Note that the average (statistical) paleoflow direction indicated by this map is from the northwest to southeast, suggesting that the sediment source area lay somewhere to the northwest. For more detailed discussion of the application of paleocurrent analysis to interpretation of sediment-dispersal patterns and basin-filling mechanisms, see Potter and Pettijohn (1977, Chapter 8).

Siliciclastic Petrofacies (Provenance) Studies

The composition of siliciclastic sediments that fill sedimentary basins is determined to a large extent by the lithology of the source rocks that furnished sediment to the basin, as well as by the climate and weathering conditions of the source area. Therefore, analysis of the particle composition of siliciclastic mineral assemblages (and rock fragments) provides a method of working backward to understand the nature of the source area. We commonly refer to such study as provenance study, where provenance is considered to include the following: (1) the lithology of the source rocks, (2) the tectonic setting of the source area, and (3) the

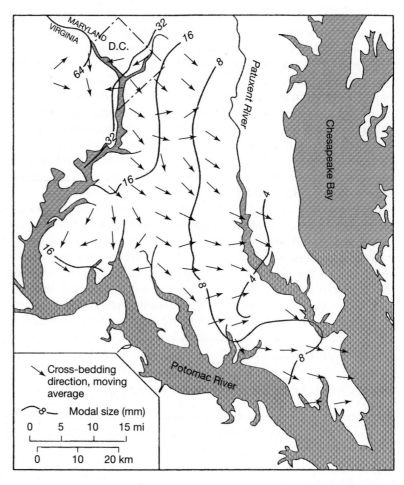

Figure 16.22
Example of the use of paleocurrent data to locate source areas, Brandywine gravel of Maryland. The contours show modal grain size (in mm). [From Potter, P. E., and F. J. Pettijohn, 1977, Paleocurrents and basin analysis, Fig. 8.9, p. 282. Reprinted by permission of Springer-Verlag, Berlin.]

climate, relief, and slope of the source area. Provenance studies provide important information about the paleoclimatology and paleogeography of the basin setting.

The lithology of the source rocks is interpreted on the basis of the kinds of minerals and rock fragments present in siliciclastic sedimentary rocks, particularly in sandstones and conglomerates. For example, the presence of abundant alkali feldspars suggests derivation from granitic source rocks; abundant volcanic rock fragments suggest derivation from volcanic source rocks; and so on. The siliciclastic mineralogy also provides information about the tectonic setting (continental block, magmatic arc, collision orogen); sediment derived from continental block settings is likely to be enriched in quartz and alkali feldspars; sediment derived from a magmatic arc is dominated by volcanic rock fragments and plagioclase feldspars; and sediment stripped from highlands along orogenic collision belts is characterized particularly by an abundance of sedimentary and metasedimentary rocks fragments. Climate, relief, and slope of the source area are more difficult to interpret from siliciclastic mineral assemblages, but some clues are provided by quartz/feldspar ratios and by degree of weathering of feldspars. The essentials of provenance analysis are discussed in greater detail in Chapter 5, Section 5.6. Zuffa (1985); Morton, Todd, and Haughton (1991), and Johnsson and Basu (1993) provide a broader and more detailed view of provenance analysis.

Geophysical Studies

Geophysical investigations, including both seismic and paleomagnetic studies of various kinds, play an important role in basin analysis. **Seismic techniques** are used to document the regional structural trends and overall basin geometry as well as to identify local structural features such as anticlines and faults that may provide traps for hydrocarbons. These seismic techniques are described in Chapter 13, which also discusses the application of seismic methods to stratigraphic interpretation, including identification of seismic sequences and facies as well as correlation of stratal reflectors. The most widespread application of **paleomagnetism** in basin analysis is the study of magnetic polarity reversals as a tool for correlation and other purposes, as described in Chapter 13. Airborne or ground-based magnetic and gravity surveys are sometimes conducted also to provide information about basin structure in addition to that provided by seismic data. Gravity surveys determine the presence of gravity anomalies (abnormally high or low gravity values), which can be related to the presence of dense (e.g., basalt) or less dense (e.g., thick sediment) rock bodies in the subsurface. These magnetic and gravity surveys furnish data that can be used to construct maps showing large-scale characteristics of basins (e.g., major basin highs and lows); however, these maps commonly do not have sufficient resolution to identify small-scale structural features, that is, features such as small anticlines or faults that might constitute petroleum traps.

16.7 APPLICATIONS OF BASIN ANALYSIS

Interpreting Geologic History

As pointed out, basin analysis is an integrated program of study that involves application of sedimentologic, stratigraphic, and tectonic principles to develop a full understanding of the rocks that fill sedimentary basins and to use this information to interpret the geologic history and evaluate the economic importance of these rocks. Thus, one major goal of basin analysis is simply to develop a better understanding of Earth history as recorded in particular depositional basins. Through analysis of sedimentary textures, structures, particle and chemical composition, fossils, and the stratigraphic characteristics of sedimentary rocks (as revealed by physical, biological, paleomagnetic, and seismic-reflection characteristics), geologists are able to interpret the important tectonic and sedimentologic events that transpired to generate and fill a particular sedimentary basin. Thus, these various kinds of basin studies, commonly involving preparation of appropriate maps and stratigraphic sections, allow geologists to interpret past tectonic, climatic, and sedimentologic events and conditions (including source-area characterization and interpretation of depositional environments) to reconstruct the paleogeography and paleogeology of Earth during specific times in the past.

A good example of the application of basin-analysis principles to interpretation of the geologic history of a particular region is provided by Clifton, Hunter, and Gardner (1988). These authors examined the influence of eustatic, tectonic, and sedimentological processes on the generation of transgressive and regressive depositional cycles in the Merced Formation near San Francisco, California. The Merced Formation consists of approximately 2000 m of shallow-marine and coastal nonmarine sediments of late Cenozoic age.

To work out the history of the Merced Formation, Clifton et al. first defined the stratigraphic boundaries (base and top) of the formation through field study and reference to previously published information. Ages of different parts of the formation were established by a combination of paleontologic data and radiometric dating of intercalated ash beds. Various depositional facies within the Merced were then characterized on the basis of contained fossils, physical and biogenic sedimentary structures, and textural characteristics (Table 16.3). Ten facies types were identified, which (from the base of the formation upward) were determined

Table 16.3   Characteristics and inferred origin of sedimentary facies in sea-cliff exposures of the Merced Formation, San Francisco, California

Inferred origin	Texture	Biota	Physical structures	Biogenic structures
Outer shelf	Sandy silt	Molluscs (*in situ* and scattered shells)	None	Intense bioturbation
Mid-shelf	Sandy silt	Molluscs	Shell lags, parallel lamination, hummocky cross-stratification, ripple lamination	Sharp-topped bioturbated intervals, between sets of laminae
Inner shelf	Very fine sand, scattered small pebbles near top	Molluscs, echinoids	Shell lags, parallel lamination, hummocky cross-stratification	Locally intense bioturbation
Nearshore	Fine to coarse sand, gravel	Molluscs, echinoids	Lenticular gravel and sand beds, high- and low-angle cross-bedding, parallel lamination	Vertical burrows, *Macaronichnus* (near top)
Foreshore	Fine to coarse sand, gravel fining-upward trend	Molluscs, echinoids	Parallel lamination, heavy mineral layers near top	*Macaronichnus* (near base)
Embayment	Mud, sand, gravel, some fining-upward cycles	Molluscs, ostracods	Fine sand-silt-clay lamination, ripples and ripple bedding (sand), cross-bedding (sand and gravel), shell lags	Bioturbation, root-rhizome structures
Backshore	Fine sand	None	Parallel lamination, low-angle cross-bedding, climbing adhesion ripple bedding, ripple lamination	Root-rhizome structures, vertebrate footprints, vertical tubes
Eolian dune	Fine sand	None	Medium- and large-scale and medium- and high-angle cross-bedding	Local mottling, vertical tubes
Alluvial	Gravel and pebbly sand	None	Indistinct stratification, lenticular bedding, medium- and small-scale trough cross-bedding	None
Pond/swamp/marsh	Mud, peat, or lignite	Freshwater diatoms, insect wings, terrestrial vertebrate bones	Flat-bedded (mud)	Root structures, burrows and intrastratal trails

Source: Clifton, Hunter, and Gardner, 1988.

to represent ten different depositional environments: pond/swamp/marsh, alluvial, eolian dune, backshore, embayment, foreshore, nearshore, inner shelf, mid-shelf, and outer shelf.

In vertical succession, these facies define alternating episodes of transgression and regression that probably reflect Pleistocene eustatic sea-level fluctuations. The transgression-regression fluctuations were matched with reasonable confidence to a Pleistocene sea-level curve, determined from oxygen-isotope data. Finally, Clifton et al. were able to calculate that the Merced Formation was deposited under shallow-marine/coastal conditions in a setting undergoing subsidence at an average rate of 1 m/1000 years, a rapid rate that must have been influenced by tectonic deformation.

This example illustrates how geologists use an armory of various sedimentological and stratigraphic approaches (i.e., basin analysis) to determine the physical and biologic characteristics of a sedimentary formation, from which the sedimentological and tectonic history of the formation can be interpreted. Several additional examples of this approach may be examined in Kleinspehn and Paola (1988).

Economic Applications

A second goal of basin analysis is to use the principles and techniques described above to evaluate the economic importance of sedimentary rocks and identify economically exploitable deposits of minerals or fossil fuels. Basin analysis finds its greatest economic application in the fields of petroleum geology and, to a lesser extent, hydrogeology. In spite of the fact that petroleum geologists have been trying for many years to locate petroleum (hydrocarbon) accumulations by geochemical analysis of surficial rocks and soil overlying such deposits, no successful method has yet been developed for direct detection of hydrocarbon deposits. To find an oil or gas deposit, geologists must (1) explore basins that have the right conditions for the formation and migration of hydrocarbons and (2) locate a suitable trap, such as a structural anticline, in which the hydrocarbons may have accumulated.

What, exactly, are the right conditions for formation and migration of hydrocarbons? Most petroleum geologists believe that petroleum originates from fine-size plant or animal organic matter through complex biochemical processes that take place during sediment burial. Because organic matter is preserved preferentially in fine-grained sediments, organic-rich shales are believed to be the principal source rocks for petroleum. Thus, the first condition for successful hydrocarbon exploration is to find a basin that contains suitable source rocks. Furthermore, the source rocks should have been buried to depths where the temperature is adequate to "crack" oil and gas from the organic matter ($\sim$90°–125°C) but not so high that the hydrocarbons are destroyed. After oil and gas are generated within the shale source rocks, they migrate, in some (poorly understood) manner, from the source rock into associated porous and permeable rocks called **reservoir rocks**. (Porosity refers to that percentage of the volume of a rock occupied by pore space; permeability refers to the ability of a rock to transmit a fluid.) Drilling experience has shown that the best reservoir rocks for hydrocarbons are sandstones and limestones. Roughly 55 percent of the world's oil and 75 percent of its natural gas occur in sandstone reservoirs; most of the remaining world reserves are in carbonate reservoirs.

Most of the pore spaces in porous sandstones and carbonate rocks in the subsurface are filled with water. When hydrocarbons migrate from source beds into these overlying or underlying (?) reservoir rocks, some of the pore water must be displaced to make room for the hydrocarbons. Because the hydrocarbons have a lower specific gravity than water, they tend to move updip to get above the water (oil floats on water). Thus, by progressive displacement of pore water, hydrocarbons will gradually move up the regional dip of a sedimentary basin. This migration will continue until the oil reaches a trap or, if no trap is encountered, until it is expelled at the surface as an oil seep.

A hydrocarbon **trap** is some kind of structural or stratigraphic feature of a rock into which oil or gas can easily migrate but from which it is difficult to escape. The three most common kinds of petroleum traps are anticlines, faults, and stratigraphic pinch-outs (see Fig. 7.25). Oil or gas will rise to the structurally highest position in an anticline, fault trap, or stratigraphic pinch-out, where it is prevented from further movement by an impermeable rock layer (e.g., shale or evaporite) that lies above the reservoir bed and seals the trap. Thus, the search for a petroleum deposit requires that a trap be located and drilled in a basin that contains suitable source and reservoir rocks. The trap, when drilled, may or may not contain petroleum. Only about 10 to 15 percent of the traps drilled in new exploration areas commonly yield commercial quantities of petroleum. Structural traps (e.g., anticlines and faults) are located primarily by reflection seismic techniques (Chapter 13). Stratigraphic traps are located through detailed analysis of stratigraphic information obtained by study of stratigraphic cross sections (based on subsurface lithologic logs and well logs) and seismic cross sections.

Basin analysis, to the petroleum geologist, thus means locating within depositional basins suitable source rocks, reservoir rocks, and traps. To do this successfully calls into play most of the principles of sedimentology and stratigraphy, as well as the various techniques for analyzing sedimentary successions and presenting data, discussed in this book. Petroleum geologists must have a thorough understanding of the physical, chemical, and biological characteristics of sedimentary rocks; must understand depositional environments and depositional systems; must be well grounded in all aspects of stratigraphy; must have a working knowledge of the principles of geophysics and structural geology; and must have a basic understanding of the principles involved in the flow of fluids through porous, subsurface rocks. No single geologist is normally capable of carrying out the many complex studies required to develop a major, successful petroleum "play." Basin assessment is commonly carried out by teams of workers, which may include stratigraphers, sedimentologists, sedimentary petrologists, paleontologists, geophysicists, and hydrologists. These teams of scientists work together to develop an understanding of the sedimentary, stratigraphic, and structural factors that control the accumulation of oil and gas in drilled basins where commercial hydrocarbon deposits have already been found. They seek then to extend the knowledge gained from such studies to exploration for new hydrocarbon deposits in untested basins.

The application of basin-assessment techniques to developing a petroleum play is described by Allen and Allen (1990). Before new exploratory wells (e.g., Fig. 16.23) are drilled in an untested basin, petroleum geologists must first work out the stratigraphic characteristics of the basin to see the kinds of potential source rocks and reservoir rocks that are present and the thickness and characteristics of these rocks. They then collect and analyze samples of potential source rocks (organic-bearing shales) to see if enough organic matter (kerogen; Chapter 7), and the right kind of kerogen, is present to generate economically significant quantities of petroleum or natural gas. Some kinds of kerogen generate liquid petroleum; others generate gas. Through outcrop study, and possibly drilling of small-diameter holes for information purposes, they determine which potential reservoir rocks have adequate porosity and permeability to make them targets for drilling.

If suitable source and reservoir rocks are present in the basin, the next step is to locate one or more traps that are large enough to hold a commercial quantity of oil or gas. As mentioned above, traps are located mainly by seismic prospecting and detailed stratigraphic analysis. An essential component of a trap is that reservoir rocks in the trap must be covered or capped by some kind of nonpermeable rock such as shale or evaporites to prevent oil in the reservoir rock from escaping upward. Finally, geologists study the fluid-flow patterns in the subsurface rocks of the basin to see if fluids are flowing with adequate pressure to move petroleum into a well bore that penetrates the reservoir rock. If all the requisite conditions of source rock, reservoir rocks, and trap are present, the well is drilled. For additional insight into application of the principles of basin analysis to petroleum exploration, see Cross et al. (1995) and Welte, Horsfield, and Baker (1997).

Figure 16.23
An exploratory gas well (Chevron Can Sup Waterton) drilling in the northwest corner of the western Canada Sedimentary Basin, Alberta, Canada. The well tested Mississippian and Devonian carbonate formations and encountered gas below a depth of 4000 m. [Photograph courtesy of Sidney Smith.]

FURTHER READING

Allen, P. A., and J. R. Allen, 1990, Basin analysis—Principles and applications: Blackwell, Oxford, 451 p.

Allen, P. A., and P. Homewood (eds.), 1986, Foreland basins: Blackwell , Oxford, 453 p.

Angevine, C. L., P. L. Heller, and C. Paola, 1990, Quantitative sedimentary basin modeling: AAPG Continuing Education Course Note Series 32, Tulsa, Okla., 133 p.

Biddle, K. T. (ed.), 1991, Active margin basins: Am. Assoc. Petroleum Geologists Mem. 52, Tulsa, Okla., 324 p.

Busby, C. J., and R. V. Ingersoll (eds.), 1995, Tectonics of sedimentary basins: Blackwell Science, Cambridge, Mass., 579 p.

Cross, T. A., R. L. Dodge, J. C. Howard, and E. S. Siraki, 1995, Basin analysis: International Human Resources Development Corp., Boston, Lass., 210 p.

Dorobek, S. L., and G. M. Ross (eds.), 1995, Stratigraphic evolution of foreland basins: SEPM Spec. Pub. 52, 310 p.

Edwards, J. D., and P. A. Santogrossi (eds.), 1990, Divergent/passive margin basins: Am. Assoc. Petroleum Geologists Mem. 48, Tulsa, Okla., 252 p.

Einsele, G., 2000, Sedimentary basins: Evolution, facies, and sediment budget, 2nd ed: Springer-Verlag, Berlin, 792 p.

Frostick, L. E., and R. J. Steel, 1993, Tectonic controls and signatures in sedimentary successions: International Association of Sedimentologists Spec. Pub. 20, Blackwell Scientific Publications, Oxford, 520 p.

Jones, S. J., and L. E. Frostick (eds.), 2002, Sediment flux to basins: Causes, controls and consequences: Geological Society Special Publication No. 191, The Geological Society, London, 284 p.

Katz, B. J. (ed.), 1990, Lacustrine basin exploration: Am. Assoc. Petroleum Geologists Mem. 50, Tulsa, Okla., 340 p.

Kleinspehn, K. L., and C. Paola (eds.), 1988, New perspectives in basin analysis: Springer-Verlag, New York, 453 p.

Leighton, M. W., D. R. Kolata, D. F. Oltz, and J. J. Eidel (eds.), 1990, Interior cratonic basins: Am. Assoc. Petroleum Geologists Mem. 51, Tulsa, Okla., 819 p.

Lerche, I., 1990, Basin analysis: Quantitative methods, vol. 2: Academic Press, Inc., San Diego, Cal., 570 p.

Macqueen, R. W., and D. A. Leckie, 1992, Foreland basins and fold belts: Am. Assoc. Petroleum Geologists Mem. 55, Tulsa, Okla., 460 p.

Miall, A. D., 2000, Principles of sedimentary basin analysis, 3rd ed.: Springer-Verlag, New York, 616 p.

Purser, B. H. (ed.), 1998, Dynamics and methods of study of sedimentary basins: Oxford & IBH Publishing Co., Pvt. Ltd, New Delhi, and Editions Technip, Paris, 392 p.

Slingerland, R., J. W. Harbaugh, and K. P. Furlong, 1994, Simulating clastic sedimentary basins: PTR Prentice Hall, Englewood Cliffs, N.J., 220 p.

Taylor, B., and J. Natland (eds.), 1995, Active margins and marginal basins of the Western Pacific: American Geophysical Union Geophysical Monograph 88, 417 p.

Welte, D. H., B. Horsfield, and D. R. Baker (eds.), 1997, Petroleum and basin evolution: Springer-Verlag, Berlin, 535 p.

Appendix A

Form and Roundness of Sedimentary Particles

Definition and Measurement of Sphericity (Form)

Numerous mathematical measures for expressing form have been proposed, but **sphericity** is probably the most widely used. The concept of sphericity was introduced by Wadell (1932), who defined sphericity mathematically as the ratio of the diameter of a sphere with the same volume as a particle to the diameter of the smallest circle that would just enclose or circumscribe the outline of the particle. Wadell determined the volumes of large particles by immersing the particle in water and measuring the volume change of the water. Krumbein (1941) modified Wadell's sphericity concept slightly to express sphericity ψ by the relationship

$$\psi = \sqrt[3]{\frac{\text{volume of the particle}}{\text{volume of the circumscribing sphere}}} \qquad \text{(A-A.1)}$$

The volume of a sphere is given by $\frac{\pi}{6}D^3$, where D is the diameter of the sphere. The approximate volume of natural particles can be calculated by assuming that the particles are triaxial ellipsoids having three diameters D_L, D_I, and D_S, where L, I, and S refer to the lengths of the long, intermediate, and short axes of the ellipsoid. By substituting appropriate values for volume into Equation A-A.1, we can express sphericity by

$$\psi_I = \sqrt[3]{\frac{\frac{\pi}{6}D_L D_I D_S}{\frac{\pi}{6}D_L{}^3}} = \sqrt[3]{\frac{D_S D_I}{D_L{}^2}} \qquad \text{(A-A.2)}$$

The sphericity of a particle determined by this relationship is called intercept sphericity and can be calculated by measuring the long, intermediate, and short axes of a particle and substituting these values into Equation A-A.2.

Sneed and Folk (1958) suggest that Krumbein's intercept sphericity does not correctly express the behavior of particles as they settle in a fluid or are acted on by fluid flow. A rod-shaped particle, for example, settles faster than a disc, although the intercept sphericity formula suggests the opposite. Particles falling through water tend to settle with maximum projection areas (the plane of the long and intermediate axes) perpendicular to the direction of motion, and small particles resting on the bottom orient themselves perpendicular to current flow direction. Sneed and Folk thus proposed a different sphericity measure called maximum projection sphericity (φ_P), which they claim better expresses particle behavior. Maximum projection sphericity is defined mathematically as the ratio between the maximum projection area of a sphere with the same volume as the particle and the maximum projection area of the particle:

$$\psi_P = \sqrt[3]{\dfrac{D_S^2}{D_L D_I}} \qquad\qquad (A\text{-}A.3)$$

Maximum projection sphericity has gained favor for expressing the shape of particles deposited by water. Conceptually, it is not necessarily more valid than intercept sphericity when applied to other modes of particle transport and deposition, e.g., transport by ice and sediment gravity flows.

Regardless of the sphericity measure used, experience has shown that particles having the same mathematical sphericity can differ considerably in their overall shape. Some additional measure or index is needed to more specifically define the form of particles. Two additional form indices that permit a more graphic representation of form are in wide use. Zingg (1935) proposed the use of two shape indices D_I/D_L and D_S/D_I to define four shape fields on a bivariate plot: oblate (disc), equant (spheres), bladed, and prolate (rollers) (see Fig. 3.9A in the text). Lines of equal intercept sphericity can be drawn on the Zingg shape fields (Fig. 3.9B), illustrating that particles of quite different form can have the same mathematical sphericity.

Sneed and Folk (1958) used two somewhat different shape indices to construct a triangular form diagram (Fig. A-A.1), in which D_S/D_L is plotted against $D_L - D_I/D_L - D_S$ to create ten form fields (e.g., compact, C; platy, P; bladed, B; elongate, E). Lines of maximum projection sphericity drawn across the field (Fig. A-A.1) again illustrate the disparity between mathematical sphericity and actual form.

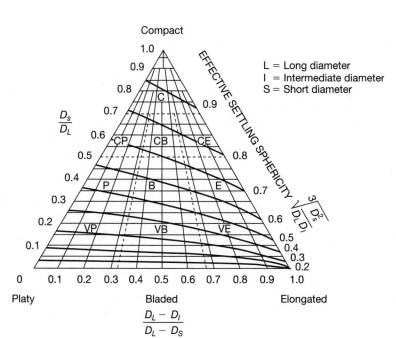

Figure A-A.1

Classification of pebble shapes, after Sneed and Folk. The symbol V refers to the adjective very (e.g., very platy, very bladed, very elongated). [After Sneed, E. D., and R. L. Folk, 1958, Pebbles in the lower Colorado River, Texas, a study in particle morphogenesis: Jour. Geology, v. 66, Fig. 2, p. 119. Reprinted by permission of University of Chicago Press.]

Definition and Measurement of Roundness

Wadell (1932) defined mathematical roundness as the arithmetic mean of the roundness of the individual corners of a grain in the plane of measurement. Roundness of individual corners is given by the ratio of the radius of curvature of the corners to the radius of the maximum-size circle that can be inscribed within the outline of the grain in the plane of measurement. The degree of Wadell roundness (R_W) is thus expressed as

$$R_W = \frac{\Sigma(r/R)}{N} = \frac{\Sigma(r)}{RN} \qquad \text{(A-A.4)}$$

where r is the radius of curvature of individual corners, R is the radius of the maximum inscribed circle, and N is the number of corners. The relationship of r to R is illustrated in Figure A-A.2.

Owing to the numerous radius measurements that must be made, it is very time-consuming to determine the Wadell roundness of large numbers of grains. Simpler roundness measures have been proposed that require only that the radius of the sharpest corner be divided by the radius of the inscribed circle (Wentworth, 1919; Dobkins and Folk, 1970); however, Wadell's roundness measure is still used by most workers. Even if the simpler roundness formula is used, measuring the radii of large numbers of small grains is a very laborious process, requiring use of either a circular protractor or an electronic particle size analyzer to measure enlarged images of grains (Boggs, 1967a). Consequently, visual comparison scales or charts consisting of sets of grain images of known roundness are often used to make rapid visual estimates of grain roundness. The visual charts of Krumbein (1941) and Powers (1953) are the most widely used of these comparison scales.

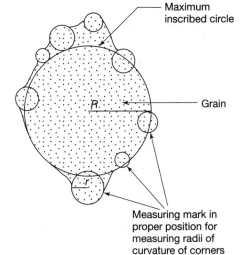

Figure A-A.2
Diagram of enlarged grain image illustrating the method of measuring the roundness of a sediment grain. (R) refers to the radius of the largest circle that can be inscribed inside the grain; (r) is the radius of curvature of the corners of the grain. [From Boggs, S., Jr., 1967, Measurement of roundness and sphericity parameters using an electronic particle size analyzer: Jour. Sed. Petrology, v. 37, Fig. 3, p. 912. Reprinted by permission of Society of Economic Paleontologists and Mineralogists, Tulsa, Okla.]

Appendix B

Paleothermometry

Estimating Diagenetic Paleotemperatures

Because temperature has a particularly significant effect on diagenetic processes, geologists are greatly interested in estimating the temperatures at which particular diagenetic reactions take place. Considerable research has been carried out to develop reliable techniques for paleotemperature analysis (e.g., Staplin et al., 1982). Tools used for determining paleotemperatures are called **geothermometers**. The principal techniques now in use for determining diagenetic paleotemperatures include the following.

Conodont Color Alteration

Conodonts are marine microfossils composed of the mineral apatite, but also containing trace amounts of organic material. They range in age from Cambrian to Triassic and occur principally in carbonate rocks and shales. Color changes in conodonts from pale yellow ($<\sim 50°-80°C$) to brown ($60°-140°C$) to black ($300°-400°$), owing to carbon fixation within trace amounts of organic material in the conodonts, as a result of increase in temperature. Extremely high temperatures may change the color to white (extreme thermal alteration) because of loss of carbon and water.

Vitrinite Reflectance

Vitrinite is structured or unstructured woody tissue plus tissue impregnations that occur as disseminated grains in sediment and as a major constituent in coals (Chapter 7). Light reflection from vitrinite (thermally altered organic grains) is measured quantitatively with a reflectance microscope. Reflection increased with increasing degree of thermal alteration (metamorphism). Up to about 240°C, percent vitrinite reflectance is related to minimum temperature by a scale that ranges from $<0.48\%$ at $<100°C$ to 2.14% at 240°C.

Graphitization Levels in Kerogen

Kerogen, which occurs principally in shales, is the disseminated organic matter of sedimentary rocks (insoluble in nonoxidizing acids, bases, and organic solvents).

Organic matter initially deposited in sediments is converted to kerogen during diagenesis by thermocatalytic processes. The carbon atoms in kerogen become increasingly well ordered (structured) with increasing levels of thermal diagenesis, a process called graphitization. Increase in the degree of ordering, or level of graphitization, as determined by X-ray diffraction methods, can thus be related to increase in diagenetic temperature. The method can be used to estimate temperatures up to about 600°C.

Analysis of Clay-Mineral Assemblages

Clay minerals (Chapter 5) are most abundant in shales but occur also as a matrix in sandstones and minor amounts in limestones. With increasing temperature and metamorphic grade, smectite clay minerals convert to illite through a mixed-layer illite/smectite series; chlorite appears, and kaolinite (as well as K-feldspar) disappears. Thus, the relative abundance of these clay minerals in rocks can be related to paleotemperatures (e.g., smectite occurs at temperatures below about 100°C; mixed-layer clays are stable up to about 200°C; illite forms at temperatures above 200°C). The useful temperature range of the method is up to about 300°C.

Analysis of Zeolite Facies Mineral Assemblages

Zeolite facies minerals form authigenically in volcaniclastic sediments through alteration of chemically reactive volcanic materials. They form in the overlapping temperature range of diagenesis and metamorphism and show a progression of mineral facies that is a reflection of temperature (and pressure) of burial. Temperature theoretically exerts a strong control on the types of zeolite minerals that occur together. For example, heulandite and analcite tend to form at temperatures below about 100°–125°C, whereas laumontite and pumpellyite form at temperatures between 100°–125°C and 175°–200°C. Thus, temperatures, up to about 200°–250°C, can be established roughly on the basis of the assemblages of zeolite minerals present in a rock.

Analysis of Fluid Inclusions

Fluid inclusions occur in a variety of sedimentary materials: geodes, vugs, and veins in sediments; sedimentary ore deposits; carbonate and quartz cements in sedimentary rocks; salt and sulfur deposits; petroleum reservoir rocks; and sphalerite (zinc-bearing ore mineral) in bituminous coal beds. Recrystallization of minerals or formation of overgrowths on minerals during diagenesis may trap fluids as minute inclusions in the crystals. Fluid inclusions commonly consist of a liquid plus a bubble of gas. Presumably, the inclusion consisted of a single fluid phase at the time of formation, which separated into two phases upon cooling. By reheating the mineral until the phase boundary between the liquid and gas can be seen to just disappear, the approximate temperature at which the inclusion formed can be established—taking into account an estimate of the pressure of formation. The method is useful at temperature ranges from about 25°C to 150°–200°C.

Oxygen Isotope Ratios

The fractionation factor α between two coexisting minerals is defined as

$$\alpha_{A-B} = \frac{(^{18}O/^{16}O)A}{(^{18}O/^{16}O)B}$$

where A and B refer to two oxygen-containing minerals. The ratio of $^{18}O/^{16}O$ in two coexisting oxygen-bearing minerals (from the same specimen) such as quartz and illite is commonly different. The amount of this difference has been shown to be a function of the maximum temperature to which the rock containing the minerals was heated during diagenesis. Therefore, the isotopic fractionation, or difference between the $^{18}O/^{16}O$ ratios of two minerals that have reached equilibrium with each other, can be used to calculate the maximum diagenetic temperature to which the rock was heated. For example, isotopic fractionation between quartz and illite pairs is greatest at low temperatures and decreases with increasing temperature. The method is useful to temperatures of about 400°C.

Detrital Thermochronology

Detrital thermochronology is a technique that permits age-dating of single mineral grains in sedimentary rocks to provide information about sediment provenance, thermal histories and exhumation of sediment source areas, and landscape evolution (Bernet and Spiegel, 2004a). Several techniques for detrital thermochronology are available, such as

1. Uranium-lead (U-Pb) dating of zircons
2. Argon-argon (^{40}A-^{39}A) dating of white micas
3. Fission-track dating of uranium-bearing minerals, particularly apatite and zircon
4. Fission-track dating of pebbles in conglomerates
5. Uranium–thorium/helium (U-Th/He) thermochronology, particularly of apatite and zircon

Application of detrital thermochronology to provenance analysis is of particular interest to sedimentologists. Estimating the ages of single mineral grains, such as zircon and apatite, provides ages of the source areas from which the minerals were derived. Such analysis makes possible correlation with specific source areas of known ages. Dating of pebbles by fission-track methods is particularly useful, because the pebbles provide information about the lithologic characteristics of the source areas as well as their ages.

Thermochemistry may also be applied to analysis of landscape evolution. Provenance information can be used to trace pathways of sediment transport and can help in the reconstruction of paleocurrent directions and identification of marine connections, which have implications for paleoecology and climatic evolution. Details of detrital thermochronologic techniques and applications may be found in Bernet and Spiegel (2004b) and references therein.

Appendix C

North American Stratigraphic Code[1]

North American Commission on Stratigraphic Nomenclature

FOREWORD

This code of recommended procedures for classifying and naming stratigraphic and related units has been prepared during a four-year period, by and for North American earth scientists, under the auspices of the North American Commission on Stratigraphic Nomenclature. It represents the thought and work of scores of persons, and thousands of hours of writing and editing. Opportunities to participate in and review the work have been provided throughout its development, as cited in the Preamble, to a degree unprecedented during preparation of earlier codes.

Publication of the International Stratigraphic Guide in 1976 made evident some insufficiencies of the American Stratigraphic Codes of 1961 and 1970. The Commission considered whether to discard our codes, patch them over, or rewrite them fully, and chose the last. We believe it desirable to sponsor a code of stratigraphic practice for use in North America, for we can adapt to new methods and points of view more rapidly than a worldwide body. A timely example was the recognized need to develop modes of establishing formal nonstratiform (igneous and high-grade metamorphic) rock units, an objective which is met in this Code, but not yet in the Guide.

The ways in which this Code differs from earlier American codes are evident from the Contents. Some categories have disappeared and others are new, but this Code has evolved from earlier codes and from the International Stratigraphic Guide. Some new units have not yet stood the test of long practice, and conceivably may not, but they are introduced toward meeting recognized and defined needs of the profession. Take this Code, use it, but do not condemn it because it contains something new or not of direct interest to you. Innovations that prove unacceptable to the profession will expire without damage to other concepts and procedures, just as did the geologic-climate units of the 1961 Code.

This Code is necessarily somewhat innovative because of: (1) the decision to write a new code, rather than to revise the old; (2) the open invitation to members of the geologic profession to offer suggestions and ideas, both in writing and orally; and (3) the progress in the earth sciences since completion of previous codes. This report strives to incorporate the strength and acceptance of established practice, with suggestions for meeting future needs perceived by our colleagues; its authors have attempted to bring together the good from the past, the lessons of the Guide, and carefully reasoned provisions for the immediate future.

Participants in preparation of this Code are listed in Appendix I, but many others helped with their suggestions and comments. Major contributions were made by the members, and especially the chairmen, of the named subcommittees and advisory groups under the guidance of the Code Committee, chaired by Steven S. Oriel, who also served as principal, but not sole, editor. Amidst the noteworthy contributions by many, those of James D. Aitken have been outstanding. The work was performed for and supported by the Commission, Chaired by Malcolm P. Weiss from 1978 to 1982.

This Code is the product of a truly North American effort. Many former and current commissioners representing not only the ten organizational members of the North American Commission on Stratigraphic Nomenclature (Appendix II), but other institutions as well, generated the product. Endorsement by constituent organizations is anticipated, and scientific communication will be fostered if Canadian, United States, and Mexican scientists, editors, and administrators consult Code recommendations for guidance in scientific reports. The Commission will appreciate reports of formal adoption or endorsement of the Code, and asks that they be transmitted to the Chairman of the Commission (c/o American Association of Petroleum Geologists, Box 979, Tulsa, Oklahoma 74101, U.S.A.).

Any code necessarily represents but a stage in the evolution of scientific communication. Suggestions for future changes of, or additions to, the North American Stratigraphic Code are welcome. Suggested and adopted modifications will be announced to the profession, as in the past, by serial Notes and Reports published in the *Bulletin* of the American Association of Petroleum Geologists. Suggestions may be made to representatives of your association or agency who are current commissioners, or directly to the Commission itself. The Commission meets annually, during the national meetings of the Geological Society of America.

1982 NORTH AMERICAN COMMISSION
ON STRATIGRAPHIC NOMENCLATURE

---

[1]Manuscript received, December 20, 1982; accepted, January 21, 1983. Copies are available at $1.00 per copy postpaid. Order from American Association of Petroleum Geologists, Box 979, Tulsa, Oklahoma 74101.

[1]Reprinted by permission from American Association of Petroleum Geologists Bulletin, v. 67, no. 5 (May 1983), pp. 841–875.

CONTENTS

PART I. PREAMBLE

BACKGROUND

PERSPECTIVE

Codes of Stratigraphic Nomenclature prepared by the American Commission on Stratigraphic Nomenclature (ACSN, 1961) and its predecessor (Committee on Stratigraphic Nomenclature 1933) have been used widely as a basis for stratigraphic terminology. Their formulation was a response to needs recognized during the past century by government surveys (both national and local) and by editors of scientific journals for uniform standards and common procedures in defining and classifying formal rock bodies, their fossils, and the time spans represented by them. The most recent Code (ACSN, 1970) is a slightly revised version of that published in 1961, incorporating some minor amendments adopted by the Commission between 1962 and 1969. The Codes have served the profession admirably and have been drawn upon heavily for codes and guides prepared in other parts of the world (ISSC, 1976, p. 104–106). The principles embodied by any code, however, reflect the state of knowledge at the time of its preparation, and even the most recent code is now in need of revision.

New concepts and techniques developed during the past two decades have revolutionized the earth sciences. Moreover, increasingly evident have been the limitations of previous codes in meeting some needs of Precambrian and Quaternary geology and in classification of plutonic, high-grade metamorphic, volcanic, and intensely deformed rock assemblages. In addition, the important contributions of numerous international stratigraphic organizations associated with both the International Union of Geological Sciences (IUGS) and UNESCO, including working groups of the International Geological Correlation Program (IGCP), merit recognition and incorporation into a North American code.

For these and other reasons, revision of the American Code has been undertaken by committees appointed by the North American Commission on Stratigraphic Nomenclature (NACSN). The Commission, founded as the American Commission on Stratigraphic Nomenclature in 1946 (ACSN, 1947), was renamed the NACSN in 1978 (Weiss, 1979b) to emphasize that delegates from ten organizations in Canada, the United States, and Mexico represent the geological profession throughout North America (Appendix II).

Although many past and current members of the Commission helped prepare this revision of the Code, the participation of all interested geologists has been sought (for example, Weiss, 1979a). Open forums were held at the national meetings of both the Geological Society of America at San Diego in November, 1979, and the American Association of Petroleum Geologists at Denver in June, 1980, at which comments and suggestions were offered by more than 150 geologists. The resulting draft of this report was printed, through the courtesy of the Canadian Society of Petroleum Geologists, on October 1, 1981, and additional comments were invited from the profession for a period of one year before submittal of this report to the Commission for adoption. More than 50 responses were received with sufficient suggestions for improvement to prompt moderate revision of the printed draft (NACSN, 1981). We are particularly indebted to Hollis D. Hedberg and Amos Salvador for their exhaustive and perceptive reviews of early drafts of this Code, as well as to those who responded to the request for comments. Participants in the preparation and revisions of this report, and conferees, are listed in Appendix I.

Some of the expenses incurred in the course of this work were defrayed by National Science Foundation Grant EAR 7919845, for which we express appreciation. Institutions represented by the participants have been especially generous in their support.

SCOPE

The North American Stratigraphic Code seeks to describe explicit practices for classifying and naming all formally defined geologic units. *Stratigraphic procedures* and principles, although developed initially to bring order to strata and the events recorded therein, are applicable to all earth materials, not solely to strata. They promote systematic and rigorous study of the compositions, geometry, sequence, history, and genesis of rocks and unconsolidated materials. They provide the framework within which time and space relations among rock bodies that constitute the Earth are ordered systematically. Stratigraphic procedures are used not only to reconstruct the history of the Earth and of extra-terrestrial bodies, but also to define the distribution and geometry of some commodities needed by society. *Stratigraphic classification* systematically arranges and partitions bodies of rock or unconsolidated materials of the Earth's crust into units based on their inherent properties or attributes.

A *stratigraphic code* or guide is a formulation of current views on stratigraphic principles and procedures designed to promote standardized classification and formal nomenclature of rock materials. It provides the basis for formalization of the language used to denote rock units and their spatial and temporal relations. To be effective, a code must be widely accepted and used; geologic organizations and journals may adopt its recommendations for nomenclatural procedure. Because any code embodies only current concepts and principles, it should have the flexibility to provide for both changes and additions to improve its relevance to new scientific problems.

Any system of nomenclature must be sufficiently explicit to enable users to distinguish objects that are embraced in a class from those that are not. This stratigraphic code makes no attempt to systematize structural, petrographic, paleontologic, or physiographic terms. Terms from these other fields that are used as part of formal stratigraphic names should be sufficiently general as to be unaffected by revisions of precise petrographic or other classifications.

The objective of a system of classification is to promote unambiguous communication in a manner not so restrictive as to inhibit scientific progress. To minimize ambiguity, a code must promote recognition of the distinction between observable features (reproducible data) and inferences or interpretations. Moreover, it should be sufficiently adaptable and flexible to promote the further development of science.

Stratigraphic classification promotes understanding of the *geometry* and *sequence* of rock bodies. The development of stratigraphy as a science required formulation of the Law of Superposition to explain sequential stratal relations. Although superposition is not applicable to many igneous, metamorphic, and tectonic rock assemblages, other criteria (such as cross-cutting relations and isotopic dating) can be used to determine sequential arrangements among rock bodies.

The term *stratigraphic unit* may be defined in several ways. Etymological emphasis requires that it be a stratum or assemblage of adjacent strata distinguished by any or several of the many properties that rocks may possess (ISSC, 1976, p. 13). The scope of stratigraphic classification and procedures, however, suggests a broader definition: a naturally occurring body of rock or rock material distinguished from adjoining rock on the basis of some stated property or properties. Commonly used properties include composition, texture, included fossils, magnetic signature, radioactivity, seismic velocity, and age. Sufficient care is required in defining the boundaries of a unit to enable others to distinguish the material body from those adjoining it. Units based on one property commonly do not coincide with those based on another and, therefore, distinctive terms are needed to identify the property used in defining each unit.

The adjective *stratigraphic* is used in two ways in the remainder of this report. In discussions of lithic (used here as synonymous with "lithologic") units, a conscious attempt is made to restrict the term to lithostratigraphic or layered rocks and sequences that obey the Law of Superposition. For nonstratiform rocks (of plutonic or tectonic origin, for example), the term *lithodemic* (see Article 27) is used. The adjective *stratigraphic* is also used in a broader sense to refer to those procedures derived from stratigraphy which are now applied to all classes of earth materials.

An assumption made in the material that follows is that the reader has some degree of familiarity with basic principles of stratigraphy as outlined, for example, by Dunbar and Rodgers (1957), Weller (1960), Shaw (1964), Matthews (1974), or the International Stratigraphic Guide (ISSC, 1976).

RELATION OF CODES TO INTERNATIONAL GUIDE

Publication of the International Stratigraphic Guide by the International Subcommission on Stratigraphic Classification (ISSC, 1976), which is being endorsed and adopted throughout the world, played a part in prompting examination of the American Stratigraphic Code and the decision to revise it.

The International Guide embodies principles and procedures that had been adopted by several national and regional stratigraphic committees and commissions. More than two decades of effort by H. D. Hedberg and other members of the Subcommission (ISSC, 1976, p. VI, 1, 3) developed the consensus required for preparation of the Guide. Although the Guide attempts to cover all kinds of rocks and the diverse ways of investigating them, it is necessarily incomplete. Mechanisms are needed to stimulate individual innovations toward promulgating new concepts, principles, and practices which subsequently may be found worthy of inclusion in later editions of the Guide. The flexibility of national and regional committees or commissions enables them to perform this function more readily than an international subcommission, even while they adopt the Guide as the international standard of stratigraphic classification.

A guiding principle in preparing this Code has been to make it as consistent as possible with the international Guide, which was endorsed by the ACSN in 1976, and at the same time to foster further innovations to meet the expanding and changing needs of earth scientists on the North American continent.

OVERVIEW

CATEGORIES RECOGNIZED

An attempt is made in this Code to strike a balance between serving the needs of those in evolving specialties and resisting the proliferation of categories of units. Consequently, more formal categories are recognized here than in previous codes or in the International Guide (ISSC, 1976). On the other hand, no special provision is made for formalizing certain kinds of units (deep oceanic, for example) which may be accommodated by available categories.

Four principal categories of units have previously been used widely in traditional stratigraphic work; these have been termed lithostratigraphic, biostratigraphic, chronostratigraphic, and geochronologic and are distinguished as follows:

1. A *lithostratigraphic unit* is a stratum or body of strata, generally but not invariably layered, generally but not invariably tabular, which conforms to the Law of Superposition and is distinguished and delimited on the basis of lithic characteristics and stratigraphic position. Example: Navajo Sandstone.

2. A *biostratigraphic unit* is a body of rock defined and characterized by its fossil content. Example: *Discoaster multiradiatus* Interval Zone.

3. A *chronostratigraphic unit* is a body of rock established to serve as the material reference for all rocks formed during the same span of time. Example: Devonian System. Each boundary of a chronostratigraphic unit is synchronous. Chronostratigraphy provides a means of organizing strata into units based on their age relations. A chronostratigraphic body also serves as the basis for defining the specific interval of geologic time, or geochronologic unit, represented by the referent.

4. A *geochronologic unit* is a division of time distinguished on the basis of the rock record preserved in a chronostratigraphic unit. Example: Devonian Period.

The first two categories are comparable in that they consist of material units defined on the basis of content. The third category differs from the first two in that it serves primarily as the standard for recognizing and isolating

materials of a specific age. The fourth, in contrast, is not a material, but rather a conceptual, unit; it is a division of time. Although a geochronologic unit is not a stratigraphic body, it is so intimately tied to chronostratigraphy that the two are discussed properly together.

Properties and procedures that may be used in distinguishing geologic units are both diverse and numerous (ISSC, 1976, p. 1, 96; Harland, 1977, p. 230), but all may be assigned to the following principal classes of categories used in stratigraphic classification (Table 1), which are discussed below:

I. Material categories based on content, inherent attributes, or physical limits,

II. Categories distinguished by geologic age:

A. Material categories used to define temporal spans, and

B. Temporal categories.

Table 1. Categories of Units Defined*

MATERIAL CATEGORIES BASED ON CONTENT OR PHYSICAL LIMITS

Lithostratigraphic (22)
Lithodemic (31)**
Magnetopolarity (44)
Biostratigraphic (48)
Pedostratigraphic (55)
Allostratigraphic (58)

CATEGORIES EXPRESSING OR RELATED TO GEOLOGIC AGE
Material Categories Used to Define Temporal Spans
Chronostratigraphic (66)
Polarity-Chronostratigraphic (83)
Temporal (Non-Material) Categories
Geochronologic (80)
Polarity-Chronologic (88)
Diachronic (91)
Geochronometric (96)

*Numbers in parentheses are the numbers of the Articles where units are defined.
**Italicized categories are those introduced or developed since publication of the previous code (ACSN, 1970).

Material Categories Based on Content or Physical Limits

The basic building blocks for most geologic work are rock bodies defined on the basis of composition and related lithic characteristics, or on their physical, chemical, or biologic content or properties. Emphasis is placed on the relative objectivity and reproducibility of data used in defining units within each category.

Foremost properties of rocks are composition, texture, fabric, structure, and color, which together are designated *lithic characteristics*. These serve as the basis for distinguishing and defining the most fundamental of all formal units. Such units based primarily on composition are divided into two categories (Henderson and others, 1980): lithostratigraphic (Article 22) and lithodemic (defined here in Article 31). A lithostratigraphic unit obeys the Law of Superposition, whereas a lithodemic unit does not. A *lithodemic unit* is a defined body of predominantly intrusive, highly metamorphosed, or intensely deformed rock that, because it is intrusive or has lost primary structure through metamorphism or tectonism, generally does not conform to the Law of Superposition.

Recognition during the past several decades that remanent magnetism in rocks records the Earth's past magnetic characteristic (Cox, Doell, and Dalrymple, 1963) provides a powerful new tool encompassed by magnetostratigraphy (McDougall, 1977; McElhinny, 1978). *Magnetostratigraphy* (Article 43) is the study of remanent magnetism in rocks; it is the record of the Earth's magnetic polarity (or field reversals), dipole-field-pole position (including apparent polar wander), the non-dipole component (secular variation), and field intensity. Polarity is of particular utility and is used to define a *magnetopolarity unit* (Article 44) as a body of rock identified by its remanent magnetic polarity (ACSN, 1976; ISSC, 1979). Empirical demonstration of uniform polarity does not necessarily have direct temporal connotations because the remanent magnetism need not be related to rock deposition or crystallization. Nevertheless, polarity is a physical attribute that may characterize a body of rock.

Biologic remains contained in, or forming, strata are uniquely important in stratigraphic practice. First, they provide the means of defining and recognizing material units based on fossil content (biostratigraphic units). Second, the irreversibility of organic evolution makes it possible to partition enclosing strata temporally. Third, biologic remains provide important data for the reconstruction of ancient environments of deposition.

Composition also is important in distinguishing pedostratigraphic units. A *pedostratigraphic unit* is a body of rock that consists of one or more pedologic horizons developed in one or more lithic units now buried by a formally defined lithostratigraphic or allostratigraphic unit or units. A pedostratigraphic unit is the part of a buried soil characterized by one or more clearly defined soil horizons containing pedogenically formed minerals and organic compounds. Pedostratigraphic terminology is discussed below and in Article 55.

Many upper Cenozoic, especially Quaternary, deposits are distinguished and delineated on the basis of content, for which lithostratigraphic classification is appropriate. However, others are delineated on the basis of criteria other than content. To facilitate the reconstruction of geologic history, some compositionally similar deposits in vertical sequence merit distinction as separate stratigraphic units because they are the products of different processes; others merit distinction because they are of demonstrably different ages. Lithostratigraphic classification of these units is impractical and a new approach, allostratigraphic classification, is introduced here and may prove applicable to older deposits as well. An *allostratigraphic unit* is a mappable stratiform body of sedimentary rock defined and identified on the basis of bounding discontinuities (Article 58 and related Remarks).

Geologic-Climate units, defined in the previous Code (ACSN, 1970, p. 31), are abandoned here because they proved to be of dubious utility. Inferences regarding climate are subjective and too tenuous a basis for the definition of formal geologic units. Such inferences commonly are based on deposits assigned more appropriately to lithostratigraphic or allostratigraphic units and may be expressed in terms of diachronic units (defined below).

Categories Expressing or Related to Geologic Age

Time is a single, irreversible continuum. Nevertheless, various categories of units are used to define intervals of geologic time, just as terms having different bases, such as Paleolithic, Renaissance, and Elizabethan, are used to designate specific periods of human history. Different temporal categories are established to express intervals of time distinguished in different ways.

Major objectives of stratigraphic classification are to provide a basis for systematic ordering of the time and space relations of rock bodies and to establish a time framework for the discussion of geologic history. For such purposes, units of geologic time traditionally have been named to represent the span of time during which a well-described sequence of rock, or a chronostratigraphic unit, was deposited ("time units based on material referents," Fig. 1). This procedure continues, to the exclusion of other possible approaches, to be standard practice in studies of Phanerozoic rocks. Despite admonitions in previous American codes and the International Stratigraphic Guide (ISSC, 1976, p. 81) that similar procedures should be applied to the Precambrian, no comparable chronostratigraphic units, or geochronologic units derived therefrom, proposed for the Precambrian

have yet been accepted worldwide. Instead, the IUGS Subcommission on Precambrian Stratigraphy (Sims, 1979) and its Working Groups (Harrison and Peterman, 1980) recommend division of Precambrian time into *geochronometric units* having no material referents.

A distinction is made throughout this report between *isochronous* and *synchronous*, as urged by Cumming, Fuller, and Porter (1959, p. 730), although the terms have been used synonymously by many. *Isochronous* means of equal duration; *synchronous* means simultaneous, or occurring at the same time. Although two rock bodies of very different ages may be formed during equal durations of time, the term *isochronous* is not applied to them in the earth sciences. Rather, isochronous bodies are those bounded by synchronous surfaces and formed during the same span of time. *Isochron*, in contrast, is used for a line connecting pints of equal age on a graph representing physical or chemical phenomena; the line represents the same or equal time. The adjective *diachronous* is applied either to a rock unit with one or two bounding surfaces which are not synchronous, or to a boundary which is not synchronous (which "transgresses time").

Two classes of time units based on material referents, or stratotypes, are recognized (Fig. 1). The first is that of the traditional and conceptually isochronous units, and includes *geochronologic units*, which are based on *chronostratigraphic units*, and *polarity-geochronologic units*. These isochronous units have worldwide applicability and may be used even in areas lacking a material record of the named span of time. The second class of time units, newly defined in this Code, consists of *diachronic units* (Article 91), which are based on rock bodies known to be diachronous. In contrast to isochronous units, a diachronic term is used only where a material referent is present; a diachronic unit is coextensive with the material body or bodies on which it is based.

A *chronostratigraphic unit*, as defined above and in Article 66, is a body of rock established to serve as the material reference for all rocks formed during the same span of time; its boundaries are synchronous. It is the referent for a *geochronologic unit*, as defined above and in Article 80. Internationally accepted and traditional chronostratigraphic units were based initially on the time spans of lithostratigraphic units, biostratigraphic units, or other features of the rock record that have specific durations. In sum, they form the Standard Global Chronostratigraphic Scale (ISSC, 1976, p. 76–81; Harland, 1978), consisting of established systems and series.

A *polarity-chronostratigraphic unit* is a body of rock that contains a primary magnetopolarity record imposed when the rock was deposited or crystallized (Article 83). It serves as a material standard or referent for a part of geologic time during which the Earth's magnetic field had a characteristic polarity or sequence of polarities; that is, for a *polarity-chronologic unit* (Article 88).

A *diachronic unit* comprises the unequal spans of time represented by one or more specific diachronous rock bodies (Article 91). Such bodies

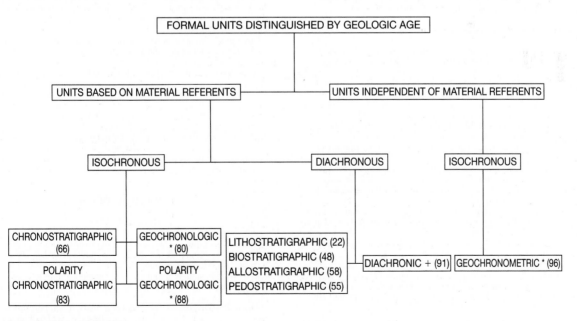

* Applicable world-wide
+ Applicable only where material referents are present
( ) Number of article in which defined

Figure 1.
Relation of geologic time units to the kinds of rock-unit referents on which most are based.

may be lithostratigraphic, biostratigraphic, pedostratigraphic, allostratigraphic, or an assemblage of such units. A diachronic unit is applicable only where its material referent is present.

A *geochronometric* (or chronometric) *unit* is an isochronous direct division of geologic time expressed in years (Article 96). It has no material referent.

Pedostratigraphic Terms

The definition and nomenclature for pedostratigraphic[2] units in this Code differ from those for soil-stratigraphic units in the previous Code (ACSN, 1970, Article 18), by being more specific with regard to content, boundaries, and the basis for determining stratigraphic position.

The term "soil" has different meanings to the geologist, the soil scientist, the engineer, and the layman, and commonly has no stratigraphic significance. The term *paleosol* is currently used in North America for any soil that formed on a landscape of the past; it may be a buried soil, a relict soil, or an exhumed soil (Ruhe, 1965; Valentine and Dalrymple, 1976).

A *pedologic soil* is composed of one or more soil horizons.[3] A *soil horizon* is a layer within a pedologic soil that (1) is approximately parallel to the soil surface, (2) has distinctive physical, chemical, biological, and morphological properties that differ from those of adjacent, genetically related, soil horizons, and (3) is distinguished from other soil horizons by objective compositional properties that can be observed or measured in the field. The physical boundaries of buried pedologic horizons are objective traceable boundaries with stratigraphic significance. A buried pedologic soil provides the material basis for definition of a stratigraphic unit in pedostratigraphic classification (Article 55), but a buried pedologic soil may be somewhat more inclusive than a pedostratigraphic unit. A pedologic soil may contain both an O-horizon and the entire C-horizon (Fig. 6), whereas the former is excluded and the latter need not be included in a pedostratigraphic unit.

The definition and nomenclature for pedostratigraphic units in this Code differ from those of soil stratigraphic units proposed by the International Union for Quaternary Research and International Society of Soil Science (Parsons, 1981). The pedostratigraphic unit, geosol, also differs from the proposed INQUA-ISSS soil-stratigraphic unit, pedoderm, in several ways, the most important of which are: (1) a geosol may be in any part of the geologic column, whereas a pedoderm is a surficial soil: (2) a geosol is a buried soil, whereas a pedoderm may be a buried, relict, or exhumed soil: (3) the boundaries and stratigraphic position of a geosol are defined and delineated by criteria that differ from those for a pedoderm; and (4) a geosol may be either all or only a part of a buried soil, whereas a pedoderm is the entire soil.

The term *geosol*, as defined by Morrison (1967, p. 3), is a laterally traceable, mappable, geologic weathering profile that has a consistent stratigraphic position. The term is adopted and redefined here as the fundamental and only unit in formal pedostratigraphic classification (Article 56).

FORMAL AND INFORMAL UNITS

Although the emphasis in this Code is necessarily on formal categories of geologic units, informal nomenclature is highly useful in stratigraphic work.

Formally named units are those that are named in accordance with an established scheme of classification; the fact of formality is conveyed by capitalization of the initial letter of the *rank* or *unit* term (for example, Morrison Formation). Informal units, whose unit terms are ordinary nouns, are not protected by the stability provided by proper formalization and recommended classification procedures. Informal terms are devised for both economic and scientific reasons. Formalization is appropriate for those units requiring stability of nomenclature, particularly those likely to be extended far beyond the locality in which they were first recognized. Informal terms are appropriate for casually mentioned, innovative, and most economic units, those defined by unconventional criteria, and those that may be too thin to map at usual scales.

Casually mentioned geologic units not defined in accordance with this Code are informal. For many of these, there may be insufficient need or information, or perhaps an inappropriate basis, for formal designations. Informal designations as beds or lithozones (the pebbly beds, the shaly zone, third coal) are appropriate for many such units.

Most economic units, such as aquifers, oil sands, coal beds, quarry layers, and ore-bearing "reefs," are informal, even though they may be named. Some such units, however, are so significant scientifically and economically that they merit formal recognition as beds, members, or formations.

Innovative approaches in regional stratigraphic studies have resulted in the recognition and definition of units best left as informal, at least for the time being. Units bounded by major regional unconformities on the North American craton were designated "sequences" (example: Sauk sequence) by Sloss (1963). Major unconformity-bounded units also were designated "synthems" by Chang (1975), who recommended that they be treated formally. Marker-defined units that are continuous from one lithofacies to another were designated "formats" by Forgotson (1957). The term "chronosome" was proposed by Schultz (1982) for rocks of diverse facies corresponding to geographic variations in sedimentation during an interval of deposition identified on the basis of bounding stratigraphic markers. Successions of faunal zones containing evolutionarily related forms, but bounded by non-evolutionary biotic discontinuities, were termed "biomeres" (Palmer, 1965). The foregoing are only a few selected examples to demonstrate how informality provides a continuing avenue for innovation.

The terms *magnafacies* and *parvafacies*, coined by Caster (1934) to emphasize the distinction between lithostratigraphic and chronostratigraphic units in sequences displaying marked facies variation, have remained informal despite their impact on clarifying the concepts involved.

Tephrochronologic studies provide examples of informal units too thin to map at conventional scales but yet invaluable for dating important geologic events. Although some such units are named for physiographic features and places where first recognized (e.g., Guaje pumice bed, where it is not mapped as the Guaje Member of the Bandelier Tuff), others bear the same name as the volcanic vent (e.g., Huckleberry Ridge ash bed of Izett and Wilcox, 1981).

Informal geologic units are designated by ordinary nouns, adjectives or geographic terms and lithic or unit-terms that are not capitalized (chalky formation or beds, St. Francis coal).

No geologic unit should be established and defined, whether formally or informally, unless its recognition serves a clear purpose.

CORRELATION

Correlation is a procedure for demonstrating correspondence between geographically separated parts of a geologic unit. The term is a general one having diverse meanings in different disciplines. Demonstration of temporal correspondence is one of the most important objectives of stratigraphy. The term "correlation" frequently is misused to express the idea that a unit has been identified or recognized.

Correlation is used in this Code as the demonstration of correspondence between two geologic units in both some defined property and relative stratigraphic position. Because correspondence may be based on various properties, three kinds of correlation are best distinguished by more specific terms. *Lithocorrelation* links units of similar lithology and stratigraphic position (or sequential or geometric relation, for lithodemic units). *Biocorrelation* expresses similarity of fossil content and biostratigraphic position. *Chronocorrelation* expresses correspondence in age and in chronostratigraphic position.

Other terms that have been used for the similarity of content and stratal succession are homotaxy and chronotaxy. *Homotaxy* is the similarity in separate regions of the serial arrangement or succession of strata of comparable compositions or of included fossils. The term is derived from *homotaxis*, proposed by Huxley (1862, p. xlvi) to emphasize that similarity in succession does not prove age equivalence of comparable units. The term *chronotaxy* has been applied to similar stratigraphic sequences composed of units which are of equivalent age (Henbest, 1952, p. 310).

Criteria used for ascertaining temporal and other types of correspondence are diverse (ISSC, 1976, p. 86–93) and new criteria will emerge in the future. Evolving statistical tests, as well as isotopic and paleomagnetic techniques, complement the traditional paleontologic and lithologic procedures. Boundaries defined by one set of criteria need not correspond to those defined by others.

PART II. ARTICLES

INTRODUCTION

Article 1.—**Purpose.** This Code describes explicit stratigraphic procedures for classifying and naming geologic units accorded formal status. Such procedures, if widely adopted, assure consistent and uniform usage in classification and terminology and therefore promote unambiguous communication.

Article 2.—**Categories.** Categories of formal stratigraphic units, though diverse, are of three classes (Table 1). The first class is of rock-material categories based on inherent attributes or content and stratigraphic position, and includes lithostratigraphic, lithodemic, magnetopolarity, biostratigraphic, pedostratigraphic, and allostratigraphic units. The second class is of material categories used as standards for defining spans of geologic time, and includes chronostratigraphic and polarity-chronostratigraphic

---

[2]From Greek, *pedon*, ground or soil.
[3]As used in a geological sense, a *horizon* is a surface or line. In pedology, however, it is a body of material, and such usage is continued here.

units. The third class is of non-material temporal categories, and includes geochronologic, polarity-chronologic, geochronometric, and diachronic units.

GENERAL PROCEDURES

DEFINITION OF FORMAL UNITS

Article 3.—Requirements for Formally Named Geologic Units. Naming, establishing, revising, redefining, and abandoning formal geologic units require publication in a recognized scientific medium of a comprehensive statement which includes: (i) intent to designate or modify a formal unit; (ii) designation of category and rank of unit; (iii) selection and derivation of name; (iv) specification of stratotype (where applicable); (v) description of unit; (vi) definition of boundaries; (vii) historical background; (viii) dimensions, shape, and other regional aspects: (ix) geologic age; (x) correlations; and possibly (xi) genesis (where applicable). These requirements apply to subsurface and offshore, as well as exposed, units.

Article 4.—Publication.[4] "Publication in a recognized scientific medium" in conformance with this Code means that a work, when first issued, must (1) be reproduced in ink on paper or by some method that assures numerous identical copies and wide distribution; (2) be issued for the purpose of scientific, public, permanent record; and (3) be readily obtainable by purchase or free distribution.

Remarks. (a) **Inadequate publication.**—The following do not constitute publication within the meaning of the Code: (1) distribution of microfilms, microcards, or matter reproduced by similar methods; (2) distribution to colleagues or students of a note, even if printed, in explanation of an accompanying illustration; (3) distribution of proof sheets; (4) open-file release; (5) theses, dissertations, and dissertation abstracts; (6) mention at a scientific or other meeting; (7) mention in an abstract, map explanation, or figure caption; (8) labeling of a rock specimen in a collection; (9) mere deposit of a document in a library; (10) anonymous publication; or (11) mention in the popular press or in a legal document.

(b) **Guidebooks.**—A guidebook with distribution limited to participants of a field excursion does not meet the test of availability. Some organizations publish and distribute widely large editions of serial guidebooks that include refereed regional papers; although these do meet the tests of scientific purpose and availability, and therefore constitute valid publication, other media are preferable.

Article 5.—Intent and Utility. To be valid, a new unit must serve a clear purpose and be duly proposed and duly described, and the intent to establish it must be specified. Casual mention of a unit, such as "the granite exposed near the Middleville schoolhouse," does not establish a new formal unit, nor does mere use in a table, columnar section, or map.

Remark. (a) **Demonstration of purpose served.**—The initial definition or revision of a named geologic unit constitutes, in essence, a proposal. As such, it lacks status until use by others demonstrates that a clear purpose has been served. A unit becomes established through repeated demonstration of its utility. The decision not to use a newly proposed or a newly revised term requires a full discussion of its unsuitability.

Article 6.—Category and Rank. The category and rank of a new or revised unit must be specified.

Remark. (a) **Need for specification.**—Many stratigraphic controversies have arisen from confusion or misinterpretation of the category of a unit (for example, lithostratigraphic vs. chronostratigraphic). Specification and unambiguous description of the category is of paramount importance. Selection and designation of an appropriate rank from the distinctive terminology developed for each category help serve this function (Table 2).

Article 7.—Name. The name of a formal geologic unit is compound. For most categories, the name of a unit should consist of a geographic name combined with an appropriate rank (Wasatch Formation) or descriptive term (Viola Limestone). Biostratigraphic units are designated by appropriate biologic forms (*Exus albus* Assemblage Biozone). Worldwide chronostratigraphic units bear long established and generally accepted names of diverse origins (Triassic System). The first letters of all words used in the names of formal geologic units are capitalized (except for the trivial species and subspecies terms in the name of a biostratigraphic unit).

Remarks. (a) **Appropriate geographic terms.**—Geographic names derived from permanent natural or artificial features at or near which the unit is present are preferable to those derived from impermanent features such as farms, schools, stores, churches, crossroads, and small communities. Appropriate names may be selected from those shown on topographic, state, provincial, county, forest service, hydrographic, or comparable maps, particularly those showing names approved by a national board for geographic names. The generic part of a geographic name, e.g., river, lake, village, should be omitted from new terms, unless required to distinguish between two otherwise identical names (e.g., Redstone Formation and Redstone River Formation). Two names should not be derived from the same geographic feature. A unit should not be named for the source of its components; for example, a deposit inferred to have been derived from the Keewatin glaciation center should not be designated the "Keewatin Till."

(b) **Duplication of names.**—Responsibility for avoiding duplication, either in use of the same name for different units (homonymy) or in use of different names for the same unit (synonymy), rests with the proposer. Although the same geographic term has been applied to different categories of units (example: the lithostratigraphic Word Formation and the chronostratigraphic Wordian Stage) now entrenched in the literature, the practice is undesirable. The extensive geologic nomenclature of North America, including not only names but also nomenclatural history of formal units, is recorded in compendia maintained by the Committee on Stratigraphic Nomenclature of the Geological Survey of Canada, Ottawa, Ontario; by the Geologic Names Committee of the United States Geological Survey, Reston, Virginia; by the Instituto de Geologia, Ciudad Universitaria, México, D.F.; and by many state and provincial geological surveys. These organizations respond to inquires regarding the availability of names, and some are prepared to reserve names for units likely to be defined in the next year or two.

(c) **Priority and preservation of established names.**—Stability of nomenclature is maintained by use of the rule of priority and by preservation of well-established names. Names should not be modified without explaining the need. Priority in publication is to be respected, but priority alone does not justify displacing a well-established name by one neither well-known nor commonly used; nor should an inadequately established name be preserved merely on the basis of priority. Redefinitions in precise terms are preferable to abandonment of the names of well-established units which may have been defined imprecisely but nonetheless in conformance with older and less stringent standards.

(d) **Differences of spelling and changes in name.**—The geographic component of a well-established stratigraphic name is not changed due to differences in spelling or changes in the name of a geographic feature. The name Bennett Shale, for example, used for more than half a century, need not be altered because the town is named Bennet. Nor should the Mauch Chunk Formation be changed because the town has been renamed Jim Thorpe. Disappearance of an impermanent geographic feature, such as a town, does not affect the name of an established geologic unit.

(e) **Names in different countries and different languages.**—For geologic units that cross local and international boundaries, a single name for each is preferable to several. Spelling of a geographic name commonly conforms to the usage of the country and linguistic group involved. Although geographic names are not translated (Cuchillo is not translated to Knife), lithologic or rank terms are (Edwards Limestone, Caliza Edwards; Formación La Casita, La Casita Formation).

Article 8.—Stratotypes. The designation of a unit or boundary stratotype (type section or type locality) is essential in the definition of most formal geologic units. Many kinds of units are best defined by reference to an accessible and specific sequence of rock that may be examined and studied by others. A stratotype is the standard (original or subsequently designated) for a named geologic unit or boundary and constitutes the basis for definition or recognition of that unit or boundary; therefore, it must be illustrative and representative of the concept of the unit or boundary being defined.

Remarks. (a) **Unit stratotypes.**—A unit stratotype is the type section for a stratiform deposit or the type area for a nonstratiform body that serves as the standard for definition and recognition of a geologic unit. The upper and lower limits of a unit stratotype are designated points in a specific sequence or locality and serve as the standards for definition and recognition of a stratigraphic unit's boundaries.

(b) **Boundary stratotype.**—A boundary stratotype is the type locality for the boundary reference point for a stratigraphic unit. Both boundary stratotypes for any unit need to be in the same section or region. Each boundary stratotype serves as the standard for definition and recognition of the base of a stratigraphic unit. The top of a unit may be defined by the boundary stratotype of the next higher stratigraphic unit.

(c) **Type locality.**—A type locality is the specified geographic locality where the stratotype of a formal unit or unit boundary was originally defined and named. A type area is the geographic territory encompassing the type locality. Before the concept of a stratotype was developed, only type localities and areas were designated for many geologic units which are now long- and well-established. Stratotypes, though now mandatory in defining most stratiform units, are impractical in definitions of many large nonstratiform rock bodies whose diverse major components may be best displayed at several reference localities.

(d) **Composite-stratotype.**—A composite-stratotype consists of several reference sections (which may include a type section) required to demonstrate the range or totality of a stratigraphic unit.

---

[4]This article is modified slightly from a statement by the International Commission of Zoological Nomenclature (1964, p. 7–9).

Table 2. Categories and Ranks of Units Defined in This Code*

A. Material Units

LITHOSTRATIGRAPHIC	LITHODEMIC	MAGNETOPOLARITY	BIOSTRATIGRAPHIC	PEDOSTRATIGRAPHIC	ALLOSTRATIGRAPHIC
Supergroup	Supersuite (Complex)				
Group	Suite (Complex)	Polarity Superzone			Allogroup
Formation	*Lithodeme*	*Polarity zone*	*Biozone* (interval, Assemblage or Abundance)	*Geosol*	*Alloformation*
Member (or Lens, or Tongue)		Polarity Subzone	Subbiozone		Allomember
Bed(s) or Flow(s)					

B. Temporal and Related Chronostratigraphic Units

CHRONO-STRATIGRAPHIC	GEOCHRONOLOGIC GEOCHRONOMETRIC
Eonothem	Eon
Erathem (Supersystem)	Era (Superperiod)
System (Subsystem)	*Period* (Subperiod)
Series	Epoch
Stage (Substage)	Age (Subage)
Chronozone	Chron

POLARITY CHRONO-STRATIGRAPHIC	POLARITY CHRONOLOGIC
Polarity Superchronozone	Polarity Superchron
Polarity Chronozone	*Polarity Chron*
Polarity Subchronozone	Polarity Subchron

DIACHRONIC	
	Episode
Diachron	Phase
Diachron	Span
	Cline

*Fundamental units are italicized.

(e) **Reference sections.**—Reference sections may serve as invaluable standards in definitions or revisions of formal geologic units. For those well-established stratigraphic units for which a type section never was specified, a principal reference section (lectostratotype of ISSC, 1976, p. 26) may be designated. A principal reference section (neostratotype of ISSC, 1976, p. 26) also may be designated for those units or boundaries whose stratotypes have been destroyed, covered, or otherwise made inaccessible. Supplementary reference sections often are designated to illustrate the diversity or heterogeneity of a defined unit or some critical feature not evident or exposed in the stratotype. Once a unit or boundary stratotype section is designated, it is never abandoned or changed; however, if a stratotype proves inadequate, it may be supplemented by a principal reference section or by several reference sections that may constitute a composite-stratotype.

(f) **Stratotype descriptions.**—Stratotypes should be described both geographically and geologically. Sufficient geographic detail must be included to enable others to find the stratotype in the field, and may consist of maps and/or aerial photographs showing location and access, as well as appropriate coordinates or bearings. Geologic information should include thickness, descriptive criteria appropriate to the recognition of the unit and its boundaries, and discussion of the relation of the unit to other geologic units of the area. A carefully measured and described section provides the best foundation for definition of stratiform units. Graphic profiles, columnar sections, structure-sections, and photographs are useful supplements to a description; a geologic map of the area including the type locality if essential.

Article 9.—**Unit Description.** A unit proposed for formal status should be described and defined so clearly that any subsequent investigator can recognize that unit unequivocally. Distinguishing features that characterize a unit may include any or several of the following: composition, texture, primary structures, structural attitudes, biologic remains, readily apparent mineral composition (e.g., calcite vs. dolomite), geochemistry, geophysical properties (including magnetic signatures), geomorphic expression, unconformable or cross-cutting relations, and age. Although all distinguishing features pertinent to the unit category should be described sufficiently to characterize the unit, those not pertinent to the category (such as age and inferred genesis for lithostratigraphic units, or lithology for biostratigraphic units) should not be made part of the definition.

Article 10.—**Boundaries.** The criteria specified for the recognition of boundaries between adjoining geologic units are of paramount importance because they provide the basis for scientific reproducibility of results. Care is required in describing the criteria, which must be appropriate to the category of unit involved.

Remarks. (a) **Boundaries between intergradational units.**—Contacts between rocks of markedly contrasting composition are appropriate boundaries of lithic units, but some rocks grade into, or intertongue with, others of different lithology. Consequently, some boundaries are necessarily arbitrary as, for example, the top of the uppermost limestone in a sequence of interbedded limestone and shale. Such arbitrary boundaries commonly are diachronous.

(b) **Overlaps and gaps.**—The problem of overlaps and gaps between long-established adjacent chronostratigraphic units is being addressed by international IUGS and IGCP working groups appointed to deal with various parts of the geologic column. The procedure recommended by the Geological Society of London (George and others, 1969; Holland and others, 1978), of defining only the basal boundaries of chronostratigraphic units, has been widely adopted (e.g., McLaren, 1977) to resolve the problem. Such boundaries are defined by a carefully selected and agreed-upon boundary-stratotype (marker-point type section or "golden spike") which becomes the standard for the base of a chronostratigraphic unit. The concept of the mutual-boundary stratotype (ISSC, 1976, p. 84–86), based on the assumption of continuous deposition in selected sequences, also has been used to define chronostratigraphic units.

Although international chronostratigraphic units of series and higher rank are being redefined by IUGS and IGCP working groups, there may be a continuing need for some provincial series. Adoption of the basal boundary-stratotype concept is urged.

Article 11.—**Historical Background.** A proposal for a new name must include a nomenclatorial history of rocks assigned to the proposed unit, describing how they were treated previously and by whom (references), as well as such matters as priorities, possible synonymy, and other pertinent considerations. Consideration of the historical background of an older unit commonly provides the basis for justifying definition of a new unit.

Article 12.—**Dimensions and Regional Relations.** A perspective on the magnitude of a unit should be provided by such information as may be available on the geographic extent of a unit; observed ranges in thickness, composition, and geomorphic expression; relations to other kinds and ranks of stratigraphic units; correlations with other nearby sequences; and the bases for recognizing and extending the unit beyond the type locality. If the unit is not known anywhere but in an area of limited extent, informal designation is recommended.

Article 13.—**Age.** For most formal material geologic units, other than chronostratigraphic and polarity-chronostratigraphic, inferences regarding geologic age play no proper role in their definition. Nevertheless, the age, as well as the basis for its assignment, are important features of the unit and should be stated. For many lithodemic units, the age of the protolith should be distinguished from that of the metamorphism or deformation. If the basis for assigning an age is tenuous, a doubt should be expressed.

Remarks.(a) **Dating.**—The geochronologic ordering of the rock record, whether in terms of radioactive-decay rates or other processes, is generally called "dating." However, the use of the noun "date" to mean "isotopic age" is not recommended. Similarly, the term "absolute age" should be suppressed in favor of "isotopic age" for an age determined on the basis of isotopic ratios. The more inclusive term "numerical age" is recommended for all ages determined from isotopic ratios, fission tracks, and other quantifiable age-related phenomena.

(b) **Calibration.**—The dating of chronostratigraphic boundaries in terms of numerical ages is a special form of dating for which the word "calibration" should be used. The geochronologic time-scale now in use has been developed mainly through such calibration of chronostratigraphic sequences.

(c) **Convention and abbreviations.**—The age of a stratigraphic unit or the time of a geologic event, as commonly determined by numerical dating or by reference to a calibrated time-scale, may be expressed in years before the present. The unit of time is the modern year as presently recognized worldwide. Recommended (but not mandatory) abbreviations for such ages are SI (International System of Units) multipliers coupled with "a" for annum: ka, Ma, and Ga[5] for kilo-annum (10^3 years), Mega-annum (10^6 years), and Giga-annum (10^9 years), respectively. Use of these terms after the age value follows the convention established in the field of C-14 dating. The "present" refers to 1950 AD, and such qualifiers as "ago" or "before the present" are omitted after the value because measurement of the duration from the present to the past is implicit in the designation. In contrast, the duration of a remote interval of geologic time, as a number of years, should not be expressed by the same symbols. Abbreviations for numbers of years, without reference to the present, are informal (e.g., y or yr for years; my, m.y., or m.yr. for millions of years; and so forth, as preference dictates). For example, boundaries of the Late Cretaceous Epoch currently are calibrated at 63 Ma and 96 Ma, but the interval of time represented by this epoch is 33 m.y.

(d) **Expression of "age" of lithodemic units.**—The adjectives "early," "middle," and "late" should be used with the appropriate geochronologic term to designate the age of lithodemic units. For example, a granite dated isotopically at 510 Ma should be referred to using the geochronologic term "Late Cambrian granite" rather than either the chronostratigraphic term "Upper Cambrian granite" or the more cumbersome designation "granite of Late Cambrian age."

Article 14.—**Correlation.** Information regarding spatial and temporal counterparts of a newly defined unit beyond the type area provides readers with an enlarged perspective. Discussions of criteria used in correlating a unit with those in other areas should make clear the distinction between data and inferences.

Article 15.—**Genesis.** Objective data are used to define and classify geologic units and to express their spatial and temporal relations. Although many of the categories defined in this Code (e.g., lithostratigraphic group, plutonic suite) have genetic connotations, inferences regarding geologic history or specific environments of formation may play no proper role in the definition of a unit. However, observations, as well as inferences, that bear on genesis are of great interest to readers and should be discussed.

Article 16.—**Subsurface and Subsea Units.** The foregoing procedures for establishing formal geologic unit apply also to subsurface and offshore or subsea units. Complete lithologic and paleontologic descriptions or logs of the samples or cores are required in written or graphic form, or both. Boundaries and divisions, if any, of the unit should be indicated clearly with their depths from an established datum.

Remarks. (a) **Naming subsurface units.**—A subsurface unit may be named for the borehole (Eagle Mills Formation), oil field (Smackover Limestone), or mine which is intended to serve as the stratotype, or for a nearby geographic feature. The hole or mine should be located precisely, both with map and exact geographic coordinates, and identified fully (operator or company, farm or lease block, dates drilled or mined, surface elevation and total depth, etc.).

(b) **Additional recommendations.**—Inclusion of appropriate borehole geophysical logs is urged. Moreover, rock and fossil samples and core and all pertinent accompanying materials should be stored, and available for examination, at appropriate federal, state, provincial, university, or museum depositories. For offshore or subsea units (Clipperton Formation of Tracey and others, 1971, p. 22; Argo Salt of McIver, 1972, p. 57), the names of the project and vessel, depth of sea floor, and pertinent regional sampling and geophysical data should be added.

(c) **Seismostratigraphic units.**—High-resolution seismic methods now can delineate stratal geometry and continuity at a level of confidence not previously attainable. Accordingly, seismic surveys have come to be the principal adjunct of the drill in subsurface exploration. On the other hand, the method identifies rock types only broadly and by inference. Thus, formalization of units known only from seismic

profiles is inappropriate. Once the stratigraphy is calibrated by drilling, the seismic method may provide objective well-to-well correlations.

REVISION AND ABANDONMENT OF FORMAL UNITS

Article 17.—**Requirements for Major Changes.** Formally defined and named geologic units may be redefined, revised, or abandoned, but revision and abandonment require as much justification as establishment of a new unit.

Remark. (a) **Distinction between redefinition and revision.**—Redefinition of a unit involves changing the view or emphasis on the content of the unit without changing the boundaries or rank, and differs only slightly from redescription. Neither redefinition not redescription is considered revision. A redescription corrects an inadequate or inaccurate description, whereas a redefinition may change a descriptive (for example, lithologic) designation. Revision involves either minor changes in the definition of one or both boundaries or in the rank of a unit (normally, elevation to a higher rank). Correction of a misidentification of a unit outside its type area is neither redefinition nor revision.

Article 18.—**Redefinition.** A correction or change in the descriptive term applied to a stratigraphic or lithodemic unit is a redefinition which does not require a new geographic term.

Remarks. (a) **Change in lithic designation.**—Priority should not prevent more exact lithic designation if the original designation is not everywhere applicable; for example, the Niobrara Chalk changes gradually westward to a unit in which shale is prominent, for which the designation "Niobrara Shale" or "Formation" is more appropriate. Many carbonate formations originally designated "limestone" or "dolomite" are found to be geographically inconsistent as to prevailing rock type. The appropriate lithic term of "formation" is again preferable for such units.

(b) **Original lithic designation inappropriate.**—Restudy of some long-established lithostratigraphic units has shown that the original lithic designation was incorrect according to modern criteria; for example, some "shales" have the chemical and mineralogical composition of limestone, and some rocks described as felsic lavas now are understood to be welded tuffs. Such new knowledge is recognized by changing the lithic designation of the unit, while retaining the original geographic term. Similarly, changes in the classification of igneous rocks have resulted in recognition that rocks originally described as quartz monzonite now are more appropriately termed granite. Such lithic designations may be modernized when the new classification is widely adopted. If heterogeneous bodies of plutonic rock have been misleadingly identified with a single compositional term, such as "gabbro," the adoption of a neutral term, such as "intrusion" or "pluton," may be advisable.

Article 19.—**Revision.** Revision involves either minor changes in the definition of one or both boundaries of a unit, or in the unit's rank.

Remarks. (a) **Boundary change.**—Revision is justifiable if a minor change in boundary or content will make a unit more natural and useful. If revision modifies only a minor part of the content of a previously established unit, the original name may be retained.

(b) **Change in rank.**—Change in rank of a stratigraphic or temporal unit requires neither redefinition of its boundaries nor alteration of the geographic part of its name. A member may become a formation or vice versa, a formation may become a group or vice versa, and a lithodeme may become a suite or vice versa.

(c) **Examples of changes from area to area.**—The Conasauga Shale is recognized as a formation in Georgia and as a group in eastern Tennessee; the Osgood Formation, Laurel Limestone, and Waldron Shale in Indiana are classed as members of the Wayne Formation in a part of Tennessee; the Virgelle Sandstone is a formation in western Montana and a member of the Eagle Sandstone in central Montana; the Skull Creek Shale and the Newcastle Sandstone in North Dakota are members of the Ashville Formation in Manitoba.

(d) **Example of change in single area.**—The rank of a unit may be changed without changing its content. For example, the Madison Limestone of early work in Montana later became the Madison Group, containing several formations.

(e) **Retention of type section.**—When the rank of a geologic unit is changed, the original type section or type locality is retained for the newly ranked unit (see Article 22c).

(f) **Different geographic name for a unit and its parts.**—In changing the rank of a unit, the same name may not be applied both to the unit as a whole and to a part of it. For example, the Astoria Group should not contain an Astoria Sandstone, nor the Washington Formation, a Washington Sandstone Member.

(g) **Undesirable restriction.**—When a unit is divided into two or more of the same rank as the original, the original name should not be used for any of the divisions. Retention of the old name for one of the units precludes use of the name in a term of higher rank. Furthermore, in order to understand an author's meaning, a later reader would have to know about the modification and its date, and whether the author is following the original or the modified usage. For these reasons, the normal practice is to raise the rank of an established unit when units of the same rank are recognized and mapped within it.

Article 20.—**Abandonment.** An improperly defined or obsolete stratigraphic, lithodemic, or temporal unit may be formally abandoned, provided that (a) sufficient justification is presented to demonstrate a concern

---

[5]Note that the initial letters of Mega- and Giga- are capitalized, but that of kilo- is not, by SI convention.

for nomenclatural stability, and (b) recommendations are made for the classification and nomenclature to be used in its place.

Remarks. (a) **Reasons for abandonment.**—A formally defined unit may be abandoned by the demonstration of synonymy or homonymy, of assignment to an improper category (for example, definition of a lithostratigraphic unit in a chronostratigraphic sense), or of other direct violations of a stratigraphic code or procedures prevailing at the time of the original definition. Disuse, or the lack of need or useful purpose for a unit, may be a basis for abandonment; so, too, may widespread misuse in diverse ways which compound confusion. A unit also may be abandoned if it proves impracticable, neither recognizable nor mappable elsewhere.

(b) **Abandoned names.**—A name for a lithostratigraphic or lithodemic unit, once applied and then abandoned, is available for some other unit only if the name was introduced casually, or if it has been published only once in the last several decades and is not in current usage, and if its reintroduction will cause no confusion. An explanation of the history of the name and of the new usage should be a part of the designation.

(c) **Obsolete names.**—Authors may refer to national and provincial records of stratigraphic names to determine whether a name is obsolete (see Article 7b).

(d) **Reference to abandoned names.**—When it is useful to refer to an obsolete or abandoned formal name, its status is made clear by some such term as "abandoned" or "obsolete," and by using a phrase such as "La Plata Sandstone of Cross (1898)". (The same phrase also is used to convey that a named unit has not yet been adopted for usage by the organization involved.)

(e) **Reinstatement.**—A name abandoned for reasons that seem valid at the time, but which subsequently are found to be erroneous, may be reinstated. Example: the Washakie Formation, defined in 1869, was abandoned in 1918 and reinstated in 1973.

CODE AMENDMENT

Article 21.—Procedure for Amendment. Additions to, or changes of, this Code may be proposed in writing to the Commission by any geoscientist at any time. If accepted for consideration by a majority vote of the Commission, they may be adopted by a two-thirds vote of the Commission at an annual meeting not less than a year after publication of the proposal.

FORMAL UNITS DISTINGUISHED BY CONTENT, PROPERTIES, OR PHYSICAL LIMITS

LITHOSTRATIGRAPHIC UNITS

Nature and Boundaries

Article 22.—Nature of Lithostratigraphic Units. A lithostratigraphic unit is a defined body of sedimentary, extrusive igneous, metasedimentary, or metavolcanic strata which is distinguished and delimited on the basis of lithic characteristics and stratigraphic position. A lithostratigraphic unit generally conforms to the Law of Superposition and commonly is stratified and tabular in form.

Remarks. (a) **Basic units.**—Lithostratigraphic units are the basic units of general geologic work and serve as the foundation for delineating strata, local and regional structure, economic resources, and geologic history in regions of stratified rocks. They are recognized and defined by observable rock characteristics; boundaries may be placed at clearly distinguished contacts or drawn arbitrarily within a zone of gradation. Lithification or cementation is not a necessary property; clay, gravel, till, and other unconsolidated deposits may constitute valid lithostratigraphic units.

(b) **Type section and locality.**—The definition of a lithostratigraphic unit should be based, if possible, on a stratotype consisting of readily accessible rocks in place, e.g., in outcrops, excavations, and mines, or of rocks accessible only to remote sampling devices, such as those in drill holes and underwater. Even where remote methods are used, definitions must be based on lithic criteria and not on the geophysical characteristics of the rocks, nor the implied age of their contained fossils. Definitions must be based on descriptions of actual rock material. Regional validity must be demonstrated for all such units. In regions where the stratigraphy has been established through studies of surface exposures, the naming of new units in the subsurface is justified only where the subsurface section differs materially from the surface section, or where there is doubt as to the equivalence of a subsurface and a surface unit. The establishment of subsurface reference sections for units originally defined in outcrop is encouraged.

(c) **Type section never changed.**—The definition and name of a lithostratigraphic unit are established at a type section (or locality) that, once specified, must not be changed. If the type section is poorly designated or delimited, it may be redefined subsequently. If the originally specified stratotype is incomplete, poorly exposed, structurally complicated, or unrepresentative of the unit, a principal reference section or several reference sections may be designated to supplement, but not to supplant, the type section (Article 8e).

(d) **Independence from inferred geologic history.**—Inferred geologic history, depositional environment, and biological sequence have no place in the definition of a lithostratigraphic unit, which must be based on composition and other lithic characteristics; nevertheless, considerations of well-documented geologic history properly may influence the choice of vertical and lateral boundaries of a new unit. Fossils may

be valuable during mapping in distinguishing between two lithologically similar, non-contiguous lithostratigraphic units. The fossil content of a lithostratigraphic unit is a legitimate lithic characteristic; for example, oyster-rich sandstone, coquina, coral reef, or graptolitic shale. Moreover, otherwise similar units, such as the Formación Mendez and Formación Velasco mudstones, may be distinguished on the basis of coarseness of contained fossils (foraminifera).

(e) **Independence from time concepts.**—The boundaries of most lithostratigraphic units may transgress time horizons, but some may be approximately synchronous. Inferred time-spans, however measured, play no part in differentiating or determining the boundaries of any lithostratigraphic unit. Either relatively short or relatively long intervals of time may be represented by a single unit. The accumulation of material assigned to a particular unit may have begun or ended earlier in some localities than in others; also, removal of rock by erosion, either within the time-span of deposition of the unit or later, may reduce the time-span represented by the unit locally. The body in some places may be entirely younger than in other places. On the other hand, the establishment of formal units that straddle known, identifiable, regional disconformities is to be avoided, if at all possible. Although concepts of time or age play no part in defining lithostratigraphic units nor in determining their boundaries, evidence of age may aid recognition of similar lithostratigraphic units at localities far removed from the type sections or areas.

(f) **Surface form.**—Erosional morphology or secondary surface form may be a factor in the recognition of a lithostratigraphic unit, but properly should play a minor part at most in the definition of such units. Because the surface expression of lithostratigraphic units is an important aid in mapping, it is commonly advisable, where other factors do not countervail, to define lithostratigraphic boundaries so as to coincide with lithic changes that are expressed in topography.

(g) **Economically exploited units.**—Aquifers, oil sands, coal beds, and quarry layers are, in general, informal units even though named. Some such units, however, may be recognized formally as beds, members, or formations because they are important in the elucidation of regional stratigraphy.

(h) **Instrumentally defined units.**—In subsurface investigations, certain bodies of rock and their boundaries are widely recognized on bore-hole geophysical logs showing their electrical resistivity, radioactivity, density, or other physical properties. Such bodies and their boundaries may or may not correspond to formal lithostratigraphic units and their boundaries. Where other considerations do not countervail, the boundaries of subsurface units should be defined so as to correspond useful geophysical markers; nevertheless, units defined exclusively on the basis of remotely sensed physical properties, although commonly useful in stratigraphic analysis, stand completely apart from the hierarchy of formal lithostratigraphic units and are considered informal.

(i) **Zone.**—As applied to the designation of lithostratigraphic units, the term "zone" is informal. Examples are "producing zone," mineralized zone," "metamorphic zone," and "heavy-mineral zone," A zone may include all or parts of a bed, a member, a formation, or even a group.

(j) **Cyclothems.**—Cyclic or rhythmic sequences of sedimentary rocks, whose repetitive divisions have been named cyclothems, have been recognized in sedimentary basins around the world. Some cyclothems have been identified by geographic names, but such names are considered informal. A clear distinction must be maintained between the division of a stratigraphic column into cyclothems and its division into groups, formations, and members. Where a cyclothem is identified by a geographic name, the word *cyclothem* should be part of the name, and the geographic term should not be the same as that of any formal unit embraced by the cyclothem.

(k) **Soils and paleosols.**—Soils and paleosols are layers composed of the in-situ products of weathering of older rocks which may be of diverse composition and age. Soils and paleosols differ in several respects from lithostratigraphic units, and should not be treated as such (see "Pedostratigraphic Units," Articles 55 et seq).

(l) **Depositional facies.**—Depositional facies are informal units, whether objective (conglomeratic, black shale, graptolitic) or genetic and environmental (platform, turbiditic, fluvial), even when a geographic term has been applied, e.g., Lantz Mills facies. Descriptive designations convey more information than geographic terms and are preferable.

Article 23.—Boundaries. Boundaries of lithostratigraphic units are placed at positions of lithic change. Boundaries are placed at distinct contacts or may be fixed arbitrarily within zones of gradation (Fig. 2a). Both vertical and lateral boundaries are based on the lithic criteria that provide the greatest unity and utility.

Remarks. (a) **Boundary in a vertically gradational sequence.**—A named lithostratigraphic unit is preferably bounded by a single lower and a single upper surfaces so that the name does not recur in a normal stratigraphic succession (see Remark b.). Where a rock unit passes vertically into another by intergrading or interfingering of two or more kinds of rock, unless the gradational strata are sufficiently thick to warrant designation of a third, independent unit, the boundary is necessarily arbitrary and should be selected on the basis of practically (Fig. 2b). For example, where a shale unit overlies a unit of interbedded limestone and shale, the boundary commonly is placed at the top of the highest readily traceable limestone bed. Where a sandstone unit grades upward into shale, the boundary may be so gradational as to be difficult to place even arbitrarily; ideally it should be drawn at the level where the rock is composed of one-half of each component. Because of creep in outcrops and caving in boreholes, it is generally best to define such arbitrary boundaries by the highest occurrence of a particular rock type, rather than the lowest.

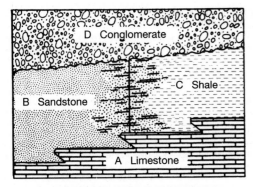

A.-- Boundaries at sharp lithologic contacts and in laterally gradational sequence.

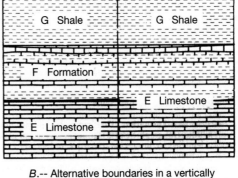

B.-- Alternative boundaries in a vertically gradational or interlayered sequence.

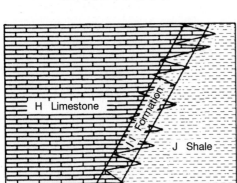

C.-- Possible boundaries for a laterally intertonguing sequence

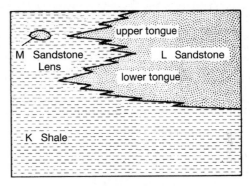

D.-- Possible classification of parts of an intertonguing sequence

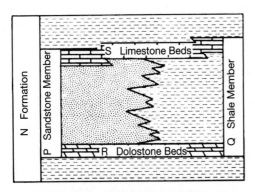

E-- Key beds, here designated the R Dolostone Beds and the S Limestone Beds, are used as boundaries to distinguish the Q Shale Member from the other parts of the N Formation. A lateral change in composition between the key beds requires that another name, P Sandstone Member, be applied. The key beds are part of each member.

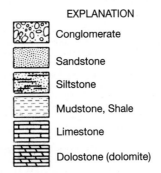

EXPLANATION

Conglomerate

Sandstone

Siltstone

Mudstone, Shale

Limestone

Dolostone (dolomite)

Figure 2.
Diagrammatic examples of lithostratigraphic boundaries and classification.

(b) **Boundaries in lateral lithologic change.**—Where a unit changes laterally through abrupt gradation into, or intertongues with, a markedly different kind of rock, a new unit should be proposed for the different rock type. An arbitrary lateral boundary may be placed between the two equivalent units. Where the area of lateral intergradation or intertonguing is sufficiently extensive, a transitional interval of interbedded rocks may constitute a third independent unit (Fig. 2c). Where tongues (Article 25b) of formations are mapped separately or otherwise set apart without being formally named, the unmodified formation name should not be repeated in a normal stratigraphic sequence, although the modified name by be repeated in such phrases as "lower tongue of Mancos Shale" and "upper tongue of Mancos Shale." To show the order of superposition on maps and cross sections, the unnamed tongues may be distinguished informally (Fig. 2d) by number, letter, or other means. Such relationships may also be dealt with informally through the recognition of depositional facies (Article 22-1).

(c) **Key beds used for boundaries.**—Key beds (Article 26b) may be used as boundaries for a formal lithostratigraphic unit where the internal lithic characteristics of the unit remain relatively constant. Even though bounding key beds may be traceable beyond the area of the diagnostic overall rock type, geographic extension of the lithostratigraphic unit bounded thereby is not necessarily justified. Where the rock between key beds becomes drastically different from that of the type locality, a new name should be applied (Fig. 2e), even though the key beds are continuous (Article 26b). Stratigraphic and sedimentologic studies of stratigraphic units (usually informal) bounded by key beds may be very informative and useful, especially in subsurface work where the key beds may be recognized by their geophysical signatures. Such units, however, may be a kind of chronostratigraphic, rather than lithostratigraphic, unit (Article 75, 75c), although others are diachronous because one, or both, of the key beds are also diachronous.

(d) **Unconformities as boundaries.**—Unconformities, where recognizable objectively on lithic criteria, are ideal boundaries for lithostratigraphic units. However, a sequence of similar rocks may include an obscure unconformity so that separation into two units may be desirable but impracticable. If no lithic distinction adequate to define a widely recognizable boundary can be made, only one unit should be recognized, even though it may include rock that accumulated in different epochs, periods, or eras.

(e) **Correspondence with genetic units.**—The boundaries of lithostratigraphic units should be chosen on the basis of lithic changes and, where feasible, to correspond with the boundaries of genetic units, so that subsequent studies of genesis will not have to deal with units that straddle formal boundaries.

RANKS OF LITHOSTRATIGRAPHIC UNITS

Article 24.—Formation. The formation is the fundamental unit in lithostratigraphic classification. A formation is a body of rock identified by lithic characteristics and stratigraphic position; it is prevailingly but not necessarily tabular and is mappable at the Earth's surface or traceable in the subsurface.

Remarks. (a) **Fundamental unit.**—Formation are the basic lithostratigraphic units used in describing and interpreting the geology of a region. The limits of a formation normally are those surfaces of lithic change that give it the greatest practicable unity of constitution. A formation may represent a long or short time interval, may be composed of materials from one or several sources, and may include breaks in deposition (see Article 23d).

(b) **Content.**—A formation should possess some degree of internal lithic homogeneity or distinctive lithic features. It may contain between its upper and lower limits (i) rock of one lithic type, (ii) repetitions of two or more lithic types, or (iii) extreme lithic heterogeneity which in itself may constitute a form of unity when compared to the adjacent rock units.

(c) **Lithic characteristics.**—Distinctive lithic characteristics include chemical and mineralogical composition, texture, and such supplementary features as color, primary sedimentary or volcanic structures, fossils (viewed as rock-forming particles), or other organic content (coal, oil-shale). A unit distinguishable only by the taxonomy of its fossils is not a lithostratigraphic but a biostratigraphic unit (Article 48). Rock type may be distinctively represented by electrical, radioactive, seismic, or other properties (Article 22h), but these properties by themselves do not describe adequately the lithic character of the unit.

(d) **Mappability and thickness.**—The proposal of a new formation must be based on tested mappability. Well-established formations commonly are divisible into several widely recognizable lithostratigraphic units; where formal recognition of these smaller units serves a useful purpose, they may be established as members and beds, for which the requirement of mappability is not mandatory. A unit formally recognized as a formation in one area may be treated elsewhere as a group, or as a member of another formation, without change of name. Example: the Niobrara is mapped at different places as a member of the Mancos Shale, of the Cody Shale, or of the Colorado Shale, and also as the Niobrara Formation, as the Niobrara Limestone, and as the Niobrara Shale.

Thickness is not a determining parameter in dividing a rock succession into formations; the thickness of a formation may range from a feather edge at its depositional or erosional limit to thousands of meters elsewhere. No formation is considered valid that cannot be delineated at the scale of geologic mapping practiced in the region when the formation is proposed. Although representation of a formation on maps and cross sections by a labeled line may be justified, proliferation of such exceptionally thin units is undesirable. The methods of subsurface mapping permit delineation of units much

thinner than those usually practicable for surface studies; before such thin units are formalized, consideration should be given to the effect on subsequent surface and subsurface studies.

(e) **Organic reefs and carbonate mounds.**—Organic reefs and carbonate mounds ("buildups") may be distinguished formally, if desirable, as formations distinct from their surrounding, thinner, temporal equivalents. For the requirements of formalization, see Article 30f.

(f) **Interbedded volcanic and sedimentary rock.**—Sedimentary rock and volcanic rock that are interbedded may be assembled into a formation under one name which should indicate the predominant or distinguishing lithology, such as Mindego Basalt.

(g) **Volcanic rock.**—Mappable distinguishable sequences of stratified volcanic rock should be treated as formations or lithostratigraphic units of higher or lower rank. A small intrusive component of a dominantly stratiform volcanic assemblage may be treated informally.

(h) **Metamorphic rock.**—Formations composed of low-grade metamorphic rock (defined for this purpose as rock in which primary structures are clearly recognizable) are, like sedimentary formations, distinguished mainly by lithic characteristics. The mineral facies may differ from place to place, but these variations do not require definition of a new formation. High-grade metamorphic rocks whose relation to established formations is uncertain are treated as lithodemic units (see Articles 31 et seq).

Article 25.—Member. A member is the formal lithostratigraphic unit next in rank below a formation and is always a part of some formation. It is recognized as a named entity within a formation because it possesses characteristics distinguishing it from adjacent parts of the formation. A formation need not be divided into members unless a useful purpose is served by doing so. Some formations may be divided completely into members; others may have only certain parts designated as members; still others may have no members. A member may extend laterally from one formation to another.

Remarks. (a) **Mapping of members.**—A member is established when it is advantageous to recognize a particular part of a heterogeneous formation. A member, whether formally or informally designated, need not be mappable at the scale required for formations. Even if all members of a formation are locally mappable, it does not follow that they should be raised to formational rank, because proliferation of formation names may obscure rather than clarify relations with other areas.

(b) **Lens and tongue.**—A geographically restricted member that terminates on all sides within a formation may be called a lens (lentil). A wedging member that extends outward beyond a formation or wedges ("pinches") out within another formation may be called a tongue.

(c) **Organic reefs and carbonate mounds.**—Organic reefs and carbonate mounds may be distinguished formally, if desirable, as members within a formation. For the requirements of formalization, see Article 30f.

(d) **Division of members.**—A formally or informally recognized division of a member is called a bed or beds, except for volcanic flow-rocks, for which the smallest formal unit is a flow. Members may contain beds or flows, but may never contain other members.

(e) **Laterally equivalent members.**—Although members normally are in vertical sequence, laterally equivalent parts of a formation that differ recognizably may also be considered members.

Article 26.—Bed(s). A bed, or beds, is the smallest formal lithostratigraphic unit of sedimentary rocks.

Remarks. (a) **Limitations.**—The designation of a bed or a unit of beds as a formally named lithostratigraphic unit generally should be limited to certain distinctive beds whose recognition is particularly useful. Coal beds, oil sands, and other beds of economic importance commonly are named, but such units and their names usually are not a part of formal stratigraphic nomenclature (Article 22g and 30g).

(b) **Key or marker beds.**—A key or marker bed is a thin bed of distinctive rock that is widely distributed. Such beds may be named, but usually are considered informal units. Individual key beds may be traced beyond the lateral limits of a particular formal unit (Article 23c).

Article 27.—Flow. A flow is the smallest formal lithostratigraphic unit of volcanic flow rocks. A flow is a discrete, extrusive, volcanic body distinguishable by texture, composition, order of superposition, paleomagnetism, or other objective criteria. It is part of a member and thus is equivalent in rank to a bed or beds of sedimentary-rock classification. Many flows are informal units. The designation and naming of flows as formal rock-stratigraphic units should be limited to those that are distinctive and widespread.

Article 28.—Group. A group is the lithostratigraphic unit next higher in rank to formation; a group may consist entirely of named formations, or alternatively, need not be composed entirely of named formations.

Remarks. (a) **Use and content.**—Groups are defined to express the natural relationships of associated formations. They are useful in small-scale mapping and regional stratigraphic analysis. In some reconnaissance work, the term "group" has been applied to lithostratigraphic units that appear to be divisible into formations, but have not yet been so divided. In such cases, formations may be erected subsequently for one or all of the practical divisions of the group.

(b) **Change in component formations.**—The formations making up a group need not necessarily be everywhere the same. The Rundle Group, for example, is widespread in western Canada and undergoes several changes in formational content. In southwestern Alberta, it comprises the Livingstone, Mount Head, and Etherington Formations in the Front Ranges, whereas in the foothills and subsurface of the adjacent plains, it comprises the Pekisko, Shunda, Turner Valley, and Mount Head Formations. However, a formation or its parts may not be assigned to two vertically adjacent groups.

(c) **Change in rank.**—The wedge-out of a component formation or formations may justify the reduction of a group to formation rank, retaining the same name. When a group is extended laterally beyond where it is divided into formations, it becomes in effect a formation, even if it is still called a group. When a previously established formation is divided into two or more component units that are given formal formation rank, the old formation, with its old geographic name, should be raised to group status. Raising the rank of the unit is preferable to restricting the old name to a part of its former content, because a change in rank leaves the sense of a well-established unit unchanged (Article 19b, 19g).

Article 29.—**Supergroup.** A supergroup is a formal assemblage of related or superposed groups, or of groups and formations. Such units have proved useful in regional and provincial syntheses. Supergroups should be named only where their recognition serves a clear purpose.

Remark. (a) **Misuse of "series" for group or supergroup.**—Although "series" is a useful general term, it is applied formally only to a chronostratigraphic unit and should not be used for a lithostratigraphic unit. The term "series" should no longer be employed for an assemblage of formations or an assemblage of formations and groups, as it has been, especially in studies of the Precambrian. These assemblages are groups or supergroups.

LITHOSTRATIGRAPHIC NOMENCLATURE

Article 30.—**Compound Character.** The formal name of a lithostratigraphic unit is compound. It consists of a geographic name combined with a descriptive lithic term or with the appropriate rank term, or both. Initial letters of all words used in forming the names of formal rock-stratigraphic units are capitalized.

Remarks. (a) **Omission of part of a name.**—Where frequent repetition would be cumbersome, the geographic name, the lithic term, or the rank term may be used alone, once the full name has been introduced; as "the Burlington," "the limestone," or "the formation," for the Burlington Limestone.

(b) **Use of simple lithic terms.**—The lithic part of the name should indicate the predominant or diagnostic lithology, even if subordinate lithologies are included. Where a lithic term is used in the name of a lithostratigraphic unit, the simplest generally acceptable term is recommended (for example, limestone, sandstone, shale, tuff, quartzite). Compound terms (for example, clay shale) and terms that are not in common usage (for example, calcirudite, orthoquartzite) should be avoided. Combined terms, such as "sand and clay," should not be used for the lithic part of the names of lithostratigraphic units, nor should an adjective be used between the geographic and the lithic terms, as "Chattanooga Black Shale" and "Biwabik Iron-Bearing Formation."

(c) **Group names.**—A group name combines a geographic name with the term "group," and no lithic designation is included; for example, San Rafael Group.

(d) **Formation names.**—A formation name consists of a geographic name followed by a lithic designation or by the word "formation." Examples: Dakota Sandstone, Mitchell Mesa Rhyolite, Monmouth Formation, Halton Till.

(e) **Member names.**—All member names include a geographic term and the word "member;" some have an intervening lithic designation, if useful; for example, Wedington Sandstone Member of the Fayetteville Shale. Members designated solely by lithic character (for example, siliceous shale member), by position (upper, lower), or by letter or number, are informal.

(f) **Names of reefs.**—Organic reefs identified as formations or members are formal units only where the name combines a geographic name with the appropriate rank term, e.g., Leduc Formation (a name applied to the several reefs enveloped by the Ireton Formation), Rainbow Reef Member.

(g) **Bed and flow names.**—The names of beds or flows combine a geographic term, a lithic term, and the term "bed" or "flow;" for example, Knee Hills Tuff Bed, Ardmore Bentonite Beds, Negus Variolitic Flows.

(h) **Informal units.**—When geographic names are applied to such informal units as oil sands, coal beds, mineralized zones, and informal members (see Articles 22g and 26a), the unit term should not be capitalized. A name is not necessarily formal because it is capitalized, nor does failure to capitalize a name render it informal. Geographic names should be combined with the terms "formation" or "group" only in formal nomenclature.

(i) **Informal usage of identical geographic names.**—The application of identical geographic names to several minor units in one vertical sequence is considered informal nomenclature (lower Mount Savage coal, Mount Savage fireclay, upper Mount Savage coal, Mount Savage rider coal, and Mount Savage sandstone). The application of identical geographic names to the several lithologic units constituting a cyclothem likewise is considered informal.

(j) **Metamorphic rock.**—Metamorphic rock recognized as a normal stratified sequence, commonly low-grade metavolcanic or metasedimentary rocks, should be assigned to named groups, formations, and members, such as the Deception Rhyolite, a formation of the Ash Creek Group, or the Bonner Quartzite, a formation of the Missoula Group. High-grade metamorphic and metasomatic rocks are treated as lithodemes and suites (see Articles 31, 33, 35).

(k) **Misuse of well-known name.**—A name that suggests some well-known locality, region, or political division should not be applied to a unit typically developed in another less well-known locality of the same name. For example, it would be inadvisable to use the name "Chicago Formation" for a unit in California.

LITHODEMIC UNITS

Nature and Boundaries

Article 31.—**Nature of Lithodemic Units.** A lithodemic[6] unit is a defined body of predominantly intrusive, highly deformed, and/or highly metamorphosed rock, distinguished and delimited on the basis of rock characteristics. In contrast to lithostratigraphic units, a lithodemic unit generally does not conform to the Law of Superposition. Its contacts with other rock units may be sedimentary, extrusive, intrusive, tectonic, or metamorphic (Fig. 3).

Remarks. (a) **Recognition and definition.**—Lithodemic units are defined and recognized by observable rock characteristics. They are the practical units of general geological work in terranes in which rocks generally lack primary stratification; in such terranes they serve as the foundation for studying, describing, and delineating lithology, local and regional structure, economic resources, and geologic history.

(b) **Type and reference localities.**—The definition of a lithodemic unit should be based on as full a knowledge as possible of its lateral and vertical variations and its contact relationships. For purposes of nomenclatural stability, a type locality and, wherever appropriate, reference localities should be designated.

(c) **Independence from inferred geologic history.**—Concepts based on inferred geologic history properly play no part in the definition of a lithodemic unit. Nevertheless, where two rock masses are lithically similar but display objective structural relations that preclude the possibility of their being even broadly of the same age, they should be assigned to different lithodemic units.

(d) **Use of "zone."**—As applied to the designation of lithodemic units, the term "zone" is informal. Examples are: "mineralized zone," "contact zone," and "pegmatitic zone,"

Article 32.—**Boundaries.** Boundaries of lithodemic units are placed at positions of lithic change. They may be placed at clearly distinguished contacts or within zones of gradation. Boundaries, both vertical and lateral, are based on the lithic criteria that provide the greatest unity and practical utility. Contacts with other lithodemic and lithostratigraphic units may be depositional, intrusive, metamorphic, or tectonic.

Remark. (a) **Boundaries within gradational zones.**—Where a lithodemic unit changes through gradation into, or intertongues with, a rockmass with markedly different characteristics, it is usually desirable to propose a new unit. It may be necessary to draw an arbitrary boundary within the zone of gradation. Where the area of intergradation or intertonguing is sufficiently extensive, the rocks of mixed character may constitute a third unit.

Ranks of Lithodemic Units

Article 33.—**Lithodeme.** The lithodeme is the fundamental unit in lithodemic classification. A lithodeme is a body of intrusive, pervasively deformed, or highly metamorphosed rock, generally non-tabular and lacking primary depositional structures, and characterized by lithic homogeneity. It is mappable at the Earth's surface and traceable in the subsurface. For cartographic and hierarchical purposes, it is comparable to a formation (see Table 2).

Remarks. (a) **Content.**—A lithodeme should possess distinctive lithic features and some degree of internal lithic homogeneity. It may consist of (i) rock of one type, (ii) a mixture of rocks of two or more types, or (iii) extreme heterogeneity of composition, which may constitute in itself a form of unity when compared to adjoining rockmasses (see also "complex," Article 37).

(b) **Lithic characteristics.**—Distinctive lithic characteristics may include mineralogy, textural features such as grain size, and structural features such as schistose of gneissic structure. A unit distinguishable from its neighbors only by means of chemical analysis is informal.

(c) **Mappability.**—Practicability of surface or subsurface mapping is an essential characteristic of a lithodeme (see Article 24d).

Article 34.—**Division of Lithodemes.** Units below the rank of lithodeme are informal.

---

[6]From the Greek *demas, -os*: "living body, frame".

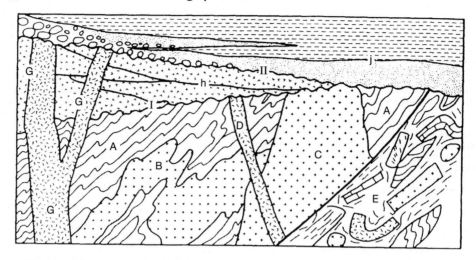

Figure 3.
Lithodemic (upper case) and lithostratigraphic (lower case) units. A *lithodeme* of *gneiss* (A) contains an *intrusion* of diorite (B) that was deformed with the gneiss. A and B may be treated jointly as a *complex*. A younger *granite* (C) is cut by a dike of *syenite* (D), that is cut in turn by unconformity I. All the foregoing are in fault contact with a *structural complex* (E). A *volcanic complex* (G) is built upon unconformity I, and its feeder dikes cut the unconformity. Laterally equivalent volcanic strata in orderly, mappable succession (h) are treated as lithostratigraphic units. A *gabbro* feeder (G'), to the volcanic complex, where surrounded by gneiss is readily distinguished as a separate lithodeme and named as a *gabbro* or an *intrusion*. All the foregoing are overlain, at unconformity II, by sedimentary rocks (j) divided into formations and members.

Article 35.—Suite. A *Suite* (metamorphic suite, intrusive suite, plutonic suite) is the lithodemic unit next higher in rank to lithodeme. It comprises two or more associated lithodemes of the same class (e.g., plutonic, metamorphic). For cartographic and hierarchical purposes, suite is comparable to group (see Table 2).

Remarks. (a) **Purpose.**—Suites are recognized for the purpose of expressing the natural relations of associated lithodemes having significant lithic features in common, and of depicting geology at compilation scales too small to allow delineation of individual lithodemes. Ideally, a suite consists entirely of named lithodemes, but may contain both named and unnamed units.

(b) **Change in component units.**—The named and unnamed units constituting a suite may change from place to place, so long as the original sense of natural relations and of common lithic features is not violated.

(c) **Change in rank.**—Traced laterally, a suite may lose all of its formally named divisions but remain a recognizable, mappable entity. Under such circumstances, it may be treated as a lithodeme but retain the same name. Conversely, when a previously established lithodeme is divided into two or more mappable divisions, it may be desirable to raise its rank to suite, retaining the original geographic component of the name. To avoid confusion, the original name should not be retained for one of the divisions of the original unit (see Article 19g).

Article 36.—Supersuite. A supersuite is the unit next higher in rank to a suite. It comprises two or more suites or complexes having a degree of natural relationship to one another, either in the vertical or the lateral sense. For cartographic and hierarchical purposes, supersuite is similar in rank to supergroup.

Article 37.—Complex. An assemblage or mixture of rocks of *two or more genetic classes*, i.e., igneous, sedimentary, or metamorphic, with or without highly complicated structure, may be named a *complex*. The term "complex" takes the place of the lithic or rank term (for example, Boil Mountain Complex, Franciscan Complex) and, although unranked, commonly is comparable to suite or supersuite and is named in the same manner (Articles 41, 42).

Remarks. (a) **Use of "complex."**—Identification of an assemblage of diverse rocks as a complex is useful where the mapping of each separate lithic component is impractical at ordinary mapping scales. "Complex" is unranked but commonly comparable to suite or supersuite; therefore, the term may be retained if subsequent, detailed mapping distinguishes some or all of the component lithodemes or lithostratigraphic units.

(b) **Volcanic complex.**—Sites of persistent volcanic activity commonly are characterized by a diverse assemblage of extrusive volcanic rocks, related intrusions, and their weathering products. Such an assemblage may be designated a *volcanic complex*.

(c) **Structural complex.**—In some terranes, tectonic processes (e.g., shearing, faulting) have produced heterogeneous mixtures or disrupted bodies of rock in which some individual components are too small to be mapped. *Where there is no doubt that the mixing or disruption is due to tectonic processes*, such a mixture may be designated as a structural complex, whether it consists of two or more classes of rock, or a single class only. A simpler solution for some mapping purposes is to indicate intense deformation by an overprinted pattern.

(d) **Misuse of "complex".**—Where the rock assemblage to be united under a single, formal name consists of diverse types of a *single class* of rock, as in many terranes that expose a variety of either intrusive igneous or high-grade metamorphic rocks, the term "intrusive suite," "plutonic suite," or "metamorphic suite" should be used, rather than the unmodified term "complex." Exceptions to this rule are the terms *structural complex* and *volcanic complex* (see Remarks c and b, above).

Article 38.—Misuse of "Series" for Suite, Complex, or Supersuite. The term "series" has been employed for an assemblage of lithodemes or an assemblage of lithodemes and suites, especially in studies of the Precambrian. This practice now is regarded as improper; these assemblages are suites, complexes, or supersuites. The term "series" also has been applied to a sequence of rocks resulting from a succession of eruptions or intrusions. In these cases a different term should be used; "group" should replace "series" for volcanic and low-grade metamorphic rocks, and "intrusive suite" or "plutonic suite" should replace "series" for intrusive rocks of group rank.

Lithodemic Nomenclature

Article 39.—General Provisions. The formal name of a lithodemic unit is compound. It consists of a geographic name combined with a descriptive or appropriate rank term. The principles for the selection of the geographic term, concerning suitability, availability, priority, etc, follow those established in Article 7, where the rules for capitalization are also specified.

Article 40.—Lithodeme Names. The name of a lithodeme combines a geographic term with a lithic or descriptive term, e.g., Killarney Granite, Adamant Pluton, Manhattan Schist, Skaergaard Intrusion, Duluth Gabbro. The term *formation* should not be used.

Remarks. (a) **Lithic term.**—The lithic term should be a common and familiar term, such as schist, gneiss, gabbro. Specialized terms and terms not widely used, such as websterite and jacupirangite, and compound terms, such as graphitic schist and augen gneiss, should be avoided.

(b) **Intrusive and plutonic rocks.**—Because many bodies of intrusive rock range in composition from place to place and are difficult to characterize with a single lithic term, and because many bodies of plutonic rock are considered not to be intrusions, latitude is allowed in the choice of a lithic or descriptive term. Thus, the descriptive term should preferably be compositional (e.g., gabbro, granodiorite), but may, if necessary, denote form (e.g., dike, sill), or be neutral (e.g., intrusion, pluton[7]. In any event, specialized compositional terms not widely used are to be avoided, as are form terms that are not widely used, such as bysmalith and chonolith. Terms implying genesis should be avoided as much as possible, because interpretations of genesis may change.

Article 41.—Suite Names. The name of a suite combines a geographic term, the term "suite," and an adjective denoting the fundamental character of the suite; for example, Idaho Springs Metamorphic Suite, Tuolumne Intrusive Suite, Cassiar Plutonic Suite. The geographic name of a suite may not be the same as that of a component lithodeme (see Article 19f). Intrusive assemblages, however, may share the same geographic name if an intrusive lithodeme is representative of the suite.

Article 42.—Supersuite Names. The name of a supersuite combines a geographic term with the term "supersuite."

MAGNESTOSTRATIGRAPHIC UNITS

Nature and Boundaries

Article 43.—Nature of Magnetostratigraphic Units. A magnetostratigraphic unit is a body of rock unified by specified remanent-magnetic properties and is distinct from underlying and overlying magnetostratigraphic units having different magnetic properties.

Remarks. (a) **Definition.**—Magnetostratigraphy is defined here as all aspects of stratigraphy based on remanent magnetism (paleomagnetic signatures). Four basic paleomagnetic phenomena can be determined or inferred from remanent magnetism: polarity, dipole-field-pole position (including apparent polar wander), the non-dipole component (secular variation), and field intensity.

(b) **Contemporaneity of rock and remanent magnetism.**—Many paleomagnetic signatures reflect earth magnetism at the time the rock formed. Nevertheless, some rocks have been subjected subsequently to physical and/or chemical processes which altered the magnetic properties. For example, a body of rock may be heated above the blocking temperature or Curie point for one or more minerals, or a ferromagnetic mineral may be produced by low-temperature alteration long after the enclosing rock formed, thus acquiring a component of remanent magnetism reflecting the field at the time of alteration, rather than the time of original rock deposition or crystallization.

(c) **Designations and scope.**—The prefix *magneto* is used with an appropriate term to designate the aspect of remanent magnetism used to define a unit. The terms "magnetointensity" or "magnetosecularvariation" are possible examples. This Code considers only polarity reversals, which now are recognized widely as a stratigraphic tool. However, apparent-polar-wander paths offer increasing promise for correlations within Precambrian rocks.

Article 44.—Definition of Magnetopolarity Unit. A magnetopolarity unit is a body of rock unified by its remanent magnetic polarity and distinguished from adjacent rock that has different polarity.

Remarks. (a) **Nature.**—Magnetopolarity is the record in rocks of the polarity history of the Earth's magnetic-dipole field. Frequent past reversals of the polarity of the Earth's magnetic field provide a basis for magnetopolarity stratigraphy.

(b) **Stratotype.**—A stratotype for a magnetopolarity unit should be designated and the boundaries defined in terms of recognized lithostratigraphic and/or biostratigraphic units in the stratotype. The formal definition of a magnetopolarity unit should meet the applicable specific requirements of Articles 3 to 16.

(c) **Independence from inferred history.**—Definition of a magnetopolarity unit does not require knowledge of the time at which the unit acquired its remanent magnetism; its magnetism may be primary or secondary. Nevertheless, the unit's present polarity is a property that may be ascertained and confirmed by others.

(d) **Relation to lithostratigraphic and biostratigraphic units.**—Magnetopolarity units resemble lithostratigraphic and biostratigraphic units in that they are defined on the basis of an objective recognizable property, but differ fundamentally in that most magnetopolarity unit boundaries are thought not to be time transgressive. Their boundaries may coincide with those of lithostratigraphic or biostratigraphic units, or by parallel to but displaced from those of such units, or be crossed by them.

(e) **Relation of magnetopolarity units to chronostratigraphic units.**—Although transitions between polarity reversals are of global extent, a magnetopolarity unit does not contain within itself evidence that the polarity is primary, or criteria that permits its unequivocal recognition in chronocorrelative strata of other areas. Other criteria, such as paleontologic or numerical age, are required for both correlation and dating. Although polarity reversals are useful in recognizing chronostratigraphic units, magnetopolarity alone is insufficient for their definition.

Article 45.—Boundaires. The upper and lower limits of a magnetopolarity unit are defined by boundaries marking a change of polarity. Such boundaries may represent either a depositional discontinuity or a magnetic-field transition. The boundaries are either polarity-reversal horizons or polarity transition-zones, respectively.

Remark. (a) **Polarity-reversal horizons and transition-zones.**—A polarity-reversal horizon is either a single, clearly definable surface or a thin body of strata constituting a transitional interval across which a change in magnetic polarity is recorded. Polarity-reversal horizons describe transitional intervals of 1 m or less; where the change in polarity takes place over a stratigraphic interval greater than 1 m, the term "polarity transition-zone" should be used. Polarity-reversal horizons and polarity transition-zones provide the boundaries for polarity zones, although they may also be contained within a polarity zone where they mark an internal change subsidiary in rank to those at its boundaries.

Ranks of Magnetopolarity Units

Article 46.—Fundamental Unit. A polarity zone is the fundamental unit of magnetopolarity classification. A polarity zone is a unit of rock characterized by the polarity of its magnetic signature. Magnetopolarity zone, rather than polarity zone, should be used where there is risk of confusion with other kinds of polarity.

Remarks. (a) **Content.**—A polarity zone should possess some degree of internal homogeneity. It may contain rocks of (1) entirely or predominantly one polarity, or (2) mixed polarity.

(b) **Thickness and duration.**—The thickness of rock of a polarity zone or the amount of time represented should play no part in the definition of the zone. The polarity signature is the essential property for definition.

(c) **Ranks.**—When continued work at the stratotype for a polarity zone, or new work in correlative rocks elsewhere, reveals smaller polarity units, these may be recognized formally as polarity subzones. If it should prove necessary or desirable to group polarity zones, these should be termed polarity superzones. The rank of a polarity unit may be changed when deemed appropriate.

Magnetopolarity Nomenclature

Article 47.—Compound Name. The formal name of a magnetopolarity zone should consist of a geographic name and the term *Polarity Zone*. The term may be modified by *Normal, Reversed,* or *Mixed* (example: Deer Park Reversed Polarity Zone). In naming or revising magnetopolarity units, appropriate parts of Articles 7 and 19 apply. The use of informal designations, e.g., numbers or letters, is not precluded.

BIOSTRATIGRAPHIC UNITS

Nature and boundaries

Article 48.—Nature of Biostratigraphic Units. A biostratigraphic unit is a body of rock defined or characterized by its fossil content. The basic unit in biostratigraphic classification is the biozone, of which there are several kinds.

Remarks. (a) **Enclosing strata.**—Fossils that define or characterize a biostratigraphic unit commonly are contemporaneous with the body of rock that contains them. Some biostratigraphic units, however, may be represented only by their fossils, preserved in normal stratigraphic succession (e.g., on hardgrounds, in lag deposits, in certain types of remanié accumulations), which alone represent the rock of the biostratigraphic unit. In addition, some strata contain fossils derived from older or younger rocks or from essentially coeval materials of different facies; such fossils should not be used to define a biostratigraphic unit.

(b) **Independence from lithostratigraphic units.**—Biostratigraphic units are based on criteria which differ fundamentally from those for lithostratigraphic units. Their boundaries may or may not coincide with the boundaries of lithostratigraphic units, but they bear no inherent relation to them.

(c) **Independence from chronostratigraphic units.**—The boundaries of most biostratigraphic units, unlike the boundaries of chronostratigraphic units, are both characteristically and conceptually diachronous. An exception is an abundance biozone boundary that reflects a mass-mortality event. The vertical and lateral limits of the rock body that constitutes the biostratigraphic unit represent the limits in distribution of the defining biotic elements. The lateral limits never represent, and the vertical limits rarely represent, regionally synchronous events. Nevertheless, biostratigraphic units are effective for interpreting chronostratigraphic relations.

Article 49.—Kinds of Biostratigraphic Units. Three principal kinds of biostratigraphic units are recognized: *interval, assemblage,* and *abundance* biozones.

Remark. (a) **Boundary definitions.**—Boundaries of interval zones are defined by lowest and/or highest occurrences of single taxa; boundaries of some kinds of assemblage

[7]Pluton—a mappable body of plutonic rock.

zones (Oppel or concurrent range zones) are defined by lowest and/or highest occurrences of more than one taxon; and boundaries of abundance zones are defined by marked changes in relative abundances of preserved taxa.

Article 50.—**Definition of Interval Zone.** An interval zone (or subzone) is the body of strata between two specified, documented lowest and/or highest occurrences of single taxa.

Remarks. (a) **Interval zone types.**—Three basic types of interval zones are recognized (Fig. 4). These include the range zones and interval zones of the International Stratigraphic Guide (ISSC, 1976, p. 53, 60) and are:
1. The interval between the documented lowest and highest occurrences of a single taxon (Fig. 4A). This is the *taxon range zone* of ISSC (1976, p. 53).
2. The interval included between the documented lowest occurrence of one taxon and the documented highest occurrence of another taxon (Fig. 4B). When such occurrences result in stratigraphic overlap of the taxa (Fig. 4B-1), the interval zone is the *concurrent range zone* of ISSC (1976, p. 55), that involves only two taxa. When such occurrences do not result in stratigraphic overlap (Fig 4B-2), but are used to partition the range of a third taxon, the interval is the *partial range zone* of George and others (1969).
3. The interval between documented successive lowest occurrences or successive highest occurrences of two taxa (Fig. 4C). When the interval is between successive documented lowest occurrences within an evolutionary lineage (Fig. 4C-1), it is the *lineage zone* of ISSC (1976, p. 58). When the interval is between successive lowest occurrences of unrelated taxa or between successive highest occurrences of either related or unrelated taxa (Fig. 4C-2), it is a kind of *interval zone* of ISSC (1976, p. 60).

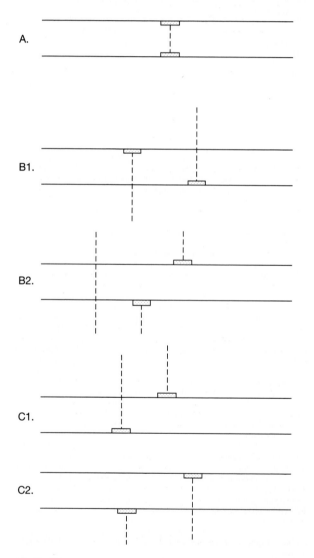

Figure 4.
Examples of biostratigraphic interval zones. Vertical broken lines indicate ranges of taxa; bars indicate lowest or highest documented occurrences.

(b) **Unfossiliferous intervals.**—Unfossiliferous intervals between or within biozones are the *barren interzones* and *intrazones* of ISSC (1976, p. 49).

Article 51.—**Definition of Assemblage Zone.** An assemblage zone is a biozone characterized by the association of three or more taxa. It may be based on all kinds of fossils present, or restricted to only certain kinds of fossils.

Remarks. (a) **Assemblage zone contents.**—An assemblage zone may consist of a geographically or stratigraphically restricted assemblage, or may incorporate two or more contemporaneous assemblages with shared characterizing taxa (*composite assemblage zones* of Kauffman, 1969) (Fig. 5c).
(b) **Assemblage zone types.**—In practice two assemblage zone concepts are used:
1. The *assemblage zone* (or cenozone) of ISSC (1976, p. 50), which is characterized by taxa without regard to their range limits (Fig. 5a). Recognition of this type of assemblage zone can be aided by using techniques of multivariate analysis. Careful designation of the characterizing taxa is especially important.
2. The *Oppel zone*, or the *concurrent range zone* of ISSC (1976, p. 55, 57), a type of zone characterized by more than two taxa and having boundaries based on two or more documented first and/or last occurrences of the included characterizing taxa (Fig. 5b).

Article 52.—**Definition of Abundance Zone.** An abundance zone is a biozone characterized by quantitatively distinctive maxima of relative abundance of one or more taxa. This is the *acme zone* of ISSC (1976, p. 59).

Remark. (a) **Ecologic controls.**—The distribution of biotic assemblages used to characterize some assemblage and abundance biozones may reflect strong local ecological control. Biozones based on such assemblages are included within the concept of ecozones (Vella, 1964), and are informal.

Ranks of Biostratigraphic Units.

Article 53.—**Fundamental Unit.** The fundamental unit of biostratigraphic classification is a biozone.

Remarks. (a) **Scope.**—A single body of rock may be divided into various kinds and scales of biozones or subzones, as discussed in the International Stratigraphic Guide (ISSC, 1976, p. 62). Such usage is recommended if it will promote clarity, but only the unmodified term *biozone* is accorded formal status.
(b) **Divisions.**—A biozone may be completely or partly divided into formally designated sub-biozones (subzones), if such divisions serve a useful purpose.

Biostratigraphic Nomenclature

Article 54.—**Establishing Formal Units.** Formal establishment of a biozone or subzone must meet the requirements of Article 3 and requires a unique name, a description of its content and its boundaries, reference to a stratigraphic sequence in which the zone is characteristically developed, and a discussion of its spatial extent.

Remarks. (a) **Name.**—The name, which is compound and designates the kind of biozone, may be based on:
1. One or two characteristic and common taxa that are restricted to the biozone, reach peak relative abundance within the biozone, or have their total stratigraphic overlap within the biozone. These names most commonly are those of genera or subgenera, binomial designations of species, or trinomial designations of subspecies. If names of the nominate taxa change, names of the zones should be changed accordingly. Generic or subgeneric names may be abbreviated. Trivial species or subspecies names should not be used alone because they may not be unique.
2. Combinations of letters derived from taxa which characterize the biozone. However, alpha-numeric code designations (e.g., N1, N2, N3 . . .) are informal and not recommended because they do not lend themselves readily to subsequent insertions, combinations, or eliminations. Biozonal systems based *only* on simple progressions of letters or numbers (e.g., A, B, C, or 1. 2, 3) are also not recommended.
(b) **Revision.**—Biozones and subzones are established empirically and may be modified on the basis of new evidence. Positions of established biozone or subzone boundaries may be stratigraphically refined, new characterizing taxa may be recognized, or original characterizing taxa may be superseded. If the concept of a particular biozone or subzone is substantially modified, a new unique designation is required to avoid ambiguity in subsequent citations.
(c) **Specifying kind of zone.**—Initial designation of a formally proposed biozone or subzone as an abundance zone, or as one of the types of interval zones, or assemblage zones (Articles 49–52), is strongly recommended. Once the type of biozone is clearly identified, the designation may be dropped in the remainder of a text (e.g., *Exus albus* taxon range zone to *Exus albus* biozone).
(d) **Defining taxa.**—Initial description or subsequent emendation of a biozone or subzone requires designation of the defining and characteristic taxa, and/or the documented first and last occurrences which mark the biozone or subzone boundaries.
(e) **Stratotypes.**—The geographic and stratigraphic position and boundaries of a formally proposed biozone or subzone should be defined precisely or characterized in one or more designated reference sections. Designation of a stratotype for each new biostratigraphic unit and of reference sections for emended biostratigraphic units is required.

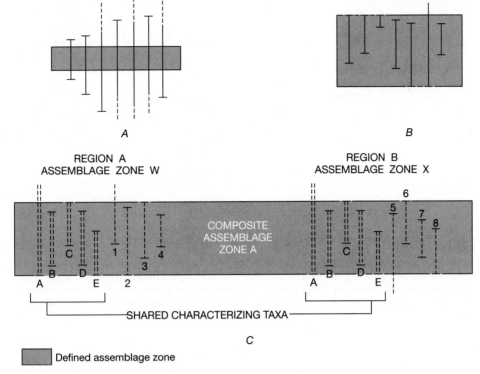

REGION A
ASSEMBLAGE ZONE W

REGION B
ASSEMBLAGE ZONE X

Figure 5.
Examples of assemblage zone concepts.

PEDOSTRATIGRAPHIC UNITS

Nature and Boundaries

Article 55.—**Nature of Pedostratigraphic Units.** A pedostratigraphic unit is a body of rock that consists of one or more pedologic horizons developed in one or more lithostratigraphic, allostratigraphic, or lithodemic units (Fig. 6) and is overlain by one or more formally defined lithostratigraphic or allostratigraphic units.

Remarks. (a) **Definition.**—A pedostratigraphic[8] unit is a buried, traceable, three-dimensional body of rock that consists of one or more differentiated pedologic horizons.

(b) **Recognition.**—The distinguishing property of a pedostratigraphic unit is the presence of one or more distinct, differentiated, pedologic horizons. Pedologic horizons are products of soil development (pedogenesis) which occurred subsequent to formation of the lithostratigraphic, allostratigraphic, or lithodemic unit or units on which the buried soil was formed; these units are the parent materials in which pedogenesis occurred. Pedologic horizons are recognized in the field by diagnostic features such as color, soil structure, organic-matter accumulation, texture, clay coatings, stains, or concretions. Micromorphology, particle size, clay mineralogy, and other properties determined in the laboratory also may be used to identify and distinguish pedostratigraphic units.

(c) **Boundaries and stratigraphic position.**—The upper boundary of a pedostratigraphic unit is the top of the uppermost pedologic horizon formed by pedogenesis in a buried soil profile. The lower boundary of a pedostratigraphic unit is the lowest *definite* physical boundary of a pedologic horizon within a buried soil profile. The stratigraphic position of a pedostratigraphic unit is determined by its relation to overlying and underlying stratigraphic units. (see Remark d).

(d) **Traceability.**—Practicability of subsurface tracing of the upper boundary of a buried soil is essential in establishing a pedostratigraphic unit because (1) few buried soils are exposed continuously for great distances, (2) the physical and chemical properties of a specific pedostratigraphic unit may vary greatly, both vertically and laterally, from place to place, and (3) pedostratigraphic units of different stratigraphic significance in the same region generally do not have unique identifying physical and chemical characteristics. Consequently, extension of a pedostratigraphic unit is accomplished by lateral tracing of the contact between a buried soil and an overlying, formally defined lithostratigraphic or allostratigraphic unit, or between a soil and two or more demonstrably correlative stratigraphic units.

(e) **Distinction from pedologic soils.**—Pedologic soils may include organic deposits (e.g., litter zones, peat deposits, or swamp deposits) that overlie or grade laterally into differentiated buried soils. The organic deposits are not products of pedogenesis, and O horizons are not included in a pedostratigraphic unit (Fig. 6); they may be classified as biostratigraphic or lithostratigraphic units. Pedologic soils also include the entire C horizon of a soil. The C horizon in pedology is not rigidly defined; it is merely the part of a soil profile that underlies the B horizon. The base of the C horizon in many soil profiles is gradational or unidentifiable; commonly it is placed arbitrarily. The need for clearly defined and easily recognized physical boundaries for a stratigraphic unit requires that the lower boundary of a pedostratigraphic unit be defined as the lowest *definite* physical boundary of a pedologic horizon in a buried soil profile, and part or all of the C horizon may be excluded from a pedostratigraphic unit.

(f) **Relation to saprolite and other weathered materials.**—A material derived by in situ weathering of lithostratigraphic, allostratigraphic, and(or) lithodemic units (e.g., saprolite, bauxite, residuum) may be the parent material in which pedologic horizons form, but is not a pedologic soil. A pedostratigraphic unit may be based on the pedologic horizons of a buried soil developed in the product of in-situ weathering, such as saprolite. The parents of such a pedostratigraphic unit are both the saprolite and, indirectly, the rock from which it formed.

(g) **Distinction from other stratigraphic units.**—A pedostratigraphic unit differs from other stratigraphic units in that (1) it is a product of surface alteration of one or more older material units by specific processes (pedogenesis), (2) its lithology and other properties differ markedly from those of the parent material(s), and (3) a single pedostratigraphic unit may be formed in situ in parent material units of diverse compositions and ages.

(h) **Independence from time concepts.**—The boundaries of a pedostratigraphic unit are time-transgressive. Concepts of time spans, however measured, play no part in defining the boundaries of a pedostratigraphic unit. Nonetheless, evidence of age, whether based on fossils, numerical ages, or geometrical or other relationships, may play an important role in distinguishing and identifying non-contiguous pedostratigraphic units at localities away from the type areas. The name of a pedostratigraphic unit should be chosen from a geographic feature in the type area, and not from a time span.

Pedostratigraphic Nomenclature and Unit

Article 56.—**Fundamental Unit.** The fundamental and only unit in pedostratigraphic classification is a geosol.

Article 57.—**Nomenclature.** —The formal name of a pedostratigraphic unit consists of geographic name combined with the term "geosol."

---

[8]Terminology related to pedostratigraphic classification is summarized on the following page.

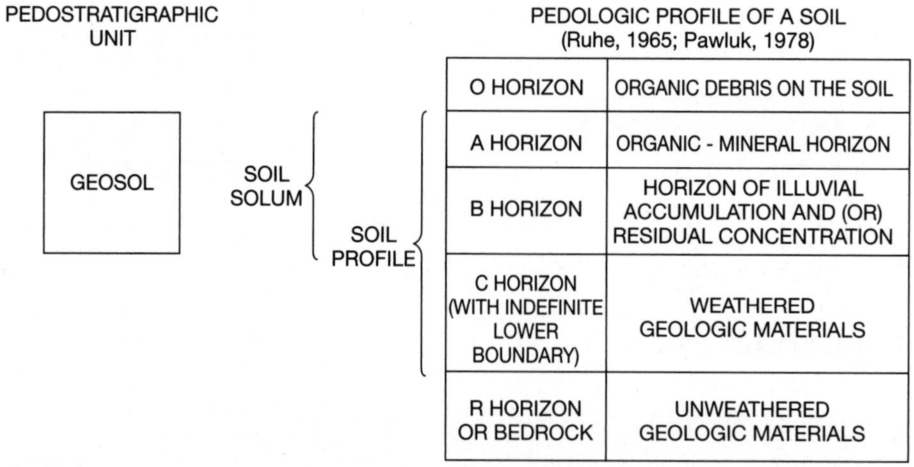

PEDOSTRATIGRAPHIC
UNIT

PEDOLOGIC PROFILE OF A SOIL
(Ruhe, 1965; Pawluk, 1978)

O HORIZON	ORGANIC DEBRIS ON THE SOIL
A HORIZON	ORGANIC - MINERAL HORIZON
B HORIZON	HORIZON OF ILLUVIAL ACCUMULATION AND (OR) RESIDUAL CONCENTRATION
C HORIZON (WITH INDEFINITE LOWER BOUNDARY)	WEATHERED GEOLOGIC MATERIALS
R HORIZON OR BEDROCK	UNWEATHERED GEOLOGIC MATERIALS

GEOSOL

SOIL SOLUM

SOIL PROFILE

Figure 6.
Relationship between pedostratigraphic units and pedologic profiles. The base of a geosol is the lowest clearly defined physical boundary of a pedologic horizon in a buried soil profile. In this example it is the lower boundary of the B horizon because the base of the C horizon is not a clearly defined physical boundary. In other profiles the base may be the lower boundary of a C horizon.

Capitalization of the initial letter in each word serves to identify formal usage. The geographic name should be selected in accordance with recommendations in Article 7 and should not duplicate the name of another formal geologic unit. Names based on subjacent and superjacent rock units, for example the super-Wilcox-sub-Claiborne soil, are informal, as are those with time connotations (post-Wilcox-pre-Claiborne soil).

Remarks.(a) **Composite geosols.**—Where the horizons of two or more merged or "welded" buried soils can be distinguished, formal names of pedostratigraphic units based on the horizon boundaries can be retained. Where the horizon boundaries of the respective merged or "welded" soils cannot be distinguished, formal pedostratigraphic classification is abandoned and a combined name such as Hallettville-Jamesville geosol may be used informally.

(b) **Characterization.**—The physical and chemical properties of a pedostratigraphic unit commonly vary vertically and laterally throughout the geographic extent of the unit. A pedostratigraphic unit is characterized by the *range* of physical and chemical properties of the unit in the type area, rather than by "typical" properties exhibited in a type section. Consequently, a pedostratigraphic unit is characterized on the basis of a composite stratotype (Article 8d).

(c) **Procedures for establishing formal pedostratigraphic units.**—A formal pedostratigraphic unit may be established in accordance with the applicable requirements of Article 3, and additionally by describing major soil horizons in each soil facies.

ALLOSTRATIGRAPHIC UNITS

Nature and Boundaries

Article 58.—**Nature of Allostratigraphic Units.** An allostratigraphic[9] unit is a mappable stratiform body of sedimentary rock that is defined and identified on the basis of its bounding discontinuities.

Remarks. (a) **Purpose.**—Formal allostratigraphic units may be defined to distinguish between different (1) superposed discontinuity-bounded deposits of similar lithology (Figs. 7, 9), (2) contiguous discontinuity-bounded deposits of similar lithology (Fig. 8), or (3) geographically separated discontinuity-bounded units of similar lithology (Fig. 9), or to distinguish as single units discontinuity-bounded deposits characterized by lithic heterogeneity (Fig. 8).

(b) **Internal characteristics.**—Internal characteristics (physical, chemical, and paleontological) may vary laterally and vertically throughout the unit.

(c) **Boundaries.**—Boundaries of allostratigraphic units are laterally traceable discontinuities (Figs. 7, 8, and 9).

(d) **Mappability.**—A formal allostratigraphic unit must be mappable at the scale practiced in the region where the unit is defined.

(e) **Type locality and extent.**—A type locality and type area must be designated; a composite stratotype or a type section and several reference sections are desirable. An

allostratigraphic unit may be laterally contiguous with a formally defined lithostratigraphic unit; a vertical cut-off between such units is placed where the units meet.

(f) **Relation to genesis.**—Genetic interpretation is an inappropriate basis for defining an allostratigraphic unit. However, genetic interpretation may influence the choice of its boundaries.

(g) **Relation to geomorphic surfaces.**—A geomorphic surface may be used as a boundary of an allostratigraphic unit, but the unit should not be given the geographic name of the surface.

(h) **Relation to soils and paleosols.**—Soils and paleosols are composed of products of weathering and pedogenesis and differ in many respects from allostratigraphic units, which are depositional units (see "Pedostratigraphic Units," Article 55). The upper boundary of a surface or buried soil may be used as a boundary of an allostratigraphic unit.

(i) **Relation to inferred geologic history.**—Inferred geologic history is not used to define an allostratigraphic unit. However, well-documented geologic history may influence the choice of the unit's boundaries.

(j) **Relation to time concepts.**—Inferred time spans, however measured, are not used to define an allostratigraphic unit. However, age relationships may influence the choice of the unit's boundaries.

(k) **Extension of allostratigraphic units.**—An allostratigraphic unit is extended from its type area by tracing the boundary discontinuities or by tracing or matching the deposits between the discontinuities.

RANKS OF ALLOSTRATIGRAPHIC UNITS

Article 59.—**Hierarchy.** The hierarchy of allostratigraphic units, in order of decreasing rank, is allogroup, alloformation, and allomember.

Remarks. (a) **Alloformation.**—The alloformation is the fundamental unit in allostratigraphic classification. An alloformation may be completely or only partly divided into allomembers, if some useful purpose is served, or it may have no allomembers.

(b) **Allomember.**—An allomember is the formal allostratigraphic unit next in rank below an alloformation.

(c) **Allogroup**—An allogroup is the allostratigraphic unit next in rank above an alloformation. An allogroup is established only if a unit of that rank if essential to elucidation of geologic history. An allogroup may consist entirely of named alloformations or, alternatively, may contain one or more named alloformation which jointly do not comprise the entire allogroup.

(d) **Changes in rank.**—The principles and procedures for elevation and reduction in rank of formal allostratigraphic units are the same as those in Articles 19b, 19g, and 28.

Allostratigraphic Nomenclature

Article 60.—**Nomenclature.** The principles and procedures for naming allostratigraphic units are the same as those for naming of lithostratigraphic units (see Articles 7, 30).

Remark. (a) **Reivison.**—Allostratigraphic units may be revised or otherwise modified in accordance with the recommendations in Articles 17 to 20.

---

[9] From the Greek *allo:* "other, different."

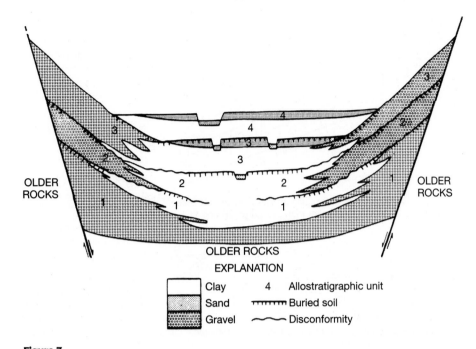

Figure 7.
Example of allostratigraphic classification of alluvial and lacustrine deposits in a graben. The alluvial and lacustrine deposits may be included in a single formation, or may be separated laterally into formations distinguished on the basis of contrasting texture (gravel, clay). Textural changes are abrupt and sharp, both vertically and laterally. The gravel deposits and clay deposits, respectively, are lithologically similar and thus cannot be distinguished as members of a formation. Four allostratigraphic units, each including two or three textural facies, may be defined on the basis of laterally traceable discontinuities (buried soils and disconformities).

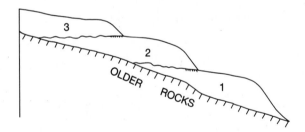

Figure 8.
Example of allostratigraphic classification of contiguous deposits of similar lithology. Allostratigraphic units 1, 2, and 3 are physical records of three glaciations. They are lithologically similar, reflecting derivation from the same bedrock, and constitute a single lithostratigraphic unit.

FORMAL UNITS DISTINGUISHED BY AGE

GEOLOGIC-TIME UNITS

Nature and Types

Article 61.—**Types.** Geologic-time units are conceptual, rather than material, in nature. Two types are recognized: those based on material standards or referents (specific rock sequences or bodies), and those independent of material referents (Fig. 1).

Units Based on Material Referents

Article 62.—**Types Based on Referents.** Two types of formal geologic-time units based on material referents are recognized: they are isochronous and diachronous units.

Article 63.—**Isochronous Categories.** Isochronous time units and the material bodies from which they are derived are twofold: geochronologic units (Article 80), which are based on corresponding material chronostratigraphic units (Article 66), and polarity-geochronologic units (Article 88), based on corresponding material polarity-chronostratigraphic units (Article 83).

Remark. (a) **Extent.**—Isochronous units are applicable worldwide; they may be referred to even in areas lacking a material record of the named span of time. The duration of the time may be represented by a unit-stratotype referent. The beginning and end of the time are represented by point-boundary-stratotypes either in a single stratigraphic sequence or in separate stratotype sections (Article 8b, 10b).

Article 64.—**Diachronous Categories.** Diachronic units (Article 91) are time units corresponding to diachronous material allostratigraphic units (Article 58), pedostratigraphic units (Article 55), and most lithostratigraphic (Article 22) and biostratigraphic (Article 48) units.

Remarks. (a) **Diachroneity.**—Some lithostratigraphic and biostratigraphic units are clearly diachronous, whereas others have boundaries which are not demonstrably iachronous within the resolving power of available dating methods. The latter commonly are treated as isochronous and are used for purposes of chronocorrelations (see biochronozone, Article 75). However, the assumption of isochroneity must be tested continually.

(b) **Extent.**—Diachronic units are coextensive with the diachronous material stratigraphic units on which they are based and are not used beyond the extent of their material referents.

Units Independent of Material Referents

Article 65.—**Numerical Divisions of Time.** Isochronous geologic-time units based on numerical divisions of time in years are geochronometric units (Article 96) and have no material referents.

CHRONOSTRATIGRAPHIC UNITS

Nature and Boundaries

Article 66.—**Definition.** A chronostratigraphic unit is a body of rock established to serve as the material reference for all rocks formed during the

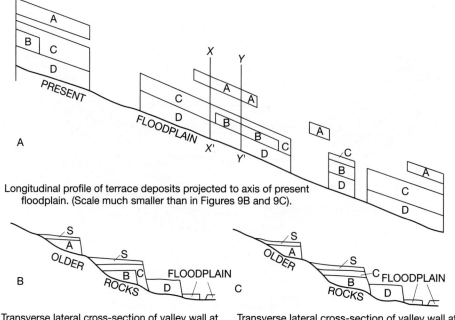

Longitudinal profile of terrace deposits projected to axis of present floodplain. (Scale much smaller than in Figures 9B and 9C).

Transverse lateral cross-section of valley wall at X-X' in Figure 9A.

Transverse lateral cross-section of valley wall at Y-Y' in Figure 9A.

Figure 9.

Example of allostratigraphic classification of lithologically similar, discontinuous terrace deposits. A, B, C, and D are terrace gravel units of similar lithology at different topographic positions on a valley wall. The deposits may be defined as separate formal allostratigraphic units if such units are useful and if bounding discontinuities can be traced laterally. Terrace gravels of the same age commonly are separated geographically by exposures of older rocks. Where the bounding discontinuities cannot be traced continuously, they may be extended geographically on the basis of objective correlation of internal properties of the deposits other than lithology (e.g., fossil content, included tephras), topographic position, numerical ages, or relative-age criteria (e.g., soils or other weathering phenomena). The criteria for such extension should be documented. Slope deposits and eolian deposits (S) that mantle terrace surfaces may be of diverse ages and are not included in a terrace-gravel allostratigraphic unit. A single terrace surface may be underlain by more than one allostratigraphic unit (units B and C in sections b and c).

same span of time. Each of its boundaries is synchronous. The body also serves as the basis for defining the specific interval of time, or geochronologic unit (Article 80), represented by the referent.

Remarks. (a) **Purposes.**—Chronostratigraphic classification provides a means of establishing the temporally sequential order of rock bodies. Principal purposes are to provide a framework for (1) temporal correlation of the rocks in one area with those in another, (2) placing the rocks of the Earth's crust in a systematic sequence and indicating their relative position and age with respect to earth history as a whole, and (3) constructing an internationally recognized Standard Global Chronostratigraphic Scale.

(b) **Nature.**—A chronostratigraphic unit is a material unit and consists of a body of strata formed during a specific time span. Such a unit represents all rocks, and only those rocks, formed during that time span.

(c) **Content.**—A chronostratigraphic unit may be based upon the time span of a biostratigraphic unit, a lithic unit, a magnetopolarity unit, or any other feature of the rock record that has a time range. Or it may be any arbitrary but specified sequence of rocks, provided it has properties allowing chronocorrelation with rock sequences elsewhere.

Article 67.—**Boundaries.** Boundaries of chronostratigraphic units should be defined in a designated stratotype on the basis of observable paleontological or physical features of the rocks.

Remark. (a) **Emphasis on lower boundaries of chronostratigraphic units.**—Designation of point boundaries for both base and top of chronostratigraphic units is not recommended, because subsequent information on relations between successive units may identify overlaps or gaps. One means of minimizing or eliminating problems of duplication or gaps in chronostratigraphic successions is to define formally as a point-boundary stratotype only the base of the unit. Thus, a chronostratigraphic unit with its base defined at one locality, will have its top defined by the base of an overlying unit at the same, but more commonly another, locality (Article 8b).

Article 68.—**Correlation.** Demonstration of time equivalence is required for geographic extension of a chronostratigraphic unit from its type section or area. Boundaries of chronostratigraphic units can be extended only within the limits of resolution of available means of chronocorrelation, which currently include paleontology, numerical dating, remanent magnetism, thermoluminescence, relative-age criteria (examples are superposition and cross-cutting relations), and such indirect and inferential physical criteria as climatic changes, degree of weathering, and relations to unconformities. Ideally, the boundaries of chronostratigraphic units are independent of lithology, fossil content, or other material bases of stratigraphic division, but, in practice, the correlation or geographic extension of these boundaries relies at least in part on such features. Boundaries of chronostratigraphic units commonly are intersected by boundaries of most other kinds of material units.

Ranks of Chronostratigraphic Units

Article 69.—**Hierarchy.** The hierarchy of chronostratigraphic units, in order of decreasing rank, is eonothem, erathem, system, series, and stage. Of these, system is the primary unit of worldwide major rank; its primacy derives from the history of development of stratigraphic classification. All systems and units of higher rank are divided completely into units of the next lower rank. Chronozones are non-hierarchical and commonly lower-rank chronostratigraphic units. Stages and chronozones in sum do not necessarily equal the units of next higher rank and need not be contiguous. The rank and magnitude of chronostratigraphic units are related to the time interval represented by the units, rather than to the thickness or areal extent of the rocks on which the units are based.

Article 70.—**Eonothem.** The unit highest in rank is eonothem. The Phanerozoic Eonothem encompasses the Paleozoic, Mesozoic, and Cenozoic

Erathems. Although older rocks have been assigned heretofore to the Precambrian Eonothem, they also have been assigned recently to other (Archean and Proterozoic) eonothems by the IUGS Precambrian Subcommission. The span of time corresponding to an eonothem is an *eon*.

Article 71.—Erathem. An erathem is the formal chronostratigraphic unit of rank next lower to eonothem and consists of several adjacent systems. The span of time corresponding to an erathem is an *era*.

Remark. (a) **Names.**—Names given to traditional Phanerozoic erathems were based upon major stages in the development of life on Earth: Paleozoic (old), Mesozoic (intermediate), and Cenozoic (recent) life. Although somewhat comparable terms have been applied to Precambrian units, the names and ranks of Precambrian divisions are not yet universally agreed upon and are under consideration by the IUGS Subcommission on Precambrian Stratigraphy.

Article 72.—System. The unit of rank next lower to erathem is the system. Rocks encompassed by a system represent a time-span and an episode of Earth history sufficiently great to serve as a worldwide chronostratigraphic reference unit. The temporal equivalent of a system is a *period*.

Remark. (a) **Subsystem and supersystem.**—Some systems initially established in Europe later were divided or grouped elsewhere into units ranked as systems. *Subsystems* (Mississippian Subsystem of the Carboniferous System) and *supersystems* (Karoo Supersystem) are more appropriate.

Article 73.—Series. Series is a conventional chronostratigraphic unit that ranks below a system and always is a division of a system. A series commonly constitutes a major unit of chronostratigraphic correlation within a province, between provinces, or between continents. Although many European series are being adopted increasingly for dividing systems on other continents, provincial series of regional scope continue to be useful. The temporal equivalent of a series is an *epoch*.

Article 74.—Stage. A stage is a chronostratigraphic unit of smaller scope and rank than a series. It is most commonly of greatest use in intra-continental classification and correlation, although it has the potential for worldwide recognition. The geochronologic equivalent of stage is *age*.

Remark. (a) **Substage.**—Stages may be, but need not be, divided completely into substages.

Article 75.—Chronozone. A chronozone is a nonhierarchical, but commonly small, formal chronostratigraphic unit, and its boundaries may be independent of those of ranked units. Although a chronozone is an isochronous unit, it may be based on a biostratigraphic unit (example: *Cardioceras cordatum* Biochronozone), a lithostratigraphic unit (Woodbend Lithochronozone), or a magnetopolarity unit (Gilbert Reversed-Polarity Chronozone). Modifiers (litho-, bio-, polarity) used in formal names of the units need not be repeated in general discussions where the meaning is evident from the context, e.g., *Exus albus* Chronozone.

Remarks. (a) **Boundaries of chronozones.**—The base and top of a *chronozone* correspond in the unit's stratotype to the observed, defining, physical and paleontological features, but they are extended to other areas by any means available for recognition of synchroneity. The temporal equivalent of a chronozone is a chron.

(b) **Scope.**—The scope of the non-hierarchical chronozone may range markedly, depending upon the purpose for which it is defined either formally or informally. The informal "biochronozone of the ammonites," for example, represents a duration of time which is enormous and exceeds that of a system. In contrast, a biochronozone defined by a species of limited range, such as the *Exus albus* Chronozone, may represent a duration equal to or briefer than that of a stage.

(c) **Practical utility.**—Chronozones, especially thin and informal biochronozones and lithochronozones bounded by key beds or other "markers," are the units used most commonly in industry investigations of selected parts of the stratigraphy of economically favorable basins. Such units are useful to define geographic distributions of lithofacies or biofacies, which provide a basis for genetic interpretations and the selection of targets to drill.

Chronostratigraphic Nomenclature

Article 76.—Requirements. Requirements for establishing a formal chronostratigraphic unit include: (i) statement of intention to designate such a unit; (ii) selection of name; (iii) statement of kind and rank of unit; (iv) statement of general concept of unit including historical background, synonymy, previous treatment, and reasons for proposed establishment; (v) description of characterizing physical and/or biological features; (vi) designation and description of boundary type sections, stratotypes, or other kinds of units on which it is based; (vii) correlation and age relations; and (viii) publication in a recognized scientific medium as specified in Article 4.

Article 77.—Nomenclature. A formal chronostratigraphic unit is given a compound name, and the initial letter of all words, except for trivial taxonomic terms, is capitalized. Except for chronozones (Article 75), names proposed for new chronostratigraphic units should not duplicate those for other stratigraphic units. For example, naming a new chronostratigraphic unit simply by adding "-an" or "-ian" to the name of a lithostratigraphic unit is improper.

Remarks. (a) **Systems and units of higher rank.**—Names that are generally accepted for systems and units of higher rank have diverse origins, and they also have different kinds of endings (Paleozoic, Cambrian, Cretaceous, Jurassic, Quaternary).

(b) **Series and units of lower rank.**—Series and units of lower rank are commonly known either by geographic names (Virgilian Series, Ochoan Series) or by names of their encompassing units modified by the capitalized adjectives Upper, Middle, and Lower (Lower Ordovician). Names of chronozones are derived from the unit on which they are based (Article 75). For series and stage, a geographic name is preferable because it may be related to a type area. For geographic names, the adjectival endings -an or -ian are recommended (Cincinnatian Series), but it is permissible to use the geographic name without any special ending, if more euphonious. Many series and stage names already in use have been based on lithic units (groups, formations, and members) and bear the names of these units (Wolfcampian Series, Claibornian Stage). Nevertheless, a stage preferably should have a geographic name not previously used in stratigraphic nomenclature. Use of internationally accepted (mainly European) stage names is preferably to the proliferation of others.

Article 78.—Stratotypes. An ideal stratotype for a chronostratigraphic unit is a completely exposed unbroken and continuous sequence of fossiliferous stratified rocks extending from a well defined lower boundary to the base of the next higher unit. Unfortunately, few available sequences are sufficiently complete to define stages and units of higher rank, which therefore are best defined by boundary-stratotypes (Article 8b).

Boundary-stratotypes for major chronostratigraphic units ideally should be based on complete sequences of either fossiliferous monofacial marine strata or rocks with other criteria for chronocorrelation to permit widespread tracing of synchronous horizons. Extension of synchronous surfaces should be based on as many indicators of age as possible.

Article 79.—Revision of units. Revision of a chronostratigraphic unit without changing its name is allowable but requires as much justification as the establishment of a new unit (Articles 17, 19, and 76). Revision or redefinition of a unit of system or higher rank requires international agreement. If the definition of a chronostratigraphic unit is inadequate, it may be clarified by establishment of boundary stratotypes in a principal reference section.

GEOCHRONOLOGIC UNITS

Nature and Boundaries

Article 80.—Definition and Basis. Geochronologic units are divisions of time traditionally distinguished on the basis of the rock record as expressed by chronostratigraphic units. A geochronologic unit is not a stratigraphic unit (i.e., it is not a material unit), but it corresponds to the time span of an established chronostratigraphic unit (Article 65 and 66), and its beginning and ending corresponds to the base and top of the referent.

Ranks and Nomenclature of Geochronologic Units

Article 81.—Hierarchy. The hierarchy of geochronologic units in order of decreasing rank is *eon, era, period, epoch,* and *age*. Chron is a non-hierarchical, but commonly brief, geochronologic unit. Ages in sum do not necessarily equal epochs and need not form a continuum. An eon is the time represented by the rocks constituting an eonothem; era by an erathem; period by a system; epoch by a series; age by a stage; and chron by a chronozone.

Article 82.—Nomenclature. Names for periods and units of lower rank are identical with those of the corresponding chronostratigraphic units; the names of some eras and eons are independently formed. Rules of capitalization for chronostratigraphic units (Article 77) apply to geochronologic units. The adjectives Early, Middle, and Late are used for the geochronologic epochs equivalent to the corresponding chronostratigraphic Lower, Middle, and Upper series, where these are formally established.

POLARITY-CHRONOSTRATIGRAPHIC UNITS

Nature and Boundaries

Article 83.—Definition. A polarity-chronostratigraphic unit is a body of rock that contains the primary magnetic-polarity record imposed when the rock was deposited, or crystallized, during a specific interval of geologic time.

Remarks. (a) **Nature.**—Polarity-chronostratigraphic units depend fundamentally for definition on actual sections or sequences, or measurements on individual rock units, and without these standards they are meaningless. They are based on material units, the polarity zones of magnetopolarity classification. Each polarity-chronostratigraphic unit is the record of the time during which the rock formed and the Earth's magnetic field had a designated polarity. Care should be taken to define polarity-chronologic units in terms of polarity-chronostratigraphic units, and not vice versa.

(b) **Principal purposes.**—Two principal purposes are served by polarity-chronostratigraphic classification: (1) correlation of rocks at one place with those of the same age and polarity at other places; and (2) delineation of the polarity history of the Earth's magnetic field.

(c) **Recognition.**—A polarity-chronostratigraphic unit may be extended geographically from its type locality only with the support of physical and/or paleontologic criteria used to confirm its age.

Article 84.—**Boundaries.** The boundaries of a polarity chronozone are placed at polarity-reversal horizons or polarity transition-zones (see Article 45).

RANKS AND NOMENCLATURE OF POLARITY-CHRONOSTRATIGRAPHIC UNITS

Article 85.—**Fundamental Unit.** The polarity chronozone consists of rocks of a specified primary polarity and is the fundamental unit of world-wide polarity-chronostratigraphic classification.

Remarks. (a) **Meaning of term.**—A polarity chronozone is the worldwide body of rock strata that is collectively defined as a polarity-chronostratigraphic unit.

(b) **Scope.**—Individual polarity zones are the basic building blocks of polarity chronozones. Recognition and definition of polarity chronozones may thus involve step-by-step assembly of carefully dated or correlated individual polarity zones, especially in work with rocks older than the oldest ocean-floor magnetic anomalies. This procedure is the method by which the Brunhes, Matuyama, Gauss, and Gilbert Chronozones were recognized (Cox, Doell, and Dalrymple, 1963) and defined originally (Cox, Doell, and Dalrymple, 1964).

(c) **Ranks.**—Divisions of polarity chronozones are designated polarity subchronozones. Assemblages of polarity chronozones may be termed polarity superchronozones.

Article 86.—**Establishing Formal Units.** Requirements for establishing a polarity-chronostratigraphic unit include those specified in Articles 3 and 4, and also (1) definition of boundaries of the unit, with specific references to designated sections and data; (2) distinguishing polarity characteristics, lithologic descriptions, and included fossils; and (3) correlation and age relations.

Article 87.—**Name.** A formal polarity-chronostratigraphic unit is given a compound name beginning with that for a named geographic feature; the second component indicates the normal, reversed, or mixed polarity of the unit, and the third component is *chronozone*. The initial letter of each term is capitalized. If the same geographic name is used for both a magnetopolarity zone and a polarity-chronostratigraphic unit, the latter should be distinguished by an -an or -ian ending. Example: Tetonian Reversed-Polarity Chronozone.

Remarks: (a) **Preservation of established name.**—A particularly well-established name should not be displaced, either on the basis of priority, as described in Article 7c, or because it was not taken from a geographic feature. Continued use of Brunhes, Matuyama, Gauss, and Gilbert, for example, is endorsed so long as they remain valid units.

(b) **Expression of doubt.**—Doubt in the assignment of polarity zones to polarity-chronostratigraphic units should be made explicit if criteria of time equivalence are inconclusive.

POLARITY-CHRONOLOGIC UNITS

Nature and Boundaries

Article 88.—**Definition.** Polarity-chronologic units are divisions of geologic time distinguished on the basis of the record of magnetopolarity as embodied in polarity-chronostratigraphic units. No special kind of magnetic time is implied; the designations used are meant to convey the parts of geologic time during which the Earth's magnetic field had a characteristic polarity or sequence of polarities. These units correspond to the time spans represented by polarity chronozones, e.g., Gauss Normal Polarity Chronozone. They are not material units.

RANKS AND NOMENCLATURE OF POLARITY-CHRONOLOGIC UNITS

Article 89.—**Fundamental Unit.** The polarity chron is the fundamental unit of geologic time designating the time span of a polarity chronozone.

Remark. (a) **Hierarchy.**—Polarity-chronologic units of decreasing hierarchical ranks are polarity superchron, polarity chron, and polarity subchron.

Article 90.—**Nomenclature.** Names for polarity chronologic units are identical with those of corresponding polarity-chronostratigraphic units, except that the term chron (or superchron, etc.) is substituted for chronozone (or superchronozone, etc.).

DIACHRONIC UNITS

Nature and Boundaries

Article 91.—**Definition.** A diachronic unit comprises the unequal spans of time represented either by a specific lithostratigraphic, allostratigraphic, biostratigraphic, or pedostratigraphic unit, or by an assemblage of such units.

Remarks. (a) **Purposes.**—Diachronic classification provides (1) a means of comparing the spans of time represented by stratigraphic units with diachronous boundaries at different localities, (2) a basis for broadly establishing in time the beginning and ending of deposition of diachronous stratigraphic units at different sites, (3) a basis for inferring the rate of change in areal extent of depositional processes, (4) a means of determining and comparing rates and durations of deposition at different localities, and (5) a means of comparing temporal and spatial relations of diachronous stratigraphic units (Watson and Wright, 1980).

(b) **Scope.**—The scope of a diachronic unit is related to (1) the relative magnitude of the transgressive division of time represented by the stratigraphic unit or units on which it is based and (2) the areal extent of those units. A diachronic unit is not extended beyond the geographic limits of the stratigraphic unit or units on which it is based.

(c) **Basis.**—The basis for a diachronic unit is the diachronous referent.

(d) **Duration.**—A diachronic unit may be of equal duration at different places despite differences in the times at which it began and ended at those places.

Article 92.—**Boundaries.** The boundaries of a diachronic unit are the times recorded by the beginning and end of deposition of the material referent at the point under consideration (Figs. 10, 11).

Remark. (a) **Temporal relations.**—One or both of the boundaries of a diachronic unit are demonstrably time-transgressive. The varying time significance of the boundaries is defined by a series of boundary reference sections (Article 8b, 8e). The duration and age of a diachronic unit differ from place to place (Figs. 10, 11).

Ranks and Nomenclature of Diachronic Units

Article 93.—**Ranks.** A diachron is the fundamental and nonhierarchical diachronic unit. If a hierarchy of diachronic units is needed, the terms episode, phase, span, and cline, in order of decreasing rank, are recommended. The rank of a hierarchical unit is determined by the scope of the unit (Article 91 b), and not by the time span represented by the unit at a particular place.

Remarks. (a) **Diachron.**—Diachrons may differ greatly in magnitude because they are the spans of time represented by individual or grouped lithostratigraphic, allostratigraphic, biostratigraphic, and(or) pedostratigraphic units.

(b) **Hierarchical ordering permissible.**—A hierarchy of diachronic units may be defined if the resolution of spatial and temporal relations of diachronous stratigraphic units is sufficiently precise to make the hierarchy useful (Watson and Wright, 1980). Although all hierarchical units of rank lower than episode are part of a unit next higher in rank, not all parts of an episode, phase, or span need be represented by a unit of lower rank.

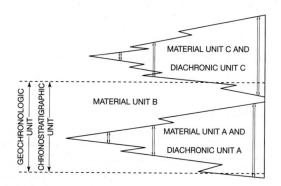

Figure 10.
Comparison of geochronologic, chronostratigraphic, and diachronic units.

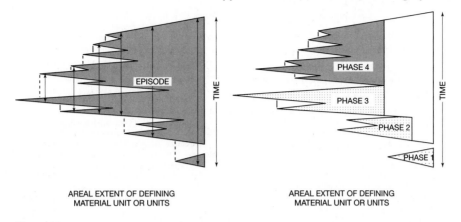

AREAL EXTENT OF DEFINING
MATERIAL UNIT OR UNITS

AREAL EXTENT OF DEFINING
MATERIAL UNIT OR UNITS

Figure 11.
Schematic relation of phases to an episode. Parts of a phase similarly may be divided into spans, and spans into clines. Formal definition of spans and clines is unnecessary in most diachronic unit hierarchies.

(c) **Episode.**—An episode is the unit of highest rank and greatest scope in hierarchical classification. If the "Wisconsinan Age" were to be redefined as a diachronic unit, it would have the rank of episode.

Article 94.—Name. The name for a diachronic unit should be compound, consisting of a geographic name followed by the term diachron or a hierarchical rank term. Both parts of the compound name are capitalized to indicate formal status. If the diachronic unit is defined by a single stratigraphic unit, the geographic name of the unit may be applied to the diachronic unit. Otherwise, the geographic name of the diachronic unit should not duplicate that of another formal stratigraphic unit. Genetic terms (e.g., alluvial, marine) or climatic terms (e.g., glacial, interglacial) are not included in the names of diachronic units.

Remarks. (a) **Formal designation of units.**—Diachronic units should be formally defined and named only if such definition is useful.

(b) **Inter-regional extension of geographic names.**—The geographic name of a diachronic unit may be extended from one region to another if the stratigraphic units on which the diachronic unit is based extend across the regions. If different diachronic units in contiguous regions eventually prove to be based on laterally continuous stratigraphic units, one name should be applied to the unit in both region. If two names have been applied, one name should be abandoned and the other formally extended. Rules of priority (Article 7d) apply. Priority in publication is to be respected, but priority alone does not justify displacing a well-established name by one not well-known or commonly used.

(c) **Change from geochronologic to diachronic classification.**—Lithostratigraphic units have served as the material basis for widely accepted chronostratigraphic and geochronologic classifications of Quaternary nonmarine deposits, such as the classifications of Frye et al (1968), Willman and Frye (1970), and Dreimanis and Karrow (1972). In practice, time-parallel horizons have been extended from the stratotypes on the basis of markedly time-transgressive lithostratigraphic and pedostratigraphic unit boundaries. The time ("geochronologic") units, defined on the basis of the stratotype sections but extended on the basis of diachronous stratigraphic boundaries, are diachronic units. Geographic names established for such "geochronologic" units may be used in diachronic classification if (1) the chronostratigraphic and geochronologic classifications are formally abandoned and diachronic classifications are proposed to replace the former "geochronologic" classifications, and (2) the units are redefined as formal diachronic units. Preservation of well-established names in these specific circumstances retains the intent and purpose of the names and the units, retains the practical significance of the units, enhances communication, and avoids proliferation of nomenclature.

Article 95.—Establishing Formal Units. Requirements for establishing a formal diachronic unit, in addition to those in Article 3, include (1) specification of the nature, stratigraphic relations, and geographic or areal relations of the stratigraphic unit or units that serve as a basis for definition of the unit, and (2) specific designation and description of multiple reference sections that illustrate the temporal and spatial relations of the defining stratigraphic unit or units and the boundaries of the unit or units.

Remark. (a) **Revision or abandonment.**—Revision or abandonment of the stratigraphic unit or units that serve as the material basis for definition of a diachronic unit may require revision or abandonment of the diachronic unit. Procedure for revision must follow the requirements for establishing a new diachronic unit.

GEOCHRONOMETRIC UNITS

Nature and Boundaries

Article 96.—Definition. Geochronometric units are units established through the direct division of geologic time, expressed in years. Like geochronologic units (Article 80), geochronometric units are abstractions, i.e., they are not material units. Unlike geochronologic units, geochronometric units are not based on the time span of designated chronostratigraphic units (stratotypes), but are simply time divisions of convenient magnitude for the purpose for which they are established, such as the development of a time scale for the Precambrian. Their boundaries are arbitrarily chosen or agreed-upon ages in years.

Ranks and Nomenclature of Geochronometric Units

Article 97.—Nomenclature. Geochronologic rank terms (eon, era, period, epoch, age, and chron) may be used for geochronometric units when such terms are formalized. For example, Archean Eon and Proterozoic Eon, as recognized by the IUGS Subcommission on Precambrian Stratigraphy, are formal geochronometric units in the sense of Article 96, distinguished on the basis of an arbitrarily chosen boundary at 2.5 Ga. Geochronometric units are not defined by, but may have, corresponding chronostratigraphic units (eonothem, erathem, system, series, stage, and chronozone).

PART III. REFERENCES[10]

American Commission on Stratigraphic Nomenclature, 1947, Note 1—Organization and objectives of the Stratigraphic Commission: American Association of Petroleum Geologists Bulletin, v. 31, no. 3, p. 513–518.

——, 1961, Code of Stratigraphic Nomenclature: American Association of Petroleum Geologists Bulletin, v. 45, no. 5, p. 645–665.

——, 1970, Code of Stratigraphic Nomenclature (2d ed.): American Association of Petroleum Geologists, Tulsa, Okla., 45 p.

——, 1976, Note 44—Application for addition to code concerning magnetostratigraphic units: American Association of Petroleum Geologists Bulletin, v. 60, no. 2, p. 273–277.

Caster, K. E., 1934, The stratigraphy and paleontology of northwestern Pennsylvania, Part 1, Stratigraphy: Bulletins of American Paleontology, v. 21, 185 p.

Chang, K. H., 1975, Unconformity-bounded stratigraphic units: Geological Society of American Bulletin, v. 86, no. 11, p. 1544–1552.

Committee on Stratigraphic Nomenclature, 1933, Classification and nomenclature of rock units: Geological Society of American Bulletin, v. 44, no. 2, p. 423–459, and American Association of Petroleum Geologists Bulletin, v. 17, no. 7, p. 843–868.

Cox, A. V., R. R. Doell, and G. B. Dalrymple, 1963, Geomagnetic polarity epochs and Pleistocene geochronometry: Nature, v. 198, p. 1049–1051.

——, 1964, Reversals of the Earth's magnetic field: Science, v. 144, no. 3626, p. 1537–1543.

---

[10]Readers are reminded of the extensive and noteworthy bibliography of contributions to stratigraphic principles, classification, and terminology cited by the International Stratigraphic Guide (ISSC, 1976, p. 111–187).

Cross, C. W., 1898, Geology of the Telluride area: U.S. Geological Survey 18th Annual Report, pt. 3, p. 759.

Cumming, A. D., J. G. C. M. Fuller, and J. W. Porter, 1959, Separation of strata: Paleozoic limestones of the Williston basin: American Journal of Science, v. 257, no. 10, p. 722–733.

Dreimanis, Aleksis, and P. F. Karrow, 1972, Glacial history of the Great Lakes-St. Lawrence region, the classification of the Wisconsin(an) Stage, and its correlatives: International Geologic Congress, 24th Session, Montreal, 1972, Section 12, Quaternary Geology, p. 5–15.

Dunbar, C. O., and John Rodgers, 1957, Principles of stratigraphy: Wiley, New York, 356 p.

Forgotson, J. M., Jr., 1957, Nature, usage and definition of marker-defined vertically segregated rock units: American Association of Petroleum Geologists Bulletin, v. 41, no. 9, p. 2108–2113.

Frye, J. C., H. B. Willman, Meyer Rubin, and R. F. Black, 1968, Definition of Wisconsinan Stage: U.S. Geological Survey Bulletin 1274–E, 22 p.

George, T. N., and others, 1969, Recommendations on stratigraphical usage: Geological Society of London, Proceedings no. 1656, p. 139–166.

Harland, W. B., 1977, Essay review [of] International Stratigraphic Guide, 1976; Geology Magazine, v. 114, no. 3, p. 229–235.

———, 1978, Geochronologic scales, *in* G. V. Cohee et al, eds., Contributions to the Geologic Time Scale: American Association of Petroleum Geologists, Studies in Geology, no. 6, p. 9–32.

Harrison, J. E., and Z. E. Peterman, 1980, North American Commission on Stratigraphic Nomenclature Note 52—A preliminary proposal for a chronometric time scale for the Precambrian of the United States and Mexico: Geological Society of America Bulletin, v. 91, no. 6, p. 377–380.

Henbest, L. G., 1952, Significance of evolutionary explosions for diastrophic division of Earth history: Journal of Paleontology, v. 26, p. 299–318.

Henderson, J. B., W. G. E. Caldwell, and J. E. Harrison, 1980, North American Commission on Stratigraphic Nomenclature, Report 8—Amendment of code concerning terminology for igneous and high-grade metamorphic rocks: Geological Society of American Bulletin, v. 91, no. 6, p. 374–376.

Holland, C. H., and others, 1978, A guide to stratigraphical procedure: Geological Society of London, Special Report 10, p. 1–18.

Huxley, T. H., 1862, The anniversary address: Geological Society of London, Quarterly Journal, v. 18, p. xl–liv.

International Commission on Zoological Nomenclature, 1964: International Code of Zoological Nomenclature adopted by the XV International Congress of Zoology: International Trust for Zoological Nomenclature, London, 176 p.

International Subcommission on Stratigraphic Classification (ISSC), 1976, International Stratigraphic Guide (H. D. Hedberg, ed.): John Wiley and Sons, New York, 200 p.

International Subcommission on Stratigraphic Classification, 1979, Magneto-stratigraphy polarity units—a supplementary chapter of the ISSC International Stratigraphic Guide: Geology, v. 7, p. 578–583.

Izett, G. A., and R. E. Wilcox, 1981, Map showing the distribution of the Huckleberry Ridge, Mesa Falls, and Lava Creek volcanic ash beds (Pearlette family ash beds) of Pliocene and Pleistocene age in the western United States and southern Canada: U.S. Geological Survey Miscellaneous Geological Investigations Map 1–1325.

Kauffman, E. G., 1969, Cretaceous marine cycles of the Western Interior: Mountain Geologist: Rocky Mountain Association of Geologists, v. 6, no. 4, p. 227–245.

Matthews, R. K., 1974, Dynamic stratigraphy—an introduction to sedimentation and stratigraphy: Prentice-Hall, New Jersey, 370 p.

McDougall, Ian, 1977, The present status of the geomagnetic polarity time scale: Research School of Earth Sciences, Australian National University, publication no. 1288, 34 p.

McElhinny, M. W., 1978, The magnetic polarity time scale; prospects and possibilities in magnetostratigraphy, *in* G. V. Cohee et al, eds., Contributions to the Geologic Time Scale, American Association of Petroleum Geologists, Studies in Geology, no. 6, p. 57–65.

McIver, N. L., 1972, Cenozoic and Mesozoic stratigraphy of the Nova Scotia shelf: Canadian Journal of Earth Science, v. 9, p. 54–70.

McLaren, D. J., 1977, The Silurian-Devonian Boundary Committee. A final report, *in* A. Martinsson, ed., The Silurian-Devonian boundary: IUGS Series A, no. 5, p. 1–34.

Morrison, R. B., 1967, Principles of Quaternary soil stratigraphy, *in* R. B. Morrison and H. E. Wright, Jr., eds., Quaternary soils: Reno, Nevada, Center for Water Resources Research, Desert Research Institute, Univ. Nevada, P. 1–69.

North American Commission on Stratigraphic nomenclature, 1981, Draft North American Stratigraphic Code: Canadian Society of Petroleum Geologists, Calgary, 63 p.

Palmer, A. R., 1965, Biomere-a new kind of biostratigraphic unit: Journal of Paleontology, v. 39, no. 1, p. 149–153.

Parsons, R. B., 1981, Proposed soil-stratigraphic guide, *in* International Union for Quaternary Research and International Society of Soil-Science: INQUA Commission 6 and ISSS Commission 5 Working Group, Pedology, Report, p. 6–12.

Pawluk, S., 1978, The pedogenic profile in the stratigraphic section, *in* W. C. Mahaney, ed., Quaternary soils: Norwich, England. GeoAbstracts, Ltd., p. 61–75.

Ruhe, R. V., 1965, Quaternary paleopedology, *in* H. E. Wright, Jr., and D. G. Frey, eds., The Quaternary of the United States: Princeton, N.J., Princeton University Press, p. 755–764.

Schultz, E. H., 1982, The chronosome and supersome—terms proposed for low-rank chronostratigraphic units: Canadian Petroleum Geology, v. 30, no. 1, p. 29–33.

Shaw, A. B., 1964, Time in stratigraphy: McGraw-Hill, New York, 365 p.

Sims, P. K., 1979, Precambrian subdivided: Geotimes, v. 24, no. 12, p. 15.

Sloss, L. L., 1963, Sequences in the cratonic interior of North America: Geological Society of American Bulletin, v. 74, no. 2, p. 94–114.

Tracey, J. L., Jr., and others, 1971, Initial reports of the Deep Sea Drilling Project, v. 8: U.S. Government Printing Office, Washington, 1037 p.

Valentine, K. W. G., and J. B. Dalrymple, 1976, Quaternary buried paleosols: A critical review: Quaternary Research, v. 6, p. 209–222.

Vella, P., 1964, Biostratigraphic units: New Zealand Journal of Geology and Geophysics, v. 7, no. 3, p. 615–625.

Watson, R. A., and H. E. Wright, Jr., 1980, The end of the Pleistocene: A general critique of chronostratigraphic classification: Boreas, v. 9, p. 153–163.

Weiss, M. P., 1979a, Comments and suggestions invited for revision of American Stratigraphic Code: Geological Society of America, News and Information, v. 1, no. 7, p. 97–99.

———, 1979b, Stratigraphic Commission Note 50—Proposal to change name of Commission: American Association of Petroleum Geologists Bulletin, v. 63, no. 10, p. 1986.

Weller, J. M., 1960, Stratigraphic principles and practice: Harper and Brothers, New York, 725 p.

Willman, H. B., and J. C. Frye, 1970 Pleistocene stratigraphy of Illinois: Illinois State Geological Survey Bulletin 94, 204 p.

Appendix D

Nomenclature of Global and Northern American Chronostratigraphic Units

Nomenclature of chronostratigraphic units generally used throughout the world. Some North American stage names are also shown, together with a chronometric scale. Source: Salvador, A., 1985, Chronostratigraphic and geochronometric scales in COSUNA stratigraphic nomenclature charts of the United States: *Am. Assoc. Petroleum Geologists Bull.*, v. 69, Fig. 1–3, p. 182–184. Reprinted by permission of AAPG, Tulsa, Okla.

ERATHEM	SYSTEMS	SERIES / STAGES			NORTH AMERICAN CHRONOSTRATIGRAPHIC UNITS SERIES / STAGES	NUMERICAL TIME SCALE (Ma)
C E N O Z O I C	QUATERNARY	HOLOCENE			NORTH AMERICAN PLEISTOCENE GLACIAL STAGES ONLY WHEN APPLICABLE AND NECESSARY	0.01
		PLEISTOCENE				1.7 to 2.8
	T E R T I A R Y — N E O G E N E	PLIOCENE		PIACENZIAN		4.6 / 5.3 / 5
				ZANCLEAN		
		M I O C E N E	UPPER	MESSINIAN		6.7
				TORTONIAN		10.8 / 10
			MIDDLE	SERRAVALLIAN		15.4 / 15
				LANGHIAN		17
			LOWER	BURDIGALIAN		20
				AQUITANIAN		23
	T E R T I A R Y — P A L E O G E N E	OLIGOCENE	UPPER	CHATTIAN	PACIFIC AREA STAGES	25 / 25
			LOWER	RUPELIAN	OR	30 / 33
					MAMMALIAN STAGES ONLY	35
		EOCENE	UPPER	PRIABONIAN	WHEN APPLICABLE	38 / 40 / 41
			MIDDLE	BARTONIAN	AND NECESSARY	
				LUTETIAN		45 / 45
			LOWER	YPRESIAN		50 / 50
						55 / 55
		PALEOCENE	UPPER	THANETIAN		60 / 62
			LOWER	DANIAN		65
						67

GLOBAL CHRONOSTRATIGRAPHIC UNITS

NORTH AMERICAN CHRONOSTRATIGRAPHIC UNITS

GLOBAL CHRONOSTRATIGRAPHIC UNITS				NORTH AMERICAN CHRONOSTRATIGRAPHIC UNITS	NUMERICAL TIME SCALE (Ma)	
ERATHEM	SYSTEMS	SERIES / STAGES		SERIES / STAGES		
M E S O Z O I C	CRETACEOUS	UPPER	MAASTRICHTIAN	SAME AS GLOBAL	67	65
			CAMPANIAN		72	70
			SANTONIAN		80	80
			CONIACIAN		85	
			TURONIAN		90	90
			CENOMANIAN		92	
		LOWER	ALBIAN		100	100
			APTIAN		108	110
			BARREMIAN		115	120
			HAUTERIVIAN		125	
			VALANGINIAN		130	130
			BERRIASIAN		135	
	JURASSIC	UPPER	TITHONIAN	SAME AS GLOBAL	140	140
			KIMMERIDGIAN		145	150
			OXFORDIAN		155	
		MIDDLE	CALLOVIAN		160	160
			BATHONIAN		165	
			BAJOCIAN		170	170
			AALENIAN		175	
		LOWER	TOARCIAN		180	180
			PLIENSBACHIAN		185	
			SINEMURIAN		190	190
			HETTANGIAN		195	
	TRIASSIC	UPPER	RHAETIAN	SAME AS GLOBAL	200	200
			NORIAN		210	210
			CARNIAN		215	220
		MIDDLE	LADINIAN		220	
			ANISIAN		230	230
		LOWER	SCYTHIAN		240	240
					245	
					250	250

GLOBAL CHRONOSTRATIGRAPHIC UNITS				NORTH AMERICAN CHRONOSTRATIGRAPHIC UNITS			NUMERICAL TIME SCALE (Ma)	
ERATHEM	**SYSTEMS**	**SERIES / STAGES**		**SERIES / STAGES**				
PALEOZOIC	PERMIAN	UPPER	TATARIAN	OCHOAN			250 / 255	250
			KAZANIAN	GUADALUPIAN				260
			KUNGURIAN					
		LOWER	ARTINSKIAN	LEONARDIAN			270 / 275	270
			SAKMARIAN	WOLFCAMPIAN				280
			ASSELIAN				285	
							290	290
	CARBONIFEROUS	UPPER	STEPHANIAN — GZHELIAN	PENNSYLVANIAN SUB-SYSTEM	VIRGILIAN			300
			STEPHANIAN — KASIMOVIAN		MISSOURIAN			
			WESTPHALIAN — MOSCOVIAN		DESMOINESIAN		310	310
		MIDDLE			ATOKAN		315	
			WESTPHALIAN — BASHKIRIAN		MORROWAN			320
			"NAMURIAN"				330	330
		LOWER	SERPUKHOVIAN	MISSISSIPPIAN SUB-SYSTEM	CHESTERIAN			
							340	340
			VISEAN		MERAMECIAN			350
					OSAGEAN		355	
			TOURNAISIAN		KINDERHOOKIAN			360
							365	
	DEVONIAN	UPPER	FAMENNIAN	CHAUTAUQUAN	CONEWANGOAN			370
					CASSADAGAN			
			FRASNIAN	SENECAN	CHEMUNGIAN		380	380
		MIDDLE	GIVETIAN	ERIAN	FINGERLAKESIAN		385	
			EIFELIAN				390	390
		LOWER	EMSIAN	ULSTERIAN	ESOPUSIAN		395	
			SIEGENIAN		DEERPARKIAN		400	400
			GEDINNIAN		HELDERBERGIAN		405	
	SILURIAN	UPPER	PRIDOLIAN	CAYUGAN				410
			LUDLOVIAN		LOCKPORTIAN		415	
		LOWER	WENLOCKIAN	NIAGARAN	CLIFTONIAN			420
					CLINTONIAN		420	
			LLANDOVERIAN	ALEXANDRIAN			425	
	ORDOVICIAN	UPPER	ASHGILLIAN	CINCINNATIAN	RICHMONDIAN			430
					MAYSVILLIAN			440
			CARADOCIAN		EDENIAN			450
							455	
		MIDDLE	SHERMANIAN / KIRKFIELDIAN / ROCKLANDIAN	BLACKRIVERIAN			460	460
			LLANDEILIAN	CHAMPLAINIAN	CHAZYAN			470
			LLANVIRNIAN		WHITEROCKIAN		475	480
			ARENIGIAN				485	
		LOWER	TREMADOCIAN	CANADIAN			490	490
	CAMBRIAN	UPPER		TREMPEALEAUAN			500	500
				FRANCONIAN				510
				DRESBACHIAN			515	
		MIDDLE						520
								530
							540	540
		LOWER						550
								560
							570	570

Appendix E

Web Sites Pertaining to Sedimentology and Stratigraphy

Web-Resource Sites

Earth Science Basic Resources

http://geology.about.com/od/activitiesbasics/
(Provides links to numerous Earth science topics, including some pertinent to sedimentology/stratigraphy)

On-Line Earth Science Journals

http://www.colostate.edu/~cwis70/journals.html
(Provides links to journal articles online; site maintained by Colorado State University)

Web Resources for Sedimentary Geologists

http://darkwing.uoregon.edu/~rdorsey/SedResources.html
(Extensive links to sites of interest to sedimentary geologists; maintained by Becky Dorsey, University of Oregon; includes links to many research sites and sedimentologists on the Web)

Selected Web Sites for Geoscientists

http://www-sul.stanford.edu/depts/branner/information/sel.geo.html
(Site maintained by Geoscience Information Society)

Organizations or Societies Promoting Sedimentology and Stratigraphy

American Association Petroleum Geologists (AAPG)

http://www.aapg.org/

American Geological Institute

http://www.agiweb.org/

American Petroleum Institute

http://api-ec.api.org/newsplashpage/index.cfm

Canadian Society of Petroleum Geologists

http://www.cspg.org/

Clay Mineral Society

http://cms.lanl.gov/

Geological Association of Canada

http://www.esd.mun.ca/~gac/ABOUT/presenti.html

Geological Society of America

http://www.geosociety.org/pubs/index.htm

Index of U.S. Geological Survey Web Sites

http://www.usgs.gov/network/

International Association of Sedimentologists (IAS)

http://www.iasnet.org/publications/index.asp

International Union of Geological Sciences (IUGS)

http://www.iugs.org/ See also Iugs Links to Iugs Bodies
http://www.iugs.org/iugs/links.htm

North American Commission on Stratigraphic Nomenclature

http://www.agiweb.org/nacsn/

Paleontological Society

http://www.paleosoc.org/

Society for Sedimentary Geology (SEPM)

http://www.sepm.org/

Soil Science Society of America

http://www.soils.org/

Publishers of Sedimentological and Stratigraphic Journals and Special Publications

AAPG Bookstore (Publications of the American Association of Petroleum Geologists)
http://bookstore.aapg.org/
(Publishes the AAPG Bulletin plus memoirs and other special papers)

American Geological Institute
http://www.agiweb.org/
(Publishes Geotimes and other materials of general geologic interest)

Blackwell Science (Publications)
http://www.blackwellpublishing.com/
(Publishes the journals "Sedimentology" and "Basin Research," as well as books dealing with sedimentology and stratigraphy)

Canadian Petroleum Society
http://www.cspg.org/
(Publishes "Bulletin of Canadian Petroleum Geology" and other works focused particularly on Canadian geology)

Elsevier Science

http://www.elsevier.com/wps/find/homepage.cws_home
(Publishes the journals "Marine Geology" and "Sedimentary Geology" and many
other journals and books of interest to sedimentologists and stratigraphers)

Geological Association of Canada Publications

http://www.esd.mun.ca/~gac/PUBLICAT/pubdesc.html
(Publishes the journal "Geoscience Canada" plus special papers and other materials)

Geological Society of America

http://www.geosociety.org/pubs/index.htm
(Publishes the "GSA Bulletin," "Geology," and various special papers and books
dealing with a variety of geological topics)

International Association of Sedimentologists (IAS)

http://www.iasnet.org/publications/index.asp
(Has produced many special publications dealing with sedimentology/stratigraphy)

Society for Sedimentary Geology (SEPM)

http://www.sepm.org/
(Publishes the "Journal of Sedimentary Petrology" plus memoirs and special
papers)

Some Sedimentology/Stratigraphy Programs of Interest

Geological Association of Canada Nuna Research Conferences

http://www.esd.mun.ca/~gac/ANNMEET/nuna.html
(Schedules research conferences to promote advancement of knowledge in special
areas such as sequence stratigraphy)

Global (Climate) Change Home Page at NASA

http://gcmd.gsfc.nasa.gov/Resources/pointers/glob_warm.html
(A directory of Earth Science and global climate change data)

Museum of Paleontology, Berkeley

http://www.ibiblio.org/expo/paleo.exhibit/paleo.html
(On-line exhibits featuring a variety of topics of stratigraphic and sedimentologic
interest)

NASA's Global Change Site Map

http://gcmd.gsfc.nasa.gov/Aboutus/sitemap.html
(Provides a search directory for a variety of topics, including climates and paleoclimates)

NOAA Paleoclimatology Program

http://www.ngdc.noaa.gov/paleo/paleo.html
(Describes NOAA research programs and paleoclimate data)

Ocean Drilling Program (ODP), Texas A&M University

http://www-odp.tamu.edu/
(Describes the mission of the Ocean Drilling Program and provides an online tour)

Western Region Coastal and Marine Geology (U.S. Geological Survey)

http://walrus.wr.usgs.gov/
(Links to USGS coastal and marine geology programs)

Other Sites of Interest

Black Smokers

http://www.amnh.org/nationalcenter/expeditions/blacksmokers/
(Short discussion of black smokers including several photographs plus an animated feature site maintained by American Museum of Natural History)

Mineral Under the Microscope

http://www.gly.bris.ac.uk/www/teach/opmin/mins.html
(Discussion of the petrographic microscope and the characteristics of minerals under the microscope; University of Bristol, England)

NOAA's Coral Reef Information Site

http://www.coris.noaa.gov/

Sequence Stratigraphy Site

http://strata.geol.sc.edu/
(Site maintained by University of South Carolina)

Earth-Science Software

Earth Sciences Products, Services, Software

http://websearch.cs.com/cs/browse?id=169316&source=CSBrowse

Geotechnical Software Directory

http://www.ggsd.com/

Litpack—a Software Package for Littoral Processes and Coastline Kinetics

//www.dhi.dk/software/litpack/litpack.htm
http://www.dhisoftware.com/litpack/

Rockware (Earth Science Software)

http://www.rockware.com/. See also http://hallogram.com/science/rockware/

Telemac Software System for Modelling Hydrodynamics, Sediment Transport, and Water Quality in Natural Environments

http://www.hrwallingford.co.uk/software/telemac.html

Bibliography

Abegg, F. E., P. M. Harris, and D. B. Loope (eds.), 2001, Modern and ancient carbonate eolianites: Sedimentology, sequence stratigraphy, and diagnesis: SEPM Special Publication 71, 207 p.

Abrahão, D., and J. E. Warme, 1990, Lacustrine and associated deposits in a rifted continental margin—Lower Cretaceous Lagoa Feia Formation, Campos Basin, offshore Brazil, in Katz, B. J. (ed.), Lacustrine basin exploration, Am. Assoc. Petroleum Geologists, Mem. 50, p. 287–305.

Adams, A. E., and W. S. MacKenzie, 1998, A color atlas of carbonate sediments and rocks under the microscope: John Wiley & Sons, New York, 180 p.

Ager, D. V., 1993a, The new catastrophism: The importance of the rare event in geologic history: Cambridge University Press, Cambridge, 231 p.

Ager, D. V., 1993b, The nature of the stratigraphical record, 3rd ed.: John Wiley & Sons, Chichester, 151 p.

Agterberg, F. P., 1990, Automated stratigraphic correlation: Elsevier, Amsterdam, 424 p.

Ahlbrandt, T. S., and S. G. Fryberger, 1981, Sedimentary features and significance of interdune deposits, in Ethridge, F. G., and R. O. Flores (eds.), Recent and ancient nonmarine depositional environments: Models for exploration: Soc. Econ. Paleontologists and Mineralogists Spec. Pub. 31, p. 293–314.

Ahlbrandt, T. S., and S. G. Fryberger, 1982, Introduction to eolian deposits, in Scholle, P. A., and D. Spearing (eds.), Sandstone depositional environments: Am. Assoc. Petroleum Geologists Mem. 31, p. 11–47.

Aigner, T., 1985, Storm depositional systems: Springer-Verlag, Berlin, 174 p.

Aissaoui, D. M., D. F. McNeill, and N. F. Hurley (eds.), 1993, Applications of paleomagnetism to sedimentary geology: SEPM (Soc. for Sedimentary Geology) Spec. Publ. 49, 216 p.

Aitken, J. D., 1967, Classification and environmental significance of cryptalgal limestones and dolomites, with illustrations from the Cambrian and Ordovician of southwestern Alberta: Jour. Sed. Petrology, v. 37, p. 1163–1179.

Alam, M. M., 1995, Tide-dominated sedimentation in the upper Tertiary succession of the Sitapahar anticline, Bangladesh, in Flemming, B. W., and A. Bartholomä (eds.), Tidal signatures in modern and ancient sediments, Internat. Assoc. of Sedimentologists Spec. Pub. 24, Blackwell Science, Oxford, p. 329–341.

Alexander, C. R., R. A. Davis, and V. J. Henry (eds.), 1998, Tidalites: Processes and products: SEPM Special Publication 61, 171 p.

Allen, J. R. L., 1963, The classification of cross-stratified units with notes on their origin: Sedimentology, v. 2, p. 93–114.

Allen, J. R. L., 1968, Current ripples: Their relation to patterns of water and sediment motion: North Holland Puslishing, Amsterdam, 433 p.

Allen, J. R. L., 1978, Studies in fluviatile sedimentation: An exploratory quantitative model for architecture of avulsion-controlled alluvial suites: Sedimentary Geology, v. 21, p. 129–147.

Allen, J. R. L., 1984, Laminations developed from upper-stage plane beds: A model based on the larger coherent structures of the turbulent boundary layer: Sed. Geology, v. 39, p. 227–242.

Allen, J. R. L., 1985, Loose-boundary hydraulics and fluid mechanics: Selected advances since 1961, in Brenchley, P. J., and B. P. J. Williams (eds.), Sedimentology: Recent developments and applied aspects: Geol. Soc., Blackwell, Oxford, p. 7–30.

Allen, J. R. L., 1994, Fundamental properties of fluids and their relation to sediment transport processes, in Pye, K. (ed.), Sediment transport and depositional processes: Blackwell Scientific Publ., Oxford, p. 25–60.

Allen, P. A., and J. R. Allen, 1990, Basin analysis—Principles and applications: Blackwell Scientific Pub., Oxford, 451 p.

623

Allen, P. A., and P. Homewood (eds.), 1986, Foreland basins: Blackwell Scientific Publications, Oxford, 453 p.

Allison, R. J., and A. S. Goudie, 1994, The effects of fire on rock weathering: An experimental study, *in* Robinson, D. A., and R. B. G. Williams (eds.), Rock weathering and landform evolution, John Wiley & Sons, Chichester, p. 1–56.

Alsharhan, A. S., and R. W. Scott (eds.), 2000, Middle East models of Jurassic/ Cretaceous carbonate systems: SEPM Special Publication 69, 364 p.

Anastas, A. S., N. P. James, C. S. Nelson, and R. W. Dalrymple, 1998, Deposition and textural evolution of cool-water limestones: Outcrop analog for reservoir potential in cross-bedded calcitic reservoirs: Am. Assoc. Petroleum Geologists Bull., v. 82, p. 160–180.

Anbar, A. D., and H. D. Holland, 1992, The photochemistry of manganese and the origin of banded iron formations: Geochimica et Cosmochimica Acta, v. 56, p. 2595–2603.

Anderson, E. J., and P. W. Goodwin, 1990, The significance of metre-scale allocycles in the quest for a fundamental stratigraphic unit: Journal of the Geological Society, London, v. 147, p. 507–518.

Anderson, J. B., and G. M. Ashley (eds.), 1991, Glacial marine sedimentation; Paleoclimatic significance: Geol. Soc. America Spec. Paper 261, 232 p.

Anderson, R. S., M. Sorensen, and B. B. Willets, 1991, A review of recent progress in our understanding of aeolian sediment transport: Acta Mechanica Supplementum 1, Springer-Verlag, Wien, New York, p. 1–19.

Angevine, C. L., P. L. Heller, and C. Paola, 1990, Quantitative sedimentary basin modeling: Am. Assoc. Petroleum Geologists Continuing Education Course Note Series 32, 133 p.

ANWR Assessment Team, 1999, The oil and gas resource potential of the Arctic National Wildlife Refuge 1002 Area, Alaska: U.S. Geological Survey Open File Report 98–34, 1999 (2 CD set).

Appel, P. W. U., and G. L. LaBerge, 1987, Precambrian iron-formations: Theophrastus Pub., S. A., Athens, Greece, 674 p.

Arvidson, R. S., F. T. Mackenzie, and M. Guidry, 2000, Ocean/atmosphere history and carbonate precipitation rates: A solution to the "dolomite problem," *in* Glen, C. R., L. Prévôt-Lucas, and J. Lucas (eds.), Marine authigenesis: From global to microbial: SEPM Spec. Pub. 66, p. 1–5.

Ashley, G. A., 1990, Classification of large-scale subaqueous bedforms: A new look at an old problem: Jour. Sed. Petrology, v. 60, p. 160–172.

Ashley, G. M., 1995, Glaciolacustrine environments, *in* Menzies, J. (ed.), Modern glacial environments: Processes, dynamics and sediments: Butterworth-Heinemann, Oxford, p. 417–444.

Asquith, G., 1982, Basic well log analysis for geologists, Methods in Exploration 3: Am. Assoc. of Petroleum Geologists, Tulsa, Okla., 216 p.

Badiozamani, K., 1973, the Dorag dolomitization model— Application to the Middle Ordovician of Wisconsin: Jour. Sed. Petrology, v. 43, p. 965–984.

Bagnold, R. A., 1941, 1954, The physics of blown sand and desert dunes: Methuen, London, 265 p.

Bagnold, R. A., 1956, The flow of cohesionless grains in fluids: Royal Soc. London, Philos. Trans., Ser. A., v. 249, p. 235–297.

Bagnold, R. A., 1962, Auto-suspension of transported sediment: Turbidity currents: Royal Soc. London Proc. (A), v. 265, p. 315–319.

Bagnold, R. A., and O. Barndorff-Nielsen, 1980, The pattern of natural size distribution: Sedimentology, v. 27, p. 199–207.

Baker, P. A., and M. Kastner, 1981, Constraints on the formation of sedimentary dolomite: Science, v. 213, p. 214–216.

Bally, A. W., and S. Snelson, 1980, Realms of subsidence, *in* Miall, A. D. (ed.), Facts and principles of world petroleum occurrence: Canadian Soc. Petroleum Geologists Mem. 6, p. 9–44.

Banerjee, I., 1991, Tidal sand sheets of the late Albian Joli Fou-Kiowa-Skull Creek marine transgression, Western Interior Seaway of North America, *in* Smith, D. G., G. E. Reinson, G. A. Zaitlin, and R. A. Rahmani (eds.), Clastic tidal sedimentology: Canadian Soc. Petroleum Geologists, Mem. 16, p. 335–348.

Barndorff-Nielsen, O., 1977, Exponentially decreasing distributions for the logarithm of particle size: Proc. Royal Society London A., v. 353, p. 401–419.

Barndorff-Nielsen, O. E., and B. B. Willets (eds.), 1991, Aeolian grain transport 1- Mechanics: Springer-Verlag, Wien, New York, 181 p.

Barndorff-Nielsen, O., K. Dalsgaard, C. Halgreen, H. Kuhlman, J. T. Moller, and G. Schou, 1982, Variation in particle size distribution over a small dune: Sedimentology, v. 29, p. 53–65.

Barnes, C., A. Hallam, D. Kaljo, E. G. Kauffman, and O. H. Walliser, 1996, Global event stratigraphy, *in* Walliser, O. H. (ed.), Global events and event stratigraphy: Springer-Verlag, Berlin, p. 319–333.

Barrett, P. J., 1980, The shape of rock particles, a critical review: Sedimentology, v. 27, p. 291–303.

Barron, J. A., 1987, Diatomite: Environmental and geologic factors affecting its distribution, *in* Hein, J. R. (ed.), Siliceous sedimentary rock-hosted ores and petroleum: Van Nostrand Reinhold, New York, p. 164–178.

Basu, A., S. W. Young, L. J. Suttner, W. C. James, and G. H. Mack, 1975, Reevaluation of the use of undulatory extinction and crystallinity in detrital quartz for provenance interpretation: Jour. Sed. Petrology, v. 45, p. 873–882.

Bates, C. C., 1953, Rational theory of delta formation: Am. Assoc. Petroleum Geologists Bull, v. 37, p. 2119–2161.

Bates, R. L., and J. A. Jackson (comps.), 1980, Glossary of geology, 2nd ed.: American Geol. Institute, Falls Church, Va., 749 p.

Bathurst, R. G. C., 1975, Carbonate sediments and their diagenesis, 2nd ed.: Developments in Sedimentology 12, Elsevier, Amsterdam, 658 p.

Baturin, G. N., 1982, Phosphorites on the sea floor: Origin, composition, and distribution: Developments in Sedimentology 33, Elsevier, Amsterdam, 343 p.

Belderson, R. H., M. A. Johnson, and N. H. Kenyon, 1982, Bedforms, *in* Stride, A. H. (ed.), Offshore tidal sands: Chapman and Hall, London, p. 27–57.

Belkaaloul, K. N., D. M. Aissaoui, M. Rebelle, and G. Sambet, 1997, Resolving sedimentological uncertainties using magnetostratigraphic correlation: An example from the Middle Jurassic of Burgundy, France: Jour. for Sedimentary Research, v. 67, p. 676–685.

Benton, M. J., 1995, Diversification and extinction in the history of life: Science, v. 268, p. 52–58.

Bentor, Y. K., 1980, Phosphorites—The unsolved problem, *in* Bentor, Y. K. (ed.), Marine phosphorites—Geochemistry, occurrence, genesis: Soc. Econ. Paleontologists and Mineralogists Spec. Pub. 29, p. 3–18.

Berg, O. R., and D. B. Wolverton (eds.), 1985, Seismic stratigraphy II, an integrated approach: Am. Assoc. Petroleum Geologists Mem. 39, 276 p.

Berger, W. H., 1974, Deep-sea sedimentation, *in* Burk, C. A., and C. L. Drake (eds.), The geology of continental margins: Springer-Verlag, New York, p. 213–241.

Berggren, W. A., and J. A. Van Couvering, 1978, Biochronology, *in* Cohee, G. V., M. F. Glaessner, and H. D. Hedberg (eds.), Contributions to the geologic time scale: Am. Assoc. Petroleum Geologists, Tulsa, Okla., p. 39–55.

Berggren, W. A., D. V. Kent, C. C. Swisher, III, and M-P. Aubry, 1995, A revised Cenozoic geochronology and chronostratigraphy, *in* Berggren, W. A., D. V. Kent, M-P Aubry, and J. Hardenbol (eds.), Geochronology, time scales and global stratigraphic correlation: SEPM (Soc. for Sedimentary Geoology) Spec. Pub. 54, p. 129–212.

Bergman, K. M., and J. W. Snedden (eds.), 1999, Isolated shallow marine sand bodies: Sequence stratigraphic analysis and sedimentologic interpretation: SEPM Special Publication 64, Tulsa, Okla., p. 13–28.

Bernasconi, S. M., 1994, Geochemical and microbial controls on dolomite formation in anoxic environments: A case study from the Middle Triassic (Ticino, Switzerland): Contributions to Sedimentology 19, Schweizerbart'sche Verlagsbuchhandlung, Stuttgart, 109 p.

Berner, R. A., 1975, The role of magnesium in crystal growth of aragonite from sea water: Geochim. et Cosmochim. Acta, v. 39, p. 489–505.

Berner, R. A., J. T. Westrich, R. Graber, J. Smith, and C. S. Martens, 1978, Inhibition of aragonite precipitation from supersaturated seawater: A laboratory and field study: Am. Jour. Sci., v. 278, p. 816–837.

Bernet, M., and C. Spiegel, 2004a, Introduction: Detrital thermochronology, *in* Bernet, M., and C. Spiegel (eds.), Detrital thermochronology: Provenance analysis, exhumation, and landscape evolution of mountain belts: Geol. Soc. America Special Paper 378, p. 1–6.

Bernet, M., and C. Spiegel (eds.), 2004b, Detrital thermochronology: Provenance analysis, exhumation, and landscape evolution of mountain belts: Geol. Soc. America Special Paper 378, 126 p.

Berry, W. B. N., 1987, Growth of a prehistoric time scale: Based on organic evolution, 2nd ed., Blackwell Scientific Pub., Palo Alto, Calif., 202 p.

Best, M. B., 2001, Some notes on the terms "deep-sea ahermatypic" and "asooxanthellate," illustrated by the coral genus *Madracis, in* Martin Willison, J. H., J. Hall, S. E. Gass, E. L. R. Kenchington, M. Butler, and P. Doherty (eds.), 2001, Proceedings of the First International Symposium on Deep-Sea Corals: Ecology Action Centre and Nova Scotia Museum, Halifax, Nova Scotia, p. 19–29.

Bhattacharya, J. P., and R. G. Walker, 1991, River- and wave-dominated depositional systems of the Upper Cretaceous Dunvegan Formation, northwestern Alberta: Canadian Petroleum Geology Bull., v. 39, p. 165–191.

Bhattacharya, J. P., and L. Giosan, 2003, Wave-influenced deltas: Geomorphological implications for facies reconstruction: Sedimentology, v. 50, p. 187–210.

Biddle, K. T. (ed.), 1991, Active margin basins: AAPG Mem. 52, Am. Assoc. Petroleum Geologists, Tulsa, Okla., 324 p.

Birkland, P. W., 1999, Soils and geomorphology, 3rd ed.: Oxford University Press, New York, 430 p.

Blair, T. C., and J. G. McPherson, 1999, Grain-size and textural classification of coarse sedimentary particles: Jour. for Sed. Research, v. 69, p. 6–19.

Blair, T. C., and J. G. McPherson, 1994a, Alluvial fans and their natural distinction from rivers based on morphology, hydraulic processes, sedimentary processes, and facies assemblages: Jour. for Sed. Research, v. A64, p. 450–489.

Blair, T. C., and J. G. McPherson, 1994b, Alluvial fan processes and forms, *in* Abrahams, A. D., and A. J. Parsons (eds.), Geomorphology of desert environments: Chapman and Hall, London, p. 354–402.

Blakey, R. C., F. Peterson, and G. Kocurek, 1988, Synthesis of late Paleozoic and Mesozoic eolian deposits of the Western Interior of the United States: Sed. Geol., v. 56, p. 3–125.

Bland, W., and D. Rolls, 1998, Weathering: An introduction to the scientific principles: Oxford University Press Inc., New York, 271 p.

Blatt, H., 1982, Sedimentary petrology: W. H. Freeman, San Francisco, 564 p.

Blatt, H., G. V. Middleton, and R. Murray, 1980, Origin of sedimentary rocks, 2nd ed.: Prentice-Hall, Englewood Cliffs, N.J., 782 p.

Boggs, S., Jr., 1967a, Measurement of roundness and sphericity parameters using an electronic particle size analyzer: Jour. Sed. Petrology, v. 37, p. 908–913.

Boggs, S., Jr., 1967b, A numerical method for sandstone classification: Jour. Sed. Petrology, v. 37, p. 548–555.

Boggs, S., Jr., 1968, Experimental study of rock fragments: Jour. Sed. Petrology, v. 38, p. 1326–1339.

Boggs, S., Jr., 1969, Relationship of size and composition in pebble counts: Jour. Sed. Petrology, v. 39, p. 1243–1247.

Boggs, S., Jr., 1974, Sand-wave fields in Taiwan Strait: Geology, v. 2, p. 251–253.

Boggs, S., Jr., 1975, Seabed resources of the Taiwan continental shelf: Acta Oceanographica Taiwanica, v. 5, p. 1–18.

Boggs, S., Jr., 1984, Quaternary sedimentation in the Japan arc-trench system: Geol. Soc. America Bull., v. 95, p. 669–685.

Boggs, S., Jr., 1992, Petrology of sedimentary rocks: Macmillan Pub. Co., New York, 707 p.

Boggs, S., Jr., D. G. Livermore, and M. G. Seitz, 1985, Humic macromolecules in natural waters: Jour. Macromolecular Science, v. C25(4), p. 599–657.

Boggs, S., Jr., W. C. Wang, and F. S. Lewis, 1979, Sediment properties and water characteristics of the Taiwan shelf and slope: Acta Oceanographica Taiwanica, v. 10, p. 10–49.

Boguchwal, L. A., and J. B. Southard, 1990, Bed configurations in steady unidirectional water flows. Part l. Scale model study using fine sands: Jour. Sed. Petrology, v. 60, p. 649–657.

Bohacs, K. M., A. R. Carroll, and J. E. Neal, 2003, Lessons from large lake systems— Thresholds, nonlinearity, and strange attractors, in Chan, M. A., and A. W. Archer (eds.), Extreme depositional environments: Mega end members in geologic time: Geol. Soc. Amer. Special paper 370, p. 75–90.

Bohacs, K. M., A. R. Caroll, J. E. Neal, and P. J. Mankiewicz, 2000, Lake-basin type, source potential, and hydrocarbon character: An integrated sequence-stratigraphic-geochemical framework, in Gierlowski-Kordesch, E. H., and K. R. Kelts (eds.), Lake basin through space and time: AAPG Studies in Geology 46, p. 3–34.

Bohrman, G., A. Abelman, R. Gersonde, H. Hubberton, and G. Kuhn, 1994, Pure siliceous ooze, a diagenetic environment for early chert formation: Geology, v. 22, p. 207–210.

Bond, G. C., M. A. Kominz, and R. E. Sheridan, 1995, Continental terraces and rises, in Busby, C. J. and R. V. Ingersoll (eds.), Tectonics of sedimentary basins: Blackwell Science, p. 149–178.

Bosellini, A., 1991, Geology of the Dolomites—An introduction: Dolomieu conference on carbonate platforms and dolomitization, Tourist Office, Ortisei, Italy, September 1991, 43 p.

Bouisset, P. M., and A. M. Augustin, 1993, Borehole magnetostratigraphy, absolute age dating, and correlation of sedimentary rocks, with examples from the Paris Basin, France: Am. Assoc. Petroleum Geologists, Bull., v. 77, p. 569–587.

Bouma, A., 1962, Sedimentology of some flysch deposits: Elsevier, Amsterdam, 168 p.

Bouma, A. H., 2000, Fine-grained, mud-rich turbidite systems: Model and comparison with coarse-grained, sand-rich systems, in Bouma, A. H., and C. G. Stone (eds.), 2000, Fine-grained turbidite systems: AAPG Memoir 72 and SEPM Special Publication No. 68, p. 9–19.

Bouma, A. H., and C. G. Stone (eds.), 2000, Fine-grained turbidite systems: AAPG Memoir 72 and SEPM Special Publication No. 68, 342 p.

Bouma, A. H., W. R. Normark, and N. E. Barnes (eds.), 1985, Submarine fans and related turbidite systems: Springer-Verlag, New York, 351 p.

Bouma, A. H., H. L. Berryhill, R. L. Brenner, and H. J. Knebel, 1982, Continental shelf and epicontinental seaways, in Scholle, P. A., and D. Spearing (eds.), Sandstone depositional environments: Am. Assoc. Petroleum Geologists Mem. 31, p. 281–327.

Bowen, R., 1988, Isotopes in the earth sciences: Elsevier Applied Sciene, London, 647 p.

Bowen, R., 1998, Radioactive dating methods, in Vértes, A., S. Nagy, and K. Süvegh (eds.), Nuclear methods in mineralogy and geology: Techniques and applications: Plenum Press, New York, p. 377–423.

Bralower, T. J., P. D. Fullagar, C. K. Paull, G. S. Dwyer, and R. M. Leckie, 1997, Mid-Cretaceous strontium-isotope stratigraphy of deep-sea sections: Geol. Soc. America Bull., v. 109, p. 1421–1442.

Braterman, P. S., A. G. Cairns-Smith, and R. W. Sloper, 1983, Photo-oxidation of hydrated Fe^{2+}—Significance for banded iron formations: Nature, v. 303, p. 163–164.

Bridge, J. S., 2003, Rivers and floodplains: Blackwell Science Ltd, Oxford, 491 p.

Bridge, J., and J. Best, 1997, Preservation of planar laminae due to migration of low-relief bed waves over aggrading, upper-stage plane beds: Comparison of experimental data with theory: Sedimentology, v. 44, p. 253–262.

Briggs, D. E. G., E. N. K. Clarkson, and R. J. Aldridge, 1983, The conodont animal: Lethaea, v. 16, p. 1–14.

Brodzikowski, K., and A. J. van Loon, 1991, Glacigenic sediments: Elsevier, Amsterdam, 674 p.

Bromley, R. G., 1996, Trace fossils: Biology, taphronomy and applications: Chapman and Hall, London, 361 p.

Bromley, R. G., S. G. Pemberton, and R. A. Rahmani, 1984, A Cretaceous woodground: The Teredolites Ichnofacies: Jour. of Paleontology, v. 58, p. 488–498.

Brookfield, M. E., 1984, Eolian sands, in Walker, R. G. (ed.), Facies models, 2nd ed.: Geoscience Canada Reprint Ser. 1, p. 91–104.

Brown, L. F., Jr., and W. L. Fisher, 1980, Seismic stratigraphic interpretation and petroleum exploration: Geophysical principles and techniques: Am. Assoc. Petroleum Geologists Continuing Education Course Notes Ser. 16, 56 p.

Buczynski, C., and H. S. Chafetz, 1993, Habit of bacterially induced precipitates of calcium carbonate: Examples from laboratory experiments and recent sediments, in Rezak, R., and D. L. Lavoie (eds.), Carbonate microfabrics: Springer-Verlag, New York, p. 105–116.

Budd, D. A., and P. M. Harris (eds.), 1990, Carbonate-siliciclastic mixtures: SEPM (Soc. for Sedimentary Geology) Reprint Series No. 14, 272 p.

Bull, P. A., 1986, Procedures in environmental reconstruction by SEM analysis, in Sieveking, G. De C., and M. B. Hart (eds.), The scientific study of flint and chert: Cambridge Univ. Press, Cambridge, p. 221–226.

Buol, S. W., F. D. Hole, R. J. McCracken, and R. J. Southard, 1997, Soil genesis and classification, 4th ed.: Iowa State University Press, Ames, 527 p.

Burchette, T. P., V. P. Wright, and T. J. Faulkner, 1990, Oolitic sandbody depositional models and geometries, Mississippian of southwest Britain: Implications for petroleum exploration in carbonate ramp settings: Sed. Geology, v. 68, p. 87–115.

Burger, H., and W. Skala, 1976, Comparison of sieve and thin-section techniques by a Monte Carlo model: Computer Geoscience, v. 2, p. 123–139.

Burnett, W. C., and P. N. Froelich (eds.), 1988, The origin of marine phosphorites: The results of the R. V. Robert D. Conrad Cruise 23-06 to the Peru shelf: Special issue of Marine Geology, v. 80, p. 181–346.

Burnett, W. C., and S. R. Riggs (eds.), 1990, Phosphate deposits of the world: Vol. 3 Neogene to modern phosphorites: Cambridge University Press, Cambridge, 464 p.

Burnett, W. C., C. R. Glenn, C. C. Yeh, M. Schultz, J. Chanton, and M. Kashgarian, 2000, U-series, ^{14}C, and stable isotope studies of recent phosphatic "protocrusts" from the Peru margin, *in* Glenn, C. R., L. Prévôt-Lucas, and J. Lucas (eds.), Marine authigenesis: From global to microbial: SEPM Spec. Pub. 66, p. 163–183.

Burst, J. F., 1965, Subaqueously formed shrinkage cracks in clay: Jour. Sed. Petrology, v. 35, p. 348–353.

Busby, C. J., and R. V. Ingersoll (eds.), 1995, Tectonics and sedimentary basins: Blackwell Science, 579 p.

Busson, G., and Schreiber, B. C. (eds.), 1997, Sedimentary deposition in rift and foreland basins in France and Spain: Columbia University Press, New York, 479 p.

Bustin, R. M., A. R. Cameron, D. A. Grieve, and W. D. Kalkreuth, 1985, Coal petrology, its principles, methods, and applications: Geol. Assoc. Canada Short Course Notes, v. 3, 230 p.

Butler, R. F., 1992, Paleomagnetism: Magnetic domains to geologic terranes: Blackwell Scientific Pub., Boston, 319 p.

Button, A., T. D. Brock, P. J. Cook, H. P. Euster, A. M. Goodwin, H. L. James, L. Margulis, K. H. Nealson, J. O. Nriagu, A. F. Trendall, and M. R. Walter, 1982, Sedimentary iron deposits, evaporites, and phosphorites, *in* Holland, H. D., and M. Schidlowski (eds.), Mineral deposits and evolution of the biosphere: Springer-Verlag, New York, p. 259–273.

Calvert, S. E., 1983, Sedimentary geochemistry of silicon, *in* Aston, R. R. (ed.), Silicon geochemistry and biogeochemistry: Acedemic Press, London, p. 143–186.

Cameron, E. M., and R. M. Garrels, 1980, Geochemical composition of some Precambrian shales from the Canadian Shield: Chem. Geology, v. 28, p. 181–197.

Camoin, G., and A. Arnaud-Vanneau, convenors, 1997, International workshop on microbial mediation in carbonate diagenesis 97, Abstract Book, International Association of Sedimentologists, 79 p.

Camoin, G. F. (ed.), 1999, Microbial mediation of carbonate diagenesis: Sed. Geology, v. 126 (special issue).

Campbell, C. V., 1967, Lamina, laminaset, bed and bedset: Sedimentology, v. 8, p. 7–26.

Canfield, D. E., 2001, Biogeochemistry of sulfur isotopes, *in* Valley, J. W., and D. R. Cole (eds.), Stable isotope geochemistry: Reviews in Mineralogy v. 43, Mineralogical Society of America, p. 607–636.

Cann, J. R., and M. R. Strens, 1989, Modeling periodic megaplume emission by black smoker systems: Jour. Geophy. Research, v. 94, p. 12,227–12,237.

Cant, D. J., 1982, Fluvial facies models and their application, *in* Scholle, P. A., and D. Spearing (eds.), Sandstone depositional environments: Amer. Assoc. Petroleum Geologists Mem. 31, p. 115–138.

Carballo, J. D., L. S. Land, and D. E. Miser, 1987, Holocene dolomitization of supratidal sediments by active tidal pumping, Sugarloaf Key, Florida: Jour. Sed. Petrology, v. 57, p. 153–165.

Carling, P. A., and M. R. Dawson (eds.), 1996, Advances in fluvial dynamics and stratigraphy: John Wiley and Sons, Chichester, 530 p.

Carney, J. L., and R. W. Pierce, 1995, Graphic correlation and composite standard databases as tools for the exploration biostratigrapher, *in* Mann, K. O., and H. R. Lane (eds.), Graphic correlation: SEPM (Soc. for Sedimetary Geology) Special Pub. 53, p. 23–43.

Carozzi, A. V., 1993, Sedimentary petrography: Prentice Hall, New Jersey, 263 p.

Castainer, S., G. Le Métayer-Levrel, and J-P. Perthuisot, 1997, Limestone genesis considered from the microbiologists point of view, *in* Camoin, G., and A. Arnaud-Vanneau, convenors, International workshop on microbial mediation in carbonate diagenesis 97, Abstract Book, International Association of Sedimentologists, p. 13–14.

Castainer, S., G. Le Métayer-Levrel, and J-P. Perthuisot, 1999, Ca-carbonates precipitation and limestone genesis—The microbiogeologists' point of view: Sedimentary Geology, v. 126, p. 9–23.

Cathcart, J. B., 1989, The phosphate deposits of Florida with a note on the deposits in Georgia and South Carolina, USA, *in* Notholt, A. J. G., R. P. Sheldon, and D. F. Davidson (eds.), Phosphate deposits of the world: Vol. 2 Phosphate rock resources: Cambridge University Press, Cambridge, p. 62–70.

Catt, J. A., 1986, Soils and Quaternary geology: A handbook for field scientists: Clarendon Press, Oxford, 267 p.

Chafetz, H. S., 1986, Marine peloids: A product of bacterially induced precipitation of calcite: Jour. Sed. Petrology, v. 56, p. 812–817.

Chafetz, H. S., 1994, Bacterially induced precipitation of calcium carbonate and lithification of microbial mats, *in* Krumbein, W. E., D. M. Paterson, and L. J. Stal (eds.), Biostabilization of sediments: Bibliotheks und Informations system der Carl von Ossietzky Universitat Oldenburg, Oldenburg, p. 148–163.

Chambers, R. L., and S. B. Upchurch, 1979, Multivariate analysis of sedimentary environments using grain size frequency distributions: Jour. Math. Geology, v. 11, p. 27–43.

Cheel, R. J., and D. A. Leckie, 1992, Coarse-grained storm beds of the Upper Cretaceous Chungo Member (Wapiabi Formation), southern Alberta, Canada: Jour. Sed. Petrology, v. 62, p. 933–945.

Cheel, R. J., and D. A. Leckie, 1993, Hummocky cross-stratification, *in* Wright, P. V. (ed.), Sedimentology Review 1, Blackwell Scientific Publications, Oxford, p. 103–122.

Chilingarian, G. V., and T. F. Yen, 1978, Bitumens, asphalts and tar sands: Elsevier, New York, 331 p.

Choquette, P. W., and L. C. Pray, 1970, Geologic nomenclature and classification of porosity in sedimentary carbonates: Am. Assoc. Petroleum Geologists Bull., v. 54, p. 207–250.

Chough, S. K., and G. J. Orton (eds.), 1995, Fan deltas: Depositional styles and controls: Sed. Geology, v. 98, 292 p.

Christie-Blick, N., G. S. Mountain, and K. G. Miller, 1990, Seismic stratigraphic record of sea-level change, *in* Revelle, R. R. et al. (eds.), Sea level change, National Research Council Studies in Geophysics, National Academy Press, Washington, D. D., p. 116–140.

Church, K. D., and A. L. Coe, 2003, Processes controlling relative sea-level change and sediment supply, *in* Coe, A. L. (ed.), The sedimentary record of sea-level change: The Open University, Cambridge University Press, Cambridge, p. 99–117.

Clarke, F. W., 1924, The data of geochemistry, 5th ed.: U.S. Geol. Survey Bull., v. 770, 841 p.

Clauer, N., 1982, The rubidium-strontium method applied to sediments: Certitudes and uncertainties, *in* Odin, G. S. (ed.), Numerical dating in stratigraphy: John Wiley & Sons, New York, p. 245–276.

Claypool, G. E., W. T. Holser, I. R. Kaplan, H. Sakai, and I. Zak, 1980, The age curves of sulfur and oxygen isotopes in marine sulfate and their mutual interpretation: Chem. Geology, v. 28, p. 199–260.

Clement, B. M., D. V. Kent, and N. D. Opdyke, 1982, Brunhes-Matuyama polarity transition in three deep-sea cores: Philosophical Transactions of Royal Society London, V. A306, p. 113–119.

Clifton, H. E., 1988, Sedimentologic relevance of convulsive geologic events, *in* H. E. Clifton (ed.), Sedimentological consequences of convulsive geologic events: Geol. Soc. America Spec. Paper 229, p. 1–5.

Clifton, H. E., R. E. Hunter, and J. V. Gardner, 1988, Analysis of eustatic, tectonic, and sedimentologic influences on transgressive and regressive cycles in the Upper Cenozoic Merced Formation, San Francisco, California, *in* Kleinspehn, K. L., and C. Paola (eds.), New perspectives in basin analysis: Springer-Verlag, New York, p. 109–128.

Cloud, P. E., 1973, Paleoecological significance of banded iron formations: Econ. Geology, v. 68, p. 1135–1143.

Cobb, J. C., and C. B. Cecil (eds.), 1993, Modern and ancient coal-forming environments: GSA Special Paper 286, 198 p.

Coe, A. L. (ed.), 2003, The sedimentary record of sea-level change: Cambridge University Press, Cambridge, 287 p.

Coe, A. L., and K. D. Church, 2003, Sequence stratigraphy, *in* Coe, A. L. (ed.), The sedimentary record of sea-level change: Cambridge University Press, Cambridge, p. 57–98.

Coleman, J. M., 1981, Deltas: Processes of deposition and models for exploration, 2nd ed.: Burgess, 124 p.

Coleman, J. M., and D. B. Prior, 1982, Deltaic environments of deposition, *in* Scholle, P. A., and D. Spearing (eds.), Sandstone depositional environments: Am. Assoc. Petroleum Geologists Mem. 31, p. 139–178.

Collinson, J. D., 1996, Alluvial sediments, *in* Reading, H. G. (ed.), Sedimentary environments: Processes, facies and stratigraphy: Blackwell Science Ltd., Oxford, p. 37–82.

Collinson, J. D., and D. B. Thompson, 1989, Sedimentary Structures, 2nd ed.: Harper Collins Academic, New York, 208 p.

Conde, K. C., 1982, Plate tectonics and crustal evolution, 2nd ed.: Pergamon, New York, 310 p.

Conglio, M., and G. R. Dix, 1992, Carbonate slopes, *in* Walker, R. G., and N. P James (eds.), Facies models: Response to sea level change: Geol. Assoc. Canada, p. 349–373.

Cook, P. J., and J. H. Shergold, 1986a, Proterozoic and Cambrian phosphorites—an introduction, *in* Cook, P. J., and J. H. Shergold (eds.), Phosphate deposits of the world: Vol. 1 Proterozoic and Cambrian phosphorites: Cambridge University Press, Cambridge, p. 1–8.

Cook, P. J., and J. H. Shergold, 1986b, Proterozoic and Cambrian phosphorites—Nature and origin, *in* Cook, P. J., and J. H. Shergold (eds.), Phosphate deposits of the world: Vol. 1 Proterozoic and Cambrian phosphorites: Cambridge University Press, Cambridge, p. 369–386.

Coplen, T. B., C. Kendall, and J. Hopple, 1983, Comparison of stable isotope reference samples: Nature, v. 302, p. 236–238.

Corliss, J. B., J. Dymond, L. I. Gordon, J. M. Edmond, R. P. von Herzen, R. D. Ballard, K. Green, D. Williams, A. Bainbridge, K. Crane, and T. H. van Andel, 1979, Submarine thermal springs on the Galápagos Rift: Science, v. 203, p. 1073–1083.

Cornelius, C. D., 1987, Classification of natural bitumens—A physical and chemical approach, *in* Meyers, R. F. (ed.), Exploration for heavy crude oil and natural bitumen: Amer. Assoc. Petroleum Geol. Studies in Geology 25, p. 165–174.

Cowen, R., 1988, The role of algal symbiosis in reefs through time: Palaios, v. 3, p. 221–227.

Cowie, J. W., and M. G. Bassett, 1989, International Union of Geological Sciences 1989 global stratigraphic chart: Supplement to Episodes, v. 12 (2).

Cox, A., 1969, Geomagnetic reversals: Science, v. 163, p. 237–245.

Cox, A., 1973, Plate tectonics and geomagnetic reversals: Introduction and reading list, *in* Cox, A. (ed.), Plate tectonics and geomagnetic reversals: W. H. Freeman, San Francisco, p. 138–153.

Cox, A., R. R. Doell, and G. B. Dalrymple, 1963, Geomagnetic polarity epochs and Pleistocene geochronometry: Nature, v. 198, p. 1049–1051.

Cressman, E. R., 1962, Nondetrital siliceous sediments: U.S. Geol. Survey Prof. Paper 440-T, 22 p.

Cross, T. A., 1990, Quantitative dynamic stratigraphy: Prentice-Hall, Englewood Cliffs, New Jersey, 615 p.

Cross, T. A., and P. W. Homewood, 1997, Amanz Gressly's role in founding modern stratigraphy: Geol. Soc. America Bull., v. 109, p. 1617–1630.

Cross, T. A., and M. A. Lessenger, 1988, Seismic stratigraphy: Ann. Rev. Earth and Planetary Science, v. 16, p. 319–354.

Cross, T. A., R. L. Dodge, J. C. Howard, and E. S. Siraki, 1995, Basin analysis: International Human Resources Development Corp., Boston, 210 p.

Cubitt, J. M., and R. A. Reyment (eds.), 1982, Quantitative stratigraphic correlation: John Wiley & Sons, New York, 301 p.

Curray, R. R., 1966, Observations of alpine mudflows in the Tenmile Range, central Colorado: Geol. Soc. America Bull., v. 77, p. 771–776.

Dalrymple, R. W., 1992, Tidal depositional systems, *in* Walker, R. G., and N. P. James (eds.), Facies models: Geol. Assoc. Canada, p. 195–238.

Dalrymple, R. W., Y. Makino, and B. A. Zaitlin, 1991, Temporal and spatial patterns of rythmite deposition on mudflats in the macrotidal, Cobequid Bay-Salmon River estuary, Bay of Fundy, *in* Smith, D. G., G. E. Reinson, B. A. Zaitlin, and R. A. Rahmani (eds.), Clastic tidal sedimentology, Canadian Soc. Petrology Geology Mem. 16, Calgary, p. 137–160.

Dalrymple, R. W., B. A. Zaitlin, and R. Boyd, 1992, Estuarine facies models: Conceptual basin and stratigraphic implications: Jour. Sed. Petrology, v. 62, p. 1130–1146.

Dalrymple, R. W., E. K. Baker, P. T. Harris, and M. G. Hughes, 2003, Sedimentology and stratigraphy of a tide-dominated foreland-basin delta (Fly River, Papua New Guinea), *in* Sidi, F. H., D. Nummedal, P. Imbert, H. Darman, and H. M. Posamentier (eds.), Tropic deltas of southeast Asia—Sedimentology, stratigraphy, and petroleum geology, SEPM Special Publ. 76, p. 147–173.

Daniels, E. J., S. P. Altaner, S. Marshak, and J. R. Eggleson, 1990, Hydrothermal alteration in anthracite in eastern Pennsylvania: Implications for the mechanisms of anthracite formation: Geology, v. 18, p. 247–250.

Darby, D. A., and Y. W. Tsang, 1987, Variation in ilmenite element composition within and among drainage basins: Implications for provenance: Jour. Sed. Petrology, v. 57, p. 831–838.

Davis, R. A., Jr., 1968, Algal stromatolites composed of quartz sandstone: Jour. Sed. Petrology, v. 38, p. 953–955.

Davis, R. A., Jr., 1983, Depositional systems: A genetic approach to sedimentary geology: Prentice-Hall, Englewood Cliffs, N.J., 669 p.

Davis, R. A., Jr., 1985, Coastal sedimentary environments, 2nd ed.: Springer-Verlag, New York, 716 p.

Davis, R. A., Jr., 1992, Depositional systems, 2nd ed.: Prentice-Hall, Englewood Cliffs, N.J., 604 p.

Davis, R. A., Jr. (ed.), 1994, Geology of Holocene barrier island systems: Springer-Verlag, Berlin, 464 p.

Davis, R. A., Jr., 1997, The evolving coast: Scientific American Library, A division of HPHLP, New York, 233 p.

de Boer, P. L., 1991, Pelagic black shale-carbonate rhythms: Orbital forcing and oceanographic response, *in* Einsele, G., W. Ricken, and A. Seilacher (eds.), Cycles and events in stratigraphy: Springer-Verlag, Berlin, p. 63–78.

de Boer, P. L., and D. G. Smith (eds.), 1994, Orbital forcing and cyclic sequences: Internat. Assoc. Sedimentologists Spec. Pub. 19, Blackwell Scientific Pub., Oxford, 559 p.

Dean, W. E., and R. Y. Anderson, 1978, Salinity cycles: Evidence for subaqueous deposition of Castile Formation and lower part of Salado Formation, Delaware Basin, Texas and New Mexico, *in* Austin, G. S., compiler, Geology and mineral deposits of Ochoan rocks in Delaware Basin and adjacent areas: New Mexico Bureau Mines Min. Res. Circ, v. 159, p. 15–20.

Dean, W. E., M. Leinen, and D. A. V. Stow, 1985, Classification of deep-sea, fine-grained sediments: Jour. Sed. Petrology, v. 55, p. 250–256.

Degens, E. T., 1965, Chemistry of sediments: Prentice-Hall, Englewood Cliffs, N.J., 342 p.

Demicco, R. V., and L. A. Hardie, 1994, Sedimentary structures and early diagenetic features of shallow marine carbonate deposits: SEPM Atlas Series No. 1, Society for Sedimentary Geology, Tulsa, Okla., 265 p.

Depaolo, D. J., and K. L. Finger, 1988, Applications of strontium isotopes in correlating Monterey Formation, California: Am. Assoc. Petroleum Geologists Bull., v. 72, p. 379.

Derbyshire, E., and L. A. Owen, 1996, Glacioaeolian processes, sediments and landforms, *in* Menzies, J. (ed.), Past glacial environments: Sediments, forms and techniques: Butterworth-Heinemann, Oxford, p. 213–237.

Desbrandes, R., 1985, Encyclopedia of well logging: Editions Technip, Paris, 584 p.

Dickin, A. P., 1995, Radiogenic isotope geology: Cambridge University Press, Cambridge, 452 p.

Dickinson, W. R., 1974, Plate tectonics and sedimentation, *in* Dickinson, W. R. (ed.), Tectonics and sedimentation: Soc. Econ. Paleontologists and Mineralogists Spec. Pub. 22, p. 1–27.

Dickinson, W. R., 1982, Composition of sandstones in circum-Pacific subduction complexes and fore-arc basins: Am. Assoc. Petroleum Geologists Bull., v. 66, p. 121–137.

Dickinson, W. R., 1993, Basin geodynamics: Basin Research, v. 5, p. 195–196.

Dickinson, W. R., 1995, Forearc basins, *in* Busby, C. J., and R. V. Ingersoll (eds.), Tectonics of sedimentary basins: Blackwell Science, p. 221–261.

Dickinson, W. R., and C. A. Suczek, 1979, Plate tectonics and sandstone composition: Am. Assoc. Petroleum Geologists Bull., v. 63, p. 2164–2182.

Dickinson, W. R., and H. Yarborough, 1976, Plate tectonics and hydrocarbon accumulation: AAPG Education Course Note Series No. 1, Am. Assoc. Petroleum Geologists, Tulsa, Okla., 34 p.

Dickinson, W. R., L. S. Beard, G. R. Brakenridge, J. L. Erjavec, R. C. Ferguson, K. F. Inman, R. A. Knepp, F. A. Lindberg, and P. T. Ryberg, 1983, Provenance of North American Phanerozoic sandstones in relation to tectonic setting: Geol. Soc. America Bull., v. 94, p. 222–235.

Dill, D. F., 1966, Sand flows and sand falls, *in* Fairbridge, R. W. (ed.), Encyclopedia of oceanography: Reinhold, New York, p. 763–765.

Dimroth, E., 1979, Models of physical sedimentation of iron formations, *in* Walker, R. G. (ed.), Facies models: Geoscience Canada Reprint Ser. 1, p. 159–174.

Dobkins, J. E., and R. L. Folk, 1970, Shape development on Tahati-Nui: Jour. Sed. Petrology, v. 40, p. 1167–1203.

Dodd, J. R., and R. J. Stanton, 1981, Paleoecology, concepts and applications: John Wiley & Sons, New York, 559 p.

Dominguez, J. M. L., 1996, The São Francisco strandplain: A paradigm for wave-dominated deltas?, *in* De Batist, M., and P. Jacobs (eds.), Geology of siliciclastic shelf seas: Geological Society Special Publication 117, p. 217–231.

Dominguez, J. M. L., L. Martin, and A. C. S. P. Bittencourt, 1987, Sea-level history and Quaternary evolution of river mouth-associated beach-ridge plains along the east-southeast Brazilian Coast: A summary, *in* Nummedal, D., O. H. Pilkey, and J. D. Howard (eds.), Sea-level fluctuations and coastal evolution: Soc. Econ. Paleontologists and Mineralogists, Spec. Pub. 41, p. 115–127.

Donselaar, M. E., 1989, The Cliff House Sandstone, San Juan Basin, New Mexico: Model for the stacking of 'transgressive' barrier complexes: Jour. Sed. Petrology, v. 59, p. 13–27.

Dorobek, S. L., and G. M. Ross (eds.), 1995, Stratigraphic evolution of foreland basins: Soc. for Sedimetary Geology Spec. Pub. 52, 310 p.

Dott, R. H., 1964, Wacke, graywacke, and matrix—what approach to immature sandstone classification: Jour. Sed. Petrology, v. 34, p. 625–632.

Dott, R. H., Jr., 1974, The geosynclinal concept, *in* Dott, R. H., and R. H. Shaver (eds.), Modern and ancient geosynclinal sedimentation: SEPM Spec. Pub. 19, p. 1–13.

Dott, R. H., Jr., and J. Bourgeois, 1982, Hummocky stratification: Significance of its variable bedding sequences: Geol. Soc. America Bull., v. 93, p. 663–680.

Dove, P. M., and J. D. Rimstidt, 1994, Silica-water interactions, *in* Heaney, P. J., C. T. Prewitt, and G. V. Gibbs (eds.), Silica: Physical behavior, geochemistry and materials applications: Mineralogical Soc. America, Rev. in Mineralogy, v 29, p. 259–308.

Dowdeswell, J. D., and J. D. Scourse (eds.), 1990, Glaciomarine environments: Processes and sediments: Geological Society of London, Special Publication 53, 423 p.

Doyle, L. J., and H. H. Roberts (eds.), 1988, Carbonate-clastic transitions: Elsevier, Amsterdam, 304 p.

Doyle, P., R. R. Bennett, and A. N. Baxter, 1994, The key to earth history: An introduction to stratigraphy: John Wiley & Sons, Chichester, 231 p.

Drever, J. L., 1971, Magnesium iron replacements in clay minerals in anoxic marine sediments: Science, v. 172, p. 1334–1336.

Drever, J. L., 1974, Geochemical model for the origin of Precambrian banded iron formations: Geol. Soc. America Bull., v. 85, p. 1099–1106.

Droz, L., and Bellaiche, G., 1985, Rhone deep-sea fan: morphostructure and growth pattern: Am. Assoc. of Petroleum Geologists, Bull., v. 69, p. 460–479.

Duke, W. L., 1985, Hummocky cross-stratification, tropical hurricanes and intense winter storms: Sedimentology, v. 32, p. 167–194.

Duke, W. L., R. W. C. Arnott, and R. J. Cheel, 1991, Shelf sandstones and hummocky cross-stratification: New insights on a stormy debate: Geology, v. 19, p. 625–628.

Dunbar, C. O., and J. Rodgers, 1957, Principles of stratigraphy: John Wiley & Sons, New York, 356 p.

Duncan, D. C., 1976, Geologic setting of oil shale deposits and world prospects, *in* Yen, T. F., and G. V. Chilingarian (eds.), Oil shale: Elsevier, New York, p. 13–26.

Dunham, R. J., 1962, Classification of carbonate rocks according to depositional textures, *in* Ham, W. E. (ed.), Classification of carbonate rocks: Am. Assoc. Petroleum Geologists Mem. 1, p. 108–121.

Durand, B. (ed.), 1980, Kerogen: Insoluble organic matter from sedimentary rocks: Editions Technip, Paris, 519 p.

Dzulynski, S., and E. K. Walton, 1965, Sedimentary features of flysch and greywackes: Developments in Sedimentology 7, Elsevier, Amsterdam, 274 p.

Easterbrook, D. J., 1982, Characteristic features of glacial sediments, *in* Scholle, P. A., and D. Spearing (eds.), Sandstone depositional environments: Am. Assoc. Petroleum Geologists Mem. 31, p. 1–10.

Easterbrook, D. J. (ed.), 1988, Dating Quaternary sediments: Geol. Soc. America Spec. Paper 227, 165 p.

Edmond, J. M., K. L. Von Damm, R. E. McDuff, and C. I. Measures, 1982, Chemistry of hot springs on the East Pacific Rise and their effluent dispersal: Nature, v. 297, p. 187–191.

Edwards, J. D., and P. A. Santogrossi (eds.), 1990, Divergent/passive margin basins: AAPG Memoir 48, Am. Assoc. Petroleum Geologists, Tulsa, Okla., 252 p.

Edwards, M., 1986, Glacial environments, *in* Reading, H. G. (ed.), Sedimentary environments and facies, 2nd ed., Blackwell Scientific Pub., Oxford, p. 445–470.

Ehrlich, R., and B. Weinberg, 1970, An exact method for characterization of grain shape: Jour. Sed. Petrology, v. 40, p. 205–212.

Eicher, D. L., 1976, Geologic time, 2nd ed.: Prentice-Hall, Englewood Cliffs, N.J., 152 p.

Einsele, G., 1992, Sedimentary basins: Springer-Verlag, Berlin, 628 p.

Einsele, G., 1998, Event stratigraphy: Recognition and interpretation of sedimentary event horizons, *in* Doyle, P., and M. R. Bennett (eds.), Unlocking the stratigraphical record: John Wiley & Sons, Ltd., p., 145–193.

Einsele, G., 2000, Sedimentary basins: Evolution, facies, and sediment budget, 2nd ed: Springer-Verlag, Berlin, 792 p.

Einsele, G., W. Ricken, and A. Seilacher, 1991a, Cycles and events in stratigraphy—Basic concepts and terms, *in* Einsele, G., W. Ricken, and A. Seilacher (eds.), Cycles and events in stratigraphy, Springer-Verlag, Berlin, p. 1–19.

Einsele, G., W. Ricken, and A. Seilacher (eds.), 1991b, Cycles and events in stratigraphy, Springer-Verlag, Berlin, 955 p.

Eisma, D., 1988, An introduction to the geology of continental shelves, *in* Postma, H., and J. J. Zijlstra, Continental shelves, Ecosystems of the World 27, Elsevier, Amsterdam, p. 39–91.

Eisma, D., P. L. de Boer, G. C. Cadée, K. Dijkema, H. Ridderinkhof, and C. Philippart, 1998, Intertidal deposits: River mouths, tidal flats, and coastal lagoons: CRC Press, Boca Raton, Fla., 524 p.

Eldredge, N., and S. J. Gould, 1972, Punctuated equilibrium: An alternative to phyletic gradualism, *in* Schopf, T. J. M. (ed.), Models in paleobiology: Freeman, Cooper, San Francisco, p. 82–115.

Eldridge, N., and S. J. Gould, 1977, Evolutionary models and biostratigraphic strategies, *in* Kauffman, E. G., and J. E. Hazel (eds.), Concepts and methods of biostratigraphy: Dowden, Hutchinson and Ross, Stroudsburg, Pa., p. 25–40.

Elliott, T., 1986, Deltas, *in* Reading, H. G. (ed.), Sedimentary environments and facies, 2nd ed., Blackwell Scientific Pub., Oxford, p. 113–154.

Embry, A. F., and J. E. Klovan, 1971, A Late Devonian reef tract on the northeastern Banks Island, N. W. T.: Canadian Petroleum Geology Bull., v. 19, p. 730–781.

Embry, A. F., and J. E. Klovan, 1972, Absolute water depth limits of Late Devonian paleoecological zones: Geol. Rundschau, v. 61, p. 672–686.

Emery, D., and K. J. Meyers (eds.), 1996, Sequence stratigraphy: Blackwell Science Ltd., Oxford, 297 p.

Emery, K. O., 1968, Relict sediments on continental shelves of the world: Am. Assoc. Petroleum Geologists Bull., v. 52, p. 445–464.

Emiliani, C., 1955, Pleistocene temperatures: Jour. Geology, v. 63, p. 538–578.

Ericson, D. B., and G. Wollin, 1968, Pleistocene climates and chronology in deep-sea sediments: Science, v. 162, p. 1227–1234.

Eswaran, H., T. Rice, R. Ahrens, B. A. Stewart, eds., 2003, Soil classification: A global desk reference: CRC Press, Boca Raton, Fla., 263 p.

Eugster, H. P., and L. A. Hardie, 1978, Saline lakes, *in* Lerman, A., (ed.), Lakes, chemistry, geology, physics: Springer-Verlag, New York, p. 237–293.

Ewers, W. E., 1983, Chemical factors in the deposition and diagenesis of banded iron-formation, *in* Trendall, A. F., and Forris, R. C. (eds.), Iron-formations: Facts and problems: Elsevier, Amsterdam, p. 491–512.

Ewing, M., and E. M. Thorndike, 1965, Suspended matter in deep-ocean water: Science, v. 147, p. 1291–1294.

Eyles, N., 1993, Earth glacial record and its tectonic setting: Earth Science Reviews, v. 35, p. 1–248.

Eyles, N., and C. H. Eyles, 1992, Glacial depositional systems, *in* Walker, R. G., and N. P. James (eds.), Facies models: Response to sea level changes, Geol. Assoc. Canada, p. 73–100.

Eyles, N., and A. D. Miall, 1984, Glacial facies, *in* Walker, R. G. (ed.), Facies models, 2nd ed.: Geoscience Canada Reprint Ser. 1, p. 15–38.

Fagerstrom, J. A., 1987, The evolution of reef communities: John Wiley & Sons, New York, 600 p.

Fairbridge, R. W., 1980, The estuary: Its definition and geodynamic cycle, *in* Olausson, E., and I. Cato (eds.), Chemistry and biochemistry of estuaries: John Wiley & Sons, New York, p. 1–35.

Fairchild, T. R., J. W. Schopf, J. Shen-Miller, E. M. Guimaraes, M. D. Edwards, A. Lagstein, X. Li, M. Pabst, and L. de S. de Melo-Filho, 1996, Recent discoveries of Proterozoic microfossils in south-central Brazil: Precambrian Research, v. 80n, p. 125–152.

Farrow, G. E., N. H. Allen, and E. B. Akpan, 1984, Bioclastic carbonate sedimentation on a high-latitude, tide-dominated shelf: Northeast Orkney Islands, Scotland: Jour. Sedimentary Petrology, v. 54, p. 373–393.

Faure, G., 1982, The marine-strontium geochronometer, *in* Odin, G. S. (ed.), Numerical dating in stratigraphy: John Wiley & Sons, Chichester, p. 73–79.

Faure, G., 1986, Principles of isotope geology, 2nd ed.: John Wiley & Sons, New York, 589 p.

Ferrusquia-Villafranca, I., R. M. Easton, L. E. Edwards, R. H. Fakundiny, and J. O. Jones, 2001, North American Commission on Stratigraphic Nomenclature Note 63—Application for amendment of the North American Stratigraphic Code concerning consistency and updating regarding electronic publishing: Amer. Assoc. Petroleum Geologists Bull., v. 85, p. 366–371.

Fieller, N. R. J., D. D. Gilbertson, and W. Olbricht, 1984, A new method for environmental analysis of particle size distribution data from shoreline sediments: Nature, v. 311, p. 648–651.

Fischer, A. G., 1984, The two Phanerozoic supercycles, *in* Berggren, W. A., and J. A. Van Couvering (eds.), Catastrophes in Earth history: Princeton University Press, p. 129–150.

Fisher, W. L., and J. H. McGowan, 1967, Depositional systems in the Wilcox Group of Texas and their relationship to occurrences of oil and gas: Gulf Coast Assoc. Geol. Soc. Trans., v. 17, p. 105–125.

Fleming, B. W., 1980, Sand transport and bedform patterns on the continental shelf between Durban and Port Elizabeth (southeast Africa continental margin): Sed. Geology, v. 26, p. 179–205.

Flügel, E., 1982, Microfacies analysis of limestones: Springer-Verlag, Berlin, 633 p.

Flügel, E., and W. Kiessling, 2002, Patterns of Phanerozoic reef crisis, *in* Kiessling, W., E. Flügel, and J. Golonka (eds.), Phanerozoic reef patterns: SEPM Special Publication No. 72, p. 691–733.

Flynn, J. J., and C. C. Swisher, III, 1995, Genozoic South American land mammal ages: Correlation to global geochronologies, *in* Berggren, W.A., D.V. Kent, M-P. Aubry, and J. Hardenbol (eds.), Geochronology, time scales and global stratigraphic correlation: SEPM Spec. Publ. 54, p. 317–333.

Folk, R. L., 1959, Practical petrographic classification of limestones: Am. Assoc. Petroleum Geologists Bull., v. 43, p. 1–38.

Folk, R. L., 1962, Spectral subdivision of limestone types, *in* Ham, W. E. (ed.), Classification of carbonate rocks: Am. Assoc. Petroleum Geologists Mem. 1, p. 62–84.

Folk, R. L., 1965, Some aspects of recrystallization in ancient limestones, *in* Pray, L. C., and R. C. Murray (eds.), Dolomitization and limestone diagenesis: Soc. Econ. Paleontologists and Mineralogists Spec. Pub. 13, p. 14–48.

Folk, R. L., 1974, Petrology of sedimentary rocks: Hemphill, Austin, Tx., 182 p.

Folk, R. L., and L. S. Land, 1975, Mg/Ca ratio and salinity: Two controls over crystallization of dolomite: Am. Assoc. Petroleum Geologists Bull., v. 59, p. 60–68.

Folk, R. L., P. B. Andrews, and D. W. Lewis, 1970, Detrital sedimentary rock classification and nomenclature for use in New Zealand: New Zealand Jour. Geol. and Geophysics, v. 13, p. 937–968.

Forgotson, J. M., 1960, Review and classification of quantitative mapping techniques: Amer. Assoc. Petroleum Geol. Bull., v. 44, p. 83–100.

Forrest, J., and N. R. Clark, 1989, Characterizing grain size distributions: Evaluation of a new approach using multivariate extension of entropy analysis: Sedimentology, v. 36, p. 711–722.

Fox, W. T., 1983, At the sea's edge, Chapter 4, Tides, p. 93–124: Prentice-Hall, Englewood Cliffs, N.J., 317 p.

Frakes, L. A., J. E. Francis, and J. I. Syktus, 1992, Climate modes of the Phanerozoic: Cambridge University Press, Cambridge, 274 p.

Frey, R. W., and S. G. Pemberton, 1985, Biogenic structures in outcrops and cores. 1. Approaches to ichnology: Canadian Petroleum Geology Bull., v. 33, p. 72–115.

Frey, R. W., and S. G. Pemberton, 1987, The *Psilonichnus* ichnocoenose and its relationship to adjacent marine and nonmarine ichnocoenoses along the Georgia coast: Canadian Petroleum Geology Bull., v. 35, p. 333–357.

Frey, R. W., and A. Seilacher, 1980, Uniformity in marine invertebrate ichnology: Lethaia, v. 13, p.183–207.

Frey, R. W., and R. A. Wheatcroft, 1989, Organism-substrate relations and their impact on sedimentary petrology: Jour. Geological Education, v. 37, p. 261–279.

Friedman, G. M., 1967, Dynamic processes and statistical parameters compared for size frequency distribution of beach and river sands: Jour. Sed. Petrology, v. 37, p. 327–354.

Friedman, G. M., 1979, Address of the retiring president of the International Association of Sedimentologists: Differences in size distribution of populations of particles among sands of various origins: Sedimentology, v. 26, p. 3–32.

Friedman, G. M., and J. E. Sanders, 1978, Principles of sedimentology: John Wiley & Sons, New York, 792 p.

Froelich, P. N., M. A. Arthur, W. C. Burnett, M. Deakin, V. Hensley, R. Jahnke, L. Kaul, K.-H Kim, K. Roe, A. Soutar, and C. Vathakanon, 1988, Early diagenesis of organic matter in Peru continental margin sediments: Phosphorite precipitation: Marine Geology, v. 80, p. 309–343.

Frostick, L. E., and R. J. Steel, 1993, Tectonic controls and signatures in sedimentary successions: International Association of Sedimentologists Spec. Pub. 20, Blackwell Scientific Pub., Oxford, 520 p.

Gains, A. M., 1980, Dolomitization kinetics: Recent experimental studies, *in* Zenger, D. H., J. B. Dunham, and R.L. Ethington (eds.), Concepts and models of dolomitization: Soc. Econ. Paleontologists and Mineralogists Spec. Pub. 28, p. 81–86.

Gale, A. S., 1998, Cyclostratigraphy, *in* Doyle, P., and M. R. Bennett (eds.), Unlocking the stratigraphical record: John Wiley & Sons, Chichester, p. 195–220.

Galehouse, J. S., 1971, Sedimentation analysis, in Carver, R. E. (ed.), Procedures in sedimentary petrology: John Wiley & Sons, New York, p. 69–94.

Galloway, W. E., 1975, Process framework for describing the morphologic and stratigraphic evolution of deltaic depositional systems, *in* Broussard, M. L. (ed.), Deltas: Models for exploration: Houston Geological Society, p. 87–98.

Galloway, W. E., 1976, Sediments and stratigraphic framework of the Copper River fan-delta, Alaska: Jour. Sed. Petrology, v. 46, p. 726–737.

Galloway, W. E., 1989, Clastic facies models, depositional systems, sequences and correlation: A sedimentologist's view of the dimensional and temporal resolution of lithostratigraphy, *in* Cross, T. A. (ed.), Quantitative dynamic stratigraphy: Prentice Hall, Englewood Cliffs, N.J., p. 459–477.

Galloway, W. E., 1998, Siliciclastic slope and base-of-slope depositional systems: Component facies, stratigraphic architecture, and classification: Am. Assoc. Petroleum Geologists Bull., v. 82, p. 569–595.

Galloway, W. E., and D. K. Hobday, 1983, Terrigenous clastic depositional systems: Springer-Verlag, New York, 423 p.

Garber, R. A., Y. Levy, and G. M. Friedman, 1987, The sedimentology of the Dead Sea: Carbonates and evaporites, v. 2, p. 43–57.

Garrels, R. M., 1987, A model for the deposition of the microbanded Precambrian iron formations: Am. Jour. Science, v. 287, p. 81–106.

Garrels, R. M., and F. T. McKenzie, 1971, Evolution of sedimentary rocks: W.W. Norton, New York, 397 p.

Garrison, R. E., R. B. Douglas, K. E. Pischotto, C. M. Isaacs, and J. C. Ingle (eds.), 1981, The Monterey Formation and related siliceous rocks of Calfornia: Soc. Econ. Paleontologists and Mineralogists, Pacific Section, Los Angeles, 327 p.

Garven, G., and R. A. Freeze, 1984, Theoretical analysis of the role of groundwater flow in the genesis of stratabound ore deposits: Am. Jour. Science, v. 284, p. 1085–1174.

Geyh, M. A., and H. Schleicher, 1990, Absolute age determination: Springer-Verlag, Berlin, 503 p.

Gierlowski-Kordesch, E., and E. Kelts (eds.), 1994a, Global geological record of lake basins, v. 1: Cambridge University Press, 427 p.

Gierlowski-Kordesch, E., and E. Kelts, 1994b, Introduction, *in* Gierlowski-Kordesch, E., and E. Kelts (eds.), Global geological record of lake basins, v. 1: Cambridge University Press, p. xvii-xxxiii.

Gierlowski-Kordesch, E. H., and E. Kelts (eds.), 2000, Lake basins through space and time: AAPG Studies in Geology 46, 648 p.

Gilette, D. A., 1999, Physics of aeolian movement emphasizing changing of the aerodynamic roughness height by saltating grains (the Owen effect), *in* Goudie, A. S., I. Livingston, and S. Stokes (eds.), John Wiley & Sons, p. 129–142.

Ginsburg, R. N. (ed.), 2001, Subsurface geology of a prograding carbonate platform margin, Great Bahama Bank: Results of the Bahamas Drilling project: SEPM Special Publication 70, 271 p.

Givens, R. K., and B. H. Wilkinson, 1987, Dolomite abundance and stratigraphic age: Constraints on rates and mechanisms of Phanerozoic dolostone formation: Jour. Sedimentary Petrology, v. 57, p. 1068–1078.

Glaister, R. P., and H. W. Nelson, 1974, Grain-size distributions, an aid to facies identifications: Canadian Petroleum Geology Bull., v. 22, p. 203–240.

Glenn, C. R., L. Prévôt-Lucas, and J. Lucas (eds.), 2000, Marine authigenesis: From global to microbial: SEPM Spec. Pub. 66, 536 p.

Glenn, C. R., and M. A. Arthur, 1988, Petrology and major element geochemistry of Peru margin phosphorites and associated diagenetic minerals: Authigenesis in modern organic-rich sediments: Marine Geology, v. 80, p. 231–267.

Glen, W., 1982, The road to Jaramillo: Critical years of the revolution in the earth sciences: Stanford University Press, Stanford, 459 p.

Glennie, K. W., 1986, Early Permian Rotliegend, *in* K. W. Glennie (ed.), Introduction to the petroleum geology of the North Sea: Blackwell, Oxford, p. 63–85.

Goldich, S. S., 1938, A study of rock weathering: Jour. Geology, v. 46, p. 17–58.

Gole, M. J., and C. Klein, 1981, Banded iron-formations through much of Precambrian time: Jour. Geology, v. 89, p. 169–183.

Goodbred, S. L., Jr., and S. A. Kuehl, 2000, The significance of large sediment supply, active tectonism, and eustasy on margin sequence development: Late Quaternary stratigraphy and evolution of the Ganges-Brahmaputra delta: Sedimentary Geology, v. 133, p. 227–248.

Gould, H. R., 1972, Environmental indicators—A key to the stratigraphic record, *in* Rigby, J. K., and W. K. Hamblin (eds.), recognition of ancient sedimentary environments: Soc. Econ. Paleontologists and Mineralogists Spec. Pub. 16, p. 1–3.

Gould, S. J., 2001, The interrelationship of speciation and punctuated equilibrium, *in* Jackson, J. B. C., S. Lidgard, and F. K. McKinney (eds.), Evolutionary patterns: Growth, form, and tempo in the fossil record, p. 196–217.

Gould, S. J., and N. Eldredge, 1993, Punctuated equilibrium comes of age: Nature, v. 366, p. 223–227.

Gournay, J., R. L. Folk, and B. L. Kirkland, 1997, Evidence for nannobacterially precipitated dolomite in Pennsylvanian carbonates, *in* Camoin, G., and A. Arnaud-Vanneau, convenors, International workshop on microbial mediation in carbonate diagenesis 97, Abstract Book, Internat. Assoc. Sedimentologists, p. 33.

Grabeau, A. W., 1904, On the classification of sedimentary rocks: Amer. Geol., v. 33, p. 228–247.

Gradstein, F. M., E. P. Agterberg, J. C. Brower, and W. S. Schwarzacher, 1985, Quantitative stratigraphy: D. Reidel Pub. Co., Dordrecht, 598 p.

Gradstein, F. M., F. P. Agterberg, J. G. Ogg, J. Hardenbol, P. van Veen, J. Thierry, and Z. Huang, 1995, A Triassic, Jurassic and Cretaceous time scale, *in* Berggren, W. A., D. V. Kent, M-P. Aubry, and J. Hardenbol (eds.), 1995, Geochronology, time scales and global stratigraphic correlation: Soc. for Sedimetary Geology Spec. Pub. 54, p. 95–126.

Green, H. G., S. H. Clarke, Jr., and M. P. Kennedy, 1991, Tectonic evolution of submarine canyons along the California continental margin, *in* Osborne, R. H., (ed.), From shoreline to abyss: Soc. for Sedimetary Geology Spec. Pub. 46, p. 231–248.

Greensmith, J. T., 1989, Petrology of the sedimentary rocks, 7th ed.: Unwin Hyman, London, 262 p.

Griffiths, C. M., 1982, A proposed geologically significant segmentation and reassignment algorithm for petrophysical borehole logs, *in* Cubitt, J. M., and R. A. Reyment (eds.), Quantitative stratigraphic correlations: John Wiley & Sons, New York, p. 287–298.

Gromet, L. P., R. F. Dymek, L. A. Haskin, and R. L. Korotev, 1984, The "North American shale composite": Its compilation and major and trace element characteristics: Geochim. et Cosmochim. Acta, v. 48, p. 2469–2482.

Gross, G. A., 1980, A classification of iron formations based on depositional environments: Canadian Mineralogist, v. 18, p. 215–222.

Gross, M. G., 1982, Oceanography, 3rd ed.: Prentice-Hall, Englewood Cliffs, N.J., 498 p.

Grotzinger, J. P., and N. P. James (eds.), 2000a, Carbonate sedimentation and diagenesis in the evolving Precambrian world: SEPM Special Publication 67, Tulsa, Okla., 364 p.

Grotzinger, J. P., and N. P. James, 2000b, Precambrian carbonates: Evolution in understanding, *in* Grotzinger, J. P., and N. P. James (eds.), Carbonate sedimentation and diagenesis in the evolving Precambrian world: SEPM Special Publication 67, p. 3–20.

Guex, J., 1991, Biochronological correlations: Springer-Verlag, Berlin, 252 p.

Gulbrandsen, R. A., and C. E. Roberson, 1973, Inorganic phosphorites in seawater: Environmental phosphorus handbook: John Wiley & Sons, New York, Chapter 5, p. 117–140.

Hails, J. R., 1976, Placer deposits, *in* Wolf, K. H. (ed.), Handbook of strata-bound and stratiform ore deposits, v. 3: Elsevier, New York, p. 213–244.

Hailwood, E. A., 1989, Magnetostratigraphy: Geological Society Special Report 19, Blackwell Scientific Pub., Oxford, 84 p.

Hallam, A., 1981, Facies interpretation and the stratigraphic record: W. H. Freeman, San Francisco, 291 p.

Hallam, A., 1984, Pre-Quaternary sea-level changes: Ann. Rev. Earth and Planetary Sciences, v. 12, p. 205–243.

Hallam, A., 1998, Interpreting sea level, *in* Doyle, P., and M. R. Bennett (eds.), Unlocking the stratigraphical record: Advances in modern stratigraphy: John Wiley & Sons, Chichester, p. 421–439.

Hallam, A., and P. B. Wignall, 1997, Mass extinctions and their aftermath: Oxford University Press, Oxford, 320 p.

Ham, W. E. (ed.), 1962, Classification of carbonate rocks: Am. Assoc. Petroleum Geologists Mem. 1, 279 p.

Ham, W. E., and L. C. Pray, 1962, Modern concepts and classifications of carbonate rocks, *in* Ham, W. E. (ed.), Classification of carbonate rocks: Am. Assoc. Petroleum Geologists Mem. 1, p. 2–19.

Hamblin, W. K., 1965, Internal structures of "homogeneous" sandstones: Kansas Geol. Survey Bull., 175, pt. 1, p. 1–37.

Hambrey, M., 1994, Glacial environments: UCL Press Ltd., London, 296 p.

Hancock, J. M., 1977, The historic development of biostratigraphic correlation, *in* Kauffman, E. G., and J. E. Hazel (eds.), Concepts and methods of biostratigraphy: Dowden, Hutchinson and Ross, Stroudsburg, Pa., p. 3–22.

Handford, C. R., 1991, Marginal marine halite: Sabkhas and salinas, *in* Melvin, J. L. (ed.), Evaporites, petroleum and mineral resources: Elsevier Sci. Pub., Amsterdam, p. 1–66.

Haney, W. D., and L. I. Briggs, 1964, Cyclicity of textures in evaporite rocks of the Lucas Formation, *in* Merriam, D. F. (ed.), Symposium on cyclic sedimentation: Kansas Geol. Survey, p. 191–197.

Hanshaw, B. B., W. Back, and R. G. Deike, 1971, A geochemical hypothesis for dolomitization by ground water: Econ. Geology, v. 66, p. 710–724.

Haq, B. U., and T. R. Worsley, 1982, Biochronology—Biological events in time resolution, their potential and limitations, *in* Odin, G. S. (ed.), Numerical dating in stratigraphy, p. 1: John Wiley & Sons, New York, p. 19–36.

Haq, B. U., J. Hardenbol, and P. R. Vail, 1987, The chronology of fluctuating sea level since the Triassic: Science, v. 235, p. 1156–1167.

Haq, B. U., J. Hardenbol, and P. R. Vail, 1988, Mesozoic and Cenozoic chronostratigraphy and eustatic cycles, *in* Wilgus, C. K., B. S. Hastings, C. G. St. C. Kendall, H. W. Posamentier, C. A. Ross, and J. C. Van Wagoner (eds.), Sea-level changes: An integrated approach: Soc. Econ. Paleontologists and Mineralogists Spec. Pub. 42, p. 71–108.

Hardie, L. A., 1984, Evaporites: marine or non-marine: Am. Jour. Science, v. 284, p. 193–249.

Hardie, L. A., 1987, Dolomitization: A critical view of some current views: Jour. Sed. Petrology, v. 57, p. 166–183.

Hardie, L. A., 1991, On the significance of evaporites: Ann. Rev. Earth and Planetary Sciences, v. 19, p. 131–168.

Hardie, L. A., and E. A. Shinn, 1986, Carbonate depositional environments modern and ancient—Part 3: Tidal flats: Colorado School of Mines Quarterly, v. 81 No. 1, 74 p.

Hardisty, J., 1990, Beaches—Form and process: Unwin Hyman, London, 324 p.

Harland, W. B., 1978, Geochronological scales, *in* Cohee, G. V., M. F. Glaessner, and H. D. Hedberg (eds.), Contributions to the geologic time scale: Am. Assoc. Petroleum Geologists, Tulsa, Okla., p. 9–32.

Harland, W. B., R. L. Armstrong, A. V. Cox, L. E. Craig, A. G. Smith, and D. G. Smith, 1990, A geologic time scale 1989, 2nd ed.: Cambridge University Press, Cambridge, 263 p.

Harms, J. C., and R. K. Fahnestock, 1965, Stratification, bed forms and flow phenomena (with examples from the Rio Grande), *in* Middleton, G. V. (ed.), Primary sedimentary structures and their hydrodynamic interpretation: Soc. Econ. Paleontologists and Mineralogists Spec. Pub. 12, p. 84–155.

Harms, J. C., J. B. Southard, and R. G. Walker, 1982, Structures and sequences in clastic rocks: Soc. Econ. Paleontologists and Mineralogists Lecture Notes for Short Course 9, variously paginated.

Harms, J. C., J. B. Southard, D. R. Spearing, and R. G. Walker, 1975, Depositional environments as interpreted from primary sedimentary structures and stratification sequences: Soc. Econ. Paleontologists and Mineralogists Short Course 2, 161 p.

Harris, P. M., C. H. Moore, and J. L. Wilson, 1985, Carbonate depositional environments modern and ancient: Part 2: Carbonate platforms: Colorado School of Mines Quarterly, v. 80, no. 4, p. 1–60.

Hasiotis, S. T., 2002, Continental trace fossils: SEPM (Society for Sedimentary Geologoy), Tulsa, Okla., 130 p.

Hatcher, B. G., 2001, What determines whether deep-water corals build reefs: Do shallow reef models apply?, *in* Martin Willison, J. H., J. Hall, S. E. Gass, E. L. R. Kenchington, M. Butler, and P. Doherty (eds.), 2001, Proceedings of the First International Symposium on Deep-Sea Corals: Ecology Action Centre and Nova Scotia Museum, Halifax, Nova Scotia, p. 6–18.

Hayes, M. O., 1975, Morphology of sand accumulations in estuaries, *in* Cronin, L. E. (ed.), Estuarine research, v. 2: Geology and engineering: Academic Press, New York, p. 3–22.

Haywick, D. W., R. M. Carter, and R. A. Henderson, 1992, Sedimentology of 40,000 year Milankovich-controlled cyclothems from central Hawke's Bay, New Zealand: Sedimentology, v. 39, p. 675–696.

Heath, G. R., 1974, Dissolved silica and deep-sea sediments, *in* Hay, W. W. (ed.), Studies in paleoceanography: Soc. Econ. Paleontologists and Mineralogists Spec. Pub. 20, p. 77–94.

Heckel, P. H., 1972, Recognition of ancient shallow marine environments, *in* Rigby, J. K., and W. K. Hamblin (eds.), Recognition of ancient sedimentary environments: Soc. Econ. Paleontologists and Mineralogists Spec. Pub. 16, p. 226–286.

Hedberg, H. D. (ed.), 1976, International Stratigraphic Guide: A guide to stratigraphical classification, terminology, and procedure: International Subcommission on Stratigraphic Classification of IUGS Commission on Stratigraphy: John Wiley & Sons, New York, 200 p.

Hein, J. R., and S. M. Karl, 1983, Comparisons between open-ocean and continental margin chert sequences, *in* Iijima, A., J. R. Hein, and R. Siever (eds.), Siliceous deposits in the Pacific region. Elsevier, Amsterdam, p. 25–43.

Hein, J. R., and J. T. Parrish, 1987, Distribution of siliceous deposits in space and time, *in* Hein, J. R. (ed.), Siliceous sedimentary rock-hosted ores and petroleum: Van Nostrand Reinhold, New York, p. 10–57.

Hein, J. R., H.-W. Yeh, and J. A. Barron, 1990, Eocene diatom chert from Adak Island, Alaska: Jour. Sed. Petrology, v. 60, p. 250–257.

Heller, F., and L. Tsungsheng, 1984, Magnetism of Chinese loess deposits: Geophys. Jour. Royal. Astron. Society, v. 77, p. 125–141.

Herring, J. R., 1995, Permian phosphorites: A paradox of phosphogenesis, *in* Scholle, P. A., T. M. Peryt, and D. S. Ulmer-Scholle (eds.), The Permian of northern Pangea. Sedimentary basins and economic resources: Springer-Verlag, Berlin, p. 292–312.

Hesse, R., 1989, Silica diagenesis: Origin of inorganic and replacement cherts; Earth-Science Reviews, v. 26, p. 253–284.

Hesse, R., 1990, Origin of chert and silica diagenesis, *in* McIlreath, I. A., and D. W. Morrow (eds.), Diagenesis: Geol. Assoc. Canada, Reprint Ser. 4, p. 227–275.

Heydari, E., 1997, Hydrotectonic models of burial diagenesis in platform carbonates based on formation water geochemistry in North American sedimentary basins, *in* Montañez, I. P., J. M. Gregg, and K. L. Shelton (eds.), Basin-wide diagenetic patterns: integrated petrologic, geochemical, and hydrologic considerations: Soc. for Sedimentary Geology Spec. Pub. 57, p. 53–79.

Hillis, L., 1991, Recent calcified Halimedacea, *in* Riding, R. (ed.), Calcareous algae and stromatolites: Springer-Verlag, Berlin, p. 167–188.

Hiscott, R. N., and G. V. Middleton, 1980, Fabric of coarse deep-water sandstones, Tourelle Formation, Quebec, Canada: Jour. Sed. Petrology, v. 50, p. 703–722.

Hofmann, H. J., K. Grey, A. H. Hickman, and R. I. Thorpe, 1999, Origin of 3.45 Ga coniform stromatolites in Warrawoona Group, Western Australia: Geol. Soc. America Bull., v. 111, p. 1256–1262.

Hollister, D. D., and A. R. M. Nowell, 1991, HEBBLE epilogue: Marine Geology, v. 99, p. 445–460.

Holmes, A., 1965, Principles of physical geology, 2nd ed., Thomas Nelson, London, 1288 p.

Holser, W. T., 1977, Catastrophic chemical events in the history of the ocean: Nature, v. 267, p. 403–408.

Holser, W. T., 1984, Gradual and abrupt shifts in ocean chemistry during Phanerozoic time, *in* Holland, H. D., and A. F. Trendall (eds.), Patterns of change in Earth evolution, Springer-Verlag, Berlin, p. 123–143.

Holser, W. T., and M. Magaritz, 1992, Cretaceous/Tertiary and Permian/Triassic boundary events compared: Geochimica et Cosmochimica Acta, v. 56, p. 3297–3309.

Holser, W. T., M. Magaritz, and J. Wright, 1986, Chemical and isotopic variations in the world ocean during Phanerozoic time, *in* Walliser, O. (ed.), Global bioevents, Lecture Notes in Earth Sciences, Vol. 8, Springer-Verlag, Heidelberg, p. 63–74.

Hooke, R. LeB., 1967, Processes on arid-region alluvial fans: Journal of Geology, v. 75, p. 438–60.

Horne, J. C., J. C. Ferm, F. T. Caruccio, and B. P. Bagnaz, 1978, Depositional models in coal exploration and mine planning in Appalachian Region: Amer. Assoc. Petroleum Geol. Bull., v. 62, p. 2379–2411.

Horowitz, A. S., and P. E. Potter, 1971, Introductory petrography of fossils: Springer-Verlag, New York, 302 p.

Horsfield, B., 1997, The bulk composition of first-formed petroleum in source rocks, *in* Welte, D. H., B. Horsfield, and D. R. Baker (eds.), Petroleum and basin evolution: Insights from petroleum geochemistry, geology, and basin modeling: Springer-Verlag, Berlin, p. 337–402.

House, M. R., and A. S. Gale (eds.), 1995, Orbital forcing timescales and cyclostratigraphy: Geol. Soc. Spec. Pub. 85, 210 p.

Hovorka, S., 1987, Depositional environments of marine-dominated bedded halite, Permian San Andres Formation, Texas: Sedimentology, v. 34, p. 1029–1054.

Howarth, M. J., 1982, Tidal currents of the continental shelf, *in* Stride, A. H. (ed.), Offshore tidal sands: Processes and deposits: Chapman and Hall, London, p. 10–26.

Howell, J. A., and S. S. Flint, 2003, Tectonic setting, stratigraphy and sedimentology of the Book Cliffs, *in* Coe, A. L. (ed.), The sedimentary record of sea-level change: Cambridge University Press, Cambridge, p. 135–157.

Hsü, K. J., 1989, Physical principles of sedimentology: Springer-Verlag, Berlin, 233 p.

Hsü, K. J., Q. He, J. A. McKenzie, H. Weissert, K. Perch-Nielson, H. Oberhänsli, K. Kelts, J. Labrecque, L. Tauxe, U. Krähenbühl, S. F. Percival, Jr., R. Wright, A. M. Karpoff, N. Peterson, P. Tucker, R. Z. Poore, A. M. Gombos, K. Pisciotto, M. F. Carman, Jr., and E. Schreiber, 1982, Mass mortality and its environmental and evolutionary consequences: Science, v. 216, p. 249–255.

Hubbard, D. K., 1992, Hurricane-induced sediment transport in open-shelf tropical systems—An example from St. Croix, U.S. Virgin Islands: Jour. Sedimentary Petrology, v. 62, p. 946–960.

Hubbard, R. J., J. Pape, and D. G. Roberts, 1985a, Depositional sequence mapping as a technique to establish tectonic and stratigraphic framework and evaluate hydrocarbon potential on a passive continental margin, *in* Berg, O. R., and D. G. Wolverton (eds.), Seismic stratigraphy II: An integrated approach: Am. Assoc. Petroleum Geologists Mem. 39, p. 79–91.

Hubbard, R. J., J. Pape, and D. G. Roberts, 1985b, Depositional sequence mapping to illustrate the evolution of a passive continental margin, *in* Berg, O. R., and D. G. Wolverton (eds.), Seismic stratigraphy II: An integrated approach: Am. Assoc. Petroleum Geologists Mem. 39, p. 93–115.

Hudson, J. H., 1985, Growth rate and carbonate production in *Halimeda opuntia*: Marquesas, Keys, Florida, *in* Toomey, D. F., and M. H. Nitecki (eds.), Paleoalgology: Springer-Verlag, Berlin, p. 257–263.

Hunt, J. M., 1979, Petroleum geochemistry and geology: W. H. Freeman, San Francisco, 617 p.

Hunt, J. M., 1996, Petroleum geochemistry and geology, 2nd ed.: W. H. Freeman, New York, 743 p.

Hunter, R. E., 1977, Basic types of stratification in small eolian dunes: Sedimentology, v. 24, p. 361–387.

Hutton, A. C., 1995, Organic petrography of oil shales, *in* Snape, C. (ed.), Composition, geochemistry and conversion of oil shales: Kluwer Academic Publishers, Netherlands, p. 17–33.

Iijima, A., H. Inagaki, and Y. Kakuwa, 1979, Nature and origin of the Paleogene cherts in the Setogawa Terrain, Shizuoka, Central Japan: Jour. Fac. Sci., University of Tokyo, v. 20, p. 1–30.

Iller, L. V., 1979, Chemistry of silica: Wiley-Interscience, New York, 866 p.

Illing, L. V., 1954, Bahaman calcareous sands: Am. Assoc. Petroleum Geologists Bull., v. 38, p. 1–95.

Inden, R. F., and C. H. Moore, 1983, Beach, *in* Scholle, P. A., D. G. Bebout, and C. H. Moore (eds.), Carbonate depositional environments: Am. Assoc. Petroleum Geologists Mem. 33, p. 211–266.

Ingersoll, R. V., 1982, Initiation and evolution of the Great Valley forearc basin of northern and central California, U.S.A., *in* Leggett, J. K. (ed.), Trench-forearc geology: modern and ancient active plate margins: Geol. Soc. Spec. Pub. 10, p. 458–467.

Ingersoll, R. V. 1988, Tectonics of sedimentary basins: Geol. Soc. America Bull., v. 100, p. 1704–1719.

Ingersoll, R. V., and C. J. Busby, 1995, Tectonics of sedimentary basins, in Busby, C. J., and R. V. Ingersoll (eds.), Tectonics of sedimentary basins: Blackwell Science, p. 1–51.

Ingersoll, R. V., S. A. Graham, and W. R. Dickinson, 1995, Remnant ocean basins, in Busby, C. J., and R. V. Ingersoll (eds.), Tectonics of sedimentary basins: Blackwell Science, p. 363–391.

Ingersoll, R. V., T. F. Bullard, R. L. Ford, J. P. Grimm, J. D. Pickle, and S. W. Sares, 1984, The effect of grain size on detrital modes: A test of the Gazzi-Dickinson point-counting method: Journal of Sedimentary Petrology, v. 54, p. 103–116.

Ingram, R. L., 1971, Sieve analysis, in Carver, R. E. (ed.), Procedures in sedimentary petrology: Wiley-Interscience, New York, p. 49–67.

Inman, D. L., and C. E. Nordstrom, 1971, On the tectonic and morphologic classification of coasts: Jour. Geology, v. 79, p. 1–21.

Isaacs, C. M., 1982, Influence of rock composition on kinetics of silica phase changes in the Monterey Formation, Santa Barbara area, California: Geology, v. 10, p. 304–308.

Isley, A. E., 1995, Hydrothermal plumes and the delivery of iron to banded iron formations: Journal of Geology, v. 103, p. 169–185.

Jackson, M. L., 1968, Weathering of primary and secondary minerals in soils: Transactions, 9th International Congress Soil Science, v. 4, p. 281–292.

Jackson, M. P. A., D. G. Roberts, and S. Snelson, 1996, Salt tectonics: A global perspective: AAPG Mem. 65, Am. Assoc. Petroleum Geologists, Tulsa, Okla., 454 p.

James, H. L., 1966, Chemistry of the iron-rich sedimentary rocks: Data of Geochemistry, 6th ed.: U.S. Geol. Survey Prof. Paper 440-W, 61 p.

James, H. L., and A. F. Trendall, 1982, Banded iron formation: Distribution in time and paleoenvironmental significance, in Holland, H. D., and M. Schidlowski (eds.), Mineral deposits and the evolution of the biosphere: Springer-Verlag, Berlin, p. 199–218.

James, N. P., 1983, Reef environment, in Scholle, P. A., D. G. Bebout, and C. H. Moore (eds.), Carbonate depositional environments: Am. Assoc. Petroleum Geologists Mem. 33, p. 345–440.

James, N. P., 1984a, Introduction to carbonate facies models, in Walker, R. G. (ed.), Facies models: Geoscience Canada Reprint Ser. 1, p. 209–212.

James, N. P., 1984b, Shallowing-upward sequences in carbonates, in Walker, R. G. (ed.), Facies models: Geoscience Canada Reprint Ser. 1, p. 213–228.

James, N. P., 1984c, Reefs, in Walker, R. G. (ed.), Facies models: Geoscience Canada Reprint Ser. 1, p. 229–244.

James, N. P., 1997, The cool-water carbonate depositional realm, in James, N. P., and J. A. D. Clarke, 1997, Cool-water carbonates: SEPM Spec. Publ. 56, Soc. for Sedimentary Geology, Tulsa, Okla., p. 1–20.

James, N. P., and P-A. Bourque, 1992, Reefs and mounds, in Walker, R. G., and N. P. James (eds.), Facies models—Response to sea level change: Geol. Assoc. Canada, p. 323–348.

James, N. P., and J. A. D. Clarke (eds.), 1997, Cool-water carbonates: SEPM Spec. Pub. 56, Soc. for Sedimentary Geology, Tulsa, Okla., 440 p.

James, N. P., and A. C. Kendall, 1992, Introduction to carbonate and evaporite facies models, in Walker, R. G., and N. P. James (eds.), Facies models—Response to sea level change: Geol. Assoc. Canada, p. 265–276.

Jenkyns, H. C., 1986, Pelagic environments, in Reading, H. G. (ed.), Sedimentary environments and facies, 2nd ed.: Blackwell, Oxford, p. 343–397.

Jervey, M. T., 1988, Quantitative geological modelling of siliciclastic rock sequences and their seismic expression, in Wilgus, C. K., B. S. Hastings, C. G. St. C. Kendall, H. W. Posamentier, C. A. Ross, and J. C. Van Wagoner (eds.), 1988, Sea-level changes: An integrated approach: Soc. Econ. Paleontologists and Mineralogists Spec. Pub. 42, Tulsa, Okla., p. 47–69.

Johansson, E. E., 1976, Structural studies of frictional sediments: Geograf. Annaler, v. 58A, p. 200–300.

Johnson, D. W., 1919, Shore processes and shoreline development: John Wiley & Sons, New York, 584 p.

Johnson, H. D., and C. T. Baldwin, 1996, Shallow clastic seas, in Reading, H. G. (ed.), Sedimentary environments: Processes, facies and stratigraphy, 3rd ed.: Blackwell Science Ltd., Oxford, p. 232–280.

Johnson, H. D. K., and B. K. Levell, 1995, Sedimentology of a transgressive, estuarine sand complex: The Lower Cretaceous Woburn Sands (Lower Greensand), southern England, in Plint, A. G. (ed.), Sedimentary facies analysis: A tribute to the research and teaching of Harold G. Reading, Internat. Assoc. Sedimentologists Spec. Pub. 22, Blackwell Science, Oxford, p. 17–46.

Johnsson, M. J., 1993, The system controlling the composition of clastic sediments, in Johnsson, M. J., and A. Basu (eds.), Processes controlling the composition of clastic sediments: Geol. Soc. America Spec. Paper 284, p. 1–19.

Johnsson, M. J., and A. Basu (eds.), 1993, Processes controlling the composition of clastic sediments: Geol. Soc. America Spec. Paper 284, 342 p.

Johnsson, M. J., R. F. Stallard, and R. H. Meade, 1988, First-cycle quartz arenites in the Orinoco River basin, Venezuela and Colombia: Jour. Geology, v. 96, p. 263–277.

Jones, B., and A. Desrochers, 1992, Shallow platform carbonates, in Walker, R. G., and N. P. James (eds.), Facies models: Response to sea level change, Geol. Assoc. Canada, p. 277–301.

Jones, D. L., and B. Murchey, 1986, Geologic significance of Paleozoic and Mesozoic radiolarian chert: Ann. Rev. Earth and Planetary Science Letters, v. 14, p. 455–492.

Jones, K. P. N., I. N. McCave, and P. D. Patel, 1988, A computer-interfaced sedigraph for modal size analysis of fine-grained sediment: Sedimentology, v. 35, p. 163–172.

Jones, S. J., and L. E. Frostick (eds.), 2002, Sediment flux to basins: Causes, controls and consequences: Geological Society Special Publication 191, The Geological Society London, 284 p.

Jones, T. A., D. E. Hamilton, and C. R. Johnson, 1986, Contouring geological surfaces with the computer: Van Nostrand Reinhold Company, New York, 314 p.

Jopling, A. V., and R. G. Walker, 1968, Morphology and origin of ripple-drift cross-lamination with examples from the Pleistocene of Massachusetts: Jour. Sed. Petrology, v. 38, p. 971–984.

Jordan, C. F., G. E. Freyer, and E. H. Hemmen, 1971, Size analysis of silt and clay by hydrophotometer: Jour. Sed. Petrology, v. 41, p. 489–496.

Jordan, T. E., 1995, Retroarc foreland and related basins, *in* Busby, C. J., and R. V. Ingersoll (eds.), Tectonics of sedimentary basins: Blackwell Science, p. 331–362.

Kadko, D., J. Baross, and J. Alt, 1995, The magnitude and global implications of hydrothermal flux, *in* Humphris, S. E., R. A. Zierenberg, L. S. Mullineaux, and R. E. Thompson, (eds.), Seafloor hydrothermal systems, Geophysical Monograph 91, American Geophysical Union, Washington, D.C., p. 446–466.

Karl, D. M., G. M. McMurtry, A. Malahoff, and M. O. Garcia, 1988, Loihi Seamount, Hawaii: A mid-plate volcano with a distinctive hydrothermal system: Nature, v. 335, p. 532–535.

Kastner, M., and J. M. Gieskes, 1983, Opal-A to opal-CT transformation: A kinetic study, *in* Iijima, A., J. R. Hein, and R. Siever (eds.), Siliceous deposits in the Pacific region: Developments in Sedimentology 36, Elsevier, Amsterdam, p. 211–228.

Katz, B. J. (ed.), 1990, Lacustrine basin exploration: Am. Assoc. Petroleum Geologists Mem. 50, 340 p.

Kauffman, E. G., 1988, Concepts and methods of high-resolution event stratigraphy: Ann. Rev. Earth and Planetary Sciences, v. 16, p. 605–654.

Kemp, T. S., 1999, Fossils and evolution: Oxford University Press, Oxford, 284 p.

Kempema, E. W., E. Reimnitz, and P. W. Barnes, 1988, Sea ice sediment entrainment and rafting in the Arctic: Jour. Sedimentary Petrology, v. 59, p. 308–317.

Kendall, A. C., 1992, Evaporites, *in* Walker, R. G., and N. P. James (eds.), Facies models—Response to sea level change: Geol. Assoc. Canada, p. 375–409.

Kendall, A. C., and G. M. Harwood, 1996, Marine evaporites: Arid shorelines and basins, *in* Reading, H. G. (ed.), Sedimentary Environments: Processes, facies and stratigraphy, Blackwell Science Ltd., Oxford, p. 281–324.

Kennedy, S. K., and J. Mazzullo, 1991, Image analysis method of grain size measurement, *in* Syvitski, J. P. M. (ed.), Principles, methods, and applications of particle size analysis: Cambridge University Press, Cambridge, p. 76–87.

Kennett, J. P., 1982, Marine geology: Prentice Hall, Englewood Cliffs, N.J., 812 p.

Kerr, R. A., 1984, Vail's sea-level curves aren't going away: Science, v. 226, p. 677–678.

Kersey, D. G., and K. J. Hsü, 1976, Energy relations and density current flows: An experimental investigation: Sedimentology, v. 23, p. 761–790.

Khramov, A. N., 1987, Paleomagnetology: Springer-Verlag, Berlin, 308 p.

Kiessling, W., 2002, Secular variations in the Phanerozoic reef ecosystem, *in* Kiessling, W., E. Flügel, and J. Golonka (eds.), Phanerozoic reef patterns: SEPM Special Publication 72, p. 625–690.

Kiessling, W. E. Flügel, and J. Golonka, 1999, Paleoreef maps: Evaluation of a comprehensive database on Phanerozoic reefs: Am. Assoc. Petroleum Geologists Bull., v. 83, p. 1552–1587.

Kiessling, W., E. Flügel, and J. Golonka (eds.), 2002, Phanerozoic reef patterns: SEPM Special Publication 72, 775 p.

Kimberley, M. M., 1994, Debate about ironstone: Has solute supply been surficial weathering, hydrothermal convection, or exhalation of deep fluids? Terra Nova, v. 6, p. 116–132.

Kingston, D. R., C. P. Dishroon, and P. A. Williams, 1983, Global basin classification system: Am. Assoc. Petroleum Geologists Bull., v. 67, p. 2175–2193.

Kirkbride, M. P., 1995, Processes of transportation, *in* Menzies, J. (ed.), Modern glacial environments: Processes, dynamics and sediments: Butterworth-Heinemann, Oxford, p. 261–292.

Kjerfve, B., and K. E. Magill, 1989, Geographic and hydrodynamic characteristics of shallow coastal lagoons: Marine Geology, v. 88, p. 187–199.

Klein, G. D., 1998, Clastic tidalites—A partial retrospective view, *in* Alexander, C. R., R. A. Davis, and V. J. Henry (eds.), 1998, Tidalites: Processes and Products: SEPM Special Publication 61, p. 5–14.

Klein, G. deV., 1963, Analysis and review of sandstone classifications in the North American geological literature, 1940–1960: Geol. Soc. America Bull., v. 74, p. 555–576.

Klein, G. deV., 1987, Current aspects of basin analysis: Sedimentary Geology, v. 50, p. 95–118.

Klein, G. deV., 1991, Basin sedimentology and stratigraphy—The basin fill, *in* Force, E. R., J. J. Eidel, and J. B. Maynard (eds.), Sedimentary and diagenetic mineral deposits: A basin analysis approach to exploration: Reviews in Economic Geology, v. 5, p. 51–89.

Klein, G. deV., 1995, Intracratonic basins, *in* Busby, C. J., and R. V. Ingersoll (eds.), Tectonics of sedimentary basins: Blackwell Science, p. 459–478.

Kleinspehn, K. L., and C. Paola (eds.), 1988, New perspectives in basin analysis: Springer-Verlag, New York, 453 p.

Kleinspehn, K. L., R. J. Steel, E. Johannessen, and A. Netland, 1984, Conglomeratic fan-delta sequences, Late Carboniferous–Early Permian, Western Spitsbergen, *in* Koster, E. H., and R. J. Steel (eds.), Sedimentology of gravels and conglomerates: Canadian Soc. Petroleum Geologists Mem. 10, p. 279–294.

Knauth, L. P., 1992, Origin and diagenesis of cherts: An isotopic perspective, *in* Clauer, N., and S. Chaudhuri (eds.), Isotopic signatures and sedimentary records: Springer-Verlag, Berlin, p. 123–152.

Knauth, L. P., 1994, Petrogenesis of chert, *in* Heaney, P. J., C. T. Prewitt, and G. V. Gibbs (eds.), Silica: Physical behavior, geochemistry and materials applications: Mineralogical Society of America, Reviews in Mineralogy, v. 29, p. 233–258.

Kneller, B. C., and M. J. Branney, 1995, Sustained high-density turbidity currents and the deposition of thick massive sands: Sedimentology, v. 42, p. 607–616.

Kocurek, G. A., 1996, Desert aeolian systems, *in* Reading, H. G. (ed.), Sedimentary environments: Processes, facies and stratigraphy, 3rd ed.: Blackwell Science, Oxford, p. 125–153.

Kocurek, G., 1999, The aeolian rock record, *in* Goudie, A. S., I. Livingston, and S. Stokes (eds.), John Wiley & Sons, p. 239–259.

Kocurek, G., 2003, Limits on extreme eolian systems; Sahara of Mauitania and Jurassic Navajo Sandstone examples, *in* Chan, M. A., and A. W. Archer (eds.), Extreme depositional environments; mega end members in geologic time: Geol. Soc. Amer. Special Paper 370, p. 43–52.

Kocurek, G., and K. G. Havholm, 1993, Eolian sequence stratigraphy—A conceptual framework, *in* Weimer, P., and H. W. Posamentier (eds.), Siliciclastic sequence stratigraphy: Recent developments and applications: Am. Assoc. Petroleum Geologists Mem. 58, p. 393–400.

Kocurek, G., and J. Nielson, 1986, Conditions favorable for the formation of warm-climate aeolian sand sheets: Sedimentology, v. 33, p. 795–816.

Kohout, F. A., H. R. Henry, and J. E. Banks, 1977, Hydrogeology related to geothermal conditions of the Floridan Plateau, *in* Smith, K. L., and G. M. Griffin (eds.), The geothermal nature of the Floridan Plateau: Florida Dept. Nat. Resources Bur. Geology Spec. Pub. 21, p. 1–34.

Kolla, V., and F. Coumes, 1985, Indus Fan, Indian Ocean, *in* Bouma, A. H., W. R. Normark, and N. E. Barnes (eds.), Submarine fans and related turbidite systems: Springer-Verlag, New York, p. 129–136.

Kolodny, Y., 1980, The origin of phosphorite deposits in the light of occurrences of Recent sea-floor phosphorites, *in* Bentor, Y. K. (ed.), Marine phosphorites: Soc. Econ. Paleontologists and Mineralogists Spec. Pub. 29, p. 249.

Komar, P. D., 1976, Beach processes and sedimentation: Prentice Hall, Englewood Cliffs, N.J., 429 p.

Komar, P. D., 1998, Beach processes and sedimentation, 2nd ed.: Prentice Hall, Upper Saddle River, N.J., 544 p.

Komar, P. D., R. H. Neudeck, and L. D. Kulm, 1972, Origin and significance of deep-water oscillatory ripple marks on the Oregon continental shelf, *in* Swift, D. J. P., D. B. Duane, and O. H. Pilkey (eds.), Shelf sediment transport processes and pattern: Dowden, Hutchinson, and Ross, Stroudsburg, p. 601–619.

Kottlowski, F. E., 1965, Measuring stratigraphic sections: Holt, Rinehart and Winston, New York, 253 p.

Krauskopf, K. B., 1959, The geochemistry of silica in sedimentary environments, *in* Ireland, H. A. (ed.), Silica in sediments: Soc. Econ. Paleontologists and Mineralogists Spec. Pub. 7, p. 4–19.

Krauskopf, K. B., 1979, Introduction to geochemistry, 2nd ed.: McGraw-Hill, New York, 617 p.

Krinsley, D., 1962, Application of electron microscopy to geology: New York Acad. Sci. Trans., v. 25, p. 3–22.

Krinsley, D., and J. Doornkamp, 1973, Atlas of quartz sand surface textures: Cambridge University Press, Cambridge, 91 p.

Krinsley, D. H., and P. Trusty, 1986, Sand grain surface textures, *in* Sieveking, G. De C., and M. B. Hart (eds.), The scientific study of flint and chert: Cambridge Univ. Press, Cambridge, p. 201–207.

Krinsley, D. H., K. Pye, S. Boggs, Jr., and N. K. Tovey, 1998, Backscattered scanning electron microscopy and image analysis of sediments and sedimentary rocks: Cambridge University Press, Cambridge, 193 p.

Krumbein, W. C., 1934, Size frequency distribution of sediments: Jour. Sed. Petrology, v. 4, p. 65–77.

Krumbein, W. C., 1941, Measurement and geological significance of shape and roundness of sedimentary particles: Jour. Sed. Petrology, v. 11, p. 64–72.

Krumbein, W. C., and F. J. Pettijohn, 1938, Manual of sedimentary petrography: Appleton-Century Crofts, New York, 549 p.

Krumbein, W. C., and L. L. Sloss, 1963, Stratigraphy and sedimentation: W. H. Freeman, San Francisco, 660 p.

Kuenen, Ph. H., 1958, Experiments in geology: Geol. Soc. Glasgow Trans., v. 23, p. 1–28.

Kuenen, Ph. H., 1959, Experimental abrasion, part 3: Fluviatile action on sand: Am. Jour. Sci., v. 257, p. 172–190.

Kuenen, Ph. H., 1960, Experimental abrasion, part 4: Eolian action: Jour. Geology, v. 68, p. 427–449.

Kwon, Y-I., and S. Boggs, Jr., 2002, Provenance interpretation of Tertiary sandstones from the Cheju Basin (NE East China Sea): A comparison of conventional petrographic and scanning cathodoluminescence techniques: Sedimentary Geology, v. 152, p. 29–43.

Kyser, K., 2000, Fluids and basin evolution: Mineralogical Association of Canada, Short Course Series, v. 28, 262 p.

LaBerge, G. L., E. I. Robbins, and T.-M. Han, 1987, A model for the biological precipitation of Precambrian iron-formation—A: Geologic evidence, *in* Appel, P. W. U., and G. L. LaBerge (eds.), Precambrian iron-formations: Theophrastus Pub., S. A., Athens, Greece, p. 69–96.

Lancaster, N., 1999, Geomorphology of desert sand seas, *in* Goudie, A. S., I. Livingstone, and S. Stokes (eds.), Aeolian environments, sediments and landforms: John Wiley & Sons, p. 49–69.

Lancaster, N., and J. T. Teller, 1988, Interdune deposits of the Namib sand sea: Sedimentary Geology, v. 55, p. 91–107.

Land, L. S., 1991, Dolomitization of the Hope Gate Formation (N. Jamaica) by seawater: Reassessment of mixing-zone model, *in* Taylor, H. P., J. R. O'Neil, and I. R. Kaplan (eds.), Stable isotope geochemistry: A tribute to Samuel Epstein: Geochem. Soc. Spec. Pub. 3, p. 121–133.

Lay, T., and T. C. Wallace, 1995, Modern global seismology: Academic Press, San Diego, 512 p.

Leckie, D. A., and R. G. Walker, 1982, Storm- and tide-dominated shorelines in Cretaceous Moosebar-Lower Gates inverval—Outcrop equivalents of deep basin gas trap in western Canada: Amer. Assoc. Petroleum Geologists Bull., v. 66, p. 138–157.

Ledesma-Vázquez, J., R. W. Berry, M. E. Johnson, and S. Gutiérrez-Sanchez, 1997, El Mono chert: A shallow-water chert from the Pliocene Infierno Formation, Baja California Sur, Mexico, *in* Johnson, M. E., and J. Ledesma-Vázquez (eds.), Pliocene carbonates and related facies flanking the Gulf of California, Baja

California, Mexico: Geol. Soc. America Spec. Paper 318, p. 73–81.

Leeder, M. R., 1993, Tectonic controls upon drainage basin development, river channel migration and alluvial architecture: Implications for hydrocarbon reservoir development and characterization: Special publication of the Geological Society of London, v. 73, p. 7–22.

Leeder, M. R., 1995, Continental rifts and proto-oceanic rift troughs, in Busby, C. J., and R. V. Ingersoll (eds.), Tectonics of sedimentary basins: Blackwell Science, p. 119–148.

Leeder, M., 1999, Sedimentology and sedimentary basins: Blackwell Science Ltd., Oxford, 592 p.

Lees, A., and A. T. Buller, 1972, Possible influences of salinity and temperature on modern shelf carbonate sedimentation: Marine Geology, v. 13, p. 1767–1773.

Leggett, J. K. (ed.), 1982, Trench-forearch geology: Sedimentation and tectonics on modern and ancient active plate margins: Geol. Soc. London Spec. Pub. 10, Blackwell, Oxford, 576 p.

Leighton, M. W., D. R. Kolata, D. F. Oltz, and J. J. Eidel (eds.), 1990, Interior cratonic basins: AAPG Memoir 51, Am. Assoc. Petroleum Geologists, Tulsa, Okla., 819 p.

Lenz, A. C., L. E. Edwards, and B. R. Pratt, 2001, North American Commission on Stratigraphic Nomenclature Note 64—Application for revision of articles 48–54, biostratigraphic units, of the North American Stratigraphic Code: Amer. Assoc. Petroleum Geologists Bull., v. 85, p. 372–375.

Lepp, H., 1987, Chemistry and origin of Precambrian iron formations, in Appel, P. W. U., and G. L. LaBerge, 1987, Precambrian iron-formations: Theophrastus Pub., S. A., Athens, Greece, p. 3–30.

Lepp, H., and S. S. Goldich, 1964, Origin of Pecambrian iron formation: Econ. Geology, v. 58, p. 1025–1061.

Lewan, M. D., 1978, Laboratory classification of very fine-grained sedimentary rocks: Geology, v. 6, p. 745–748.

Lewin, J. C., 1961, The dissolution of silica from diatom walls: Geochimica et Cosmochimica Acta, v. 21, p. 182–198.

Light, M. P. R., M. P. Maslanyj, R. J. Greenwood, and N. L. Banks, 1993, Seismic sequence stratigraphy and tectonics offshore Nambia, in Williams, G. D., and A. Dobb (eds.), Tectonics and seismic sequence stratigraphy, The Geological Society, Bath, U.K. p. 163–191.

Lindholm, R., 1987, A practical approach to sedimentology: Allen and Unwin, London, 276 p.

Link, M. H., and R. H. Osborne, 1978, Lacustrine facies in the Pliocene Ridge Basin Group: Ridge Basin, California, in Matter, A., and M. E. Tucker, eds., Modern and ancient lake sediments: International Association Sedimentologists, Special Publication 2, Blackwell, Oxford, p. 169–187.

Liu, J. T., R. T. Hsu, J-S. Huang, and S-Y. Chao, 2003, Sediment dispersal pattern off an eroding delta on the west coast of Taiwan, in Sidi, F. H., D. Nummedal, P. Imbert, H. Darman, and H. W. Posamentier (eds.), Tropic deltas of southeast Asia—Sedimentology, stratigraphy, and petroleum geology: SEPM Spec. Publ. 76, Society for Sedimentary Geology, Tulsa, Okla., p. 45–70.

Livingston, D. A., 1963, Data of geochemistry, Chapter G., Chemical composition of rivers and lakes: U.S. Geol. Survey Prof. Paper 440-G, 64 p.

Logan, B. W., R. Rezak, and R. N. Ginsburg, 1964, Classification and environmental significance of algal stromatolites: Jour. Geology, v. 72, p. 68–83.

Lomando, A. J., and P. M. Harris (eds.), 1991, Mixed carbonate-siliciclastic sequences: Soc. Econ. Paleontologists and Mineralogists Core Workshop 15, 568 p.

Longman, M. W., 1981, A process approach to recognizing facies of reef complexes, in Toomey, D. F. (ed.), European fossil reef models: Soc. Econ. Paleontologists and Mineralogists Spec. Pub. 30, p. 9–40.

López-Gómez, J., and A. Arche, 1993, Architecture of the Cañizar fluvial sheet sandstones, early Triassic, Iberian ranges, eastern Spain, in Marzo and Puigdefábregas (eds.), Alluvial Sedimentation: International Association of Sedimentologists Special Publ. 17: Blackwell Scientific Publ., p. 363–381.

Lowe, D. R., 1976, Subaqueous liquefied and fluidized sediment flows and their deposits: Sedimentology, v. 23, p. 285–308.

Lowe, D. R., 1982, Sedimentary gravity flows: II. Depositional models with special reference to the deposits of high-density turbidity currents: Jour. Sed. Petrology, v. 52, p. 279–297.

Lowe, D. R., and R. D. LoPiccolo, 1974, The characteristics and origin of dish and pillar structures: Jour. Sed. Petrology, v. 44, p. 484–501.

Lumsden, D. N., and R. V. Lloyd, 1997, Three dolomites: Jour. Sed. Research, v. 67, p. 391–396.

Lundegard, P. D., and N. D. Samuels, 1980, Field classification of fine-grained rocks: Jour. Sed. Petrology, v. 50, p. 781–786.

MacFadden, B. J., and R. M. Hunt, Jr., 1998, Magnetic polarity stratigraphy and correlation of the Arikaree Group, Arikareean (late Oligocene–early Miocene) of northwestern Nebraska, in Terry, D. O., H. E. LaGarry, and R. M. Hunt, Jr. (eds.), Depositional environments, lithostratigraphy and biostratigraphy of the White River and Arikaree Groups (late Eocene to early Miocene, North America): Geol. Soc. America Spec. Paper 325, p. 143–165.

Machel, G.-G., and E. W. Mountjoy, 1986, Chemistry and environments of dolomitization—A reappraisal: Earth Science Rev., v. 23, p. 175–222.

Macintyre, I. G., and R. P. Reid, 1992, A comment on the origin of aragonite needle mud: A picture is worth a thousand words: Jour. Sed. Petrology, v. 62, p. 1095–1097.

Macintyre, I. G., and R. P. Reid, 1995, Crystal alteration in a living calcareous alga (Halimeda): Implications for studies in skeletal diagenesis: Jour. Sedimentary Petrology, v. A65, p. 143–153.

Mackenzie, F. T., and R. Gees, 1971, Quartz synthesis at Earth-surface conditions: Science, v. 172, p. 533–535.

Macqueen, R. W., and D. A. Leckie, 1992, Foreland basins and fold belds: American Association Petroleum Geologists, Mem. 55, Tulsa, Okla., 460 p.

Magaritz, M., 1991, Carbon isotopes, time boundaries and evolution: Terra Nova, v. 3, p. 251–256.

Maguregui, J., and N. Tyler, 1991, Evolution of middle Eocene tide-dominated deltaic sandstones, Lagunillas Field, Maracaibo Basin, western Venezuela, *in* Miall, A. D., and N. Tyler (eds.), Three-dimensional facies architecture of terrigenous clastic sediments and its implications for hydrocarbon discovery and recovery: Concepts in sedimentology and paleontology 3, Soc. for Sedimetary Geology, Tulsa, Okla., p. 233–244.

Mahaney, W. C. (ed.), 1984, Quaternary dating methods: Elsevier, New York, 431 p.

Maizels, J., 1995, Sediments and landforms of modern proglacial terrestrial environments, *in* Menzies, J. (ed.), Modern glacial environments: Processes, dynamics and sediments: Butterworth-Heinemann, Oxford, p. 365–417.

Maliva, R. G., and R. Siever, 1988a, Pre-Cenozoic nodular cherts: Evidence for opal-CT precursors and direct quartz replacement: Am. Jour. Science, v. 288, p. 798–809.

Maliva, R. G., and R. Siever, 1988b, Diagenetic replacement controlled by force of crystallization: Geology, v. 16, p. 688–691.

Maliva, R. G., and R. Siever, 1989, Chertification histories of some late Mesozoic and middle Paleozoic platform carbonates: Sedimentology, v. 36, p. 907–926.

Manfrino, C., and R. N. Ginsburg, 2001, Pliocene to Pleistocene depositional history of the upper platform margin, *in* Ginsburg, R. N. (ed.), Subsurface geology of a prograding carbonate platform margin, Great Bahama Bank: Results of the Bahamas Drilling Project: SEPM Special Publication 70, p. 17–39.

Manley, P. L., and R. D. Flood, 1988, Cyclic sediment deposition within Amazon deep-sea fan: Am. Assoc. Petroleum Geologists Bull., v. 72, p. 912–925.

Mann, C. J., 1981, Stratigraphic analysis: Decades of revolution (1970–1979) and refinements (1980–1989), *in* Merriam, D. F. (ed.), Computer applications in earth sciences—An update of the 70s: Plenum, New York, p. 211–242.

Mann, K. O., and H. R. Lane (eds.), 1995, Graphic correlation: Soc. for Sedimetary Geology Spec. Pub. 53, 259 p.

Manspeizer, W., (ed.), 1988, Triassic-Jurassic rifting, 2 vols: Elsevier, New York, 998 p.

Markevich, V. P., 1960, The concept of facies: Internat. Geol. Rev., v. 2, p. 376–379, 498–507, 582–604.

Marsaglia, K. M., 1995, Interarc and backarc basins, *in* Busby, C. J., and R. V. Ingersoll (eds.), Tectonics of sedimentary basins: Blackwell Science, p. 299–329.

Marsaglia, K. M., and R. V. Ingersoll, 1992, Compositional trends in arc-related, deep-marine sand and sandstone: A reassessment of magmatic-arc provenance: Geological Society of America Bull., v. 104, p. 1637–1649.

Marsaglia, K. M., S. Boggs, Jr., P. Clift, A. Seyedolali, and R. Smith, 1995, Sedimentation in Western Pacific backarc basins: New Insights from Recent ODP drilling, *in* Taylor, B., and J. Natland (eds.), Active margins and marginal basins of the Western Pacific: Am. Geophy. Union Geophysical Monograph 88, p. 291–314.

Marshall, J. R. (ed.), 1987, Clastic particles: Scanning electron microscopy and shape analysis of sedimentary and volcanic clasts: Van Nostrand Reinhold Co., New York, 346 p.

Martin Willison, J. H., J. Hall, S. E. Gass, E. L. R. Kenchington, M. Butler, and P. Doherty (eds.), 2001, Proceedings of the First International Symposium on Deep-Sea Corals: Ecology Action Centre and Nova Scotia Museum, Halifax, Nova Scotia, 231 p.

Martin, L., Suguio, K., J.-M. Flexor, J. M. L. Dominguez, and A. C. D. S. P. Bittencourt, 1987, Quaternary evolution of the central part of the Brazilian coast: The role of relative sea-level variation and shoreline drift, *in* Quaternary coastal geology of West Africa and South America, UNESCO reports in Marine Science, 43, p. 97–145.

Martini, I. P., M.E. Brookfield, and S. Sadura, 2001, Principles of glacial geomorphology and geology: Prentice Hall, Upper Saddle River, N.J., 379 p.

Martinsen, O. J., 1990, Fluvial, inertia-dominated deltaic deposition in the Namurian (Carboniferous) of northern England: Sedimentology, v. 37, p. 1099–1113.

Marzo, M., and C. Puigdefábregas (eds.), 1993, Alluvial Sedimentation: Special Publication 17, International Association of Sedimentologists: Blackwell Science, Oxford, 586 p.

Marzolf, J. E., 1988, Controls on late Paleozoic and early Mesozoic eolian deposition of the western United States: Sedimentary Geology, v. 56, p. 167–191.

Mason, B., 1966, Principles of geochemistry: John Wiley & Sons, New York, 329 p.

McArthur, J. M., 1994, Recent trends in strontium isotope stratigraphy: Terra Nova, v. 6, p. 331–358.

McArthur, J. M., 1998, Strontium isotope stratigraphy, *in* Doyle, P., and M. R. Bennett (eds.), Unlocking the stratigraphical record: Advances in modern stratigraphy: John Wiley & Sons, p. 221–241.

McBride, E. F., 1963, A classification of common sandstones: Jour. Sed. Petrology, v. 33, p. 664–669.

McBride, E. F., R. G. Shepard, and R. A. Crawley, 1975, Origin of parallel, near-horizontal laminae by migration of bed forms in a small flume: Jour. Sed. Petrology, v. 45, p. 132–139.

McCabe, P. J., 1984, Depositional environments of coal and coal-bearing strata, *in* Rahmani, R. A., and R. M. Flores (eds.), Sedimentology of coal and coal-bearing sequences: International Assoc. Sedimentologists Spec. Pub. 7, p. 13–42.

McCave, I. N., R. J. Bryant, H. F. Cook, and C. A. Coughanowr, 1986, Evaluation of a laser-diffraction-size analyzer for use with natural sediments: Jour. Sed. Petrology, v. 56, p. 561–564.

McClellan, G. H., and S. J. Van Kauwenbergh, 1990, Mineralogy of sedimentary apatites, *in* Northolt, A. J. G., and I. Jarvis (eds), Phosphorite research and development: The Geological Society Special Publication 52, Bath, U.K. p. 23–31.

McCubbin, D. J., 1982, Barrier-island and strand-plain facies, *in* Scholle, P. A., and D. Spearing (eds.), Sandstone depositional environments: Am. Assoc. Petroleum Geologists Mem. 31, p. 247–279.

McDonald, K. C., K. B. F. N. Spiess, and R. D. Ballard, 1980, Hydrothermal flux of the "black smoker" vents on the East Pacific Rise: Earth and Planetary Sci. Letters, v. 48, p. 1–7.

McDougall, I., 1979, The present status of the geomagnetic polarity time scale, in McElhinny, M. W. (ed.), The Earth: Its origin, structure, and evolution: Academic Press, London, p. 543–566.

McDougall, I., and T. M. Harrison, 1988, Geochronology and thermochronology by the $^{40}Ar/^{39}Ar$ method: Oxford University Press, New York, 212 p.

McElhinny, M. W., and P. L. McFadden, 2000, Paleomagnetism: Continents and oceans: Academic Press, San Diego, 386 p.

McEwan, I. K., and B. B. Willets, 1993, Sand transport by wind: A review of the current conceptual model, in Pye, K. (ed.), The dynamics and environmental context of aeolian sedimentary systems: Geological Society Spec. Publ. 72, p. 7–16.

McKee, E. D., 1965, experiments on ripple lamination, in Middleton, G. V. (ed.), Primary sedimentary structures and their hydrodynamic interpretation: Soc. Econ. Paleontologists and Mineralogists Spec. Pub. 12, p. 66–83.

McKee, E. D., 1982, Sedimentary structures in dunes of the Namib Desert, Southwest Africa: Geol. Soc. America Spec. Paper 188, 64 p.

McKee, E. D., and G. W. Weir, 1953, Terminology for stratification and cross-stratification in sedimentary rocks: Geol. Soc. America Bull., v. 64, p. 381–390.

McKee, E. D., J. R. Douglass, and S. Rittenhouse, 1971, Deformation of lee-side laminae in eolian dunes: Geol. Soc. America Bull., v. 82, p. 359–378.

McKelvey, V. E., J. S. Williams, R. P. Sheldon, E. R. Cressman, and T. M. Channey, 1959, The Phosphoria, Park City and Shedhorn formations in the Western Phosphate Field: U.S. Geol. Survey Prof. Paper 313-A.

Mellere, D., and R. J. Steel, 1996, Tidal sedimentation in Inner Hebrides half grabens, Scotland: the Mid-Jurassic Bearreraig Sandstone Formation, in De Batist, M., and P. Jacobs (eds.), Geology of siliciclastic shelf seas: Geological Society Special Publication 117, p. 49–79.

Melnik, Y. P., 1982, Precambrian banded iron-formations: Developments in Precambrian Geology 5, Elsevier, Amsterdam, 310 p.

Melvin, J. L. (ed.), 1991, Evaporites, petroleum and mineral resources: Elsevier, Amsterdam, 555 p.

Menning, M., G. Katzung, and Lutzner, H., 1988, Magnetostratigraphic investigations in the Rotliegendes (300–252 Ma) of central Europe: Z. geol. Wiss., v. 16, p. 81–102.

Menzies, J., 1995, The dynamics of ice flow, in Menzies, J. (ed.), Modern glacial environments: Processes, dynamics and sediments: Butterworth-Heinemann, Oxford, p.101–196.

Mero, J. L., 1965, The mineral resources of the sea: Elsevier, New York, 312 p.

Merriam, D. F. (ed.), 1981, Computer applications in the earth sciences—An update of the 70s: Plenum, New York, 385 p.

Merrill, R. T., M. W. McElhinny, and P. L. McFadden, 1996, The magnetic field of the Earth: Academic Press, San Diego, 531 p.

Meyer, R. F. (ed.), 1987, Exploration for heavy crude oil and natural bitumens: AAPG Studies in Geology 25, Amer. Assoc. Petroleum Geologists, Tulsa, Okla., 731 p.

Meyer, R. F., and W. De Witt, Jr., 1990, Definition and world resources of natural bitumens: U.S. Geol. Survey Bull. 1944, 14 p.

Miall, A. D., 1984, Deltas, in Walker, R. G. (ed.), Facies models, 2nd ed: Geoscience Canada Reprint Ser. 1, p. 105–118.

Miall, A. D., 1986, Eustatic sea-level changes interpreted from seismic stratigraphy: A critique of the methodology with particular reference to the North Sea Jurassic record: American Association Petroleum Geol. Bull., v. 70, p. 131–137.

Miall, A. D., 1990, Principles of sedimentary basin analysis, 2nd ed.: Springer-Verlag, New York, 668 p.

Miall, A. D., 1991, Exxon global cycle chart: An event for every occasion? Geology, v. 20, p. 797–790.

Miall, A. D., 1992, Alluvial deposits, in Walker, R. G., and N. P. James (eds.), Facies models: Response to sea level changes, Geol. Assoc. Canada, p. 119–142.

Miall, A. D., 1994, Sequence stratigraphy and chronostratigaphy: Problems of definition and precision in correlation, and their implication for global eustasy: Geoscience Canada, v. 21, p. 1–26.

Miall, A. D., 1996, The geology of fluvial deposits: Springer-Verlag, Berlin, 582 p.

Miall, A. D., 1997, The geology of stratigraphic sequences: Springer-Verlag, Berlin, 433 p.

Miall, A. D., 2000, Principles of sedimentary basin analysis, 3rd edition: Springer-Verlag, Berlin, 616 p.

Michels, K. H., H. R. Kudrass, C. Hübscher, A. Suchow, and M. Wiedicke, 1998, The submarine delta of the Ganges-Brahmaputra: Cyclone-dominated sedimentation patterns: Marine Geology, v. 149, p. 133–154.

Middleton, G. V., 1973, Johannes Walther's Law of the Correlation of Facies: Geol. Soc. America Bull., v. 84, p. 979–988.

Middleton, G. V., 1991, Mechanics of sediment gravity flows (unpublished manuscript).

Middleton, G. V., 1993, Sediment deposition from turbidity currents: Ann. Rev. Earth and Planetary Science, v. 21, p. 89–114.

Middleton, G. V., and M. A. Hampton, 1976, Subaqueous sediment transport and deposition by sediment gravity flows, in Stanley, D. J., and D. J. P. Swift (eds.), Marine sediment transport and environmental management: John Wiley & Sons, New York, p. 197–218.

Middleton, G. V., and J. B. Southard, 1984, Mechanics of sediment transport: SEPM Short Course 3, 2nd ed., SEPM, Tulsa, Okla., variously paginated.

Middleton, G. V., and P. R. Wilcock, 1994, Mechanics in the earth and environmental sciences: Cambridge University Press, Cambridge, 459 p.

Miller, A. I., 2000, Conversations about Phanerozoic global diversity, in Erwin, D. H., and S. L. Wing (eds.), Deep time: Paleobiology's perspective: The Paleontolgtical Society, Lawrence, Kans., p. 53–73.

Miller, F. X., 1977, Graphic correlation method in biostratigraphy, *in* Kauffman, E. G., and E. Hazel (eds.), Concepts and methods in biostratigraphy: Dowden, Hutchinson and Ross, Stroudsburg, Pa., p. 165–186.

Mitchell, A. H. C., and H. G. Reading, 1986, Sedimentation and tectonics, *in* Reading, H. G. (ed.), Sedimentary environments and facies, 2nd ed., Blackwell, p. 471–519.

Mitchum, R. M., Jr., and P. R. Vail, 1977, Seismic stratigraphic interpretation procedures, *in* Payton, C. E. (ed.), Seismic stratigraphy—Applications to hydrocarbon exploration: Am. Assoc. Petroleum Geologists Mem. 26, p. 135–143.

Mitchum, R. M., Jr., P. R. Vail, and J. B. Sangree, 1977, Seismic stratigraphy and global change of sea level, Part 6: Stratigraphic interpretation of seismic reflection patterns in depositional sequences, *in* Payton, C. E. (ed.), Seismic stratigraphy—Applications to hydrocarbon exploration: Am. Assoc. Petroleum Geologists Mem. 26, p. 117–133.

Mitchum, R. M., Jr., P. R. Vail, and S. Thompson, III, 1977, Seismic stratigraphy and global change of sea level. Part 2: The depositional sequence as a basic unit for stratigraphic analysis, *in* Payton, C. E. (ed.), Seismic stratigraphy—Applications to hydrocarbon exploration: Am. Assoc. Petroleum Geologists Mem. 26, p. 53–62.

Monty, C. L. V., D. W. J. Bosence, P. H. Bridges, and B. R. Pratt (eds.), 1995, Carbonate mud-mounds—Their origin and evolution: International Association Sedimentologists, Spec. Pub. 32, Blackwell Science Ltd., Oxford, 537 p.

Moore, D. M., 1978, A sample of the Purington Shale prepared as a geochemical standard: Jour. Sed. Petrology, v. 48, p. 995–998.

Moore, R. C., 1949, Meaning of facies: Geol. Soc. America Mem. 39, p. 1–34.

Morey, G. W., R. O. Flournier, and J. J. Rowe, 1962, The solubility of quartz in water in the temperature interval from 25°C to 300°C: Geochim. et Cosmochim. Acta, v. 26, p. 1029–1043.

Morey, G. W., R. O. Flournier, and J. J. Rowe, 1964, The solubility of amorphous silica at 25°C: Jour. Geophys. Research, v. 69, p. 1995–2002.

Morris, R. C., 1993, Genetic modeling for banded iron formation of the Hamersley Group, Pibara Craton, Western Australia: Precambrian Research, v. 60, p. 243–286.

Morrow, D. W., and H. J. Abercrombie, 1994, Rates of dolomitization: The influence of dissolved sulphate, *in* Purser, B., M. Tucker, and D. Zenger (eds.), Dolomites: A volume in honor of Dolomieu: Internat. Assoc. Sedimentologists, Spec. Pub. 21, Blackwell Scientific Pub, Oxford, p. 377–386.

Morse, J. W., and F. T. Mackenzie, 1990, Geochemistry of sedimentary carbonates: Elsevier, Amsterdam, 707 p.

Morse, J. W., J. S. Hanor, and S. He, 1997, The role of mixing and migration of basinal waters in carbonate mineral mass transport, *in* Montañez, I. P., J. M. Gregg, and K. L. Shelton (eds.), Basin-wide diagenetic patterns: Integrated petrologic, geochemical, and hydrologic considerations: SEPM Spec. Pub. 57, Soc. for Sed. Geology, Tulsa, Okla., p. 41–50.

Morton, A. C., and C. R. Hallsworth, 1999, Processes controlling the composition of heavy mineral assemblages in sandstones: Sedimentary Geology, v. 124, p. 3–29.

Morton, A. C., S. P. Todd, and P. D. W. Haughton (eds.), 1991, Developments in sedimentary provenance studies: Geological Society Spec. Publ. 57, Geol. Soc. London, 370 p.

Morton, R. A., 1988, Nearshore responses to great storms, *in* Clifton, H. E. (ed.), Sedimentologic consequences of convulsive geologic events: Geological Society of America Special Paper 229, p. 7–22.

Mount, J., 1985, Mixed siliciclastic and carbonate sediments: A proposed first-order textural and compositional classification: Sedimentology, v. 32, p. 435–442.

Mountjoy, E. W. and J. E. Amthor, 1994, Has burial dolomitization come of age? Some answers from the Western Canada Sedimentary Basin, *in* Purser, B., M. Tucker, and D. Zenger (eds.), Dolomites: A volume in honor of Dolomieu: Internat. Assoc. Sedimentologists, Spec. Pub. 21, Blackwell Scientific Pub., Oxford, p. 203–229.

Mucci, A., and J. W. Morse, 1983, The incorporation of Mg^{2+} and Sr^{2+} into calcite overgrowths: Influence of growth rates and solution composition: Geochim. et Cosmochim. Acta, v. 47, p. 217–233.

Muerdter, D. R., J. P. Dauphin, and G. Steele, 1981, An interactive computerized system for grain size analysis of silt using electro-resistance: Jour. Sedimentary Petrology, v. 51, p. 647–650.

Mullins, H. T., and H. E. Cook, 1986, Carbonate apron models: Alternatives to the submarine fan model for paleoenvironmental analysis and hydrocarbon exploration: Sed. Geology, v. 48, p. 37–79.

Multer, H. G., 1988, Growth rate, ultrastructure and sediment contribution of *Halimeda incrassata* and *Halimeda monile*, Nonsuch and Falmouth Bays, Antigua, W.I.: Coral Reefs, v. 6, p. 179–186.

Murray, R. W., D. L. Jones, and M. R. Bucholtz ten Brink, 1992, Diagenetic formation of bedded chert: Evidence from chemistry of chert-shale couplet: Geology, v. 20, p. 271–274.

Mutti, E., 1985, Turbidite systems and their relations to depositional sequences, *in* Zuffa, G. G. (ed.), Provenance of arenites: D. Riedel, Dordrecht, p. 65–93.

Mutti, E., 1992, Turbidite sandstones: Agip, Instituto di Geologia, Università di Parma, Milan, 275 p.

Nahon, D. B., 1991, Introduction to the petrology of soils and chemical weathering: John Wiley & Sons, New York, 313 p.

Nardin, T. R., B. D. Edwards, and D. S. Gorsline, 1979, Santa Cruz Basin, California borderland: Dominance of slope processes in basin sedimentation, *in* Doyle, L. J., and O. H. Pilkey (eds.), Geology of continental slopes: Soc. Econ. Paleontologists and Mineralogists Spec. Pub. 27, p. 209–221.

Nathan, Y., 1984, The mineralogy and geochemistry of phosphorites, *in* Nriagu, J. O., and P. B. Moore (eds.), Phosphate minerals: Springer-Verlag, Berlin, p. 275–291.

Nelson, C. S., 1988, An introductory perspective on nontropical carbonates: Sed. Geology, v. 60, p. 3–12.

Nemec, W., 1990, Deltas—Remarks on terminology and classification, *in* Colella, A., and D. B. Prior (eds.), Coarse-grained deltas: International Association of Sedimentologists Special Pub. 10, Blackwell Scientific Publications, Oxford, p. 3–12.

Nemec, W., and R. J. Steel (eds.), 1988a, What is a fan delta and how do we recognize it, *in* Nemic, W., and R. J. Steel (eds.), Fan deltas: Sedimentology and tectonic settings: Blackie, Glasgow and London, p. 3–13.

Nemec, W., and R. J. Steel (eds.), 1988b, Fan deltas: Sedimentology and tectonic settings: Blackie, Glasgow and London, 444 p.

Nettleton, L. L., 1940, Geophysical prospecting for oil: McGraw-Hill, New York, 444 p.

Neumann, A. C., and L. S. Land, 1975, Lime mud deposition and calcareous algae in the Bight of Abaco, Bahamas: A budget: Jour. Sed. Petrology, v. 45, p. 763–786.

Nichols, M. N., and J. D. Boon, 1994, Sediment transport processes in coastal lagoons, *in* Kjerfve, B. (ed.), 1994, Coastal lagoon processes: Elsevier, Amsterdam, p. 157–219.

Nickling, W. G., 1994, Aeolian sediment transport and deposition, *in* Pye, K. (ed.), Sediment transport and depositional processes: Blackwell Scientific Pub., Oxford, p. 293–350.

Nilsen, T. H., 1982, Alluvial fan deposits, *in* Scholle, P. A., and D. Spearing (eds.), Sandstone depositional environments: Am. Assoc. Petroleum Geologists Mem. 31, p. 49–86.

Nilsen, T. H., and A. G. Sylvester, 1995, Strike-slip basins, *in* Busby, C. J., and R. V. Ingersoll (eds.), Tectonics of sedimentary basins: Blackwell Science, p. 425–457.

Nittrouer, A. A., and L. D. Wright, 1994, Transport of particles across continental shelves: Rev. Geophysics, v. 32, p. 85–113.

Normark, W. R., 1978, Fan valleys, channels, and depositional lobes on modern submarine fans: Characters for recognition of sandy turbidite environments: Am. Assoc. Petroleum Geologists Bull., v. 62, p. 912–931.

Normark, W. R., and D. J. W. Piper, 1991, Initiation process and flow evolution of turbidity currents: Implications for the depositional record, *in* Osborne, R. H. (ed.), From shoreline to abyss: Soc. for Sed. Geology Spec. Pub. 46, p. 207–230.

North American Commission on Stratigraphic Nomenclature, 1983, North American Stratigraphic Code: Am. Assoc. Petroleum Geologists Bull., v. 67, p. 841–875.

Notholt, A. J. G., R. P. Sheldon, and D. F. Davidson (eds.), 1989, Phosphate deposits of the world: Vol. 2 Phosphate rock resources: Cambridge University Press, Cambridge, 566 p.

NOVA, 2003, Magnetic storm: Earth's invisible shield: WGBH Boston Video, 60 minutes.

Odin, G. S. (ed.), 1982, Numerical dating in stratigraphy: John Wiley & Sons, New York, Pt. 1, p. 1–630, Pt. II, p. 631–1040.

Odin, G. S., and M. H. Dodson, 1982, Zero isotopic ages of glauconites, *in* Odin, G. S. (ed.), Numerical dating in stratigraphy: John Wiley & Sons, New York, p. 277–306.

Odin, G. S., M. Renard, and C. V. Grazzini, 1982, Geochemical events as a means of correlation, *in* Odin, G.

S. (ed.), Numerical dating in stratigraphy: John Wiley & Sons, New York, p. 37–72.

Odin, G. S., D. Curry, N. H. Gale, and W. J. Kennedy, 1982, The Phanerozoic time scale in 1981, *in* Odin, G. S. (ed.), Numerical dating in stratigraphy: John Wiley & Sons, New York, p. 957–960.

Ogg, J. G., 1995, Magnetic polarity time scale of the Phanerozoic, *in* Ahrens, T. J. (ed.), Global Earth physics—A handbook of physical constants: AGU Reference Shelf 1, Am. Geophy. Union, Washington, D.C., p. 240–270.

Okada, H., 1971, Classification of sandstones: Analysis and proposal: Jour. Geology, v. 79, p. 509–525.

Olea, R. A., 1988, Correlator—An interactive computer system for lithostratigraphic correlation of wireline logs: Kansas Geological Survey, Lawrence, Kansas, Petrophysical Series 4, 85 p.

Ollier, C., and C. Pain, 1996, Regolith, soils and landforms: John Wiley & Sons, New York, 316 p.

Open University Course Team, 1989, Ocean circulation: Open University, Milton Keynes, Pergamon Press, Oxford, 238 p.

Orton, G. J., 1988, A spectrum of Middle Ordovician fan deltas and braid plain deltas, North Wales: A consequence of varying fluvial clastic input, *in* Nemic, W., and R. J. Steel (eds.), Fan deltas: Sedimentology and tectonic setting: Blackie, Glasgow and London, p. 23–49.

Orton, G. J., and H. G. Reading, 1993, Variability of deltaic processes in terms of sediment supply, with particular emphasis on grain size: Sedimentology, v. 40, p. 475–512.

Osleger, D., and J. F. Read, 1991, Relation of eustasy to stacking patterns of meter-scale carbonate cycles, Late Cambrian, U.S.A.: Jour. Sedimentary Petrology, v. 61, p. 1225–1252.

Otto, G. H., 1938, The sedimentation unit and its use in field sampling: Jour. Geology, v. 46, p. 569–582.

Palmer, M. R., and J. M. Edmond, 1989, The strontium isotope budget of the modern ocean: Earth and Planetary Science Letters, v. 92, p. 11–26.

Pantin, H. M., 1979, Interaction between velocity and effective density in turbidity flow: Phase plane analysis with criteria for autosuspension: Marine Geology, v. 31, p. 59–99.

Parkash, B., and G. V. Middleton, 1970, Downcurrent textural changes in Ordovician turbidite graywackes: Sedimentology, v. 14, p. 259–293.

Parker, G., 1982, Conditions for the ignition of catastrophic erosive turbidity currents: Marine Geology, v. 46, p. 307–327.

Parrish, J. T., 1990, Paleooceanographic and paleoclimatic setting of the Miocene phosphogenic episode, *in* Burnett, W. C., and S. R. Riggs (eds.), Phosphate deposits of the world: Vol. 3 Neogene to modern phosphorites: Cambridge University Press, Cambridge, p. 223–240.

Parrish, R., and J. C. Roddick, 1985, Geochronology and isotope geology for the geologist and explorationist: Geol. Assoc. Canada, Cordilleran Section, Short Course 4, variously paginated.

Parson, L. M., C. L. Walker, and D. R. Dixon (eds.), 1995, Hydrothermal vents and processes: The Geol. Society, London, 411 p.

644   Bibliography

Passega, R., 1964, Grain size representation by CM patterns as a geological tool: Jour. Sed. Petrology, v. 34, p. 830–847.

Passega, R., 1977, Significance of CM diagrams of sediments deposited by suspension: Sedimentology, v. 24, p. 723–733.

Payton, C. E. (ed.), 1977, Seismic stratigraphy—Applications to hydrocarbon exploration: Am. Assoc. Petroleum Geologists Mem. 26, 516 p.

Pemberton, S. G., J. A. MacEachern, and R. W. Frey, 1992, Trace fossil facies models: Environmental and allostratigraphic significance, in Walker, R. G., and N. P. James (eds.), Facies models—Response to sea level change, Geol. Assoc. Canada, p. 47–72.

Perillo, G. M. E ., 1995a, Definitions and geomorphic classification of estuaries, in Perillo, G. M. E. (ed.), Geomorphology and sedimentology of estuaries, Developments in Sedimentology 53, Elsevier Science B.V., Amsterdam, p. 17–47.

Perillo, G. M. E. (ed.), 1995b, Geomorphology and sedimentology of estuaries, Developments in Sedimentology 53: Elsevier Science B. V., Amsterdam, 471 p.

Peterson, M. N. A., 1966, Calcite: Rates of dissolution in a vertical profile in the central Pacific: Science, v. 154, p. 1542–1544.

Peterson, M. N., and C. C. von der Borch, 1965, Chert: Modern inorganic deposition in a carbonate-precipitating locality: Science, v. 149, p. 1501–1503.

Pettijohn, F. J., 1963, Chemical composition of sandstones—excluding carbonate and volcanic sands, in Data of geochemistry, 6th ed.: U.S. Geol. Survey Prof. Paper 440S, 19 p.

Pettijohn, F. J., 1975, Sedimentary rocks, 3rd ed.: Harper & Row, New York, 628 p.

Pettijohn, F. J., P. E. Potter, and R. Siever, 1987, Sand and sandstone, 2nd ed.: Springer-Verlag, New York, 553 p.

Piazzola, J., and V. V. Cavaroc, 1991, Comparison of grain-size-distribution statistics determined by sieving and thin-section analyses: Jour. Geological Education, v. 39, p. 364–367.

Picard, M. D., 1971, Classification of fine-grained sedimentary rocks: Jour. Sed. Petrology, v. 41, p. 179–195.

Picard, M. D., and L. R. High, Jr., 1981, Physical stratigraphy of ancient lacustrine deposits, in Ethridge, F. G., and R. M. Flores (eds.), Recent and nonmarine depositional environments: Models for exploration: Soc. Econ. Paleontologists and Mineralogists Spec. Pub. 31, p. 233–259.

Pickering, K. T., R. N. Hiscott, and F. J. Hein, 1989, Deep marine environments: Clastic sedimentation and tectonics: Unwin-Hyman, London, 416 p.

Pickering, K. T., R. N. Hiscott, N. H. Kenyon, F. Ricci-Lucchi, and R. D. A. Smith (eds.), 1995, Atlas of deep water environments: Architectural style in turbidite systems: Chapman and Hall, London, 333 p.

Piper, D. J. A., 1987, Paleomagnetism and the continental crust: The Open University Press, Milton Keynes, 434 p.

Piper, D. J. W., P. Cochonat, and M. L. Morrison, 1999, The sequence of events around the epicenter of the 1929 Grand Banks earthquake: Inititation of debris flows and turbidity currents inferred from sidescan sonar: Sedimentology, v. 46, p. 79–97.

Piper, D. J. W., A. N. Shor, and J. E. H. Clark, 1988, The 1929 "Grand Banks" earthquake, slump, and turbidity current, in Clifton, H. E. (ed.), Sedimentologic consequenes of convulsive geologic events: Geol. Soc. America Spec. Paper 229, p. 77–92.

Piper, D. J . W., D. A. V. Stow, and W. R. Normark, 1984, The Laurentian Fan; Sohm Abyssal Plain: Geo-Marine Letters, v. 3, p. 141–146.

Pirrie, D., 1998, Interpreting the record: Facies analysis, in Doyle, P., and M. R. Bennett (eds.), Unlocking the stratigraphical record: Advances in modern stratigraphy: John Wiley & Sons, p. 395–420.

Pitman, W. C., III, 1978, Relationship between eustasy and stratigraphic sequences of passive margins: Geol. Soc. America Bull., v. 89, p. 1389–1403.

Plint, A. G., N. Eyles, C. H. Eyles, and R. G. Walker, 1992, Control of sea level change, in Walker, R. G., and N. P. James (eds.), Facies models—Response to sea level change: Geological Association of Canada, p. 15–25.

Plummer, P. S., and V. A. Gostin, 1981, Shrinkage cracks: Desiccation or synaeresis? Jour. Sed. Petrology, v. 51, p. 1147–1156.

Poppe, L. J., A. H. Eliason, and J. J. Fredricks, 1985, APSAS—An automated particle size analysis system: U.S. Geological Survey Circ. 963, 77 p.

Posamentier, H. W., and G. P. Allen, 1999, Siliciclastic sequence stratigraphy—Concepts and applications: SEPM Concepts in sedimentology and Paleontology, Tulsa, Okla., 210 p.

Posamentier, H. W., M. T. Jervey, and P. R. Vail, 1988, Eustatic controls on clastic deposition I - Conceptual framework, in Wilgus, C. K., B. S. Hastings, C. G. St. C. Kendall, H. W. Posamentier, C. A. Ross, and J. C. Van Wagoner (eds.), Sea-level changes: An integrated approach, Soc. Econ. Paleontologists and Mineralogists Spec Pub. 42, p. 109–124.

Potter, P. E., and F. J. Pettijohn, 1977, Paleocurrents and basin analysis, 2nd ed.: Springer-Verlag, Berlin, 413 p.

Potter, P. E., J. B. Maynard, and W. A. Pryor, 1980, Sedimentology of shale: Springer-Verlag, New York, 306 p.

Powell, R., and D. Domack, 1995, Modern glaciomarine environments, in Menzies, J. (ed.), Modern glacial environments: Processes, dynamics and sediments: Butterworth-Heinemann, Oxford, p. 445–486.

Powers, M. C., 1953, A new roundness scale for sedimentary particles: Jour. Sed. Petrology, v. 23, p. 117–119.

Pratt, B. R., and N. P. James, 1986, The St. George Group (Lower Ordovician) of western Newfoundland: Tidal flat island model for carbonate sedimentation in shallow epeiric seas: Sedimentology, v. 33, p. 323–343.

Prior, D. B., and B. D. Bornhold, 1990, The underwater development of Holocene fan deltas, in Colella, A., and D. B. Prior (eds.), Coarse-grained deltas: International Assoc. Sedimentologists Spec. Pub. 10, Blackwell Scientific Pub., p. 75–90.

Prospero, J. M., 1981, Eolian transport to the world ocean, in Emiliani, C. (ed.), The oceanic lithosphere: The sea, v. 7, John Wiley & Sons, New York, p. 801–874.

Pulham, A. J., 1989, Controls on internal structure and architecture of sandstone bodies within Upper Carboniferous fluvial-dominated deltas, County Clare, western Ireland, *in* Whateley, M. K. G., and K. T. Pickering (eds.), Deltas: Sites and traps for fossil fuels: Geol. Soc. Spec. Pub. 41, Blackwell Scientific Pub., Oxford, p. 179–203.

Purser, B., M. Tucker, and D. Zenger (eds.), 1994a, Dolomites: A volume in honor of Dolomieu: International Assocation of Sedimentologists, Special Publication 21, Blackwell Scientific Publications, Oxford, 451 p.

Purser, B., M. Tucker, and D. Zenger, 1994b, Problems, progress, and future research concerning dolomites and dolomitization, *in* Purser, B., M. Tucker, and D. Zenger (eds.), Dolomites: A volume in honor of Dolomieu: Internat. Assoc. Sedimentologists, Spec. Pub. 21, Blackwell Scientific Pub., Oxford, p. 3–20.

Pye, K., and H. Tsoar, 1990, Aeolian sand and sand dunes: Unwin Hyman, London, 396 p.

Raup, D. M., 1991, Extinction: Bad genes or bad luck? W. W. Norton and Co., New York, 210 p.

Rautman, C. A., and R. H. Dott, Jr., 1977, Dish structures formed by fluid escape in Jurassic shallow marine sandstones: Jour. Sed. Petrology, v. 47, p. 101–106.

Ravnås, R., and R. J. Steel, 1998, Architecture of marine rift-basin successions: Am. Assoc. Petroleum Geologists Bull., v. 82, p. 110–146.

Raymo, M. E., W. F. Ruddiman, J. Backman, B. M. Clement, and D. G. Martinson, 1989, Late Pliocene variation in Northern Hemisphere ice sheets and North Atlantic deep water circulation: Paleooceanography, v. 4, p. 413–446.

Read, J. F., 1982, Carbonate platforms of passive (extensional) continental margins: Types, character and evolution: Tectonophysics, v. 81, p. 195–212.

Read, J. F., 1985, Carbonate platform facies models: Am. Assoc. Petroleum Geologists Bull., v. 69, p. 1–21.

Reading, H. G., and J. D. Collinson, 1996, Clastic coasts, *in* Reading, H. G. (ed.), Sedimentary environments: Processes, facies and stratigraphy, 3rd ed.: Blackwell Science, Oxford, p. 154–231.

Reading, H. G., and B. K. Levell, 1996, Controls on the sedimentary rock record, *in* Reading, H. G. (ed.), Sedimentary environments: Processes, facies and stratigraphy: Blackwell Science, Oxford, p. 5–36.

Reading, H. G., and M. Richards, 1994, Turbidite systems in deep-water basin margins classified by grain size and feeder systems: Am. Assoc. Petroleum Geologists Bull., v. 78, p. 792–822.

Reed, W. R., R. LeFever, and G. J. Moir, 1975, Depositional environment interpreted from settling velocity (psi) distributions: Geol. Soc. America Bull., v. 86, p. 1321–1328.

Reeder, R. J. (ed.), 1983, Carbonates: Mineralogy and chemistry: Rev. in Mineralogy, v. 11, 394 p.

Reineck, H. E., and I. B. Singh, 1980, Depositional sedimentary environments, 2nd ed.: Springer-Verlag, Berlin, 549 p.

Reinhardt, J., and W. R. Sigleo (eds.), 1988, Paleosols and weathering through geologic time: Principles and applications: Geol. Soc. America Spec. Paper 216, 181 p.

Reinson, G. E., 1992, Transgressive barrier island and estuarine systems, *in* Walker, R. G., and N. P. James (eds.), Facies models, Geol. Assoc. Canada, p. 179–194.

Retallack, G. J., 1988, Field recognition of paleosols, *in* Reinhardt, J., and W. R. Sigleo (eds.), Paleosols and weathering through geologic time: Geol. Soc. Amer. Spec. Paper 216, p. 1–20.

Retallack, G. J., 1990, Soils of the past: Unwin Hyman, Boston, 520 p.

Retallack, G. J., 1992, Paleozoic paleosols, *in* Martini, I. P. and W. Chesworth (eds.), Weathering, soils & paleosols: Elsevier, Amsterdam, p. 543–564.

Retallack, G. J., 1997, A colour guide to paleosols: John Wiley & Sons, Chichester, 175 p.

Revelle, R., (ed.), 1990, Sea level change: National Research Council, Studies in Geophysics: National Academy Press, Washington, D.C., 234 p.

Ricci-Lucchi, F., 1995, Sedimentographica: A photographic atlas of sedimentary structures, 2nd ed: Columbia University Press, New York, 255 p.

Rich, J. L., 1951, Three critical environments of deposition and criteria for recognition of rocks deposited in each of them: Geol. Soc. America Bull., v. 62, p. 1–20.

Ridderinkhof, H., 1998, Sediment transport in intertidal areas, *in* Eisma, D., P. L. de Boer, G. C. Cadée, K. Dijkema, H. Ridderinkhof, and C. Philippart, Intertidal deposits: River mouths, tidal flats, and coastal lagoons: CRC Press, Boca Raton, Fla., p. 363–381.

Rider, M. H., 1986, The geological interpretation of well logs: John Wiley and Sons, New York, 175 p.

Robbin, L. L., and P. L. Blackwelder, 1992, Biochemical and ultrastructural evidence for the origin of whitings: A biologically induced calcium carbonate precipitation mechanism: Geology, v. 20, p. 464–468.

Robbins, L. L., Y. Tao, and C. A. Evans, 1997, Temporal and spatial distribution of whitings on Great Bahama Bank and a new lime mud budget: Geology, v. 25, p. 947–950.

Roberts, H. H., and J. Sydow, 2003, Late Quaternary stratigraphy and sedimentology of the offshore Mahakam Delta, East Kalimantan (Indonesia), *in* Hanson, S. F., D. Nummendal, P. Imbert, H. Darman, and H. W. Posamentier (eds.), Tropical deltas of Southeast Asia; Sedimentology, stratigraphy, and petroleum geology: Soc. for Sed. Geology, Spec. Publ. 76, p. 125–145.

Robinson, J. E., 1982, Computer applications in petroleum geology: Hutchinson Ross Pub. Co., New York, 164 p.

Rodgers, J., 1959, The meaning of correlation: Am. Jour. Sci., v. 257, p. 684–691.

Roehler, H. W., 1992, Correlation, composition, areal distribution, and thickness of Eocene stratigraphic units, greater Green River Basin, Wyoming, Utah, and Colorado: U.S. Geological Survey Professional paper 1506-E, 49 p.

Rogers, M. A., J. D. McAlary, and N. J. L. Bailey, 1974, Significance of reservior bitumens to thermal-maturation studies, western Canada basin: Amer. Assoc. Petroleum Geol. Bull., v. 58, p. 1806–1824.

Ronov, A. B., and A. A. Migdisov, 1971, Geochemical history of the crystalline basement and sedimentary cover of the Russian and North American platforms: Sedimentology, v. 16, p. 137–185.

Ronov, A. B., V. E. Khain, A. N. Balukhovsky, and K. B. Seslavinsky, 1980, Quantitative analysis of Phanerozoic sedimentation: Sed. Geology, v. 25, p. 311–325.

Rouse, H., and J. W. Howe, 1953, Basic mechanics of fluids: John Wiley & Sons, New York, 245 p.

Ruddiman, W. F., M. E. Raymo, D. G. Martinson, B. M. Clement, and J. Backman, 1989, Pleistocene evolution: Northern Hemisphere ice sheets and North Atlantic Ocean: Paleooceanography, v. 4, p. 353–412.

Russell, P. L., 1990, Oil shales of the world: Their origin, occurrence and exploitation: Pergamon Press, Oxford, 736 p.

Russell, R. D., and R. E. Taylor, 1937, Roundness and shape of Mississippi River sands: Jour. Geology, v. 45, p. 225–267.

Rust, B. R., 1981, Alluvial deposits and tectonic style: Devonian and Carboniferous successions in eastern Gaspé, in Miall, A. D. (ed.), Sedimentation and tectonics in alluvial basins: Geological Association of Canada Special Paper 23, p. 49–76.

Ryder, R. T., T. D. Fouch, and J. H. Elison, 1976, Early Tertiary sedimentation in the western Uinta Basin, Utah: Geological Society of America, Bull, v. 87, p. 496–512.

Ryer, T. A., and A. W. Langer, 1980, Thickness change involved in peat-to-coal transformation for a bituminous coal of Cretaceous age in central Utah: Jour. Sed. Petrology, v. 50, p. 987–992.

Sagoe, K-M. O., and G. S. Visher, 1977, Population breaks in grain-size distributions of sand—A theoretical model: Jour. Sedimentary Petrology, v. 47, p. 285–310.

Saller, A. H., P. M. Harris, B. L. Kirkland, and S. J. Mazullo (eds.), 1999, Geologic framework of the Capitan Reef: SEPM Special Publication No. 65, 224 p.

Salvador, A., 1985, Chronostratigraphic and geochronometric scales in COSUNA Stratigraphic Correlation Charts of the United States: Am. Assoc. Petroleum Geologists Bull., v. 69, p. 181–189.

Salvador, A. (ed.), 1994, International stratigraphic guide: A guide to stratigraphic classification, terminology, and procedure, 2nd ed.: Internat. Union of Geol. Sciences and Geol. Soc. America, Inc., Trondheim, Norway, 214 p.

Sandberg, P. A., 1983, An oscillating trend in Phanerozoic non-skeletal carbonate mineralogy: Nature, v. 305, p. 19–22.

Saxov, S., and J. K. Nieuwenhuis (eds.), 1982, Marine slides and other mass movements: NATO Conference Series IV: Marine Sciences, v. 6, Plenum, New York, 353 p.

Scheltema, R. S., 1977, Dispersal of marine invertebrate organisms: Paleobiogeographic and biostratigraphic implications, in Kauffman, E. G., and J. E. Hazel (eds.), Concepts and methods of biostratigraphy: Dowden, Hutchinson and Ross, Stroudsburg, Pa., p. 73–108.

Schieber, J., W. Zimmerle, and P. S. Sethi (eds.), 1998, Shales and mudstones: E. Schweizerbartsche Verlagsbuchhandlung, Stuttgart, Vol. I, 384 p., Vol. II, 296 p.

Scholle, P. A., 1978, A color illustrated guide to carbonate rock constituents, textures, cements, and porosities: Am. Assoc. Petroleum Geologists Mem. 27, 241 p.

Scholle, P. A., and D. S. Ulmer-Scholle, 2003, A color guide to the petrography of carbonate rocks: Grains, textures, porosity, diagenesis: AAPG Memoir 77, American Association of Petroleum Geologists, Tulsa, Okla., 474 p.

Scholle, P. A., M. A. Arthur, and A. A. Ekdale, 1983, Pelagic environment, in Scholle, P. A., D. G. Bebout, and C. H. Moore (eds.), Carbonate depositional environments: Am. Assoc. Petroleum Geologists Mem 33, p. 620–691.

Schopf, J. M., 1956, A definition of coal: Econ. Geology, v. 51, p. 521–527.

Schopf, T. J. M., 1980, Paleoceanography: Harvard University Press, Cambridge, Mass., 341 p.

Schreiber, B. C., 1988, Subaqueous evaporite deposition, in Schreiber, B. C. (ed.), Evaporites and hydrocarbons: Columbia University Press, New York, p. 182–255.

Schreiber, B. C., M. E. Tucker, and R. Till, 1986, Arid shorelines and evaporites, in Reading, H. G. (ed.), Sedimentary environments and facies, Blackwell, p. 189–228.

Schubel, K. A., and B. M. Simonson, 1990, Petrography and diagenesis of cherts from Lake Magadi, Kenya: Jour. Sed. Petrology, v. 60, p. 761–776.

Schwab, W. C., H. J . Lee, and D. C. Twichell, 1993, Submarine landslides: Selected studies of the U.S. Exclusive Economic Zone: USGS Bull., v. 2002, U.S. Govt. Print. Office, 204 p.

Schwan, J., 1988, The structure and genesis of Weichselian to early Holocene aeolian sheets in Western Europe: Sedimentary Geology, v. 55, p. 197–232.

Schwarzacher, W. J., 1975, Sedimentation models and quantitative stratigraphy: Developments in Sedimentology 19, Elsevier, Amsterdam, 382 p.

Schwennicke, T., H. Siegmund, and C. Jehl, 2000, Marine phosphogenesis in shallow-water environments: Cambrian, Tertiary, and Recent examples, in Glen, C. R., L. Prévôt-Lucas, and J. Lucas (eds.), Marine authigenesis: From global to microbial: SEPM Spec. Pub. 66, p. 481–498.

Schwarzacher, W., 1993, Cyclostratigraphy and the Milankovich theory: Elsevier, Amsterdam, 225 p.

Sedimentation Seminar, 1981, Comparison of methods of size analysis for sands of the Amazon-Solimtões rivers, Brazil and Peru: Sedimentology, v. 28, p. 123–128.

Seliacher, A., 1964, Biogenic structures, in Imbrie, J., and N. D. Newell (eds.), Approaches to paleoecology: John Wiley & Sons, New York, p. 296–315.

Seilacher, A., 1967, Bathymetry of trace fossils: Marine Geology, v. 5, p. 413–428.

Seilacher, A., 1992, Event stratigraphy: A dynamic view of the sedimentary record, in Brown, G. C., C. J. Hawkesworth, and R. C. L. Wilson (eds.), Understanding the Earth: Cambridge University Press, Cambridge, p. 375–385.

Sellwood, B. W., 1986, Shallow-marine carbonate environments, in H. G. Reading (ed.), Sedimentary environments and facies, 2nd ed.: Blackwell Scientific Publications, Oxford, p. 283–342.

Sengör, A. M.C., 1995, Sedimentation and tectonics of fossil rifts, in Busby, C. J., and R. V. Ingersoll (eds.), Tectonics of sedimentary basins: Blackwell Science, p. 53–117.

Sepkoski, J. J., Jr., 1995, Patterns of Phanerozoic extinction: A perspective from global data bases, *in* Walliser, O. H. (ed.), Global events and event stratigraphy in the Phanerozoic: Springer-Verlag, Berlin, p. 35–51.

Seyedolali, A., D. H. Krinsley, S. Boggs, Jr., P. F. O'Hara, H. Dypvik, and G. G. Goles, 1997, Provenance interpretation of quartz by scanning electron microscope-cathodoluminescence fabric analysis: Geology, v. 25, p. 789–790.

Seymour, R. J., 1990, Autosuspending turbidity flows, *in* LeMéhauté, B., and D. M. Hanes (eds.), The sea, Vol. 9, Ocean Engineering Science, John Wiley & Sons, New York, p. 919–940.

Shackleton, N. J., 1967, Oxygen isotope analyses and paleotemperatures reassessed: Nature, v. 215, p. 15–17.

Shane, P. A. R., T. M. Black, B. V. Alloway, and J. A. Westgate, 1996, Early to middle Pleistocene tephrochronology of North Island, New Zealand: Implications for volcanism, tectonism, and paleoenvironments: Geol. Soc. America Bull., v. 108, p. 915–925.

Shanley, K. W., and P. J. McCabe, 1994, Perspectives on sequence stratigraphy of continental strata: Am. Assoc. Petroleum Geology Bull., v. 78, p. 544–568.

Sharp, R. F., and L. H. Nobles, 1953, Mudflow of 1941 at Wrightwood, southern California: Geol. Soc. America Bull., v. 64, p. 547–560.

Shaw, A. B., 1964, Time in stratigraphy: McGraw-Hill, New York, 365 p.

Shaw, B. R., 1982, A short note on the correlation of geologic sequences, *in* Cubitt, J. M., and R. A. Reyment (eds.), Quantitative stratigraphic correlation: John Wiley & Sons, New York, p. 7–12.

Shaw, B. R., and J. M. Cubitt, 1978, Stratigraphic correlation of well logs: An automated approach, *in* Gill, C., and D. F. Merriam (eds.), Geomathematical and petrophysical studies in sedimentology: Pergamon, Oxford, p. 127–148.

Shaw, D. M., 1956, Geochemistry of pelitic rocks. Part III: Major elements and general geochemistry: Geol. Soc. America Bull., v. 67, p. 919–934.

Shearman, D. J., 1978, Evaporites of coastal sabkhas, *in* Dean, W. E., and B. C. Schreiber (eds.), Marine evaporites: Soc. Econ. Paleontologists and Mineralogists Short Course Notes 4, p. 6–42.

Sheldon, P. R., 1996, Plus ça change—A model for stasis and evolution in different environments: Palaeogeography, Palaeoclimatology, and Palaeoecology, v. 127, p. 209–227.

Sheldon, R. P., 1989, Phosphorite deposits of the Phosphoria Formation, western United States, *in* Nothold, A. J. G., R. P. Sheldon, and D. F. Davidson (eds.), Phosphate deposits of the world: Vol. 2 Phosphate rock resources: Cambridge University Press, Cambridge, p. 53–61.

Shepard, F. P., 1932, Sediments on the continental shelves: Geol. Soc. America Bull., v. 43, p. 1017–1039.

Shepard, F. P., 1961, Deep-sea sand: 21st Internat. Geol. Cong. Rept., p. 23, p. 26–42.

Shepard, F. P., 1973, Submarine geology, 3rd ed.: Harper & Row, New York, 551 p.

Shepard, F. P., 1979, Currents in submarine canyons and other types of sea valleys, *in* Doyle, L. J., and O. H. Pilkey (eds.), Geology of continental slopes: Soc. Econ. Paleontologists and Mineralogists Spec. Pub. 27, p. 85–94.

Shepard, F. P., and R. F. Dill, 1966, Submarine canyons and other sea valleys: Rand McNally, Chicago, 381 p.

Sheriff, R. E., 1980, Seismic stratigraphy: International Human Resources Development Corp., Boston, 227 p.

Shiki, T., Chough, S. K., and Einsele, G. (eds.), 1996, Marine sedimentary events and their records: Sedimentary Geology, v. 104 (special issue).

Shinn, E. A., and D. M. Robbin, 1983, Mechanical and chemical compaction in fine-grained shallow-water limestones: Jour. Sed. Petrology, v. 53, p. 595–618.

Shinn E. A., C. W. Holmes, and M. Marot, 2000, Abstract short-lived isotopes and the investigation of microbially precipitated calcium carbonate; a new approach to the "whiting problem": Geological Society of America, 2000 Annual Meeting, Abstract with programs 32, no. 7, p. 279.

Shinn, E. A., R. P. Steinen, B. H. Lidz, and P. K. Swart, 1989, Whitings, a sedimentologic dilemma: Jour. Sed. Petrology, v. 59, p. 147–161.

Sibley, D. F., and J. M. Gregg, 1987, Classification of dolomite rock textures: Jour. Sed. Petrology, v. 57, p. 967–975.

Siever, R., 1983, Evolution of chert at active and passive continental margins, *in* Iijima, A., J. R. Hein, and R. Siever (eds.), Siliceous deposits in the Pacific region: Developments in Sedimentology 36, Elsevier, Amsterdam, p. 7–24.

Siever, R., 1992, The silica cycle in the Precambrian: Geochimica et Cosmochimica Acta, v. 56, p. 3265–3272.

Simo, J. A. T., R. W. Scott, and J-P. Masse (eds.), 1993, Cretaceous carbonate platforms: AAPG Memoir 56, Am. Assoc. Petroleum Geologists, Tulsa, Okla., 479 p.

Simons, D. B., and E. V. Richardson, 1961, Forms of bed roughness in alluvial channels: Am. Soc. Civil Engineers Proc., Jour. Hydraulics Div., v. 87 (HY3), p. 87–105.

Simonson, B. M., 1985, Sedimentological constraints on the origins of Precambrian iron-formations: Geol. Soc. America Bull., v. 96, p. 244–252.

Simonson, B. M., 2003, Origin and evolution of large Precambrian iron formations, *in* Chan, M. A., and A. W. Archer (eds.), Extreme depositional environments: Mega end members in geologic time: Geological Society of America Special Paper 370, p. 231–244.

Simonson, B. M., and S. W. Hassler, 1996, Was the deposition of large Precambrian iron formations linked to major marine transgressions?: Jour. Geology, v. 104, p. 665–676.

Singer, J. K., J. B. Anderson, M. T. Ledbetter, I. N. McCave, K. P. N. Jones, and R. Wright, 1988, An assessment of analytical techniques for size analysis of fine-grained sediments: Jour. Sed. Petrology, v. 58, p. 534–543.

Sisodia, M. S., and D. S. Chauhan, 1990, The influence of magnesium ions during the formation of stromatolitic phosphorites of Udaipur, Rajasthan, India, *in* Northolt, A. J. G., and I. Jarvis (eds.), Phosphorite research and development: The Geological Society Special Publication 52, Bath, U.K., p. 313–320.

Slansky, M., 1986, Geology of sedimentary phosphates: North Oxford Academic Pub., Essex, Great Britain, 210 p.

Slingerland, R., J. W. Harbaugh, and K. P. Furlong, 1994, Simulating clastic sedimentary basins: PTR Prentice Hall, Englewood Cliffs, 220 p.

Sloss, L. L., 1963, Sequences in the cratonic interior of North America: Geol. Soc. America Bull., v. 74, p. 93–114.

Sloss, L. L., 1979, Global sea level changes: A view from the craton, *in* Watkins, J. S., L. Montadert, and W. Dickerson (eds.), Geological and geophysical investigation of continental margins: American Association Petroleum Geologists Memoir 29, p. 461–468.

Sloss, L. L., 1982, The Midcontinent Province: United States, *in* Palmer, A. R. (ed.), Perspectives in regional geological syntheses—Planning for The Geology of North America: D-Nag Special Publication 1, Geological Society of America, p. 27–39.

Sloss, L. L., 1996, Sequence stratigraphy on the craton: Caveat emptor, *in* Witzke, B. J., G. A. Ludvigson, and J. Day (eds.), 1996, Paleozoic sequence stratigraphy: Views from the North American craton: GSA Special Paper 306, p. 425–434.

Smith, D. G., G. E. Reison, B. A. Zaitlin, and R. A. Rahmani (eds.), 1991, Clastic tidal sedimentology: Canadian Soc. Petroleum Geologists Mem. 16, 307 p.

Smith, M. A., 1990, Lacustrine oil shale in the geologic record, *in* Katz, B. J. (ed.), Lacustrine basin exploration: Am. Assoc. Petroleum Geologists Mem. 50, p. 43–60.

Smith, N. D., and J. Rogers, 1999, Fluvial Sedimentology VI, Special Publication 28, International Association of Sedimentologists: Blackwell Science, Oxford, 478 p.

Smoot, J. P., and T. K. Lowenstein, 1991, Depositional environments of non-marine evaporites, *in* Melvin, J. L. (ed.), Evaporites, petroleum and mineral resources: Elsevier Sci. Pub., Amsterdam, p. 189–347.

Snedden, J. W., and R. W. Dalrymple, 1999, Modern shelf sand ridges: From historical perspective to a unified hydrodynamic and evolutionary model, *in* Bergman, K. M., and J. W. Snedden (eds.), Isolated shallow marine sand bodies: Sequence stratigraphic analysis and sedimentologic interpretation: SEPM Special Publication No. 64, p. 13–28.

Snedden, J. W., D. Nummedal, and A. F. Amos, 1988, Storm- and fair-weather combined flow on the central Texas continental shelf: Jour. Sedimentary Petrology, v. 58, p. 580–595.

Sneed, E. D., and R. L. Folk, 1958, Pebbles in the lower Colorado River, Texas, a study in particle morphogenesis: Jour. Geology, v. 66, p. 114–150.

Snelling, N. J., (ed.), 1985a, The chronology of the geological record: Geological Society Memoir 10, Blackwell Scientific Pub., Oxford, 343 p.

Snelling, N. J., 1985b, Geochronology and the geological record, *in* Snelling, N. J. (ed.), The chronology of the geological record: Geol. Soc. London Mem. 10, Blackwell Scientific Pub., Oxford, p. 3–9.

Soil Survey Staff, 1999, Soil taxonomy: A basic system of soil classification for making independent soil surveys, 2nd ed.: U.S. Department of Agriculture, Natural Resources Conservation Service, Agricultural Handbook 436, 869 p.

Sonnenfeld, P., and G. C. St. C. Kendall, convenors, 1989, Marine evaporites: Genesis, alteration, and associated deposits: Penrose Conference Report, Geology, v. 17, p. 573–574.

Soudry, D., 1992, Primary bedded phosphorites in the Campanian Mishash Formation, Negev, southern Israel: Sedimentary Geology, v. 80, p. 77–88.

Southard, J. B., and L. A. Boguchwal, 1990, Bed configurations in steady unidirectional water flows. Part 2. Synthesis of flume data: Jour. Sedimentary Petrology, v. 60, p. 658–679.

Sperling, C. H. B., and R. U. Cooke, 1980, Salt weathering in an arid environment: Experimental investigations of the relative importance of hydration and recrystallization processes. II. Laboratory studies: Bedford College, London, Papers in Geography 9, 53 p.

Stanistreet, I. G., and T. S. McCarthy, 1993, The Okavango Fan and the classification of subaerial fan systems: Sedimentary Geology, v. 85, p. 115–133.

Stanley, G. D., Jr., and S. D. Cairns, 1988, Constructional azooxanthellate coral communities: An overview with implications for the fossil record: Palaios, v. 3, p. 233–242.

Stanley, G. D., Jr., and J. A. Fagerstrom (eds.),1988, Ancient reef ecosystems: Palaios, v.3, p. 111–254.

Stanley, S. M., 1979, Macroevolution, pattern and process: W. H. Freeman, San Francisco, 332 p.

Stanley, S. M., 1985, Rates of evolution: Paleobiology, v. 11, p. 13–26.

Stanley, S. M., and L. A. Hardie, 1998, Secular oscillations in the carbonate mineralogy of reef-building and sediment-producing organisms driven by tectonically forced shifts in seawater chemistry: Palaeogeography, Palaeoclimatology, Palaeoecology, v. 144, p. 3–19.

Stanley, S. M., and L. A. Hardie, 1999, Hypercalcification: Paleontology links plate tectonics and geochemistry to sedimentology: GSA Today, v. 9, p. 1–7.

Staplin, F. L., W. G. Dow, C. W. D. Milner, D. I. O'Connor, S. A. J. Pocock, P. van Gijzel, D. H. Welte, and M. A. Yükler, 1982, How to assess maturation and paleotemperatures: Soc. Econ. Paleontologists and Mineralogists Short Course 7, 298 p.

Stein, R., 1985, Rapid grain-size analyses of clay and silt fraction by Sedigraph 5000D: Comparison with Coulter counter and Atterberg methods: Jour. Sed. Petrology, v. 55, p. 590–593.

Stein, S., and M. Wysession, 2003, An introduction to seismology, earthquakes, and earth structure: Blackwell Pub., Malden, Ma., 498 p.

Stewart, F. H., 1963, Marine evaporites, *in* Fleischer, M. (ed.), Data of geochemistry: U.S. Geol. Survey Prof. Paper 440-Y, 54 p.

Stockman, K. W., R. N. Ginsburg, and E. A. Shinn, 1967, The production of lime mud by algae in south Florida: Jour. Sed. Petrology, v. 37, p. 633–648.

Stokes, S., C. S. Nelson, and T. R. Healy, 1989, Textural procedures for the environmental discrimination of late Neogene coastal sand deposits, southwest

Auckland, New Zealand: Sedimentary Geology, v. 61, p. 135–150.

Stone, W. N., and R. Siever, 1996, Quantifying compaction, pressure solution and quartz cementation in moderately- and deeply-buried quartzose sandstones from the greater Green River Basin, Wyoming, *in* Crossey, L. J., R. Loucks, and M. W. Totten (eds.), 1996, Siliciclastic diagenesis and fluid flow: Soc. for Sed. Geology Spec. Pub. 55, p. 129–150.

Stonecipher, S. A., 2000, Applied sandstone diagenesis—Practical petrographic solutions for a variety of common exploration, development, and production problems: SEPM Short Course Notes 50, SEPM, Tulsa, Okla., 143 p.

Stopes, M. C., 1919, On the four visible ingredients in banded bituminous coal: Studies in the composition of coal: Royal Soc. London Proc., Ser. B., v. 90, p. 470–487.

Stopes, M. C., 1935, On the petrology of banded bituminous coal: Fuel, London, v. 14, p. 4–13.

Stow, D. A. V., 1994, Deep sea processes of sediment transport and deposition, *in* Pye, K. (ed.), Sediment transport and depositional processes: Blackwell Scientific Pub., Oxford, p. 257–291.

Stow, D. A V., and D. J. W. Piper, 1984a, Deep-water fine-grained sediments: Facies models, *in* Stow, D. A. V., and D. J. W. Piper (eds.), Fine-grained sediments: Deep-water processes and facies: The Geological Society, Blackwell, Oxford, p. 611–646.

Stow, D. A. V., and D. J. W. Piper (eds.), 1984b, Fine-grained sediments: Deep-water processes and facies: Geol. Soc. Spec. Pub. 15, Blackwell, Oxford, 659 p.

Stow, D. A. V., J-C. Faugères, A. Viana, and E. Gonthier, 1998, Fossil contourites: A critical review: Sedimentary Geology, v. 115, p. 3–31.

Stow, D. A. V., H. G. Reading, and J. D. Collinson, 1996, Deep seas, *in* Reading, H. G. (ed.), Sedimentary environments: Processes, facies and stratigraphy, Blackwell Science Ltd., Oxford p. 395–453.

Stride, A. H., R. H. Belderson, N. H. Kenyon, and M. A. Johnson, 1982, Offshore tidal deposits: Sand sheet and sand bank facies, *in* Stride, A. H. (ed.), Offshore tidal sands: Processs and deposits: Chapman and Hall, London, p. 95–125.

Sudgen, D. E., and B. S. John, 1976, Glacier and landscape: A geomorphical approach: Edward Arnold, London, 376 p.

Swift, D. J. P., 1975, Tidal sand ridges and shoal retreat massifs: Marine Geology, v. 18, p. 105–134.

Swift, D. J. P., and J. A. Thorne, 1991, Sedimentation on continental margins, I: A general model for shelf sedimentation, *in* Swift, D. J. P., G. F. Oertel, R. W. Tillman, and J. A. Thorne (eds.), Shelf sand and sandstone bodies—Geometry, facies and sequence stratigraphy: Internat. Assoc. Sedimentologists Spec. Pub. 14, Blackwell Scientific Pub., Oxford, p. 3–31.

Swift, D. J. P., G. Han, and C. E. Vincent, 1986, Fluid processes and sea-floor response on a modern storm-dominated shelf: Middle Atlantic shelf of North America. Part I: The storm-dominated regime, *in* Knight, R.

J., and J. R. McLean (eds.), Shelf sands and sandstones: Canadian Soc. Petroleum Geologists Mem. 11, p. 99–119.

Swift, D. J. P., J. R. Schubel, and R. E. Sheldon, 1972, Size analysis of fine-grained suspended sediments: A review: Jour. Sed. Petrology, v. 42, p. 122–134.

Swift, D. J. P., D. J. Stanley, and J. R. Curray, 1971, Relict sediments on continental shelves: A recommendation: Jour. Geology, v. 79, p. 322–346.

Sylvester, A. G., 1988, Strike-slip faults: Geol. Soc. America Bull., v. 100, p. 1666–1703.

Syvitski, J. P. M., 1991, Principles, methods, and applications of particle size analysis: Cambridge University Press, Cambridge, 368 p.

Taira, A., and P. A. Scholle, 1979, Discrimination of depositional environments using setting tube data: Jour. Sed. Petrology, v. 49, p. 787–800.

Tarling, D. H., 1983, Paleomagnetism: Principles and applications in geology, geophysics and archaeology: Chapman and Hall, London, 379 p.

Tauxe, L., 2002, Paleomagnetic principles and practice: Kluwer academic Publishers, Dordrecht, 299 p.

Taylor, G., and R. A. Eggleton, 2001, Regolith geology and geomorphology: John Wiley & Sons, Chichester, 375 p.

Taylor, J. M., 1950, Pore-space reduction in sandstones: Am. Assoc. Petroleum Geologists Bull., v. 34, p. 701–716.

Teichert, C., 1958, Concepts of facies: Am. Assoc. Petroleum Geologists Bull., v. 42, p. 2718–2744.

Tetzlaff, D. M., and J. W. Harbaugh, 1989, Simulating clastic sedimentation: Van Nostrand Reinhold, New York, 202 p.

The Open University Team, 1989, Waves, tides and shallow-water processes: Pergamon Press, Oxford, 187 p.

Ting, F. T. C., 1982, Coal macerals, *in* Meyers, R. A. (ed.), Coal structure: Academic Press, Inc., p. 8–49.

Tissot, B. P., and D. H. Welte, 1984, Petroleum formation and occurrence, 2nd ed.: Springer-Verlag, Berlin, 699 p.

Tourtelot, H. A., 1960, Origin and use of the word "shale": Am. Jour. Sci., Bradley Volume, v. 258–A, p. 335–343.

Trappe, J., 2001, A nomenclature system for granular phosphate rock according to depositional texture: Sed. Geology, v. 145, p. 135–150.

Trendall, A. F., 1983, Introduction, *in* Trendall, A. F., and R. C. Morris (editors), 1983, Iron-formation facts and problems: Developments in Precambrian Geology 6, Elsevier, Amsterdam, p. 1–12.

Trendall, A. F., and R. C. Morris (eds.), 1983, Iron-formation facts and problems: Developments in Precambrian Geology 6, Elsevier, Amsterdam, 558 p.

Tucker, M. E., 1993, Carbonate diagenesis and sequence stratigraphy: Sedimentology Review 1, p. 51–72.

Tucker, M. E., and V. P. Wright, 1990, Carbonate sedimentology: Blackwell Scientific Pub., Oxford, 482 p.

Tucker, M. E., J. L. Wilson, P. D. Crevello, J. R. Sarg, and J. F. Read (eds.), 1990, Carbonate platforms: Facies, sequences and evolution: International Association Sedimentologists Spec. Pub. 9, Blackwell Scientific Pub., Oxford, 328 p.

Tucker, R. W., and H. L. Vacher, 1980, Effectiveness of discriminating beach, dune, and river sands by moments and cumulative weight percentages: Jour. Sed. Petrology, v. 50, p. 165–172.

Udden, J. A., 1898, Mechanical composition of wind deposits: Augustana Library Pub. 1, 69 p.

Underwood, M. B., and G. F. Moore, 1995, Trenches and trench-slope basins, *in* Busby, C. J., and R. V. Ingersoll (eds.), Tectonics of sedimentary basins: Blackwell Science, p. 179–219.

Usdowski, E., 1994, Synthesis of dolomite and geochemical implications, *in* Purser, B., M. Tucker, and D. Zenger (eds.), Dolomites: A volume in honor of Dolomieu: International Assocation of Sedimentologists, Special Publication 21, Blackwell Scientific Pub., Oxford, p. 345–360.

Vacher, H. L., and T. M. Quinn (eds.), 1997, Geology and hydrology of carbonate islands: Elsevier Science B.V., Amsterdam, 948 p.

Vail, P. R., 1987, Seismic stratigraphy interpretation procedure, *in* Bally, A. W. (ed.), Atlas of seismic stratigraphy: Am. Assoc. Petroleum Geology Studies in Geology 27, Vol. 1, p. 1–10.

Vail, P. R., F. Audemard, S. A. Bowman, P. N. Eisner, and C. Perez-Cruz, 1991, The stratigraphic signatures of tectonics, eustasy, and sedimentology—An overview, *in* Einsele, G., W. Ricken, and A. Seilacher (eds.), Cycles and events in stratigraphy: Springer-Verlag, Berlin, p. 617–659.

Vail, P. R., J. Hardenbol, and R. G. Todd, 1984, Jurassic unconformities, chronostratigraphy, and sea-level changes from seismic stratigraphy and biostratigraphy, *in* J. S. Schlee (ed.), Interregional unconformities and hydrocarbon accumulation: Am. Assoc. Petroleum Geologists Mem. 36, Tulsa, Okla., p. 129–144.

Vail, P. R., R. M. Mitchum, Jr., and S. Thompson, III, 1977a, Seismic stratigraphy and global change of sea level. Part 3: Relative changes of sea level from coastal onlap, *in* Payton, C. E. (ed.), Seismic stratigraphy—Applications to hydrocarbon exploration: Am. Assoc. Petroleum Geologists Mem. 26, p. 63–81.

Vail, P. R., R. M. Mitchum, Jr., and S. Thompson, III, 1977b, Seismic stratigraphy and global change of sea level. Part 4: Global cycles of relative changes of sea level, *in* Payton, C. E. (ed.), Seismic stratigraphy—Applications to hydrocarbon exploration: Am. Assoc. Petroleum Geologists Mem. 26, p. 83–97.

Valentine, J. W., 1977a, General patterns of metazoan evolution, *in* Hallam, A. (ed.), Patterns of evolution as illustrated by the fossil record: Elsevier, New York, p. 27–57.

Valentine, J. W., 1977b, Biogeography and biostratigraphy, *in* Kauffmann, E. G., and J. E. Hazel (eds.), Concepts and methods of biostratigraphy: Dowden, Hutchinson and Ross, Stroudsburg, Pa., p. 143–162.

Valley, J. W., and D. R. Cole (eds.), 2001, Stable isotope geochemistry: Reviews in mineralogy and geochemistry, v. 43, Mineralogical Society of America, 662 p.

Van der Leeden, F., 1975, Water resources of the world—Selected statistics: Water Information Centre, Point Washington, N.Y., 568 p.

Van der Voo, R., 1993, Paleomagnetism of the Atlantic, Tethys, and Iapetus Oceans: Cambridge University Press, New York, 411 p.

Van der Voo, R., C. R. Scotese, and N. Bonhommet (eds.), 1984, Plate reconstruction from Paleozoic paleomagnetism: Geodynamics Series v. 12, American Geophysical Union, Washington, D.C., 136 p.

Van Houten, F. B., 2000, Ironstone ooids and phosphorites—A comparison from a statigrapher's view, *in* Glen, C. R., L. Prévôt-Lucas, and J. Lucas (eds.), Marine authigenesis: From global to microbial: SEPM Spec. Pub. 66, p. 127–132.

Van Houten, F. B., and D. P. Bhattacharya, 1982, Phanerozoic oolitic ironstone: Geologic record and facies models: Ann. Rev. Earth and Planetary Sci., v. 10, 441–457.

van Lith, Y., R. Warthmann, G. Vasconcelos, and J.A. McKenzie, 2003, Microbial fossilization in carbonate sediments: A result of the bacterial surface involvement in dolomite precipitation: Sedimentology, v. 50, p. 237–245.

Van Valen, L., 1973, A new evolutionary law: Evolution theory, v. 1, p. 1–30.

Van Wagoner, J. C., H. W. Posamentier, R. M. Mitchum, P. R. Vail, J. F. Sarg, T. S. Loutit, and J. Hardenbol, 1988, An overview of the fundamentals of sequence stratigraphy and key definitions, *in* Wilgus, C. K., B. S. Hastings, C. G. St. C. Kendall, H. W. Posamentier, C. A. Ross, and J. C. Van Wagoner (eds.), Sea-level changes: An integrated approach: Soc. Econ. Paleontologists and Mineralogists Spec. Pub. 42, p. 39–45.

Van Wagoner, J. C., R. M. Mitchum, K. M. Campion, and V. D. Rahmanian, 1990, Siliciclastic sequence stratigraphy in well logs, cores, and outcrops: AAPG Methods in Exploration Series 7, Am. Assoc. Petroleum Geologists, Tulsa, Okla., 55 p.

Vandenberghe, N., 1975, An evaluation of CM patterns for grain-size studies of fine grained sediments: Sedimentology, v. 22, p. 615–622.

Vasconcelos, C., and J. A. McKenzie, 1995, Microbial mediation as a possible mechanism for natural dolomite formation at low temperature: Nature, v. 377, p. 220–222.

Vasconcelos, C., and J. A. McKenzie, 1997, Microbial mediation of modern dolomite precipitation and diagenesis under anoxic conditions (Lagoa Vermelha, Rio De Janeiro, Brazil): Jour. Sedimentary Research, v. 67, p. 378–390.

Veizer, J., 1989, Strontium isotopes in seawater through time: Annual Review Earth and Planetary Sciences, v. 17, p. 141–167.

Viau, C., 1983, Depositional sequences, facies and evolution of the Upper Devonian Swan Hills Reef buildup, central Alberta, Canada, *in* Harris, P. M. (ed.), carbonate buildups—A core workshop: SEPM Core Workshop 4, Soc. Econ. Paleontologists and Mineralogists, Tulsa, Okla., p. 112–143.

Vincent, C. E., 1986, Processes affecting sand transport on a storm-dominated shelf, *in* Knight, R. J. and J. R. McLean (eds.), Shelf sands and sandstones: Canadian Soc. Petroleum Geologists Mem. 11, p. 121–132.

Vincent, S. J., D. M. Macdonald, and P. Gutteridge, 1998, Sequence stratigraphy, *in* Doyle, P., and M. R. Bennett (eds.), Unlocking the stratigraphical record: Advances in modern stratigraphy: John Wiley & Sons, Chichester, p. 299–350.

Vine, F. H., and D. H. Matthews, 1963, Magnetic anomalies over oceanic ridges: Nature, v. 199, p. 947–949.

Visher, G. S., 1969, Grain size distributions and depositional processes: Jour. Sed. Petrology, v. 39, p. 1074–1106.

Von Damm, K. L., 1990, Seafloor hydrothermal activity: Black smoker chemistry and chimneys: Ann. Rev. Earth and Planetary Sciences, v. 18, p. 173–204.

Wadell, H., 1932, Volume, shape and roundness of rock particles: Jour. Geology, v. 40, p. 443–451.

Wagner, G. A., and P. Van den Haute, 1992, Fission track dating: Kluwer, Dordrecht, Boston, 285 p.

Walker, R. G., 1978, Deep-water sandstone facies of ancient submarine fans: Models for exploration for stratigraphic traps: Am. Assoc. Petroleum Geologists Bull, v. 62, p. 932–966.

Walker, R. G., 1990, Facies modeling and sequence stratigraphy: Jour. Sedimentary Petrology, v. 60, p. 777–786.

Walker, R. G., 1992, Facies, facies models and modern stratigraphic concepts, in Walker, R. G., and N. P. James (eds.), Facies models—Response to sea level changes: Geol. Assoc. Canada, p. 1–14.

Walker, R. G., and D. J. Cant, 1979, Facies models 3. Sandy fluvial systems, in Walker, R. G. (ed.), Facies models: Geoscience Canada Reprint Ser. 1, p. 23–31.

Walker, R. G., and N. P. James (eds.), 1992, Facies models—Response to sea level changes: Geol. Assoc. of Canada, 407 p.

Walker, R. G., and A. G. Plint, 1992, Wave- and storm-dominated shallow marine systems, in Walker, R. G., and N. P. James (eds.), Facies models—Response to sea level changes: Geol. Assoc. Canada, p. 219–238.

Walliser, O. H. (ed.), 1996, Global events and event stratigraphy: Springer-Verlag, Berlin, 333 p.

Ward, C. R. (ed.), 1984, Coal geology and coal technology: Blackwell, Melbourne, 345 p.

Ward, W. C., C. G. St. C. Kendall, and P. M. Harris, 1986, Upper Permian (Guadalupian) facies and their association with hydrocarbons—Permian Basin, West Texas and New Mexico: Am. Assoc. Petroleum Geologists Bull., v. 70, p. 239–262.

Warren, J. K., 1989, Evaporite sedimentology: Prentice-Hall, Englewood Cliffs, N.J., 285 p.

Warren, J. K., 1991, Sulfate dominated sea-marginal and platform evaporative settings: Sabkhas and salinas, mudflats and salterns, in Melvin, J. L. (ed.), Evaporites, petroleum and mineral resources: Elsevier Sci. Pub., Amsterdam, p. 69–187.

Warren, J. K., and G. C. St. C. Kendall, 1985, Comparison of marine sabkhas (subaerial) and salina (subaqueous) evaporites: Modern and ancient: Am. Assoc. Petroleum Geologists Bull., v. 69, p. 843–858.

Warren, J., 2000, Dolomite: Occurrence, evolution, and economically important associations: Earth Sciences Reviews, v. 52, p. 1–81.

Warthmann, Y. van Lith, C. Vasconcelos, J. A. McKenzie, and A. M. Karpoff, 2000, Bacterially induced dolomite precipitation in anoxic culture experiments: Geology, v. 28, p. 1091–1094.

Watkins, N. D., 1972, A review of the development of the geomagnetic polarity time scale and discussion of prospects for its finer definition: Geol. Soc. America Bull., v. 83, p. 551–574.

Watson, A., 1992, Desert soils, in Martini, I. P., and W. Chesworth (eds.), Weathering, soils & paleosols: Elsevier, Amsterdam, p. 225–260.

Watts, A. B., 1981, The U.S. Atlantic continental margin: Subsidence history, crustal structure and thermal evolution, in Bally, A. W., et al., Geology of passive continental margins: Am. Assoc. Petroleum Geologists Education Course Note Series 19, p. 2-i to 2-ii and 2-1 to 2-75.

Weaver, M., and S. W. Wise, Jr., 1974, Opaline sediments of the southeastern coastal plain and Horizon A.: Biogenic origin: Science, v. 184, p. 899–901.

Wei, W., 1993, Calibration of upper Pliocene—lower Pleistocene nannofossil events with oxygen isotope stratigraphy: Paleooceanography, v. 8, p. 85–99.

Weimer, P., 1989, Sequence stratigraphy, facies geometries, and depositional history of the Mississippi Fan, Gulf of Mexico: Am. Assoc. of Petroleum Geologists, Bull., v. 74, p. 425–453.

Weise, B. R., 1980, Wave-dominated deltaic systems of the Upper Cretaceous San Miguel Formation, Maverick Basin, south Texas: Texas Bureau of Economic Geology, Report of Investigations 107, 39 p.

Weller, J. M., 1958, Stratigraphic facies differentiation and nomenclature: Am. Assoc. Petroleum Geologists Bull, v. 42, p. 609–639.

Weller, J. M., 1960, Stratigraphic principles and practices: Harper and Brothers, New York, 725 p.

Welte, D. H., B. Horsfield, and D. R. Baker (eds.), 1997, Petroleum and basin evolution: Springer-Verlag, Berlin, 535 p.

Wendt, J., Z. Belka, B. Kaufmann, R. Kostrewa, and J. Hayer, 1997, The world's most spectacular carbonate mud mounds (Middle Devonian, Algerian Sahara): Jour. Sed. Research, v. 67, p. 424–436.

Wentworth, C. K., 1919, A laboratory and field study of cobble abrasion: Jour. Geology, v. 27, p. 507–521.

Wentworth, C. K., 1922, A scale of grade and class terms for clastic sediments: Jour. Geology, v. 30, p. 377–392.

Whalen, M. T., 1995, Barred basins: A model for eastern ocean basin carbonate platforms: Geology, v. 23, p. 625–628.

Wheeler, H. E., 1958, Time stratigraphy: American Association of Petroleum Geologists Bull., v . 42, p. 1049–1063.

Wheeler, H. E., and V. S. Mallory, 1956, Factors in lithostratigraphy: Am. Assoc. Petroleum Geologists Bull, v. 40, p. 2711–2723.

Whitaker, F. F., and P. L. Smart, 1990, Active circulation of saline ground water in carbonate platforms: Evidence from the Great Bahama Bank: Geology, v. 18, p. 200–203.

Whittaker, A., 1998, Principles of seismic stratigraphy, in Doyle, P., and M. R. Bennett (eds.), Unlocking the stratigraphical record: Advances in modern stratigraphy, John Wiley & Sons, Chichester, p. 275–298.

Whittaker, A., J. C. W. Cope, J. W. Cowie, W. Gibbons, E. A. Hailwood, M. R. House, D. G. Jenkins, P. F. Rawson, A. W. A. Rushton, D. G. Smith, A. T. Thomas, and W. A. Wimbledon, 1991, A guide to stratigraphical procedure: Jour. Geological Society, London, v. 148, p. 813–824.

Wilgus, C. K., B. S. Hastings, C. G. St. C. Kendall, H. W. Posamentier, C. A. Ross, and J. C. Van Wagoner (eds.), 1988, Sea-level changes: An integrated approach: Soc. Econ. Paleontologists and Mineralogists Spec. Publ. 42, 407 p.

Wilkinson, B. H., R. M. Owen, and A.R. Carroll, 1985, Submarine hydrothermal weathering, global eustasy, and carbonate polymorphism in Phanerozoic marine oolites: Jour. Sedimentary Petrology, v. 55, p. 171–183.

Wilkinson, B. R., 1979, Biomineralization, paleooceanography, and evolution of calcareous marine organisms: Geology, v. 7, p. 524–527.

Wilkinson, B. R., 1982, Cyclic cratonic carbonates and Phanerozoic calcite seas: Jour. Geol. Education, v. 30, p. 189–203.

Williams, D. F., I. Lerche, and W. Full, 1988, Isotope chronostratigraphy: Theory and methods: Academic Press, San Diego, 345 p.

Williams, H., F. J. Turner, and C. M. Gilbert, 1982, Petrography, 2nd ed.: W.H. Freeman, San Francisco, 626 p.

Williams, J., 1996, Turbulent flow in rivers, *in* Carling, P. A., and M. R. Dawson (eds.), Advances in fluvial dynamics and stratigraphy: John Wiley and Sons, Chichester, p. 1–32.

Williams, L. A., G. A. Parks, and D. A. Crerar, 1985, Silica diagenesis, I Solubility controls: Jour. Sed. Petrology, v. 55, p. 301–311.

Willis, B. J., J. P. Bhattacharya, S. L. Gabel, and C. D. White, 1999, Architecture of a tide-influenced river delta in the Frontier Formation of central Wyoming, USA: Sedimentology, v. 46, p. 667–688.

Wilson, C., 1992, Sequence stratigraphy: An introduction, *in* Brown, G. C., C. J. Hawkesworth, and R. C. L. Wilson (eds.), Understanding the Earth, Cambridge University Press, Cambridge, p. 388–414.

Wilson, I. G., 1972, Aeolian bedforms—Their development and origins: Sedimentology, v. 19, p. 173–210.

Wilson, J. L., 1975, Carbonate facies in geologic history: Springer-Verlag, Berlin, 471 p.

Wilson, J. L., and C. Jordan, 1983, Middle shelf, *in* Scholle, P. A., D. G. Bebout, and C. H. Moore (eds.), Carbonate depositional environments: Am. Assoc. Petroleum Geologists Mem. 33, p. 297–344.

Wilson, J. T., 1966, Did the Atlantic close and then reopen? Nature, v. 211, p. 676–681.

Witzke, B. J., G. A. Ludvigson, and J. Day (eds.), 1996, Paleozoic sequence stratigraphy: Views from the North American craton: GSA Special Paper 306, 446 p.

Woodruff, F., and S. M. Savin, 1991, Mid-Miocene isotope stratigraphy in the deep sea: High-resolution correlations, paleoclimatic cycles, and sediment preservation: Paleooceanography, v. 6, p. 755–806.

Worden, R. H., and S. D. Burley, 2003, Sandstone diagenesis: The evolution of sand to stone, *in* Burley, S. D. and R. H. Worden (eds.), Sandstone diagenesis: Recent and ancient: Blackwell Publishing, p. 3–44.

Wright, D. T., 2000, Benthic microbial communities and dolomite formation in marine and lacustrine environments—a new dolomite model, *in* Glen, C. R., L. Prévôt-Lucas, and J. Lucas (eds.), Marine authigenesis: From global to microbial: SEPM Spec. Pub. 66, p. 7–20.

Wright, L. D., 1977, Sediment transport and deposition at river mouths: A synthesis: Geol. Soc. America Bull., v. 88, p. 857–868.

Wright, L. D., 1995, Morphodynamics of inner continental shelves: CRC Press, Boca Raton, Fla., 241 p.

Wright, V. P., 1992, A revised classification of limestones: Sed. Geol., v. 76, p. 177–185.

Wright, V. P., and T. P. Burchette, 1996, Shallow-water carbonate environments, *in* Reading, H. G. (ed.), Sedimentary environments: Processes, facies and stratigraphy: Blackwell Science Ltd., Oxford, p. 325–394.

Wuellner, E. E., L. R. Lehtonen, and W. C. James, 1986, Sedimentary-tectonic development of the Marathon and Val Verde Basins, West Texas, U.S.A.: A Permo-Carboniferous migrating foredeep, *in* Allen, P. A., and P. Homewood (eds.), Foreland Basins, Internat. Assoc. Sedimentologists Spec. Pub. 8, Blackwell, Oxford, p. 347–368.

Wyrwoll, K.-H., and G. K. Smyth, 1985, On using the log-hyperbolic distribution to describe the textural characteristics of eolian sediments: Jour. Sedimentary Petrology, v. 55, p. 471–478.

Yanov, E. N., 1978, Classification of sandstones and siltstones by composition of grains: Lithology and Mineral Resources, v. 12, p. 466–472.

Yates, K. K., and L. L. Robbins, 2001, Microbial lime-mud production and its relation to climate change, *in* Gearhard, L. C., W. E. Harrison, and B. M. Hanson (eds.), Geological perspectives of global climate Change: AAPG Studies in Geology 47, p. 267–283.

Yen, T. F., and G. V. Chilingarian (eds.), 1976, Oil shales: Elsevier, New York, 292 p.

Young, F. G., and G. E. Reinson, 1975, Sedimentology of Blood Reserve and adjacent formations (Upper Cretaceous), St. Mary River, southern Alberta, *in* Shawa, M. S. (ed.), Guidebook to selected sedimentary environments in southwestern Alberta, Canada: Canadian Soc. Petroleum Geologists Field Conference, p. 10–20.

Young, T. P., 1989, Phanerozoic ironstones: An introduction and review, *in* Young, T. P., and W. E. G. Taylor (eds.), Phanerozoic ironstones. Geol. Soc. London Spec. Pub. 46, p. ix–xxv.

Young, T. P., and W. E. G. Taylor (eds.), 1989, Phanerozoic ironstones: Geological Society Special Publication 46, The Geol. Soc., London, 251 p.

Zempolich, W. C. and H. E. Cook (eds.), 2002, Paleozoic carbonates of the Commonwealth of Independent States (CIS): Subsurface reservoirs and outcrop analogs: SEPM Special Publication 74, 245 p.

Zenger, D. H., F. G. Bourrouilh-Le Jan, and A. V. Carozzi, 1994, Dolomieu and the first description of dolomite, *in* Purser, B., M. Tucker, and D. Zenger (eds.), Dolomites: A volume in honor of Dolomieu: Internat. Assoc. Sedimentologists, Spec. Pub. 21, Blackwell Scientific Pub., Oxford, p. 21–28.

Zingg, Th., 1935, Beiträge zur Schotteranalyse: Schweiz. Mineralog. Petrog. Mitt., v. 15, p. 39–140.

Zuffa, G. G. (ed.), 1985, Provenance of arenites: D. Reidel Publishing Co., Dordrecht, 408 p.

Zuffa, G. G., 1980, Hybred arenites: Their composition and classification: Jour. Sed. Petrology, v. 50, p. 21–29.

Index